SURVEY SAMPLING

SURVEY SAMPLING

Leslie Kish

Professor of Sociology
Program Director, Survey Research Center
Institute for Social Research
The University of Michigan

Wiley Classics Library Edition Published 1995

A Wiley-Interscience Publication
JOHN WILEY & SONS, INC.
New York • Chichester • Brisbane • Toronto • Singapore

This text is printed on acid-free paper.

Wiley Classics Library Edition Published 1995.

Library of Congress Cataloging in Publication Data:

Library of Congress Catalog Card Number: 65-19479

ISBN 0-471-48900-X

ISBN 0-471-10949-5 (Classics Edition)

Printed in the United States of America

10 9 8 7 6 5

Preface

This is designed to be a simple book on sampling methods, with emphasis on and illustrations from surveys of human populations. Although I have designed samples for a variety of nonhuman populations, my experience has been mostly in social applications; the book emphasizes these areas, and it draws most of its examples from them. However, essentially the same methods can be used for sampling animals, plants, minerals, physical products, accounts, and inventories. Sampling specialists find that their skills possess broad generality and transferability; thus this book can also be useful to medical men, biologists, chemists, engineers, and accountants.

Sampling plays a vital role in research design involving human populations; it commands increasing attention from social scientists and practitioners. I include not only the disciplines of economics, sociology, anthropology, psychology, and political science, but also the professions of public health, biostatistics, education, social work, public administration, and business administration. Sampling problems are equally material to practitioners engaged in marketing, commerce, and industry.

While writing this book, I tried to keep a definite audience in mind and to maintain a dialogue with them. My first consideration was for quantitatively oriented students in the social sciences and allied fields. I believe that sampling methods should become roughly the third course of statistics for students in the social sciences. It should follow a course in the fundamentals of statistical reasoning, and another devoted to major statistical tools. I hope that this book can serve as a text for professors of social statistics who are not specialists in sampling.

Second, I have also aimed to provide a reference book on sampling methods for professors, researchers, and officials who want to understand survey sampling without necessarily becoming sampling specialists. Improving their methods is crucial, since most samples in survey research represent occasional small or medium-sized efforts by researchers who are not specialists.

I have not written this for my colleagues in sampling, but chiefly for my fellow social scientists, whose specialties differ from mine. Reading one book will not turn anyone into an expert sampling statistician; but the careful reader should become able to design and execute valid samples of moderate dimensions and difficulty, to avoid selection biases, and to achieve reasonable efficiency. He should also become more adept at evaluating the sample results he encounters, to judge their validity, their limits of inference, applicability, and precision.

The reader will be able to design satisfactory rather than optimum samples—to *satisfice* (as Herbert Simon says), rather than to optimize. I paid less attention than others to some technical efforts for getting the last 2 percent efficiency out of samples. These efforts may be worthwhile for large, complicated samples—where it may be best to consult an experienced sampling statistician. In that case, the researcher with sufficient knowledge to meet him half-way will be better able to utilize the scarce time of the sampling consultant.

Social research places special emphasis on the comparative and analytical uses of samples. Hence, I have stressed statistics of subclasses in the sample, rather than focusing all attention on the entire sample. Subclasses are treated for each of the major designs; and the comparison of subclass means receives frequent attention. Other issues relating to analytical statistics from complex samples are also discussed. These discussions frequently utilize research developed under a grant (G-7571) from the National Science Foundation.

The book is oriented toward providing a working knowledge of practical sampling methods, with an understanding of their theoretical background. The necessary working formulas are given. I have tried to state clearly the assumptions underlying the formulas, and to outline the potentialities and limitations inherent in the assumptions. I often use illustrations to explain in detail the meaning of formulas and definitions. I also provide a variety of examples, with computations laid out in painstaking detail.

The book contains many practical procedures, the "domestic arts" of sampling along with its science; the valuable "tricks" that are usually learned only in apprenticeship. Perhaps the principal task of any book or course on methods should be the drastic shortening of the period of apprenticeship in a discipline. Throughout the book the reader can discover my attempts to synthesize complete sets of coherent rules for conduct ordinarily explained—if at all—with *ad hoc* rules-of-thumb.

My efforts are perhaps most evident in methods for dealing with frame problems. They receive a great deal of emphasis, especially in an early summary (2.7) of frame defects and techniques for treating them, and in a detailed treatment in Sections 11.1 to 11.4. Throughout the book the reader's attention is called to possible frame defects and their effects on sample design. Problems invite the reader to participate in finding and overcoming defects he is likely to encounter in research situations.

Sampling methods are not developed here for their own sake, but as means to ends originating in substantive research problems, especially in the social sciences and their applications. They are designed to provide valid, scientific, and economical tools for those research problems. Hence, the approach is directed less toward delivering a neat theoretical package than toward practical research projects in all their complexity, while preserving a fundamental simplicity of presentation.

To achieve simplicity I frequently resort to useful approximations. Briefly, this book attempts to preach what reasonable sampling statisticians actually practice, rather than an arid dogma. For example, the factor $(N - 1)/N$, if inconvenient, can be ignored when the population size N is large. The book emphasizes estimated variances that the researcher actually computes, rather than their expected values. And so on.

I did not try to make this book "self-contained"—an illusory aim. For the algebraic steps of derivations the interested reader is directed, by frequent references, to relevant passages in several fine textbooks. In these references I sought neither to discover original sources, nor the prestige of the most elegant mathematical treatment. Instead, I sought the most relevant, readable, and available sources, preferably textbooks; there the reader can find further references.

However, I am glad that, on the urging of friends, I abandoned my original intention to avoid derivations completely. My derivations are few and simple, yet they cover the essentials of sampling theory adequately for readers who understand certain fundamental concepts of statistical theory. Sampling theory was condensed by eliminating some repetition and detail. The basic derivations are in Sections 2.8, 4.5, 4.6, 5.6, 6.6, 8.5, and in scattered remarks. These and other technical portions carry a (*); continuity can be maintained without them.

My original intention was to formalize and disseminate the notes that my students and I have evolved from teaching a one-semester course. This restricted but vital core is presented in the simple, nontechnical, carefully organized and related sections of Chapters 1 to 8 in Part I: Fundamentals of Survey Sampling. To complete the course, I add

Sections 9.1 to 9.5 on area sampling, plus most of Chapter 13 on non-sampling errors. Brief selections may also be added from Chapters 11 and 12, which deal with special techniques.

These two chapters and the technical portions of others have served as the base for advanced courses in special problems. However, they are designed primarily to serve as sources of reference for individual, technical problems. These sections are mostly self-contained, rather than inter-connected. I hope that researchers will often find in one of them ready answers to specific technical problems.

Those who can avoid the complexities of cluster sampling will find a self-contained treatment of element sampling in Chapters 1 to 4. To this can be added a modest treatment of the essentials of cluster sampling: 6.4B, 6.5A and B, 7.2, 7.3, and 9.1 to 9.3. These could comprise a third of a course, incorporated into a full year course in statistics.

For a briefer introduction to sampling, teachers have used the following sections of 81 pages: 1.0–1.7, 2.6, 2.7, 3.1, 3.4, 3.5, 5.1, 5.4, 6.4B, 6 5A, 7.2, 7.3, 9.1–9.3.

I hope that readers will especially like the problems. They should give much credit to my students, who eliminated the worst, stimulated the best, and sharpened their phrasing; also some credit to my many colleagues and clients, whose questions and research projects gave birth to most of these problems. I tried hard to make them life-like; hence, the student's crucial task is to discover what the principal issues are, then pose appropriate questions. When this is well done, the answers come rather smoothly. The focus is on principal issues, rather than on fine nuances. This was made possible by eliminating cumbersome details, while retaining the essence of actual problems. In addition to testing the student's ability to apply the book's methods, these problems perform a double task. They deepen understanding of those methods by extending them to new and varied areas. Furthermore, the extensions themselves represent valuable methodological tools. Work on these problems should yield half the value the student can extract from this book. Of a subject devoted to methods, we can say with Sophocles: "One must learn by doing the thing; for though you think you know it, you have no certainty, until you try."

It is with genuine pleasure that I acknowledge my debt to many friends and colleagues. Since Yates' book in 1948, six fine books on sampling have appeared in English, and I have learned and taught from each in turn. Deming, Cochran, Hansen, and Hurwitz have been personal teachers, friends, and consultants. I was fortunate to have, at different times, Roe Goodman, Benjamin Tepping, and Irene Hess as colleagues in the Sampling Section of the Survey Research Center. Angus Campbell and Rensis Likert, in creating and directing that Center and the Institute for Social Research, have provided valuable support and opportunity. Questions, problems, and issues posed daily by the social scientists at the Center provide a uniquely stimulating atmosphere. I also benefited from daily contact with colleagues in the Sociology and Mathematics Departments. Of the many statisticians who have given me personal stimulation, support, and criticism, I must single out L. J. Savage, Bruce Hill, and Howard Raiffa; William Ericson gave me valuable criticism on much of the final draft, tightening many a loose phrase or formulation. They stimulated and helped me to make this book more suitable as a text in departments of statistics than I first planned. V. K. Sethi and C. T. Tharakan eliminated many technical errors. Above all, I was fortunate to find in my wife, Rhea, an expert and devoted editor, who has greatly improved the book's style.

Primary and essential stimulation came from the many students in about forty courses in sampling I have given in fifteen years. They represent a good cross section of researchers, teachers, practitioners, and students from the social sciences, mathematics, and allied fields. The students in the seventeen yearly Summer Institutes on Survey Techniques have been mostly officials, scientists, and professors; they came from all over the globe, representing yearly an average of ten countries. Their experience, searching questions, suggestions, and arguments have contributed substantially to the origin, form, and content of this book.

Leslie Kish

Ann Arbor, Mich.
June 1965

Contents

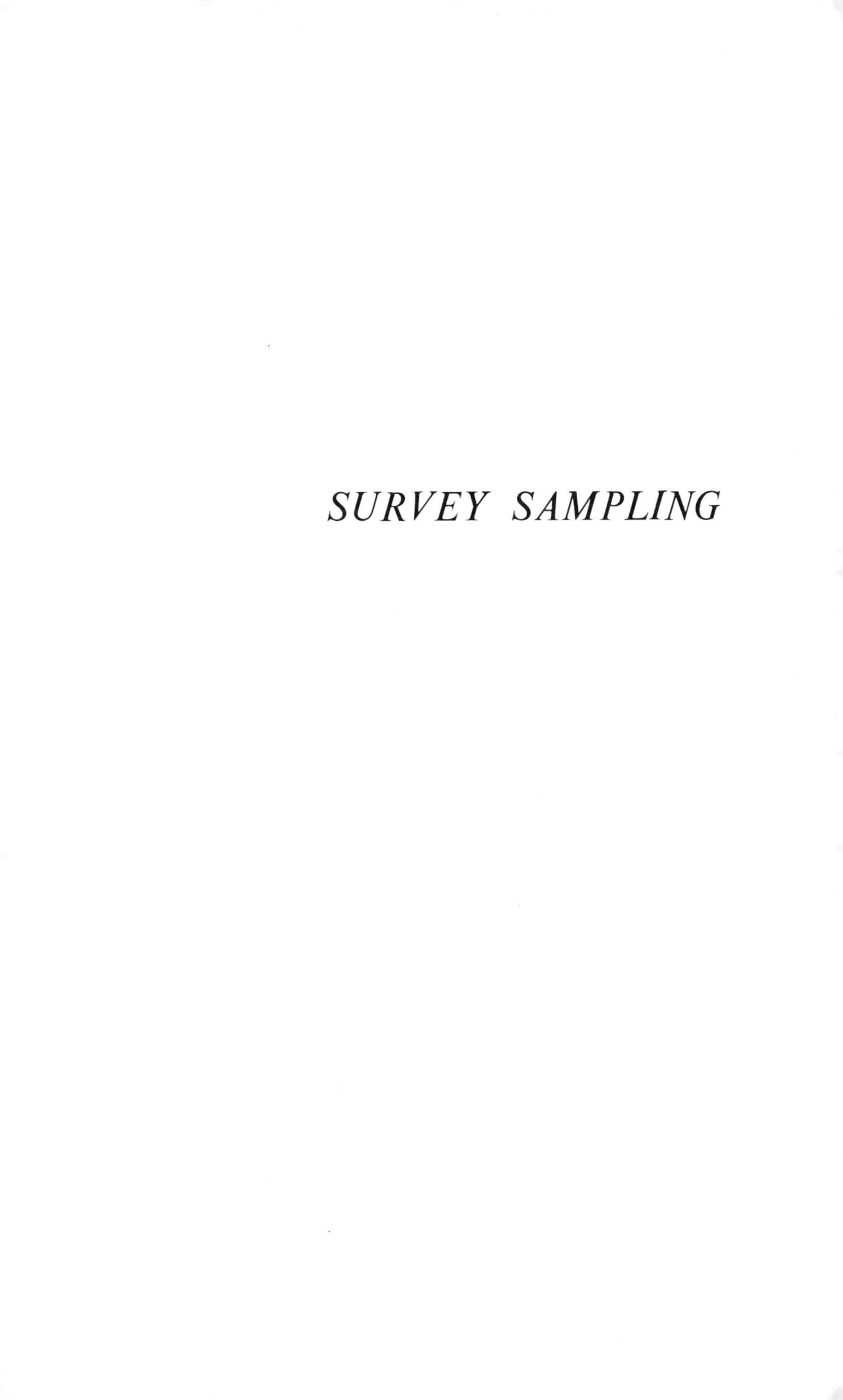

SURVEY SAMPLING

PART I
FUNDAMENTALS OF SURVEY SAMPLING

1
Introduction

1.0 GUIDE TO THE INTRODUCTION

This chapter differs from the others—as introductions to technical books often do. The rest of this book is devoted to specific methods of sampling, and the detailed treatments should serve the reader as clear guides to new methods. This chapter, however, gives only a brief coverage of important matters that lie outside the narrow province of sampling. But since they are adjacent to it, mutual understanding of these ideas, based on a common vocabulary, is essential. Readers with various backgrounds may find some sections obvious and others too condensed. They can skim over the former and pursue the references to further reading in the latter. Generally, it would be best to read this chapter once before and once after the rest of the book—and parts of it in between.

This introduction provides the *why*, *what*, and *whence* for the book: some rationale and motivation for the uses of "good" sampling methods; a bird's-eye view of the basic problems and methods for meeting them; and an indication of how the approach of population sampling fits into survey methods and into the general quest for scientific knowledge.

I attempt here a systematic classification of the units, tools, methods, problems, and goals of population sampling. Where good definitions are available I use them; for new definitions I try to conform to good usage and to reason. Where I had to coin new terms, I warn the reader to distinguish these novelties from the tried and trusted standard terms.

Because this chapter deals with fundamentals rather than techniques, and because it deals with them briefly yet wholeheartedly, it is also unavoidably more controversial than the rest of the book.

1.1 SURVEY DESIGN AND SAMPLE DESIGN

Sample design has two aspects: a *selection process*, the rules and operations by which some members of the population are included in the sample; and an *estimation process* (or estimator) for computing the sample statistics, which are *sample estimates* of population values. The overall design of surveys includes other important aspects that can be called jointly the *survey objectives:*

(*a*) *The definition of the survey variables* should specify the nature of the characteristics, the rules of classification categories, and the units for expressing them. It must also specify the extent and content of the *survey population.*

(*b*) The *methods of observation* (measurement), including both data collecting and data processing, give operational meaning to the survey variables and determine the nature of the survey data.

(*c*) The *methods of analysis*, statistical and substantive, reduce the survey data to results that can be comprehended and utilized.

(*d*) The *utilization of survey results* may sometimes take the form of specified decisions, based on the survey results and other relevant information. More frequently, the results become part of the public fund of knowledge, and the researcher has only vague understanding of the future use of his results.

(*e*) *The desired precision* of survey results may be clearly stated for samples designed for a specified statistical decision. More often the survey aims are many and vaguely stated, yet the researcher can find some broad limits of desired precision. Commonly, however, instead of specifying precision, the researcher must work from a reasonable allowed expenditure, and adjust accordingly the aims and the scope of the survey. This occurs in the design of surveys with many objectives, none of which is of predominant importance.

The survey objectives should determine the sample design; but the determination is actually a two-way process, because the problems of sample design often influence and change the survey objectives. We shall encounter examples of the ways in which survey objectives and sample design interact to produce overall survey designs. A dialogue between the researcher and the sampler must occur before any aspect of the survey design is "frozen," because a change in one aspect may dictate a change in others. Instead of a dialogue, the decisions may involve a larger cast:

sampler, researcher, and "consumer"; and the last, perhaps the grantor of the project, may feel behind him the silent pressure of the "ultimate consumers" of the data—the members of a profession or, perhaps, a wider public. The dialogue may occur silently within one head, if the researcher and sampler are one; but the dialogue should nevertheless take place.

Most samples are prepared by statisticians and other researchers who are not primarily sampling specialists. Nevertheless, it is helpful, although sometimes difficult, to separate sampling design from the related activities involved in survey research. The sample design covers the tasks of selection and estimation for making inference from sample value to the population value. Beyond this are the problems of making inferences from the survey population to another and generally broader population, with measurements free from error.

Imagine a file of N cards, each of which contains the desired measurement (Y_i) or measurements (X_i, Y_i, Z_i, ...) for one of the population elements. Computations on the entire file would produce the constant population value. Sampling would consist of selecting a fraction of the cards, then computing statistics from the sample to estimate the population value. For example, the sample mean $\bar{y}$ can be computed for estimating the population mean $\bar{Y}$. Moreover, the standard error of $\bar{y}$ can also be computed to estimate the average variability (the root-mean-square error) of $\bar{y}$ from $\bar{Y}$. The actual deviation $(\bar{y} - \bar{Y})$ is unknown; only the average fluctuation, the standard error, can be estimated. Different sampling designs would result in different standard errors, and choosing the design with the smallest error is the principal aim of sampling design.

The sampling fluctuation of $(\bar{y} - \bar{Y})$ must be distinguished from the measurement bias that can affect the population value $\bar{Y}$, based on a complete census of the file. The average accuracy of the measurements of the sample elements is equal to the average accuracy of the cards in the entire file. Comparing the sample value to the population value of the entire file, based on a *complete equal coverage* [Deming, 1960, p. 50], separates sampling errors from the nonsampling errors of measurement. It compares the sample value to an imaginary population value, obtained under *similar essential survey conditions*.

In practice the file does not exist, and the measurements are made only after selection and only on the sample. But this difference is not crucial for the distinction that places the measurement process outside the sampling design proper. Obtaining the element values (Y_i, etc.) and placing them on the cards belong to the measurement process (collecting and processing). If several complete files were prepared under similar essential survey conditions, their population values would differ slightly from each other, due to errors in the individual measurements. Even the number (N) of cards in the file can be subject to errors of coverage. The population values would fluctuate slightly around a common expected value, determined by the essential survey conditions. To include the appropriate amount of this variable nonsampling error in the computed total survey error is sometimes possible, although difficult (13.8).

Finally, imagine several complete files, each prepared under different essential conditions, so that the accuracy of the cards differs from file to file. But the cost of obtaining access to the files differs also. These differences are not strictly within the sampler's special domain, but he should not disregard them. The overall survey design should consider the problems of choosing the best file, together with the best sample design from the chosen file.

The economic design of surveys requires the joint consideration and planning of sampling and nonsampling errors (Chapter 13). Since the sources and magnitudes of measurement errors in any science are subjects as deep as the entire science, to treat the errors of measurement in specific detail would be an impossible task. Hence, knowledge about measurement errors must come from specialists in the substantive field(s) involved in the observations.

The researcher, on the other hand, may not know enough sampling to design a valid and economic sample, and may consult a specialist in survey sampling. Survey sampling is a specialized skill within statistics because deep knowledge of the art and science requires full-time professional preparation and attention. Furthermore, knowledge of sampling design can be applied equally well to many fields of scientific, public, and business research.

The sampler also concerns himself with other aspects of the survey objectives, and participates in their definitions and clarifications. In some cases he may even possess expert knowledge in some of those aspects. Conversely, the researcher may have ideas about the design of samples. Nevertheless, it is useful to make a separation between the two functions. Knowledge of measurement errors must come from specialists in the different fields involved in the observations. The sampler's own statistical skills allow him to raise the right questions at the proper time, but not to answer them. He qualifies for Tukey's definition of an expert as somebody "who thinks with other people's brains."

1.2 A TAXONOMY OF SURVEY UNITS AND CONCEPTS

The elements of a population are the units for which information is sought; they are the *individuals*, the *elementary units* comprising the population about which inferences are to be drawn. *They are the units of analysis, and their nature is determined by the survey objectives.*

The *population* is defined jointly with the elements: *the population is the aggregate of the elements*, and the elements are the basic units that comprise and define the population. The population must be defined in terms of (1) content, (2) units, (3) extent, and (4) time. For example, in the design of a survey of consumer expectations we may desire to specify: (1) all persons, (2) in family units, (3) in the United States and its territories, (4) in 1965. Often the desired population must be redefined to obtain a practicable *survey population*. For example, the above might be redefined as: (1) all persons above 18 years of age living in private dwelling units, (2) in spending units, (3) in the continental United States without Alaska, Hawaii, and the territories, (4) on January 1, 1965.

The *survey population* actually achieved may differ somewhat from the desired target population. The chief difference frequently arises from nonresponses and noncoverage. Strictly speaking, only the survey population is represented in the sample. But this may be difficult to describe exactly, and it is easier to write about the defined target population.

One survey may yield information about several diverse populations. *Different contents* may be covered by the same survey. For example, a survey of home accidents may yield information about separate accidents, persons injured in accidents, families incurring accidents, and homes with accidents. *Different units* may be formed from the same data. Consumer data may be presented in terms of persons, spending units, families, or households. *Different extents* usually appear in the form of subclasses for which survey results are commonly prepared. The divisions may be geographic, as for regional data, or not, as with occupational or age subclasses. A *subclass* is a portion of the sample regarded as a sample from the corresponding portion of the survey population. Domains denote subclasses specifically planned for in the sample design. *Different times* may be represented in a survey when information is obtained about two or more periods. These, for example, could be current and past month, or past month and past year.

We avoid using *universe* as a synonym for population. It denotes a hypothetical infinite set of elements generated by a theoretical model. This may be an ideal operation repeated endlessly, such as the endless tossing of a perfect coin. Behind every survey population stands some hypothetical universe, explicit or implicit, definite or indefinite. For example, a statistic about the U.S. adult population on a fixed date is expected to have relevance for other dates and, often hopefully, for other cultures. The sample provides statistical inference to the survey population. Inference to hypothetical universe is more complex and not statistical (at least, not in the same sense).

Characteristics of population elements are transformed to *variables* Y_i by the survey operations of measurement. Some literature deals directly with the statistical population of the variables Y_i. But I prefer to say that the ith *element has the variable* Y_i. This permits us to talk of the many variables (Y_i, X_i, Z_i, W_i, P_i, etc.) of the same element. We can also consider relationships between variables of an element, changes of variables, and accuracy of measurements of variables. A statistic based on the variables found in a sample results in a random variable that we call a *variate* [Kendall and Buckland, 1957].

Sampling units contain the elements, and they are used for selecting elements into the sample. In element sampling, each sampling unit contains only one element; but in cluster sampling, any sampling unit called *cluster* may contain several elements. For example, a sample of students may be obtained from a sample of classrooms, or a sample of dwellings from a sample of blocks. The same survey may use different kinds of sampling units, and in multistage sampling a hierarchy of sampling units or clusters is used, so that the element belongs uniquely to one sampling unit at each stage. For example, a sample of persons in a state may be taken by successively selecting counties, townships, segments, dwellings, and finally persons. The population is also an aggregate of the sampling units, specified for each stage.

To improve the selection, the population is commonly divided into subpopulations called *strata*. The aggregate of the strata comprises the population.

Listing units (briefly, *listings*) are used to identify and select sampling units from *lists*. Sometimes detailed procedures are needed to convert listings into sampling units; for example, converting listed addresses into dwellings and households. The problems may be serious if the elements fail to have unique identification with the listings. For example, a sample of families from telephone listings may involve serious difficulties (2.7).

Observational units are the units from which the observations are obtained. In interview surveys they are called respondents. Observational units are often the elements, as in surveys of attitudes. But the two may be different. The household head may give data about all persons in the household, or about all children; in the former case the respondents belong, with others, to the population; in the latter they do not belong. In a survey of school children the respondent may be the teacher.

Hansen, Hurwitz, and Madow [1953, II, Chs. 1 and 2] define and develop well the relationships of several survey units.

1.3 POPULATION VALUES AND STATISTICS

Empirical research may be performed in different ways: by haphazard observations, controlled observations, experiments, or surveys. This book concentrates on sampling for surveys. Survey research is aimed at estimating specified population values. *A population value* is a numerical expression that summarizes the values of some characteristic(s) for all N elements of an entire population; it is a summary measure of some feature of the distribution of the variable(s) in the defined population. The basic example in survey sampling is the *population mean* $\bar{Y} = \Sigma Y_i/N$, where Y_i is the value for some variable of the ith element in the population. Population values closely allied to the mean are proportions, medians, and other quantiles, and the population total or aggregate. Still other population values measure relationships; the most common of these is the difference between two means; further examples are coefficients of regression and correlation. Any population value is determined by four factors which appeared in the first three survey objectives: (*a*) the defined survey population; (*b*) the nature of the survey variable(s), and their distributions in some cases; (*c*) the methods of observation; and (*d*) the mathematical expression for deriving the population value from the individual element values.

Both *population value* and *true value* refer to numerical expressions derived from the entire population. The difference between them arises from errors of observation. The true value would be obtained from all population elements, if the observations were not subject to error. The population value, also a function of all the observations, is subject to the same nonsampling errors as its sample estimate; it is a value that would be obtained if the entire population—rather than just a sample—were designated for observation under the actual survey conditions. Even that value is subject to fluctuation, since repeated measurements would obtain different values. But this variability of the population value is ordinarily small compared to the sampling variance, when the sample is only a small fraction of the population. Hence the population value is usually regarded as an unknown constant, neglecting its measurement error. We avoid the word "parameter" because its use in statistics is not clearly defined; in different contexts it may stand either for the true value or the population value; in mathematics it means something else again.

The *sample value*, or *statistic*, is an *estimate* computed from the n elements in the sample. An important example is the mean of the sample elements $\bar{y} = \Sigma y_j/n$, where y_j is the observed value of the jth element in the sample. It is a *variate*, or *random variable*, that depends on the sample design and on the particular combination of elements which happen to be selected. Hence the particular estimate is only one among the many possible estimates which could have been obtained by the same sample design. On the contrary, the population value depends on all N values in the population. This is a constant independent of the vagaries of selection, although its value is usually unknown.

Statistics are subject to both nonsampling errors and sampling errors. The latter arise because only part of the total population is designated for observation in the sample. The nonsampling errors occur because the procedures of observation are imperfect. Their contribution to the total error of the survey should be considered jointly with the sample design, and they are treated in Chapter 13. Most of this book, and sampling design in general, deals chiefly with the variability of statistics around the population value.

The information that the mean $\bar{y}$ of a specific sample is, say, 14.7 units, has no intrinsic practical worth; its value lies in what it may tell us about the population mean $\bar{Y}$. We know that the unknown $\bar{Y}$ should differ from the known $\bar{y}$, but we do not know by how much. The sample mean depends on which sample of n elements happened to be selected, since different samples, though of the same size n, and taken with the same design from the same population, would result in different sample means. The deviation of any single mean from the population mean is unknown; it may be plus or minus, large or small. The only statistical way of regarding sampling variability is in probability terms: what size deviations are likely to occur in the long run? We can ask: Given a sample design and size, what values of the sample mean $\bar{y}$ are possible, and what is the probability of occurrence of each of those values? This array of possible values of $\bar{y}$, each with its probability of occurrence, is the *sampling distribution* of the possible sample means $\bar{y}$ for a fixed population, sample design, and size.

"Sample mean" here denotes the sample estimate of the population mean; it is not necessarily the simple mean of the sample cases. We shall see examples where other estimators are preferred; for example, a weighted mean or a ratio mean. Furthermore, the discussions in this section are generally relevant not only to the mean but also to other statistics.

Imagine that a sample design has been specified, including the sample size, and the selection and the estimation procedures; that we apply it to a fixed population, drawing sample after sample; and that we compute the mean for each sample and then tabulate and plot these values as the distribution of their relative frequencies. As the number of plotted values increases, the shape of the distribution becomes more stable and gradually approaches the true sampling distribution. This occurs as the "relative frequencies" for the different values of $\bar{y}$ approach the true probabilities.

of their occurrence: the theoretical result of an infinitely large number of drawings. Thus the *sampling distribution* of an estimate (mean) is the theoretical distribution of all possible values of the estimate ($\bar{y}_c$), each with its probability of occurrence (P_c). The possible values and their probabilities depend on the sample design (size, selection, and estimation) applied to a fixed population of characteristics.

The mean of the sampling distribution is the expected value of the estimate: thus,

$$E(\bar{y}) = \sum_c P_c \bar{y}_c. \tag{1.3.1}$$

This mean value $E(\bar{y})$ may or may not be equal to the population value $\bar{Y}$. The difference between the two we call the *sampling bias* $= E(\bar{y}) - \bar{Y}$. *A sample design is called unbiased if* $E(\bar{y}) = \bar{Y}$. Note that this is not a property of a single sample, but of the entire sampling distribution, and that it belongs neither to the selection nor the estimation procedure alone, but to both jointly. Examples of unbiased estimates are the mean and the variance obtained from simple random samples. Many estimates are not unbiased: the standard deviation of a simple random sample, the simple mean of a systematic sample of elements, and the ratio mean based on a random selection of clusters of unequal sizes; but in all these cases the sampling bias becomes negligibly small with increasing sample size.

The standard deviation of the sampling distribution is called the standard error; it is the square root of the variance of the sampling distribution, which is the mean squared deviation around the mean $E(\bar{y})$:

$$\text{Var}(\bar{y}) = \sum_c P_c[\bar{y}_c - E(\bar{y})]^2$$

and

$$\text{SE}(\bar{y}) = \sqrt{\text{Var}(\bar{y})}. \tag{1.3.2}$$

The sampling distribution represents the random fluctuation of $\bar{y}_c$ due to the specific sample design, and this variability is measured by the standard error. This definition is fundamental and general. But it cannot be used directly to obtain the standard error because in practical situations we know only a single point in the whole sampling distribution. However, using this concept, statisticians have developed specific formulas for the variances of sample means for many practical sample designs. Different specific expressions appear for the different sample designs. For example, for the mean of an unrestricted sample (a simple random sample with replacement), the standard error is equal to the familiar $\sigma_y/\sqrt{n}$. But this formula does not apply for other sample designs.

However, knowing the population value of the standard error, denoted SE ($\bar{y}$), is still of no practical use, because its value depends on the entire population, hence cannot be computed from the sample. For example,

$\sigma_y^2 = \sum_i^N (Y_i - \bar{Y})^2/N$ cannot be computed from the sample. But statisticians have also developed practical formulas for computing variance estimates from the sample data. For example, se $(\bar{y}) = s_y/\sqrt{n}$ is the estimated standard error of an unrestricted sample where $s_y^2 = \sum_j^n (y_j - \bar{y})^2/(n-1)$. Formulas for computing appropriate sample estimates of the standard

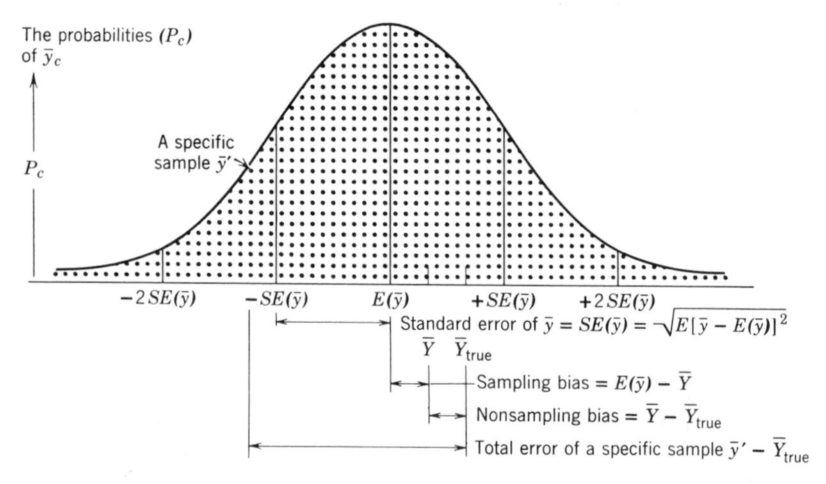

FIGURE 1.3.I Schematic View of a Sampling Distribution
We can think of each dot as a possible sample value, or as a group of equal numbers of values with similar values. The horizontal scale represents the possible values $\bar{y}_c$ of the statistics, and the vertical scale represents the probability P_c of occurrence of those values. The normal bell-shape is approached with moderate sized samples for most statistics, and for most variables ordinarily encountered, although these are typically not normal. The standard deviation of the sampling distribution is the standard error of the statistic. Arbitrary points were chosen to illustrate the population value $\bar{Y}$, the true value $\bar{Y}_{true}$, and a specific sample value $\bar{y}'$. The standard error may more properly be called "variable error" if it includes variable errors from nonsampling sources.

errors form important parts of the sections devoted to specific designs. These are denoted as var $(\bar{y})$ and se $(\bar{y})$ for sample estimates of the variance and the standard error, respectively, of the estimator $\bar{y}$. Subscripts under $\bar{y}$ often distinguish the kind of sample designs for which specific formulas are appropriate.

To be precise, an *estimator* should be distinguished from a particular *estimate* (or statistic, or sample value) for a specific sample. A *sample design* specifies the selection and estimation methods; for example, the mean of a simple random sample. A sample design with specified sample size, applied to a population of characteristics defined by survey operations defines an estimator and gives rise to the *sampling distribution of an estimator*. The mean and the standard deviation of the distribution are the expected value and standard error of the estimator. Both of these are properties of the estimator ($\bar{y}$), and not of any specific sample estimate ($\bar{y}_c$). Strictly speaking, then, se ($\bar{y}$) is the "estimated standard error of the estimator $\bar{y}$." But we shall often say briefly, "standard error of the estimate" or "computed standard error."

The mean square error of the sampling distribution is related to its variance, but with the deviations taken around the population value $\bar{Y}$:

$$\text{MSE}(\bar{y}) = \sum_c P_c(\bar{y}_c - \bar{Y})^2 = \text{Var}(\bar{y}) + [E(\bar{y}) - \bar{Y}]^2. \tag{1.3.3}$$

The added term is the square of the sampling bias, and it vanishes for unbiased sample designs. When not zero, the sampling bias is small in most well-designed samples, and it tends to diminish with increasing sample size. This property resembles the consistency property in mathematical statistics.

In many practical situations the sampling distribution of the estimated mean is approximately normally distributed. Just how good that approximation is depends on the underlying distribution of the characteristics in the population and on the sample design; and the approximation improves with increasing sample size.

This approach to normality of the sampling distribution of large samples is not based on the normality of distribution of the elements in the population. On the contrary, the distributions of survey characteristics in the population are usually far from normal. They are more often of the kinds pictured in Fig. 1.3.II, and the distributions of elements in the samples selected with equal probability tend to resemble that in the population. Nevertheless, the sampling distributions of means of even moderate size samples, perhaps 100 or perhaps 1000 cases, will often be close enough to normal for the practical purpose of making statistical inference from sample estimates to population values. Furthermore, the approach to normality of sampling distributions for large samples occurs not only for the mean but for most estimators commonly used to present survey results. This problem is the concern of Central Limit Theorems.

The term bias covers three concepts that have diverse origins and different practical consequences. Although, formally, each denotes a difference $[E(\bar{y}) - \bar{Y}_{\text{true}}]$ between the expected value of a statistic and its "true" population value, they differ in important ways. First, the mathematical statistician thinks of biases due to using n for $n - 1$, or s to estimate σ, or the median to estimate the median of a skewed distribution, or y/x to estimate Y/X. These may have academic interest, but they are of little concern for survey samples and will be treated only in a few specific remarks.

Second, large selection biases frequently result from the improper use of imperfect selection frames. These errors can mostly be avoided by any skilled sampler, yet they are amazingly common. This kind of common and avoidable selection bias will be our chief concern throughout the book, and especially in Sections 2.8 and 11.1 to 11.5.

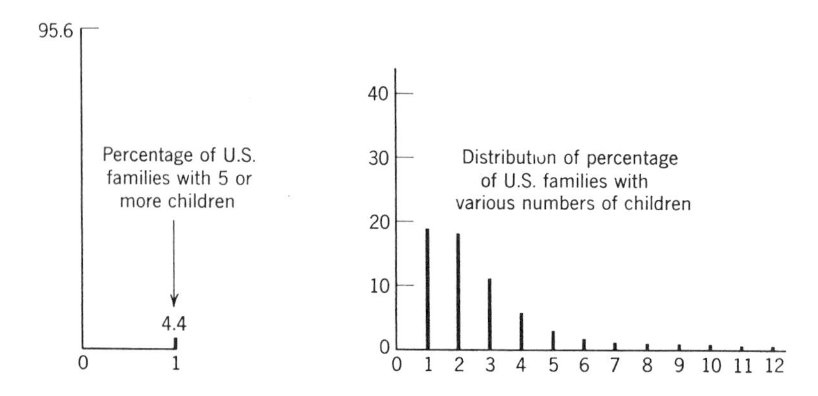

FIGURE 1.3.II Two Illustrations of Population Distributions
The cases in the sample will tend to have similar nonnormal distributions. But the sampling distribution of statistics based on moderate or large samples (for example, sample means based on a few hundred cases) will be approximately normal.

Third, many surveys have large biases of observation that are difficult either to reduce or to assess. The sampler cannot remain indifferent to this central problem of research. However, the problem does not lie strictly within his competence and is treated separately in Chapter 13.

1.4 STATISTICAL INFERENCE IN SURVEYS

The theory of sample surveys has concentrated on standard errors of estimates, especially of means and aggregates, for complex sample designs. That is also a central concern of this book. This concentration is justified because statistical inference is based on standard errors. Statistical inference from surveys typically takes the form $[\bar{y} \pm t_p \operatorname{se}(\bar{y})]$. Briefly, this denotes the statement that the population mean $\bar{Y}$ is within the interval $[\bar{y} - t_p \operatorname{se}(\bar{y})]$ to $[\bar{y} + t_p \operatorname{se}(\bar{y})]$; and the probability P of that statement is a function (usually and approximately normal or Student) of the chosen constant t_p.

In Example 3.3*a* the mean is $\bar{y} = 0.20$, and the standard error se $(\bar{y}) = 0.017$. Then, choosing $t_p = 2.6$, we may say, with about $P = 0.99$ confidence, that the population mean is between the limits

$$0.20 \pm 2.6(0.017) = 0.20 \pm 0.044 = 0.156 \text{ to } 0.244.$$

If we are satisfied with $P = 0.95$ confidence and use $t_p = 2$, then we can use the narrower limits $0.20 \pm 2(0.017) = 0.20 \pm 0.034 = 0.166$ to 0.234.

1	$P = 2P' - 1$; two sides	0.50	0.68	0.80	0.90	0.95	0.99	0.9973	0.999
2	$P' = 1/2 + P/2$; one side	0.75	0.84	0.90	0.95	0.975	0.995	0.9986	0.9995
3	t_p normal (df $= \infty$)	0.67	1.00	1.28	1.64	1.96	2.58	3.00	3.29
4	t_p Student (df $= 30$)	0.67	1.03	1.31	1.70	2.04	2.75	3.27	3.65
5	t_p Student (df $= 10$)	0.70	1.05	1.37	1.81	2.23	3.17	3.96	4.59

TABLE 1.4.I Selected Values of Normal and "Student" Deviates

Probability levels of two-sided (1) and one-sided (2) intervals. Normal deviates (3) and "Student" deviates with 30 degrees (4) and with 10 degrees (5) of freedom.

Frequently, we may prefer to construct one-sided intervals. Thus we may state that $\bar{Y}$ is greater than $\bar{y} - t_p$ se $(\bar{y})$; or, alternatively, that $\bar{Y}$ is less than $\bar{y} + t_p$ se $(\bar{y})$. For these one-sided intervals the probabilities are $P' = 0.5 + 0.5P$ (or $P = 2P' - 1$), where P denotes the probability for two-sided intervals corresponding to t_p. In the above example, we may say with $P' = 0.5 + 0.5(0.95) = 0.975$ confidence *either* that the population mean is greater than 0.166, *or* that it is less than 0.234.

The length of the interval depends on the standard error and, through t_p, on the probability level (P or P'). Choosing any two factors determines the third. The reader is assumed to have acquired the necessary statistical knowledge to use and to choose proper values of the standard error.

Statisticians agree that it is good survey practice to publish standard errors together with survey results. This permits the reader to construct intervals and to make inferences according to his needs. Statisticians are well agreed also on the approximate amount of information conveyed by standard errors computed from survey samples, and on the practical procedures for using them to construct intervals.

There is considerable disagreement about the meaning of those intervals. I avoid any adjective for the intervals t_p se $(\bar{y})$, and the reader will utilize them according to his knowledge. If he wants confidence intervals, he should know that for most statistics based on complex samples the standard errors yield only approximate, not exact, confidence intervals.

Others will prefer to think of credible or likelihood or fiducial intervals (14.3).

The interval $[\bar{y} \pm t_p \text{ se } (\bar{y})]$ includes symbols for the sample mean, but its place can be taken by many other statistics. We could write

$$[\bar{v} \pm t_p \text{ se } (\bar{v})],$$

where $\bar{v}$ stands for some unspecified statistic and se $(\bar{v})$ is its computed standard error. We should also note that se $(\bar{v})$ is the estimate of the standard error of the entire sampling distribution, and not of the specific $\bar{v}$ obtained from the sample. Typically, se $(\bar{v})$ is computed from the same data as $\bar{v}$; but this is not a mathematical requirement, and exceptions do arise in practice. For example, se $(\bar{v})$ may be better known or estimated from other sources; or it may be estimated from only part of the entire sample.

Using the normal deviates t_p for constructing the interval $[\bar{y} \pm t_p \text{ se } (\bar{y})]$ involves the assumption that the normal distribution is a good approximation for the sampling distribution of the statistic $\bar{y}$. The normality of the sampling distribution depends on Central Limit Theorems for statistics based on large samples. (The approach to normality for distributions of statistics based on large samples *does not* require normality for the variables in the population distributions.) To the degree that the sampling distribution departs from the normal, the interval fails to possess the intended probabilities. I believe that for most sample estimates encountered in practical survey research, assumptions of normality lead to errors that are small compared to other sources of inaccuracy.

For two reasons complex probability samples are typically large, from several hundred to thousands of cases. First, the organization and operations needed to obtain complex samples of large populations are so ponderous and expensive that only large samples can justify their economical utilization. Second, most complex survey samples are aimed at measuring effects that are too small for detection with small samples.

Complex probability samples are essentially tools for large samples. Small samples from small populations can be selected with simple random sampling, and then the vast available theory, parametric and nonparametric, can be resorted to. Small samples from widespread populations may be too difficult to select with either complex or simple random sampling. Then purposive selection can replace probability sampling, which is essentially a tool for large samples (1.7 and 14.4).

Nevertheless, serious departures from normality can occur. When normality cannot be assumed, remedies may be sought in the statistical literature. Perhaps a *transformation* of the variables will bring the statistics closer to normal. Or a nonnormal distribution can be found and applied to the sampling distribution of the statistics. Or a distribution-free, or nonparametric, method can be applied. But most of these are not easily applicable, because they assume simple random selection (14.2).

Criteria for judging the seriousness of departures from the normal would be difficult to define. Searching for clues of serious departures is also difficult, but several factors may be mentioned:

1. Departures from normality are proportionately greater at the higher probability levels, corresponding to the tails of the sampling distribution. An artificial example of a typical situation should clarify our meaning. At $t_p = 3.29$ instead of the expected $P = 0.999$, nonnormality may give $P = 0.995$ for an increase factor of $0.005/0.001 = 5$ in the error rate. At $t_p = 1.96$, a change from $P = 0.95$ to 0.945 would mean an increase factor of $0.055/0.05 = 1.1$ in the error rate.

2. The approach to normality of the sampling distribution of the statistic is faster for some variables than for others. Faster approach to normality means a better approximation of the sampling distribution for a given sample size, or a desired approximation reached with a smaller sample size. Even for small samples, the approximation is good for many common variables. For example, 30 to 200 element selections suffice for proportions between 20 and 80 percent [Cochran, 1963, p. 57]. Contrariwise, skewed distributions can result in serious departures from normality even for moderate size samples. For example, special methods are needed for analyzing income and wealth, and for most statistics of business and other establishments and organizations (11.4B).

3. Subclasses even from large samples can present problems. First, suppose that the subclass consists of a few primary clusters, each of which contains a fair number of elements. The analysis then depends on rough normality for the sample values of the individual clusters. Then the values of "Student's" t_p are used if the number of primary clusters is small. Degrees of freedom refer to numbers of primary clusters (less the number of strata) if the sample sizes within the clusters do not differ greatly (8.6D). Second, suppose that the subclass consists of a few elements spread individually over the entire breadth and width of the sample. Then, one can conjecture that the sample resembles a random choice, and refer to the vast literature for random selection.

1.5 A TAXONOMY OF SELECTION PROCEDURES

Any 100 percent census can be regarded as a sample for two reasons. To the degree that it is subject to errors of observation, the population value of a census is only one of many that could have resulted from essentially the same operations. Second, the particular population is arbitrarily specified from a universe of interest that is usually greater as to time, space, and perhaps other dimensions. For example, the decennial census singles one day out of the 10 × 365 days of a 10-year period. Nevertheless, we should separate these problems from the province of sampling proper, and consider the population from which a sample is drawn as a specified, entire population.

Survey sampling, or population sampling, deals with methods for selecting and observing a part (sample) of the population in order to make inferences about the whole population. A sample can have several advantages over a complete census: (1) economy; (2) speed and timeliness; (3) feasibility (if the observation is "destructive," a census is not practical); (4) quality and accuracy (in some situations money simply cannot buy the trained personnel and supervisors for a good census, or even a large sample).

On the other hand, complete censuses possess special advantages in some situations. (1) Data for small units can be obtained. (2) Public acceptance is easier to secure for complete data. (3) Public compliance and response may be better secured. (4) Bias of coverage *may* be easier to check and reduce. (5) Sampling statisticians are not required [Yates, 1960, p. 2; Zarkovich, 1961, Chs. 1 and 2].

An important field of sampling that lies outside the scope of survey sampling is quality control. This deals with acceptance sampling by lot inspection; that is, sampling each lot to judge if it conforms to the quality specified for the entire universe. For this, frequent use is made of sequential sampling, where the sample size depends on the results of successive selections. But the collecting, processing, and analyzing procedures of surveys are generally lengthy and cumbersome, ill-suited to sequential sampling (8.4D). Survey sampling concentrates on the study of one-step probability samples for estimating population values.

We frequently make inferences about populations from arbitrary and informal samples: we judge a basket of grapes by tasting one of them; a buyer accepts a shipment after inspecting a few items he picks haphazardly. Much research in the physical and biological sciences is based on items picked in a haphazard manner. The researchers assume, vaguely and implicitly, that "typical items" were chosen. They hope that the important characteristics are distributed either uniformly or randomly in the population. These are simple examples of *model sampling*, which we define generally as sampling based on broad assumptions about the distribution of the survey variables in the population. Several forms of model sampling can be distinguished.

1. *Haphazard* or *fortuitous* samples form the bases of most research in many fields. Samples of *volunteer subjects* should be included here. Archaeology, history, and often medicine draw conclusions from whatever items come to hand. Astronomy, experimental physics, and chemistry are leading sciences which display little care about the representativeness of their specimens.

2. *Expert choice* is a form of *purposive* or *judgment sampling* used by experts to pick "typical" or "representative" specimens, units, or portions. For example, consider the judgment sampling of "representative" individuals or specimens for experiments; or the accounting practice of choosing typical weeks for auditing ledgers; or the common practice of picking a typical city or village to represent a national urban or rural population. Experts often hold differing views on the best way to choose representative specimens, or to decide which are the most representative units.

Sometimes the researcher asks that, instead of a real population, a hypothetical universe be postulated as the parent of the sample. Some believe that a hypothetical universe suffices for theoretical models. But I think that inference from empirical data to a hypothetical universe does not lead to useful results if the gap between that universe and any real population is too great (14.4).

3. *Quota sampling* is a form of purposive sampling widely used in opinion, market, and similar surveys. The enumerators are instructed to obtain specified quotas from which to build a sample roughly proportional to the population, on a few demographic variables. Within the quotas, the enumerators are supposed to obtain representative individuals. The nature of the controls and instructions depends on the expert judgment of the practitioner (13.7).

4. *Sampling of mobile populations* often depends on "capture-tag-recapture" methods. The total population is estimated from the proportion in the recapture of individuals (insects, fish, deer) which have been previously captured and tagged. Ingenious theoretical models are used to state explicitly the assumptions of the method (Remark 11.1.I).

These different types of model sampling vary greatly as to degree and area of justification. However, they have in common a heavy dependence on the validity of broad assumptions about the distributions of the survey variables in the population. On the contrary, from the results of ideal probability sampling, the inferences to the population can be made entirely by statistical methods, without assumptions regarding the population distributions. The need for assumptions of randomization of the population is by-passed by introducing randomization into the selection procedures. Similarly, in card playing and in lotteries, instead of relying on the dealer's judgment for a fair distribution, we insist on open and thorough shuffling.

In probability sampling, every element in the population has a known nonzero probability of being selected. This probability is attained through some mechanical operation of randomization (1.7). Its value is determined in accord with the demands of the sample design. Probability samples are usually designed to be *measurable;* that is, so designed that statistical inference to population values can be based on *measures of variability*, usually standard errors, *computed from the sample data.*

The desired and idealized properties of probability samples can only be approached, because many imperfections creep into the actual

I. *Epsem:* equal probability for all elements (*a*) Equal probabilities at all stages (*b*) Equal overall probabilities for all elements obtained through compensating unequal probabilities at several stages	*Unequal probabilities* for different elements; ordinarily compensated with inverse weights (*a*) Caused by irregularities in selection frames and procedures (*b*) Disproportionate allocation designed for optimum allocation
II. *Element Sampling:* single stage, sampling unit contains only one element	*Cluster Sampling:* sampling units are clusters of elements (*a*) One-stage cluster sampling (*b*) Subsampling or multistage sampling (*c*) Equal clusters (*d*) Unequal clusters
III. *Unstratified Selection:* sampling units selected from entire population	*Stratified Sampling:* separated selections from partitions, or strata, of population
IV. *Random Selection* of individual sampling units from entire stratum or population	*Systematic Selection* of sampling units with selection interval applied to list
V. *One-Phase Sampling:* final sample selected directly from entire population	*Two-Phase (or Double) Sampling:* final sample selected from first-phase sample, which obtains information for stratification or estimation

TABLE 1.5.I A Taxonomy of Probability Selection Methods. Five Alternatives That May Be Combined.

execution of practical samples. Hence, inference from sample values to population values also involves assumptions regarding the possible effects of imperfections; but often we can investigate, successively reduce, and put some limits on the possible or likely effects of imperfections. Therefore, I believe that separating probability sampling from the different types of model sampling is justified and useful.

Simple random sampling (srs) is the basic selection process, and all other procedures can be viewed as modifications of it, introduced to provide more practical, economical, or precise designs. There are five major types of modifications and, since any of these can occur with any other, a large variety of possible designs appear immediately. This variety increases rapidly in multistage samples, where the choices must be made at each stage of selecting sampling units; for example, counties, blocks, dwellings. To these varieties must be added the possible diversity in estimating procedures. The great flexibility provided by different sampling methods is further increased with the choice of different kinds of frames and practical procedures:

1. *Epsem* (equal probability of selection method) *sampling* describes any *sample in which the population elements have equal probabilities of selection*. Note that this is a special type of probability sampling and, in turn, *srs* is a special type of epsem. *Epsem is used widely because it usually leads to self-weighting samples*, where the simple mean of the sample cases is a good estimate of the population mean (2.8.3). Epsem sampling can result either from equal probability selection throughout, or from variable probabilities that compensate each other through the several stages of multistage selection. In contrast to epsem selection, some designs—we call them *loaded*—call for variable probabilities of selection for different kinds of elements.

2. In *element sampling* (Chapters 2 to 4) the elements are also the only sampling units. *Cluster sampling*, on the contrary, involves the selection of groups, called *clusters* of elements, as sampling units. *Subsampling* of the clusters results in *multistage sampling*: where the selection of the elements results from selection of sampling units in two or more stages (Chapters 5 to 10).

3. *Stratification* denotes selection from several subpopulations, called *strata*, into which the population is divided.

4. *Systematic selection*, an alternative to random choice, denotes the selection of sampling units in sequences separated on lists by the interval of selection. Briefly, we select every kth sampling unit.

5. *Two-phase sampling*, or *double sampling*, refers to the subselection of the final sample from a preselected larger sample, that provides information for improving the final selection. *Multiphase sampling* refers to the possibility of more than two phases of selection.

1	2	3	4	5	6	7	8
$y_1y_2y_3$	y	Σy_j^2	var (y)	$y - fY$	$y - fY$	$y - fY$	$y - fY$
013	4	10	3.5	−8			−8
015	6	26	10.5	−6			
016	7	37	15.5	−5			
019	10	82	36.5	−2			
035	8	34	9.5	−4			
036	9	45	13.5	−3	−3	−3	
039	12	90	31.5	0	0		
056	11	61	15.5	−1	−1		
059	14	106	30.5	2	2		
069	15	117	31.5	3			
135	9	35	6	−3			
136	10	46	9.5	−2	−2		
139	13	91	26	1	1		
156	12	62	10.5	0	0		
159	15	107	24	3	3	3	
169	16	118	24.5	4			
356	14	70	3.5	2			
359	17	115	14	5			
369	18	126	13.5	6			
569	20	142	6.5	8			8
Mean	12	76	16.8	0	0	0	0
Var $(y) = E(y - fY)^2$				16.8	3.5	9	64

TABLE 1.5.II Numerical Illustration of a Sampling Distribution

1. The *population* of $N = 6$ elements have values $Y_i = \{0, 1, 3, 5, 6, 9\}$. The population total is $Y = \Sigma Y_i = 24$ and the population mean is $Y/N = 24/6 = 4$. From $\Sigma Y_i^2 = 152$, the element variance is $S_y^2 = (\Sigma Y_i^2 - Y^2/N)/(N-1) = (152 - 24^2/6)/5 = 56/5 = 11.2$.
2. *A simple random sample* of $n = 3$ elements results in $f = n/N = 3/6 = 0.5$. The *sample total* $y = \Sigma y_j$ is the estimator of $fY = nY/N = n\bar{Y}$. It has the variance $\text{Var}(y) = (1 - n/N)nS_y^2 = (1 - 0.5)3(11.2) = 16.8$. The *sampling distribution* contains $\binom{N}{n} = \binom{6}{3} = (6 \times 5 \times 4)/(1 \times 2 \times 3) = 20$ possible samples, in column 1. Each of the elements appears in $f = 0.5$ of the samples.

   The *sample estimates* of y in column 2 are seen to vary. Their *average or expected value* is 12, illustrating that $E(y) = fY$, that y is an unbiased estimate of fY. Then y/f is an unbiased estimate of Y, and y/n of $\bar{Y}$. Note in column 3 that the expected value of Σy_j^2 is 76, equal to $f\Sigma Y_i^2$, hence also an unbiased estimate.

   The computed variances in column 4 vary greatly, but their expected value is 16.8; thus $E[\text{var}(y)] = \text{Var}(y)$ is another unbiased estimate. The deviations $(y - fY)$

1.6 CRITERIA OF SAMPLE DESIGN

Survey sampling deals mainly with departures from simple random sampling. Departures from simplicity must be justified by strong considerations. A good sample design requires the judicious balancing of four broad criteria:

1. *Goal orientation.* The entire design, both selection and estimation, should be oriented to the research objectives, tailored to the survey design, and fitted to the survey conditions. These considerations should influence the choice and definitions of the population, the measurement, and the sampling procedures. Although these admonitions seem vague, they are not trivial. For example, too often a neat sample of a single unit is used, such as a town, or a tract, or a class of students, when the research objectives would be better met with a less neat sample of a larger population (14.4).

2. *Measurability* denotes designs which allow the computation, from the sample itself, of valid estimates or approximations of its sampling variability. This is ordinarily expressed with standard errors in surveys, but sometimes other expressions of the likelihood function or the sampling distribution may serve. This is the necessary basis for statistical inference, serving as the objective, scientific bridge between the sample result and the unknown population value. Other methods for judging the adequacy of samples depend on personal judgment. A nonmeasurable sample, such as a quota sample or a typical city, may represent a population well or poorly; but statistical theory alone does not suffice for judging its precision. The theory of statistical inference depends on probability

---

in the sampling distribution are shown in column 5, also that $E[y - E(y)] = 0$, because y is an unbiased estimate. Their variance $E[y - E(y)]^2 = 16.8$, gives the variance of the estimate directly.

3. Columns 6, 7, and 8 contain the sampling distribution of three different epsem samples, each with $f = n/N = 0.5$. Each of them represents a *restriction* on simple random selection, a *control* of the selection that suppresses some of the possible combinations. Note that in each design, every element occurs in $f = 0.5$ of the samples.

   In column 6 appears the proportionate *stratified* sample, one selection from the three strata (0, 1), (3, 5), (6, 9). There are $2^3 = 8$ possible samples. In column 7 a *systematic* sample was taken with the interval $1/f = F = 2$. There are $F = 2$ possible samples only. Column 8 shows a *clustered* sample selected with the fraction $f = 1/2$ from the $A = 2$ clusters (0, 1, 3), (5, 6, 9).

   Each of the epsem methods yields unbiased estimates y of fY, and $E(y - fY) = 0$ for each. But note that the variances of y, computed directly from the sampling distributions, show that $\text{Var}(y) = E[y - E(y)]^2$ is different for the diverse designs.

samples, that is, where the probability of selection of every element of the population is known.

While only probability samples are objectively measurable, probability sampling does not guarantee measurability automatically. Randomized replication is needed, with variance computations that follow faithfully the replicated design. Generally, selecting and identifying at least two sampling units at random from each stratum permits the valid computation of standard errors. This book is devoted to measurable designs, and the reader will also be warned against nonmeasurable probability samples. Three examples are: selecting a single cluster, a systematic sample from a population with periodic variation, and any cluster sample in which the primary clusters are not identified.

3. *Practicality* refers to problems in accomplishing the design essentially as intended. A probability sample cannot be created by assumption, nor will it be "given," as in theoretical problems. The dictum of quota samplers to their interviewers, "go out and get a random sample," is most impractical; neither the interviewer nor his dispatcher can do it. Care is needed to translate the theoretical selection model into a set of instructions for the office and the field. These instructions should be *simple*, *clear*, *practical*, and *complete*. For example, to identify a sample segment, the interviewer should not be asked to locate a long arbitrary straight line marked on a map; his duties should be confined to locating streets and addresses. Moreover, clear, simple, practical, and complete instructions represent aims, because practical work, particularly in the field, cannot be done without errors. Simplicity should always be among the aims, because it reduces the risks of errors and it is worth the sacrifice of some theoretical efficiency. Simplicity must be sought not in the abstract, but within the actual operating conditions. Each actual design represents an adaptation of sampling theory to the problems and resources at hand (8.4).

Practical work consists in good part of guessing what irregularities, where, and how much, one can afford to tolerate. The art of sampling involves making the practical design conform well, even if not perfectly, to a model. Its central problem concerns the proper construction and use of selection frames or population lists.

4. *Economy* concerns the fulfillment of survey objectives with minimum cost (effort) and the degree of achieving that aim. Efficiency denotes this criterion in some sampling books; but in general statistical literature efficiency measures the number of sample elements for a fixed precision. That number, although important, is not ordinarily the only cost factor in survey sampling (8.3).

Research objectives are commonly stated in terms of the *precision*, the inverse of the variance of survey estimates. *Accuracy* is the inverse of the total error, including bias as well as the variance. If important biases, especially nonsampling errors, are present and distinguishable, accuracy is a better measure of survey objectives than precision alone (13.1).

A sample is too small if its results are not precise enough to make appreciable contributions to decisions. For example, it would be generally useless to forecast that one of two parties in an election will win from 40 to 60 percent of the votes. On the other hand, a sample is too large if its results are more precise than is warranted by their likely uses; or if the nonsampling errors overwhelm the sampling precision. This suggests some rational decision function that considers for any degree of precision the cost of the sample against its value for the survey objectives, in the light of information from other sources. A complete and formal statement of this function is often too difficult. Instead, we may be able to *fix the desired precision* that the sample result should have, and then determine the sample design and size that would yield it *for a minimum cost.* In most survey problems the research aims are many and indefinite; hence, only vague ideas surround the precision required from the sample results, but the available funds may be fixed between rigid bounds. Then the sample design should aim to obtain *maximum precision* (minimum variance) for the *fixed allowed cost.* These two basic approaches give similar numerical results, hence discussing precision or economy amounts to the same thing. A sample is *economical* if the precision per unit cost is high (the variance for unit cost is low), or the cost per unit of variance is low. Comparisons of precision are confined to samples that are measurable, because precision has been defined only for them.

The four criteria frequently conflict, and the sampler must balance and blend them to obtain a good sample design. No unique definition exists for a good or desirable sample. The following favorable adjectives for samples are often used and confused. Yet they all possess distinct meanings, and no single one is both necessary and sufficient to denote a good sample.

Probability samples require nonzero *known probabilities* of selection.

Measurable samples are probability samples, so designed that they permit estimating the sampling variability, ordinarily the standard errors, of the estimates.

Epsem samples, special kinds of probability samples, require *equal probabilities* for the elements.

Area samples use area segments as sampling units. It often serves as a procedure of probability sampling, when the selection frame consists of segments.

Unbiased samples denote *designs* (selection plus estimation) for which the average or *expected value is equal to the population value.*

Precise samples have low standard errors. The precision must be judged against the requirements of survey objectives.

Accurate samples have low total errors, including biases and non-sampling errors, together with sampling variability.

Economic samples have low unit costs for fixed unit precision (or variance).

Efficient samples denote high precision (low variance) per element in general statistical literature, but in many sampling books the word *efficient* has the same meaning as *economic* does here.

We avoid the term *random sampling* because of the different meanings it has in different contexts. In most statistics it refers to simple random sampling, usually with replacement. In sampling literature it often denotes probability sampling. In everyday language it stands for haphazard choice. We let the term *random choice* mean simple random selection among the sampling units without replacement, unless replacement is specified.

Representative sampling is a term easier to avoid because it is disappearing from the technical vocabulary. At different times it has been used for random sampling, proportionate sampling, quota sampling, and purposive sampling. In general, it often denotes the aims of representing a population well with a sample; and this is the sense of the terms *population sampling* and *survey sampling* in our vocabulary.

Sampling is synonymous in this book with survey sampling, but sometimes it appears in the common sense of selecting. Some samplers restrict the term to probability sampling. In statistical literature it is generally synonymous with random sampling, and both refer to unrestricted random sampling. Contrariwise, in common parlance it means vaguely taking any portion.

1.7 MECHANICAL SELECTION, RANDOMIZATION, AND FRAMES

Probability sampling requires that the actual selection of units into the sample be made by a *mechanical procedure* that assigns the desired probabilities. This *randomization* process requires a practical physical operation which is exactly or reasonably congruent with a probability model. Statistical pictures of "tossing perfect coins" or "drawing perfect balls from a completely mixed urn" describe intuitively the nature of the needed physical operation. But this naive picture needs modification because it is impractical to mix, toss, shuffle, or draw the kind of elements which compose actual survey populations. Instead, we could assign numbers to the elements, list these numbers, then mix and draw uniform objects bearing the identification numbers of the elements—as in a lottery. But it is difficult to construct and operate a fair lottery. Its equivalent has been performed by careful technicians who put their results in *tables of random numbers*, our practical equivalent, twice removed, of mixing the elements. Thus the mechanical operation of selection, indispensable for probability sampling, is achieved in practice as follows: a set of numbers, properly selected from a good table of random numbers, identifies a set of numbers on a list of sampling units; from the selected listing units identification is made to a set of physical units which will comprise the actual sample.

Hence the key function of lists arises eventually from the need for statistical inference through a chain in which each of the links depends on the next link: statistical inference → measurability → probability sampling → mechanical selection → lists or frames.

The word *frame* is used widely for sampling lists because when the actual physical listing of all sampling units in the population is too difficult, an equivalent procedure can be substituted for it. The nature of available or feasible frames is an important consideration in the sample design. Relevant factors include the types of sampling units, extent of coverage, the accuracy and completeness of the list, and the amount and quality of auxiliary information on the list. This last factor can be useful, as we shall see, for stratification, for measures of size, and in the estimation process. All these factors influence the sample design and the selection procedures.

Practical problems arise in identifying elements associated with selected listings. For example, identifying members of a household, or of a family, requires some care. Problems of field procedure must be solved with clear, simple, and practical instructions, not merely assumed. Some tasks may be easy, such as the identification of employees from a payroll list, or students from a class roll. Other tasks, such as identifying dwellings from block listings, may call for skill and detailed instructions—and mistakes may still occur. Frame problems are discussed in many sections, especially in 2.7 and 11.1 to 11.5.

Area sampling is an important type of frame widely used in social research, because of the relative dependability of identifying every member of a human population with one, and only one, dwelling. Dwellings, in turn, can be identified uniquely with area segments. Area sampling is also used for many other types of surveys, for crops and other flora, grocery stores, rocks and soil types (9.1).

Probability sampling is important for three reasons: (1) Its measurability leads to objective statistical inference, in contrast to the subjective inference from judgment sampling. (2) Like any scientific method, it permits *cumulative* improvement through the separation and objective appraisal of its sources of errors. (3) When simple methods fail, researchers turn to probability sampling; this explains, it seems to me, both the recent interest in probability sampling in the social sciences and the lack of it in physical and biological sciences.

Probability selection demands randomized selection. When randomization is both simple and important, disregarding it amounts to carelessness or ignorance. These vices are easy to deride, and objects of derision abound all around us.

More challenging are the many situations where randomization is expensive. Then its value must be balanced against the sacrifices it entails in higher cost, and often in decreased control over the measurements and experimental variables (14.4). Hence, in most fields, probability sampling has been avoided as long as possible by persons operating behind three lines of defense. First, sampling is unimportant if the elements are uniform for example, all hydrogen atoms of weight 1 can be considered equal. Second, lacking uniformity, sampling may still be avoided if the predictor variables can be measured and controlled; for example, it is easy to control for sex in selecting individuals. Third, an uncontrollable variable, if randomly distributed in the population, yields a random sample under any selection design.

Such automatic population randomization was, and is, the hope and unstated assumption behind the many past and current examples of purposive and judgment samples. Since the sample selections are not uniform, uncontrolled variability is perceived, but it is hoped that the variation is randomly distributed in the population. Sometimes this hope is buttressed with such statements as: "I know of no reasons for doubting randomness." But ignorance of reasons would prove the absence of causes only for an omniscient being.

Great advances of the most successful sciences—astronomy, physics, chemistry—were, and are, achieved without probability sampling. Statistical inference in these researches is based on subjective judgment about the presence of adequate, automatic, and natural randomization in the population. This is relied on, not only for avoiding serious biases, but also to provide the model for the distribution of sampling variability. Scientific research abounds in successful results of assumptions of natural randomization in the population.

On the other hand, the literature of some sciences—especially the social, medical, and biological sciences—is replete with cases where assumptions of natural randomization failed openly and badly. There must be many more failures that remain undetected.

Only after the fallacies of random assumptions are revealed by observed selection biases in a specific research area, do some researchers manifest interest in probability sampling. In many situations the assumption of random distribution of variables in the population seems to work well enough. For example, for most physical and chemical experiments the selection of specimens does not seem to require special care. In the biological sciences the picture is mixed. At the other extreme, in the social sciences the distribution of characteristics is typically far from random. It is in these fields that probability sampling is most needed, as well as most developed.

No clear rule exists for deciding exactly when probability sampling is necessary, and what price should be paid for it. The decision involves scientific philosophy and research strategy, discussed in 14.4. The relationship of sample size to probability selection is merely mentioned here. If a research project must be confined to a single city in the United States, I would rather use my judgment to choose a "typical" city than select one at random. Even for a sample of 10 cities, I would rather trust my knowledge of U.S. cities than a random selection. But I would raise the question of enlarging the sample to 30 or 100 cities. For a sample of that size a probability selection should be designed and controlled with stratification. A classical example is Neyman's [1934] proposal for a stratified random selection of about 1400 communes, instead of a purposive judgment selection of 29 large districts as a representative 15 percent sample of Italy's 1921 census.

Probability sampling for randomization is not a dogma, but a strategy, especially for large numbers. Just what a large number is depends greatly on circumstances, and it may begin with 4 or 10 or 100. Sample size enters the decision to balance uncontrolled random variation against the likely bias of judgment selection. Serious biases of subjective selections have been demonstrated time and again [Yates, 1960, 2.4; Kendall and Stuart, I, 1958, 9.1–9.5]. Personal judgment has been shown inadequate for selecting random samples of integers, or stones from a pile, or plants from a field, or people on streets or in homes.

Remark 1.7.1 Several good tables of random numbers exist. Their defects, if any, are likely to be negligible compared to other sources of survey errors. Descriptions of tables, of methods, and of problems of construction appear in statistic books [Kendall and Stuart, 1958, 9.11–9.18] and in the introduction to *A Million Random Digits* [RAND, 1955]. The *Cambridge Tables for Random Numbers* [Kendall and Babington Smith, 1954] is an inexpensive set of 100,000 numbers, from which a page is reproduced in Appendix D. Several statistical tables also contain a few pages of random numbers. To attain good results the tables must be used properly, because careless selection can easily lead to mistakes. Proper usage can be learned easily. But I think that rather than with abstract concepts, instruction is best imparted with the practice of selection and with problems (see Problem 1.13).

PROBLEMS

1.1. Comment briefly on changes in sampling concepts since the following appeared in a report by Jensen at the 1926 meeting of the International Statistical Institute.

"We have briefly described the two methods which are comprehended under the common term of 'The Representative Method.' It will be seen that the difference between the two methods lies in the—in principle—quite different methods of selecting the sample.

"In the first, which we designate *Inquiry by Random Selection*, we may place complete dependence on Bernoulli's laws and their more modern development, subject only to the difficulties of carrying out in practice the strict rules which the application of these laws demands. The essential condition of this method is that we are concerned with a defined population or universe of persons or things to all of which we have access, and that we make a selection at random of some of these persons or things, in such a way that every unit in the universe has an equal chance of being selected, while the method of selection is completely independent of the characteristics to be examined. The degree of accordance to be expected from the results obtained from the examination of the sample with the corresponding quantities in the universe is then only a question of pure mathematical probability. Any precision required can be attained by including a sufficiently large number of units in the sample.

"In the second method, which we call the *Method of Purposive Selection*, we do not select individuals, but groups or districts. Here we have no definite mathematical guidance, but have rather to depend on sound judgment and general knowledge of the circumstances. Districts or groups are selected which together yield the same (or very nearly the same) averages or proportions as the whole country or population in respect of those quantities or qualities which are already a matter of statistical knowledge (e.g., from a population census). It is then assumed that in other respects the districts or groups taken together will be typical of the whole."

1.2. Comment briefly on this statement: "Only to the extent that the sample is truly representative of the population can the researcher be sure that the sample average is the same as the population average. Since he can *never* be absolutely certain that the sample is truly representative (for example, urban voters may be more Democratic than rural voters), the researcher allows a narrow range for error."

1.3. The population mean $\bar{Y}$ of the numbers of children per family is estimated with the self-weighting mean $\bar{y}$ of the numbers of children in a sample of $n = 1000$ families selected with epsem (equal probabilities) from a population of $N = 1{,}000{,}000$ families. Consider, in turn, three distributions:

(1) the distribution of $n = 1000$ families of the sample actually selected; (2) the distribution of the $N = 1{,}000{,}000$ elements in the population; (3) the sampling distribution of all possible sample estimates that could have been obtained with the same estimator (or sample design). For the means of each of the three distributions, state whether it is *exactly* equal to $\bar{Y}$: (*a*) necessarily and always; (*b*) in most cases; (*c*) seldom; (*d*) never; (*e*) sometimes, depending on (complete sentence); or (*f*) do not know (explain). Give the numbers of the three means (1, 2, and 3) and the code letters of the answer you choose for each.

1.4. State for each of the three distributions in 1.3 whether it is: (*g*) exactly normal; (*h*) close enough to normal for practical purposes and to pass eye examination; (*i*) so far from normal that it is obvious from a glance; or (*j*) do not know (explain). Again, give the numbers of the three distributions and the code letters of the answers you choose. You may explain briefly any of your choices.

1.5. Assume in 1.3 that the variance per family has been computed as

$$s^2 = \frac{1}{1000}\left[\sum_j^{1000} y_j^2 - \frac{\left(\sum_j^{1000} y_j\right)^2}{1000}\right] = 1.024$$

and that the researcher computed the standard error of the mean as

$$\text{se}\,(\bar{y}) = s/\sqrt{n} = \sqrt{1.024/1000} = 0.032.$$

What questions or objections would you raise?

1.6. "I know of scarcely anything so apt to impress the imagination as the wonderful form of cosmic order expressed by the 'Law of Frequency of Error.' The law would have been personified by the Greeks and deified, if they had known of it. It reigns with serenity and in complete self-effacement amidst the wildest confusion. The huger the mob, (and) the greater the apparent anarchy, the more perfect its sway. It is the supreme law of Unreason. Whenever a large sample of chaotic elements are taken in hand and marshalled in the order of their magnitude, an unsuspected and most beautiful form of regularity proves to have been latent all along."

Thus Tippett [1956, p. 135] quotes Galton in connection with a sampling distribution of means for samples of size $n = 10$ from a rectangular distribution ($Y_i = 1, 2, \ldots, 10$; $P_i = \frac{1}{10}$). Comment on the appropriateness of Galton's words to describe: (*a*) sampling distributions in general, and Tippett's in particular; (*b*) the distribution of the characteristics of *elements* in very large probability samples in general, and simple random samples in particular.

1.7. Some interviewers on national probability samples have complained that the sample blocks assigned to them do not properly represent their cities. They complain that their sample blocks (3 to 10 in number) are poorer than the "typical" blocks, that they overrepresent the poor and underrepresent the better parts of the city. Not one has ever complained that his sample has overrepresented the better blocks. Assume that the interviewers are not only honest but also observant; assume also that a new selection of blocks in each complaining city would probably improve the representation from that city. Describe the effect of such a strategy of reselection on the national sample.

1.8. Attack and defend, or explain and reconcile, these three statements: (a) A physicist, when asked if the distribution of heavy hydrogen was random among all hydrogen atoms, replied that he thought it was, because he could not readily think of a reason to the contrary. He could quote Jeffries [1948, 6.12]: "Variation is random until the contrary is shown; .... We saw before that progress was possible only by testing hypotheses in turn, at each stage treating the outstanding variation as random; .... The charge 'you have not considered all possible variations' is not an admissible one; the answer is, 'The onus is on you to produce *one*'." (*b*) "When the selections are made by judgment, inferences may be made only by judgment, not by the theory of probability." [Deming, 1960, p. 28.] (*c*) "Only by using probability methods can objective numerical statements be made concerning the precision of the of the results of the survey. It is necessary to be sure that the conditions imposed by the use of probability methods are satisfied." [Hansen, Hurwitz, and Madow, 1953, Vol. II, 1.6.]

1.9. In the two designs below discuss the decisions and whether the two decisions are inconsistent or not. (*a*) The research had to be confined to three blocks in one city. The blocks were chosen with judgment sampling. (*b*) A sample consists of three blocks each in 100 cities. Both cities and blocks within cities were selected with probability sampling.

1.10. "After spending January there, I found Cleveland winters much milder than reputed; therefore, I changed my opinion and plans and decided to move there." Comment on the statistical basis of this man's decision.

1.11. A sample of $n = 1000$ cases (elements) yields the sample totals $\Sigma y_j = 2000$, and $\Sigma y_j^2 = 40{,}000$. How large is the standard error of the mean $\bar{y}$ of the sample? What assumptions did you make? How might your answers change if the assumptions do not fit the actual situation?

1.12. A probability sample of 1500 families yields a mean of $\bar{y}_1 = 560$ with a standard error of $s_{\bar{y}} = 10$.

(*a*) You design a similar sample for one month later. How large do you think the standard error of the difference $(\bar{y}_1 - \bar{y}_2)$ is? What assumptions did you need to give that answer? How would your answer change if the assumptions were not met?

(*b*) If the second sample is designed for only 500 families, how large would the standard error of the difference $(\bar{y}_1 - \bar{y}_2)$ be? What assumptions did you need? How would your answer change if your assumptions were not met?

(*c*) From the first survey of 1500, two subclasses (*a* and *b*) are compared, each comprising $\frac{1}{8}$ of the entire sample. What is the standard error of the difference $(\bar{y}_a - \bar{y}_b)$ between the two subclasses? What assumptions did you make? If these were not met, how would your answer change?

1.13. In each case you want to assign an equal probability (epsem) to each listing indicated in parentheses, by drawing a random number *r* from a good table. Indicate for each procedure whether it does (*Y*) or does not (*N*) obtain epsem.

(*a*) (1–112): First select *r* from 1–56; then select *r* from 1–2 to decide between *r* and 56 + *r*.

(*b*) (1–112): First select *r* from 1–2 to decide between two groups 1–100 and 101–112; then select final *r* to select one *r* from the selected group.

(*c*) (1–1109): Select *r* from 1–10,000. If first digit is even, take the number to 1000 indicated by last three digits; if first digit is odd and the last three digits are between 001 and 109, take the number indicated between 1001 and 1109; if first digit is odd and the last three numbers are between 110 and 1000, discard it and select a new *r*.

(*d*) (67,084–68,192): Select *r* from 1–1109 and add 67,083.

(*e*) (67,084–68,192): Select *r* from 1–2000. If *r* is from 0084 to 1192, add 67,000 and take indicated number; otherwise, discard it and select new *r*.

(*f*) (1109 numbers spread between 61,000 and 68,000): Select four-digit *r* and add 60,000. If indicated number exists on the list, take it; otherwise, discard it and select new *r*.

(*g*) (1–17): Select *r* 1–100 and divide by 20. Use remainder if 1–17; otherwise, discard it and select new *r*.

(*h*) (1–17): Select *r* from 1–100 and divide by 17. Use remainder.

(*i*) You need one *r* from each of two populations: (1–17) and (1–63). Select *r* from (1–63). If selected *r* is (1–17), take the indicated number from first population, then select another (1–63) from second. If selected *r* is (18–63), take the indicated number from the second population, then select another (1–17) from the first.

(*j*) Again you need both (1–17) and (1–63). Select from (1–63). If selected *r* is (1–17), take indicated number from each group. If selected *r* is (18–63), take the indicated number from the second population, then select another (1–17) from the first.

1.14.

j	1	2	3	4	5	6	7	8	9	10	11
y_{1j}	5	6	3	1	1	5	3	2	4	1	2
y_{2j}	0	1	2	3	4	5	6	7	8	9	10

Compute:

(*a*) $y_1 = \sum_j y_{1j}$ and $\bar{y}_1 = \dfrac{y_1}{n}$ (*b*) $\sum_j y_{1j}^2$ (*c*) $\left(\sum_j y_{1j}\right)^2$

(*d*) $\dfrac{n}{n-1} \sum_j (y_{1j}^2 - \bar{y}_1^{\,2})$ (*e*) $\dfrac{1}{n-1}\left[n \sum_j y_{1j}^2 - \left(\sum_j y_{1j}\right)^2\right]$

(*f*) $\dfrac{n}{n-1}\left(\sum_j y_{1j}^2 - n\bar{y}_1^{\,2}\right)$ (*g*) $\sum_j y_{1j} \sum_j y_{2j}$ (*h*) $\sum_j y_{1j}y_{2j}$

(*i*) $\sum_i \sum_j y_{ij}$ (*j*) $\sum_j \sum_i y_{ij}$ (*k*) $\dfrac{\sum_j y_{1j}}{\sum_j y_{2j}}$

(*l*) $\sum_j \dfrac{y_{1j}}{y_{2j}}$ (*m*) $\sum_j 5y_{1j}$ (*n*) $\sum_j ky_{1j}$

(*o*) $\left(\sum_j ky_{1j}\right)^2$ (*p*) $\sum_i^2 \sum_j^{11} x_i y_{ij}$, where $x_1 = 2$ and $x_2 = 3$

(*q*) $r = \dfrac{n \sum_j y_{1j}y_{2j} - \sum y_{1j} \sum y_{2j}}{\sqrt{n \sum y_{1j}^2 - (\sum y_{1j})^2}\sqrt{n \sum y_{2j}^2 - (\sum y_{2j})^2}}$

1.15. Compute parts (*a*) to (*f*) of 1.14 for y_{2j}.

2

Basic Concepts of Sampling

2.0 SOME BASIC SYMBOLS

The basic symbols defined in this section are appropriate not only to simple random sampling, but also to most other designs. They provide a key to the system used throughout. Capital letters refer to population values and lower case letters denote corresponding sample values. A bar ($^-$) over a symbol indicating a total (aggregate) value denotes a mean value. The number of elements in the sample is n. For equal probability of selection methods (epsem) f is the uniform overall sampling fraction for the elements. Note that $f = n/N$; this holds exactly when both f and n are fixed; in other epsem samples it holds on the average.

Some Population Values

N = number of elements in the population.

Y_i = value of the "y" variable for the ith population element. It is also used to denote the "y" variable as a general element.

$Y = \sum_i^N Y_i$ = the *population total* for the "y" variable. The subscript i stands for the listing number of the population element. The sign $\sum_i^N$ denotes summation over all the population elements ($i = 1, 2, \ldots, N$).

$\bar{Y} = \dfrac{Y}{N} = \dfrac{1}{N}\sum_i^N Y_i$ = the *population mean* per element of the Y_i variable.

$$S_y^2 = \frac{1}{N-1}\sum_i^N (Y_i - \bar{Y})^2$$

or

$$\sigma_y^2 = \frac{1}{N}\sum_i^N (Y_i - \bar{Y})^2$$

The variance of population elements has two definitions. σ_y^2 is more traditional, but S_y^2 simplifies many formulas. Since $S_y^2 = \frac{N}{N-1}\sigma_y^2$, the difference between the two disappears for large populations.

S_y or σ_y = the standard deviation of the population elements as distinguished above.

Some Sample Values or Statistics

y_j = value of the Y_i variable for the jth sample element.

$y = \sum_j^n y_j$ = simple *sample total* for the Y_i variable.

$\bar{y} = \frac{y}{n} = \frac{1}{n}\sum_j^n y_j$ = simple *sample mean* per element of the Y_i variable. The sample count number "j" runs from 1 to n. The sign $\sum^n$ denotes summation over all the sample elements ($j = 1, 2, \ldots, n$).

In many, but not all, designs of "epsem" sampling, $\bar{y}$ is used to estimate $\bar{Y}$. Similarly, $N\bar{y}$ may be used as an estimate of Y. Note that $N\bar{y} = \frac{N}{n}y = \frac{y}{f}$.

$s_y^2 = \frac{1}{n-1}\sum_j^n (y_j - \bar{y})^2$ = variance of the sample elements. This is easier to compute as $\frac{1}{n-1}\left[\sum^n y_j^2 - n\bar{y}^2\right]$, or as $\frac{1}{n-1}\left[\sum^n y_j^2 - \frac{y^2}{n}\right]$. In simple random samples s_y^2 is an unbiased estimate of S_y^2. In other epsem selections, while not unbiased, s_y^2 is often a good estimate of S_y^2. Survey situations generally call for large samples, and it is not important whether $n - 1$ or n is used as the divisor. Also, s_y = the standard deviation of the sample elements. Usually we can omit the subscript y without causing confusion.

2.1 PROCEDURES FOR SIMPLE RANDOM SELECTION (SRS)

An *operational definition* may read: From a table of random digits select with equal probability n *different* selection numbers, corresponding to n of the N *listing numbers* of the population elements. The n listings selected from the list, on which each of the N population elements is represented separately by exactly one listing, must identify uniquely n different elements. At each of the n successive drawings, every *unselected* element has an equal probability of selection, but previously selected numbers are disregarded. Each of the n selections identifies a specific listing number and leads to a different specified population element.

The N listing numbers need not be consecutive. Instead, they may be scattered among $N + B$ numbers, of which B are "blanks" not corresponding to population elements. On each draw disregard all previously selected numbers and the B blanks that do not correspond to population elements.

For example, suppose we want an srs selection of $n = 400$ out of the $N = 10{,}000$ employees on the payroll list of a firm. If they are numbered from 1 to 10,000, select 400 different random four-digit numbers (0000 stands for 10,000). On any draw, selecting any of the previously unselected numbers from 1 to 10,000 brings another element into the sample.

If the number of employees is $N = 8000$, the last 2000 four-digit numbers are disregarded as blanks. If, for some reason, we go to the trouble of using five-digit numbers, then the last 92,000 numbers are blanks to be disregarded.

Suppose now that 8000 employees have five-digit payroll identification numbers scattered among the 100,000 possible numbers. We could assign to them new consecutive listing numbers, but this is not necessary. Instead, we can select 400 different random five-digit numbers, each of which must correspond to an unselected payroll number. We disregard the 92,000 blanks scattered through the 100,000 five-digit numbers. (See Example 2.2*a*.)

The words "different" and "unselected" in the definition mean that, on any later draw, a previously selected element cannot be reselected and must be disregarded. This denotes sampling *without replacement*: the already selected elements are not placed in the pool again for possible further selection. In practical survey sampling, we usually avoid including elements more than once.

But *in sampling with replacement* the selected elements are placed in the selection pool again and *may be reselected* on subsequent draws. Striking the words "different" and "unselected" from the definition describes this procedure. Tossing a coin is sampling with replacement from a population consisting of two elements, head and tail. The sample size n cannot exceed the population size N when sampling without replacement, but n can be any size when sampling with replacement. The drawing of balls from urns provides a clear illustration of sampling either with or without replacement; indeed, these terms have their origin in urn problems.

Briefly, we shall refer to *simple random sampling* (*srs*) when sampling *without* replacement; and to *unrestricted sampling* when sampling *with* replacement. Selection without replacement represents a restriction because it suppresses the combinations with repeated appearances of elements, combinations which can appear in sampling with replacement.

We shall also restrict the term "simple random sampling" to situations where the elements are selected individually, hence the elements are also the sampling units. This differs from cluster sampling, where the sampling units are clusters containing several elements. If "sampling units" is substituted for "elements" in the definition, it defines the *random selection* of a out of A sampling units, regardless of whether these are elements or clusters.

If all the elements, or their listing numbers, are thoroughly mixed and are put in a random order, then any predesignated segment can be considered an srs sample. For example, we can take simply the first n elements. But this procedure, however valid, could rarely serve as a simple procedure for obtaining an srs. On the other hand, without mixing, the natural order of elements on most lists cannot safely be assumed to be that of a "well-mixed urn." Hence this mixing must be replaced by randomization from a table of random numbers.

In practical survey sampling we seldom actually use an srs design. Why, then, does srs loom so large in sampling theory? First, because of its simple mathematical properties, most statistical theories and techniques assume simple random selection of elements, though usually from an infinite population or with unrestricted selection. Second, all probability selections may be viewed as restrictions on simple random selection (1.5), which suppress some combinations of population elements, whereas srs permits all possible combinations. Third, the relatively simple srs computations are often used on data obtained by more complex selections. This procedure leads to good approximation in situations where the distribution of the variable in the population is effectively random. But this assumption of random distribution is often wrong and leads to gross mistakes (5.4). Fourth, srs computations can often be used as a convenient base, then adjusted for the "design effect" of the sample design actually used (8.2).

Remark 2.1.1 Suppose the list contains $N = M + B$ listings for M population elements and B blanks. If we specify a sample of size m and continue drawing until exactly m different random elements are selected from the M population elements, disregarding all blanks, the result is a strictly srs sample of m elements.

But suppose that with a fixed f one selects $n = fN$ listings from the list containing M elements and B blanks. This is an epsem because every element has a selection probability of f. However, it is not strictly an srs because the sample size is not fixed at m, but varies around its expected value of fM. This can be treated as a *subclass* (Remark 2.2.II).

Now suppose that, wishing to estimate the proportion M/N, we select elements at random from the list until we obtain a specified number m of "successes." Then the number of selected elements n becomes a random variable (called negative binomial). The sample mean m/n is the maximum likelihood estimate, although an unbiased estimate is $(m - 1)/(n - 1)$. In samples of even moderate size the difference is negligible.

Remark 2.1.II Suppose that a sample had to be selected with replacement, and that one can distinguish in the sample the n different selections plus d duplications. What can he do with this knowledge? The variance is reduced by discarding the duplications and retaining only the n different elements in the sample. This follows because the original sample is equivalent to selecting n elements without replacement and then duplicating a random portion d of that sample. But this duplication of part of the sample must increase the variance (Section 11.7B). The sample of n elements has a smaller variance than the sample of $n + d$ selections containing d duplicates. A sample of $n + d$ different elements would have a still smaller variance.

Remark 2.1.III A *theoretical definition of srs* would be: Each possible combination of n different elements out of N has the same probability of being selected for the sample. There are

$$\binom{N}{n} = \frac{N!}{(N-n)!n!} = \frac{N(N-1)(N-2)\cdots(N-n+1)}{n(n-1)\cdots(2)(1)}$$

different combinations of elements, any one of which may constitute the sample. Their probabilities of selection are all equal to $1\bigg/\binom{N}{n}$.

Simple random selection may be viewed as a restriction on *unrestricted* sampling with replacement. The additional samples permitted with unrestricted sampling are those with duplicate selections of some elements. Altogether there are N^n equally possible unrestricted samples (distinguishing orders of appearance).

Simple random selection is a special type of epsem selection, because elements have the same fixed selection probability of n/N. All other epsem designs may be viewed as restrictions that prevent the selection of some of the $\binom{N}{n}$ combinations possible under srs.

The equal selection probability for each of the N elements is common to all epsem samples. But srs is distinct among them, because each of the $\binom{N}{2}$ possible pairs also have the same selection probability of $n(n-1)/N(N-1)$, and selection equality also holds in srs for any triple, quadruple, or any combination of r elements, where r is any number from 1 to n. In other epsem selections most of these combinations are excluded.

Remark 2.1.IV The equal selection probability of n/N can be computed in detail by considering the probability of the selection for a specific element in each of the draws, which are "mutually exclusive events."

In unrestricted sampling *with* replacement the probability is $1/N$ for any specific element on each of the n draws; the expectation for an element is

$$1/N + 1/N + 1/N + \cdots = n \times 1/N = n/N.$$

The result of any draw is truly independent of the results of other draws.

In srs *without* replacement the probability of selecting a specific element on the first draw is $1/N$. On the second draw, its probability is $1/(N-1)$, *conditional* on the probability $(N-1)/N$, that it went unselected on the first draw; thus the probability of its selection on the second draw is

$$[(N-1)/N] \times [1/(N-1)] = 1/N.$$

On the third draw, it can be selected with probability $1/(N-2)$, conditional on the probability $[(N-1)/N] \times [(N-2)/(N-1)] = (N-2)/N$ that it went unselected on both the first and second draws; and the joint probability is again $1/N$. And so on for n draws. The total probability is the sum of n probabilities, each of which comes to $1/N$:

$$\frac{1}{N} + \frac{N-1}{N} \cdot \frac{1}{N-1} + \frac{N-1}{N} \cdot \frac{N-2}{N-1} \cdot \frac{1}{N-2} + \cdots$$
$$= \frac{1}{N} + \frac{1}{N} + \frac{1}{N} + \cdots = n\frac{1}{N} = \frac{n}{N}.$$

Thus in srs the selection probability of $1/N$ is maintained on each draw, although the results of different draws are not independent (Section 2.8C).

In srs sampling there are $\binom{N}{n}$ equally probable combinations. Among these there are $\binom{N-1}{n-1}$ which contain any specified element. Hence its probability of selection is $\binom{N-1}{n-1} \Big/ \binom{N}{n} = n/N$, as before.

2.2 MEAN AND VARIANCE OF SRS

The simple mean of the sample of an srs selection is the *srs mean*, and we distinguish it with the subscript 0:

$$\bar{y}_0 = \frac{y}{n} = \frac{1}{n}\sum_j^n y_j = \frac{1}{n}[y_1 + y_2 + \cdots + y_n]. \tag{2.2.1}$$

The results of an srs selection may be used for other estimators also, for example, with post-stratification (3.4*C*) or with a ratio estimator (6.6).

But we treat those separately as other designs. Simple random sampling is a sample design specifying both the *srs* selection and the simple mean estimate. The variance of the *srs* mean $\bar{y}_0$ is computed as

$$\text{var}(\bar{y}_0) = (1 - f)\frac{s^2}{n}, \tag{2.2.2}$$

where $s^2 = \frac{1}{n-1}\sum_j^n (y_j - \bar{y})^2 = \frac{1}{n-1}\left[\sum_j^n y_j^2 - \frac{y^2}{n}\right] = \frac{n\sum^n y_j^2 - y^2}{n(n-1)}$.

The standard error of $\bar{y}_0$ is the square root of its variance:

$$\text{se}(\bar{y}_0) = \sqrt{\text{var}(\bar{y}_0)} = \sqrt{1-f}\frac{s}{\sqrt{n}}. \tag{2.2.3}$$

Sometimes we may want to estimate $Y = N\bar{Y}$, the aggregate or total of the Y_i variable in the population. A simple estimator of Y is $N\bar{y}_0$ and its standard error is estimated by

$$\text{se}(N\bar{y}_0) = N\text{se}(\bar{y}_0) = N\sqrt{1-f}\frac{s}{\sqrt{n}}. \tag{2.2.4}$$

We can also point out that the expected value of s^2 in srs is

$$E(s^2) = S^2 = \frac{N}{N-1}\sigma^2. \tag{2.2.2'}$$

This is shown in Section 2.8B; also that the expected value of the sample estimate of the variance of the mean is

$$E\left(\frac{1-f}{n}s^2\right) = \frac{1-f}{n}S^2 = \frac{N-n}{N-1}\frac{\sigma^2}{n}. \tag{2.2.2''}$$

Remark 2.2.1 For the *difference* $(\bar{y} - \bar{x})$ *of two means*, the variance is simply the sum of the two variances if the two samples are independent. But if the two means are not independent, a covariance term must be subtracted from the sum of the variances (2.8.9'): $\text{var}(\bar{y} - \bar{x}) = \text{var}(\bar{x}) + \text{var}(\bar{y}) - 2\,\text{cov}(\bar{y}, \bar{x})$. For *n pairs* of values, each pair selected with srs, the difference has the variance:

$$\text{var}(\bar{y} - \bar{x}) = \frac{1-f}{n}(s_x^2 + s_y^2 - 2s_{yx}). \tag{2.2.5}$$

Note the use of the *covariance* of the two variables. This statistic resembles the variance, but contains cross-product terms instead of the squared terms of the variance

$$\text{cov}(\bar{y}_0, \bar{x}_0) = (1-f)\frac{s_{yx}}{n}, \quad \text{where} \quad s_{yx} = \frac{1}{n-1}\left[\sum_j^n y_j x_j - \frac{yx}{n}\right]. \tag{2.2.5'}$$

Note also that for the pairs of elements

$$(\bar{y} - \bar{x}) = \frac{\sum_j^n y_j}{n} - \frac{\sum_j^n x_j}{n} = \sum_j^n \frac{y_j - x_j}{n} = \sum_j^n \frac{d_j}{n}.$$

Hence we may treat this as the mean of a sample of n elements $(y_j - x_j) = d_j$. The variance can also be computed as

$$\text{var}\left(\frac{1}{n}\sum_j^n d_j\right) = \frac{1-f}{n} s_d^2, \quad \text{where} \quad s_d^2 = \frac{1}{n-1}\left[\sum_j^n d_j^2 - \frac{(\sum d_j)^2}{n}\right], \tag{2.2.5''}$$

which is numerically equal to (2.2.5). The covariance is absent for two independent samples, but present for two overlapping samples (12.4).

The variance of the difference becomes more complicated if the two samples are neither completely independent nor completely overlapping. The problems of *partially overlapping* samples are treated in Section 12.4A.

Remark 2.2.II The subclass mean $\bar{y}_m = \sum_j^m y_j/m$ from an srs of n elements can be treated as an srs of m elements. That is, we consider the variance of the sample *conditional* on obtaining a sample of m elements:

$$\text{var}(\bar{y}_m) = \frac{1-f}{m(m-1)}\sum_j^m (y_j - \bar{y}_m)^2 = \frac{1-f}{m} s_m^2. \tag{2.2.6}$$

We can use $f = m/M$ if we know M, the population size of the subclass. If we do not know M, we can use $f = n/N$, neglecting the difference from m/M.

But if we want to estimate $Y = M\bar{Y}$, the population total of Y_i for the subclass, then knowledge of the subclass size M becomes important. If M is known, then $(M\bar{y}_m)$ has the variance $M^2 \text{ var}(\bar{y}_m)$. If we do not know M and use $(mN/n)\bar{y}_m$ to estimate Y, then the element variance s_y^2 is increased to $[s_m^2 + (1 - \bar{m})\bar{y}_m^2]$ (see 11.8).

Example 2.2a The population of $N = 270$ blocks in Appendix E is numbered from 232 to 772, with 271 blanks scattered among the 541 listing numbers. With three-digit random numbers, the following srs of $n = 20$ was selected.

j Sample No.	Block Listing Numbers	x_j Total *Du's*	y_j Rented *Du's*	j Sample No.	Block Listing Numbers	x_j Total *Du's*	y_j Rented *Du's*
1	689	5	3	11	701	29	17
2	537	9	5	12	566	31	14
3	545	18	5	13	680	5	0
4	420	68	52	14	735	2	0
5	436	32	21	15	528	4	2
6	385	48	34	16	541	102	54
7	575	11	3	17	564	20	11
8	727	1	0	18	380	15	11
9	753	1	0	19	730	1	0
10	451	4	0	20	376	29	23

$x = 435 \quad \sum^n x_j^2 = 22{,}239 \quad y = 255 \quad \sum^n y_j^2 = 8545$

The sample means for the two variates are, respectively,

$$\bar{y}_0 = \frac{y}{n} = \frac{255}{20} = 12.75 \quad \text{and} \quad \bar{x}_0 = \frac{x}{n} = \frac{435}{20} = 21.75.$$

The sample variances of the two means are

$$\text{var}(\bar{y}_0) = \left(1 - \frac{20}{270}\right)\frac{278.62}{20} = 12.90, \quad \text{where } s_y^2 = \frac{1}{19}\left[8545 - \frac{255^2}{20}\right] = 278.62,$$

and

$$\text{var}(\bar{x}_0) = \left(1 - \frac{20}{270}\right)\frac{672.51}{20} = 31.14, \quad \text{where } s_x^2 = \frac{1}{19}\left[22{,}239 - \frac{435^2}{20}\right] = 672.51,$$

and the standard deviations are

$$s_y = \sqrt{278.62} = 16.7 \quad \text{and} \quad s_x = \sqrt{672.51} = 25.9.$$

The standard errors of the means are

$$\text{se}(\bar{y}_0) = \sqrt{12.90} = 3.59 \quad \text{and} \quad \text{se}(\bar{x}_0) = \sqrt{31.14} = 5.58.$$

The estimates of the total rented dwellings and of the total dwellings of all kinds in the entire population of $N = 270$ blocks are, respectively,

$$N\bar{y}_0 = 270(12.75) = 3442.50 \quad \text{and} \quad N\bar{x}_0 = 270(21.75) = 5872.50.$$

The standard errors of these estimates are

$$\text{se}(N\bar{y}_0) = 270(3.59) = 969.30 \quad \text{and} \quad \text{se}(N\bar{x}_0) = 270(5.58) = 1506.60.$$

This example serves only to illustrate methods of computation. The small size of the population is unrealistic and usually the factor n/N is much smaller. Another disturbing note is that both the X_i and Y_i variables have distributions that are highly skewed. This may be guessed from the several large values that deviate from the average by large multiples of the standard deviations. Skewed distributions are encountered in practical survey work. Means based on samples of only 20 might be severely affected, with serious doubts about normal approximation. This problem becomes less acute with larger samples, say several hundred blocks, in this kind of distribution. The problem can be eased with stratification and other techniques (11.4).

2.3 THE "fpc"; SAMPLING WITHOUT REPLACEMENT

The formulas of 2.2 are familiar to most readers, with the possible exception of the factor $(1-f) = (1 - n/N)$. This factor is usually called the *finite population correction*, briefly the fpc. When sampling without replacement, it appears as a correction factor to the main portion of variance terms, which is s^2/n for srs.

If we think of a fixed sample size n being applied to larger and larger populations, the sampling fraction $f = n/N$ tends to zero, and the factor $(1-f)$ approaches 1. Multiplication by one has no effect, and the fpc can be omitted when the population is much larger than the sample. For an "infinite population" the factor disappears from the variance formula; hence, its name. Also, when selecting *with* replacement, the factor $(1-f)$ becomes 1 and disappears. The effect is similar to selection from an infinite population (2.8*B*).

Actual populations of physical objects are finite, hence, for selections without replacement, the fpc is less than one. However, in most practical examples the population is so much larger than the sample that, neglecting the fpc $(1 - n/N)$ leads to negligible overestimation of the variance. The effect on the computed standard error is approximately $(1-f/2)$; that this is negligible in most cases can be judged from this table:

$f = n/N$	0.10	0.04	0.01	0.001	0.0001
$\sqrt{1-f}$	0.9487	0.9798	0.9950	0.9995	0.99995
$1 - f/2$	0.9500	0.9800	0.9950	0.9995	0.99995

For a sample of 1000 homes, consider the effect of $(1-f)$ on the variance if the population is one of the following: city ward of 4000 homes; city of 10,000 homes; county of 25,000 homes; metropolitan area of 100,000 homes; state of 1,000,000 homes; nation of 10,000,000 homes. Note that the fpc has some effect on the sample of the ward, little on the sample of the city, and it makes no practical difference whether the sample is taken from the county, metropolitan area, state, or nation.

The variances for other sample designs contain components different from s^2/n, but the above statements about the fpc will have analogous meaning. In particular, for many epsem designs, the fpc remains $(1 - n/N)$. Because this factor is usually small, we should look past the fpc to the important parts of the variance formulas.

The sampling fraction is usually small, because the population is large. The aims of research generally concern inferences about large populations or infinite theoretical universes. When, for financial reasons, the sample is confined to a small population, this often is hopefully considered a "sample" for making inferences about some much larger actual population or theoretical universe. In these situations the fpc is best forgotten.

But censuses aimed specifically at small populations do occur, and sometimes these run into large sampling fractions of 10 percent and more. For example, inventories for legal or accounting purposes; also, the strata of large items in disproportionate sampling (3.5). In these rare cases the fpc is needed.

Note that the variance can be written as var $(\bar{y}) = s^2/n'$, where $n' = n/(1 - n/N) = nN/(N - n)$. From this we easily find that $n = n'/(1 + n'/N)$. In other words, the effect of $(1 - n/N)$ is to increase the "effective sample size" from n to n'. It might be convenient to write all the variance formulas with this convention. However, I refrained from doing so in order to conform to common usage in sampling literature.

2.4 PROPORTIONS: THE MEANS OF BINOMIALS

In many cases the survey result is presented as the proportion of population elements that belong to a defined class, or possess a defined attribute. A proportion is the mean of a dichotomous variable, when members of a class receive the value $Y_i = 1$, and nonmembers the value $Y_i = 0$. We sometimes call this a binomial variable, or simply a *binomial.*

The results of a multiple classification, or polytomy, can similarly be called a *multinomial* variable. Many survey results appear as multinomial variables, sometimes called "counted data" or "classifications" or "nominal scale." Multinomial variables may occur because of the nature of data, as with occupation classes; they may result from the observation process, as with some attitudes; or be due to classification in the coding and machine process, as with income classes. Much of the analysis and presentation takes the statistical form of a sequence of proportions.

Because of their prevalence, it is worth presenting the special simple form that the variance of a proportion takes when the design is srs. For stratified element sampling, some modifications are necessary, and for cluster sampling, they are entirely inapplicable. In a binomial the assigned value of the variable is 1 if the element belongs to the defined class, and 0 if it does not belong. $NP = Y = \sum^{N} Y_i$ is the number of elements in the population that belong to the class, and $P = \bar{Y} = Y/N$ is their proportion among the population elements. Then $Q = 1 - P$ is the population proportion of those that do not belong, and $NQ = N(1-P) = N - NP$ is their number in the population.

$p = \bar{y} = y/n$ is the *proportion of sample elements* that belong to the defined class; and $q = 1 - p$ is the proportion that does not belong.

$np = y = \sum^{n} y_j$ is the *number of sample elements* that belong to the class; and $nq = n(1-p) = n - np$ is the number that do not belong.

For srs samples of proportions, the usual formula of the variance (2.2.2) can also be computed by a simple procedure from the sample proportion p:

$$\text{var}(p) = (1-f)\,\frac{p(1-p)}{n-1}. \tag{2.4.1}$$

The standard error is

$$\text{se}(p) = \sqrt{\text{var}(p)} = \left[(1-f)\,\frac{p(1-p)}{n-1}\right]^{1/2}. \tag{2.4.2}$$

This is the popular $\sqrt{pq/n}$ formula, and the factor $(1-f)$ is seldom important. Furthermore, for moderate to large samples, it matters little whether n or $n-1$ is used in the denominator. The choice between the two divisors involves technical arguments, but we use $n-1$ here to be consistent with sampling textbooks and with formula (2.2.2). The distribution of p for a simple random sample is called a *binomial distribution* when selecting with replacement, and a *hypergeometric distribution* when selecting without replacement.

Note for computing ease that

$$\left[\frac{p(1-p)}{n-1}\right]^{1/2} = \frac{1}{n}\left[\frac{y(n-y)}{n-1}\right]^{1/2}. \tag{2.4.3}$$

Example 2.4a Assume that in a sample of 400 employees selected with srs from a payroll of 10,000 employees, the number responding "no" to an attitudinal question was 80. The sample proportion with the negative attitude is 80/400 = 0.20 or 20 percent. The standard error is

$$\begin{aligned}\text{se}(p) &= \left[\left(1-\frac{400}{10{,}000}\right)\left(\frac{0.20\times 0.80}{399}\right)\right]^{1/2} \doteq 0.98\,\frac{0.40}{20}\\ &= 0.98(0.0200) = 0.0196 = 1.96\times 10^{-2} = 1.96 \text{ percent.}\end{aligned}$$

Or we can use (2.4.3) to compute

$$\frac{1}{400}\left(\frac{80\times 320}{399}\right)^{1/2} = 0.0200.$$

Although the formula is in terms of proportions, the standard error can be computed from percentages as

$$\text{se}(p_0) = \left[\left(1-\frac{400}{10{,}000}\right)\left(\frac{20\times 80}{399}\right)\right]^{1/2} \doteq 0.98\,\frac{40}{20} = 0.98(2.00) = 1.96.$$

This procedure reduces the number of mistakes due to misplaced decimal points that occur with proportions. We can also estimate that 10,000 × 20 percent = 2,000 employees have this negative attitude, and that the standard error of that estimate is 10,000 × 1.96 per cent = 196.

Remark 2.4.I The consistency of these formulas for proportions with the general srs variance formulas of 2.2 can be noted:

$$\sigma^2 = P(1\text{-}P), \qquad S^2 = \frac{N}{N-1} P(1\text{-}P), \tag{2.4.5}$$

and $s^2/n = p(1\text{-}p)/(n-1)$ is the unbiased estimate of $S^2/n = NP(1\text{-}P)/(N-1)n$. Thus (2.4.1) comes from replacing S^2 with s^2 in

$$\text{Var}(p) = \frac{N-n}{N-1}\frac{\sigma^2}{n} = (1\text{-}f)\frac{S^2}{n} = (1\text{-}f)\left(\frac{N}{N-1}\right)\frac{P(1\text{-}P)}{n}. \tag{2.4.6}$$

All these are simply results of the fact that when np elements have the value 1 and all others 0, then

$$\sum^n y_j^2 = np \qquad \text{and} \qquad \sum^n y_j^2 - n\bar{y}^2 = np - np^2 = np(1\text{-}p).$$

Similarly,

$$\sum^N Y_i^2 = NP \qquad \text{and} \qquad \sum^N Y_i^2 - N\bar{Y}^2 = NP - NP^2 = NP(1\text{-}P).$$

Remark 2.4.II The statistical analysis of multinomials may involve the exclusion of some of the classes. For example, from a sample divided into three classes, a, b, c, we may want the proportion $a/(a+b)$ and disregard class c. The analysis can regard the class c as blanks and treat $a/(a+b)$ as a proportion based on the *subclass* $a+b$. As noted in Remark 2.2.II, the subclass mean can be treated with the usual formulas, but estimates of the aggregate for the subclass present other problems.

2.5 RELATIVE ERROR

In some situations it is useful to consider some *relative* measures instead of the *absolute* measures of the variation. The absolute measures, the standard deviation and the standard error, appear in the units of measurement of the variable, and this causes difficulties in some comparisons. Common relative measures are the *coefficients of variation*, in which the unit of measurement is canceled by dividing with the mean. *The element coefficient of variation* is derived from the standard deviation:

$$C_y = \frac{S_y}{\bar{Y}}, \qquad \text{estimated by } c_y = \frac{s_y}{\bar{y}}. \tag{2.5.1}$$

The coefficient of variation of the mean ($\bar{y}$) is derived similarly from the standard error:

$$\text{CV}(\bar{y}) = \frac{\text{SE}(\bar{y})}{\bar{Y}}, \qquad \text{estimated by cv}(\bar{y}) = \frac{\text{se}(\bar{y})}{\bar{y}}. \tag{2.5.2}$$

The squares of these quantities correspond, respectively, to the variances of the element and of the mean:

$$C_y^2 = \frac{S_y^2}{\bar{Y}^2}, \qquad \text{estimated by } c_y^2 = \frac{s_y^2}{\bar{y}^2},$$

is the *element relvariance* and

$$\text{CV}^2(\bar{y}) = \frac{\text{Var}(\bar{y})}{\bar{Y}^2}, \qquad \text{estimated by cv}^2 = \frac{\text{var}(\bar{y})}{\bar{y}^2},$$

is the *relvariance* of the mean ($\bar{y}$).

Coefficients of variation are useful for variables that are always or mostly positive; these occur frequently in surveys, especially as "count data." Comparisons of the variability of these items often becomes more meaningful when expressed in relative terms. For example, in comparing the "income spread" in two countries, the use of the two standard deviations would be confused by the different monetary units as well as by different standards of living; but coefficients of variation may provide a reasonable comparison in terms of average income. Also in Example 2.2*a*, we can say that the total dwellings per block seem to be more variable than rented dwellings, because $s_x = 25.9$ while $s_y = 16.7$. After all, the means are 21.75 and 12.75, respectively, and common sense tells us that comparing the standard deviations alone does not tell the full story. The values of the coefficient of variation per element are roughly equal: $25.9/21.75 = 1.19$ and $16.7/12.75 = 1.31$.

Reasonable constancy of C values can often be found in groups of related variables, even when both the means and standard deviations differ widely. Using estimates or guesses of C values, when S values are difficult to guess directly, facilitates the design of samples.

The coefficients of variation are the same for the mean and for the aggregate:

$$\text{cv}(N\bar{y}) = \frac{N\,\text{se}(\bar{y})}{N\bar{y}} = \frac{\text{se}(\bar{y})}{\bar{y}} = \text{cv}(\bar{y}). \tag{2.5.3}$$

These general expressions hold for different sample designs. Specifically, for srs samples we can use

$$\text{cv}(\bar{y}_0) = \text{cv}(N\bar{y}_0) = \sqrt{1-f}\,\frac{c_y}{\sqrt{n}} = \sqrt{1-f}\,\frac{s_y}{\bar{y}_0\sqrt{n}}. \tag{2.5.4}$$

In some situations the coefficients of variation should be used only with caution, or not at all. (1) If the mean of the variable is close to zero, the cv's are large and *unstable*. Measures of change usually fall in this class; for example, yearly changes in individual income may have a large S, but a mean near zero. (2) For binomial variables, the element variance is the same $P(1-P)$ for both P and $Q = 1 - P$; but the coefficients of variation differ, depending on the arbitrary decision of which side of the binomial cut is regarded as P and which as Q. That is:

$$C_y^2 = \frac{P(1-P)}{P^2} = \frac{1-P}{P} \quad \text{and} \quad \text{cv}^2(p) = \text{cv}^2(Np) = (1-f)\frac{1-p}{p(n-1)}. \tag{2.5.5}$$

The *element relvariance* $C_y^2 = (1 - P)/P = 1$ for $P = 0.5$. It increases rapidly for small values of P; it is 9 for $P = 0.10$, and it is nearly $1/P$ for very small values of P.

2.6 THE DESIGN OF ECONOMIC SAMPLES

The sampler is sometimes faced with this kind of question: "What percent sample do I need?" or "Will a five percent sample give me adequate precision?" These questions must be drastically reformed before they can be treated reasonably.

First, the sample size n is the important factor in S^2/n; but the sampling fraction n/N as such usually has only a negligible effect on the fpc $=$ $(1 - n/N)$. Second, the variance of the sample depends not only on its size n, but also on the sample design. In complex designs it depends on the magnitudes of other components of variation. The present discussion assumes an srs design; hence, it applies strictly only to it. But the general principles and even the techniques used here have much broader implications that will be developed in 8.1.

When designing a sample, we must also consider several issues not entirely covered by sampling theory, but closely related to it. First, the survey objectives usually require several statistics; considered separately, each could lead to a different design; their conflicting claims must be reconciled and compromised (14.5). Second, the magnitude of nonsampling errors also affects sample design (13.1). Third, to be sensible, the discussion of "needed" or "adequate" precision should be preceded by some formulation of the survey objectives in cost and utility terms.

For brevity we shall assume first that the acceptable precision can be stated as V^2, the variance of the mean. Second, that the cost of the sample can be stated in terms of n, the sample size. We can then answer one of two questions:

(1) "For a fixed sample size n, what variance shall we expect for $\bar{y}$?" The answer comes directly from (2.2.2):

$$V^2 = \frac{(1 - n/N)S^2}{n} = \frac{S^2}{n'}, \quad \text{where } n' = \frac{n}{(1 - n/N)}. \tag{2.6.1}$$

(2) "For a desired variance V^2 of $\bar{y}$, how large should the sample be?" The answer comes from solving for n in (2.6.1). But it is easier to do this in two steps. Ignoring $(1 - n/N)$, first compute

$$n' = \frac{S^2}{V^2}, \quad \text{and then} \quad n = \frac{n'}{1 + n'/N}. \tag{2.6.2}$$

The proof consists simply in noting that

$$\frac{1}{n'} = \left(1 - \frac{n}{N}\right)\frac{1}{n} = \frac{1}{n} - \frac{1}{N}; \quad \text{hence,} \quad \frac{1}{n} = \frac{1}{n'} + \frac{1}{N} = \frac{1 + n'/N}{n'}.$$

The two steps yield not an approximation but an exact solution. They also permit us to note separately the effect of the fpc. Usually this is negligible, in view of much larger errors involved in the design.

Notice that we omit here, as in many other places, the subscript y in $S = S_y$, $C = C_y$, etc.

More generally, instead of fixing either V^2 or n, we note that V^2 is inversely proportional to n'. And n' equals n, except if n begins to approach the population size N.

Often laymen are surprised to hear that precision depends only on the size of the sample and not on the population size. But population size affects only the factor $(1 - n/N)$, and this can usually be ignored in designing the sample.

With fixed S^2 and fixed desired precision $V^2 = S^2/n = S^2/Nf$, the sample size n is also fixed. Thus the sampling fraction f varies inversely with the population size, since $f = (1/N)(S^2/V^2)$. For example, with $S^2 = PQ = 0.25$ and a desired $V = 0.01$ the needed $n = 0.25/0.0001 = 2500$ for either a city of 1 million or a country of 100 million. Hence, the sampling fraction would be 1/400 for the city and 1/40,000 for the country.

When dealing with *relative variances* instead of absolute variances, we can write

$$\text{CV}^2 = \left(1 - \frac{n}{N}\right)\left(\frac{C^2}{n}\right), \tag{2.6.1$'$}$$

or

$$n' = \frac{C^2}{\text{CV}^2} \quad \text{and then} \quad n = \frac{n'}{1 + n'/N}. \tag{2.6.2$'$}$$

These formulas are also directly applicable to estimating aggregates of the type $N\bar{y}$, because the relative variances are the same for the aggregate $N\bar{y}$, as for the mean $\bar{y}$.

More unusual would be trying to fix the absolute variance of the aggregate $N\bar{y}$ which would be $V_t^2 = \text{Var}(N\bar{y}) = N^2S^2/n = NS^2/f$. Thus the needed n would be proportional to N^2, since $n = N^2(S^2/V_t^2)$; and the sampling fraction would vary directly with N, since $f = N(S^2/V_t^2)$. If in the example above, somebody would specify a fixed $V_t = 50{,}000$, that would mean $S^2/V_t^2 = (0.5/50{,}000)^2 = 1/100{,}000^2$. This indicates a sample size of $n = N^2/100{,}000^2$, resulting in a sample of 100 for the city of 1 million, but a sample of $n = 1{,}000{,}000$ for the country of 100 million. The sampling fraction would be 1/10,000 for the city but 1/100 for the country. The blame for the seeming absurdity of these results should be placed on the unreasonable requirement of equal absolute precision for populations of very unequal sizes. It is more likely, perhaps, that the tolerable errors in the aggregate should be *relative* to the size of the estimate. Then $\text{CV}^2 = N^2S^2/nN^2\bar{Y}^2 = (S^2/\bar{Y}^2)/n$. For fixed coefficients of variation, the sample size n is unaffected by population size but for $(1-f)$.

In these computations the key term is S^2, the element variance. In practice this is unknown, and it must be estimated or guessed. Although it may seem paradoxical, we must make this guess regarding the variable that is the subject of the sample design. What are the sources for those guesses?

1. We should search for data from *past surveys* of similar variables, or seek the advice of an expert survey statistician, with his knowledge of past surveys, and his ability to unearth relevant aspects of past surveys. The statistician can ask the questions that will elicit the relevant data from specialists in the subject matter concerned. Supported by that knowledge, we may construct a model of the population distribution, its shape, and its probable limits, and deduce S^2 from it.

2. If we know the variance var* $(\bar{y}_0)$ of a past srs of size n^*, then we use $S^{*2} = \text{var}^*(\bar{y}_0) \times n^*/(1-f)$. If the sample is not srs, the "design effect" may be used to adjust the variance (see 8.2).

3. Often, instead of S, we can more easily guess $C = S/\bar{Y}$, because C is less variable than S; hence, this facilitates borrowing data from the results of similar variables. With an estimate of C and also of $\bar{Y}$, we can estimate $S = C\bar{Y}$.

4. The frequent problem of estimating proportions is relatively easy to treat. The variance $P(1-P)$ *is not sensitive* to changes in the middle range of $P = 0.2$ to $P = 0.8$, and generally a reasonable guess of P can be made. Of course, the "safe" choice is the maximum $\sigma^2 = 0.25$, corresponding to $P = 0.5$.

5. To design efficiently a large sample in an unknown field, a *pilot study may be conducted prior to the survey, to gain information for designing the survey*. But most studies are too small and too hurried to support a large enough pilot study to produce *serviceable* estimates of S^2. If the pilot study is too small, its results are useless, because they are less dependable than the expert guesses we can obtain without it.

Sometimes the sample size can be adjusted more closely to the survey's demands. This adjustment is best made by building flexibility into the design. First, collect a basic sample of a reasonable minimum size that might meet the demands. Then compute the results and, if the demands are not met, collect a supplementary sample of desired size. This two-step procedure can be used to obtain either a desired variance or sample size. However, it cannot be used for surveys with rigid time schedules that preclude using this two-step procedure (8.4D).

Naturally the guesses about S^2 are subject to error, and the actual variance of the sample mean may be smaller or greater than planned. But these errors do not affect the validity of variances computed from actual sample values, which are not influenced at all by the guessed values of S^2. Uncertainty about S^2 leads to a range of requisite sample sizes, from which one can be chosen. The choice may be based on principles of personal decision theory [Schlaifer, 1961].

Moreover, the errors in estimates of S^2 are generally exceeded by the margin of ignorance about several related issues. First, we often know even less about the cost factors than about S^2. Second, the statement of an "adequate" or "desired" variance is generally subject to more vagueness than is S^2. This is especially true when the survey has several objectives, with conflicting demands on the desirable sample size. Third, our knowledge about the effects of nonsampling errors is generally less adequate than about S^2.

Example 2.6a From the population of $N = 10{,}000$ employees of a factory we want to take an srs sample to ascertain attitudes, measured in proportions which will run anywhere from 10 percent to 60 percent. Hence the variances

$$S^2 = P(1-P) \text{ will vary from } 0.09 \text{ for } P = 0.1 \text{ to } 0.25 \text{ for } P = 0.5.$$

Question 1. "For $n = 400$ questionnaires, what will be the standard errors of the estimated proportions?" From (2.6.1) for items near $P = 0.10$,

$$V^2 = \left(1 - \frac{400}{10{,}000}\right)\frac{900 \times 10^{-4}}{400} = 2.16 \times 10^{-4},$$

and
$$V = \sqrt{2.16 \times 10^{-4}} = 1.47 \times 10^{-2}.$$

For items near $P = 0.5$,

$$V^2 = \left(1 - \frac{400}{10{,}000}\right)\frac{2500 \times 10^{-4}}{400} = 6.00 \times 10^{-4},$$

and
$$V = \sqrt{6.00 \times 10^{-4}} = (2.45) \times 10^{-2}.$$

The standard errors will be 1.5 percent for P near 10 percent, and 2.5 percent for P near 50 percent.

Question 2. "What size sample would we need for standard errors no greater than 2 percent for any proportion?" The critical items are proportions near 0.5 for which the variance is the maximum $S^2 = 0.25$, and from (2.6.2) we have

$$n' = \frac{0.25}{(0.02)^2} = \frac{0.25}{0.0004} = 625, \quad \text{and} \quad n = \frac{625}{1 + 625/10{,}000} = 588.$$

2.7 FRAME PROBLEMS

The sampling frame or list is the keystone around which the selection process must be designed. Appraisal of the available or obtainable frames must dominate the search for good selection procedures and the choice among several alternatives.

We saw the need for a list as the basis for statistical inference in probability sampling (1.7). *Frame* is a more general concept: it includes physical lists and also procedures that can account for all the sampling units without the physical effort of actually listing them. For example, in area sampling, the frame consists of maps, but the frame can often be constructed without mapping the entire population (9.1). A frame for school children consists of school districts containing schools, their classes, and finally children. The design can be carried through in several stages without obtaining a complete list of all the children.

"The frame consists of previously available descriptions of the material in the form of maps, lists, directories, etc., from which sample units may be constructed and a set of units selected. The specification of the frame should define the geographical scope of the survey and the categories of material covered; also the date and source of the frame. Frames that are originally available often require emendation, particularly in the later stages of multi-stage sampling, before they may be considered adequate; at times a frame may need to be constructed *ab initio*. In such cases the method of emendation or construction should be described." [*United Nations*, 1950.]

In our terms (1.7), the frame contains listings, but we want to select *sampling units*. For simplicity's sake, this discussion centers on the selection of *elements*, where each sampling unit consists of a single element. But it holds equally well for selecting sampling units that are clusters of elements. Substituting the term *sampling unit* for *element* in the discussion will generally suffice.

The frame is perfect if every element appears on the list separately, once, only once, and nothing else appears on the list. In other words: every element must appear in a listing, and in only one listing; also, every listing must contain an element, and only one element. But perfect frames are rare, and we must often use frames with serious deficiencies that must be detected and remedied. Before undertaking a selection, the sampler must probe thoroughly for possible faults in the frame. He can discover some faults by skillful questioning of persons with specialized knowledge of the lists. Other faults may have to be found by empirical investigations. Recognizing the faults may permit dealing with them adequately, economically, and practically.

Actual frame problems exhibit a bewildering variety and their recognition constitutes an important part of the sampler's art. Examples of problems

$L - S$	No problem; one-to-one correspondence of listing and element.
(A) $O - S$	*Missing elements, noncoverage, incomplete frame*: contradict rule that every element must *appear* once, to be *present* in some one listing.
(B) $L < \begin{matrix} S \\ S \end{matrix}$	*Clusters* of elements *together* in one listing: contradict rule that elements must appear *separately*, so each listing contains *only one* element.
(C) $L - O$	*Blanks or foreign elements*: contradict rule that list contain *nothing else*, so each listing contains *at least one element*.
(D) $\begin{matrix} L \\ L \end{matrix} > S$	*Duplicate listings*: contradict rule that element appear *once only*, so that each element appears *in only one listing*.

TABLE 2.7.I Four Basic Frame Problems

appear widely in the literature and good examples are given by Yates [1960, Sections 4.1–4.28] and Hansen [1953, 2.3–2.8]. I attempt to introduce some order and systematic treatment for frame problems by reducing the many varieties to four basic categories. Each represents one of four possible contradictions in the basic requirements of a one-to-one correspondence between listing (frame unit) and element (sampling unit). Thus each interferes with our aim of selecting a single element with epsem, when a single listing is selected with epsem.

Before giving specific treatments for the four basic problems, we mention *three general ways of avoiding the problems:*

1. *Ignore and disregard the problem* if it is known to be small compared to other errors, and if correcting it would be too costly. For example, a small proportion of families have two dwellings, and two chances of selection; but except for summer cottages, this is not a big problem in

small samples. From external evidence or from previous investigations the proportion of missing elements and of duplicate listings may be known to be small. A statement about the magnitude of the ignored problem should be added to the description of the sample.

2. *Redefine the population to fit the frame.* This should be avoided if the orientation of the sample would be seriously deflected from its goal. But it can be used if the result of the redefinition is trivial or preferred. For example, a firm's payroll list may exclude recently hired employees; but these may be few and the researcher may prefer to exclude them. Another example: for a sample of customers of a store (or library) we may accept for elements the individual visits instead of the customer (or library user). This gives each customer selection probabilities in proportion to his visits and may lead to better analysis.

We may accept a population reduced by redefinition, if the error caused by the missed portion is outweighed by the cost of adding it. We then *explicitly* exclude the missing elements from the population, preferably with a statement about the magnitude of the excluded portion. For example, for a sample of 200 dwellings of a city we might accept a directory list, only 95 percent complete. Many area samples of the U.S. population include only persons living in private dwellings, because the inclusion of the institutional and transient populations would be too costly.

3. *Correct the entire population list.* This means finding all missing elements, splitting each cluster, and eliminating all blanks, foreign elements, and duplicate listings. The clerical correction of thousands or tens of thousands of simple cards or line entries may be less expensive than the skill required for a more sophisticated treatment; and machine methods may extend the limits of routine treatment. But individual treatment of lists running into millions may be so costly that routine must be replaced by skill.

Suppose now that we can neither ignore the faults of the frame, nor redefine the population to fit the frame, nor correct the entire frame. If the problem is great and the task is not, we should introduce some remedies to compensate for the faults in the frame. The remedies may involve modifications of the original selection design; they may introduce some clustering or some variation in the sample size. It is wise to have some foreknowledge of the magnitude of these problems and to adjust the sample design accordingly. For example, the selection rate should anticipate the ratio of elements to listings in the frame.

The remedies proposed below have the virtue of restoring the equal probabilities that the faults tend to destroy. Two of the remedies do accept the inequalities, but compensate for them with proper weights. Separate remedies for each of the four kinds of problems are presented, after a brief statement for each. They are treated in more detail in Chapter 11.

A. Missing elements, also called *noncoverage and incomplete frame*, frequently present important practical problems (see 11.5 and 13.3).

1. *A supplement in a separate stratum* for the missed elements may be formed to provide for their separate selection. For example, institutional and transient populations are included with special procedures by the Bureau of the Census, as a supplement to its sample of private dwellings. A firm may supplement its payroll list of older employees with a separate list of its newly hired employees.

2. *Linking procedures* attach uniquely the selection of the missed elements to specified listings. This is a useful device in many situations where a separate stratum is too costly, and the missed elements are scattered individually or in small clusters. This last condition is intended to guarantee against the appearance of giant clusters of missed elements. For example, on the payroll list we can prespecify certain unique positions, such as the employee listed last in each section; whenever he is included, any new employee in that section is also to be listed and included. Note that the missed elements receive the same probability of selection as the prespecified unique listings.

This procedure is also called the *half-open interval*, which means that the listings are defined to include the interval up to, but not including, the next listing. The procedure must fit the missed elements into the interval in a clear, practical, and unique manner. The actual inspection of the population needs to be done only within the selected intervals. For example, on the payroll listing, if and only if the last name is selected in a work section, can the procedure specify an interval for listing any employee newly hired into that section. Another example is provided by occasional dwellings missed in block listings; a half-open interval can tell the interviewer to include with the sample dwelling any dwelling up to but not including the next listed dwelling (9.4C). For either a supplement or a linking procedure, we may specify the same sampling fraction as for the main frame. But if the cost per element is much greater for these procedures, we may introduce a smaller sampling fraction, according to the principles of disproportionate allocation (3.5).

B. Clusters of elements can appear together, associated with single listings. If this occurs often, and the average cluster size is not small, we should consider it fully and formally as a cluster sampling problem, discussed in Chapters 5, 6, and 7. (See also 11.3.)

1. *Include all the elements that occur with each selected listing*, if the clusters are rare and small. For example, entries of dwellings prepared

by the enumerator occasionally contain two or perhaps three dwellings; all dwellings occurring with single selected entry should be included in the sample, thus receiving the probabilities assigned to the entries. Similarly, in a small proportion of dwellings, two or even three families are found; all such families found in selected dwellings should be included. With this procedure the elements receive the proper selection probabilities assigned to the listing. This is not commonly understood and one often observes the misguided procedure of the random selection of one element, without recognition of the unequal probabilities incurred. This is one of those situations where "common sense" can fail as a guide.

2. *Select one element* from the cluster, at random, *and weight* it up with the number of elements in the cluster. This weighting compensates for the fact that the probability received in the subselection is inversely proportional to the cluster size.

3. *Relist a larger sample* and then select from it an epsem of elements. For example, if it is known that practically no household contains more than five adults, one may select an epsem sample of dwellings designed to contain five times the number of adults needed. Since dwellings contain an average of about two adults, a sample of 500 dwellings will yield a list of about 1000 adults. A sampling rate of 1/5 applied to this list can produce an epsem of 200 adults without taking more than one adult from any dwelling.

C. ***Blanks or foreign elements*** occur in many frames when some listings contain no elements of the target population. If purifying the frame before selection is too costly, we face problems treated in detail in 11.1.

The selected *blanks must be rejected* and omitted, because they contribute no element to the sample. The question arises: Can the blanks be discovered after selection, but before making costly observations, such as interviews? Such an expensive screening procedure justifies some thought and perhaps double sampling.

Another question is whether the sample size remains fixed, or is permitted to vary as we select from a list containing M members plus B blanks. With simple random sampling, we may specify the selection of exactly m member elements.

For more complex selections, the *expected* sample size is $m = fM$, but the actual sample size will vary as an estimate of the proportion $M/(M + B)$. Thus, by simply omitting the blanks, we obtain the elements as a *subclass*, selected with epsem from the entire list. Treatments for subclasses appear in sections devoted to the different designs.

Avoid the common fallacy of substituting for a blank the element that follows next on the list. This procedure actually increases the probability of selecting any element in proportion to the number of blanks that precede the element on the list. Usually these numbers are unequal and, if they are not compensated, or if they are unknown, they can lead to biases.

Think of the proportion of elements to listings as the density of elements on the list. By selecting the "nearest" element, or any fixed number of sample elements, at points selected with equal probability, the elements are unwittingly given probabilities inversely proportional to the densities where they occur. To the degree that the densities are associated with a variable, its estimate will be biased.

D. Duplicate listings give each element a selection probability proportional to its number P_i of listings (11.2).

1. *Weighting* each case by the inverse of its number of listings can compensate for the selection probabilities proportional to that number. If the element has P_i listings, assign the weight $1/P_i$ to its selection and include duplicate selections. But formal weighting may be replaced with corrective elimination of selections in the ratio $(P_i - 1)/P_i$, and retaining them with the probability $1/P_i$. The relative advantages of weighting versus elimination are discussed in 11.7B; both involve some losses due to the unequal probabilities of selections. One of these losses must be accepted if the P_i can be established only after an expensive observation, e.g., during the interview.

2. *Unique identification* of a single listing for each element may be established at the time of the selection. For example, we may define the *first* (or last) listing as the unique selector; this is particularly easy if the listing is clearly ordered and if the duplicates follow each other. The *selection is confined to the unique listings and all other listings become blanks to be rejected* and treated as in *C* above. For example, in a card file of a university, some students may have more than one card, one for each school in which they are registered. But one card may be designated uniquely as his principal identification. Furthermore, if the duplicate cards can be removed cheaply before selection, then the problem of blanks and duplications can be avoided altogether.

If the ordering of duplicates is not simple, some unique feature may still be designated. For example, on a list of farm parcels, each farmer may be associated uniquely with his *largest* parcel. In the mailing file of insurance policies of a company, the *oldest* policy can be clearly and uniquely identified for each policy holder. These unique mailing cards could be separated mechanically and cheaply, and a sample of policyholders confined to them.

If unique identification seems difficult, but the duplicates can be found, then one of them can be *designated at random* to be the selector listing.

3. *Correction for the sample selections involves removal of duplications from the entire population.* This only restates the procedure of unique identification, but is especially useful when the population consists of several separate frames, in which some elements may be duplicated.

For example, each school of a university may have its own file of students and some students appear in more than one file. We designate an ordering of the frames, and the first frame for each element designates a set of unique selectors. Hence, for each sample case a search of all preceding frames must be made, and the selection discarded if a preceding duplicate is found.

Some further words of caution against common fallacies. First, avoid substitutions for the discarded duplicates. Second, it is a naive belief that the problem is solved by removing duplicates from the sample selections only. This falls far short of correcting for the increased chances of duplicate selection. For example, with simple random sampling of n out of N listings, an element with two listings has an expectation of $2(n/N)$ in the sample, but only $(n/N)(n-1)/(N-1)$ chances of appearing twice in the sample.

*2.8 BASIC FORMULAS AND DERIVATIONS

2.8A Some Fundamentals

This brief development of fundamental formulas serves chiefly to define the most basic concepts and their symbols. These serve as central references for the subsequent development of other important formulas. The development is too brief to prepare fully those readers who have never seen these formulas before. It assumes that the reader is acquainted with the algebra of *expected values*, and with the concepts of *variance* and *covariance*. If he needs them, the reader can refresh his memory in his favorite statistics book, or in one of the texts on sampling.

Consider a population of N elements and the variable Y_i. In the population the total is $Y = \Sigma Y_i$ and the mean is $\bar{Y} = Y/N$. If y_j is the value for a single element selected at random, its *expected value* is the mean of the population distribution:

$$E(y_j) = \sum_i^N \frac{1}{N} Y_i = \bar{Y}. \tag{2.8.1}$$

More generally, *a sampling distribution* has C components y_c, the values possible under the sample design, and the probability of sample value y_c is P_c. The sample value y_c is a statistic, an estimate, and its distribution depends on both the selection and estimation procedures. The expected value of the sample value y_c is the mean value of its sampling distribution:

$$E(y_c) = \sum_c^C P_c y_c. \tag{2.8.2}$$

If the expected value coincides exactly with the corresponding *population value* Y, then y_c is its *unbiased estimate.* Otherwise $E(y_c) - Y = Bias$. The random selection of a single element is a special case with $C = N$, and with $P_c = 1/N$. The sampling distribution of medians or standard deviations based on a simple random sample are instances when $E(y_c) \neq Y$ generally. The sampling distribution of the mean is unbiased for simple random sampling and some other selection designs, but not for all.

The *variance* of the sample value y_c is the mean square deviation from its expected value:

$$\text{Var}(y_c) = \sum P_c[y_c - E(y_c)]^2 = E[y_c - E(y_c)]^2 = E(y_c^2) - [E(y_c)]^2. \tag{2.8.3}$$

The variance is a special case of covariance, $\text{Var}(y_c) = \text{Cov}(y_c, y_c)$, and the *covariance* of two statistics y_c and x_c is defined as

$$\text{Cov}(y_c, x_c) = E\{[y_c - E(y_c)][x_c - E(x_c)]\}. \tag{2.8.3'}$$

The standard deviation of the sampling distribution, the positive square root of its variance, is the *standard error* of the statistic y_c: $\text{SE}(y_c) = \sqrt{\text{Var}(y_c)}$. The *mean square error* is

$$\begin{aligned} \text{MSE}(y_c) &= E(y_c - Y)^2 = E[y_c - E(y_c) + E(y_c) - Y]^2 \\ &= \text{Var}(y_c) + \text{Bias}^2. \end{aligned} \tag{2.8.4}$$

The addition of the constant $[E(y_c) - Y]$ has no effect on the variance, and the unbiased $(y_c - \text{Bias})$ has the same variance as y_c. Thus Bias^2 measures the increase of the mean square error due to the bias. Good designs of large samples usually result in bias that is either absent or negligible compared to the variance.

The addition of a constant has no effect on the variance, and multiplication with constant factors has simple effect. Thus,

$$\begin{aligned} &\text{Var}(y_c + K) = \text{Var}(y_c), \\ &\text{Var}(Wy_c) = W^2\,\text{Var}(y_c), \\ &\text{Var}(Wy_c + K) = W^2\,\text{Var}(y_c), \\ &\text{SE}(Wy_c) = |W|\,\text{SE}(y_c), \\ &\text{Cov}(W_1y_1, W_2y_2) = W_1W_2\,\text{Cov}(y_1, y_2). \end{aligned} \tag{2.8.5}$$

The sum of n random variables has this variance:

$$\begin{aligned}\text{Var}\left(\sum_j^n y_j\right) &= \sum_j^n \text{Var}(y_j) + \sum_{j\neq k}^n \text{Cov}(y_j, y_k)\\ &= \sum_j^n \text{Var}(y_j) + 2\sum_{j<k}^n \text{Cov}(y_j, y_k).\end{aligned} \qquad (2.8.6)$$

This is the sum of all the terms in the $n \times n$ *covariance matrix*, in which the variances are the diagonal terms. When the n variables are uncorrelated, the covariances vanish, resulting in

$$\text{Var}\left(\sum y_c\right) = \sum \text{Var}(y_c). \qquad (2.8.7)$$

A single element with value y_j, selected at random from a population of N elements with values Y_i, has the expected value $\bar{Y}$, and variance σ_y^2, equal to the variance of the population distribution of elements:

$$\text{Var}(y_j) = E(y_j - \bar{Y})^2 = \sum \frac{1}{N}(Y_i - \bar{Y})^2 = \sigma_y^2. \qquad (2.8.8)$$

If we select at *random with replacement* n elements, we obtain an *unrestricted random sample*. The variance of the sample sum $y = \Sigma y_j$ becomes

$$\text{Var}(y) = \text{Var}\left(\sum y_j\right) = \sum \text{Var}(y_j) = \sum \sigma_y^2 = n\sigma_y^2. \qquad (2.8.9)$$

To obtain the sample mean $\bar{y} = y/n$ from the sample sum y involves merely the known constant factor $1/n$; this is true for an unrestricted sample and for other designs with fixed sample size n. The variance of $\bar{y}$ is obtained by applying (2.8.5) to (2.8.9):

$$\text{Var}(\bar{y}) = \text{Var}\left(\frac{y}{n}\right) = \frac{1}{n^2}n\sigma_y^2 = \frac{\sigma_y^2}{n}. \qquad (2.8.10)$$

To estimate the population aggregate Y with $N\bar{y}$, using the known population size N, involves again only (2.8.5):

$$\text{Var}(N\bar{y}) = N^2\,\text{Var}(\bar{y}), \quad \text{and} \quad \text{SE}(N\bar{y}) = N\,\text{SE}(\bar{y}). \qquad (2.8.11)$$

When N is unknown, the inverse of the sampling fraction $F = 1/f$ may be used for the estimate Fy of the aggregate; then

$$\text{Var}(Fy) = F^2\,\text{Var}(y), \quad \text{and} \quad \text{SE}(Fy) = F\,\text{SE}(y).$$

The covariance of the sums y and x, based on pairs of measures (y_j and x_j) on the n elements of an unrestricted random sample, is similar to the variance in (2.8.7) and (2.8.9):

$$\text{Cov}(y, x) = \text{Cov}\left(\sum y_j, \sum x_j\right) = \sum \text{Cov}(y_j, x_j) = \sum \sigma_{yx} = n\sigma_{yx}, \qquad (2.8.9')$$

where $\sigma_{yx} = \frac{1}{N} \Sigma (Y_i - \bar{Y})(X_i - \bar{X})$ is the covariance of the two variables Y_i and X_i for the N elements in the population distribution. The covariance of the sample means $\bar{y} = y/n$ and $\bar{x} = x/n$ is obtained similarly to the variance (2.8.10):

$$\text{Cov}(\bar{y}, \bar{x}) = \text{Cov}\left(\frac{y}{n}, \frac{x}{n}\right) = \frac{1}{n^2} n\sigma_{yx} = \frac{\sigma_{yx}}{n}. \qquad (2.8.10')$$

2.8B Variance in Simple Random Sampling Without Replacement

In (2.8.9) and (2.8.10) we found variances for unrestricted sampling, that is, simple random sampling with replacement. We now derive the variance of the sample sum y for a *simple random sample*, which is understood to be *without* replacement. This sample sum y is shown later to be an unbiased estimate of $n\bar{Y}$, that is, $E(y) = n\bar{Y}$. Hence, the variance is defined for y as

$$\begin{aligned}\text{Var}(y) &= E\left\{\left(\sum_j^n y_j - n\bar{Y}\right)^2\right\} = E\left\{\left[\sum_j^n (y_j - \bar{Y})\right]^2\right\} \\ &= E\left\{\sum_{j,k}^n (y_j - \bar{Y})(y_k - \bar{Y})\right\} \\ &= E\left\{\sum_j^n (y_j - \bar{Y})^2 + \sum_{j \neq k}^n (y_j - \bar{Y})(y_k - \bar{Y})\right\}.\end{aligned}$$

Squaring the n terms results in an $n \times n$ matrix in which we separate the n variances on the diagonal terms from the other $n(n-1)$ covariance terms. In unrestricted sampling only the first term, the variance $n\sigma_y^2$, remains. The covariance terms vanish due to the independence between all pairs, j and k, of selections. But sampling without replacement lacks that complete independence; for each of the $n(n-1)$ sample covariance pairs, the expected value is the average value among the $N(N-1)$ population covariance pairs. Similarly, for each of the n variance terms, the expected value is the variance of population elements. Thus,

$$\begin{aligned}\text{Var}(y) &= \frac{n}{N}\sum_i^N (Y_i - \bar{Y})^2 + \frac{n(n-1)}{N(N-1)}\left[\sum_{i \neq h}^N (Y_i - \bar{Y})(Y_h - \bar{Y})\right] \\ &= n\sigma_y^2 + \frac{n(n-1)}{N(N-1)}\left[\left\{\sum_i^N (Y_i - \bar{Y})\right\}^2 - \sum_i^N (Y_i - \bar{Y})^2\right].\end{aligned}$$

The N population terms squared result in the $N \times N$ matrix $\{\Sigma (Y_i - \bar{Y})\}^2$, and subtracting the N variance terms on the diagonal leaves the needed covariance terms. Note now that the term in the { } braces vanishes because $\Sigma (Y_i - \bar{Y}) = 0$; also that $\Sigma (Y_i - \bar{Y})^2 = N\sigma_y^2 = (N - 1)S_y^2$. Thus,

$$\text{Var}(y) = n\sigma_y^2 - \frac{n(n-1)}{N-1}\sigma_y^2 = \frac{N-n}{N-1}n\sigma_y^2$$

$$= \frac{N-n}{N}nS_y^2 = (1-f)nS_y^2, \tag{2.8.12}$$

where $f = n/N$. Then, because the sample size n is a fixed known constant, we get for the sample mean $\bar{y} = y/n$:

$$\text{Var}(\bar{y}) = \text{Var}\left(\frac{y}{n}\right) = \frac{1}{n^2}\text{Var}(y) = \frac{1-f}{n}S_y^2 = \left(\frac{1}{n} - \frac{1}{N}\right)S_y^2. \tag{2.8.13}$$

Also,

$$\text{Var}(N\bar{y}) = N^2\,\text{Var}(\bar{y}) = (1-f)\frac{N^2}{n}S_y^2. \tag{2.8.14}$$

Similarly to the variances, we have for the covariances:

$$\text{Cov}(y, x) = (1-f)nS_{yx} \quad \text{and} \quad \text{Cov}(\bar{y}, \bar{x}) = \frac{1-f}{n}S_{yx},$$

where
$$S_{yx} = N\sigma_{yx}/(N-1). \tag{2.8.15}$$

Finally, we show that

$$s_y^2 = \sum_j^n (y_j - \bar{y})^2/(n-1) = \left(\sum_j^n y_j^2 - n\bar{y}^2\right)\Big/(n-1)$$

is an unbiased estimate of S_y^2: that $E(s_y^2) = S_y^2$. Because $\text{Var}(\bar{y}) = E(\bar{y} - \bar{Y})^2 = E(\bar{y}^2) - \bar{Y}^2$, we have

$$E\left(\sum_j^n y_j^2 - n\bar{y}^2\right) = \sum_j^n E(y_j^2) - nE(\bar{y}^2) = \frac{n}{N}\sum_i^N Y_i^2 - n[\text{Var}(\bar{y}) + \bar{Y}^2]$$

$$= n\left(\frac{1}{N}\sum_i^N Y_i^2 - \bar{Y}^2\right) - n\left(\frac{1}{n} - \frac{1}{N}\right)S_y^2$$

$$= n\sigma_y^2 + \frac{n}{N}S_y^2 - S_y^2 = n\left(\frac{N-1}{N} + \frac{1}{N}\right)S_y^2 - S_y^2$$

$$= (n-1)S_y^2.$$

Hence,

$$E(s_y^2) = S_y^2, \quad \text{where } s_y^2 = \frac{1}{n-1}\left(\sum_j^n y_j^2 - n\bar{y}^2\right). \tag{2.8.16}$$

For unrestricted sampling we had $\text{Var}(\bar{y}) = \sigma_y^2/n$, and the above derivation gives

$$E\left(\sum y_j^2 - n\bar{y}^2\right) = n\sigma_y^2 - n\sigma_y^2/n = (n-1)\sigma_y^2.$$

Hence $E(s_y^2) = \sigma_y^2$. Thus, either with or without replacement, we have these formulas, except that $f = 0$ when sampling with replacement.

$$E\,[\text{var}\,(y)] = \text{Var}\,(y), \qquad \text{where} \quad \text{var}\,(y) = (1-f)ns_y^2, \tag{2.8.17}$$

and
$$E\,[\text{var}\,(\bar{y})] = \text{Var}\,(\bar{y}), \qquad \text{where} \quad \text{var}\,(\bar{y}) = (1-f)s_y^2/n. \tag{2.8.18}$$

2.8C The Expectations of Some Important Types of Samples

Suppose that on each of n draws a single element is selected and that $1/N$ is the probability of selection for each of the N population elements on each draw. Then the expected value of the sample sum is

$$E(y) = E(\sum y_j) = \sum E(y_j) = \sum \bar{Y} = n\bar{Y} = \frac{n}{N}\,Y = fY. \tag{2.8.19}$$

The central point is that

$$E(y_j) = \sum^{N} P_i Y_i = \sum \frac{1}{N}\,Y_i = \bar{Y},$$

because the probability of selection is $P_i = 1/N$ for every element ($i = 1, 2, \ldots, N$) on every draw ($j = 1, 2, \ldots, n$). This is obvious in unrestricted random sampling, where each element is drawn separately from the entire population of N elements, in which the previous draw has been replaced. Although less obvious, it can also be shown for simple random sampling without replacement (Remark 2.2.II). In either case, for any of the N population elements the expected appearance in a sample of size n is equal to $n(1/N) = n/N = f$, the sampling fraction.

Unrestricted and simple random selections are only two special cases of a general situation that holds true for all epsem selections. Epsem denotes *any* method of selection that guarantees for N population elements an equal and known selection probability, denoted by f. Hence in an epsem selection the expected appearance in the sample sum y equals the same value f for all N values Y_i in the population. Then the sample sum y of any epsem may be regarded as having been selected in a single draw with equal probability from a sampling distribution, for each of whose components it is true that

$$E(y) = fY \qquad \text{and} \qquad E\left(\frac{y}{f}\right) = Y = N\bar{Y}. \tag{2.8.20}$$

This can be proven with auxiliary random variables δ_i associated with each of the population elements ($i = 1, 2, \ldots, N$), and whose expected value in the sample is f for every element. These variables need not be generally independent, as they are with unrestricted random sampling.

When the sample size is not fixed, these variables show that the expected sample size is $E(n) = E\left(\sum_i \delta_i\right) = E(\Sigma f) = fN$.

When sampling without replacement, this variable is $\delta_i = 1$ if the ith element appears in a particular sample, and $\delta_i = 0$ if it does not appear. Hence it represents the probability as well as the expected contribution of the ith element to a sample. The expected contribution of the ith population element to any one sample is the expectation for the product of its constant value Y_i, with the random variable that represents its appearance in the sample: $E(\delta_i Y_i) = Y_i E(\delta_i) = Y_i f$. This follows from formula (1.3.1) for expected values:

$$E(\delta_i Y_i) = \sum_c Y_i \, \delta_{ic} P_c = Y_i \sum_c \delta_{ic} P_c = Y_i f,$$

the summation being taken over all possible samples in the sampling distribution. The sample total $y = \sum_j y_j$ represents the sum of contributions from all elements which appear in the sample. Therefore, the expected value of the sample sum y is the sum of individual expectations for all N population elements:

$$E\left(\sum_j y_j\right) = E\left(\sum_i Y_i f\right) = fY.$$

When sampling with replacement, $\delta_i = 0$ if the element does not appear in the sample, $\delta_i = 1$ if it appears once, $\delta_i = 2$ if it appears twice, and $\delta_i = k$ if it appears k times. Its expected value over all samples is f, and it is a random variable that represents the number of appearances of the ith element in the sample. If an element appears k times in the sample, it gets counted k times. Proof that $E(y) = fY$ follows as before.

Thus y/f is an unbiased estimate of Y, when y is the sample sum of an epsem selected with a known fixed constant f. When the sample size is also a fixed constant, the sample mean $\bar{y} = y/n$ is also an unbiased estimate of $\bar{Y}$, because $n = fN$ necessarily when n and N are both fixed constants; hence,

$$E(\bar{y}) = E\left(\frac{y}{n}\right) = \frac{1}{n} E(y) = \frac{f}{n} Y = \bar{Y}. \qquad (2.8.21)$$

These results have wide utility. It is usually easy to state for a sampling design if it is epsem and if n is fixed. Then the above relations follow immediately, without having to derive them separately for the many kinds of epsem designs that are the bases of most survey samples. Epsem with fixed n occurs in different kinds of element sampling, and in the sampling and subsampling of equal clusters; for all of these, both the

$\bar{y}$ and y/f are unbiased. Epsem with variable n occurs with unequal sized clusters, and when dealing with subclasses; for these, y/f is still unbiased, whereas $\bar{y}$ is unbiased for some designs, but not for others. However, even when it has a small bias in epsem selections, $\bar{y}$ is usually employed and preferred. With a fixed sampling fraction $E(n/f) = N$, and we have

$$\frac{E(y/f)}{E(n/f)} = \frac{Y}{N} = \bar{Y}.$$

But if n is not fixed, we may still have $E(y/n) \neq \bar{Y}$, and lack an unbiased sample mean.

If all the elements have an equal selection probability f' which is *not* known, then (2.8.20) cannot be applied directly. This can occur in situations where the population size is unknown. For example, we may select a fixed number n of balls from an urn with equal probability f' without knowing either f' or the population size. This kind of problem occurs with "capture-recapture" techniques of sampling fish and other free animals. Also, consider a frame with $N = M + B$ listings with B blanks and the number M of population elements unknown; if a sample with fixed probability $f = n/N$ is selected, the sample size m will not be fixed, but a variate (see Remark 2.2.II). Alternatively, if a fixed sample size m is selected, the selection probability f' will not be fixed, but a variate.

When the probability is unknown, the selection is not probability sampling according to its definition of known and nonzero probability. Hence, it is not epsem either, if that term is restricted to mean *equal and known probabilities.* When the probabilities are equal but not known, we call the selection an *equal chance* selection method. An important case is sampling with equal probabilities from an infinite hypothetical universe. If the population size and the equal selection probability are both entirely unknown, then the population aggregate Y cannot be estimated from the sample alone. If a fairly good estimate f' of the selection probability is available, then y/f' can be used; it may or may not be an unbiased estimate of Y. The sample mean $\bar{y} = y/n$ would be unbiased if the selection method were unrestricted or simple random. With other equal chance selections, $\bar{y}$ may not be unbiased, but would likely be used.

The preceding discussion was based on equal selection probabilities. But the concepts should and can be generalized to selections when the probabilities vary between the elements. Assume now that the selection probability of element i with value Y_i is fP_i, where the known and nonzero P_i vary between the elements, and f is a known common selection constant. This describes *probability sampling*, with known nonzero selection probabilities for all population elements, and *epsem* is only its special case with all $P_i = 1$. Now if the elements in the sample are weighted inversely to the selection probabilities, then for the sample sum y_{pw} we obtain the generalization of (2.8.20):

$$E(y_{pw}) = E\left(\sum_j \frac{y_j}{p_j}\right) = fY. \tag{2.8.20'}$$

The proof here is an extension of that of (2.8.20). The expected contribution of the ith element to the sample is $E(\delta_i Y_i) = Y_i E(\delta_i) = Y_i f P_i$. The value of Y_i varies between elements, but is a constant for a fixed element; hence, $E(\delta_i Y_i / P_i) = fY_i$. For the sample total we sum these element expectations over the population; hence,

$$E(y_{pw}) = E\left(\sum_i \frac{\delta_i Y_i}{P_i}\right) = \sum_i E\left(\frac{\delta_i Y_i}{P_i}\right) = \sum_i f Y_i = fY. \tag{2.8.19'}$$

The sample size represents the sample total for the count variable $Y_i = 1$; hence, $E\left(\sum_i \delta_i / P_i\right) = fN$. Consequently, this important property of the sample sum holds for all probability sampling. In some situations the selection probabilities may be $f'P_i$, and the relative values P_i known and nonzero, but with f' an unknown constant of proportionality. Examples are similar to those given above for urns of unknown size and hypothetical infinite universes. We shall denote as *randomized selection* this generalization of probability sampling; similarly, when all $P_i = 1$, we used the term equal chance selection for the generalization of epsem selection.

For a sample mean corresponding to $\bar{y} = y/n$, we have, for unequal probabilities,

$$\bar{y}_{pw} = \frac{\sum y_j / p_j}{\sum 1/p_j}. \tag{2.8.21'}$$

This estimate is not unbiased; because its denominator is not fixed, but a random variable, it is a ratio mean (6.6). Nevertheless it will usually be the preferred estimate, as $\bar{y} = y/n$ is the preferred estimate for epsem selections.

Note the vital fact that *subclasses inherit* from the entire sample any of the four broad types of selection we discussed: epsem, equal chance, probability, or randomized selection. Fixed sample size is not inherited by subclasses. But for this exception, simple random sampling or unrestricted sampling are also inherited by subclasses. On the other hand, other selection types—such as stratified, systematic, or equal sized clusters—are not generally inherited by subclasses.

The nature of Y_i was not specified. It may represent not only a simple variable of the ith population element, but some function of it, or even a function of several variables. For example, if $Y_i = X_i^n$ or $Y_i = X_i^n Z_i^m$, the unbiased nature of higher moments of one or two variables are established for the four types of randomized selections. Furthermore, linear combinations of unbiased estimates of the functions will also be unbiased. Many statistics computed from the unbiased sample functions will be, if not unbiased, consistent estimators [Kish, 1964].

The importance of this principle can be illustrated by obtaining a much-needed result: estimates of the population variance σ_y^2 from any epsem or other probability sample. From the sample we construct n, $y = \Sigma\, y_j$, and $\Sigma\, y_j^2$, either self-weighted or properly weighted. Since $E(n) = fN$, and $E(y) = fY$, and $E(\Sigma\, y_j^2) = f\Sigma\, Y_i^2$, we get

$$E\left(\sum_j^n y_j^2 - \frac{y^2}{n}\right) = E\left[\left(\sum_j^n y_j^2 - \frac{fY^2}{N}\right) - \left(\frac{y^2}{n} - \frac{f^2Y^2}{fN}\right)\right]$$
$$= f\left(\sum_i^N Y_i^2 - \frac{Y^2}{N}\right) - E\left(\frac{y^2}{n} - \frac{f^2Y^2}{fN}\right).$$

Thus
$$E(nv_y^2) = fN\sigma_y^2 - E\left(\frac{y}{n}\,y - \frac{f^2Y^2}{fN}\right), \tag{2.8.22}$$

where $v_y^2 = (\Sigma\, y_j^2/n - \bar{y}^2) = (n-1)s^2/n$. We should like also to express the expectation of this element variance in the sample. When n is fixed at fN for the sample, we have directly that

$$E(v_y^2) = \sigma_y^2 - E\frac{(y^2 - f^2Y^2)}{(fN)^2} = \sigma_y^2 - \text{Var}\,(\bar{y}),$$

and

$$E[v_y^2 + \text{var}\,(\bar{y})] = \sigma_y^2\,, \text{ when } E[\text{var}\,(\bar{y})] = \text{Var}\,(\bar{y}). \tag{2.8.23}$$

When n is not actually fixed, the analysis may be made *conditional* on a fixed n, and arrive at essentially the same result. I believe that these will be the preferred statistics from some modern points of view (14.3). Furthermore, it can be shown with methods of 6.6B that the bias is bound to be usually small for nv_y^2/n considered as a ratio mean. Hence,

$$E(v_y^2) \doteq \frac{E(nv_y^2)}{E(n)} = \sigma_y^2 - E\left(\frac{y}{n}\cdot\frac{y}{fN} - \frac{Y}{N}\cdot\frac{Y}{N}\right)$$
$$\doteq \sigma_y^2 - R_{\frac{y}{n}\cdot\frac{y}{fN}}\cdot\sigma_{\frac{y}{n}}\cdot\sigma_{\frac{y}{fN}}\,. \tag{2.8.23'}$$

The second term becomes Var (y/n) for fixed n, and it should approach the mean square error $\sigma_{\bar{y}}^2$ of $\bar{y}$ when n is *not* fixed. Generally, then, $v_y^2 + \sigma_{\bar{y}}^2$ computed from the sample will be a good estimate of σ^2 (or S^2) among the population elements.

Var $(\bar{y})$ is *roughly* σ^2/n for many designs, and then $s_y^2 = v_y^2 n/(n-1) = (\Sigma\, y_j^2 - y^2/n)/(n-1)$ may be employed to estimate σ_y^2. The result in

(2.8.16) is the special case for simple random sampling when Var $(\bar{y}) = (1 - f)S_y^2/n$.

Similarly v_{yx} + cov $(\bar{y}, \bar{x})$ estimates σ_{yx} generally, and s_{yx} is a reasonably good estimate of S_{yx}, for all probability samples. Hence $v_{yx}/v_y v_x$ will also estimate $R_{yx} = \sigma_{yx}/\sigma_x\sigma_y$.

The criterion of unbiasedness has limited usefulness, I think. One reason for its appearance here is that I am anxious to preserve ties to the literature of sampling. Although unbiasedness is given a prominent role, it is usually abandoned for the most important designs of survey sampling, such as unequal clusters. This seems to indicate the implicit acceptance of more important criteria. The criterion of unbiasedness is also criticized by the *neo-Bayesian* school of statisticians, and by others who emphasize the likelihood principle (see 14.3).

2.8D Variances for Linear Combinations

We can utilize (2.8.5) and (2.8.6) to obtain variances for some more complicated linear combinations that we shall need later. The sum of H random variables, weighted by the constant factors W_h has the variance:

$$\text{Var}\,(\sum W_h y_h) = \sum_h W_h^2 \,\text{Var}\,(y_h) + 2\sum_{h<g} W_h W_g \,\text{Cov}\,(y_h, y_g). \tag{2.8.24}$$

A common example is the sum or difference of two random variables y_1 and y_2, when $W_1 = 1$ and W_2 is either 1 or -1:

$$\text{Var}\,(y_1 \pm y_2) = \text{Var}\,(y_1) + \text{Var}\,(y_2) \pm 2\,\text{Cov}\,(y_1, y_2). \tag{2.8.25}$$

The covariance vanishes if y_1 and y_2 are uncorrelated. Another important special case occurs when all H variates are uncorrelated, because they are based on independent samples from H strata (Section 3.1). Then all the covariances vanish and

$$\text{Var}\,(\sum W_h y_h) = \sum W_h^2 \,\text{Var}\,(y_h). \tag{2.8.26}$$

If all the $W_h = 1$, then this becomes (2.8.7).

Similarly, we can consider the covariance of the sums $\Sigma\, W_h y_h$ and $\Sigma\, V_h x_h$ of two sets of random variables, again assuming independence between the H sets; for example, these could be pairs of measurements on H independently selected elements:

$$\text{Cov}\,(\sum W_h y_h, \sum V_h x_h) = \sum W_h V_h \,\text{Cov}\,(y_h, x_h). \tag{2.8.27}$$

When the constants W_h and V_h are all unity, we have the simple case similar to (2.8.9′):

$$\text{Cov}\,(\sum y_h, \sum x_h) = \sum \text{Cov}\,(y_h, x_h). \tag{2.8.28}$$

The formulas for variances and covariances of linear combinations were developed for population values, written with capital letters as Var and Cov. But they apply also to their sample estimates, which we write with lower case letters as var and cov. The property of unbiasedness is inherited in these formulas for linear combinations (2.8.5, 2.8.24–2.8.28). Summation of estimated variances and covariances for sample totals within strata is simple and frequently needed. Therefore, we employ the brief notation dy_h^2, dx_h^2, and $dy_h dx_h$. When the y_h and x_h represent two variates for selections that are independent between strata, we have

$$\text{var}\left(\sum y_h\right) = \sum \text{var}(y_h) = \sum dy_h^2,$$

$$\text{var}\left(\sum x_h\right) = \sum \text{var}(x_h) = \sum dx_h^2,$$

$$\text{cov}\left(\sum y_h x_h\right) = \sum \text{cov}(y_h, x_h) = \sum dy_h dx_h. \tag{2.8.29}$$

PROBLEMS

2.1. This is a practical problem and its first purpose is to teach basic symbols and computations. The results are also used later for basic problems in stratified and cluster sampling.

You can compute in one set of operations the five needed terms Σx, Σy, Σx^2, Σy^2, and $2\Sigma xy$. This is done by entering the quantities $(10^k x + y)$ on the keyboard of a desk computer, squaring them, and cumulating the 20 squares. The three product terms will remain distinct if $k \geq 6$. The choice of k depends on the size of the computer and the numbers x, y, and n. This operation is particularly useful in many problems of computing survey variances and covariances.

The population consists of the $N = 270$ blocks in Appendix E. Select an srs of $n = 20$ blocks by drawing three-digit random numbers until 20 different block numbers are selected. Disregard any random number denoting either a blank (no corresponding block number) or a number already selected. (For a class project, it is convenient if the students use identical selections.) For the selected 20 blocks, copy the pairs of numbers denoting the numbers of x_j = total dwellings and y_j = rented dwellings.

(*a*) Compute the mean number of dwellings per block : $\bar{x} = x/n = \sum^n x_j/n$.

(*b*) Compute the element variance $s_x^2 = \left(\sum^n x_j^2 - x^2/n\right) \Big/ (n - 1)$.

(*c*) Compute $\text{var}(\bar{x}) = (1 - f)s_x^2/n$ and $\text{se}(\bar{x}) = \sqrt{\text{var}(\bar{x})}$.

(*d*) Estimate the total number of dwellings in the population $N\bar{x}$ and $\text{se}(N\bar{x}) = N\,\text{se}(\bar{x})$.

(*e*) Compute the coefficients of variation

$$\text{cv}(\bar{x}) = \text{se}(\bar{x})/(\bar{x}) = \text{se}(N\bar{x})/N\bar{x} = \text{cv}(N\bar{x}).$$

(*f*) Compute the mean number of rented dwellings per block $\bar{y} = y/n = \sum^{n} y_j/n$ and the total number of rented dwellings $N\bar{y}$.

(*g*) Compute the element variance s_y^2.

(*h*) Compute var $(\bar{y})$ and se $(\bar{y})$; also se $(N\bar{y})$ and cv $(\bar{y})$ = cv $(N\bar{y})$.

(*i*) Compute the element covariance term

$$2s_{xy} = 2\left(\sum^{n} x_j y_j - xy/n\right) \Big/ (n - 1)$$

and the covariance of the means 2cov $(\bar{x}, \bar{y}) = (1 - f)2s_{xy}/n$.

(*j*) Compute the variance of the difference $(\bar{x} - \bar{y})$ of the two means: var $(\bar{x} - \bar{y})$ = var $(\bar{x})$ + var $(\bar{y})$ − 2cov $(\bar{x}, \bar{y})$ and se $(\bar{x} - \bar{y}) = \sqrt{\text{var}(\bar{x} - \bar{y})}$. Note that the same values can be obtained from the $n = 20$ values of $d_j = (x_j - y_j)$, computing $\bar{d} = (\bar{x} - \bar{y})$ and se $(\bar{d})$.

(*k*) If the sample size were increased from $n = 20$ to $n = 45$, what would be the estimated standard error of $\bar{x}$?

(*l*) What sample size n is needed to make the standard error 3?

2.2. For the sample of $n = 20$ blocks of Example 2.2*a*, compute also the variance of the difference of the means for total and rented dwellings: var $(\bar{x}_0 - \bar{y}_0)$ = var $(\bar{x}_0)$ + var $(\bar{y}_0)$ − 2 cov $(\bar{x}_0, \bar{y}_0)$, This is the same as the variance of the sample mean $\bar{d}$ for the 20 variables $d_j = (x_j - y_j)$, the numbers of nonrented, that is owned, dwellings in the blocks. Use this as your check.

2.3. In Example 2.2*a* the element coefficient of variation was $c_y = s_y/\bar{y} = 16.7/12.75 = 1.31$ for rented dwellings. This coefficient, obtained in one ward, is expected to hold roughly for the entire city of $N = 1111$ blocks. (*a*) For a sample of $n = 100$ blocks in the city, what coefficient of variation, cv $(\bar{y})$, for the city do you expect? (*b*) Suppose that error is too large and a cv $(\bar{y})$ of about 0.05 is desired; how many blocks are needed for that sample?

2.4. For a health survey of a large population, estimates are wanted for two proportions, each measuring the yearly incidence of a disease. For designing the sample, we guess that one occurs with a frequency of 50 percent and the other with a frequency of only 1 percent. To obtain the same standard error of $\frac{1}{2}$ percent, how large an srs is needed for each disease? The large difference in the needed n causes a re-evaluation of the requirements. Now the same coefficient of variation of 0.05 is declared desirable for each disease; how large a sample is needed for each disease? Write a few sentences about how else we might formulate the required precision.

2.5. The following values of y_j were noted in an srs of $n = 40$ from a population of $N = 4000$.

10, 8, 6, 5, 3,	3, 8, 5, 0, 9,	9, 0, 4, 3, 1,	2, 3, 4, 0, 6,
9, 5, 0, 8, 9,	0, 4, 10, 8, 0,	10, 5, 6, 1, 3,	3, 1, 5, 5, 4.

Compute the mean and the aggregate $N\bar{y}$ and their standard errors. You may use (2.2.1) and (2.2.2), either directly or with the grouped formulas where f_g is the number of sample cases with the value y_g:

$$y = \sum^n y_j = \sum f_g y_g \quad \text{and} \quad \sum^n y_j^2 = \sum f_g y_g^2.$$

In the above

f_g	6	3	1	6	4	6	3	0	4	4	3
y_g	0	1	2	3	4	5	6	7	8	9	10

Note: These numbers seem to be rather *uniformly distributed* between 0 and 10. Actually, they are 40 numbers from 0 to 10 taken consecutively from a table of random numbers. Thus they come from a rectangular distribution with $\bar{Y} = 5$, an expectation of $f_g = 40/11$ for each y_g and $S^2 = 10$ (see 8.2).

2.6. Using the s^2 computed in Problem 2.5 as the estimate of S^2, how large an n do you need to obtain for the mean $\bar{y}$ a standard error of 0.25, from populations of: (*a*) $N = 8000$, (*b*) $N = 800{,}000$, (*c*) $N = 320$?

2.7. In another population of $N = 8000$, the mean is guessed to be about $\bar{Y} = 10$, the distribution roughly uniform from 0 to 20, and the coefficient of variation $C = S/\bar{Y}$, about the same as in Problem 2.5. How large a sample do you need to obtain a standard error of 0.25 for the mean?

2.8. The six faces of a die represent the population $Y_i = 1, 2, 3, 4, 5$, and 6; $N = 6$; $\bar{Y} = 3.5$; and $\sigma^2 = 35/12$. Throwing a pair of good dice represents *unrestricted* selection of $n = 2$ from that population.

(*a*) How many equally probable samples are there? Write them down.

(*b*) Write down all values in the sampling distribution for the sample total y and mean $\bar{y} = y/2$. Check that the expected value $E(\bar{y}) = \bar{Y}$. Would this hold for any numbers on the faces of the die?

(*c*) The variance of the sample total is Var $(y) = 2\sigma^2$, and the variance of the mean is Var $(y/2) = \frac{1}{4}(2\sigma^2) = \sigma^2/2$. Check Var (y) by computing the mean square error of its sampling distribution.

2.9. Excluding all "doubles" from the throws of a pair of good dice represents a *simple random selection* of $n = 2$ from the population of $N = 6$ above.

(*a*) Write down all $\binom{6}{2}$ equally probable combinations.

(*b*) Write down all values in the sampling distributions of the sample total y and $\bar{y} = y/n$.

(*c*) The variance of the sample total is Var $(y) = (1 - n/N)nS^2 = n^2$ Var $(\bar{y})$. Check Var (y) by computing the mean square error of its sampling distribution.

(*d*) Note that the Var ($\bar{y}$) for the simple random sample is smaller by the factor $(N - n)/(N - 1)$ than Var ($\bar{y}$) for the unrestricted sample in Problem 2.8.

2.10. Simple random sampling in Problem 2.9 was a restriction on the unrestricted sample of Problem 2.8. Now impose further restrictions, while maintaining the epsem selection fraction $f = 2/6$. Note also that each of the three designs uses a subset of all possible designs of the srs.

(*a*) After dividing the population into the *strata* (1, 2, 3) and (4, 5, 6), one element can be selected from each stratum. Write down the possible samples and compute the mean square error for the distribution of the sample total y.

(*b*) A *systematic* sample can be selected with the interval 3 after a random start from 1 to 3. For the three possible values of y, compute the mean square error.

(*c*) After forming the *clusters* (1, 2), (3, 4), and (5, 6), one cluster can be selected at random. Compute the mean square error for the three possible values of y.

2.11. This shows how we can speed up the selections from a table of random numbers.

(*a*) The population of $N = 1111$ is numbered from $i = 100{,}001$ to 101,111; select five random numbers from the 1,000,000 six-digit numbers.

(*b*) Consider only the numbers 0001 to 1111, and select five numbers from the 10,000 four-digit numbers.

(*c*) Consider that each element can be selected by any of the five numbers i, $i + 2000$, $i + 4000$, $i + 6000$, $i + 8000$. With the 1111 elements thus confined to 2000 numbers, select five four-digit numbers.

2.12. Sometimes the frame for a large population does not separately list the elements, but only clusters, with counts of elements for each cluster. But an srs can be selected from such a frame if lists of elements can be prepared later for each selected cluster.

The 270 blocks of Appendix E contain 6786 dwellings. Select 10 different random numbers from 1 to 6786 and put these *selection numbers* in ascending order. Now add the total dwellings (column 2) cumulatively. Each time the addition of a block reaches or passes a selection number, that block is chosen.

The selection numbers also indicate the specific dwellings to be taken from the lists of dwellings in the chosen blocks. Alternatively, we could make a new random selection of a dwelling from each chosen block.

Assume that rented dwellings precede owned dwellings in each block, and note whether each dwelling is rented (1) or owned (0). Compute the proportion of rented dwellings in your sample of 10 dwellings.

2.13. Suppose that each of the samples below was simple random, selected with the fraction $f = 1/4$. State whether you think the factors $(1-f) = 3/4$ should or should not be used in computing the variances (Yes or No).

(*a*) The current population count of a town is estimated from a sample census.
(*b*) The prevalence of habitual smokers is estimated in a "typical" town, considered "a random sample of all similar towns of its kind."
(*c*) The change $(t_2 - t_1)$ in dollar volume of a store's inventory is estimated from samples taken on December 30 of years 1 and 2.
(*d*) The difference $(P_a - P_b)$ in proportions of habitual smokers in two countries (*a* and *b*) is estimated from samples selected from a town chosen to be "typical" in each country.
(*e*) The "null hypothesis" of $(P_a - P_b) = 0$ is tested: that is, that the means of the two towns are random samples from the same population.

3

Stratified Sampling

3.1 DEFINITION AND PURPOSES

This chapter develops general properties common to various types of stratified sampling and then applies these properties to *element* sampling. In broad terms, stratified sampling consists of the following steps: (*a*) The entire population of sampling units is divided into *distinct subpopulations*, called *strata*. (*b*) *Within each stratum a separate sample is selected* from all the sampling units composing that stratum. (*c*) From the sample obtained in each stratum, a separate stratum mean (or other statistic) is computed. These stratum means are properly *weighted* to form a *combined estimate* for the entire population. (*d*) The variances are also computed separately within each stratum and then properly weighted and added into a combined estimate for the population.

Detailed explanations and elaborations of each step are presented later. Furthermore, we shall encounter modifications in which some of these steps are altered or omitted. In proportionate sampling (3.4), step (*c*) can be by-passed for computing a self-weighting mean; but step (*d*) must still be performed. Step (*b*) can be performed without step (*a*) (3.6*B*). The method of post-stratification (3.4*C*) by-passes steps (*a*) and (*b*). In the method of double sampling (12.1), steps (*a*) and (*b*) are not performed on the population, but only on a sample basis. In spite of these modifications, which can often be introduced fruitfully, the above four principal steps constitute a sound basis for presenting the fundamentals of stratification.

In some situations the sorting of the sampling units into strata can be costly. Step (*a*) can be omitted, yet (*b*), (*c*), and (*d*) performed completely, if the stratum weights W_h are available, and if the strata can be identified for all selected sampling units. Then each selection is made at random from the entire population, but only the specified number n_h is accepted from the hth stratum. The n_h selections from the hth stratum represent a random selection, and for that purpose we can disregard as "blanks" the selections from other strata.

There are three principal reasons for resorting to stratification:

1. *Stratification may be used to decrease the variances of the sample estimates.* In *proportionate* sampling, the sample size selected from each stratum is made proportionate to the population size of the stratum. The variance is decreased to the degree that the stratum means diverge and that homogeneity exists within strata.

On the contrary, in *disproportionate or optimal allocation*, different sampling rates are used deliberately in the different strata. The variance (per unit cost) can be decreased by increasing the sampling fractions in strata having higher variation or lower sampling cost.

2. *Strata may be formed to employ different methods and procedures within them.* Different sampling procedures or different methods of observation and data collection may be needed in several portions of the population. Why should one abandon simplicity and employ several competing procedures in one survey?

(*a*) If the *physical distribution* of parts of the population differ radically, it may be useful to tailor different procedures to the several parts. For example, in selecting a sample of people, separate selection procedures may be employed for persons living in private dwellings, for those in institutions of various kinds, for transients, and for those in military service. These subpopulations are dealt with in separate strata. Actually, for interview surveys of the people in the United States, all but the largest samples exclude the latter strata and are confined to private dwellings that contain most of the people (9.1). For another example, consider the selection of an interview sample of the employees of a widespread electric power company. The home office and plant employees might be sampled as individuals. But the employees scattered in substations, and the repair crews, may be better selected as *clusters* to save travel costs.

Stratification by size of clusters is valuable for controlling the sample size, when the cluster size varies. Similarly, it can be employed for reducing the variation in *subsample* sizes (7.1).

(*b*) There may be *differences in the lists* available, or preferred, for different parts of the population. For example, we may use a city directory to select most of the dwellings within a city, then supplement it with an area sample for dwellings missed by the directory (11.5). Thus the population of dwellings in the city is divided into two strata: the directory listing and the supplement. The supplement should include all those missed, but none of those listed by the directory. Supplementing an incomplete list is a frequent operation in selection.

(*c*) The diverse *nature of the elements* in parts of the population may call for different procedures. For example, in a study of a firm's employees we may prefer written questionnaires for the "white collar" workers, but personal interviews for the "blue collars." This is done conveniently by separating the two groups into two strata. Another example: for a sample of young people between 13 and 17 years of age, schools may provide both good lists and favorable situations of observation for the majority who are students; those who are not at school can then be added in a separate stratum with other procedures, perhaps an area sample.

3. *Strata may be established because the subpopulations within them are also designated as domains of study.* A *domain* is a part of the population for which separate estimates are *planned* in the sample design. "Any sub-division about which the enquiry is planned to supply numerical information of known precision may be termed a domain." [U.N., 1950.] For example, the results of national surveys are often published separately for its component regions; therefore, it helps to treat the regions as strata with separate selection from each. In some domains the sampling fraction may have to be increased to produce the required precision.

On the other hand, most subclasses cannot be selected in separate strata, because the information needed for separating them is not readily available before selection. For example, the results for the subclasses of age, of sex, of occupation from area samples of dwellings are often presented without deliberately planning for them in the design, hence without their being made domains.

3.2 THE WEIGHTED MEAN AND ITS VARIANCE

First we examine the properties of a weighted mean as a general concept. This will allow us to develop later in detail the special formulas appropriate to diverse methods of stratified sampling. We want to develop a sample estimate for a weighted population mean, $\bar{Y}_w$:

$$\bar{Y}_w = \sum W_h \bar{Y}_h = W_1 \bar{Y}_1 + W_2 \bar{Y}_2 + \cdots + W_h \bar{Y}_h + \cdots + W_H \bar{Y}_H. \tag{3.2.1}$$

That is, the population mean is equal to the sum of the H strata means $\bar{Y}_h$, each multiplied by its proper weight W_h, where $\Sigma W_h = 1$. The weighted sample mean is

$$\bar{y}_w = \sum W_h \bar{y}_h. \tag{3.2.2}$$

The sample mean is obtained separately and *independently* for each stratum, and it is then multiplied by the weight of the stratum. These products are summed over the H strata to obtain the weighted sample mean. The variance of this weighted mean is obtained by combining the separate variances of the stratum means: the variance of each stratum mean is multiplied by the square of the stratum weight and the products are added over the H strata:

$$\text{var}(\bar{y}_w) = \sum W_h^2 \text{ var}(\bar{y}_h). \tag{3.2.3}$$

A sample has to be taken in each stratum to estimate its stratum mean; and the sample from each stratum must contain at least two sampling units to permit the computation of the variance in the stratum. The processes of selection and estimation are performed *separately* and *independently* within each stratum. Note that nothing was said in this discussion about the sample design within the strata. Thus, within the several strata, different sampling fractions may be used, as well as different methods of selection, estimation, and observation. This general formula is discussed in 2.8D.

The sum of the elements contained in all H strata equals the totality of the N elements in the entire population, because each element of the population occurs in one, and only one, stratum:

$$N = \sum N_h = N_1 + N_2 + \cdots + N_h + \cdots + N_H.$$

The number of elements selected into the sample from the hth stratum is denoted by n_h. The number of elements in the entire sample is

$$n = \sum n_h = n_1 + n_2 + \cdots + n_h + \cdots + n_H.$$

Typically, $\bar{Y}_h$ is the mean of the N_h elements in the hth stratum; that is,

$$\bar{Y}_h = \frac{1}{N_h} \sum_i^{N_h} Y_{hi} = \frac{Y_h}{N_h},$$

where Y_{hi} is the value of the ith element in the hth stratum, and Y_h is their sum in the hth stratum. The weights frequently, but not always, represent the proportions of the population elements in the strata and $W_h = N_h/N$. Then the weighted mean is equal to the ordinary mean per element of the population:

$$\bar{Y}_w = \sum \frac{N_h}{N} \bar{Y}_h = \frac{1}{N} \sum Y_h = \frac{Y}{N} = \bar{Y}, \tag{3.2.1$'$}$$

and then

$$\sum W_h = \sum \frac{N_h}{N} = \frac{1}{N} \sum N_h = \frac{N}{N} = 1.$$

The weight W_h of the stratum is generally the *proportion of the population* contained in that stratum, and so $\Sigma W_h = 1$. This can be reinforced now by permitting N_h to be any arbitrary measure of the size of the stratum, no longer restricting it to a count of elements. If we denote with $N = \Sigma N_h$ the sum of these arbitrary measures, $W_h = N_h/N$, we obtain

$$\bar{y}_w = \frac{1}{N} \sum N_h \bar{y}_h, \tag{3.2.2$'$}$$

and

$$\text{var}(\bar{y}_w) = \frac{1}{N^2} \sum N_h^2 \text{var}(\bar{y}_h). \tag{3.2.3$'$}$$

We may want to estimate the *population total* $N\bar{Y}_w$, where N is a *constant*, known from outside sources and independent of the sample results. This can be estimated with $N\bar{y}_w$, and the variance and standard error of this estimate are

$$\text{var}(N\bar{y}_w) = N^2 \text{var}(\bar{y}_w), \qquad \text{and} \qquad \text{se}(N\bar{y}_w) = N \text{se}(\bar{y}_w). \tag{3.2.4}$$

For computing aggregates, the weights can be the stratum totals rather than the proportions. For example, instead of (3.2.4) we get

$$N\bar{y}_w = N \sum \frac{N_h}{N} \bar{y}_h = \sum N_h \bar{y}_h,$$

and

$$\text{var}(N\bar{y}_w) = \sum N_h^2 \text{var}(\bar{y}_h). \tag{3.2.4$'$}$$

The weights can also be incorporated into the stratum estimates. For example, we can estimate $Y = \Sigma Y_h$ with $y = \Sigma y_h$, where $E(y_h) = Y_h$. Then

$$\text{var}(y) = \text{var}(\sum y_h) = \sum \text{var}(y_h). \tag{3.2.4$''$}$$

The general problem of assigning proper weights to strata, that is, how the different parts of the population should be summed, occurs *in the same form* whether we deal with a complete population census or a sample; hence, it is not strictly a sampling problem. Nevertheless, we should examine a few situations in which the weights represent not a simple count of the elements, but of some other units.

1. A sample of the mothers of city A has been taken, and the number of children born per mother has been estimated separately for several age classes of the mothers, called "age specific birth rates." To compare the overall birth rate of city A with other cities, we can assign some *standard* age distribution of mothers as weights to the age strata, instead of the age distribution of the sample city.

2. The entire equipment of a large company was sorted into strata of different types of equipment, from each of which a separate sample of items

was selected. The items in the sample were observed, and a measure of "depreciation ratio" was assigned to each of them. The aim was to produce a single, overall estimate of the depreciation ratio for the company's entire equipment. The weights for the strata were not based on the number of items in the strata, because of very large variation in the dollar values per item in the different strata; the weights were based on the original dollar values within strata.

3. A sample of families in a city was obtained by selecting a sample of blocks from a map on which all the blocks have been numbered and divided into strata. The sample estimates, made separately within the strata, must be combined to estimate the characteristics of the entire city. The families are the elements in this case, but their population numbers cannot be used for weights because they are not known. If we cannot find good estimates for the stratum sizes, we must use the known probabilities of selection within each stratum as weights. If a uniform sampling rate has been used, the sample is self-weighting. Otherwise, the weights are inversely proportional to the known probabilities of selection within the strata: if f_h is the probability of selection of each element in the hth stratum, then $1/f_h$ can be assigned as a weight to each sample element selected from that stratum. Thus the relative size of the hth stratum is estimated as

$$w_h = \frac{n_h/f_h}{\sum n_h/f_h} \qquad \text{and} \qquad \sum w_h = 1.$$

Note, however, that here the weights are not known constants, but only estimates, subject to sampling variation. Hence, the formulas, such as (3.2.2) and (3.2.3), do not hold exactly, but only as approximations. These problems arise when the number of elements within the sampling units are not known. This occurs in cluster and multistage sampling with sampling rates applied to clusters of unequal and unknown sizes. It also appears in the sampling of elements from a list containing blanks and in dealing with subclasses of the sample (4.5).

3.3 MEAN AND VARIANCE FOR STRATIFIED ELEMENT SAMPLING

We now treat a basic class of stratified samples: those with *random* selections of *elements* within each stratum. They are samples of *elements* because the elements are selected individually and separately, rather than in clusters. They are stratified because the selection is carried on separately and independently within each stratum. They are *random* because the n_h sample elements are selected with simple random sampling. In this section we present general fundamentals and formulas which can be used for *any* stratified random sample of elements. They can be used for both disproportionate designs and for proportionate samples, although simpler formulas are also available for the latter. The simple mean of the elements in the hth stratum is

$$\bar{y}_{h0} = \frac{1}{n_h} \sum_i^{n_h} y_{hi}.$$

This is the mean of the hth stratum, and selected with srs, as the subscript 0 denotes. For combining the different strata we use (3.2.2) and obtain for the mean of any stratified random sample of elements:

$$\bar{y}_{w0} = \sum_h^H W_h \bar{y}_{h0} = \sum_h^H W_h \frac{1}{n_h} \sum_i^{n_h} y_{hi}. \tag{3.3.1}$$

The variance of the simple random sample of n_h elements in the hth stratum is [directly from (2.2.2)]

$$\text{var}(\bar{y}_{h0}) = (1 - f_h) \frac{s_h^2}{n_h}, \qquad \text{where } s_h^2 = \frac{1}{n_h - 1} \left(\sum_i^{n_h} y_{hi}^2 - \frac{y_h^2}{n_h} \right).$$

We combine the variances of the stratum means according to (3.2.3) and obtain the variance of the sample mean ($\bar{y}_{w0}$) as

$$\text{var}(\bar{y}_{w0}) = \sum W_h^2 (1 - f_h) \frac{s_h^2}{n_h}. \tag{3.3.2}$$

If the weights are based on the proportions of population elements in the strata, then (since $W_h = N_h/N$ and $n_h = f_h N_h$) (3.3.1) and (3.3.2) can also be expressed, respectively, as

$$\bar{y}_{w0} = \frac{1}{N} \sum_h^H N_h \frac{1}{n_h} \sum_i^{n_h} y_{hi} = \frac{1}{N} \sum_h^H \frac{1}{f_h} \sum_i^{n_h} y_{hi}, \tag{3.3.1'}$$

$$\text{var}(\bar{y}_{w0}) = \frac{1}{N^2} \sum_h^H (1 - f_h) \frac{N_h^2}{n_h} s_h^2 = \frac{1}{N^2} \sum_h^H \frac{1 - f_h}{f_h} N_h s_h^2. \tag{3.3.2'}$$

Disregarding the factors $(1 - f_h)$, the variance is inversely proportional to n_h in (3.3.2) and to f_h in (3.3.2′). Thus, for all stratified element sampling, as for simple random sampling, multiplying all sampling fractions by k results in a proportional reduction of the combined variance.

The standard error for the aggregate ($N\bar{y}_w$) equals N [se ($\bar{y}_w$)]. If the weights are $W_h = N_h/N$, then the variance can also be written

$$\text{var}(N\bar{y}_{w0}) = \sum (1 - f_h) \frac{N_h^2}{n_h} s_h^2 = \sum \frac{1 - f_h}{f_h} N_h s_h^2. \tag{3.3.3}$$

If the $\bar{y}_{w0}$ denotes a proportion $p_{w0} = \Sigma W_h p_h$, its variance may be written in an equivalent form which is easier to compute:

$$\text{var}(p_{w0}) = \sum W_h^2(1 - f_h)\frac{p_h(1 - p_h)}{n_h - 1}. \tag{3.3.4}$$

Example 3.3a For the numerical illustration in 3.4, the sample mean is

$$P_{w0} = \sum w_h \bar{y}_{h0} = 0.50(7) + 0.30(15) + 0.20(60) = 20 \text{ percent}$$

And the variance of that mean is

$$\text{var}(p_{w0}) = \left[0.50^2 \frac{24}{25}\frac{7(93)}{199} + 0.30^2\frac{24}{25}\frac{15(85)}{119} + 0.20^2\frac{24}{25}\frac{60(40)}{79}\right] \times 10^{-4}$$
$$= 2.88 \times 10^{-4}.$$

The standard error of the mean $p_w = 0.20$ is

$$\sqrt{2.88 \times 10^{-4}} = 1.70 \times 10^{-2} = 1.70 \text{ percent}.$$

And the standard error of the estimate $Np_w = 10{,}000 \times 0.200 = 2000$ is $10{,}000 \times 0.0170 = 170$.

Subclass means and their variances can also be handled with formulas (3.3.1) and (3.3.2) but *only if the stratum weights W_h are known separately for the subclasses*. This would occur in domains for which distinct strata are used. But for many subclasses the stratum weights are not known; then the formulas developed in 4.5 are necessary.

Comparison of two means presents no problems if the two samples are independent: then the variance of the difference equals the sum of the variances. This holds for comparisons of two domains that come from separate strata. But for comparing two subclasses that cut across strata the difference has a complicated variance, also shown in 4.5.

3.4 PROPORTIONATE SAMPLING OF ELEMENTS

3.4A The Mean and Variance of Proportionate Samples

This is perhaps the most widely recognized method of selection. It is what people generally and vaguely mean by talking of "representative sampling," of samples which are "miniatures of the population," and by the notion that the "different parts of the population should be appropriately represented in the sample."

Suppose we select a sample of $n = 400$ employees out of the $N = 10{,}000$ employed in a factory. (See accompanying table.) We use the several departments of the factory as strata, because it can be done easily, and because we suspect that there may be large differences among the departments for the survey variables. The data used in planning the sample appear on lines *a*, *b*, *c* and the results for one variable on lines *d* and *e*.

In proportionate samples, the *sampling fraction in each stratum is made equal to the sampling fraction for the population as a whole.* That is, n_h/N_h is made equal to n/N for every h. In terms of sampling fractions, we have $f_1 = f_2 = f_3 = f_h = f$, which is the overall sampling fraction. In our example, after the total sample size $n = 400$ has been decided, it follows that the sampling fraction in each stratum should be $400/10{,}000 = 1/25$. This fraction is applied in turn to each of the N_h to obtain the values

		Symbol	Assembly	Foundry and Machine	Office and Miscellaneous	Entire Factory
	Stratum Number	h	1	2	3	Total
a	Population Size	N_h	5000	3010	1990	10,000
b	Stratum Weight	W_h	0.50	0.30	0.20	1.00
c	Sample Size	n_h	200	120	80	400
d	Number of "yes" answers	y_h	14	18	48	80
e	The percentage of "yes" answers	$\bar{y}_h$	7	15	60	20

Proportionate Sample of a Population Divided into Three Strata

of the $fN_h = n_h$, the sample sizes to be selected from the various strata. The rounding of n_h to the nearest integer introduces slight departures in the values of the actual sampling fractions n_h/N_h from their common value of $\frac{1}{25}$. This trivial departure is common and can usually be neglected.

Another view of a proportionate sample is that *it represents all strata among the sample cases in the ratios* of the strata in the population: that is, $n_h/n = N_h/N = W_h$ for all h. For example, the proportion of the sample that falls into the first stratum is $200/400 = 5{,}000/10{,}000 = 0.50$. Thus the proportionate n_h can also be obtained by multiplying n by each of the W_h in turn; for example, the n_1 above would be obtained as $nW_1 = 400 \times 0.50 = 200$. The selection of the n_h sample elements from the N_h in each stratum is made with random choices. The full descriptive title of the selection process is proportionate stratified random sample of elements; but we shall refer to it briefly as *proportionate sample.* This design yields a *self-weighting* sample, and the population mean can be estimated with the simple mean of the sample cases, the sample total divided by the number of cases in the sample:

$$\bar{y}_{\text{prop}} = \frac{y}{n} = \frac{1}{n}\sum^{n} y_j. \tag{3.4.1}$$

In our example, we compute simply $\frac{80}{400} = 20$ percent. The same result was obtained in 3.3 through weighting. The *self-weighting* formula in effect automatically assigns the weights n_h/n to each of the $\bar{y}_h$, whereas the weighted computations use N_h/N as weights. Since $n_h/n = N_h/N$ in proportionate sampling, the two results must be equal within rounding errors. This can be shown in symbols by substituting f for the f_h and n for Nf in (3.3.1′):

$$\bar{y}_{\text{prop}} = \frac{1}{N}\sum_h^H \frac{1}{f_h}\sum_i^{n_h} y_{hi} = \frac{1}{Nf}\sum_h^H \sum_i^{n_h} y_{hi} = \frac{1}{n}\sum_j^n y_j.$$

The double summation simply means that after the sum of all values in each stratum is obtained, these partial sums must be added over the H strata to obtain the total for the entire sample. But this summation can be done in one step for all the $\sum^H n_h = n$ cases in the sample. Thus the mean $\bar{y}_{\text{prop}}$ of a proportionate sample can be estimated without sorting the elements into different strata.

The elements must be sorted into separate strata for computing the variance properly. The computing may follow (3.3.2), good for any stratified random sample of elements; but it can also be modified in the following two ways. Because the factors $(1-f_h)$ remain constant for all the strata, they may be moved outside the summation, and, substituting in (3.3.2′), we get another form of the variance of a proportionate sample:

$$\text{var}\,(\bar{y}_{\text{prop}}) = \frac{1-f}{n}\left[\sum W_h s_h^2\right] = \frac{1-f}{n}s_w^2. \tag{3.4.2}$$

The sample value s_w^2 estimates $\Sigma\, W_h S_h^2 = S_w^2$, the *within stratum element variance* that plays the same role here as S^2, the element variance in the population, does in srs variance. The sample variance may be computed by substituting n_h/n for the W_h:

$$\text{var}\,(\bar{y}_{\text{prop}}) = \frac{1-f}{n^2}\sum_h^H n_h s_h^2 = \frac{1-f}{n^2}\sum_h^H \frac{n_h}{n_h - 1}\left[\sum_i^{n_h} y_{hi}^2 - \frac{y_h^2}{n_h}\right]. \tag{3.4.2′}$$

In proportionate sampling *equal allocation* occurs when all $n_h = c = n/H$ are equal, hence all weights are also equal $W_h = 1/H$, and

$$\text{var}\,(\bar{y}_{\text{prop}}) = \frac{1-f}{n}\left[\frac{1}{H(c-1)}\sum_h^H\left(\sum_i^c y_{hi}^2 - \frac{y_h^2}{c}\right)\right]. \tag{3.4.3}$$

Inside the brackets we find the *within stratum element variance*, called the "mean squared error" in traditional analysis of variance tables. When the weights are elements, then each stratum contains $N_h = c/f = cF$ elements, with cFH elements in the entire population.

Note also that the variance of the *sample total* y for a proportionate sample is

$$\text{var}(y) = (1 - f)\sum_h^H \frac{n_h}{n_h - 1}\left[\sum_i^{n_h} y_{hi}^2 - \frac{y_h^2}{n_h}\right]. \tag{3.4.4}$$

The variance of *proportions* from a proportionate sample (3.4.2) can also be computed as

$$\begin{aligned}\text{var}(p_{\text{prop}}) &= \frac{1-f}{n}\sum^H W_h \frac{n_h}{n_h - 1} p_h(1 - p_h)\\ &= \frac{1-f}{n^2}\sum^H \frac{n_h^2}{n_h - 1} p_h(1 - p_h) = \frac{1-f}{n^2}\sum^H \frac{y_h(n_h - y_h)}{n_h - 1}.\end{aligned} \tag{3.4.5}$$

In our example we get

$$\begin{aligned}\text{var}(\bar{y}_{\text{prop}}) &= \frac{1 - 1/25}{400}\left[0.50\,\frac{200}{199}(7)(93) + 0.30\,\frac{120}{119}(15)(85)\right.\\ &\qquad\left. + 0.20\,\frac{80}{79}(60)(40)\right] \times 10^{-4}\\ &= \frac{24/25}{400}[1199] \times 10^{-4} = 2.88 \times 10^{-4}.\end{aligned}$$

This value is equal to that obtained in 3.3 because the general formula (3.3.2) and the special formula (3.4.2) yield the same correct value for a proportionate sample.

Remark 3.4.1 *Systematic selection*, instead of true random, is often used for proportionate sampling of elements. This consists of taking every kth individual after a random start from 1 to k (4.1). For example, suppose that a payroll listing of the 10,000 employees of a factory is prepared and sorted by department. Then the interval 10,000/400 = 25 is applied after a random start from 1 to 25. If the random number is 8, then listings numbered 8, 33, 58, 83, . . . , 9983 become selected. The sample will contain the correct proportions of elements from each department (stratum), within a fraction of an element.

If the payroll cards of each department had been shuffled thoroughly before listing, the systematic sample would be equivalent to a proportionate stratified random sample. For the latter, the shuffling is not necessary, because the procedure of n_h independent choices within each stratum accomplishes the equivalent of a shuffling process. With the regular intervals of systematic selection this shuffling is lacking. But many lists contain haphazard arrangements that can be considered to give results similar to random selection within the strata. In such situations the formulas of proportionate sampling are appropriate. Details and precautions in the use of systematic selection are given in 4.1 and 4.2.

3.4B The Design of Proportionate Samples

If the stratified nature of the selection is ignored and the srs formula is used for the variance, instead of the stratified formula, the result would be

$$(1-f)\frac{p(1-p)}{n-1} = \left(1-\frac{1}{25}\right)\frac{20 \times 80}{399}10^{-4}$$

$$= \frac{24/25}{399}1600 \times 10^{-4} = 3.85 \times 10^{-4}.$$

This computation would lead to the wrong level of probability; for example, instead of a desired level of $P = 0.95$, it would lead to an actual level of $P = 0.98$ (see 8.2). However, it can serve as a good estimate of the variance that *would* have been obtained *had* a simple random sample of the same size, $n = 400$, been taken. The ratio of the variances of the two kinds of designs is $2.88/3.85 = 0.75$. Taking the srs variance as standard at 100 percent, we note that in this case proportionate sampling reduced the variance by 25 percent.

We may also view this "efficiency" in terms of the number of elements needed for an srs design to have the same precision, or variance, as the proportionate sample of 400 elements. This comes nearly to $400 \times 3.85/2.88 = 535$ elements. To obtain a fixed precision with the proportionate sample, we need only $400/535 = 75$ percent as many elements as we would need for an srs design.

This discussion can be generalized. By comparing (3.4.2) with (2.2.2) we find that, as measured by their variances, the "efficiency" of proportionate sampling over srs is

$$\frac{\text{Var}(\bar{y}_{\text{prop}})}{\text{Var}(\bar{y}_0)} = \frac{\sum^{H} W_h S_h^2}{S^2} = \frac{S_w^2}{S^2}. \tag{3.4.6}$$

The numerator S_w^2 is the average element variance *within* the strata, while the denominator is the *total* element variance in the population (4.6). For estimating proportions this ratio is neglecting factors $N_h/(N_h - 1)$

$$\frac{\text{Var}(p_{\text{prop}})}{\text{Var}(p_0)} = \frac{\sum W_h P_h(1-P_h)}{P(1-P)}. \tag{3.4.6$'$}$$

Unbiased estimators oi S_w^2 are $\Sigma W_h s_h^2$, and $\Sigma W_h n_h p_h(1 - p_h)/(n_h - 1)$, respectively. If the values of n_h are large enough, the factors $n_h/(n_h - 1)$

become negligible. In our example, $s_w^2 = \Sigma W_h s_h^2$ was computed as 1199×10^{-4}.

The element variance s^2 computed from stratified samples is a good estimate of the element variance S^2 (2.8.23). For proportions $s^2 = p(1-p)[n/(n-1)]$, but the correction $n/(n-1)$ is usually negligible. Thus with $p(1-p) = 1600 \times 10^{-4}$, we obtain for the ratio of the variances $S_w^2/S^2 = 1199/1600 = 75$ percent. This reduction of 25 percent in within stratum variance over the total variance is the source of similar results in the preceding computations.

Now, compute the simple random sample necessary to obtain the same variance as the 2.88×10^{-4} obtained with proportionate sampling in the illustration of 3.4. Using the methods of 2.6 and the estimate of $p(1-p) = 1600 \times 10^{-4}$, for S^2 we get

$$n' = \frac{1600}{2.88} = 556 \quad \text{and} \quad n = \frac{556}{1 + 556/10{,}000} = \frac{556}{1.0556} = 528.$$

The trivial difference between 528 and the 535 computed earlier is the effect of the change in the finite population correction, which results in the factor 1.0556 instead of 1.04.

But now pretend to design a proportionate sample to obtain the variance 2.88×10^{-4}. Suppose you can guess that the element variance in the population is $S^2 = 1600 \times 10^{-4}$, and also that proportionate sampling should reduce the variance by a factor of about $S_w^2/S^2 = 0.75$. Then you estimate $S_w^2 = 0.75 \times 1600 \times 10^{-4} = 1200 \times 10^{-4}$ and compute

$$n' = \frac{1200}{2.88} = 416 \quad \text{and} \quad n = \frac{416}{1 + 416/10{,}000} = 400.$$

Suppose that, for another variable in the same sample or in another survey, the reduction in the variance due to proportionate sampling is guessed to be only 10 percent. If $S^2 = 1600 \times 10^{-4}$, then

$$S_w^2 = 0.9 \times 1600 \times 10^{-4} = 1440 \times 10^{-4},$$

and if the variance is specified again to be 2.88×10^{-4}, then

$$n' = \frac{1440}{2.88} = 500 \quad \text{and} \quad n = \frac{500}{1 + 500/10{,}000} = 476.$$

These examples illustrate that, for designing proportionate samples, methods similar to those developed for simple random samples in 2.6 can be used. The variance and the sample size stand in the relationship $\text{Var}(\bar{y}_{\text{prop}}) = (1-f)S_w^2/n$. How can we estimate $S_w^2 = \Sigma W_h S_h^2$? In some situations, reasonable estimates or guesses can be made of the S_h^2, stratum by stratum. For proportions $S_h^2 = P_h(1 - P_h)$, and these can be estimated fairly well even with rough guesses about P_h. In other situations, it may be easier to estimate S^2 and then reduce it to $S_w^2 = S^2 \times (S_w^2/S^2)$ by also making an estimate or guess about the reduction (S_w^2/S^2) that proportionate sampling may yield.

Remember that poor guesses for S_w^2 do not invalidate the computations of the mean and standard error from the sample results. They will merely cause the precision of the sample to be somewhat different from what was guessed at the time of the design. The gains in precision arise because the sample represents proportionately each stratum in the population. This method eliminates from the variance of the mean that component of the population variance which is due to the differences of the means between strata. Success in choosing strata with great *heterogeneity among their means* determines the gains made by proportionate sampling. Or, looking at the other side of this same coin, we note that the variability of proportionate samples comes only from sampling *within* the strata. To the degree that we can form *homogeneous* strata, making the variance within the strata less than in the population at large, just so far will proportionate sampling reduce the variance (4.6).

It is useful to consider S_w^2 the *element variance*, the equivalent in proportionate sampling of S^2 in srs. Proportionate sampling may be considered as equivalent to srs selection from an artificial population in which the element variance has been reduced from S^2 to S_w^2. The ratio S_w^2/S^2 of reduction is called the *design effect* (8.2).

Generally we obtain only small or moderate gains from proportionate sampling of elements, because the variables available for stratification, such as age and sex, do not separate the population into very homogeneous strata. Variables with the high relationships necessary for large gains are rarely available for stratification. Since stratification yields only modest gains, little effort should be spent on it for proportionate samples. Knowledge of the subject matter will usually enable the researcher to make good choices from available variables.

However, the wide use of proportionate sampling can be justified for several good reasons. First, it often yields some modest gains in reduced variances. Furthermore, in cluster sampling its gains are often greater (5.5). Second, it is safe because the variances cannot be greater than for an unstratified sample of the same size (4.6.6). Third, it can typically be done simply and easily. Fourth, it results in self-weighting means. The last three arguments do not hold for disproportionate sampling, which is neither safe nor simple nor self-weighting, although it can yield much larger gains in favorable situations.

Typical gains are particularly modest when the sample estimate is a proportion p. The gain of 25 percent shown in our example is not common and it results from the remarkable differences among the proportions (p_h) in the three strata: 7, 15, and 60 percent, respectively. But suppose that for another characteristic the proportions were 52, 40, and 60 percent, which are more common results. In estimating the overall proportion ($p = 0.50$), the gain from proportionate sampling would amount to only 2 percent, or 8 interviews in 400 (using 3.4.6). A basic reason for the small gains in estimating proportions is that variances $p_h(1 - p_h)$ of elements are insensitive to moderate differences in the p_h induced by stratification, especially in the central region from 20 to 80 percent. For very small or very large p_h, the changes in $p_h(1 - p_h)$ are more rapid and the chances are better for finding situations with larger gains.

The usual modest gains from proportionate sampling sharply contrast with the exaggerated notions prevalent about this method. Many believe it to be necessary for good sample design, but it is far from that. The small gains it typically yields could be obtained instead with only a modest increase in the size of a simple random sample. Note that without any stratification, in most cases a simple random sample tends to obtain nearly the correct proportions from each stratum. In our example, the proportionate sample allocates 200 out of the 400 sample elements to the first stratum: the simple random sample would obtain some number between 190 and 210 two-thirds of the time, and between 180 and 220 in 95 percent of possible samples. This is computed by noting that with srs the number of elements selected from the first stratum becomes a random variable with mean $= nW_h = 400 \times 0.50 = 200$. The standard error of that 200 is $\sqrt{nW_h(1 - W_h)} = \sqrt{400 \times 0.5 \times 0.5} = 10$.

Remark 3.4.II Means based on *subclasses* of the sample are often important. It is shown in 4.5 that the gains of proportionate sampling become lost as the ratio of subclass to entire sample diminishes. So for a subclass that represents 10 percent of the sample, only 10 percent of the gains of proportionate sampling are left. Can the full gains be recovered for subclasses too?

The gains of proportionate sampling are obtained for any subclass if it is made a *domain* and its members are specifically selected proportionately from the strata. This requires knowing the correct stratum weights for the subclasses, and the capacity for classifying them on the population list. For example, in the sample of employees, five age classes can be designated as domains; then for each age class, correct proportions for the three strata are computed and selected.

Frequently it becomes too difficult to control the selection simultaneously for many subclasses. But if the correct weights of the subclasses are known, then *post-stratification* can adjust for random variations in the sample sizes. This will introduce most of the gains of proportionate sampling for the subclass.

For many subclasses the correct stratum weights are unknown. Then the gains of proportionate sampling must be sacrificed; and the variance of small subclasses becomes similar to simple random samples. (4.5A)

3.4C Post-Stratification or Stratification after Selection

Proportionate sampling produces a sample in which each stratum is represented in its proper proportion (W_h) among the sample cases. Suppose, however, that we select, without stratification, a simple random sample of n cases from the N elements in the population. Furthermore, suppose that we can classify the sample cases into strata and also obtain the proportions W_h of these strata in the population. How can we use this information to improve the sample estimate?

The sample cases should be sorted into the proper strata, the stratum means computed, and then weighted by W_h (according to 3.3.1) to form the estimate of the combined mean. The estimate of the variance is given by the formula (3.3.2). If the original sample was not srs, but stratified, then post-stratification can be used to establish further "strata" within the actual strata. In the illustration of Section 3.4, the sample was proportionately stratified according to departments; post-stratification of each department according to sex and age may be introduced into the estimate, if these are known for each department. But the "strata" established by the weighting must not be too small: the expected sample size nW_h should not be less than 10. This caution guards against very unequal weights.

This is *post-stratification* or *stratification after selection* by weighting. Having failed to make the sample proportionate by selection, the effect of stratification is introduced into the estimation process. Post-stratification is an example of improving the estimator by the proper utilization of *ancillary* sources of information; other and more dramatic examples are discussed elsewhere (12.1–12.3). How effective is this procedure in improving the sample estimate? The variance of the post-stratified design is approximately [Hansen, Hurwitz, Madow, 1953, II, 5.13]

$$\text{Var}(\bar{y}) \doteq \frac{1-f}{n}\sum W_h S_h^{\,2} + \frac{1-f}{n}\sum W_h(1-W_h)\frac{S_h^{\,2}}{n_h}. \qquad (3.4.7)$$

The first term here equals the variance of a proportionate sample. The second term is smaller by the factor $(1 - W_h)/n_h$ and becomes negligible for moderately large n_h values. This term arises because the sample sizes obtained in the different strata are subject to variation around the expected proportionate numbers n_h. Thus the variance of post-stratification can approach that of a proportionate sample of the same size. But it cannot be less; and for similar reasons it seldom results in large gains in precision.

Post-stratification requires: (*a*) information on the proportions W_h of the population in the several strata; and (*b*) information for classifying the sample cases into the same strata. It must be stressed that the criteria of classification must be the same for (*a*) and (*b*), and if they differ, the procedure is biased.

Post-stratification does not require, as proportionate selection does, that every member of the population be classified and sorted into its stratum before selection. These considerations help to answer the question: When should post-stratification be used instead of a proportionate selection?

1. The stratifying variable may be *unavailable* for classifying and sorting each element. For example, in a sample of a factory, suppose that union membership is confidential and not available for all workers; suppose also that the respondents reveal their membership, and that the union will give the data on the total number of members. Then union membership can be used for post-stratification, but not for stratification in the selection. *Unavailable* may also simply mean that it is too expensive to stratify the entire population. When the total vote for each party is available after an election, this variable can be used for post-stratification, but not for selection. In this example, as in the preceding, it is important that the self-classification of respondents correspond closely to the behavior (voting and union membership) used to compute the weights W_h.

2. The stratifying variable, although available, may not be used. Perhaps at the time of selection the sampler overlooked it, or for other reasons failed to use it. Perhaps there were too many variables available, and he chose some others instead.

3. Post-stratification may be used on the *subclasses* even if a proportionate sample of the entire population has been selected. The effect of proportionate sampling becomes lost on small subclasses; the sample sizes vary in the strata, and the variance of the mean approaches that of simple random sampling. But introducing the proper weights, the number of subclass members in each stratum, can restore most of the gains of proportionate sampling. In our example, the departments are represented proportionately in the entire sample; but if we separate workers over fifty years of age, their representation among the departments is subject to sampling fluctuation. If for these older workers the proper stratum weights are available, their use can yield means with reduced variances.

Thus post-stratification is an "adjustment" or "correction" of the mean.

It is based on the assumption of random selection of the sample; the adjustments are generally small, because random selection alone tends to result in nearly proportionate stratum sample sizes. These facts are in sharp contrast with adjustments sometimes used to force conformity to population proportions, on data obtained in some haphazard fashion. The latter may have proportions drastically different from the population; hence the corrections may be drastic, and they may bring the estimate much closer to the population mean. But if the data are not based on a probability sample, statistical theory will not bridge the gulf from sample mean to population mean, and the conjecture must be made by expert judgment. This also must be the basis for deciding whether the adjustment did more good than harm.

The initial error can be written as $\Sigma w_h \bar{y}_h - \Sigma W_h \bar{Y}_h$; the w_h are sample results corresponding to the W_h. *Post-stratification* adjusts this to $\Sigma W_h(\bar{y}_h - \bar{Y}_h)$, and the remaining error is free from bias when the selection is random. But when the adjustment is applied to data selected by judgment, much bias can remain in the differences $(\bar{y}_h - \bar{Y}_h)$.

3.5 DISPROPORTIONATE SAMPLING OR OPTIMUM ALLOCATION

This method of using stratification to increase the precision of the sample mean is in contrast to proportionate sampling. It involves the deliberate use of widely different sampling rates for the various strata. The designation *optimum allocation* refers to the *aim* of assigning sampling rates to the strata in such a way as to achieve the *least variance for the overall mean* per unit of *cost*.

Sampling rates are increased in strata where the variance among the elements is large—at the expense of the sampling rates in strata with smaller element variances. If *allowed a fixed number* $n = \Sigma n_h$ of elements, the *variance* of $\bar{y}_w$ can be made a *minimum if the sampling rate within each stratum is made proportional to the standard deviation within the stratum.* That is, for fixed $n = \Sigma n_h$, the minimum var $(\bar{y}_w)$ occurs if the H values of f_h are chosen so that

$$f_h = \frac{n_h}{N_h} = kS_h. \tag{3.5.1}$$

S_h is the standard deviation per element in the hth stratum, and k is a constant of proportionality. If, instead of fixing n, we specify the variance, then allocating the n_h into the various strata according to (3.5.1) will yield a specified variance for *the smallest n*. Generally, the variance decreases as n increases and for any point on this curve of relationship, (3.5.1) provides the optimum allocation.

Next, consider the cost per element in the different strata: if these exhibit large differences, it pays to increase the sampling rates in strata with low costs and decrease them in strata with high costs. To be more exact, denote the cost per sample element in the hth stratum by J_h, so that the total of this cost factor for the entire sample is $\Sigma\, n_h J_h$. Again we can view economy two ways: to obtain the lowest variance for a fixed expenditure; or to obtain a specified variance for the least expenditure. More generally, we consider the decrease of variance with increasing costs. For any of these requirements the "optimum allocation" is reached when the *sampling rates* within the strata are made *directly proportional to the standard deviations* within the strata and *inversely proportional to the square roots of the costs per element* within the strata. That is, the f_h are to be designated so that

$$f_h = \frac{n_h}{N_h} = K \frac{S_h}{\sqrt{J_h}}, \tag{3.5.2}$$

where K is a constant of proportionality. We may confine our attention to (3.5.2) because (3.5.1) is merely its special form for situations when element costs J_h are treated as equal for all strata. In other words, if the cost per element is the same in all strata, the fixing (or minimizing) of the summed cost for elements is the same as fixing (or minimizing) of the number of elements in the sample.

However, not all the cost factors vary simply with the number of observations in the strata, as assumed above. Whereas proportionate sampling is self-weighting, disproportionate sampling is not; hence cost for weighting should be added for any allocation which departs from proportionality. This cost should include either the expense of employing weights in the computations or that incurred with duplicate cards (11.7). Furthermore, it should cover the extra trouble of keeping accounts, and the mistakes that can occur with unequal weights.

The constants of proportionality k and K depend on the factors W_h, S_h, J_h; see formulas (4.6.9). But the values of n_h can be computed easily without computing the constants. For example, suppose that you aim at minimum variance for a fixed cost of $C = \Sigma\, J_h n_h$. First, assume a reasonable *provisional* value for k. Next, compute the provisional values for n_h' and $\Sigma\, J_h n_h'$. Then use the ratio of $C/\Sigma\, J_h n_h'$ to multiply each n_h'. A similar method will also work to minimize cost for fixed variance, if the effects of the factors $(1 - f_h)$ can be neglected.

The optimum allocation applies to the variances of both the mean $\bar{y}_w$ and the estimate $N\bar{y}_w$ of the aggregate Y.

The allocation $n_h = KN_h S_h/\sqrt{J_h}$ refers to the common situation when the weights denote proportions of population elements, $W_h = N_h/N$. More generally for any set of constant weights, it is shown in 8.5 that optimum allocation occurs when

$$n_h = \frac{K'W_hS_h}{\sqrt{J_h}}. \tag{3.5.3}$$

For computing means, aggregates, and their variances the formulas of Section 3.3 are applied. In the variance formula (3.3.2) it may be noted again that increasing all the allocations n_h proportionally by the factor k decreases the overall variance by the same factor—if changes in f_h are negligible.

But optimum allocation is precisely the method that may lead to situations where the values of $(1 - f_h)$ in some strata are too large to be negligible. Moreover, sometimes in one or more of the extreme strata the formula (3.5.2) may point to values of f_h close to 1 or even over 1. In such cases, all elements of these extreme strata should be taken into the sample. Then the sample size available for the remaining strata is computed and reallocated among them. The extreme strata, if sampled completely, do not contribute to the variance of the combined estimate.

If the differences among the factors $S_h/\sqrt{J_h}$ are large, optimum allocation may yield large gains, much larger than proportional allocation. This can occur for characteristics which are distributed with great inequality in the population, often in highly skewed distributions, if good information is available for separation into strata. Examples of such situations, where small proportions of the population account for large proportions of the survey characteristic and its variance, are frequent in sampling of establishments; farm production; and in industrial and commercial applications (see Problems 3.4 and 3.5).

The variance is insensitive to small or even moderate changes in the allocation. For guidance I suggest that using sampling ratios (f_h) anywhere between half and twice the optimum ratios will generally yield variances within 10 percent of the optimum variance [Cochran, 1963, 5A.1]. This insensitivity leads to several valuable rules:

1. Dc not resort to disproportionate allocation unless there are substantial differences among the factors $S_h/\sqrt{J_h}$ in various strata. Otherwise, the gain over proportionate sampling may be consumed by the extra costs of weighting and special care. Generally, differences of several-fold are required to make disproportionate sampling worthwhile. Therefore, if the several values of $S_h/\sqrt{J_h}$ are roughly equal (say, within a factor of 2), use a proportionate sample.

2. It follows then that disproportionate allocation is not usually economical for estimating proportions, because the standard deviations

are equal to $\sqrt{P_h(1 - P_h)}$. These are insensitive (lying between 0.3 and 0.5) to fluctuations of the values of P_h between 0.10 and 0.90. (See Chart 8.2.I and Example 3.5a.)

3. In applying disproportionate sampling rates it is practical to avoid complicated sampling fractions. Choose, instead, some nearby convenient numbers; for example, integers for intervals, or a set of intervals which are integral multiples of each other.

4. Most of the potential gain can often be secured by using only a few different rates. Sometimes, merely using two rates will extract much of the gain readily: a low sampling rate for most elements and a high rate for a special stratum containing only the large elements. Sometimes, this "stratum of large items" can be designated for selection with certainty ($f_h = 1$), eliminating it entirely as a source of sampling error (11.4B).

Example 3.5a In Section 3.4 proportionate sampling reduced the sample size necessary for a variance of 2.88×10^{-4} by the ratio $400/535 = 0.75$. Would optimum allocation yield substantial further reductions in the needed sample size?

	h	1	2	3	Total
Stratum weight	W_h	0.50	0.30	0.20	1.00
Proportion of *yes* answers	$\bar{y}_h = p_h$	0.07	0.15	0.60	0.20
Variance s_h^2	$p_h(1 - p_h)$	0.0651	0.1275	0.2400	0.1600
Standard deviation	s_h	0.255	0.357	0.490	0.400

Application of formula (4.6.10) shows that the variance 2.88×10^{-4} could be obtained with a sample size of $n = 369$ by using optimum allocation. This would be allocated into the three strata as 141, 119, 109. The proportionate sample of 400 was allocated as 200, 120, 80. Thus in this case, a further saving of 31 elements, or 7 percent of the $n = 400$, can be obtained with optimum allocation. But the actual saving would be even less, because we cannot guess the S_h exactly. In most situations, this would not justify the extra expense and trouble of departing from proportionate sampling. Furthermore, conflicting losses for other survey variables would probably arise. This example illustrates that optimum allocation seldom yields large gains for estimating proportions. In the above example, these show large differences (7, 15, and 60), but the standard deviations $\sqrt{p_h(1 - p_h)}$ differ much less, being in the ratio 1, 3/2, and 2. Important gains need larger differences in the allocations. The low percentages of rare items may bring that about, if they are segregated in strata, as in Problem 3.13c. But large gains typically arise only when the variable takes on large values in a skewed distribution.

Remark 3.5.I Sometimes what is wanted is not the sum of two (or more) subpopulations, but the difference of their means. To minimize the variance *for the difference* $(\bar{y}_1 - \bar{y}_2)$, *the optimum allocation is reached when the sample sizes* (not the sampling rates) *are made proportional to the quantities* $S_h/\sqrt{J_h}$. Thus the minimum variance is obtained when the n_h are such that

$$n_h = k \frac{S_h}{\sqrt{J_h}}. \tag{3.5.4}$$

In other words, for the difference $(\bar{y}_1 - \bar{y}_2)$ to have the smallest variance, make

$$\frac{n_1}{n_2} = \frac{S_1/\sqrt{J_1}}{S_2/\sqrt{J_2}}. \tag{3.5.4'}$$

Often $S_1/\sqrt{J_1}$ is about equal to $S_2/\sqrt{J_2}$, and then we should *make the sample sizes equal* for the two subpopulations.

Remark 3.5.II Exact values are never available for the S_h and J_h in (3.5.2), and we must use estimates or rough guesses. Where can we get these? The alternative sources are similar to those discussed in 2.6: a pilot study, past surveys of similar character, the experience of experts, and intelligent guesses based on some knowledge of the distribution of the characteristic. Among the last, note in particular that $\sigma_h = \sqrt{P_h(1 - P_h)}$ is insensitive to moderate changes in P_h, which the researcher can usually guess fairly well. In some situations we can get good estimates of the stratum means $\bar{Y}_h$, and also assume that the S_h are roughly proportional to the $\bar{Y}_h$; that is, assume that the coefficients of variation $C_h = S_h/\bar{Y}_h = k'$ roughly. Then make the sampling rates proportional to the stratum means:

$$n_h/N_h = (kk')\bar{Y}_h \qquad \text{and} \qquad n_h = (kk')Y_h, \tag{3.5.5}$$

since $N_h\bar{Y}_h = Y_h$, the *total* y in the stratum. This rule says: make the sample size proportional to the "y content" of the stratum. Remember that if our guesses of the S_h are incorrect, the efficiency of the design is less than optimum; however, the validity of the conclusions drawn from the sample is not affected.

Remark 3.5.III The formulas (3.5.1) and (3.5.2) yield the optimum allocation for one variate at a time. What can we do when the survey must produce several, perhaps many, variates? This calls for caution before resorting to disproportionate sampling. If the optimum allocations for several variates are close to each other, a good compromise can be worked out; this follows because of the insensitivity of the optimum. But if they are far apart, the optimum allocation for some variates will produce large increases in the variances for other survey variates. The variance for the latter may be increased not only in comparison with optimum or with proportionate sampling, but also with that of simple random sampling. How can we avoid this undesirable situation?

Before deciding on disproportionate sampling, we should make at least a rough estimate of the variances of all the important items—or of a well selected subset. If the different answers point toward the same design, there is no great problem. But if the answers are contradictory, the choice involves good judgment and theory that is beyond this book—and largely beyond existing theoretical developments. In deciding among them we should keep in mind not only the relative importance of the item, but also the relative importance of its variances under different designs. An item may be important, yet a lesser precision may still be acceptable. Some problems of general survey design are treated in 14.5.

Remark 3.5.IV Conflict can arise when equal precision is specified for *domains* if the domain sizes N_h differ greatly. Proportionate sampling is often preferred for the overall mean, and equal sampling fractions, $f_h = r$, mean $n_h = rN_h$. But equal precision for the means of domains requires equal $n_h = C$ for each domain, hence $f_h = C/N_h$. When the domain sizes N_h are roughly equal, no conflict need arise between the aims of designing for the population and for the domains. But when some domains are much smaller than others, the separate estimates can strain the survey's resources.

It is preferable therefore to define domains roughly equal in size. Sometimes the definitions of the domains can be changed to decrease the grossest inequalities, perhaps by merging small domains with others. Or we may leave them distinct and allow greater errors for the smaller domains; perhaps without publishing separate estimates for them.

In Table 3.5.I assume that the standard deviations S_h are about the same within the different strata, as they often are. Note that equal allocation obtains equal errors of the mean (and equal coefficients of variation), but it results in

Description	Sample Sizes	Sampling Rates	*Relative* Values of Domain Error, Omitting Constants	
			SE of Mean and CV	SE of Aggregate
Equal allocation	$n_h = C$	$f_h = C/N_h$	S_h	$S_h N_h$
Proportionate sampling	$n_h = rN_h$	$f_h = r$	$S_h/\sqrt{N_h}$	$S_h\sqrt{N_h}$
Rates rising with N_h	$n_h = rN_h^2$	$f_h = rN_h$	S_h/N_h	S_h
Optimum Allocations Varying with Differences in Domain S_h				
Equal domain errors	$n_h = CS_h^2$	$f_h = CS_h^2/N_h$	Constant	N_h
Optimum for mean	$n_h = rN_hS_h$	$f_h = rS_h$	$\sqrt{S_h/N_h}$	$\sqrt{S_hN_h}$
Optimum for comparison	$n_h = rS_h$	$f_h = rS_h/N_h$	$\sqrt{S_h}$	$N_h\sqrt{S_h}$

TABLE 3.5.I Effects of Several Allocations on the Errors of Domain Statistics

greater standard errors for the aggregates of the larger domains, in the proportions of N_h. Rates of $f_h = rN_h$ have opposite effects. *The equal sampling fractions of proportionate sampling may often provide a compromise solution.* The standard error of the mean and the coefficients of variation decrease as $1/\sqrt{N_h}$ for the larger domains; but the standard errors of the aggregates increase with $\sqrt{N_h}$. Between these two effects proportionate sampling is a compromise. It also minimizes the variance of the entire sample.

When the differences in the S_h are important, the discussion above can be generalized by considering N_hS_h instead of N_h. The last three lines compare three optimum allocation designs. Equal standard errors for the mean (and equal coefficients of variation) for the domains require n_h proportional to $N_hS_h^2$. But optimum allocation for the overall mean requires n_h proportional to N_hS_h. Finally, optimum comparison of domains would require n_h proportional to S_h. In all of this we assume that domain differences in the cost per element J_h and the factors $(1-f_h)$ can be neglected; otherwise the discussion can be generalized to include them.

3.6 FORMING THE STRATA

These remarks have general significance for all types of stratified designs and for both element and cluster sampling. Of the many aspects that might be discussed under this title, the following seem to be of practical importance, also subject to misunderstanding, but amenable to brief treatment. Points *A* to *D* present procedures for obtaining valid samples, whereas points *E* to *L* deal with ways to increase the efficiency of the sample.

3.6A Classifying Sampling Units

Every sampling unit must be classified distinctly into one of the strata. Hence, for any variable used for stratification, *information must be available on all of the sampling units in the population.* Information available for only a small part of the sampling units is not generally useful for stratification. This strict rule of universal information can be relaxed in several ways. (1) If the information is missing for a small proportion of the sampling units, these may be simply dropped into a "miscellaneous" stratum. (2) Sometimes no unique variable is either available or preferred for all the sampling units in the population. But *some relevant variable for each portion* of the entire population may be found and employed efficiently (see *E* below). (3) At each stage of a multistage sample, we need to subdivide internally into strata only those sampling units that have been selected into the sample in the previous stage. (4) If stratifying the entire population is too expensive, it can be done on a sample basis through the method of *double sampling* (12.1).

3.6B Stratification with Random "Quotas"

Typically, stratification consists of sorting the units into strata before selection. But in some situations the *sorting of all units can be too expensive*, although the stratifying variable is readily available for all units and the stratum weights W_h are also known. Then the desired sample sizes n_h *should be specified and filled by drawing selections at random from the entire population.* The n_h selections from the hth stratum represent a random selection; this can be understood by regarding all the selections from other strata as "blanks" within the hth stratum (see Remark 2.1). Draw random selections from the entire population until the "quotas" of n_h are filled within all strata. The n_h can be specified for either proportionate or disproportionate sampling. This procedure is exactly equivalent to stratification after sorting the population.

In other situations, although the weights W_h are available, the units of the population cannot be classified until after full observation. Then post-stratification can still recover most of the gains of proportionate sampling.

3.6C Errors in Sorting

Sorting a few sampling units into the wrong strata does not greatly decrease the efficiency of the stratification. Similarly, minor inaccuracies in the stratifying variables cause little damage.

If after selection a few units are discovered to have been sorted into the "wrong" strata, it is generally best to leave them in their sorted strata. This will merely decrease slightly the efficiency of the stratification. But it will not bias the selection, because on the average the sample will contain merely its portion of wrong sortings. Contrariwise, correcting only the sample can lead to bias.

3.6D Overlapping Lists

If the population frame is defined as the sum of several lists from each of which a sample is selected separately, then each list becomes a stratum. The appearance of some sampling units on two or more lists results in duplicate listings; these can be handled by one of three alternative procedures (2.7*D* and 11.2).

3.6E Objectivity and Regularity Unnecessary

Neither objectivity nor regularity are needed for sorting sampling units into strata. (1) On the contrary, subjective sorting may be superior to rigid procedures for creating homogeneous strata. In the entire selection procedure this is the area *par excellence* for the exercise of personal judgment, based on expert knowledge of the list and subject matter. (2) There is no need for regularity and uniformity in the stratification process. For one thing, the limits of class boundaries of the stratifying variables may be changed at will to fit circumstances. For another, the classes need not be used symmetrically throughout the selection process. The stratifying variables should be used where they are meaningful, denoting important sources of variation, but not where those conditions are absent. For example, in one sample of United States counties, the northern counties were stratified by percentage of Democratic voters, and the southern counties by percentage of nonwhites. In another national sample, the urban counties were stratified by percentage of the labor force employed in manufacturing industries, whereas the rural counties were stratified by the average size of farm.

3.6F Homogeneity within Strata

For large reductions of the variance, we need stratifying variables closely related to the main survey objectives. The aim is to form strata within which the sampling units are relatively *homogeneous* in survey variables. Their variances are reduced to the extent that the variation among sampling units within the strata is less than their variation in the entire population. Hence, we strive to increase and maximize the *homogeneity of the sampling units within strata.* For a given population of sampling units, this is equivalent to increasing the *differences*, or *heterogeneity, among the means of the strata* (see 4.6*A*). This aim suffices for proportionate sampling and is also relevant to the design of disproportionate stratified samples.

However, for the design of *disproportionate* samples, another aim must be added: *to increase the heterogeneity of the standard deviations* of the sampling units among the strata (3.5 and 4.6*B*).

3.6G Utilizing Available Variables

How can we best utilize several variables available for stratification? It may not be clear just which variables yield more gain. Furthermore, the diverse aims of the survey may point to the use of different stratifying variables and lead to compromise choices.

Neither a long search for the best variables, nor elaborate sorting procedures are likely to be justified for small or moderate samples. The potential gains from stratification tend to be moderate in many situations and especially in proportionate sampling. The researcher generally knows enough about the subject to make a satisfactory choice among the variables available for stratification. Several practical hints are in order.

(1) Generally, more gain accrues from the use of coarser divisions of several variables than from the finer divisions of one. This is related to point *J* below. (2) When using several variables, there is no need for completeness and symmetry in forming the cells; smaller and less important cells may be combined. (3) Stratifying variables *unrelated to each other* (but related to the survey variables) should be preferred. Contrariwise, if two stratifying variables are highly correlated, using either one will give as much gain as using both. This resembles the problem of regression from correlated variables. (4) We may have information on two variables, one a qualitative attribute and the other quantitative, perhaps continuous and metric. The former can be used for stratified selection only. The latter can be used for estimation in a way not readily available for the former; for example, in a ratio or regression estimate (see 12.1–12.3).

3.6H Elaborate Controls Unnecessary

Elaborate controls for several stratifying variables would result in too many strata. When each variable leads to a number of divisions, the total number of cells thus formed equals the product of the divisions. Too many cells can also be created if stratification is attempted not only for the entire population, but also for several domains, perhaps of different kinds.

The number of strata must be limited by the effort that can be profitably spent on stratification, and ultimately by the number of sampling units comprising the sample. At least two sampling units must be selected from each stratum to compute an unbiased estimate of the variance. Hence, the number of strata should not be greater than half the number of sampling units the study can afford. But for element sampling, fewer strata are often preferable. First, because there is a loss of one "degree of freedom" for each stratum in computing the variance. Second, because too many strata increase the difficulties in the analysis of subclasses.

Deep stratification may be justified for large-scale cluster samples, where the gains of stratification are larger. Selecting only two primary clusters per strata is common practice (4.3). Often only one primary cluster is selected from each stratum, with "collapsing" of strata for the variance computations. Indeed, the number of stratum cells may exceed the number of selections. Procedures for doing this are called "multiple stratification," "two-way stratification," "deep stratification," "controlled selection," or "Latin Square" designs (12.8).

3.6I Number of Strata

How many strata should be formed from any single stratifying variable? The choice exists when the stratifying variable is continuous, e.g., income in dollars; or when its division can be continued for a long time, e.g., the geographic division of the U.S. into regions, then states, then counties, etc. Generally it is not advisable to carry the division very far. (1) Very small strata—the odd "fringes" of the population—contribute little to the gains from stratification; these gains are proportional to the stratum weight W_h (4.6). (2) The formation of only a few strata will typically yield most of the possible gains from a variable. Further subdivisions of these would result in only small additional gains. Perhaps between three and ten strata suffice for any single variable.

*3.6J Effect of Increasing Numbers of Strata

The effect of increasing the number of strata from H may be represented by the model $R^2/H^2 + (1 - R^2)$. Here R^2 is the portion of the variance affected by the stratification and corresponds to the relationship between the stratifying and survey variables. This portion decreases with the square of the number of strata. But the portion $(1 - R^2)$ of variance unrelated to the stratifying variable is unaffected by increasing the strata. Thus the variance approaches this level after the creation of a moderate number of strata. For example, $R^2 = 0.64$ represents a strong correlation $R = 0.8$ between the stratifying and survey variables. Yet $H = 6$ strata reduces the variance to $0.018 + 0.36 = 0.378$; doubling the number of strata to $H' = 12$ would further decrease it only to $0.004 + 0.36 = 0.364$. This model has been developed for optimum allocation with linear regression on the stratifying variable [Dalenius and Gurney, 1951; Dalenius, 1957, Ch. 8; Cochran, 1961; Cochran, 1963, 5A.7]. It also seems to perform well in tests we are making with nonmetric variables and for proportional allocation.

3.6K Equal Allocation

By equal allocation we denote the selection of a constant $n_h = c$ number of sampling units from each stratum. The sampling fractions are $f_h =$

$1/F_h = c/cF_h$, where the c units are selected from $cF_h = N_h$ units in the hth stratum. With a constant number c of elements selected at random from each stratum, the variance of the mean (3.2.2) becomes

$$\text{var}(\bar{y}_w) = \frac{1}{c(c-1)} \sum_h^H W_h^2(1-f_h)\left(\sum_i^c y_{hi}^2 - \frac{y_h^2}{c}\right). \tag{3.6.1}$$

Equal allocation is especially practical when the stratum boundaries are not rigidly fixed. Then the samples can divide the population into implicit strata, or *zones* (4.4).

The special case of proportionate allocation occurs when all strata contain the same number $F = 1/f$ of units and all stratum weights are $W_h = 1/H$; this leads to the variance formula given under proportionate sampling:

$$\text{var}(\bar{y}_{\text{prop}}) = \frac{(1-f)}{n}\left[\frac{1}{H(c-1)} \sum_h^H \left(\sum_i^c y_{hi}^2 - \frac{y_h^2}{c}\right)\right]. \tag{3.4.3}$$

For example, a list of names or a file of cards can be divided into $H = n/c$ zones, each of size cF, and then a sample of c selected at random from each zone. The map of a city can be divided into blocks; after grouping the blocks into $a/2$ zones, two blocks are selected from each zone.

Equal allocation can also be simple for disproportionate sampling. The number of selected units remains constant at c, and the size of the stratum is fixed at $cF_h = c/f_h$ to fit deliberately any desired sampling fraction f_h. The sampling fractions can be expressed as $f_h = f/k_h$, where the k_h are factors inversely proportional to the sampling fractions and $f = 1/F$, a constant basic rate. The k_h should generally be kept as a few simple "raising factors," often integers. Then the zone is created from cFk_h units, from which c are selected at random. The weights of the strata are $W_h = cFk_h/\Sigma\, cFk_h = k_h/\Sigma\, k_h$, and these can be kept simple too.

We assume in this discussion that the weights represent numbers of units, $W_h = N_h$. But the procedures can be applied to measuring stratum sizes $F_h = Fk_h$ in other units also. The simplicity and flexibility of the design can be exploited in several ways. Section 4.3 deals with paired selections that, with $c = 2$, permit the creation of many strata. Section 4.4 presents replicated sampling, where the c selections can be "collapsed" across strata to present the sample in the simple form of c independent replications. Moreover, if the strata are also domains, then having equal n_h tends to benefit their estimates and comparisons.

Beside simplicity, equal allocation is often efficient. If the standard deviations S_h are rather uniform over the population, and proportionate sampling is used, then a constant sample size from equal-sized strata may have intuitive appeal. Theory shows that constant n_h is also efficient in optimum allocation when $n_h = W_h S_h$ is desired. This means creating strata with equal values of $W_h S_h$. Why this is an efficient way of dividing the population is discussed next.

*3.6L Optimum Stratification

Methods of *choosing the best boundaries for strata* were denoted as *optimum stratification* by Dalenius in several papers [1957, 1959]. He deals with determining the best boundary points $y_h (h = 1, 2, \ldots, H - 1)$ to create H strata. The sample is to be optimally allocated with $n_h = kW_h S_h$, and the boundary points are to provide the minimum variance for a fixed number (H) of strata. He assumes the variable Y_i to be continuous, and investigates the theoretical situation when the survey variable itself is the stratifying variable.

Distributions of Y_i often have long tails, particularly in the skewed distributions that give rise to problems of optimum allocation. Two contrasting methods for establishing stratum boundaries can be dismissed on intuitive grounds. If the range of the distribution is divided into equal intervals, so that the ranges $(y_h - y_{h-1})$ of the strata are equal for all h, then the relative sizes W_h of the strata become too unequal, those on the extreme tail becoming too small. On the other hand, if the area of the population distribution is divided into equal relative sizes W_h for the strata, then the ranges on the extremes become too great. For example, the stratum of large elements on the right tail of a skewed distribution will have a long range $(y_H - y_{H-1})$. In these extreme strata the standard deviation S_h will be too large. Some compromise between equal proportions W_h and equal ranges $(y_h - y_{h-1})$ is called for. A similar compromise is often employed to present classes of economic, demographic, and Census data. For example, the distributions of incomes, or city sizes (in thousands), often appear in classes with boundaries such as 1, 2.5, 5, 10, 25, 50, 100, 250, 500, 1000, etc.

These intuitive compromises may help in understanding the boundaries provided by theory, which are compromises of a similar kind. The solution given directly by theory is that the best boundary y_h (between strata h and $h + 1$) should satisfy the relationship

$$\frac{(y_h - \bar{Y}_h)^2 + S_h^2}{S_h} = \frac{(y_h - \bar{Y}_{h+1})^2 + S_{h+1}^2}{S_{h+1}}. \tag{3.6.2}$$

This optimal solution is not practical, because all five parameters in the equation depend on y_h. However, several rules have been advanced for establishing strata, permitting us to approach the optimum.

Equal values of $W_h S_h$ for strata will lead to boundaries near the optimal. This rule ($W_h S_h = W_{h+1} S_{h+1}$) does not provide a simple procedure for establishing directly a set of optimum boundaries, because its parameters depend on the boundaries. But for data given in classes, the rule can help us to combine some classes and perhaps divide others to attain good boundaries. Equal values of $W_h S_h$ also mean equal sample sizes n_h in optimum allocation. Hence, *optimum stratification with optimum allocation also tends to result in equal allocation samples* ($n_h = n_{h+1}$).

Similarly, it has been shown that creating strata with *equal values of* $W_h(y_h - y_{h-1})$ yields good approximations to the optimal boundaries. This also can serve as a rule for combining and dividing classes, but not for a direct solution of optimal boundaries. Note their relationship to the intuitive speculations above.

The relative weights W_h for the strata represent proportions of the relative frequency curve for the variable Y_i. Its height is f_y at the value y. If the curve is approximated with a histogram of rectangles of height f_y and width d, then $W_h = \Sigma f_y d$, summed over the range $(y_h - y_{h-1})$ for the stratum. The rule of equal $W_h(y_h - y_{h-1})$ requires creating strata of equal summations of $f_y d^2$.

The square root of the last quantity leads to a rule that has been found both practical and efficient: compute the values of $\sqrt{f_y}$, then *cumulate* $\sqrt{f_y}$, *and divide the cumulation into approximately equal parts*. The corresponding values of y_h will be approximately the best boundaries. In practice, the stratifying variable comes in class intervals, and rough approximations may be necessary. The above rule assumes equal widths d for the class intervals, and it must be modified when the class intervals have variable widths d_y: *cumulate the values* $\sqrt{d_y f_y}$ *and divide the cumulation into approximately equal parts.*

The cumulated $\sqrt{f_y}$ rule has been shown to give excellent results for a variety of distributions. Cochran [1961; and 1963, 5A.6] presents convincing motivations for the three approximate rules above, and evidence of their good performance on empirical data. Sethi [1963, 1964] investigating some theoretical distributions in detail, also found the three rules successful; and obtained optimal boundaries for proportionate sampling and for equal allocation. Both authors found poorer performance for the simple rule of *equal aggregate stratum sizes;* that is, equal values of $W_h Y_h$ in the strata [Hansen, Hurwitz, and Madow, 5.9–5.11]. This would perform well to the extent that the coefficient of variation $S_h/\bar{Y}_h$ were about equal in strata. Then equal $W_h \bar{Y}_h$ would imply equal $W_h S_h$, which does yield close to the best solution.

The rules for optimum stratification can serve as practical guides for forming strata, but seldom as precise procedures for optimal solutions. More evidence is needed for the common situation when the correlation between the stratifying and survey variables is not very high. Two further modifications are necessary, when stratifying variables are nonmetric (qualitative), and when several stratifying variables can achieve more than the best single variable.

PROBLEMS

3.1. A population of $N = 1111$ blocks of a city is divided into 4 strata. The variable X_i denotes total dwellings and Y_i rented dwellings. To complete the problem, enter either (*a*) the results of Problem 2.1; or (*a'*) the results of Example 2.2*a*; or (*a''*) $\bar{x}_h = 21$, $\bar{y}_h = 14$, var $(\bar{x}_h) = 26$, and var $(\bar{y}_h) = 19$. (*b*) Compute the sampling fraction $f_h = n_h/N_h$ for each stratum. Compute also the number $n_h \doteq 80\ W_h$ for a proportionate sample of 80 elements. In a proportionate sample, the sample sizes are also equal if . . . (complete the sentence). (*c*) Compute $\bar{x}_w$ and se $(\bar{x}_w)$. (*d*) Compute $\bar{y}_w$ and se $(\bar{y}_w)$. (*e*) Compute the total $N\bar{x}_w$, se $(N\bar{x}_w)$, and var $(N\bar{x}_w)$.

Stratum	Wards	N_h	W_h	n_h	$\bar{x}_h$	$\bar{y}_h$	var $(\bar{x}_h)$	var $(\bar{y}_h)$	W_h^2
1	1	270	0.243	20					0.0590
2	2, 3, 4	260	0.234	20	34	25	43	40	0.0548
3	5, 6, 7	227	0.204	20	42	34	58	53	0.0416
4	8, 9	354	0.319	20	22	13	15	11	0.1018
	Total	1111	1.000	80					

3.2. Suppose that in 3.1 you may increase the sample size in the first stratum to $n_1 = 100$. (*a*) Estimate and explain briefly the variance var $(\bar{x}_1)$ you would expect. (*b*) Compute the var $(\bar{x}_w)$ you expect with $n_1 = 100$; note that this variance is only slightly smaller than in 3.1(*c*), although the total sample size has been doubled from $n = 80$ to $n = 160$. (*c*) Compute also the var $(\bar{x}_w)$ you would expect with $n_1 = n_2 = n_3 = n_4 = 40$. To save time, use the approximation that this variance is less than the result in 3.1(*c*) by the factor $(20/40)(1111 - 160)/(1111 - 80) = 0.461$. Note that this variance is much less than in 3.2(*b*), although they both have the same total size $n = 160$.

3.3. (*a*) From data in 3.1, compute the standard error of the difference se $(\bar{x}_2 - \bar{x}_3)$. This was based on $n_2 = n_3 = 20$; under what circumstances is this an "optimum allocation"? (*b*) Compute se $(\bar{x}_2 + \bar{x}_3)$.

(*c*) Compute the weighted mean $(N_2\bar{x}_2 + N_3\bar{x}_3)/(N_2 + N_3)$ and its standard error. (*d*) Compute the standard error of the difference $(N_2\bar{x}_2 - N_3\bar{x}_3)$ of the two stratum aggregates.

3.4. The farms of a small nation were divided into 7 strata based on their areas reported in the last census. The yields of the chief cash crop were estimated from the results of a sample survey, and rounded to convenient numbers.

(*a*) Design a sample to yield minimum variance for *about* $\Sigma n_h = n = 3000$ sample farms. All of the 200 farms of strata 6 and 7 fall into the sample; design the sample for 2800 cases from the other 5 strata.

(*b*) What is the variance, var $(\bar{y}_w)$, of this sample?

(*c*) Compare this variance to the variance of a proportionate sample of size $n = 3200$. For this comparison we suppose that the cost of each element is $J_h = \$3$, and that the cost of weighting the disproportionate sample, \$600, would buy 200 more elements for the proportionate sample.

h	N_h	$100W_h$	$\bar{Y}_h$	S_h^2	S_h
1	50,000	50.00	0.13	0.25	0.5
2	23,000	23.00	0.72	2.89	1.7
3	20,000	20.00	3.34	72.25	8.5
4	5,300	5.30	18.03	1,225	35
5	1,500	1.50	68.85	9,025	95
6	120	0.12	786	40,000	200
7	80	0.08	434	28,900	170
Total	100,000	100.00	$Y = 417{,}730$ $\bar{Y} = 4.1773$	$\Sigma W_h S_h^2 = 286.66$ $S^2 = 1250.57$	

3.5. Suppose that $J_1 = J_2 = \$2$, $J_3 = J_4 = \$3$, $J_5 = \$4$, and $J_6 = J_7 = \$6$. Suppose also that all 200 cases of strata 6 and 7 remain in the sample. Compute the variance of a sample allocated for minimum variance for cost $\Sigma J_h n_h = \$10{,}000$ (about same as 3.4*a*).

3.6. (*a*) For a city population of about $N = 130{,}000$ families it is desired to estimate the mean number $\bar{y}$ of children per family. Compute the necessary sample size n, assuming a simple random sample; assuming also that the desired standard error is set at SE $(\bar{y}) = 0.025$; that on several diverse previous studies the coefficient of variation per family was found to be around $C = S/\bar{Y} = 0.3$; and that experts guess that in this city $\bar{Y} = 3$, roughly.

(*b*) Suppose that previous surveys have shown that proportionate stratification yields standard errors about 5 percent less than would srs with the same sample size n. Using the proportionate design, how large an n will you need?

(*c*) Another agency wants to estimate $\bar{y}$ for the entire state of $N = 2{,}600{,}000$ families, and for a town of $N = 2600$ families. Making the same assumptions as in (*a*), what size samples are needed for the state and for the town?

(*d*) Suppose that the survey results show that the last two assumptions in (*a*) were wrong and that the standard error was actually about 30 percent greater than expected. How will this affect the survey results? Is the computed standard error a valid estimate of the sampling variation?

3.7. When listing dwellings within sample blocks for a national survey, the interviewer assigned economic ratings L, M, and H (for low, medium, and high). From data of one year's survey we have the following data about dwellings in the three strata:

Stratum	Proportion of Entire Sample	Mean Liquid Assets	Variance of Liquid Assets	Standard Deviation of L.A.	Proportion Who Own Less Than $2,000 in L.A.
		In Thousands of Dollars			
L	0.40	0.7	3.2	1.8	70
M	0.50	1.9	16.5	4.1	48
H	0.10	4.8	79.7	8.9	26

(*a*) For a sample of a city with $N = 800{,}000$, assume that the distribution of ratings and wealth is about the same. You now want a stratified subsample so allocated as to produce minimum variance for the mean liquid assets holdings in the city. Assume that the sampling rate for the stratum rated L is to be 1/1000. Choose an answer among (1) through (8) below, which represents adequate approximations for the optimum sampling rates in strata L, M, and H.

(1) 1/1000, 1/5000, and 1/25,000
(2) 1/1000, 1/2000, and 1/5000
(3) 1/1000, 2/3000, and 1/3000
(4) 1/1000, 2/1000, and 5/1000
(5) 1/1000, 1/1000, and 1/1000
(6) 1/1000, 3/1000, and 7/1000
(7) 1/1000, 5/1000, and 25/1000
(8) 1/800, 1/800, and 1/800.

(*b*) The rates above would yield about how many interviews? Suppose that your budget allows for only 1000. What adjustment of the sampling rates you chose in (*a*) should you make?

(*c*) What allocation should be used to estimate, with a minimum variance, the proportion of spending units in the city that have less than \$2000 in liquid assets? Choose an answer among (1) through (8) above.

3.8. If the results of a large srs are sorted into six groups, then reweighted according to the *true* stratum sizes, the process is called *post-stratification*. It is resorted to at times because in some situations (answer True or False for each of the following statements):
(*a*) It can produce results more precise than simple random sampling.
(*b*) It can produce results more precise than proportionate sampling.
(*c*) It can produce results more precise than optimum allocation.
(*d*) The stratifying variable is not available at the time of selection.
(*e*) The sampler did not use the strata for selecting the sample.
(*f*) It reduces the variance *about* as much as true proportionate stratified random sampling.

3.9. In a metropolitan area, the central city has about 400,000 dwellings and the suburban area 100,000. The cost per sample dwelling is about the same in the city as in the suburb. Money for a sample of 600 dwellings is available for a survey with many objectives. Some of these involve estimating means for the entire metropolitan area, others the estimation of the difference of city and suburban means $(\bar{y}_1 - \bar{y}_2)$. All involve estimating proportions, so it can be assumed that the element variances are about the same for the suburbs as for the city: $\sigma_1^2 = \sigma_2^2 = \sigma^2 = PQ$. Disregard the finite population correction, and differences in effects of clustering and stratification.
(*a*) What sampling rates in the city and suburbs (f_1 and f_2) would produce the minimum variance for the mean of the metropolitan area? About what sample sizes (n_1 and n_2) would result in city and suburb?
(*b*) What sampling rates in the city and suburbs would produce the minimum variance var $(\bar{y}_1 - \bar{y}_2)$ of the difference of suburban and city means?
(*c*) Compare the variance of the mean for the entire metropolitan area obtained with the rates of (*a*) and (*b*), respectively.
(*d*) Compare the variance of the city versus suburb difference—var $(\bar{y}_1 - \bar{y}_2)$—obtained with the rates of (*a*) and (*b*), respectively.
(*e*) A sample of $n_1 = 400$ and $n_2 = 200$ would yield $f_1 = 1/1000$ and $f_2 = 2/1000$. Compute the variance of the combined mean for the entire metropolitan area. Compute also var $(\bar{y}_1 - \bar{y}_2)$. Compare these results with those of (*c*) and (*d*), and draw conclusions.

3.10. In a metropolitan area the central city contains 200,000 dwellings and its suburban area 200,000 also. It is assumed that for most survey variables the variances per element (dwellings) are equal in the two strata ($\sigma_1^2 = \sigma_2^2 = \sigma^2$). Three sampling plans are proposed, each involving $n = 600$ dwellings.
(*a*) Take $n_1 = 300$ in the city and $n_2 = 300$ in the suburbs; that is, $f_1 = f_2$.
(*b*) Take $n_1 = 400$ and $n_2 = 200$; that is, $f_1 = 2f_2$.
(*c*) Take $n_1 = 480$ and $n_2 = 120$; that is, $f_1 = 4f_2$.

Compare these three variances for the mean of the entire metropolitan area, and draw some conclusions. The factors $(1 - f_h)$ are negligible.

3.11. Suppose that you want a proportionate stratified sample of elements from a population of $N = 10{,}000$. The critical variates are proportions, guessed to be around 45 percent. Proportionate stratification is expected to reduce the element variance by 15 percent. (*a*) What standard error do you expect with $n = 250$? (*b*) What n is needed for obtaining a standard error of 2 percent? (*c*) If the actual gain from stratification is only 5 percent, instead of the expected 15 percent, what is the effect of this mistake on the validity and magnitudes of the sample results obtained in (*a*) and (*b*)?

3.12. A large population is divided into three strata (with weights W_h) in which the proportions (P_h) of a survey variable are guessed to vary as follows: $W_1 = 0.5$, $P_1 = 0.52$; $W_2 = 0.3$, $P_2 = 0.40$; $W_3 = 0.2$, $P_3 = 0.60$. What size proportionate sample can be expected to yield the same variance as a simple random sample of $n = 600$?

3.13. The relative gain from proportionate element sampling can be written as $\Sigma W_h(\bar{Y}_h - \bar{Y})^2/\sigma^2$, and this becomes $\Sigma W_h(P_h - P)^2/P(1-P)$ for estimating the proportion P. (*a*) Suppose $P = 0.5$, based on two strata in which $W_1 = W_2 = 0.5$. How large must the deviations $(P_h - P)$ be to yield an appreciable gain of 20 percent? Give P_1 and P_2. (*b*) Repeat (*a*) for $P = 0.2$. (*c*) Suppose $W_1 = 0.05$, $P_1 = 45$ percent; $W_2 = 0.20$, $P_2 = 5$ percent; $W_3 = 0.75$, $P_3 = 1$ percent. This represents a rather high concentration of a fairly rare characteristic; $P = 4$ percent. How much is the gain of proportionate over simple random sampling? (*d*) Optimum allocation would yield a variance of $(\Sigma W_h\sqrt{P_hQ_h})^2 = 0.0204$; what is its relative gain over proportionate and over simple random sampling?

3.14. On the payroll list of a company, all 20,000 employees are designated with a *single* and *distinct* five-digit payroll number. These are printed in order, but without the 80,000 blank numbers.

The seven departments of the factory have the following number of employees, respectively: 1000, 1000, 2000, 2000, 4000, 5000, 5000. The first digit of the payroll number denotes department. Within the department the allotment of the last four digits of the payroll numbers is haphazard; and after an inquiry it is decided that the distribution of employees within the departments may be considered random.

It is desired to select a sample of 500, giving each employee an equal probability of being selected. Assume that the selection will be made by an honest, willing, and intelligent person who is not a statistician, but who can use a table of random numbers. Write brief (3 to 5 sentences) and clear instructions for each of the following:

(*a*) A simple random sample of employees.

(*b*) A proportionate stratified sample of employees, where the departments are the strata.

(*c*) A systematic selection applied to the printed payroll list.

(*d*) Does it make a difference to any of the three methods whether the selection is applied to the 100,000 potential numbers, or to a compact list of 20,000 ordered payroll numbers?

3.15. From the sample of 3.14 you want to estimate the proportion of all employees of the factory who hold certain attitudes. Your first tabulations show that there are important differences in attitudes among the 7 departments. A code for each respondent identifies the department to which he belongs. To use this information for stratification, what would you need to do in each of the following situations?

(*a*) If the selection were simple random.

(*b*) If the selection were a proportionate stratified random.

(*c*) If the selection were systematic applied to the 20,000 listings.

To estimate the overall proportion P (choose either 1 or 2 for each design): (1) we *may* use the simple mean, $\frac{1}{500}\sum_j^{500} y_j$; or (2) we *must* use a weighted mean, $\sum_h^7 W_h p_h$.

The proper estimate of the variance of p is (choose either 3, 4, or 5 for each design):

$$(3)\ \frac{p(1-p)}{500} \qquad (4)\ \frac{39}{40}\frac{p(1-p)}{499} \qquad (5)\ \frac{39}{40}\sum_h^7 W_h^2 \frac{p_h(1-p_h)}{n_h - 1}.$$

3.16. A sample of $n = 5000$ persons is selected with simple random selection and each is assigned a *sample number* in the order they are selected, the rth draw being assigned number r. The persons are identified with *payroll numbers* of a large company, the first digits identifying factories and departments. Furthermore, each interview is also assigned an *interview number*, denoting its order of arrival in the office. It is desired to select for further intensive research (perhaps a reinterview) a subsample of 500 from the 5000. Six alternative plans are proposed:

(*a*) Take the lowest 500 of the 5000 *sample numbers* selected.

(*b*) Take the lowest 500 of the 5000 *interview numbers* selected.

(*c*) Take 500 of the 5000 *interview numbers* with simple random sampling.

(*d*) Order the interviews by *payroll numbers* and take the lowest 500 payroll numbers.

(*e*) Form packs of 5 interviews with the previous ordering, then take every tenth pack, after a random start.

(*f*) Take every *tenth* payroll number after a random start.

For each of the alternative plans (*a*) to (*f*) state whether it is true (T) or false (F) that the proposed selection is:

(1) a probability sample of the entire population,

(2) an epsem (equal probability of selection method),

(3) an exact equivalent of a simple random sample (srs),

(4) although not an exact equivalent of a simple random sample, it is probably an acceptable substitute for it.

(*g*) Now think of the subselection of 500 elements as a practical problem where you want the best sample you can get for further research. Can you discuss *briefly* one or two selection procedures you might prefer to any of the above six?

3.17.

h	W_h	$\bar{y}_h$	S_h^2	S_h	P_h	P_hQ_h	$\sqrt{P_hQ_h}$
1	0.35	3.1	4	2.0	0.54	0.2484	0.50
2	0.55	3.9	11	3.3	0.39	0.2379	0.49
3	0.10	7.8	128	11.3	0.24	0.1824	0.43
Total	1.00	4.0	21.98		0.43	0.2451	0.50

$$\sum W_hS_h^2 = 20.25 \qquad \sum W_hP_hQ_h = 0.2360$$

This is a table of personal incomes in \$1000, and of the proportions P_h who have incomes under \$3000, based on a sample of the U.S. urban population. Dwellings were separated into three strata according to interviewer ratings (11.4*C*). Ignore the finite population corrections $(1-f_h)$.

(*a*) Estimate variances of $\bar{y}$ and p for simple random samples of $n = 1000$.

(*b*) Estimate variances of $\bar{y}$ and p for proportionate samples of $n = 1000$. How much is the relative gain over (*a*)?

(*c*) Allocate a sample of $n = 1000$ for minimum variance of $\bar{y}$. Compute variance of $\bar{y}$ and p and compare to results of (*a*) and (*b*).

(*d*) Allocation for minimum variance of p would be very close to proportionate. What part of the table is the basis for this judgment? The median income was near \$3000, and the variance of its estimate depends on the proportion p having this income (12.9).

4

Systematic Sampling; Stratification Techniques

4.1 PROCEDURES AND USES OF SYSTEMATIC SELECTION

4.1A Uses of Systematic Samples

Systematic sampling is perhaps the most widely known selection procedure. It is commonly used and simple to apply; it consists of taking every kth sampling unit after a random start. It provides an alternative for random and independent choice of sampling units and is sometimes called a "pseudo-random" selection. It is often used jointly with stratification and with cluster sampling. The following illustration deals with the selection of n out of N elements, but the same procedure can also be applied directly to the selection of a out of A clusters.

Suppose that the population size N is an integral multiple of the desired sample size n. Then the desired sampling *interval* $k = N/n$ is also an integer. If a number is drawn at random from 1 to k, this *random start* r for the interval determines the unit to be selected in each of n *implicit strata* or zones.

The interval k divides the population into n *zones* of k units each. One unit gets selected from each zone and has the same location in each zone. Since the first number is drawn at random from 1 to k, each unit gets the same probability $1/k$ of selection. The order of selection—1, 2, . . . , g, $(g + 1)$, . . . , $(n - 1)$, n—reflects the order of numbering of the population elements. We use the subscript g for systematic selections to indicate the importance of the selection order for computing variances.

The symbol k commonly denotes the selection interval in systematic sampling. Since we use $f = 1/F$ for selection fractions, we note that $k = F = N/n$. The interval $k = F$ selects one unit from each zone of $k = F$ units. The random start from 1 to k imparts to each unit the selection probability of $1/k = 1/F = f$.

The prime reason for using systematic sampling is that its application is easy, "foolproof," and flexible. For example, the clerical tasks of selecting every kth line from listing sheets or every kth block from numbered maps can be done more easily than in a corresponding random selection. Furthermore, it is easier to check the clerical application of intervals than of random selections.

These advantages seem even more pertinent when the field worker is entrusted with the selection of sampling units. For example, his instructions may call for listing the dwellings in a block (or students of a class, or

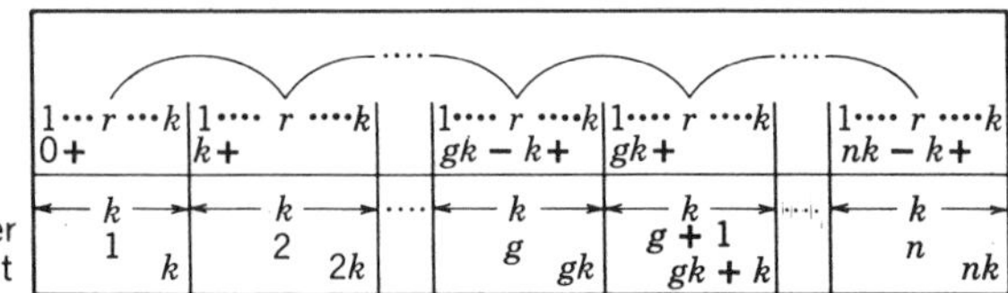

FIGURE 4.1.I Systematic selection of n from population of $N = nk$ units. Same location r is selected from n "*implicit strata*" or zone of k units each. The length of the line represents the $N = nk$ units. The arcs point to the n selections. After drawing the random start r, the interval is laid off $(n - 1)$ times.

members of a club, or employees of a work section) and then selecting units numbered r, $r + k$, $r + 2k$, etc. The assignment of a set of random numbers is more difficult, and asking the field worker to select random numbers is risky and not easy to check. The application of intervals has an added advantage in the common situation where cluster size is not known beforehand; it can be simply carried on by the field or office worker as long as needed, until the cluster size is exceeded. It can also be applied readily to sampling with probability proportional to size (7.4).

Another advantage of systematic sampling is that it can easily yield a proportionate sample, if we take advantage of its even spread over the population through a corresponding ordering of the latter. For example, a sample of every kth dwelling of a block will be spread around its sides, which may differ considerably in characteristics. Or a systematic sample over an alphabetical list of names will yield about the same proportion of names from each letter.

Whatever stratification exists in the ordering of the population list, the sample will reflect it. Of course, definite limits to the amount of stratification are set by the selection of a single unit from each set of k consecutive units. Fractions of k units yield either one or zero selection, and the proportionality has an absolute error of less than one unit for any stratum. The relative error for a stratum of ck units is less than $1/c$, hence is small for large strata. It also follows that if we have flexibility in ordering the population list, we can try making each set of k units relatively homogeneous. Or each set of $2k$ units can be made homogeneous, to facilitate variance computations as described below.

Both advantages, simplicity and proportionality, may be present in the type of bureaucratic situation where the accessions to the list get numbered as they arrive in a haphazard order. Examples of this might be social security numbers, hospital admissions, arrivals in stores and travel stations, bills and invoices in files, etc. A 10 percent, or 1 percent, or 0.1 percent sample can be produced easily by designating for selection respectively a single-digit, two-digit, or three-digit ending. We may take several endings to obtain other sampling fractions. For example, four endings of three digits produce a sample of 0.4 percent. This can become a particularly simple routine task for clerical administration. It also provides incidental stratification over the order of accessions—perhaps over the diurnal, weekly, seasonal, or secular variations. The sampler should investigate the order and nature of accessions before accepting it as similar to random or stratified random selection. If in doubt, he can select at random several (five, ten, or twenty) selection numbers; comparing these will allow him to guard against the undesirable effects of regularities (4.4).

4.1B Problems with Intervals

If the population size N is not an integral multiple of k, a problem arises. It can be solved in several ways and the sampler should choose the most convenient:

1. *Permit the sample size to be either n or* $(n + 1)$. Choose k so that N is greater than nk, but less than $(n + 1)k$. Then the random start will determine whether the sample size will be n or $(n + 1)$. You can imagine having added enough *blanks* to make the list exactly $nk + k$ long. Typically the difference of a single element is negligible compared to other sources of variation, such as nonresponse. The probability of selection is $1/k$ for each unit. This can be regarded as applying the interval k to a population of $k(n + 1)$ elements, larger than N by the number of blanks in the last zone of k units. The proportion of these blanks in the population is less than $1/(n + 1)$.

A modification of the above would keep the sample size constant at n, by omitting one element at random if $(n + 1)$ were selected. This procedure is not epsem. For example, applying $k = 3$ to $N = 7$ for a final sample of $n = 2$, the units (2, 5) and (3, 6) receive the proper probability of 1/3; but the units (1, 4, and 7) receive only the probability $1/3 \times 2/3 = 2/9$. The overall probability of 2/7 would be made up of $W_h = 4/7$ of the units with $P_r = 1/3$, and of $W_h = 3/7$ of the units with $P_r = 2/9$; thus $(4/7)(1/3) + (3/7)(2/9) = 2/7$. This departure from epsem may be trivial in most situations, where the ordering of the units has little effect and where n is large. But if n is large, we can also ordinarily accept $(n + 1)$.

2. *Eliminate with epsem* enough units to reduce the listings to exactly nk before selection with the interval k. The probability of selection over the two procedures is n/N. Instead of elimination, it may be more convenient to select some listings with epsem, then add these duplicates to the end of the list.

3. *Consider the list to be circular*, so that the last unit is followed by the first. Choose a random start from 1 to N. Now add the intervals k until exactly n elements are chosen, going to the end of the list and then continuing to the beginning. Any convenient interval k will result in an epsem of n elements selected with the probability n/N. Generally, the integer k closest to the ratio N/n will be most suitable. This procedure has great flexibility and can be applied to many situations. It can be used to apply an interval separately to many strata. It is especially useful for applying an interval to many clusters in multistage sampling. It is hardly worth using for selecting a single sample.

4. *Using fractional intervals* is simple with a decimal fraction. For example, suppose that to select a sample of $n = 100$ units from a population of $N = 920$ units, the interval $k = N/n = 920/100 = 9.2$ is applied. Select a random start from 1 to 92 and add the interval 92 successively. Then round down these numbers by eliminating the last digits, obtaining exactly $n = 100$ numbers. This is done easily on desk computers and the selection numbers can be written down without the decimals.

Remark 4.1.1 That the last procedure yields the probability of selection $1/k = 1/9.2$ for all units can be easily shown. Unit 1 would be selected by any of the 10 random starts from 10 through 19; unit 2 by any of the 10 from 20 through 29. Unit 9 would be selected by 90, 91, 92 plus the 7 from 01 through 07; and unit 10 by the 10 numbers from 08 through 17. Note that a random number from 01 through 09 will yield the first selection, either 9 or 10, only after the interval is added. Thus any unit will be selected by 10 of the 92 possible random starts, hence with probability $10/92 = 1/9.2$.

The population list may contain many blanks or foreign elements scattered throughout. In some cases renumbering would be too difficult and the blanks must be tolerated. Moreover, when treating a subclass, the nonmembers appear as blanks on the list. Nonresponse may also be regarded as blanks. Consider the list of $nk = N = M + B$ units composed of M

members and B blanks. The interval k will select m units, with probability $1/k$. However m is not fixed, but a variate, and its expected value is M/k. (Its standard error is $\sqrt{n\bar{M}(1 - \bar{M})}$, with $\bar{M} = M/N$, if the members and blanks are distributed at random; but probably less if they are distributed irregularly but not truly at random.)

If the population size M is known, we can compute $k = M/m^*$ from the desired sample size m^*, and obtain an actual m not far from the desired m^*. If M is not known, the sample size is subject to the degree of our ignorance. A preliminary sample of m can be drawn and reduced to size m^*, by eliminating from it a selection with the interval $m/(m-m^*)$, if the selected sample is greater than desired ($m > m^*$). If the selected sample is smaller than desired ($m < m^*$), we can add another selection with the interval $km/(m^* - m)$; the two samples together should be about size m^*. These procedures are not strictly epsem, but can be used in many situations if m^* is large or if we have enough knowledge about the listing order. Otherwise, we may abandon systematic selection in favor of a random selection, perhaps after stratification.

4.1C Variances for Systematic Samples

A systematic sample selected with the interval k after a random start yields an epsem because each element has a probability $1/k$ of being selected. Therefore, the sample mean is an unbiased estimate of the population mean, if the sample size was fixed at n (2.8C). If n was not fixed, then the mean is not technically unbiased, but it will usually be a good estimate.

Theoretically, variances for systematic samples present a formidable problem. To be valid under all conditions, variance computations require at least two random selections per stratum. But systematic samples are composed of single selections from each implicit stratum; furthermore these selections are not independent between strata. The interval k divides the population into k "clusters" of n elements each, and the random start from 1 to k selects one of these clusters. Hence, strictly speaking, a systematic selection of units is not "measurable," because the variance cannot be computed from the sample alone.

We shall see later that for computing the overall variance of the sample only the summary values of primary selections (PS) are used ordinarily; the computations do not depend on whether systematic or random choices are used in later stages. Hence, the simplicity of systematic selection for the field work of second and later stages can be employed without complicating the variance computations. Therefore, we shall concentrate on the effect of using systematic selection, rather than random choice, for the primary selections. Element sampling is treated specifically below, and cluster sampling in the appropriate sections.

Computing the variance for a systematic selection requires assumptions about the distribution in the population on the list, and for these the sampler should have good theoretical and empirical bases. If he is not satisfied that the problems raised in 4.2 can be avoided, he should use some random choice in place of systematic selection. Some statisticians prefer to avoid systematic sampling in the primary stage. Others advise against its use by any but the trained and painstaking sampling specialist. Often, however, one of the following four models can be assumed—after investigation—with reasonable confidence.

Simple Random Model. If the population units were thoroughly shuffled or mixed before they were ordered on the list, the systematic sample would be equivalent to a simple random sample. Though short of this ideal, in some populations the overwhelming proportion of the variance of survey variables is irregular and haphazard, and the result of a systematic sample can be accepted for practical purposes as a good approximation for random choice. In these situations we can use simply the proper random selection formula for the kind of sampling units selected (see 2.2 for elements; 5.2 or 5.3 for equal clusters; and 6.2 or 6.3 for unequal clusters). For example, suppose that a small systematic sample is selected from an alphabetical list of names. For most lists, this would approximate an srs except for the mild proportionate stratification effect associated with the starting initials. This effect may be negligible, particularly in element sampling, and especially for estimates of proportions.

Stratified Random Model. In many situations, in addition to much irregular variation, there is some tendency for nearby units to resemble each other, to deviate similarly from the population mean. Chronologically kept files are subject to variation along the time dimension. A list of employees may reflect the partial homogeneities of work groups. The mailing list of members of a society, subscribers, or customers of a company may be in geographical order. Numbering the blocks of a city, district after district, establishes a list with geographical divisions.

Most files have some meaningful divisions, and the variance may be decidedly lower than random selection formulas would indicate, because of the proportionate stratification effect induced by the systematic selection. This effect is likely to have more of an impact on cluster samples than on element sampling. In such situations, use the proportionate stratification formula appropriate for the sampling unit (see 3.4 for element sampling, 5.6 for equal clusters, and 6.4 for unequal clusters).

Formally, this model assumes the division of the population into strata and thorough mixing of the sampling units within the strata. The sampler must investigate the nature of the population list (and not the sample) to determine the strata. For example, on an alphabetical list these might follow the initial letters. Determining these boundaries may prove a subjective and worrisome task.

Paired Selections Model. Frequently, it is a reasonable assumption that each successive pair of selections was drawn at random, *two from each implicit stratum* or *zone*. This model assumes that within each zone the $2k$ elements were thoroughly shuffled before the 2 selections were drawn from it. The pairs of selections compared are determined by the order of selection; the first selection is contrasted with the second, third with fourth, and so on until $(n - 1)$ with n. The number of zones is $H = n/2$. The pairs may be denoted with the subscripts ha and hb, as h takes on the values, $1, 2, \ldots, n/2$. The $n/2$ paired differences yield the estimated variance. Appropriate variance formulas are developed in 4.3 for elements and in 6.5*B* for clusters. For element sampling, the variance of the mean is

$$\text{var}(\bar{y}) = \frac{1-f}{n^2} \sum_h^{n/2} (y_{ha} - y_{hb})^2. \tag{4.1.1}$$

A small problem arises if n is odd. Then select one of the elements at random and use it twice, making $(n + 1)/2 = m'$ pairs. Then

$$\text{var}(\bar{y}) = \frac{1-f}{n(2m')} \sum_h^{m'} (y_{ha} - y_{hb})^2. \tag{4.1.1'}$$

Successive Difference Model. This modifies the last model and uses all the $(n - 1)$ successive differences (1-2), (2-3), (3-4), . . . , ($[n - 1]$-n). Thus the precision of the computed variance is somewhat increased. The "degrees of freedom" are more than $n/2$, but less than $(n - 1)$. The variance is computed as

$$\text{var}(\bar{y}) = \frac{1-f}{2n(n-1)} \sum_g^{n-1} (y_g - y_{g+1})^2. \tag{4.1.2}$$

Aggregates for the population can be estimated simply as $N\bar{y}$, and its standard error as $N \,\text{se}(\bar{y})$, when the population size N is known. But when the list contains blanks, then the population size M may be unknown, and the simple expansion $y/f = yk$ may have to be used (for this case, see 11.8).

Example 4.1a Suppose a systematic sample of $n = 40$ elements out of a population of $N = 4000$ elements results in the following sample, presented in the order they were drawn:

10,8, 6,5, 9,8, 8,5, 9,9, 9,10, 4,3, 1,2, 3,4, 0,6,

3,5, 0,3, 0,0, 4,0, 8,0, 10,5, 6,1, 3,3, 1,5, 5,4.

The mean of the sample is $\bar{y} = y/n = 185/40 = 4.625$.

Using the successive difference formula (4.1.2), the variance is

$$\text{var}(\bar{y}) = \frac{0.99}{2 \times 40 \times 39} 540 = 0.99 \times 0.173 = 0.171 = 0.41^2,$$

where $(10 - 8)^2 + (8 - 6)^2 + (6 - 5)^2 + (5 - 9)^2 + (9 - 8)^2 + \cdots$
$+ (1 - 5)^2 + (5 - 5)^2 + (5 - 4)^2 = 540.$

If the variance were computed with a simple random formula, disregarding the order of selection, it would come to $0.265 = 0.515^2$. This would disregard the stratifying effect of the systematic selection applied to the listing order. Actually, the 40 values are merely the values of Problem 2.5, but rearranged to have some stratification effect.

4.2 PROBLEMS OF SYSTEMATIC SELECTION

The sampler must be alert to two types of departure from randomness in the ordering of units on the population list. The presence of either of these tendencies may greatly increase the variance of a systematic sample. Worse still, the computed variance may fail to include these sources of variation, thus resulting in underestimation of the true variance, and leading to exaggerated confidence in the results.

A *monotonic trend* may exist in the ordered population list. For example, suppose that the 20,000 mortgages (loans) granted by a bank to home buyers were numbered in the order they were granted over 15 years. There is a tendency for these mortgages to increase from the lowest to the highest number on the list, because the cost of homes and the mortgages granted have been rising over the years. This tendency is much greater for the debt remaining unpaid, because the mortgages are reduced gradually by the monthly payment of the home buyers. Hence, for a study of mortgage debts, the mean of a systematic sample of 1 in 100 could depend greatly on the choice of a random number: a sample consisting of the first element in each zone (1, 101, 201, etc.) could have a much smaller mean than a sample of the last elements (100, 200, 300, etc.). The monotonic trend induces a variation among the 100 possible sample means.

Periodic fluctuations are potentially more dangerous, but also rarer, and easier to find and avoid. For example, taking the same digits for selecting house or telephone numbers may result in poor samples, because of the propensity of some people to choose certain digits (say, 0 and 00) and to avoid others (perhaps 13). This problem is present when the population members have some freedom to choose their own listing numbers.

For another example, consider a housing development composed of identical structures of eight dwellings each. The dwellings in each structure vary in type, numbers of rooms, numbers of occupants and children; hence, probably also in related characteristics. A systematic sample over the development of every eighth dwelling would result in a sample composed of similar types of families; and any integral multiple of 8 would have the same disadvantage. The possible samples would differ greatly, and the variance computed from the relatively homogeneous sample could badly underestimate the true variance of the sample design. Some time series data can have regular periodicities. Sometimes lists, ledgers, and files may have tendencies to peculiar regularities. For example, the first entries on every page of a long list may tend to differ from later entries.

Generally, whenever fluctuations of period k are present in the population, a sampling interval of k or any integral multiple of k (or even simple fractions such as $k/2$ or $k/3$) should be avoided. Fortunately, such regularities are seldom present in population lists, and the alert sampler can usually discover and avoid them.

The problems induced by either a monotonic trend or a periodic variation are revealed if we regard a systematic sample as the result of choosing a single cluster from a population of k clusters. The sample cluster consists of n elements; consider that each has a random error component e_{rg} and a deviation d_r that is constant for all the n elements in the cluster ($r = 1, 2, \ldots, k$). The random components tend to cancel out for large n, but d_r is a systematic deviation as large for the sample mean as for the individual element. This resembles a systematic bias of measurement, and its effect on a single large sample can be similar. It is not formally called a bias because it cancels over all k possible samples.

The constant deviation d_r appears greater for large samples with relatively small errors. Furthermore, the constant deviation remains unmeasured, excluded from the variance computations. Thus a systematic sample, although a probability selection, is not strictly "measurable," because the sample itself does not provide all the information for computing the variance. We must also provide a model for the nature of variation over the listing order. This is not needed for truly random selection, hence systematic sampling is called "pseudo-random."

Because ordinarily we cannot be *absolutely* certain of having avoided all danger, some statisticians prefer to avoid systematic sampling altogether. I, and many others, are cautious about the pitfalls of systematic selection, but take advantage of its convenience and simplicity in many practical selection situations. In most practical situations after investigating, we can dismiss the dangers both of a monotonic trend and of a periodic fluctuation coinciding with the selection interval. When dangers are not negligible, we have several alternatives:

1. *Randomizing the population ordering* through some shuffling procedure would permit the application of systematic selection. But this is seldom practical and would destroy trends beneficial for stratification.
2. *Random selection* can be introduced in several ways: simple random selection, random selections within strata or pairs of random selections within implicit strata (zones) of size $2k$ units each (4.3).
3. *Changing the random start* several times reduces the effect of any single start, yet it can retain most of the practical convenience. For example, when selecting cards from a long file or entries from a long list, the clerk can apply a regular interval for, say, 20 selections to the first 20 k zones, after one random start (e.g., 3, 13, 23, . . . , 193). Then another random start yields the next 20 selections (e.g., 208, 218, . . . , 398). And so on.
4. *Replicated selection* of c different samples permits the use of systematic selection for each. Instead of one systematic interval of k, we can apply the interval ck c times. Consider the example of the bank's list of mortgages with a strong monotonic trend: we can select 10 random starts each from 1 to 1000 and apply the interval 1000 with each. In other words, the clerk needs to look for all accounts ending in any of the 10 selected three-digit numbers. This is slightly more difficult than a single two-digit ending, but easier clerically than finding a set of random numbers.

When considering the problems of systematic sampling for a multistage sample, we must distinguish between the primary selections and the later stages of selection. Systematic selection in the primary stage can entail serious hazards in some situations. Those hazards can be avoided with alternate designs; for example, with paired selections per stratum. But within the selected primary units, the smaller units can be drawn with systematic selection to exploit its practical advantages. For example, we may select first a stratified random sample of 50 banks; then within each bank a systematic sample of 1/1000 can be designated with a three-digit end number. Applying a different random number in each bank does not complicate the clerical task of selection within each bank.

Remark 4.2.1 If we can find and safely define a strong and regular trend, monotonic or periodic, in the listing order, we may extract special gains from it. This may be accomplished with a sample design fitted to the situation, that takes advantage, either in the selection or the estimation, of the known regular trend in the listing order. For example, if the length of a regular period is known to be eight units, then an interval of either seven or nine will cover well the period of variation. For strong linear trends, "centrally located samples" can be used in the selection, or its equivalent in the estimation with "end corrections." These technical matters are described well by Cochran [1946, 1963], Yates [1948, 1960], Madow [1944], and Milne [1959]. But regular and securely known trends occur rarely in surveys, especially in social surveys. Those that do occur can be treated also with stratification.

Remark 4.2.II That the sampling distribution of a systematic selection consists only of k possible samples strains the assumption of its normality. This theoretical problem has a theoretical approach: consider the actual listing order as but one chance situation among many potential permutations under the prevailing causal system [Cochran, 1963, p. 214].

Remark 4.2.III If $N \neq nk$, the use of the interval k leads to samples which vary by a single unit; if $nk = N - c$, then c of the possible random starts would result in $(n + 1)$ units and $(k - c)$ in n. Formally, then, the usual sample mean is not unbiased. But this has negligible effect in practical situations, where n is large enough and the correlation between sample size and the survey variables is nil. This technical problem arises because the average value of all possible samples is $\Sigma\, \bar{y}_r/k = \Sigma\,(y_r/n_r)/k$; this, because n_r varies from n to $(n + 1)$, is not exactly equal to the population mean, $\bar{Y} = \Sigma\, Y_j/N$.

4.3 PAIRED SELECTIONS

Designs based on two random selections from each stratum have many applications and two main advantages. First, for a fixed number of selections, it is the farthest we can take stratification and still compute the variance from the sample data alone. The population is divided into $n/2$ strata; the two selections yield a single comparison from each stratum; hence, the entire sample yields $n/2$ comparisons for computing the variance. Second, these comparisons permit simple formulas for computing the variance.

The method's wide applicability and great advantages for selecting pairs of clusters is described in 7.3. Here we describe its use for selecting n elements in pairs from $n/2$ strata. The sample sum consists of the $n/2$ pairs $(y_a + y_b)$ of elements, and the sample mean is

$$\bar{y} = \frac{y}{n} = \frac{\sum^{n} y_j}{n} = \frac{1}{n} \sum_{h}^{n/2} (y_{ha} + y_{hb}). \tag{4.3.1}$$

The variance of the mean is

$$\text{var}\,(\bar{y}) = \frac{1 - f}{n^2} \sum_{h}^{n/2} (y_{ha} - y_{hb})^2 = \frac{1-f}{n^2} \sum D^2 y_h, \tag{4.3.2}$$

where $D^2 y_h = (y_{ha} - y_{hb})^2$. First, the variance of a single pair of selections is shown to be var $(y_{ha} + y_{hb}) = (1 - f)(y_{ha} - y_{hb})^2$ (4.3.7). Second, the variance of the sum of the $n/2$ independent pairs is simply the sum of the variances (2.8.7′). The fixed sample size n appears squared in the variance. The finite population correction $(1 - f)$ is often trivial. The sampling fraction is $n/N = f = 1/F = 2/2F$, when 2 selections are made from each stratum, or zone, of $2F$ units. *Zones* here denote strata of equal size created primarily to facilitate the selection [Deming, 1963]. With $n/2$ zones, we have $2F(n/2) = Fn = n/f = N$ elements in the population.

When the population size N is not an integral F multiple of the desired sample size n, we face a minor problem. We can create $nF < N$ listings by first eliminating at random the right number $(N - nF)$ of elements; or, to create $nF > N$ listings, we can choose $nF - N$ listings at random, and add them to the end. Or a few of the strata can be made one unit larger or smaller than $2F$ to create the $n/2$ strata. If these procedures seem undesirable, we can choose among the several alternatives for creating integral zones for systematic selection (4.1*B*).

Two random numbers are needed for each of the $n/2$ strata of $2F$ numbers. We can choose random numbers from 1 to N until two are found for each stratum. However, to save time, we can first draw $n/2$ pairs of random numbers from 1 to $2F$; then add 0 to the first pair, $2F$ to the second pair, $4F$ to the third pair, and so on, until reaching

$$2F(n/2 - 1)$$

for the last pair.

If the aggregate Y is estimated with yF, we have

$$\text{var}\,(Fy) = F^2(1 - f) \sum D^2 y_h. \tag{4.3.3}$$

Further, this method can be extended readily to selections with varying sampling fractions in different strata. One can increase the sampling fraction by decreasing the number of units composing the zone, so that $f_h = 2/2F_h = 1/F_h$. Then the weighted mean is

$$\bar{y}_w = \frac{\sum F_h(y_{ha} + y_{hb})}{2\sum F_h}, \tag{4.3.4}$$

and its variance is

$$\text{var}\,(\bar{y}_w) = \frac{\sum F_h^2(1-f_h)D^2 y_h}{(2\sum F_h)^2}. \tag{4.3.5}$$

The numerators are estimates of the aggregate and its variance. In the denominator $2\Sigma F_h = N$, and this becomes $2\Sigma F = 2Fn/2 = nF$ when F is constant over all strata.

The simplicity of the design depends on its symmetry, and this also can be its weakness. Nonresponses can destroy the symmetry. Although in some situations nonresponses can be replaced with substitute selections from the same strata, in other situations they may have to be omitted. Suppose that among n' responses there remains a large enough number

m' of paired selections, and that these can be assumed to be a fair sample. Since each of the m' pairs estimates $2s_y^2$, we can estimate the variance of the sample mean $\bar{y} = y/n'$ with

$$\text{var}(\bar{y}) = \frac{1-f}{n' \times 2m'} \sum_h^{m'} (y_{ha} - y_{hb})^2. \tag{4.3.6}$$

But the missing units can become too numerous if the list contains many blanks. This becomes a major problem in analyzing subclasses. Perhaps when subclasses are important, paired selections should not be used for element sampling. But paired selections are especially useful for sampling large clusters, where the subclasses in each are treated as clusters of unequal sizes.

If paired selection of elements results in many blanks, we can choose between two alternative treatments for computing the variance. Suppose that a subclass comprising half the sample is represented by the y entries, and that the x entries represent nonmembers or blanks:

$$(yx)\,(xy)\,(xx)\,(yy)\,(xx)\,(yx)\,(yx)\,(yy)\,(yy)\,(yx).$$

Ten zones are represented from a larger sample. Note that only $m' = 3$ pairs of those 10 pairs would be available for computing variances by the preceding method. That proportion decreases for small subclasses.

The *collapsed stratum* method treats all selections from a group of zones as random selections from one stratum. In our example, the 11 subclass members that appeared in the 10 zones form one collapsed stratum. The variance of the mean of stratified sampling is applied to the elements in the collapsed stratum. By disregarding the finer strata, the computed variance is subject to some overestimation. This effect, however, becomes negligible for small subclasses (4.5*A*). This method can be used for any subclass, and for the entire sample, but it is particularly suitable for small subclasses. The size of the collapsed stratum should be large enough to expect about 5 to 10 subclass members in each.

With the *combined stratum* method, take one random selection from each pair and treat it as one combined selection; then the other selection in each pair is treated as the second combined selection from the same stratum. In our example, suppose that the selections within each pair are in random order; note that the first combined selection contains 7 subclass members and the second contains 4 subclass members. These can be treated as paired selections of unequal clusters, and the variances computed as in 6.5*B*. The combined selection should be planned to contain no less than 5 to 10 selections, hence 10 to 20 (or more) per stratum. The number of combined strata should preferably be 20 or more to yield a precise enough estimate of the variance. Thus this method requires 200 or more elements in the subclass. It also requires a more difficult computation than the collapsed stratum method. But it preserves the gains of stratification built into the design, and it may be preferred for larger subclasses.

Example 4.3a Suppose now that the data of Problem 4.1*a* came as paired selections from 20 strata. The mean of the sample is still $\bar{y} = 4.625$. The variance, using (4.3.2), is

$$\text{var}(\bar{y}) = \frac{0.99}{40^2} 215 = 0.99 \times 0.134 = 0.133 = 0.36^2,$$

where $(10 - 8)^2 + (6 - 5)^2 + (9 - 8)^2 + \cdots + (1 - 5)^2 + (5 - 4)^2 = 215.$

Remark 4.3.I The form $(y_a - y_b)^2$ has convenient properties for computing the variance of the sum $(y_a + y_b)$ of 2 random selections; also the variance of their difference $(y_a - y_b)$. First note that for $n = 2$ selections the general formula of the element variance becomes

$$s_y^2 = \frac{1}{n-1}\sum^{n} (y_j - \bar{y})^2 = \left(y_a - \frac{y_a + y_b}{2}\right)^2 + \left(y_b - \frac{y_a + y_b}{2}\right)^2 = \tfrac{1}{2}(y_a - y_b)^2.$$

Since $Es_y^2 = S_y^2$ for any $n \geq 2$, we get

$$E(y_a - y_b)^2 = E(2s_y^2) = 2S_y^2.$$

For the sum $y = (y_a + y_b)$ of two random selections from N units we have (from 2.8.12) that Var $(y_a + y_b) = (1 - 2/N)2S_y^2$. Of this, we have an unbiased estimate in

$$\text{var}(y_a + y_b) = (1\text{-}f)(y_a - y_b)^2. \tag{4.3.7}$$

Here $f = 2/N$, and $f = 2/2F = 1/F$ when there are $2F$ units in the stratum. Typically, we deal with the sum of many independent pairs, and the variance of that sum is the sum of the variances (2.8.7′).

This relationship can also be shown with methods used in the derivation of (2.8.12). Note that the variance of $(y_a + y_b)$ is

$$\begin{aligned} E[(y_a - \bar{Y}) + (y_b - \bar{Y})]^2 &= E(y_a - \bar{Y})^2 + E(y_b - \bar{Y})^2 + 2E(y_a - \bar{Y})(y_b - \bar{Y}) \\ &= 2\sigma_y^2 - \frac{2\sigma_y^2}{N-1} = 2\frac{N-1-1}{N-1}\sigma_y^2 = 2\frac{N-2}{N-1}\sigma_y^2 \\ &= 2\frac{N-2}{N}S_y^2. \end{aligned}$$

Similarly, we can also derive the variance of the difference $(y_a - y_b)$:

$$\begin{aligned} E[(y_a - \bar{Y}) - (y_b - \bar{Y})]^2 &= E(y_a - \bar{Y})^2 + E(y_b - \bar{Y})^2 - 2E(y_a - \bar{Y})(y_b - \bar{Y}) \\ &= 2\sigma_y^2 + \frac{2\sigma_y^2}{N-1} = 2\frac{N-1+1}{N-1}\sigma_y^2 = 2\frac{N}{N-1}\sigma_y^2 \\ &= 2S_y^2. \end{aligned}$$

Thus $2S_y^2$ is the variance of the difference of two random selections.

For the difference of two selections we can also derive generally and directly that the variance of the sampling distribution is simply

$$\text{Var}\,(y_a - y_b) = E[(y_a - y_b) - E(y_a - y_b)]^2 = E(y_a - y_b)^2,$$

whenever $E(y_a - y_b) = 0$. Hence, we have the unbiased estimate for the difference of two selections,

$$\text{var}\,(y_a - y_b) = (y_a - y_b)^2, \tag{4.3.8}$$

and this holds even if the two selections are correlated, so long as their expected values are equal.

Finally, we observe that if the two selections are drawn unrestricted with replacement, the covariances vanish and we have

$$\text{Var}\,(y_a + y_b) = \text{Var}\,(y_a - y_b) = 2\sigma_y^2 = E(y_a - y_b)^2. \tag{4.3.9}$$

****Remark 4.3.II*** We saw that for the difference $Dy = (y_a - y_b)$ of two random selections $E(D^2y/2) = S_y^2$. Hence, from any simple random sample of n selections one can choose a pair at random and use $D^2y/2$ to estimate s_y^2. To get more precision one can choose many pairs and use their average. Altogether there are $n(n-1)/2$ pairs, and the average of the values $(y_a - y_b)^2$ can be shown to be exactly equal to $2\Sigma\,(y_j - \bar{y})^2/(n-1) = 2s_y^2$.

4.4 REPLICATED SAMPLING

We discussed in 4.2 the problem of a systematic selection of every 100th listing from a serially numbered list of the $N = 20{,}000$ mortgages of a bank. This means simply selecting all numbers with a specified two-digit ending. Suppose, instead, that we draw 10 different three-digit random numbers. All mortgages ending in any of these 10 numbers are taken into the sample. This procedure also permits rather simple instructions and checking. The entire sample consists of $c = 10$ replications, each an epsem with a sampling fraction of $1/cF = 1/(10 \times 100)$, and each a systematic sample selected with the interval of $cF = 1000$. Each replication consists of $n_\gamma = n/c = N/cF = 20{,}000/1000 = 20$ selections, one from each zone. The entire sample contains $n = cN/cF = N/F = 20{,}000/100 = 200$ cases. The overall sampling fraction is $n/N = c/cF = 1/F = f = 1/100 = 10/1000$.

If the population size N is not an integral multiple of the desired zone size cF, it is possible to deal with a fractional zone size. But it is more convenient to create an artificial population size N', so that both the zone size cF and the number of zones $n_\gamma = N'/cF$ are integers. Two simple procedures are available. (1) Add enough blanks to the last zone to increase N to N'. Then some of the replications can contain one less actual selection than others. This source of variation in n_γ is generally smaller than those introduced by nonresponses, by blanks on the list, and especially by the use of subclasses. (2) If exact equality of the sample sizes n_γ for all c replications is needed, before the selection we can create exactly N' units with epsem elimination of $(N - N')$ units, or with epsem duplication and addition of $(N' - N)$ units.

The bare essentials of the method consist of four steps:

(1) Compute the zone size cF from the desired number of replications c, and from $N/n = cF/c = F$, based on the desired sample size $n = cn_\gamma$. There will be $n/c = N/cF$ zones. It is convenient to make cF an integer.

(2) Within each zone of size cF, make c selections that are independent of each other and similar in design. Each replication consists of all its n_γ selections, one from each of N/cF zones.

(3) Compute the desired statistics x_γ for each of the c replications. Each x_γ is designed to be an estimate of the corresponding population value.

(4) Use (4.4.1) and (4.4.2) to compute the sample statistics and their variances:

$$\bar{x} = \frac{1}{c}\sum_{\gamma}^{c} x_\gamma, \tag{4.4.1}$$

and

$$\operatorname{var}(\bar{x}) = \frac{1-f}{c(c-1)}\sum (x_\gamma - \bar{x})^2 = \frac{1-f}{c(c-1)}\left[\sum x_\gamma^2 - \frac{(\sum x_\gamma)^2}{c}\right]. \tag{4.4.2}$$

These essentials are few and simple, and the method has great flexibility. The *flexibility of selection* method is implied by step 2 above, requiring only that the c replications be independent, and similar in design. All c replications must use the same sampling design and fractions, but we can choose any convenient and efficient sample design. We have great latitude in the composition of the n_γ selections of each replication. They need not be independent between the zones. In our initial example, each replication was a systematic sample applied with the interval 1000. This systematic procedure illustrates the possibility of using simple selection procedures that can result in cheap clerical operations. There may be monotonic or other trends within each replication, but our confidence comes from having c replications. Thus large samples can be drawn with routine operations. Furthermore, each selection within the zones may consist of clusters of elements linked in any specified manner.

The *flexibility of statistics* that the method can produce is implied by steps 3 and 4. To estimate the population mean $\bar{Y}$, we form its estimate $x_\gamma = \bar{y}_\gamma$ with each replication, and then produce the overall sample mean $\bar{y}$ and its variance. To estimate the population total Y, the replicated statistic can be its estimate $x_\gamma = cFy_\gamma$, based on the sample total y_γ in each replication. Thus the two formulas above can serve any statistic based on c replicated statistics x_γ, each of which alone exists as an estimate of the target population value. The samples should be large enough so that $\bar{x}$, the mean of the c replications x_γ, is approximately normally distributed around that population value. The simplicity of variance computations is the key to the flexibility and utility of this method. Although survey sampling deals chiefly with means and totals, replicated methods can be extended to other statistics. For example, x_γ may stand for the difference of two means, for regression coefficients, etc. This flexibility of replicated sampling should lead to its application in many new areas. This can be greatly facilitated by the growth of computer technology.

The flexible nature of the method permits the introduction of many variations. Stratification can be readily applied. The zones themselves are implicit strata, and we can simply order stratum after stratum and divide the entire list into zones. But we can also create H distinct strata and apply the method separately to each. If c_h replicated means $\bar{y}_{h\gamma}$ are taken independently in the hth stratum, then the application of the general stratified formulas (3.2.2) and (3.2.3) follow directly, with c_h replications instead of n_h elements, and $\bar{y}_{h\gamma}$ instead of y_{hi}. Furthermore, if the $\bar{y}_{h\gamma}$ are self-weighting, either because they are epsem or because they have been properly weighted, then the formulas of (3.3.1) and (3.3.2) apply also.

Disproportionate sampling rates may be used in different strata if needed. It is possible to maintain the simplicity of a constant $c = c_h$ and yet vary the sampling fractions f_h by varying the zone size as cF_h. The larger the zone size cF_h, the smaller the sampling fraction f_h. This is called "thinning of the zone" and is illustrated by Deming [1960].

Deming [1960] has many examples that exploit the method with extreme symmetry. All his H zones are of the same size c (generally 10), and the random selections within the different zones are independent of each other. Hence the zones can be used as H strata with $H(c - 1)$ degrees of freedom; or they can be "collapsed" into c replications; or, between those two extremes, any degree of "thickening of the zones" may be employed.

Further advantages are available when the number of replications c is small. First, some variable nonsampling errors can be automatically included with the variance computations by randomizing them properly over the replications. For example, the variability between coders and between interviewers can be included in the variance by randomizing coders and interviewers over the replications; see 13.8, Deming [1960, p. 249], Mahalanobis [1946], and Lahiri [1954]. Second, the visual evidence present in the consistent results x_γ of the c replications may have

Zone: 1	$1 \cdots \gamma \cdots c_1$	$- \cdots -$	$1 \cdots \gamma \cdots c_h$	$- \cdots -$	$1 \cdots \gamma \cdots c_H$
2	$1 \cdots \gamma \cdots c_1$		$1 \cdots \gamma \cdots c_h$		$1 \cdots \gamma \cdots c_H$
3	$1 \cdots \gamma \cdots c_1$				$1 \cdots \gamma \cdots c_H$
4	$1 \cdots \gamma \cdots c_1$				
n_γ	$1 \cdots \gamma \cdots c_1$				
Stratum data	$\bar{1} \cdots \bar{\gamma} \cdots \bar{c}_1$ Stratum 1	$- \cdots -$	$\bar{1} \cdots \bar{\gamma} \cdots \bar{c}_h$ Stratum h	$- \cdots -$	$\bar{1} \cdots \bar{\gamma} \cdots \bar{c}_H$ Stratum H

TABLE 4.4.I. The Relationship between Selection Zones and Ordinary Strata

The first column marked "Stratum 1" represents the simple notion of c_1 replicates through the entire population. There are c_1 independent samples, each based on the vertical summation across zones. The variance has $(c_1 - 1)$ degrees of freedom. The zones may be genuine strata; but they need not be, as in the case of systematic selection across zones for each replicate.

Ordinary stratified sampling may be represented along the bottom row, with c_h selections from the hth stratum. These numbers can vary from stratum to stratum. The c_h selections are random within the stratum and independent between strata. The variance has $(\Sigma\, c_h - H)$ degrees of freedom.

A combination of the two can be represented by the cells of the table. Within each stratum the number of replications c_h is constant for all zones. But c_h may vary between strata. Also the number of zones may vary between strata. Each replication is an independent random selection at the stratum level, but members of a replication may be connected across the zones.

strong intuitive appeal for nonstatisticians, who may be skeptical about variance computations they fail to understand. Third, the standard error can be estimated without computing the sum of squares, by using the range thus:

$$\text{se}(\bar{x}) = \frac{\text{range}(x_\gamma)}{(R/\sigma)_c\sqrt{c}}, \tag{4.4.3}$$

where range = maximum (x_γ) − minimum (x_γ), and $(R/\sigma)_c$ is the ratio range/σ for samples of size c from a normal population. It is convenient to note in Appendix C that for c between 3 and 13 the values of $(R/\sigma)_c$ are closely equal to $\sqrt{c}$ itself; hence range/c will estimate the standard error.

The precision of this estimate is lower than the root mean square error using (4.4.2). Also note that the application of formulas (4.4.3) requires that the x_γ be approximately normally distributed.

To use (4.4.2) when c is large, we need to assume only that the mean of the c replications be normally distributed. When c is small, we should use the *Student t* tables with $(c - 1)$ degrees of freedom. This also requires the assumption of normal distribution for the replicates x_γ, but is not very sensitive to mild departures. As the sample sizes on which they are based increase, the values of x_γ tend to normality.

How many replications should be used for a fixed sample size? *Designing few replications has several advantages.* (1) It may simplify the selection procedures. (2) It permits better stratification, when the zones are utilized for strata. (3) The computation of variances and the visual display of replications is easy. (4) Randomization of nonsampling variables (coders and interviewers) is feasible. (5) The replicate results x_γ, based on larger samples, more closely approach normality.

But designing many replications also has advantages. (6) The variance has greater precision. This effect appears in the size of the *Student t* needed for constructing intervals. With few replications this effect may more than cancel the gains of increased stratification. (7) These effects are aggravated, and the dangers of having too few replications are greater, when the selection is exposed to regularities, such as periodic variations in systematic selection. (8) The normality of $\bar{x}$ can be based on more replications, even when the separate replicates x_γ lack it.

Mahalanobis [1946] and Lahiri [1958] have frequently employed 4 replicates for the first five reasons. Tukey and Deming [1960] have often used 10 replicates, and Deming's Chapters 6–10 contain rich illustrations and extension of the method. Jones [1956, pp. 64–66] presents reasons and rules for using 25 to 50 replicates. Generally, I too favor a large number, perhaps between 20 and 100.

When the population is first broken into H strata, fine zoning with few replicates can be used within each stratum. By combining the $(c_h - 1)$ degrees of freedom over the H strata, adequate precision (enough degrees of freedom) can be secured for the overall variance. On the other hand, introducing stratification into the variance computation sacrifices some of the method's simplicity.

Precision for the variance needs a fairly large number of replicates. But some advantages of the method require large sample sizes for the separate replicates. If both are needed, the method requires a large number, hundreds or thousands, of primary selections. These are often available for element sampling, but seldom available if the primary selections are large clusters. When the design is confined to a few score primary selections, the method of paired selections (4.3 and 7.4*B*) is usually preferable.

Remark 4.4.I The presence of "blanks" among the listings results in *unequal sample sizes* n_γ for the different replications. Nonresponses also have this effect. The problem becomes more serious when dealing with small subclasses. Suppose that the replicated statistics are means $\bar{y}_\gamma = y_\gamma/n_\gamma$ of subclasses with unequal n_γ's.

Formula (4.4.1) obtains a mean of the ratios y_γ/n_γ, but we may prefer

$$\bar{y}_r = \frac{y}{n} = \frac{\sum y_\gamma}{\sum n_\gamma} = \frac{1}{c}\sum_\gamma^c \frac{n_\gamma}{n/c}\bar{y}_\gamma. \tag{4.4.4}$$

This is a *ratio mean* developed in detail in Chapter 6. It treats the c replications as so many *clusters with unequal sizes* n_γ. Note that, in contrast to (4.4.1), this mean weights each replication with its relative size $n_\gamma/(n/c)$. Its variance can be estimated with

$$\begin{aligned}\text{var}(\bar{y}_r) &= \frac{1-f}{c(c-1)}\sum\left[\frac{n_\gamma}{n/c}(\bar{y}_\gamma - \bar{y}_r)\right]^2 \\ &= \frac{(1-f)c}{n^2(c-1)}\sum(y_\gamma - \bar{y}_r n_\gamma)^2 \\ &= \frac{(1-f)c}{n^2(c-1)}\left(\sum y_\gamma^2 + \bar{y}_r^2\sum n_\gamma^2 - 2\bar{y}_r\sum y_\gamma n_\gamma\right).\end{aligned} \tag{4.4.5}$$

The first form demonstrates that the deviations are weighted by their relative sizes before squaring and adding. The other two are convenient computing formulas. If the replications are large, the n_γ may have relatively small variation, and it makes little difference whether (4.4.1) or (4.4.4) is computed. But when the variation in the n_γ is considerable, (4.4.4) and (4.4.5) may be preferable. They too can be combined into stratified forms. For these, see the formulas for stratified samples of unequal clusters (6.4).

4.5 SUBCLASS MEANS, TOTALS, AND COMPARISONS IN STRATIFIED SAMPLES

4.5A Subclass Means, Totals, and their Variances

Presenting statistics based on subclasses of the sample is common practice in survey analysis. Subclasses can also result from the presence of foreign elements in the selection frame, the target population being a subclass of the listed elements. In the presence of nonresponse, the respondents can also be considered as a subclass of the entire population.

In simple random sampling the means of subclasses can be readily treated. When we come to cluster sampling, the general formulas for unequal clusters apply also to subclasses. But for stratified element sampling we need to consider the treatment of subclasses separately.

Suppose we select a stratified random sample, consisting of n_h elements from the N_h elements in the hth stratum ($h = 1, 2, \ldots, H$). Instead of dealing with the variable X_{hi} of the entire population of N elements, we need to treat only a subclass of M elements. We can avoid new problems and apply the standard stratified formulas to a subset of all H strata, if the subclass boundaries coincide with stratum boundaries; that is, if in each stratum either all or none of the elements belong to the subclass. Of course, the subclass can contain several of these "clean" strata. We should try to create such "clean" strata for important domains of analysis. But we may lack the necessary information for separating the subclasses into strata. When a great deal of information is available, we may be unable to use all of it, because that would result in too many strata. Thus we must often deal with subclasses which cut across strata so that only part of the stratum, only M_h of the population of N_h elements, and only m_h of the sample of n_h elements, belong to the subclass. Although n_h is fixed, the number m_h of subclass members in the sample is a random variable, and that causes new problems.

We can still deal easily with "unclean" strata if we know the numbers M_h of the subclass population within each stratum. Within the strata, the simple random samples of n_h elements result in simple random samples of the m_h subclass members. Thus, when the weights $W_h = M_h/M$ are known constants, the general stratified random formulas (Section 3.3) for the variance can be applied to the mean $\bar{y}_w = \sum_h^H (M_h/M) \sum_i^{m_h} y_{hi}/m_h$. The gains of proportionate sampling would be maintained approximately, for reasons similar to those of post-stratification. The gains of disproportionate sampling would depend on the relationship of subclass weights and variances to those of the entire population. We must understand that using the population weights N_h/N in the place of the correct subclass weights M_h/M could result in large biases if the two sets of weights differ considerably.

Here we treat the frequent problem that arises because we do not know the constants M_h, but only the random variable m_h of subclass members that are found among the n_h elements of the random sample selected in the hth stratum. The proportion of subclass members, $\bar{m}_h = m_h/n_h$, is the binomial mean of the count variable M_{hi} that takes on the values

$m_{hi} = n_{hi} = 1$ for the m_h subclass members in the sample,

and

$m_{hi} = 0$ for the $(n_h - m_h)$ nonmembers.

The average value of $\bar{m}_h$ is $\bar{M}_h = M_h/N_h$. Its variance $\bar{M}_h(1 - \bar{M}_h)/n_h$ causes the increase of the variance (4.5.4) due to using subclasses. The computations and derivations for subclasses involve a similar auxiliary variable Y_{hi} for the variable X_{hi}:

$y_{hi} = x_{hi}$ for the m_h members of the subclass,

and

$y_{hi} = 0$ for the $(n_h - m_h)$ nonmembers.

Note also the definition of the stratum total $y_h = \sum_i^{n_h} y_{hi} = \sum_i^{m_h} y_{hi}$ and the stratum mean:

$$\bar{y}_h = \frac{y_h}{\bar{m}_h n_h} = \frac{y_h}{m_h} = \frac{\sum^{m_h} y_{hi}}{\sum^{m_h} m_{hi}}.$$

We want to estimate the *subclass total* $Y = \sum_h^H \sum_i^{M_h} Y_{hi}$ for the M subclass members in the population, and the *subclass mean* $\bar{Y} = Y/M$. The *subclass total* can be estimated by

$$\tilde{Y}_w = \sum N_h \frac{y_h}{n_h} = \sum F_h y_h. \tag{4.5.1}$$

The subclass sample mean is

$$\bar{y}_w = \frac{\sum N_h y_h/n_h}{\sum N_h m_h/n_h} = \frac{\sum F_h y_h}{\sum F_h m_h} = \sum w_h \bar{y}_h, \tag{4.5.2}$$

where $w_h = F_h m_h/\Sigma F_h m_h = F_h m_h/m_w$, the symbol m_w defining the weighted sample count. These weights of the stratum means are not constants, but random variables. We shall use $N_h/n_h = F_h = 1/f_h$ throughout this section, assuming that N_h, the population count, is the proper measure of each stratum's importance. But other weights, if desired, can be substituted for N_h in the formulas. For proportionate samples $F_h = F$ is constant, and the weights become $w_h = m_h/\Sigma m_h$; the estimated total becomes $\tilde{Y} = F\Sigma y_h$ and the mean $\bar{y} = \Sigma y_h/\Sigma m_h$.

For computing variances we shall need *the element variances within the strata:*

$$v_h^2 = \frac{1}{m_h} \sum_i^{m_h} (y_{hi} - \bar{y}_h)^2 = \frac{1}{m_h} \sum_i^{m_h} y_{hi}^2 - \bar{y}_h^2. \tag{4.5.3}$$

The form v_h^2 involves division by m_h, and it results in simpler formulas than would the form $s_h^2 = v_h^2 m_h/(m_h - 1)$. For each stratum we shall also need *the element variance taken around the sample mean* $\bar{y}_w$:

$$t_h^2 = \frac{1}{m_h} \sum_i^{m_h} (y_{hi} - \bar{y}_w)^2 = v_h^2 + (\bar{y}_h - \bar{y}_w)^2. \tag{4.5.3'}$$

The variance for the sample mean $\bar{y}_w$ can be computed with

$$\text{var}(\bar{y}_w) \doteq \sum_h^H (1 - f_h) \frac{w_h^2}{m_h'} [v_h^2 + (1 - \bar{m}_h)(\bar{y}_h - \bar{y}_w)^2], \tag{4.5.4}$$

where $m_h' = m_h(n_h - 1)/n_h$.

The variance for the subclass mean resembles the ordinary variance for a stratified mean, but with $[v_h^2 + (1 - \bar{m}_h)(\bar{y}_h - \bar{y}_w)^2]$ replacing the element variance s_h^2. Approximately $(\bar{y}_h - \bar{y}_w)^2$ represents the between-strata component and the effect of proportionate stratification on the element variance. We should also note an equivalent form for the variance of subclass means:

$$\text{var}(\bar{y}_w) \doteq \sum_h^H (1 - f_h) \frac{W_h^2}{m_h'} [t_h^2 - \bar{m}_h(\bar{y}_h - \bar{y}_w)^2]. \qquad (4.5.5)$$

For proportionate sampling this becomes, with $m_w = \Sigma\, m_h$,

$$\text{var}(\bar{y}_w) = \frac{1-f}{m_w^2} \sum_h^H \frac{n_h}{n_h - 1} \left[\sum_i^{m_h} (y_{hi} - \bar{y}_w)^2 - m_h \bar{m}_h (\bar{y}_h - \bar{y}_w)^2 \right]. \qquad (4.5.5')$$

When $\bar{m}_h$ is so small that the second term within the brackets can be neglected, and when $n_h/(n_h - 1) \doteq 1$, this is essentially the variance of a simple random sample, s^2/n. It becomes pq/n when Y_i is a binomial variable.

From the above formulas we may deduce several valuable rules:

1. For large subclasses $(1 - \bar{m}_h)$ is small. Hence the ordinary stratified variance formulas, neglecting $(1 - \bar{m}_h)(\bar{y}_h - \bar{y}_w)^2$, would underestimate only slightly the actual variance.
2. However, for small subclasses, the use of ordinary stratified formulas that neglect the addition of $(1 - \bar{m}_h)(\bar{y}_h - \bar{y}_w)^2$ can lead to gross underestimation of the actual variance. *The gains due to proportionate stratification tend to vanish for small subclasses.* The formula (4.5.5) using t_h^2 shows that computing the variance as if it were not stratified may lead to acceptable approximations for small subclasses.
3. The gains (or losses) of disproportionate stratification, which are additional to those of proportionate sampling, tend to persist if the weights w_h and standard deviations v_h for the subclass are similar to those of the entire sample.

The variance for the difference $(\bar{y}_{wa} - \bar{y}_{wb})$ of two subclass means involves two variances, plus a covariance term that tends to vanish because it involves the factor $\bar{m}_{ha}\bar{m}_{hb}$. The variance of the comparison can then be well approximated with

$$\text{var}(\bar{y}_{wa} - \bar{y}_{wb}) \doteq \sum (1 - f_h) \left[\left(\frac{W_h^2 t_h^2}{m_h'} \right)_a + \left(\frac{W_h^2 t_h^2}{m_h'} \right)_b \right]. \qquad (4.5.6)$$

The variance for a subclass total $\tilde{Y}_w = \Sigma\, F_h y_h$ is

$$\text{var}(\tilde{Y}_w) = \sum (1 - f_h) F_h^2 \frac{n_h}{n_h - 1} \left[\sum^{m_h} y_{hi}^2 - \frac{y_h^2}{n_h} \right]$$

$$= \sum (1 - f_h) \frac{N_h^2}{n_h - 1} \bar{m}_h [v_h^2 + (1 - \bar{m}_h) \bar{y}_h^2]. \qquad (4.5.7)$$

The latter form shows that the element variance is increased by about $(1 - \bar{m}_h)\bar{y}_h^2$ beyond its value in ordinary stratified sampling. Note that, if $\bar{y}_h$ does not vanish, the element relvariance $v_h^2/\bar{y}_h^2$ is increased by $(1 - \bar{m}_h)$. For a large subclass this increase may be small, but when dealing with small subclasses this increase approaches the value of 1.

*4.5B Derivations of Variances for Stratified Subclasses

The variances for subclasses can be derived from the stratified random variances of Section 3.3, by employing zero values for the variables of the $(n_h - m_h)$ nonmembers of the subclass. Actually, by using $(n_h - m_h)$ zeros in a total deck of n_h machine cards, the variance can be computed with the ordinary stratified formulas. This may be cumbersome, however. Furthermore, insight is gained by expressing the variance in terms of only the m_h subclass values. Begin with the variance of the total $\tilde{Y}_w = \Sigma F_h y_h = \Sigma N_h y_h/n_h$ of a stratified sample: $\text{var}(\tilde{Y}_w) = \Sigma (1 - f_h)N_h^2 s_h^2/n_h$. Now with both $y_h = \Sigma y_{hi}$ and Σy_{hi}^2 having only m_h nonzero values, we get

$$(n_h - 1)s_h^2 = \sum^{m_h} y_{hi}^2 - \frac{y_h^2}{n_h} = \sum^{m_h} y_{hi}^2 - \frac{y_h^2}{m_h} + \frac{y_h^2}{m_h} - \frac{y_h^2}{n_h}$$

$$= m_h v_h^2 + m_h \bar{y}_h^2 - m_h \bar{m}_h \bar{y}_h^2 = m_h[v_h^2 + (1 - \bar{m}_h)\bar{y}_h^2]. \tag{4.5.8}$$

Since the variance of a stratified total is $\Sigma (1 - f_h)N_h^2 s_h^2/n_h$, we readily obtain (4.5.7) from (4.5.8). In these substitutions of entirely equivalent forms the properties of ordinary stratified estimates are preserved. With similar procedures we can see that (4.5.7) is an unbiased estimate of

$$\text{Var}(\tilde{Y}_w) = \sum_h^H (1 - f_h)\frac{N_h^2}{n_h}\left(\frac{N_h}{N_h - 1}\right)\bar{M}_h[\sigma_{yh}^2 + (1 - \bar{M}_h)\bar{Y}_h^2]. \tag{4.5.9}$$

Hence, when we estimate the aggregate for a subclass rather than the entire sample, we increase the element variance from σ_{yh}^2 to $[\sigma_{yh}^2 + (1 - \bar{M}_h)\bar{Y}_h^2]$, where $\sigma_{yh}^2 = \Sigma (\bar{Y}_{hi} - \bar{Y}_h)^2/M_h$, and $\bar{Y}_h = Y_h/M_h$.

The subclass mean is a ratio of two random variables:

$$\bar{y}_w = \frac{y_w}{m_w} = \frac{\sum F_h y_h}{\sum F_h m_h} = \frac{\sum_h^H (N_h/n_h) \sum_i^{n_h} y_{hi}}{\sum_h^H (N_h/n_h) \sum_i^{n_h} m_{hi}}. \tag{4.5.10}$$

A general formula for stratified ratio means appears later in (6.4.3). We accept that general formula here, to derive from it the special formula

(4.5.5) which provides both insight and easier computations. We begin with the general form:

$$\operatorname{var}(\bar{y}_w) \doteq \frac{1}{m_w^2}\sum_h^H (1-f_h)F_h^2 \frac{n_h}{n_h-1}\left[\sum_i^{n_h} z_{hi}^2 - \frac{\left(\sum^{n_h} z_{hi}\right)^2}{n_h}\right], \tag{4.5.11}$$

where $z_{hi} = (y_{hi} - \bar{y}_w m_{hi})$. Among the n_h values, only m_h of the z_{hi} and the z_{hi}^2 have nonzero values. Also, for these m_h values $z_{hi} = (y_{hi} - \bar{y}_w)$, because $m_{hi} = 1$. Now, remembering the definition of t_h^2 we obtain

$$\sum_i^{m_h} z_{hi}^2 = \sum_i^{m_h} (y_{hi} - \bar{y}_w)^2 = m_h t_h^2.$$

Since

$$\sum_i^{m_h} z_{hi} = \sum_i^{m_h} y_{hi} - \sum^{m_h} \bar{y}_w = m_h(\bar{y}_h - \bar{y}_w),$$

we obtain

$$\frac{\left(\sum_i^{m_h} z_{hi}\right)^2}{n_h} = m_h \bar{m}_h (\bar{y}_h - \bar{y}_w)^2.$$

Thus

$$\left[\sum_i^{m_h} z_{hi}^2 - \frac{\left(\sum_i^{m_h} z_{hi}\right)^2}{n_h}\right] = m_h[t_h^2 - \bar{m}_h(\bar{y}_h - \bar{y}_w)^2]. \tag{4.5.12}$$

Combining (4.5.11) with (4.5.12), we obtain the equation (4.5.5). These are estimates of the population value,

$$\operatorname{Var}(\bar{y}_w) \doteq \frac{1}{M^2}\sum_h^H (1-f_h)\frac{N_h}{N_h-1}\frac{N_h^2}{n_h}\bar{M}_h[T_h^2 - \bar{M}_h(\bar{Y}_h - \bar{Y})^2]$$

$$= \sum_h^H (1-f_h)\frac{N_h}{N_h-1}\frac{W_h'^2}{n_h \bar{M}_h}[T_h^2 - \bar{M}_h(\bar{Y}_h - \bar{Y})^2], \tag{4.5.13}$$

that can be derived similarly, but with summations going to M_h instead of m_h, and with $W_h' = M_h/M$. Here $T_h^2 = \Sigma (Y_{hi} - \bar{Y})^2/M_h$, the variance of the M_h stratum elements around the population mean. In terms of the within-stratum variance

$$[\sigma_{yh}^2 + (1 - \bar{M}_h)(\bar{Y}_h - \bar{Y})^2] = [T_h^2 - \bar{M}_h(\bar{Y}_h - \bar{Y})^2].$$

This bracketed quantity takes the place of the σ_{yh}^2 that we would find in the formula $\Sigma (1 - f_h)W_h^2 S_{yh}^2/m_h$, for the variance of a sample stratified specifically for the subclass of ΣM_h members.

For the difference of two means, the variance can be formulated first as

$$\operatorname{var}(\bar{y}_{wa} - \bar{y}_{wb}) = \operatorname{var}(\bar{y}_{wa}) + \operatorname{var}(\bar{y}_{wb}) - 2\operatorname{cov}(\bar{y}_{wa}, \bar{y}_{wb}).$$

Now the variance of $\bar{y}_w$ can be stated as

$$\text{var}\left(\frac{y_w}{m_w}\right) \doteq \sum_h^H (1 - f_h)\frac{F_h^2 n_h^2}{n_h - 1}\left[\frac{\bar{m}_h t_h^2}{m_w^2} - \frac{\bar{m}_h^2(\bar{y}_h - \bar{y}_w)^2}{m_w^2}\right]. \tag{4.5.14}$$

With the methods that led to (4.5.12) we can also obtain the covariance. Note that the first term is $\Sigma\, z_{ahi} z_{bhi} = 0$ for mutually exclusive subclasses a and b, whereas the second term is

$$\frac{-(\sum z_{ahi})(\sum z_{bhi})}{n_h} = -n_h[\bar{m}_h(\bar{y}_h - \bar{y}_w)]_a[\bar{m}_h(\bar{y}_h - \bar{y}_w)]_b\,.$$

Hence we obtain

$$2\,\text{cov}\,(\bar{y}_{wa}, \bar{y}_{wb}) = \sum_h^H (1\text{-}f_h)\frac{F_h^2 n_h^2}{n_h - 1}\left[0 - 2\left\{\frac{\bar{m}_h(\bar{y}_h - \bar{y}_w)}{m_w}\right\}_a\left\{\frac{\bar{m}_h(\bar{y}_h - \bar{y}_w)}{m_w}\right\}_b\right].$$

Combining this with the second terms in the brackets of (4.5.14) for both a and b we obtain

$$-\left\{\left[\frac{\bar{m}_h(\bar{y}_h - \bar{y}_w)}{m_w}\right]_a - \left[\frac{\bar{m}_h(\bar{y}_h - \bar{y}_w)}{m_w}\right]_b\right\}^2.$$

In most cases this term is bound to be negligible compared to the two main variance terms. Neglecting it leaves (4.5.6), and we can use its comparatively simple form for the variance of the difference.

Similarly, the variance of the difference of two totals can be computed as

$$\text{var}\,(\tilde{Y}_{wa} - \tilde{Y}_{wb}) = \sum_h^H (1\text{-}f_h)\frac{N_h^2}{(n_h - 1)n_h}$$

$$\times\left[\left(\sum^{m_h} y_{hi}^2\right)_a + \left(\sum^{m_h} y_{hi}^2\right)_b - \frac{1}{n_h}(y_{ha} - y_{hb})^2\right]. \tag{4.5.15}$$

*4.5C Optimum Allocation for Subclasses

The variance formulas for both the total (4.5.9) and the mean (4.5.13) can be put in the form

$$\Sigma\left(1 - \frac{n_h}{N_h}\right)\frac{N_h^2}{n_h}B_h^2 = \Sigma\frac{N_h^2 B_h^2}{n_h} - \Sigma\frac{N_h^2 B_h^2}{N_h}\,.$$

The methods of optimum allocation used in 3.5 and 4.6 for ordinary stratified samples can also be applied to subclasses. Optimum allocation is reached when the sampling fractions $f_h = n_h/N_h$ are made proportional to

$B_h/\sqrt{J_h}$. For subclass means the population value B_h^2 and its estimate b_h^2 are

$$B_h^2 = \bar{M}_h[T_h^2 - \bar{M}_h(\bar{Y}_h - \bar{Y})^2] = \bar{M}_h[\sigma_{yh}^2 + (1 - \bar{M}_h)(\bar{Y}_h - \bar{Y})^2],$$

and

$$b_h^2 = \bar{m}_h[t_h^2 - \bar{m}_h(\bar{y}_h - \bar{y}_w)^2] = \bar{m}_h[v_h^2 + (1 - \bar{m}_h)(\bar{y}_h - \bar{y}_w)^2].$$

Thus the B_h^2 depend mostly on the products $\bar{M}_h T_h^2$ of the subclass sizes $\bar{M}_h$, with the element variances T_h^2 taken as deviations from the population mean. Instead of $\bar{m}_h$, $m_h/(n_h - 1)$ may be better [Kish, 1961a].

For estimates $\Sigma F_h y_h$ of totals, the values of B_h^2 and b_h^2 are

$$B_h^2 = \bar{M}_h[\sigma_{yh}^2 + (1 - \bar{M}_h)\bar{Y}_h^2] \quad \text{and} \quad b_h^2 = \bar{m}_h[v_{yh}^2 + (1 - \bar{m}_h)y_h^2].$$

To obtain realistic cost factors J_h, we should consider not only the cost factor c_h for the m_h subclass elements, but also the factor d_h for the $n_h - m_h$ eliminated elements. We can write the aggregate cost, proportioned to the number n_h of elements in a stratum, as

$$J_h n_h = c_h m_h + d_h(n_h - m_h) = n_h[c_h \bar{m}_h + d_h(1 - \bar{m}_h)].$$

The bracketed term expresses J_h, the combined cost per element in the hth stratum.

*4.6 PRECISION OF PROPORTIONATE AND DISPROPORTIONATE SAMPLES

4.6A Precision of Proportionate Samples

Here we compare the variance of a proportionate stratified random sample of elements to an srs of the same size n. The gains of proportionate sampling can be realized either in reduced variance for fixed n, or in reduced n for a specified variance.

The variance of proportionate sampling, for which (3.4.2) provides the sample estimate, is

$$\operatorname{Var}(\bar{y}_{\text{prop}}) = \frac{1 - f}{n}[\Sigma W_h S_h^2] = \frac{1 - f}{n} S_w^2 . \tag{4.6.1}$$

$S_h^2 = \dfrac{1}{N_h - 1} \sum_i^{N_h} (Y_{hi} - \bar{Y}_h)^2$ is the variance of the N_h elements *within* the hth stratum, and $S_w^2 = \Sigma W_h S_h^2$ is the *average within stratum variance*. Consider the result of analyzing the variance of a simple random sample of size n into two components—the *between-stratum* and *within-stratum* components:

$$\frac{1 - f}{n} S^2 \doteq \frac{1 - f}{n} [\Sigma W_h S_h^2 + \Sigma W_h(\bar{Y}_h - \bar{Y})^2]. \tag{4.6.2}$$

On the right side the first term denotes the variance remaining in proportionate sampling, and the second term denotes the amount eliminated by proportionate sampling; that is, the variance of a proportionate sample is less than that of an srs of the same n by an amount which represents the weighted deviations of the stratum means from the population mean:

$$\text{Var}(\bar{y}_{\text{prop}}) \doteq \text{Var}(\bar{y}_0) - \frac{1-f}{n} \sum W_h(\bar{Y}_h - \bar{Y})^2. \tag{4.6.3}$$

We now express the two components of the variance as ratios of the basic srs standard:

$$1 \doteq \frac{\sum W_h S_h^2}{S^2} + \frac{\sum W_h(\bar{Y}_h - \bar{Y})^2}{S^2}. \tag{4.6.4}$$

Hence, the ratio of the total srs *variance remaining* in a proportionate sample is

$$\frac{\text{Var}(\bar{y}_{\text{prop}})}{\text{Var}(\bar{y}_0)} = \frac{S_w^2}{S^2} = \frac{\sum W_h S_h^2}{S^2} \doteq 1 - \frac{\sum W_h(\bar{Y}_h - \bar{Y})^2}{S^2}. \tag{4.6.5}$$

The *relative gain* due to proportionate sampling is expressed as the portion of the srs variance it eliminated:

$$\frac{\text{Var}(\bar{y}_0) - \text{Var}(\bar{y}_{\text{prop}})}{\text{Var}(\bar{y}_0)} \doteq \frac{\sum W_h(\bar{Y}_h - \bar{Y})^2}{S^2} = 1 - \frac{S_w^2}{S^2}. \tag{4.6.6}$$

From these expressions we derive several important consequences for likely gains from proportionate sampling:

1. The magnitude of the *gain depends on the heterogeneity among the sample means.* Because the gain $\Sigma W_h(\bar{Y}_h - \bar{Y})^2$ cannot be negative, *we cannot lose by proportionate sampling.* The variance of proportionate sampling cannot be greater than that of srs (except for a trivial factor).

2. This also means that *the greater the homogeneity within the strata, the smaller the variance* of proportionate sampling. This homogeneity depends on the relationship between the characteristic being estimated and the stratifying variate. If this relationship can be expressed by the coefficient R of *linear* correlation, then we may say that the gain is directly proportional to R^2, and the remaining variance to $(1 - R^2)$. This occurs because the proportion of the variance "explained" by the stratifying variable is $R^2 = \Sigma W_h(\bar{Y}_h - \bar{Y})^2$, if the relationship between the Y_i and the stratifying variable is a linear correlation. However, frequently the relationship is not linear, perhaps not even numerical. For example, the strata may be based on geographical or other attributes. The components of the variance will measure these relationships too, but R^2 will not.

3. The portion of the gain due to each stratum is proportional to the product of its squared deviation from the population mean $(\bar{Y}_h - \bar{Y})^2$ with its relative size W_h. Hence, even very unusual subpopulations fail to be useful for proportionate stratification if they are too small.

4. The relative gain is inversely proportional to the population variance. If this is large, we get small gains, even when the differences between stratum means appear large and "significant" both socially and statistically. Usually it is difficult to find strata where the difference of means $\sum W_h(\bar{Y}_h - \bar{Y})^2$ is large compared to the unit variance S^2 in the population. Therefore, the gains from proportionate sampling of elements are often modest.

In estimating proportions (P), the gains are usually small because $S^2 = P(1 - P)$ is relatively large. This means that it is difficult to find "clean" sorting of the 0 and 1 element values Y_{hi} into the strata to obtain large enough differences among the stratum means. Thus the average within-stratum variance $S_w^2 \doteq \sum W_h P_h Q_h$ remains relatively large even after stratification. We shall see (5.6) that when we use clusters instead of elements as sampling units, we can usually obtain larger *relative* gains through the use of stratification, because the denominator, the variance among the cluster means, is smaller.

Equation (4.6.2) can be derived by breaking the deviation of each element from the population mean into two components; the deviation of the element from its stratum mean, and the deviation of the stratum mean from the population mean:

$$(Y_{hi} - \bar{Y}) = (Y_{hi} - \bar{Y}_h) + (\bar{Y}_h - \bar{Y}). \tag{4.6.7}$$

Squaring these terms and summing them for all elements, we get

$$\sum_h^H \sum_i^{N_h} (Y_{hi} - \bar{Y})^2 = \sum_h^H \sum_i^{N_h} (Y_{hi} - \bar{Y}_h)^2 + \sum_h^H N_h(\bar{Y}_h - \bar{Y})^2 + 2\sum_h^H (\bar{Y}_h - \bar{Y}) \sum_i^{N_h} (Y_{hi} - \bar{Y}_h).$$

The cross product term at the end is zero. The other terms divided by N become

$$\sigma^2 = \sum W_h \sigma_h^2 + \sum W_h(\bar{Y}_h - \bar{Y})^2,$$

and these terms equal the terms in

$$\frac{N-1}{N} S^2 = \sum W_h \frac{N_h - 1}{N_h} S_h^2 + \sum W_h(\bar{Y}_h - \bar{Y})^2$$
$$= \sum W_h S_h^2 + \sum W_h(\bar{Y}_h - \bar{Y})^2 - \sum \frac{W_h}{N_h} S_h^2. \tag{4.6.8}$$

When $W_h = N_h/N$, the last factor is $\Sigma S_h^2/N$, and is ordinarily negligible. Hence the factors that we ignored in (4.6.2) to (4.6.6) for the sake of simplicity, are usually negligible. To a good approximation then, $\Sigma W_h S_h^2$ is bound to be less than S^2; hence the variance of a proportionate sample is less than that of a simple random sample of the same size. To contradict this, we would need small strata, and so small a variation $\Sigma W_h(\bar{Y}_h - \bar{Y})^2$ among the stratum means that it could not cancel the effects of the $(N_h - 1)/N_h$. This situation can occur only if the variation among stratum means is less than one could expect from random sorting. It corresponds to a negative intraclass correlation among the strata (5.5) that can rarely occur.

4.6B The Precision of Optimum Allocation

To aim at "optimum" results, the n_h of a sample should be allocated (4.4.2) so that $n_h/N_h = KS_h/\sqrt{J_h}$, when $\Sigma J_h n_h$ represents the relevant survey costs. These are proportional to $n = \Sigma n_h$ if the cost per element within the hth stratum is considered the same for all strata, and the optimum allocation of the n_h becomes simply $n_h/N_h = kS_h$

This allows us to allocate the sample rather simply. First, compute some preliminary values for the n_h, or for the f_h, based on some assumed reasonable value of k or K. Then, compare either the desired variance or the allowed cost (computed either in terms of n or $\Sigma J_h n_h$) with what the preliminary sample would yield. Finally, adjust the sampling rates to conform to the required fixed quantity.

Alternatively, we may obtain the n_h or f_h directly by choosing the proper constant of proportionality from (4.6.9). The four forms correspond to the four different conditions we can impose as "fixed" (derived in 8.5.8, with k and K substituted for K'/N).

For J_h same for all strata	For J_h variable among strata
For fixed $n = \Sigma n_h$	For fixed $\Sigma J_h n_h$
$k = \dfrac{n}{N \sum W_h S_h}$	$K = \dfrac{\sum J_h n_h}{N \sum W_h S_h \sqrt{J_h}}$
For fixed $\text{Var}_f(\bar{y})$	For fixed $\text{Var}_f(\bar{y})$
$k = \dfrac{\sum W_h S_h}{N \cdot \text{Var}_f(\bar{y}) + \sum W_h S_h^2}$	$K = \dfrac{\sum W_h S_h \sqrt{J_h}}{N \cdot \text{Var}_f(\bar{y}) + \sum W_h S_h^2}$

(4.6.9)

However, note that: (1) Instead of fixing either costs or variances, we can examine their changing relation to each other. For this, too, the basic proportionality of optimum rates would hold. (2) If any of the sampling rates f_h would exceed 1, use $f_h = 1$ for those strata and omit them from a recomputed allocation.

Using estimates of S_h, J_h, and W_h in one of the four forms of (4.6.9), one can obtain the sample sizes n_h and fractions f_h. Then we can compute either the expected variance, or cost of the design. These two steps can be done algebraically to obtain

$$\text{Var}(\bar{y}_{\text{opt}}) + \frac{1}{N}\sum W_h S_h^2 = \frac{(\sum W_h S_h)^2}{n}, \qquad \text{for } J_h \text{ constant,} \tag{4.6.10}$$

$$\text{Var}(\bar{y}_{\text{opt}}) + \frac{1}{N}\sum W_h S_h^2 = \frac{(\sum W_h S_h \sqrt{J_h})^2}{\sum J_h n_h}, \qquad \text{for } J_h \text{ variable.} \tag{4.6.10$'$}$$

Entering these equations with fixed variance, we solve for either n or $\Sigma J_h n_h$. Or entering with fixed n or $\Sigma J_h n_h$, we find the variance of optimum allocation. The term $(1/N)\Sigma W_h S_h^2$ is negligible, unless the sampling rates are large.

We can obtain insight from a generalized formula for the gains of optimum allocation over proportionate sampling:

$$\frac{\text{Var}(\bar{y}_{\text{opt}})}{\text{Var}(\bar{y}_0)} \doteq \left(1 - \frac{\sum W_h(\bar{Y}_h - \bar{Y})^2}{S^2}\right) - \frac{\sum W_h(S_h - \bar{S})^2}{S^2}. \tag{4.6.11}$$

This equation from Cochran [1963, 5.6] is an approximation that neglects terms in $1/N_h$. Here $\bar{S} = \Sigma W_h S_h$, the average standard deviation. J_h is assumed to be the same for all strata. For a fixed set of J_h values use a "cost-weighted" mean: $S_h' = S_h\sqrt{J_h n_h / \Sigma J_h n_h}$. The second term indicates the gain over srs sampling arising from the variance among stratum means; this holds good for both proportionate and optimal sampling. The additional gain of optimal over proportionate sampling appears in the last term as a function of the variations among the values of S_h. Hence, to obtain large gains by optimal allocation, we must be able to establish strata with great differences in the values of S_h.

Optimal allocation rarely gives appreciable gains for estimating proportions because the $S_h = \sqrt{P_h(1 - P_h)}$ are insensitive to the differences in P_h that we may expect in practice. On the contrary, for some highly skewed variables we can obtain very large gains. These arise from the separation of the high values of S_h into a few strata.

Formulas (4.6.10) and (4.6.11) point approximately to the *possible* gains obtainable by a sample allocated optimally. They would be obtained only if we could secure exact values for S_h and J_h. To the degree that the estimates miss their marks, the gains will generally be less. We may even incur losses instead of gains—as compared with proportionate or even with simple random sampling. The last possibility does not exist in proportionate sampling. Furthermore, realistic comparisons of optimum with proportionate sampling should include an appraisal of the costs, worries, and risks that may result from departing from a self-weighting sample.

The optimum formulas $n_h = KN_hS_h/\sqrt{J_h}$ may result in the impossible allocation $n_h > N_h$ and $f_h > 1$. This can occur when a rather large sample is needed from a skewed population of establishments or inventories. In the extreme strata the values of S_h may be so large that the formula leads to the impossible preliminary result. This solution lies in the mathematical region where the contribution of the extreme strata to the sum of variances $\sum_h W_h^2 \operatorname{Var}(\bar{y}_h)$ would be negative. The first practical step is to make these variances zero by making $f_h = 1$ and $n_h = N_h$. In practice we may decide to take this step for any stratum where $f_h > 0.5$, let us say. Denote these "complete" strata with the summation sign $\sum_c$.

Having "increased" the sum of variances in $\sum_c$ from an imaginary negative value to zero, the sum of variances in the rest of the strata thereby becomes greater.

Denote these "incomplete" strata with the summation $\sum_i$, so that $\sum_i + \sum_c = \sum_h$. In these remaining incomplete strata the sample sizes n_h are too small to satisfy the requirement originally intended. However, these n_h are in the correct proportions to each other for optimum allocation. Hence we need only increase them proportionately by the factor that will satisfy the original requirement.

The original requirement may have been a fixed number of elements $n_f = \Sigma n_h$, or a fixed cost $C_f = \sum_h J_h n_h$, or a fixed variance

$$\operatorname{Var}_f(\bar{y}) = \sum_h W_h^2 S_h^2/n_h - \sum_h W_h^2 S_h^2/N_h.$$

The n_h in each stratum should be increased accordingly either by the ratio $(n_f - \sum_c n_h)/\sum_i n_h$, or by $(C_f - \sum_c J_h n_h)/\sum_i J_h n_h$, or by

$$\left(\sum_i W_h^2 S_h^2/n_h\right)/\left(\operatorname{Var}_f(\bar{y}) + \sum_h W_h^2 S_h^2/N_h\right).$$

PROBLEMS

4.1. Suppose that the 40 values of Problem 2.5 came in that same order from a systematic selection. Compute the variance of the mean, based on the 39 successive differences. This value corresponds closely to the $0.265 = 0.52^2$ obtained with the simple random formula.

4.2. Suppose that the 40 values of Problem 2.5 came as paired selections from 20 implicit strata (zones). Compute the variance of the mean.

4.3. From the list of 270 blocks in Appendix E, select a sample of 10 blocks with the interval 27, after a random start. Note the number of total dwellings x_j and rented dwellings y_j in each block. (*a*) Estimate the means per block and their standard errors. (*b*) Estimate the population totals ΣY_i and ΣX_i and their standard errors. (*c*) Estimate the mean of the block ratios Y_i/X_i and its standard error. Explain why this is not the same as the proportion of rented dwellings for the population of dwellings.

4.4. In Appendix E, the block numbers 232–772 constitute a list of 541 numbers. From this list draw a systematic sample with the interval 27, after a random start. Note the numbers of total dwellings x_j and rented dwellings y_j in the selected blocks. (*a*) What can you say about the number of actual selections with this method? (*b*) How do substitutions for blank selections cause bias? (*c*) Estimate the means per block and their standard errors. (*d*) Estimate the population totals ΣX_i and ΣY_i and their standard errors, first without and then with the knowledge that $N = 270$ is the number of blocks in the population. This problem (unlike 4.3 and 2.1) leaves unresolved problems.

4.5. Divide the population of 270 blocks in Appendix E into 5 strata of 54 consecutive blocks. Select a paired selection from each stratum by drawing 5 pairs of random numbers 1 to 54. Note the number of dwellings x_j and rented dwellings y_j in each block. (*a*) Estimate the means per block and their standard errors. (*b*) Estimate the population totals ΣX_i and ΣY_i and their standard errors.

4.6. The assessed valuations of two classes (R and C) of land parcels were estimated from a sample of pages of tax records. Ten independent replications were selected, each comprising a sample of pages. Each R sample consisted of all R parcels on every 40th page throughout the 23,279 pages that contained a mixture of C and R classes; the overall sampling fraction was 10/40. One part of the C sample consisted of all C parcels on every 9th page of the R sample; hence, the R and C samples were not independent. Another part of the C sample consisted of 10 selections, each one 1/360 of the 21,406 pages of volumes known to contain only C parcels. Thus both parts were sampled with 10/360 for the C parcels. The 10 replications yielded the following estimates $10Fy$ of the aggregate assessed valuations in thousands of dollars:

Subsample	R Parcels	C Parcels
1	18,932	291,744
2	20,760	327,024
3	27,256	392,796
4	14,176	380,772
5	18,568	442,728
6	25,980	389,412
7	30,576	398,096
8	21,220	417,568
9	25,060	488,340
10	25,532	482,364
Mean	22,806	401,084

(*a*) Compute the standard errors of the aggregate R and C valuations.
(*b*) The sample was designed, after a pilot study, to yield coefficients of variation at approximately 5 percent; how well were these achieved?
(*c*) Compute the combined aggregates for the R and C parcels and its standard error. Should the variance of $R + C$ equal the sum of variances of R and C? You may leave unresolved problem caused by different *fpc*'s (Remark 11.7.I).
(*d*) Compute the standard error of R and C aggregates from figures rounded to the closest million dollars (19, 21, 27, etc.). How does this compare to (*a*)?
(*e*) Estimate the standard error by using the range and compare with those in (*a*).

4.7.

Replications γ	1	2	3	4	5	6	7	8	9	10	Total
Class size y_γ	16	20	18	17	14	15	19	13	19	17	168
Sample size x_γ	94	85	85	80	82	79	86	82	93	88	854
Proportions $p_\gamma = y_\gamma/x_\gamma$	0.170	0.236	0.212	0.213	0.171	0.190	0.221	0.159	0.204	0.193	0.197

These 10 replications were selected from a file of motel owners [Deming, 1960, p. 107]. Each replication represents a simple random sample of 1/70. The list contains many blanks; this causes the variation in the number of motels x_γ in the samples. The numbers y_γ denote motel owners who said that people "frequently" asked them to make reservations.
(*a*) Estimate the total number of motel owners and its standard error.
(*b*) Estimate the total number who say "frequently" and its standard error.
(*c*) Estimate the mean proportion who say "frequently" and its standard error.
(*d*) If you know that the total number of motel owners is 5988, can you use this number to estimate the number who say "frequently" and its standard error? Explain the difference between this and (*b*).

(*e*) Estimate the four standard errors by using the range and compare with (*a*) to (*d*).

4.8. Explain why a replicated design for computing variances is particularly well-adapted to periodic surveys which repeatedly gather similar statistics. Describe an example.

4.9. You want to select an epsem sample of "social scientists" for a mail survey. For practical purposes they become defined as: any member of any one or more of four professional societies. You decide, after some thought, that a systematic sample of the alphabetical list of each of the four societies will provide a reasonable equivalent of a simple random sample of that society. The number of members in the four societies are respectively 6000, 8000, 9000, and 21,000. (*a*) Describe briefly the major problems of selection and your proposed selection method. (*b*) What is your estimator (p) of the proportion among the social scientists having some characteristic? Let p_h = the proportion in the sample taken from one society having some characteristic; use numbers where you can, and define any other term you use. (*c*) What is your estimator of the standard error of this proportion? Discuss briefly any unusual problems you see, your proposed solutions, and their justification. (*d*) Do you need any modification of your sampling plan to produce separate estimates for each of the four societies?

5

Cluster Sampling and Subsampling

5.1 NATURE OF CLUSTERS

When individual selection of elements seems too expensive, survey tasks can be facilitated by selecting *clusters*; that is, sampling units containing several elements. For example, employees of a firm may be selected in work groups, a school's students in designated classes, or the dwellings of a city in blocks. Cluster sampling contrasts to element sampling where the elements are also the sampling units, so that each sampling unit contains only a single element, or none in the case of "blanks."

Clustering, or cluster sampling, denotes methods of selection in which the sampling unit, the unit of selection, contains more than one population element; hence the sampling unit is a cluster of elements. Each element must be *uniquely identified* with one, and only one, sampling unit. The sampler must investigate the frame and, if necessary, design procedures to overcome confusion in identification.

Suppose that we need an epsem sample of about $n = 400$ of the $N = 10{,}000$ estimated dwellings of a city. A sample of individual dwellings would be difficult to select because a good list of all dwellings does not exist and would be too costly to prepare. Instead, we can obtain a sample of dwellings by selecting a sample of blocks. This can be done by dividing the entire area of the city's map into blocks and then selecting $1/25$ of the blocks into the sample. The dwellings located within the boundaries of the sample blocks comprise the sample. The probability of selection for any dwelling in the city is the probability of selection of its block; this has been set at $1/25$, to correspond to the desired sampling rate of

400/10,000. Since the number of dwellings per block varies, the actual sample size depends on the blocks which happen to be selected. The number of sample blocks would depend on the *average* number of dwellings per block. The dwellings are the elements determined by the objectives of

Population	Variables	Elements	Clusters, or sampling units
(1) City *A*	Characteristics of household	Dwellings	Blocks
(2) City *B*	Clothing purchases	Persons	Dwellings
(3) Airport	Travel information	Departing passengers	Plane load
(4) High School	Career plans	Students	Homeroom classes
(5) Village people	Social attitudes	Adults	Villages
(6) Annual traffic over bridge	Origin and destination	Motor vehicles	Intervals of 40 minutes
(7) File of land holdings in city	Tax information	Land holdings	Pages in ledgers
(8) File of health insurance	Medical data	Cards	Groups of 10 consecutive cards

TABLE 5.1.I Illustrations of Possible Clusters

Dwellings are elements in (1), but are clusters in (2). Perhaps in city *B* a good list of dwellings is available; otherwise they will have to be selected in larger sampling units, perhaps blocks. In (3) and (4) the entire clusters of elements can readily be induced to fill out a questionnaire. "Homeroom classes" in (4) provide a unique identification for each student, whereas ordinary classes would not. A list of villages in (5) provides a complete list of people who live only in villages. In (6) a time interval is employed to provide a cluster for a convenient work-load. The land holdings in (7) are identified by the appearance of their "initial lines" on selected pages; these are unequal clusters of initial lines. From a file of cards in (8) equal clusters of 10 consecutive cards are selected to facilitate the work and checking.

the analysis. However, the sampling units are blocks, selected from a complete list: the numbered map. The selection of sample blocks determines the selection of the dwellings that are clearly and uniquely located within them.

If we compare a cluster sample with an element sample comprised of the same number of elements, typically we will find that in cluster sampling: (1) *the cost per element is lower*, due to the lower cost of listing, or of locating, or both (see 8.3); (2) *the element variance is higher*, resulting from the usual, though irregular, *homogeneity* of elements in clusters (5.4); (3) the costs and problems of *statistical analysis* are greater. Clustering should be preferred over individual selection when the lower cost per element more than compensates for its two disadvantages. This occurs often in large, widespread samples.

Just what makes a desirable cluster is a matter of practical expediency that depends on the survey's situation and resources. The individual elements are determined by the survey objectives. The sampler must then decide whether he can use them also as unique sampling units, or whether he must designate clusters of the specified elements as sampling units. In some studies the household is regarded as a cluster of persons; but in other studies entire cities may be used as elements. The population of the United States may alternatively be regarded as an aggregate of units which are entire counties; or of cities, towns, and townships; or of area segments and blocks; or of dwellings; or, finally, as individual persons. Indeed all those sampling units are employed in turn for area samples of the United States. The proper sampling units are defined to conform with the requirements of a practical, economical sampling design applied to the physical distribution of the population and to its selection frame. The choice of clusters is the recognition in the sampling design of some features in the physical distribution of the population, and in the nature of the frame.

The number of elements in a cluster is called the *size* of the cluster. The clusters in most populations are of *unequal size;* for example, dwellings in blocks, persons in households, employees in sections, etc. Clusters of equal size seldom occur in sampling design, but they provide a simple introduction to the theory of cluster sampling.

Clusters of *equal size* are often the result of planned conditions such as manufacturing; for example, cigarettes in packs, in cartons, in cases, in carloads. They rarely exist in nature or society, but they occur in special sampling situations. They result from the *planned uniformity of a population;* for example, dwellings in large housing developments, some work units, or army units. *The sampler can create equal clusters* from a frame; for example, equal size packs from a file of cards, as in Example 5.2*a*. *Equal subsamples* can be obtained from unequal sized clusters, especially by selecting with "probabilities proportional to size" (Chapter 7).

Even designs that begin with equal clusters often end in actual sample clusters which are unequal due to: (1) imperfect frames, (2) nonresponse, and (3) the use of subclasses for analysis. But minor imperfections in cluster sizes, due to the frame and nonresponse, may be repaired. Having equal clusters could greatly facilitate some complex statistical analysis, if needed. A "missing case" may be provided by duplicating a randomly selected element from the cluster; or by selecting a new "substitute" from the cluster, when subsampling. In most practical situations, clusters sizes are unequal, selection is improved with stratification, and two selections per stratum is both efficient and simple. The impatient reader can turn directly to the *paired selection of clusters* in 6.4*B*.

5.2 RANDOM CHOICE OF CLUSTERS

Suppose that from a population of A clusters, a sample clusters are selected with epsem (equal probability). In the selected clusters, all B elements are included in the sample which consists of $a \times B = n$ elements. The equal probability of selection of any of the $N = AB$ population elements is

$$\frac{a}{A} = \frac{a}{A} \cdot \frac{B}{B} = \frac{n}{N} = f. \qquad (5.2.1)$$

The factor $B/B = 1$ symbolizes the selection of any element, whenever its cluster is selected; but that selection is *conditional on the probability* a/A *of selecting the cluster.*

The *sample mean* of the n elements in the sample typically serves to estimate the population mean $\bar{Y}$. It is also the mean of the a cluster means:

$$\bar{y} = \frac{y}{n} = \frac{1}{n}\sum_{j}^{n} y_j = \frac{1}{aB}\sum_{\alpha}^{a}\sum_{\beta}^{B} y_{\alpha\beta} = \frac{1}{aB}\sum_{\alpha}^{a} y_\alpha = \frac{1}{a}\sum_{\alpha}^{a} \bar{y}_\alpha. \qquad (5.2.2)$$

The selection is epsem, and the sample size is fixed at $n = a \times B$; hence, the sample mean is an unbiased estimate of $\bar{Y}$ (2.8*C*). The selection of the sample clusters may be systematic, stratified, or further clustered, so long as it remains epsem. But assume further that the *a clusters* are selected with *simple random* choice; that is, from the list of A population numbers a different random numbers are selected from a good table, to designate the a sample clusters. For this design we can state the variance as

$$\text{var}(\bar{y}) = (1-f)\frac{s_a^2}{a}, \qquad \text{where } s_a^2 = \frac{1}{a-1}\sum_{\alpha}^{a}(\bar{y}_\alpha - \bar{y})^2. \qquad (5.2.3)$$

This formula resembles the variance of simple random sampling (2.2.2). In both cases, the variance of the mean is *directly proportional to the variance between sampling units and inversely proportional to the number of sampling units*. The unit variances are respectively s^2 and s_a^2; and the sample sizes are, respectively, n and a. Similarly, the cluster sample mean (5.2.2) is the mean of a sampling units selected at random. The values

Elements \ Clusters	1	2 ···	α ···	A
1	Y_{11}	Y_{21}	$Y_{\alpha 1}$	Y_{A1}
2	Y_{12}	Y_{22}	$Y_{\alpha 2}$	Y_{A2}
·	·	·	·	·
·	·	·	·	·
·	·	·	·	·
β	$Y_{1\beta}$	$Y_{2\beta}$	$Y_{\alpha\beta}$	$Y_{A\beta}$
·	·	·	·	·
·	·	·	·	·
·	·	·	·	·
B	Y_{1B}	Y_{2B}	$Y_{\alpha B}$	Y_{AB}
Cluster Totals	Y_1	$Y_2 \cdots$	$Y_\alpha \cdots$	Y_A
Cluster Means	$\bar{Y}_1$	$\bar{Y}_2 \cdots$	$\bar{Y}_\alpha \cdots$	$\bar{Y}_A$

TABLE 5.1.II A population of N elements divided into A clusters of B elements each, so that $N = A \times B$.

These columns can represent $A = 4000$ stacks of $B = 10$ cards each, into which a population of $N = 40{,}000$ cards has been divided. The stacks of 10 cards each serve as clusters, and $\bar{Y}_\alpha$ would be the mean value of the αth cluster. The mean of the "typical" αth cluster is

$$\bar{Y}_\alpha = \frac{1}{B}(Y_{\alpha 1} + Y_{\alpha 2} \cdots + Y_{\alpha\beta} \cdots + Y_{\alpha B}) = \frac{1}{B}\sum_\beta^B Y_{\alpha\beta} = \frac{Y_\alpha}{B},$$

where Y_α is the total for the αth cluster. The *population mean* is still defined as the mean of the N elements in the population; but it is also equal to the mean of the A cluster means in the population:

$$\bar{Y} = \frac{Y}{N} = \frac{1}{N}\sum_i^N Y_i = \frac{1}{AB}\sum_\alpha^A\sum_\beta^B Y_{\alpha\beta} = \frac{1}{A}\left[\frac{1}{B}\sum_\alpha^A Y_\alpha\right] = \frac{1}{A}\sum_\alpha^A \bar{Y}_\alpha.$$

For the first stage units, the subscript α runs from 1 to A in the population and 1 to a in the sample. For the second stage, the subscript β runs from 1 to B for equal sized clusters. For equal subsamples, the subscript β runs from 1 to b. When a third stage exists, the subscript γ runs from 1 to C in the population, and to c in the sample.

$\bar{y}_\alpha$ of the sample clusters enter without sampling error, since they are based on the values of the entire population of B elements within the clusters. The variance of the sample mean arises entirely from variance between the cluster means.

The derivation of (5.2.3) follows directly from the variance of a simple random sample (2.2.2). Consider the cluster mean as the mean of a simple random sample of a elements selected from a population of A elements comprised of the cluster means $\bar{Y}_\alpha$.

Whereas the mean $\bar{y}$ may be computed simply from the entire sample, computing var $(\bar{y})$ requires the separation of the a cluster values; hence, cluster identification should appear on the data and on the machine cards. Often more convenient than (5.2.3) would be its equivalent:

$$\text{var}(\bar{y}) = \frac{1-f}{a}\frac{s_y^2}{B^2}, \qquad \text{where } s_y^2 = B^2 s_a^2 = \left[\frac{1}{(a-1)}\left(\sum_\alpha^a y_\alpha^2 - \frac{y^2}{a}\right)\right]. \tag{5.2.3'}$$

The form s_a^2 denotes *unit variance* between the cluster *means* $\bar{y}_\alpha$; the form $s_y^2 = B^2 s_a^2$ is the unit variance between the cluster *totals* y_α. Computing with the y_α we can avoid the separate divisions by B.

The factor $(1-f)$ becomes negligible for small f, and it disappears for selection *with* replacement, when clusters are permitted to appear more than once in the sample.

The simple and unbiased estimate of the *population aggregate* Y *is* $N\bar{y} = (N/n)y = y/f$. Its standard error is se $(N\bar{y}) = N$ se $(\bar{y}) = \dfrac{N\sqrt{1-f}\,s_a}{\sqrt{a}}$.

In practical situations a stratified design is preferred to a simple random choice of clusters. The latter serves, like simple random sampling, as an introduction to clustering, and sometimes as an acceptable approximation for other selections actually used.

Example 5.2a A newspaper has 39,800 subscribers served by carrier routes. There is a card for each subscriber; in the file the cards of each carrier route are kept together in geographical order, and neighboring routes follow each other. The number of cards per carrier varies between 50 and 200. The chief purpose of the survey is to find out how many of the subscribers own their homes. An interview survey of about 400 subscribers is wanted, in clusters of 10 subscribers each. These save travel time, because an interviewer can generally obtain the 10 interviews in one neighborhood in half a day.

The sampler regards the $N = 39{,}800$ cards as a frame of $A = 3980$ clusters of $B = 10$ each. A few of the clusters will be split between two routes. He selects $a = 40$ different random numbers, from 1 to 3980. Each random number r denotes the selection of 10 cards numbered from $(10r - 9)$ to $10r$; e.g., number 179 will select cards 1781–1790.

The results in the 40 clusters follow, in terms of y_α, the number of homeowners in each cluster of 10 households:

10, 8, 6, 5, 9, 8, 8, 5, 9, 9, 9, 10, 4, 3, 1, 2, 3, 4, 0, 6

3, 5, 0, 3, 0, 0, 4, 0, 8, 0, 10, 5, 6, 1, 3, 3, 1, 5, 5, 4.

We need to compute only two numbers: $y = \sum^{n} y_j = \sum^{a} y_\alpha = 185$; and $\sum^{a} y_\alpha^2 = 1263$. The sample mean is simply (5.2.2):

$$\bar{y} = \frac{y}{n} = \frac{185}{400} = 0.4625 = 46.2 \text{ percent.}$$

The estimated variance of the sample mean is (5.2.3′)

$$\begin{aligned}
\text{var}(\bar{y}) &= \frac{1-f}{a} s_a^2 = \frac{1-f}{a}\left[\frac{1}{(a-1)B^2}\left(\sum^{a} y_\alpha^2 - \frac{y^2}{a}\right)\right] \\
&= \frac{0.99}{40}\left[\frac{1}{39 \times 100}\left(1263 - \frac{185^2}{40}\right)\right] \\
&= \frac{0.99}{40}\frac{(1263 - 855.6)}{3900} = \frac{0.99}{40}\frac{407.4}{3900} = \frac{0.99}{40} 0.1045 \\
&= 0.99(26.03) \times 10^{-4} = 25.85 \times 10^{-4}.
\end{aligned}$$

The standard error is $\sqrt{25.85} \times 10^{-2} = 5.084 \times 10^{-2} = 5.1$ percent. The *total number* of subscribers who own their own home is estimated as $N(\bar{y}) = 39{,}800 \times 0.4625 = 18{,}408 \doteq 18{,}400$, with a standard error of $N\text{se}(\bar{y}) = 39{,}800 \times 0.05084 = 2023 \doteq 2000$.

In most actual situations some departures from the above simple design would arise. First, suppose that the population size N were not an exact multiple of the cluster size B; say there were 39,803 cards in the file. A preliminary random elimination of the excess 3 cards may be the easiest solution; but several others mentioned in 4.1 can also be used. Second, a stratified or systematic procedure may be preferred for selecting the clusters of cards. Third, inequalities in cluster sizes introduced by nonresponse could perhaps be controlled with substitutions.

5.3 SUBSAMPLING OR MULTISTAGE SAMPLING

5.3A *The Aims of Subsampling*

When comparing a cluster sample with a sample of elements of the same size n, we can expect a larger variance but a smaller cost for the cluster sample. The greater spread of a sample of elements over the population usually obtains greater precision, but at a price. Although this situation is not universal, it is common and may lead to subsampling. Subsampling results from the search for a good compromise between the two conflicting effects of clustering on the economy of the design: decreasing the degree of clustering may decrease the variance greatly, without incurring a proportional increase in cost. A detailed treatment in Chapter 8 considers several cost factors and variance components; the basis for comparing designs is shifted from sample sizes to survey costs.

First selecting clusters, and then elements, requires two *stages of selection*. This method can readily be extended to more stages. *Multistage sampling* consists of "a hierarchy of different types of units, each first-stage unit being divided, or potentially divisible, into second-stage units, etc. A frame will be required at each stage for the units that have been selected at that stage. Initially, a frame is required by which first-stage units may be defined and selected. For the second stage of selection a frame is required by which second-stage units may be defined within the first-stage units which have been selected. One of the advantages of multistage sampling is that second-stage frames are only required for selected first-stage units and so on" [U.N., 1950].

The first stage is particularly important, especially in variance computations; and the sampling units of the first stage are called *Primary Sampling Units*, or *PSU's*. The subsequent stages are called second-stage (or secondary), third-stage, etc. units and selections. We shall denote the sample obtained from the selected PSU's as *Primary Selections* or PS's. In the example below, the blocks are the PSU's, and dwellings selected into the sample in the sample blocks comprise the PS's. Primary selections have also been called *ultimate clusters*, to describe the result of the series of operations required to obtain ultimately the subsample from each primary unit. "The term *ultimate cluster* is used to denote the aggregate of units included in the sample from a primary unit" [Hansen, 1953, p. 242].

Suppose that from the population of $N = 40{,}000$ dwellings in $A = 4000$ blocks of a city we want to select a sample of about $n = 400$ dwellings. We can select a sample of individual dwellings (elements) or a sample of blocks (clusters), using the same overall sampling fraction of $1/100$ in either case. The sample of 400 elements would appear scattered over the city, but the cluster sample would be confined to 40 blocks. Clustering could decrease the costs of listing and locating the dwellings, but increase the variance of survey variates. A subsampling design might use sampling fractions of $1/50$ for blocks and $1/2$ for the dwellings in the sample blocks; or it might use $1/20$ for blocks and $1/5$ for dwellings. The former would yield a sample of 80 blocks with 5 dwellings per block, and the latter 200 blocks with 2 dwellings per block, with the same $1/100$ sampling fraction.

The subsampling procedure may call either for the selection of compact segments into which the sample blocks were divided; or for subsampling the lists of individual dwellings prepared for the sample blocks. The effort of listing for the second stage, either segmenting the blocks or listing dwellings, is confined to the sample blocks only. Either procedure aims to divide the blocks into smaller and preferable subsamples.

Generally, subsampling is used to divide larger clusters into smaller clusters. But why resort to subsampling instead of creating and selecting the ultimate clusters as primary selections, directly in a single stage? There are four reasons that justify subsampling, in preference to the direct creation of smaller clusters and their selection in one-stage cluster sampling:

(1) *Natural clusters* may exist as convenient sampling units, yet larger than the desired economic size of PS. For example, the boundaries of city blocks can be identified rather clearly and easily; but they often contain too many dwellings, and subsampling yields preferable PS's. Similarly, we may conveniently use classes of students as sampling units, and then subsample smaller PS's from them. (2) We can *avoid the cost of creating smaller clusters* in the entire population and confine it to the selected sampling units. For example, the creation of small area segments for an entire city or state could be prohibitively expensive. (3) *The effect of clustering* expressed as *rho* in 5.4 is often less in larger clusters. For example, a compact cluster of 4 dwellings from a city block may bring into the sample similar dwellings, perhaps from one building; but 4 dwellings selected separately can be spread around the dissimilar sides of the block. (4) The sampling of *compact clusters may present practical difficulties.* For example, independent interviewing of all members of a household may seem impractical.

5.3B Features of Simple Replicated Subsampling

Because each stage of sampling permits variations, a multiplicity of designs becomes possible. To present only one or two designs would be too restrictive. To present many of them in detail would be too exhausting. But a variety of useful designs can be covered with a few formulas by assuming the following reasonable common features. Modifications of these features are considered later.

1. The *overall uniform sampling fraction*, and the equal probability of selecting any of the $N = AB$ population elements into the sample of $n = ab$ elements is

$$f = \frac{n}{N} = \frac{a}{A} \cdot \frac{b}{B} = f_a \cdot f_b.$$

2. *Random choice with replacement* is used to select the sample of a primary sampling units from the population of A PSU's. The more complicated, and more usual, cases of stratified or systematic selection of PSU's are treated later as elaborations of the basic formulas.

3. *Epsem* selection is used to *subsample without replacement* a PS of b out of the B elements from each selected PSU. The theory is neat for random choice; but one simple variance formula covers all epsem methods. Among the practical alternatives are systematic sampling, or a multistage sample after subdividing each primary unit into two or more stages of sampling units. This flexibility is the basis for the method's wide utility. If the PSU is selected twice, two separate subsamples (PS) must be selected, each containing b different elements.

4. In most situations a is fairly large, A very large, and a/A, the sampling fraction of PSU's, is small. On the other hand, the cluster size B is often small; the sampling fraction b/B may or may not be small.

The mean of the αth sample PS is

$$\bar{y}_\alpha = \frac{1}{b} y_\alpha = \frac{1}{b} \sum_\beta^b y_{\alpha\beta} = \frac{1}{b} [y_{\alpha 1} + y_{\alpha 2} + \cdots + y_{\alpha\beta} + \cdots + y_{\alpha b}]. \quad (5.3.1)$$

The sample mean $\bar{y}_\alpha$ is an unbiased estimate of the population mean

$$\bar{y}_\alpha{}^* = \frac{1}{B} \sum_\beta^B y_{\alpha\beta}$$

of the αth PSU in the sample. We generally omit the asterisk in $\bar{y}_\alpha{}^*$ when the risk of confusion is small, and use it only when the PSU mean $\bar{y}_\alpha{}^*$ must be distinguished from the sample mean of the PS.

The simple mean of the sample is

$$\bar{y} = \frac{y}{n} = \frac{1}{n} \sum_j^n y_j = \frac{1}{ab} \sum_\alpha^a \sum_\beta^b y_{\alpha\beta} = \frac{1}{a} \sum_\alpha^a \bar{y}_\alpha$$

$$= \frac{1}{a} [\bar{y}_1 + \bar{y}_2 + \cdots + \bar{y}_\alpha + \cdots + \bar{y}_a]. \quad (5.3.2)$$

This mean is based on an epsem with fixed sample size n; hence, it is an unbiased estimator of the population mean $\bar{Y}$ (2.8C). It is the simple mean of the sample elements; and it also equals the simple mean of PS means.

The variance of the sample mean can be stated as

$$\text{var}(\bar{y}) = (1-f) \frac{s_a{}^2}{a}, \qquad \text{where } s_a{}^2 = \frac{1}{a-1} \sum_\alpha^a (\bar{y}_\alpha - \bar{y})^2$$

$$= \frac{1}{a-1} \left(\sum_\alpha^a \bar{y}_\alpha{}^2 - a\bar{y}^2 \right). \quad (5.3.3)$$

This is proven in 5.6. It may also be given as

$$\text{var}(\bar{y}) = (1-f) \frac{s_y{}^2}{ab^2}, \qquad \text{where } s_y^2 = \frac{1}{a-1} \left(\sum_\alpha^a y_\alpha{}^2 - \frac{y^2}{a} \right). \quad (5.3.3')$$

This gives an unbiased estimate of the variance for the design described above. Note that the mean can be computed directly from the elements, but the variance demands the separation of the PS values. The formula is similar to those of simple clusters in Section 5.2; they even seem identical, but they are not. The formulas of 5.2 involve the population values of the clusters, $\bar{y}_\alpha{}^*$ and B; here in 5.3 we use instead the sample values $\bar{y}_\alpha$ and b; that is, while the *forms* of the two variance formulas are similar, their *contents* differ because they represent different sampling designs.

Observe that from the values of the *primary variates* y_α alone, we can compute the entire variance, including both the "between" cluster and "within" cluster *components of variation*. However, if we need to estimate separately the components of the variance, we must compute them with methods discussed in 5.5.

Example 5.3a Suppose that the 40 values of y_α of Example 5.2*a* came from the following design. The population is imagined as divided into 398 clusters of 100 elements each. Then $a = 40$ random numbers, from 1 to 398, are drawn with replacement; that is, with repetition allowed. Number 179, for example, would define the cluster of elements 17,801–17,900. From the selected clusters of $B = 100$ elements each, $b = 10$ different elements are drawn with a specific epsem procedure. If the cluster is selected twice (or k times), then two (or k) epsem selections of 10 elements are made, without replacement.

Given the same 40 values of y_α, we would get the same estimates as in 5.2*a*. The overall uniform sampling fraction is

$$f = \frac{a}{A} \cdot \frac{b}{B} = \frac{40}{398} \cdot \frac{10}{100} = \frac{400}{39{,}800}.$$

The mean is $\bar{y} = y/n = 185/400 = 0.4625$, and the variance is

$$\text{var}(\bar{y}) = \frac{1-f}{a}\left[\frac{1}{(a-1)b^2}\left(\sum^{a} y_\alpha{}^2 - \frac{y^2}{a}\right)\right] = 0.002585.$$

Here we used the same values of $\bar{y}_\alpha$ as in the sample of compact clusters to emphasize the similarity of the computing forms. But the content and expected values of the two procedures are not equivalent, and the *expected variances* typically differ. For example, we would generally scatter the $b = 10$ elements throughout the clusters of $B = 100$ listings, perhaps by taking a systematic sample of every tenth. This would tend to decrease the variance, because the *homogeneity* (measured as *rho*) is typically less in larger clusters. Then we could expect a greater variance from the design for compact clusters described in Example 5.2*a* than from the design of 5.3*a* for a subsample of the same n dispersed into larger clusters.

The expected variance depends on the subsampling design employed. In the extreme case, the design could call for drawing 10 successive elements (for example, 17,801–17,810); this design and its expected variance would be equivalent to the sample of compact clusters in Example 5.2*a*.

Remark 5.3.I The design has great flexibility because it requires only that the subsampling (b from B) be epsem. This permits a great variety of selection designs, and the flexibility can be exploited to choose designs as needed.

If b out of B elements are chosen with *simple random sampling*, the well-known formulas of random multistage sampling hold. These are discussed in Section 5.6. But instead of simple random selection, stratification or clustering may seem preferable. Note that two or more random selections per primary unit are needed, if we must separate the variance components between and within the primary units (8.6A and 8.6E).

Systematic selection can often attain stratification by taking advantage of the ordering of elements in the primary units. Furthermore, it often has an advantage in simpler clerical operations.

Clustering may be introduced to subsample the elements. For example, in 5.3a, subsamples of $b = 10$ from $B = 100$ dwellings were needed; these could also be selected as pairs of clusters of 5 consecutive dwellings each. Such *compact segments* are often used for selecting dwellings from sample blocks.

Multistage sampling can also be utilized, and the overall sampling probability appears as the product of the selection probabilities at each stage. For a three-stage sample,

$$f = \frac{n}{N} = \frac{a}{A} \cdot \left[\frac{b}{B} \cdot \frac{c}{C}\right] = f_a \cdot [f_b \cdot f_c].$$

We may extend this to any number of stages:

$$f = \frac{n}{N} = \frac{a}{A} \cdot \left[\frac{b}{B} \cdot \frac{c}{C} \cdots \frac{g}{G}\right].$$

For example, imagine drawing a sample of A carloads of cigarettes when each carload holds B crates; each crate contains C boxes; each box, D cartons; each carton, E packs; and each pack, F cigarettes. Such uniform organization of people is rare, but equal clusters can sometimes be created out of files or lists of people.

The brackets enclose the subsampling rates, separated from the primary stage. Their final result forms the basis for each value y_α of primary selections that are required for variance computations. We assume that the last-stage selection is *without* replacement, which is generally preferable. Selection in the preceding stages should theoretically be *with* replacement if one wants (5.3.3) to be an unbiased estimate of the variance (8.6A).

If the selection in the last stage, and in all other stages too, were made *with* replacement, then the fpc $= (1-f)$ would become 1 and vanish from the variance formula. Typically $(1-f)$ is negligible even when the last stage is selected without replacement.

Remark 5.3.II What if the selection rates of the units in the different stages are *not* epsem? A few simple rules can cover most practical situations, although complex problems may need special technical treatment. First, when the different rates are confined to separate strata, the stratified formulas will generally handle them. Second, when the primary units are selected with epsem (a/A), but the subsampling rates vary, we should still be able to produce the properly *weighted primary variates* y_α, with which both means and variances may be computed. Third, selection with probabilities proportional to size is introduced in Chapter 7, where uniform overall probabilities f and uniform subsample sizes b are obtained from unequal sized clusters. This design also can be covered with (5.3.3).

Remark 5.3.III The model calls for sampling in the primary stage *with* replacement. But sampling *without* replacement has distinct practical advantages. The difference between the two procedures generally has little effect, because the primary sampling fraction (a/A) is small. We can choose between three alternatives.

1. In most cases I would select the PSU's *without replacement*, because it is more efficient. With it, I would use the simple variance formula (5.3.3). This design would tend to *slightly overestimate the actual variance.* It disregards a term that in most cases is positive and small (5.6.7). The term vanishes when the homogeneity within clusters is zero; negative values are possible but rare. It decreases with small sampling fractions a/A. Furthermore, if the fpc $(1-f)$ is omitted from the formula (5.3.3), this will always overestimate the variance by only s_a^2/A. A slight overestimate of the actual variance often seems acceptable.
2. We can select the PSU's *without replacement*, and then compute the two components and the unbiased estimate of the variance with (5.6.5) or Table 5.6.I.
3. Select the PSU's *with replacement*, and compute the unbiased estimate of the variance with (5.3.3). This follows the features of the simple model and accepts a slight increase in the actual variance.

Remark 5.3.IV The variance (5.3.3) for simple subsampling can be proven by reducing it to the variance (5.2.3) of complete clusters. We create artificial and arbitrary clusters as we desire, then select them in one imaginary stage. Imagine any uniform procedure for dividing all the A primary units into smaller clusters, each of b elements. Altogether, AB/b clusters could be thus formed if B/b is integral. Then we could select at random, without replacement, a of these AB/b artificial clusters, with a sampling fraction of $a/(AB/b) = ab/AB$. This leads directly to the formula (5.2.3) for complete clusters.

The actual division of the entire population is laborious and unnecessary. Its equivalent is a specified selection procedure. This provides the frame for the selection of a clusters from a population of AB/b artificial clusters. The a primary selections, each of b elements, comprise the sample.

The division of the B elements into small clusters presents problems if B/b is not an integer. These problems have little practical importance because a/A is so small that multiple selection of PSU's is rare; and because the ratio B/b is large enough to permit duplicate selections. But this problem can be handled completely with an imaginary distribution of all possible samples (8.6*A*).

Remark 5.3.V It is helpful to distinguish and to relate several measures of variances. Their role in sample design is further exploited in 8.2. From $\text{var}(\bar{y}) = (1-f)s_a^2/a$ we extract the *unit variance* $s_a^2 = \Sigma(\bar{y}_\alpha - \bar{y})^2/(a-1)$, the variance among unit means. Note that $s_a^2 = a[\text{var}(\bar{y})/(1-f)]$.

The *element variance* is defined as

$$v_v^2 = n[\text{var}(\bar{y})/(1-f)] = (n/a)s_a^2 = bs_a^2.$$

The ratio of the element variance for a particular design to the element variance of an srs of the same n is denoted the *design effect*:

$$\text{deff} = \frac{\text{var}(\bar{y})}{(1-f)s^2/n} = \frac{s_a^2/a}{s^2/n} = \frac{v_v^2}{s^2}.$$

This ratio for cluster samples is denoted $[1 + \text{roh}(b - 1)]$ in (5.4.2). The variance of cluster totals, $s_y^2 = (\Sigma y_\alpha^2 - y^2/a)/(a - 1)$, is related to the above: $s_y^2 = b^2 s_a^2 = bv_v^2$. Note also that the variance of the sample total y is $(1-f)as_y^2 = \text{var}(y) = n^2 \text{var}(\bar{y})$.

These concepts apply also to stratified cluster samples, provided s_a^2, s_y^2, and roh refer to variations within strata and are so defined. Thus if $y = \sum_h y_h$, we have $as_y^2 = \sum_h \sum_\alpha (y_{h\alpha}^2 - y_h^2/a_h)a_h/(a_h - 1)$ and a similar expression for s_a^2.

5.4 EFFECTS OF CLUSTERING; INTRACLASS CORRELATION, ROH

Among diverse selection procedures, clustering often has by far the greatest effect on both the variance and the cost. Compare two samples: one of n independently selected elements, the other containing n elements within a selected clusters. The number of independent choices is reduced in the cluster sample from n to a; the sample is confined to a clusters, rather than spread in n independent points in the population. The effects of clustering on the variance come from two sources. First, the selection consists of actual clusters of the physical distribution of the population. Second, the distribution of the population in those clusters is generally not random. Instead, it is characterized by some *homogeneity* that tends to increase the variance of the sample. The degree of homogeneity and its effect are often small, but sometimes great. The measure of that homogeneity is *roh, the coefficient of intraclass correlation.* The word stands for ρ, commonly used for this measure; but we can remember it as the initials for *rate of homogeneity*.

Example 5.4a In Example 5.2*a*, the proportion of home owners in the sample of $n = 400$ subscribers was 46.25 percent. If this were a simple random sample, we could estimate the variance as:

$$\text{var}(p_0) = (1-f)\frac{pq}{n - 1} = (0.99)\frac{46.25 \times 53.75}{399} \times 10^{-4} = 6.168 \times 10^{-4}.$$

In practice, we can omit trivial refinements and simply compute $pq/n = 6.21 \times 10^{-4}$. The variance of the cluster sample was computed properly in 5.2*a* as 25.85×10^{-4}. But, if by mistake (commonly made) we should use the srs estimate, we would underestimate the variance by a ratio of $25.85/6.17 = 4.19$;

and the standard error by the ratio of $\sqrt{4.19} = 2.05$. For example, desiring to use $t = 1.96$ standard errors, we would actually be using $t = 1.96/2.05 = 0.96$ standard errors; that is, $1.96 \times \sqrt{6.17} = 1.96 \times 2.48 = 4.86$; and 4.86 is only 0.96 of the properly computed standard error $\sqrt{25.85} = 5.08$. Therefore, instead of making correct statements at the $P = 0.95$ level, we would be making correct statements at the $P = 0.66$ level, corresponding to $t = 0.96$.

It is useful to compute the ratio of the two variance estimates: the proper estimate divided by the estimate of an srs of the same n. This ratio (4.19 in the example) measures the *design effect*, or *deff* (8.2). We may think of $400/4.19 = 95$ as the *effective n* of the above design, *as compared to the standard of an srs.* Furthermore, we must note that the ratio estimates to a good approximation (5.6.17) the effect of clustering in terms of the measure of homogeneity, roh:

$$\text{deff} = \frac{s_a^2/a}{s^2/n} = [1 + \text{roh}\,(B - 1)]. \tag{5.4.1}$$

In the example above, the number of elements per cluster was $B = 10$ and $[1 + \text{roh}\,(10\text{–}1)] = 4.19$; hence,

$$\text{roh} = \frac{\text{deff} - 1}{B - 1} = \frac{4.19 - 1}{10 - 1} = 0.35.$$

The formula provides a concise expression for the effect of clustering on sample design. The expression also has great flexibility. If subsamples of b are selected randomly from equal clusters, we may use

$$\text{deff} = \frac{s_a^2/a}{s^2/n} = [1 + \text{roh}\,(b - 1)]. \tag{5.4.2}$$

Similarly, in unequal sized clusters, the average size $\bar{n} = n/a$ in place of b generally yields serviceable approximations. When the selection of the b elements is simply random, the roh in (5.4.2) also denotes the roh of the entire parent cluster of B elements.

Often the subsampling is not simply random. Systematic, stratified, and other designs may be employed to obtain the b elements. Quite generally one may regard (5.4.2) as measuring the ratio of homogeneity *roh* in the primary selections of size b as created by the sample design. Moreover, for stratified random samples it may be regarded as measuring the effect of clustering within the strata (see also 8.1).

Note also that since $n/a = b$, we have in

$$\frac{s_a^2}{s^2} = \frac{[1 + \text{roh}\,(b - 1)]}{b} = \left[\text{roh} + \frac{1 - \text{roh}}{b}\right] \tag{5.4.3}$$

a measure of precision, relative to the precision of elements, for the created sampling units of size b.

These serviceable working rules are subject to approximations. First, we use the variance among the sample elements to estimate S^2 (2.8.23). Second, the formulas (5.4.1) and (5.4.2) involve slight approximations. Third, the measure of homogeneity depends on the model of variation we adopt; in 5.6 we shall consider a slightly different version, rho.

When roh is positive, the design effect exceeds 1, because the cluster sample has a greater variance than an srs with the same n. The highest possible value of roh is $+1$. This corresponds to complete segregation of the variable within clusters: all elements comprising any cluster have the same value. In this case the design effect $[1 + \text{roh}\,(b - 1)] = b$. Also $s_a^2/s^2 = [1 + \text{roh}\,(b - 1)]/b = 1$, so that the variance of cluster units is as great as of single elements.

The lowest possible value for roh is $-1/(b - 1)$. This corresponds to $[1 + \text{roh}\,(b - 1)] = 0$, and to zero variance between cluster means. Negative values of roh are rare; they occur when cluster means are more uniform than would be produced by random sorting.

If the variable is distributed completely at random among the clusters, then we expect a roh of zero, and the design effect $[1 + \text{roh}\,(b - 1)] = 1$. Also, since $s_a^2/s^2 = [1 + \text{roh}\,(b - 1)]/b = 1/b$, the expected variance for cluster units is $1/b$ as great as for single elements. For example, the expected variance of n balls from a "well-mixed urn" will be the same whether we select them individually or in clusters of b. Our discussion concerns expected variances, and actual samples are subject to sampling variations around them.

Generally, population variables are not "well-mixed": they are not randomly distributed in groups and clusters. In practical survey clusters, roh tends to be greater than zero, sometimes by much, often by little. But even a relatively small positive roh can have a large effect, $[1 + \text{roh}\,(b - 1)]$, on the variance, if the sample cluster b is large.

The variance of cluster samples, particularly in social research, is typically greater than for a comparable sample of elements. This is not a logical necessity, but a generalization based on research with groups of many kinds. In most groups roh tends to be positive; the individuals within groups tend to resemble each other. The homogeneity of groups is greater than if individuals were assigned to them at random. The homogeneity may be due to selective factors in grouping, to joint exposure to similar influences, to the effects of mutual interaction, or to some combination of these three sources. Regardless of source, roh measures the homogeneity in terms of the *portion of the total element variance* that is due to group membership. Sampling units employed as convenient clusters typically possess some group homogeneity.

The homogeneity, roh, has distinct values for diverse variables and for different populations. For a specified variable and population, its value depends on the nature and size of the clusters. It is a characteristic of the clusters, and affects the variance when the clusters serve as sampling units. The precision of a cluster sample cannot be given merely in terms of the number of sample elements, because the variance depends on the numbers of sample units at each stage.

The increase in the variance, $[1 + \text{roh}\,(b - 1)]$, for fixed n and specified size b of the sample cluster depends on roh. Since homogeneity within sample clusters increases the variance, the sampler wants to reduce it, to increase its heterogeneity. First, he may spread his subsample b into a cluster with greater size B and with less homogeneity. Larger clusters tend to have smaller roh, and this is one reason for subsampling, rather than confining the sample into compact clusters. The decrease of roh with increasing cluster size has been investigated [Hansen, Hurwitz, and Madow, 1953, 6.27; Cochran, 1963, 9.5]. For many distributions we can quite simply assume that "neighboring" units tend to resemble each other. Then it follows that the resemblance decreases with increasing distance over the distribution.

Second, the sampler may try to assemble dissimilar units to form the sample clusters. For example, when selecting pairs of dwellings from sample blocks, or pairs of students from sample classes, the sampler may try to link contrasting elements (dwellings or students) to form the sample clusters (the pairs). Since formal composition of clusters is difficult, existing groups generally serve as clusters. Nevertheless, the sampler may decrease the homogeneity of the sample clusters with judicious subsampling procedures. For example, with systematic selection a sample of four dwellings can be spread around the block; this should produce less homogeneity than a cluster of four adjacent dwellings.

5.5 STRATIFIED CLUSTER SAMPLING

Cluster samples are generally selected with stratification, because stratification has more advantages for cluster than for element sampling. First, stratification is typically easier, because the sorting involves fewer units and because more information is available for them. For example, compare the tasks of stratifying the United States into 3000 counties and into 55 million families.

Second, the relative gains of proportionate stratification are greater for clusters than for element sampling from the same set of strata. For example, suppose the regions of the United States comprise the strata from which we might select families, either directly as elements or in clusters of counties. For either element sampling or cluster sampling, the reduction of the sampling unit variance due to proportionate sampling equals $\Sigma W_h(\bar{Y}_h - \bar{Y})^2$, the weighted squared deviations of the stratum means. But that equal reduction comes from S_a^2 for cluster sampling and S^2 for element sampling. Hence the *relative gain* is $\Sigma W_h(\bar{Y}_h - \bar{Y})^2/S^2$ for element sampling and $\Sigma W_h(\bar{Y}_h - \bar{Y})^2/S_a^2$ for cluster sampling. From (5.6.17) we see that

$$S_a^2/S^2 = [1 + \text{Roh}\,(B-1)]/B = [(1 - \text{Roh})/B + \text{Roh}];$$

and for commonly low values of Roh S_a^2 is much less than S^2. Hence, the relative gains are much greater for sampling clusters than elements.

Third, the problems of selecting clusters without replacement also favor stratification.

The formulas of stratified cluster sampling appear complex due to multiple subscripts and summations. But the formulas below represent the direct application of basic stratification notions (3.2) to the subsampling formulas of 5.3. (The complete clusters of 5.2 appear merely as the special case when $b_h = B_h$.) We require *equal clusters* and epsem within strata, so that $f_h = (a_h/A_h) \times (b_h/B_h) = n_h/N_h$. But we may allow differences in the sampling fractions between strata; also in the number of sampling units, both in the population and in the sample.

Application of the general stratification formula to subsampling of equal clusters results in

$$\bar{y} = \sum W_h \bar{y}_h = \sum W_h \frac{y_h}{n_h} = \frac{1}{N} \sum \frac{y_h}{f_h}. \tag{5.5.1}$$

The last form holds only when $W_h = N_h/N$. For proportionate sampling ($f_h = f$), this reduces to the simple mean of the sample y/n, where $n = \Sigma\, n_h$. The variance of the mean is

$$\text{var}\,(\bar{y}) = \sum_h^H W_h^2 \frac{1 - f_h}{a_h} \left[\frac{1}{(a_h - 1)b_h^2}\left(\sum_\alpha^{a_h} y_{h\alpha}^2 - \frac{y_h^2}{a_h}\right)\right]. \tag{5.5.2}$$

This can be simplified for important special cases. For proportionate sampling, when $W_h = N_h/N$, and with equal allocation, $a_h = a_c$, we have

$$\text{var}\,(\bar{y}) = \frac{1-f}{n^2(a_c - 1)}\left[\sum_h \left(a_c \sum_\alpha y_{h\alpha}^2 - y_h^2\right)\right]. \tag{5.5.3}$$

The special case of *paired selections* ($a_c = 2$) deserves attention. The procedures resemble those developed in 4.3. The variance (5.5.3) becomes simply

$$\text{var}\,(\bar{y}) = \frac{1-f}{n^2} \sum (y_{h1} - y_{h2})^2. \tag{5.5.3'}$$

For disproportionate sampling with paired selections (5.5.2) becomes

$$\text{var}(\bar{y}) = \sum \frac{W_h^2(1-f_h)}{n_h^2}(y_{h1} - y_{h2})^2. \tag{5.5.2$'$}$$

Example 5.5a Suppose now that the 40 cluster totals of example 5.2a were the result of paired selections from 20 strata. Assume that the $A = 3980$ clusters had been divided into 20 strata of 199 clusters each. Two different random numbers are drawn from each stratum: two from 1–199, two from 200–398, two from 399–597, etc. Successive pairs of the data represent successive cluster totals (y_{h1} and y_{h2}). Then substituting these in (5.5.3′) yields

$$\sum (y_{h1} - y_{h2})^2 = (10 - 8)^2 + (6 - 5)^2 + \cdots + (5 - 4)^2 = 215,$$

and
$$\text{var}(\bar{y}) = \frac{0.99}{400^2} 215 = 13.30 \times 10^{-4}.$$

Compared to 5.2a, this shows a marked reduction of the variance due to stratification. Note that $\text{var}(\bar{y}) = [(1-f)/a]s_a^2$; hence, $s_a^2 = 215/40(10^2) = 0.05375$ represents the variance of cluster means within strata.

The simple random variance was computed (5.4a) as 6.168×10^{-4}; hence, the design effect is $13.30/6.168 = 2.15$.

*5.6 COMPONENTS OF THE VARIANCE

5.6A Components of Two-Stage Random Sampling

The element variance is $\sigma^2 = \sum_i^N (Y_i - \bar{Y})^2/N = \sum_\alpha^A \sum_\beta^B (Y_{\alpha\beta} - \bar{Y})^2/AB$. It is instructive to analyze the variance into components corresponding to the several stages of selection. For two stages of equal clusters there are two components, $\sigma^2 = \sigma_a^2 + \sigma_b^2$, defined by

$$\frac{1}{AB}\sum_\alpha^A \sum_\beta^B (Y_{\alpha\beta} - \bar{Y})^2 = \frac{1}{A}\sum_\alpha^A (\bar{Y}_\alpha - \bar{Y})^2 + \frac{1}{AB}\sum_\alpha^A \sum_\beta^B (Y_{\alpha\beta} - \bar{Y}_\alpha)^2. \tag{5.6.1}$$

The *between clusters component* σ_a^2 denotes the variance of the cluster means $\bar{Y}_\alpha$ around the population mean $\bar{Y}$. The *within clusters component* σ_b^2 denotes the variance of elements $Y_{\alpha\beta}$ around their cluster means $\bar{Y}_\alpha$. The left side of (5.6.1) can be shown to equal the right side by first inserting two steps:

$$\frac{1}{AB}\sum_\alpha^A \sum_\beta^B (\bar{Y}_\alpha - \bar{Y} + Y_{\alpha\beta} - \bar{Y}_\alpha)^2$$

$$= \frac{1}{AB}\sum_\alpha^A \sum_\beta^B [(\bar{Y}_\alpha - \bar{Y})^2 + 2(\bar{Y}_\alpha - \bar{Y})(Y_{\alpha\beta} - \bar{Y}_\alpha) + (Y_{\alpha\beta} - \bar{Y}_\alpha)^2].$$

After noting that the middle term sums to zero, sum the first and third terms to get the right side of (5.6.1).

For estimating the components from samples, the S^2 terms below will appear more convenient and we reformulate (5.6.1) into its equivalent:

$$\frac{N-1}{N} S^2 = \frac{A-1}{A} S_a^{\,2} + \frac{B-1}{B} S_b^{\,2}, \tag{5.6.2}$$

where $S^2 = \dfrac{N}{N-1}\sigma^2, \qquad S_a^{\,2} = \dfrac{A}{A-1}\sigma_a^{\,2}, \qquad S_b^{\,2} = \dfrac{B}{B-1}\sigma_b^{\,2}.$

Impose on this clustered population a *two-stage selection*, a from A clusters and b from B elements. The selection is *random and without replacement* in both stages. This design occupies a basic position in sampling literature, and other designs can be regarded as its modifications. Now we define two useful sample statistics $s_a^{\,2}$ and $s_b^{\,2}$. First, the variance of sample elements within clusters is

$$s_b^{\,2} = \frac{1}{a(b-1)} \sum_\alpha^a \sum_\beta^b (y_{\alpha\beta} - \bar{y}_\alpha)^2. \tag{5.6.3}$$

Then $E(s_b^{\,2}) = S_b^{\,2}$, and $S_b^{\,2} = \dfrac{1}{A(B-1)} \displaystyle\sum_\alpha^A \sum_\beta^B (Y_{\alpha\beta} - \bar{Y}_\alpha)^2.$

The mathematical expectation of $s_b^{\,2}$ is $S_b^{\,2}$, but the relationship between $s_a^{\,2}$ and $S_a^{\,2}$ is more complex. We had already defined

$$s_a^{\,2} = \frac{1}{a-1}\sum_\alpha^a (\bar{y}_\alpha - \bar{y})^2 \qquad \text{and} \qquad S_a^{\,2} = \frac{1}{A-1}\sum_\alpha^A (\bar{Y}_\alpha - \bar{Y})^2.$$

Then
$$E(s_a^{\,2}) = \tilde{S}_a^{\,2} = S_a^{\,2} + \left(1 - \frac{b}{B}\right)\frac{S_b^{\,2}}{b},$$

and
$$E\left[s_a^{\,2} - \left(1 - \frac{b}{B}\right)\frac{s_b^{\,2}}{b}\right] = \tilde{S}_a^{\,2} - \left(1 - \frac{b}{B}\right)\frac{S_b^{\,2}}{b} = S_a^{\,2}. \tag{5.6.4}$$

With the expectations above for $s_a^{\,2}$ and $s_b^{\,2}$ it is easy to show that

$$E[\text{var}(\bar{y})] = \text{Var}(\bar{y}),$$

where var $(\bar{y})$ is defined as

$$\text{var}(\bar{y}) = \left(1 - \frac{a}{A}\right)\frac{s_a^{\,2}}{a} + \left(1 - \frac{b}{B}\right)\frac{a}{A}\frac{s_b^{\,2}}{ab} = \frac{s_a^{\,2}}{a} - \frac{1}{A}\left[s_a^{\,2} - \left(1 - \frac{b}{B}\right)\frac{s_b^{\,2}}{b}\right], \tag{5.6.5}$$

and
$$\text{Var}(\bar{y}) = \left(1 - \frac{a}{A}\right)\frac{S_a^{\,2}}{a} + \left(1 - \frac{b}{B}\right)\frac{S_b^{\,2}}{ab}. \tag{5.6.5'}$$

This is the population variance for the mean when selection in both stages is made at random and without replacement. This is derived in Remark 5.6.I, together with the expectations of s_a^2 and s_b^2. The statistic (5.6.5) is an unbiased estimate of the variance. The formula requires the separate computation of the within-component s_b^2; hence computing the variance s_a^2 among the primary clusters alone does not suffice. But observe the change from a to A in the denominator of s_b^2; when a/A is small, all but s_a^2/a becomes negligible.

Note two special cases. If complete clusters are selected without subsampling, then $b = B$, $E(s_a^2) = S_a^2$, and $\text{var}(\bar{y}) = (1 - a/A)s_a^2/a$. The formula for s_b^2 has $(B - 1)$ for $(b - 1)$, and it estimates S_b^2 (5.6.3).

On the contrary, if all clusters are selected $(a = A)$, then formula (5.6.5) loses the first term and becomes the variance of proportionate stratified sampling within A strata.

From (5.6.4) and (5.6.5) we can see that

$$E\left(\frac{s_a^2}{a}\right) = \frac{S_a^2}{a} + \left(1 - \frac{b}{B}\right)\frac{S_b^2}{ab} = \text{Var}(\bar{y}) + \frac{S_a^2}{A}. \tag{5.6.6}$$

Hence, the use of s_a^2/a would overestimate the variance by S_a^2/A. Generally, A is large compared to a, and the overestimation is not important. Similarly, we can show that

$$E\left[\left(1 - \frac{ab}{AB}\right)\frac{s_a^2}{a}\right] = \text{Var}(\bar{y}) + \frac{S_a^2}{A}\left(1 - \frac{b}{B}\right)\left(1 - \frac{S_b^2}{BS_a^2}\right). \tag{5.6.7}$$

In most cases $(1 - S_b^2/S_a^2B)$ lies between 0 and 1; then the formula on the left side would lead to a slight overestimation of the variance. This appears reasonable since the formula $(1 - f)s_a^2/a$ becomes the variance when the clusters are selected with replacement (5.3.3). The extra term in (5.6.7) arises when the clusters are selected without replacement; this generally leads to a slight reduction of the variance.

If the number of primary units A in an actual or hypothetical population is so large that a/A is considered zero, then

$$\text{Var}(\bar{y}) = E\left(\frac{s_a^2}{a}\right) = \frac{S_a^2}{a} + \left(1 - \frac{b}{B}\right)\frac{S_b^2}{ab}, \tag{5.6.8}$$

and $\text{var}(\bar{y}) = s_a^2/a$ is an unbiased estimate of the variance. This also holds when b/B is considered zero. This corresponds to the classical "random model" of the analysis of variance.

The second-stage units need not be elements. The preceding analysis will also serve situations where each primary cluster contains B secondary clusters from which b are selected with epsem. Each secondary cluster contains an equal number C of elements that are included in the sample. However, if an epsem of c out of C must be subsampled, the above methods can be further developed in an analogous manner for three or more stages. The variance of the mean obtained by sampling at random and without replacement in three stages can be estimated as

$$\text{var}(\bar{y}) = \left(1 - \frac{a}{A}\right)\frac{s_a^2}{a} + \left(1 - \frac{b}{B}\right)\frac{a}{A}\frac{s_b^2}{ab} + \left(1 - \frac{c}{C}\right)\frac{ab}{AB}\frac{s_c^2}{abc}. \quad (5.6.9)$$

The components are unbiased estimates of similar population values [Cochran, 1963, 10.8].

Remark 5.6.1 The variance *within* any specified αth cluster, due to selecting b from its B elements with random choice, is

$$\text{Var}(\bar{y}_\alpha) = \left(1 - \frac{b}{B}\right)\frac{S_{b\alpha}^2}{b}, \quad \text{where } S_{b\alpha}^2 = \frac{1}{B-1}\sum_\beta^B (Y_{\alpha\beta} - \bar{Y}_\alpha)^2.$$

This is merely simple random sampling within the small population of the cluster. We also have the unbiased estimates within the fixed αth cluster:

$$\text{var}(\bar{y}_\alpha) = \left(1 - \frac{b}{B}\right)\frac{s_{b\alpha}^2}{b}, \quad \text{where } s_{b\alpha}^2 = \frac{1}{b-1}\sum_\beta^b (y_{\alpha\beta} - \bar{y}_\alpha)^2.$$

The expectations for all random samples of b out of the B elements within the αth cluster are

$$E_\beta(s_{b\alpha}^2) = S_{b\alpha}^2 \quad \text{and} \quad E_\beta[\text{var}(\bar{y}_\alpha)] = \text{Var}(\bar{y}_\alpha).$$

The mean value of $S_{b\alpha}^2$ over all A clusters is S_b^2, defined in (5.6.3). For a single randomly selected cluster, the expectation over all A values of $S_{b\alpha}^2$ is also S_b^2. Hence, the double expectation of $s_{b\alpha}^2$ from a random cluster is also S_b^2; that is, $\underset{\alpha}{E}[\underset{\beta}{E}(s_{b\alpha}^2)] = \underset{\alpha}{E}[S_{b\alpha}^2] = S_b^2$. Hence, for a randomly selected cluster, the expected within variance is

$$\underset{\alpha}{E}[\text{Var}(\bar{y}_\alpha)] = \left(1 - \frac{b}{B}\right)\frac{S_b^2}{b}.$$

For a random sample of a clusters, the mean of the within component $s_{b\alpha}^2$ is s_b^2; and its expected value must be S_b^2. For the mean of a clusters, the *within-variance component* equals the variance of a single cluster divided by a:

$$E\left\{\frac{1}{a^2}\sum_\alpha^a [\text{Var}(\bar{y}_\alpha)]\right\} = \left(1 - \frac{b}{B}\right)\frac{S_b^2}{ab}, \text{estimated with } \left(1 - \frac{b}{B}\right)\frac{s_b^2}{ab}. \quad (5.6.10)$$

Now suppose we select a random clusters from A clusters and measure the true cluster means $\bar{y}_\alpha^*$ without subsampling. The result is a simple random sample of a values. The variance of single clusters is

$$s_a^{*2} = \frac{1}{a-1}\sum^a (\bar{y}_\alpha^* - \bar{y}^*)^2 \quad \text{and} \quad E(s_a^{*2}) = S_a^2. \quad (5.6.11')$$

The variance of the mean $\bar{y}^* = \sum_\alpha^a \bar{y}_\alpha^*/a$ of a random sample of complete clusters is

$$\text{Var}(\bar{y}^*) = \left(1 - \frac{a}{A}\right)\frac{S_a^2}{a}, \quad \text{estimated by } \text{var}(\bar{y}^*) = \left(1 - \frac{a}{A}\right)\frac{s_a^{*2}}{a}. \quad (5.6.11)$$

For *two-stage* random sampling, both components are present independently: the *within-clusters* component (5.6.10) and the *between-clusters* component (5.6.11). Hence,

$$\text{Var}(\bar{y}) = \left(1 - \frac{a}{A}\right)\frac{S_a^2}{a} + \left(1 - \frac{b}{B}\right)\frac{S_b^2}{ab}. \tag{5.6.5'}$$

The variance of a single random subsample, from (5.6.10) and (5.6.11), has an unbiased estimate in

$$s_a^2 = \frac{1}{a-1}\sum(\bar{y}_\alpha - \bar{y})^2; \quad \text{and} \quad E(s_a^2) = S_a^2 + \left(1 - \frac{b}{B}\right)\frac{S_b^2}{b}. \tag{5.6.4}$$

Note that this combines the independent variances of a single complete cluster and its within-component. Now we see that the expectation of (5.6.5) is (5.6.5'):

$$\left(1 - \frac{a}{A}\right)\frac{1}{a}\left[s_a^2 - \left(1 - \frac{b}{B}\right)\frac{s_b^2}{b}\right] + \left(1 - \frac{b}{B}\right)\frac{s_b^2}{ab}$$

$$= \left(1 - \frac{a}{A}\right)\frac{s_a^2}{a} + \left(1 - \frac{b}{B}\right)\frac{s_b^2}{ab}\left(1 - 1 + \frac{a}{A}\right) = \left(1 - \frac{a}{A}\right)\frac{s_a^2}{a} + \left(1 - \frac{b}{B}\right)\frac{s_b^2}{Ab}. \tag{5.6.5}$$

This then is an unbiased estimate of (5.6.5′) because

$$E(s_b^2) = S_b^2 \quad \text{and} \quad E\left[s_a^2 - \left(1 - \frac{b}{B}\right)\frac{s_b^2}{b}\right] = S_a^2.$$

The details of derivations for two-stage sampling made us resort to several auxiliary definitions. These also occur in the literature for reasons on which the following relations may shed some light.

$$E(s_a^2) = \tilde{S}_a^2 = S_a^2 + \left(1 - \frac{b}{B}\right)\frac{S_b^2}{b} = S_a^2 + \frac{S_b^2}{b} - \frac{S_b^2}{B} = S_u^2 + \frac{S_b^2}{b}.$$

$$E\left(s_a^2 - \frac{s_b^2}{b}\right) = S_u^2 = S_a^2 - \frac{S_b^2}{B}$$

$$S_\cdot^2 = S_a^2 + S_b^2 - \frac{S_b^2}{B} = S_u^2 + S_b^2 \qquad \left(\text{when } \frac{A-1}{A} \doteq 1 \doteq \frac{N-1}{N}\right).$$

5.6B The Intraclass Correlation

The coefficient of *intraclass correlation measures the homogeneity* of the elements within clusters. It can be defined in terms of the variance components:

$$\text{Rho} = \frac{\sigma_a^2 - \sigma_b^2/(B-1)}{\sigma^2} = \frac{\dfrac{A-1}{A}S_a^2 - \dfrac{1}{B}S_b^2}{\dfrac{N-1}{N}S^2}. \tag{5.6.12}$$

Knowing that $\sigma^2 = \sigma_a^2 + \sigma_b^2$, we can also express this relationship by using only one of the two components:

$$\text{Rho} = \frac{B}{B-1} \cdot \frac{\sigma_a^2}{\sigma^2} - \frac{1}{B-1} = \left[\frac{A-1}{A} \cdot \frac{N}{N-1}\right] \cdot \frac{B}{B-1} \cdot \frac{S_a^2}{S^2} - \frac{1}{B-1}$$

$$= 1 - \frac{S_b^2}{S^2} \cdot \frac{N}{N-1}. \tag{5.6.12'}$$

We shall find two other expressions useful:

$$\sigma_a^2 = \frac{\sigma^2}{B}[1 + \text{Rho}(B-1)]$$

or

$$\left(\frac{A-1}{A} \cdot \frac{N}{N-1}\right) S_a^2 = \frac{S^2}{B}[1 + \text{Rho}(B-1)], \tag{5.6.13}$$

and $\sigma_b^2 = \dfrac{B-1}{B}\sigma^2(1 - \text{Rho})$ or $\left(\dfrac{N}{N-1}\right) S_b^2 = S^2(1 - \text{Rho})$.

The word "correlation" comes from its original definition:

$$\text{Rho} = \frac{1}{\sigma^2}\left[\frac{1}{A}\sum_{\alpha}^{A} \frac{2}{B(B-1)} \sum_{\beta<\gamma}^{B} (Y_{\alpha\beta} - \bar{Y})(Y_{\alpha\gamma} - \bar{Y})\right], \tag{5.6.14}$$

which can be shown to be equivalent to (5.6.12). The bracket contains the mean product of deviations from the population mean. The products are formed for each of the $B(B-1)/2$ different pairs of elements within each of the A clusters. To the degree that the various elements of the same cluster tend, on the average, to deviate in the same direction from the population mean, the average of the products, hence Rho, tend to be positive. Because neither σ_a^2 nor σ_b^2 may be negative, the limits of Rho are easily determined:

1. *Complete homogeneity* within clusters, when $\sigma_b^2 = 0$ and $\sigma_a^2 = \sigma^2$, leads to Rho $= 1$.

2. *Extreme heterogeneity* within clusters, when $\sigma_b^2 = \sigma^2$ and $\sigma_a^2 = 0$, leads to Rho $= -1/(B-1)$.

3. When the homogeneity equals that of *random* sorting of elements into clusters, then Rho $= -1/(N-1) \doteq 0$.

In many situations we may consider $(A-1)/A$, hence also $(N-1)/N$, as equal to one. We may do this either because the large size of A justifies the approximation, or because the population can properly be viewed as a random sample from an indefinitely large universe. Under these assumptions the relationship $\sigma^2 = \sigma_a^2 + \sigma_b^2$ becomes

$$S.^2 = S_a^2 + \frac{B-1}{B} S_b^2. \tag{5.6.15}$$

This leads to somewhat simpler formulas, when we substitute for Rho (5.6.12) the slightly different

$$\text{Roh} = \frac{S_a^2 - S_b^2/B}{S.^2} = \frac{B}{B-1} \cdot \frac{S_a^2}{S.^2} - \frac{1}{B-1} = 1 - \frac{S_b^2}{S.^2}. \quad (5.6.16)$$

Instead of (5.6.13) we get

$$S_a^2 = \frac{S.^2}{B}[1 + \text{Roh}\,(B-1)] \quad \text{and} \quad S_b^2 = S.^2[1 - \text{Roh}]. \quad (5.6.17)$$

Generally, it is needless to worry about small corrections like $(A-1)/A$, because the variance components are subject to rather large errors. We should conserve our attention for the chief issues, such as computing sample values of the components.

When *complete clusters of B* are sampled, then s_a^2 and s_b^2 are unbiased estimates of S_a^2 and S_b^2, and they can be used directly in (5.6.15) to (5.6.17). The formulas for roh lead to good estimates of Roh.

Epsem subsampling of b elements from randomly selected clusters has been described for a simple and flexible model (5.3). We may accept the clusters formed by the subsampling design and conduct the analysis as if dealing with randomly selected complete clusters of size b. Then s_b^2 and s_a^2 estimate $\tilde{S}_b^2$ and $\tilde{S}_a^2$ for the clusters as formed by the design. Generally, $\tilde{S}_b^2$ can differ from the S_b^2 for the complete clusters, though probably little. However, $\tilde{S}_a^2$ can differ considerably from S_a^2 for the complete cluster. $\tilde{S}_a^2$, based on compact clusters of b elements, tends to be greater than S_a^2, based on B elements. The estimates of Roh from formulas (5.6.16) and (5.6.17) will estimate $\widetilde{\text{Roh}}$ for the cluster as formed by the subsampling design. This may depart from Roh for the entire cluster to the degree that the subsampling departs from simple random. If the subsampling is clustered, roh may overestimate Roh. If we manage, perhaps with systematic selection, to introduce stratification into the subsampling, then roh (and $\widetilde{\text{Roh}}$) may be lower than Roh. This may be clarified by noting what happens in random subsampling.

Random subsampling formulas were developed in 5.6*A*. In this case, s_b^2 estimates S_b^2 (5.6.3). But s_a^2 estimates $\tilde{S}_a^2$ for the sample cluster, and $E(s_a^2) = S_a^2 + (1 - b/B)S_b^2/b$; hence we compute $[s_a^2 - (1 - b/B)s_b^2/b]$ to estimate S_a^2. If we use this in (5.6.16), note that

$$\text{roh} = \frac{[s_a^2 - (1 - b/B)s_b^2/b] - s_b^2/B}{\hat{s}^2} = \frac{s_a^2 - s_b^2/b}{\hat{s}^2}. \quad (5.6.18)$$

In the last form, the random subsamples take the place of complete clusters. This value of roh equals the first form, based on the original clusters of B elements and, therefore, it estimates Roh. Hence, the relationships (5.6.17) also involve the Roh of the original clusters.

Similar reasoning also leads to the conclusion that under all three conditions we may estimate $S_{.}^2$ as

$$\hat{s}^2 = s_a^2 + \frac{b-1}{b} s_b^2. \tag{5.6.19}$$

When selecting complete clusters of B, the results follow directly. For any epsem subsampling the same result follows if we consider the subsamples as clusters of size b. Specifically for random subsampling we note that, as required,

$$E(\hat{s}^2) = E\left(s_a^2 + \frac{b-1}{b} s_b^2\right) = S_a^2 + \frac{S_b^2}{b} - \frac{S_b^2}{B} + S_b^2 - \frac{S_b^2}{b}$$

$$= S_a^2 + \frac{B-1}{B} S_b^2 = S_{.}^2$$

5.6C Tables for Analysis of Variance

The "analysis of the variance" of the mean into the components of the total variance is often conveniently displayed in tables. The components of (5.6.5) can be seen in Table 5.6.I.

Source of Variation	Degrees of Freedom	Sums of Squares (SQ)	Mean Squares	Expected Values of Mean Squares
Between clusters	$a-1$	$\frac{1}{b}\sum_\alpha y_\alpha^2 - \frac{y^2}{ab}$	bs_a^2	$\left(\frac{B-b}{B}\right)S_b^2 + bS_a^2$
Within clusters	$a(b-1)$	$\sum_\alpha\sum_\beta y_{\alpha\beta}^2 - \frac{1}{b}\sum_\alpha y_\alpha^2$	s_b^2	S_b^2
Total	$ab-1$	$\sum_\alpha\sum_\beta y_{\alpha\beta}^2 - \frac{y^2}{ab}$	s^2	

TABLE 5.6.I Analysis of Variance for Two Stages of Random Selections

In the language of the analysis of variance the above would often be called a *finite* model of a *nested* or *hierarchal* situation. Note that when the *between* and *within mean squares* are equal, then $s_a^2 - s_b^2/b = 0$. In expected values we have $S_a^2 - S_b^2/B = 0$, whether we have complete clusters ($b = B$) or subsampling ($b < B$). These are the conditions for Roh $= 0$ (5.6.16). Hence the usual test to decide whether the ratio $F = bs_a^2/s_b^2 = 1$ or >1, also serves as a test to decide whether Roh $= 0$ or >0. Less frequently one may find genuine cases of $F < 1$, hence, negative Roh.

Source of Variation	Degrees of Freedom	Sums of Squares (SQ)	Mean Squares	Expected Values of Mean Squares
Between strata	$H - 1$	$\frac{1}{a_c b}\sum_h^H y_h{}^2 - \frac{y^2}{Ha_c b}$	$a_c b s_h{}^2$	$\left(\frac{B-b}{B}\right)S_b{}^2 + \left(\frac{A-a_c}{A}\right)bS_a{}^2 + ba_c S_h{}^2$
Between clusters within strata	$H(a_c - 1)$	$\frac{1}{b}\sum_h^H\sum_\alpha^{a_c} y_{h\alpha}^2 - \frac{1}{a_c b}\sum_h^H y_h{}^2$	$bs_a{}^2$	$\left(\frac{B-b}{B}\right)S_b{}^2 + bS_a{}^2$
Within clusters	$Ha_c(b - 1)$	$\sum_h^H\sum_\alpha^{a_c}\sum_\beta^b y_{h\alpha\beta}^2 - \frac{1}{b}\sum_h^H\sum_\alpha^a y_{h\alpha}^2$	$s_b{}^2$	$S_b{}^2$
Total	$Ha_c b - 1$	$\sum_h^H\sum_\alpha^{a_c}\sum_\beta^b y_{h\alpha\beta}^2 - \frac{y^2}{Ha_c b}$	s^2	

TABLE 5.6.II Analysis of Variance of Two-Stage Random Sampling within H Strata

In the last subsection we observed that $s_a^2 + (b-1)s_b^2/b$ estimates $S.^2$. If we denote the three sums of squares of $\text{SQ}_a + \text{SQ}_b = \text{SQ}_t$, we can compute the estimate of the element variance $S.^2$ in the population as

$$\hat{s}^2 = \frac{\text{SQ}_a}{(a-1)b} + \frac{\text{SQ}_b}{ab} = \frac{\text{SQ}_t}{ab} + \frac{\text{SQ}_a}{ab(a-1)} = \frac{\text{SQ}_t}{ab} + \frac{s_a^2}{a}$$

$$= s^2 + \left(\frac{s_a^2}{a} - \frac{s^2}{ab}\right), \qquad \text{where } s^2 = \frac{\text{SQ}_t}{ab-1} = \frac{\text{SQ}_t}{ab} + \frac{\text{SQ}_t}{ab(ab-1)}. \tag{5.6.20}$$

When b denotes numbers of elements, then s^2 is the variance of the sample elements. With the correction in parentheses it becomes an unbiased estimate of $S.^2$. The correction, amounting to $[(s^2/ab)\,\text{roh}\,(b-1)]$, is generally negligible compared to errors in estimating the variance. This illustrates the general point made in (2.8.23).

An important special case occurs when $b = 2$, representing either two elements or two second-stage clusters selected within each primary unit. Then the within sum of squares can be computed as

$$\text{SQ}_b = \sum_\alpha^a (y_{\alpha 1} - y_{\alpha 2})^2/2 \quad \text{and} \quad s_b^2 = \sum_\alpha^a (y_{\alpha 1} - y_{\alpha 2})^2/2a.$$

In the simplest "random models" of the analysis of variance it is assumed that $b/B = 0$, as well as $(A-1)/A = 1$. The lack of a finite B for cluster size resembles sampling with replacement from finite clusters. The expected value of the between mean squares bs_a^2 becomes $S_b^2 + bS_a^2$. Observe that if $s_b^2 > bs_a^2$, the estimate of S_a^2 becomes negative. This is the consequence of the linear model $Y_{\alpha\beta} = \bar{Y}_\alpha + e_{\alpha\beta}$ on which the analysis is based.

The analysis can be carried further to three or more stages, and the interested reader may consult Anderson and Bancroft [1952, Ch. 22] or Kempthorne [1952, Ch. 6]. Table 5.6.II may be interpreted as selecting randomly in three stages H primary units, a_c from A secondary units, and b from B third-stage units. However, the notation aims at a more common problem: to select in each of H strata a_c/A clusters and then b/B of secondary units. In this case we estimate $\sigma_h^2 = \dfrac{H-1}{H} S_h^2$, the between strata variance, because the strata are not selected. For stratified samples the ratio $(A_h - 1)/A_h$ may not be negligible, and estimating $\sigma_a^2 = S_a^2(A_h - 1)/A_h$ in formulas (5.6.12) and (5.6.13) may be preferable to formulas (5.6.15) to (5.6.17).

If the between stratum component is not needed, one may disregard the first line. The rest of the table resembles the two-stage sampling in Table 5.6.I, but the between and within cluster sums of squares do not add to the total; that is, $SQ_a + SQ_b \neq SQ_t$. This partial analysis yields the cluster variance reduced by the stratum variance.

Example 5.6a We can apply the methods of Table 5.6.I to an analysis of the variance of Example 5.2*a*. The sample contained $a = 40$ randomly selected complete clusters of $b = B = 10$ elements each. All sample estimates will be based on three sample results:

$$y = 185, \qquad \sum^{a} y_\alpha^2 = 1263, \qquad \sum^{a}\sum^{B} y_{\alpha\beta}^2 = 185.$$

The last would have to be computed for continuous variables, but for binomials it equals y. Therefore, $y^2/aB = 185^2/400 = 85.56$, and the total mean square is $SQ_t = 185 - 85.56 = 99.44$. Generally, for binomials, $SQ_t = npq = y - y^2/n$. The "between mean square" is $SQ_a = 1263/10 - 85.56 = 40.74$. The "within mean square" is computed as $SQ_b = SQ_t - SQ_a$.

Source of Variation	Degrees of Freedom	Sum of Squares (SQ)	Mean Square	Estimate of
Between clusters	39	$126.3 - 85.56 = 40.74$	1.0446	$10S_a^2$
Within clusters	360	$185 - 126.3 = 58.70$	0.1631	S_b^2
Total	399	$185 - 85.56 = 99.44$	0.2492	

If we disregard $A/(A - 1) = 3980/3979$ and $N/(N - 1) = 39800/39799$, we may compute the estimate of S^2 employing (5.6.19):

$$\hat{s}^2 = s_a^2 + [(B - 1)/B]s_b^2 = 0.1045 + (9/10)0.1631$$
$$= 0.1045 + 0.1468 = 0.2513.$$

The "naive estimate" is $s^2 = 0.2492$ and differs by less than 1 percent. That difference is negligible in view of greater errors in the estimate. For a binomial variable, S^2 cannot be greater than 0.2500; and that is its population value, given in Problem 2.5.

We may compute roh with formula (5.6.18):

$$\text{roh} = \frac{s_a^2 - s_b^2/B}{\hat{s}^2} = \frac{0.1045 - 0.0163}{0.2513} = 0.35.$$

We can get the same result with the more usual procedure, of simply employing (5.6.17), without computing s_b^2:

$$\text{roh} = \frac{\left(\dfrac{Bs_a^2}{s^2} - 1\right)}{(B - 1)} = \frac{\left(\dfrac{10 \times 0.1045}{0.2492} - 1\right)}{(10 - 1)} = \frac{(4.19 - 1)}{9} = 0.35.$$

Example 5.6b In Example 5.5*a*, the data from 5.2*a* were recognized and treated as a stratified sample. The 40 clusters represent $H = 20$ strata, with $a_c = 2$ clusters selected from each. Then the complete clusters of $B = 10$ are included in the sample.

In Example 5.6*a*, the analysis of the same data was conducted as if the clusters had been drawn at random without stratification. The only additional value we need is $\Sigma\, y_h^2 = \Sigma\,(y_{h1} + y_{h2})^2 = (10 + 8)^2 + (6 + 5)^2 \cdots (5 + 4)^2 = 2311$, the sum of the squares of the stratum totals. With it, the mean square *between clusters within strata* can be computed as

$$\frac{1263/10 - 2311/20}{20} = \frac{10.75}{20} = 0.5375.$$

This is $10s_a^2$; hence, $s_a^2 = 0.05375$. In 5.6*a* this was computed as

$$\sum (y_{h1} - y_{h2})^2 = 215 = Ha_cb^2s_a^2;$$

hence, $s_a^2 = 215/20 \times 2 \times 10^2 = 215/4000 = 0.05375$. This quantity suffices to compute the variance of the mean.

Source of Variation	Degrees of Freedom	Sum of Squares (SQ)	Mean Square	Estimate of
Between strata	19	$\frac{2311}{20} - \frac{185^2}{400} = 29.99$	1.5789	$10\frac{197}{199}S_a^2 + 20S_h^2$
Between clusters within strata	20	$\frac{1263}{10} - \frac{2311}{20} = 10.75$	0.5375	$10S_a^2$
Within clusters	360	$185 - \frac{1263}{10} = 58.70$	0.1631	S_b^2
Total	399	$185 - \frac{185^2}{400} = 99.44$	0.2492	

We can compute the following estimates:

$$S_b^2 = 0.1631 \text{ and } \sigma_b^2 = \frac{9}{10}\,0.1631 = 0.1468$$

$$S_a^2 = 0.05375 \text{ and } \sigma_a^2 = \frac{198}{199}\,0.05375 = 0.05348$$

$$S_h^2 = 0.05234 \text{ and } \sigma_h^2 = 0.04972.$$

The variance of clusters in the whole population is $\sigma_h^2 + \sigma_a^2 = 0.04972 + 0.05348 = 0.1032$. Note that the "naive estimate" obtained in 5.6*a* was $S_a^2 = 0.1045$, a negligible difference. Similarly, the variance of the elements in the population, estimated as $\sigma_h^2 + \sigma_a^2 + \sigma_b^2 = 0.04972 + 0.05348 + 0.1468 = 0.2500$ is close to the "naive estimate" 0.2492.

	Cluster Sampling	Element Sampling	Clusters / Elements
Stratified	$B\sigma_a^2 = 0.5348$	$\sigma_a^2 + \sigma_b^2 = 0.2003$	$\frac{0.5348}{0.2003} = 2.67$
Unstratified	$B(\sigma_h^2 + \sigma_a^2) = 1.032$	$\sigma^2 = \sigma_h^2 + \sigma_a^2 + \sigma_b^2 = 0.2500$	$\frac{1.032}{0.2500} = 4.13$
Stratified / Unstratified	$\frac{0.5348}{1.032} = 0.519$	$\frac{0.2003}{0.2500} = 0.801$	$\frac{0.5348}{0.2500} = 2.14$

The gain from proportionate stratification is $1 - 0.519 = 48$ percent for cluster sampling, but only 20 percent for element sampling. The increase in variance due to clustering is 4.13 for unstratified sampling, but only 2.67 for stratified sampling.

The overall "design effect" of the stratified clustered sample compared to a simple random sample is $B\sigma_a^2/\sigma^2 = 2.14$. In Example 5.5*a* we computed easily, without the analysis of variance, the simple estimate $(s_a^2/a)/(s^2/n) = 2.15$; that approximation is surely satisfactory. The design effect yields the synthetic roh of $(2.14 - 1)/(10 - 1) = 0.127$, representing the homogeneity within stratified clusters.

The natural homogeneity can be measured by (5.6.12) as

$$\text{rho} = [(\sigma_a^2 + \sigma_h^2) - \sigma_b^2/(B - 1)]/\sigma^2 = (0.1032 - 0.0163)/0.2500 = 0.348,$$

where $(\sigma_a^2 + \sigma_h^2)$ is the cluster variance including the between strata component. On the other hand, within strata the element variance is 0.2003; hence, the rho $= (0.0535 - 0.0163)/0.2003 = 0.186$; this is the homogeneity of clusters within strata.

The increase of the variance due to selecting clusters is $[1 + \text{rho}(B - 1)]$. For unstratified sampling, this is $1 + 0.348(9) = 4.13$; and for proportionate sampling, $1 + 0.186(9) = 2.67$.

PROBLEMS

5.1.

5 1 2 7 3 6 3 0 2 10

6 7 9 4 1 2 3 4 1 1 .

(*a*) These y_α values represent the number of elements that possess an attribute in subsamples of $b = 10$ elements each. Assume that the $a = 20$ clusters were selected at random with replacement, and the elements with epsem without replacement. Compute the element mean $\bar{y}$ and its standard error. The sampling fraction is $f = ab/AB = n/N = 200/200{,}000 = 1/1000$. Say in a few words what difference it would make if the clusters were selected without replacement. Estimate the variance you would expect if the sample had $a = 40$ clusters of $b = 10$ each.

(*b*) Since the mean is a proportion p, you easily estimate the variance of the simple random sample of the same size $n = 200$. Compare with the variance in (*a*), discuss the difference between them, and estimate *roh*.

(*c*) Estimate now the variance you would expect from $a = 40$ clusters, if 5 elements were subselected at random from each subsample of 10 elements.

(*d*) Suppose now that all you know is that these 20 values are the result of 20 replications, each selected with the sampling fraction 1/20,000. Estimate the population aggregate Y, and the standard error of this estimate. Check this against the results in (*a*).

5.2. Repeat the questions of 5.1, supposing that the y represent 10 paired selections from that many strata.

5.3. Repeat the questions of 5.1 for this sample of 20 selections, each representing subsamples of $b = 20$ elements with $f = 1/20{,}000$.

12	1	10	12	5	13	6	1	13	18
9	13	10	6	6	4	4	10	1	2.

5.4. Repeat the questions of 5.1 on the data for 5.3, supposing that these represent 10 paired selections of $b = 20$ each.

5.5.* (*a*) Estimate approximately the components S_a^2 and S_b^2 of the element variance in 5.1. (*b*) Estimate the variance of a mean of 78 random clusters.

5.6.* Repeat the questions of 5.5 with the data of 5.3.

5.7. An Army Corps comprises 400 infantry companies of about 100 enlisted men each. A sample of 10 companies was selected at random, and all men completed questionnaires. The *percentages* of "yes" answers to a question in the 10 companies were: 25, 33, 12, 32, 17, 24, 26, 23, 37, 21. (*a*) Estimate the proportion p of enlisted men in the Army Corps who hold this attitude, and the standard error of p. (*b*) Estimate the variance that a simple random sample of $n = 1000$ would have. Compute the ratio of the actual variance to the simple random variance; also estimate roh, the intraclass correlation. (*c*) Estimate the variance of a sample of 40 companies. (*d*) Estimate the variance of a sample of $n = 1000$ men, based on subsamples $b = 25$ men from 40 companies.

5.8. In 5.7, estimate S_a^2 and S^2, then $S_b^2 = (S^2 - S_a^2)B/(B - 1)$. Use these to answer (*c*) and (*d*) in 5.7.

5.9.* In 5.1, estimate the components of the variances S_a^2 and S_b^2, supposing that $a/A \doteq 0$, $S_u^2 + S_b^2 = S^2, S_b^2 = S^2(1 - \text{roh})$, and $S^2 \text{roh} = S_a^2 - S_b^2/B = S_u^2$. Use the components to answer 5.1(*c*).

5.10.* In 5.3, estimate the components of the variance S_a^2 and S_b^2 and use them to answer 5.3(*c*).

---

* Components refer to clusters created by subsampling design.

5.11. Compare the following three samples, each based on $n = 3600$, drawn from the same population of $N = 1{,}800{,}000$ elements. (1) A simple random sample of $n = 3600$ had $\bar{y} = 513$ and se $(\bar{y}) = 3.3$. (2) A random sample of $a = 180$ from $A = 90{,}000$ clusters, each containing $B = 20$ elements, had $\bar{y} = 524$ and se $(\bar{y}) = 10.1$. (3) Random subsamples of $b = 4$ from the $B = 20$ elements of $a = 900$ clusters, selected at random from the same population of $A = 90{,}000$ clusters, had $\bar{y} = 509$ with se $(\bar{y}) = 5.2$.

Note that the estimated standard error for the second sample is about three times as large as for the first sample. This indicates that [Answer true (T) or false (F)]:

(*a*) The intraclass correlation of the 90,000 clusters is greater than zero.

(*b*) All of the elements within each cluster are equal ($Y_{ij} = Y_{ik}$ for all j and k).

(*c*) The standard error of the second design can be reduced to that of the first by a large increase in the number of sample clusters (a) in the second design.

(*d*) Elements tend to resemble other elements in the same cluster.

(*e*) If the sample of the first design were cut to 1200 elements, it would have about the same standard error as the second design.

(*f*) Any sample selected by the first design would have a mean closer to the population mean of the 1,800,000 elements than any mean selected by the second design.

5.12. Note in 5.11 that the estimated standard error of the third design is markedly smaller than that of the second design. Answer true (T) or false (F).

(*g*) Although its effect is different on the two designs, the true (population) value of the coefficient of intraclass correlation is the same for both designs.

(*h*) That sample (3) has a smaller standard error than sample (2) is not unusual, since sample (3) contains more clusters and the same number of elements.

(*i*) If $a = 720$ clusters were drawn by the second design, its standard error would become about equal to that of the third sample.

(*j*) If the third design were altered to $b = 2$ instead of 4, and $a = 1800$ instead of 900, the standard error of the mean would remain unchanged.

5.13. Try now (either in your mind or on paper) to superimpose graphs representing *roughly* the *sampling distributions* of the three designs in 5.11. Answer true (T) or false (F).

(*k*) The means of the three sampling distributions coincide exactly.

(*l*) The standard deviation of curve (1) is the smallest and of curve (2) the largest.

(*m*) The three sampling distributions will be roughly normal if the population distribution is normal; but if this is skewed, the three sampling distributions will be just as skewed.

5.14. A sample of compact area segments contains about four farms each. In a second sample two farms were selected at random from each segment. These half segments proved quite efficient, since the reduction by half in number of farms increased the variance by only about 20 percent. (*a*) Estimate roh, S_u^2 and S_b^2, if $S^2 = S_u^2 + S_b^2 = 1$. (*b*) Make similar estimates for another variable for which the increase in variance was 50 percent. (Note: $S_u^2 = S_a^2 - S_b^2/B$.)

5.15. A country is divided into a set of administrative areas; you have been asked if they can be used as compact sampling units, which—if selected—are to be completely included in the sample. (*a*) What do you need to know about the areas to decide if they will serve well as the only sampling units? List the desired features, preferably with some imaginary numbers, or other specific details. (*b*) For two or three features describe some supposed inadequacies and corresponding changes that you would recommend in the sample design.

6
Unequal Clusters

6.1 PROBLEMS OF UNEQUAL CLUSTERS

In most clustered samples, we must deal with clusters of unequal sizes. Natural clusters of both human and nonhuman populations contain unequal numbers of elements. Dwellings in blocks, households in villages, passengers in airplanes, and students in classes are all of unequal sizes.

Moreover, even a sample designed and initiated with equal clusters may often end up with unequal clusters. Inequality may exist in the *actual sizes* of clusters that were designed with equal *measures of sizes*. Second, nonresponses also introduce inequalities into the final cluster results. Third, members of subclasses that cut across the clusters are distributed unequally in the clusters.

The population of $N = X$ elements is grouped in A clusters, with $N_\alpha = X_\alpha$ elements in the (typical) αth cluster, so that $\sum_\alpha^A X_\alpha = X = N$. Then $Y_{\alpha\beta}$ is the value of the variable Y_i of the βth element in the αth cluster; and the population of Y_i values may be represented in:

$$\begin{aligned}
X_1\bar{Y}_1 &= Y_1 = Y_{11} + Y_{12} + Y_{13} + \cdots + Y_{1\beta} + \cdots + Y_{1X_1} \\
X_2\bar{Y}_2 &= Y_2 = Y_{21} + \cdots + Y_{2X_2} \\
&\cdot \\
&\cdot \\
&\cdot \\
X_\alpha\bar{Y}_\alpha &= Y_\alpha = Y_{\alpha 1} + Y_{\alpha 2} + Y_{\alpha 3} + \cdots + Y_{\alpha\beta} + \cdots + Y_{\alpha X_\alpha} \\
&\cdot \\
&\cdot \\
&\cdot \\
X_A\bar{Y}_A &= Y_A = Y_{A1} + \cdots + Y_{A\beta} + \cdots + Y_{AX_A}.
\end{aligned} \qquad (6.1.1)$$

The total Y_i variable in the population is

$$Y = \sum_{\alpha}^{A} Y_\alpha = \sum_{\alpha}^{A} \sum_{\beta}^{X_\alpha} Y_{\alpha\beta} = \sum_{i}^{N} Y_i. \tag{6.1.2}$$

The population mean per element is

$$\bar{Y} = R = \frac{Y}{X} = \frac{1}{X} \sum_{\alpha}^{A} \sum_{\beta}^{X_\alpha} Y_{\alpha\beta} = \frac{1}{X} \sum_{\alpha}^{A} Y_\alpha = \frac{\sum_{\alpha}^{A} Y_\alpha}{\sum_{\alpha}^{A} X_\alpha}. \tag{6.1.3}$$

The population mean can be denoted variously as $Y/N = \bar{Y} = R = Y/X$. The corresponding sample values are $y/n = \bar{y} = r = y/x$. The complexity of the notation is annoying and merits apologies, especially the use of x for counting elements: X for N in the population, and x for n in the sample. This usage occurs in the sampling literature for unequal clusters and for ratio estimates generally; Yates [1960], Cochran [1963], and Sukhatme [1954] use y/x; Deming [1960] and Hansen, Hurwitz, and Madow [1953] use x/y.

This notation effectively serves several purposes. Using x instead of n for the sample size stresses the fact that the variance of y/x is not simply var $(y)/x^2$, but a more complex statistic. The notation has the flexibility required for the diverse uses of the ratio mean, and for dealing with unequal cluster sizes. For merely *counting the elements in clusters*, the variable $X_{\alpha\beta}$ is 1 for every element in the population, and X_α represents the number of elements in the cluster. But the formulas for the ordinary mean and variance of the sample also serve other important aims. First, the $X_{\alpha\beta} = 1$ may count members of any *subclass* of the population, vanishing for nonmembers; then X_α denotes the number of subclass members in the cluster. Second, the $X_{\alpha\beta}$ need not be restricted to counting variables; rather, *it can denote the values of some auxiliary variable introduced to improve the estimate;* then X_α denotes the cluster total in the auxiliary variable. For example, $(y/x)X$ may be an improved estimate of the population aggregate Y (11.8). If $X_{\alpha\beta}$ is not a counting variable, then we should use N_α instead of X_α as subscripts and limits of summations in (6.1.1) to (6.1.3).

Selecting unequal clusters creates several problems. First, the size of the sample is not fixed; it becomes a random variable, depending on the chance selection of larger or smaller clusters. This augments the uncertainties in planning the cost and variance of the sample. Second, the ratio mean, though typically a good and practical estimate, is not an unbiased estimate of the population mean. Third, practical variance formulas are not unbiased estimates of the true variances, but they too are good approximations in well designed samples. Fourth, the variance formulas appear complicated, although this problem is considerably eased with the development of simpler computing forms.

These problems are discussed later in detail. Generally, we find it neither practical nor necessary to control completely the sample size x down to a prefixed constant. However, a reasonable control over the variability of x is both practical and desirable, and we shall see that all these problems can be diminished by controlling the coefficient of variation of x. This coefficient can be reduced by selecting a large number of clusters. For a fixed number of clusters it can be reduced with several selection techniques: stratification according to size, dividing and combining natural clusters to create more equal artificial clusters, subsampling with probabilities proportional to cluster size. These techniques are described in 7.1.

Paired selection is the most practical method for selecting clusters, and the impatient reader may turn directly to 6.4B.

6.2 RANDOM SELECTION OF UNEQUAL CLUSTERS, EPSEM SUBSAMPLING

For selecting complete primary units without subsampling, we can select at random without replacement a from the A clusters in the population. This can be regarded as the special case of subsampling when the subsampling fraction is unity. The sampling fraction $f = a/A$ is then also the selection probability of any of the N population elements.

For selecting samples in two or more stages a large variety of alternative designs is possible. We introduce some reasonable restrictions to keep the presentation and formulas simple. The restrictions admit many practical designs; they are similar to those introduced for equal clusters in 5.3.

1. The first stage consists of selecting, at random, a from the A primary units. The selection probability here is the sampling fraction $f_a = a/A$. If the selection is *without* replacement, the simple variance formula involves an approximation; this generally results in an overestimate that is negligible for large A (see 5.6.7). If the selection is *with* replacement, some primary units may be selected twice (seldom more). Within such units two separate selections are made without replacement, and each selection constitutes a separate subsample (PS). These ideas are extended in 6.4 to stratified random selection of PSU's.

2. Within each selected primary unit we subsample with the same probability, denoted by the sampling fraction f_b. The selection probability of any of the N population elements is f_b, conditional on the initial

probability f_a. Hence, we have an epsem sample with $f = f_a \cdot f_b$, the fixed probability for all elements.

3. The variance computations are based on the *a primary selections* (PS). Each of these consists of x_α elements selected from the total of x_α^* elements in the αth primary unit. The subsample size x_α equals $f_b \cdot x_\alpha^*$, either exactly or approximately. The lack of exactitude results from the chance fluctuations of the subsampling procedures.

4. In equal sized clusters, a constant subsampling rate f_b also yields equal subsamples b. In unequal clusters, one of these must be sacrificed. When the primary units are selected with equal probability f_a, selecting a fixed size subsample would result in the unequal combined probabilities $f_a \cdot b/x_\alpha^*$. That is, elements would be selected with probabilities inversely proportional to the size of the primary unit. This would destroy the epsem nature of the selection, and would be neither practical nor efficient in most situations.

We emphasize maintaining the fixed f_b, allowing the subsample sizes x_α to vary. The elements are selected with the probability f_b and without replacement. A large variety of subsampling procedures is possible, and this flexibility allows us to choose and develop efficient procedures for specific situations.

(*a*) Select systematically with the interval $F_b = 1/f_b$, after a random start from 1 to F_b, drawn separately for each primary unit. This can yield the gains of proportionate sampling, if it takes advantage of the ordering present in the cluster or introduced into it. Clerical instructions are simple and especially suited to field situations where the cluster total x_α^* is not known beforehand. For example, the enumerator can be sent to list dwellings in sample blocks, and then to apply the interval F_b to the lists. Note that this systematic procedure can be applied either to a list of elements or to clusters of elements. For example, in sample blocks we can select either from lists of dwellings or of segments.

(*b*) Compute $x_\alpha' = f_b x_\alpha^*$, the expected sample size. If this is a fraction, some scheme, probably randomized, must be introduced to obtain an integral x_α. Then select x_α elements at random. Similar procedures can be designed for clusters of elements, such as segments in blocks. Note that x_α^* must be known beforehand, and the instructions become more difficult than for systematic selection.

(*c*) Divide each primary unit into F_b secondary units, and select one at random. Or divide into $2F_b$ units, and select two at random, to permit computing the within PSU variance. Or divide into kF_b units, and select k at random. By introducing heterogeneity into the units, the variance may be reduced.

(*d*) Select the subsample in two or more stages, using any combination of probabilities whose product is f_b. Multistage sampling is developed in Chapter 10.

(*e*) Different procedures may be specified to suit different types of PSU's.

The sample mean $\bar{y} = y/n = y/x$ is here denoted by

$$r = \frac{y}{x} = \frac{1}{x}\sum_{\alpha}^{a} y_\alpha = \frac{\sum y_\alpha}{\sum x_\alpha}, \tag{6.2.1}$$

where $x = \sum_\alpha x_\alpha$, is the "size" of the sample. Ordinarily this is merely n, the count of elements. Here x_α is the size of the αth primary selection. Similarly, $y = \sum_\alpha y_\alpha$ is the sample total for the Y_i variable; $y_\alpha = \sum_\beta y_{\alpha\beta}$ is its total for the αth primary selection, and $y_{\alpha\beta}$ is the value of the βth element in it.

Both y and x are random variables. Based on epsem selections, y is an unbiased estimate of fY and x of fX (2.8*C*). However, since x is not fixed, but a random variable, r is not an unbiased estimate of R; it is a *ratio estimate* (6.5). The ratio mean may also be regarded as the weighted mean of the a selection means:

$$r = \frac{y}{x} = \frac{1}{a}\sum_{\alpha}^{a} \frac{x_\alpha}{x/a}\,\bar{y}_\alpha, \quad \text{where} \quad \bar{y}_\alpha = \frac{y_\alpha}{x_\alpha}. \tag{6.2.1'}$$

The selection means are weighted with their relative sizes, $x_\alpha/(x/a)$, where x/a is the average size. When the selections have a constant size $n/a = b$, the above mean reduces to the mean of equal clusters. The ratio mean estimates the population mean, which is a similar weighted mean:

$$\bar{Y} = \frac{Y}{X} = \frac{1}{A}\sum_{\alpha}^{A} \frac{X_\alpha}{(X/A)}\,\bar{Y}_\alpha, \quad \text{where} \quad \bar{Y}_\alpha = \frac{Y_\alpha}{X_\alpha}.$$

All the formulas in this chapter are equally valid for sampling *complete clusters;* but then we employ y_α^* and x_α^*, the cluster totals, instead of the estimates y_α and x_α. We omit the (*) when this causes no confusion.

Subsampling with the constant rate $f_b = 1/F_b$ signifies that $F_b y_\alpha$ estimates y_α^*, and $F_b x_\alpha$ estimates x_α^*. Moreover, with the constant overall expansion factor $F = 1/f$, we note also that Fy and Fx are unbiased estimates of Y and X (2.8*C*). If so desired, the primary variates y_α and x_α may incorporate any constant factor, such that Fky and Fkx are unbiased estimates of Y and X.

Unequal sampling rates may also be introduced, if compensated with proper weighting. The weighted values of $y_\alpha = \sum_\beta w_{\alpha\beta} y_{\alpha\beta}$ and $x_\alpha = \sum_\beta w_{\alpha\beta} x_{\alpha\beta}$ must be such that for some constant F_b' we have $F_b' y_\alpha$ and $F_b' x_\alpha$ as unbiased estimates of y_α^* and x_α^*. With unequal weights, the simple value of $(1\text{-}f)$ in the variance computation is lost, but often this is a negligible quantity. Again, we must have some overall expansion factor F', such that Fy and Fx (also $F_a' y_\alpha$ and $F_a' x_\alpha$) be unbiased estimates of Y and X.

The X_i variable may represent the elements of the entire population or of a subclass. Or it may represent an auxiliary variable, so that $y_{\alpha\beta}$ and $x_{\alpha\beta}$ are two observed variables on each sample element. The same formulas for ratio means serve our needs in all these situations.

6.3 VARIANCES FOR RANDOM UNEQUAL CLUSTERS

The variance for the ratio mean $r = y/x$ is

$$\text{var}(r) = \frac{1}{x^2}[\text{var}(y) + r^2 \text{var}(x) - 2r \text{cov}(y, x)]. \qquad (6.3.1)$$

This general expression for the variance of the ratio of two random variables is derived in(6.6A).There we also find some details and cautions about the kind of approximation that the formula (6.3.1) represents. A simple, valuable rule is: do not permit a large coefficient of variation of x. As a rule of thumb, be sure se $(x)/x < 0.20$ before using (6.3.1)—see 7.1.

The general expression above must be made specific by developing formulas for the three variance terms in it. For a sample of a simple random selections, with $y = \Sigma y_\alpha$ and $x = \Sigma x_\alpha$, we have from (2.8.17):

$$\text{var}(y) = (1\text{-}f)as_y^2 = (1\text{-}f)\frac{a}{a-1}\left(\sum y_\alpha^2 - \frac{y^2}{a}\right),$$

and similar expressions for var (x) and cov (y, x). Hence, for r based on a random selections,

$$\begin{aligned} \text{var}(r) &= \frac{1-f}{x^2} a[s_y^2 + r^2 s_x^2 - 2rs_{yx}] \qquad (6.3.2) \\ &= \frac{1-f}{x^2}\frac{a}{a-1}\left[\left(\sum y_\alpha^2 - \frac{y^2}{a}\right) + r^2\left(\sum x_\alpha^2 - \frac{x^2}{a}\right) \right. \\ &\qquad \left. - 2r\left(\sum y_\alpha x_\alpha - \frac{yx}{a}\right)\right]. \end{aligned}$$

The second part of the three terms sum to zero ($y^2 + r^2x^2 - 2ryx = 0$ because $r = y/x$), and we get here (but not in stratified formulas)

$$\text{var}(r) = \frac{1-f}{x^2}\frac{a}{a-1}\left[\sum y_\alpha^{\,2} + r^2 \sum x_\alpha^{\,2} - 2r \sum y_\alpha x_\alpha\right]. \qquad (6.3.3)$$

With desk machines we can conveniently compute all three cumulated products by using the form

$$\sum^{a} (10^k y_\alpha + x_\alpha)^2 = 10^{2k} \sum y_\alpha^{\,2} + 10^k\, 2 \sum y_\alpha x_\alpha + \sum x_\alpha^{\,2}.$$

The multipliers, 10^{2k} and 10^k, separate the three sums on the machine dial.

The above formulas for the variance also yield its three components separately, and knowing the components may help. That knowledge is sacrificed by the z formulas below. But these have the advantages, especially in complex computations, of simplicity and amenability to easy checking.

Define the auxiliary variate

$$z_\alpha = y_\alpha - rx_\alpha.$$

Since $\Sigma z_\alpha^{\,2} = \Sigma (y_\alpha^{\,2} + r^2 x_\alpha^{\,2} - 2ry_\alpha x_\alpha)$, by substituting in (6.3.3) we get its equivalent:

$$\text{var}(r) = \frac{1-f}{x^2}\frac{a}{a-1}\sum z_\alpha^{\,2} = \frac{1-f}{x^2}\frac{a}{a-1}\sum (y_\alpha - rx_\alpha)^2. \qquad (6.3.4)$$

This needs simply the sum of squares of the z_α; these are easy to check against gross errors, because they sum to zero:

$$z = \sum z_\alpha = \sum (y_\alpha - rx_\alpha) = 0.$$

Remark 6.3.1 This simplifies formula (6.3.4), because $\Sigma z_\alpha^{\,2} = \Sigma\, z_\alpha^{\,2} - z^2/a$. A notation that we can use later is

$$s_z^{\,2} = (s_y^{\,2} + r^2 s_x^{\,2} - 2rs_{yx}) = \frac{1}{a-1}\left(\sum z_\alpha^{\,2} - \frac{z^2}{a}\right) = \frac{1}{a-1}\sum z_\alpha^{\,2}$$

$$= \frac{1}{a-1}\sum (y_\alpha - rx_\alpha)^2. \qquad (6.3.5)$$

This denotes the variance of the z variate for a single selection. With this unit, the variance of r based on a random selections is

$$\text{var}(r) = \frac{(1-f)}{x^2} a s_z^{\,2}. \qquad (6.3.5')$$

Of course, all these forms of var (r) are equal to each other, and one more will be enlightening. The unit variances $s_z^{\,2}$, $s_y^{\,2}$, $s_x^{\,2}$, and s_{yx} are all variances for the totals of primary selections. Thus $s_z^{\,2}$ corresponds to $s_y^{\,2} = b^2 s_a^{\,2}$ for equal clusters, where $b = n/a$ is the cluster size. For computing the variances of unequal clusters, the variates z_α take the place of the variates y_α for equal clusters.

We obtain the variance corresponding to s_a^2 in

$$s_r^2 = \frac{1}{(x/a)^2} s_z^2 = \frac{1}{a-1} \sum \frac{z_\alpha^2}{(x/a)^2} = \frac{1}{a-1} \sum \left[\frac{1}{x/a} (y_\alpha - r x_\alpha) \right]^2. \qquad (6.3.6)$$

Then

$$\text{var}(r) = \frac{1-f}{a} s_r^2 = \frac{1-f}{a(a-1)} \sum_\alpha^a \left[\frac{x_\alpha}{x/a} (\bar{y}_\alpha - r) \right]^2, \qquad (6.3.6')$$

where $\bar{y}_\alpha = y_\alpha / x_\alpha$. The variance involves deviations of the $\bar{y}_\alpha$ from the sample mean, as with equal clusters (5.3.2). For unequal clusters the deviations are weighted with $x_\alpha/(x/a)$, the relative size of the cluster.

If the cluster sizes vary only slightly, we may use the selection means $\bar{y}_\alpha$ in the unweighted formula (5.3.2) as an approximation. For example, slight variations in cluster sizes due to nonresponse could be tolerated in the approximation. But the cluster totals y_α should not be used directly in that formula, because that would tend to overestimate the variance.

Example 6.3a In Example 2.2*a*, a simple random sample of 20 blocks was selected; the blocks served both as elements and as sampling units. We now take the same selection of blocks, but consider them as *clusters* of dwellings, which are the elements defined for the present problem. We want to estimate the proportion of renter-occupied dwellings. The blocks constitute clusters of unequal sizes, serving as sampling units. The symbols for blocks change with the definitions; the number of blocks is $a = 20$ in the sample, and $A = 270$ in the population; the sampling fraction is $f = a/A = 20/270 = 0.074$. Despite the change of units, we can utilize many computations of Problem 2.2*a* for our present problem.

The proportion of renter-occupied dwellings is estimated by

$$r = \frac{\sum y_\alpha}{\sum x_\alpha} = \frac{y}{x} = \frac{255}{435} = 0.5862.$$

We can compute the variance by any of three procedures:

1. Although (6.3.2) is not the most convenient, it has heuristic value in expressing var (r) in terms of the unit variances:

$$\begin{aligned} \text{var}(r) &= \frac{0.926}{435^2} 20[278.62 + 0.5862^2 \times 672.51 - 0.5862 \times 839.13] \\ &= \frac{0.926}{9461}[278.62 + 231.09 - 491.90] = \frac{0.926}{9461}(17.81) \\ &= 17.43 \times 10^{-4}. \end{aligned}$$

2. Formula (6.3.3) fails to obtain the separate variance terms, but it is more direct:

$$\begin{aligned} \text{var}(r) &= \frac{0.926}{435^2} \frac{20}{19}[8545 + 0.5862^2 \times 22{,}239 - 0.5862 \times 27{,}036] \\ &= 0.05151 \times 10^{-4}[8545 + 7642 - 15{,}849] = 0.05151(338) \times 10^{-4} \\ &= 17.41 \times 10^{-4}. \end{aligned}$$

3. Formula (6.3.4) uses the auxiliary variates $z_\alpha = y_\alpha - rx_\alpha$, serviceable in many cluster designs. The 20 values of z_α are

0.069, −0.274, −5.548, 12.152, 2.248, 5.872, −3.446, −0.586, −0.586, −2.344, 0.006, −4.166, −2.930, −1.172, −0.344, −5.772, −0.720, 2.210, −0.586, 6.006.

Thus, $\text{var}(r) = \dfrac{0.926}{435^2}\dfrac{20}{19}[338.2] = 17.42 \times 10^{-4}$.

The variance is 17.42×10^{-4} for all three procedures, within rounding errors. The standard error of $r = 0.586$ is

$$\text{se}(r) = \sqrt{\text{var}(r)} = \sqrt{17.42 \times 10^{-4}} = 4.17 \times 10^{-2} = 4.17 \text{ percent.}$$

Note that the coefficient of variation of the sample size x can be easily computed as

$$\frac{\sqrt{\text{var}(x)}}{x} = \frac{\sqrt{.926(20)(672.51)}}{435} = \frac{111.60}{435} = 0.257$$

To design other samples it is worth noting that the variance of the mean of a single block, s_r^2, may be computed from (6.3.6) as

$$\text{var}(r) = \frac{(1-f)s_r^2}{a} = \frac{0.926s_r^2}{20} = 17.42 \times 10^{-4}.$$

Hence the block variance is estimated to be

$$s_r^2 = \frac{20(17.42 \times 10^{-4})}{0.926} = 0.03765.$$

To obtain an estimate for the variance of a simple random sample of the same number of elements, $n = 435$, we can compute $(1-f)s^2/n$ directly from the sample. Since the mean is a proportion ($p = 0.586$), this is easy:

$$\frac{(1-f)p(1-p)}{n-1} = \frac{0.926 \times 0.586 \times 0.414}{434} = 5.176 \times 10^{-4}.$$

Due to the homogeneity of rented dwellings in blocks, the design effect is $17.42/5.176 = 3.37$. With a simple random formula we underestimate the standard error by the factor $\sqrt{3.37} = 1.84$. For example, instead of $P = 0.95$ for 2 standard errors, we would operate with $2/1.84 = 1.09$ standard errors; hence, at the $P = 0.72$ level.

6.4 STRATIFIED SAMPLING OF UNEQUAL CLUSTERS

6.4A Any Number (a_h) of Clusters

Clusters are selected generally with stratification, because this may considerably reduce the variance. Stratified sampling and subsampling of unequal clusters can be accomplished in many ways. To trace in detail the varieties and complexities of many designs would be too burdensome.

However, a relatively simple treatment will cover many actual situations. This is an extension of the simple ratio mean to stratified samples, and it can be denoted as

$$r = \frac{y}{x} = \frac{1}{x}\sum_h^H \sum_\alpha^{a_h} y_{h\alpha} = \frac{\sum_h^H \sum_\alpha^{a_h} y_{h\alpha}}{\sum_h^H \sum_\alpha^{a_h} x_{h\alpha}} = \frac{\sum_h^H y_h}{\sum_h^H x_h}. \tag{6.4.1}$$

The mean is the ratio of the variates y and x. Each of these is the simple sum of the stratum totals y_h and x_h. These in turn are simple totals of the primary variates $y_{h\alpha}$ and $x_{h\alpha}$. The $y_{h\alpha}$ and $x_{h\alpha}$ are sample values for a_h random selections from the hth stratum. We impose restrictions on the design, similar to those described for the unstratified random sample (6.2). Within these restrictions we still possess wide flexibility for choosing suitable sampling designs. *Paired selection*, with $a_h = 2$, is the simplest and most common method for selecting clusters (6.4*B*).

The simple additions for y and x indicate that the $y_{h\alpha}$ and $x_{h\alpha}$ have been properly weighted. First, if there is an overall uniform sampling fraction $f = 1/F$, the sample is self-weighting; then $Fa_hy_{h\alpha}$ and $Fa_hx_{h\alpha}$ are unbiased estimates of Y_h and X_h. The sampling rates, f_{ha} for the primary stage and f_{hb} for subsampling, may be varied as long as the uniform sampling rate, $f = f_{ha}f_{hb}$, is maintained. Moreover, the methods for applying f_{hb} may vary between strata; and so may the number a_h of primary selections.

Second, if the sampling rates f_h are uniform within the strata, they may differ between strata, provided that $y_{h\alpha}$ and $x_{h\alpha}$ are properly weighted to compensate for unequal sampling fractions. This means that there is a uniform expansion factor F' such that $F'a_hy_{h\alpha}$ and $F'a_hx_{h\alpha}$ are unbiased estimates of Y_h and X_h. Then all our formulas for the ratio mean and its variance still hold.

Third, we may need to depart from uniform sampling fractions within strata. This is permissible, provided the elements are properly weighted; that is, for a uniform expansion factor F' we have $F'a_hy_{h\alpha}$ and $F'a_hx_{h\alpha}$ as unbiased estimates of Y_h and X_h. Lacking a uniform f_h, the factor $(1 - f_h)$ is not determined; but this factor is often negligible.

In addition to determining the sampling fractions, we have a wide choice among selection methods, within the restrictions of *simple replicated subsampling*, developed in 5.2 and 6.2. The last stage is assumed to be without replacement, as commonly required in practice. The primary units may be selected with replacement, to suit the variance formulas. But often we may select them without replacement and accept a mild overestimate of the variance. That is, by selecting the primary units with replacement, the actual variance will probably be made a little lower than the formula shows.

For the variance of the ratio mean $r = y/x$ we begin with the general form (6.3.1). Note that $y = \Sigma y_h$ and $x = \Sigma x_h$, where the H values are the results of independent selections in the separate strata. Hence (2.8.7), $\text{var}(y) = \text{var}(\Sigma y_h) = \Sigma \text{var}(y_h)$, with similar expressions for var (x) and for cov (y, x). Thus the variance of the stratified ratio mean becomes

$$\text{var}(r) = \frac{1}{x^2}[\sum \text{var}(y_h) + r^2 \sum \text{var}(x_h) - 2r \sum \text{cov}(y_h, x_h)]$$

$$= \frac{1}{x^2}[\sum d^2 y_h + r^2 \sum d^2 x_h - 2r \sum dy_h\, dx_h]. \tag{6.4.2}$$

These d symbols (2.8.29) denote concisely the variance terms of the sample sums within strata, each based on a_h random primary selections. According to the development of the preceding sections, they can be computed as

$$d^2 y_h = \frac{(1 - f_h)}{a_h - 1}\left(a_h \sum_{\alpha}^{a_h} y_{h\alpha}^2 - y_h^2\right), \qquad d^2 x_h = \frac{(1 - f_h)}{a_h - 1}\left(a_h \sum_{\alpha}^{a_h} x_{h\alpha}^2 - x_h^2\right),$$

$$dy_h\, dx_h = \frac{(1 - f_h)}{a_h - 1}\left(a_h \sum_{\alpha}^{a_h} y_{h\alpha} x_{h\alpha} - y_h x_h\right). \tag{6.4.3}$$

The three terms can be combined into one with the auxiliary variates $z_{h\alpha} = y_{h\alpha} - r x_{h\alpha}$. By simple expansion it can be seen that the equivalent of (6.4.2) is

$$\text{var}(r) = \frac{1}{x^2}\sum_h^H d^2 z_h, \quad \text{where } d^2 z_h = \frac{(1 - f_h)}{a_h - 1}\left[a_h \sum_{\alpha}^{a_h} z_{h\alpha}^2 - z_h^2\right]. \tag{6.4.4}$$

Here $z_h = \sum_{\alpha}^{a_h} z_{h\alpha} = \sum_{\alpha}^{a_h} y_{h\alpha} - r \sum_{\alpha}^{a_h} x_{h\alpha}$, and the equality of these forms provides a check on the computations. Another check is also available, because $z = \sum_h^H z_h = \sum_h^H \sum_{\alpha}^{a_h} z_{h\alpha} = 0$.

Simplified computing formulas may be developed for several special cases that occur frequently. Suppose that for all strata we have both the same sampling fraction ($f_h = f$) and the same number of primary selections ($a_h = a_c$); this is *proportionate sampling with equal allocation*. Then,

$$\sum_h^H dy_h\, dx_h = \frac{1 - f}{a_c - 1}\left[a_c \sum_h^H \sum_{\alpha}^{a_c} y_{h\alpha} x_{h\alpha} - \sum_h^H y_h x_h\right], \tag{6.4.5}$$

with $\Sigma\, d^2y_h$ and $\Sigma\, d^2x_h$ taking similar forms. Or we may have the equivalent of these three forms by computing

$$\sum_h^H d^2z_h = \frac{1-f}{a_c-1}\left[a_c \sum_h^H \sum_\alpha^{a_c} z_{h\alpha}^2 - \sum_h^H z_h^2\right]. \qquad (6.4.6)$$

Remark 6.4.I We define several measures of variances for unequal clusters. They are analogous to those for equal clusters in Remark 5.3.V, but based on the average subsample size $x/a = n/a$ instead of the constant subsample size b. The *unit variance* is $s_r^2 = a[\text{var}\,(r)/(1\text{–}f)]$.

The *element variance* is $v_v^2 = n[\text{var}\,(r)/(1\text{–}f)] = (n/a)s_r^2$. The ratio of the element variance to the variance of an *srs* of the same n is the *design effect*:

$$\text{deff} = \frac{\text{var}\,(r)}{(1\text{–}f)s^2/n} = \frac{s_r^2/a}{s^2/n} = \frac{v_v^2}{s^2}\,.$$

We may conceive of a sample total z with the variance $\text{var}\,(z) = n^2\,\text{var}\,(r) = (1\text{–}f)as_z^2$. Now note that $s_z^2 = (n/a)^2s_r^2 = (n/a)v_v^2$.

For random selection of clusters $s_z^2 = \Sigma\, z_\alpha^2/(a-1)$, where $z_\alpha = (y_\alpha - rx_\alpha)$, and s_r^2 is defined accordingly (6.3.5 and 6.3.6). For stratified clusters $s_z^2 \doteq \Sigma\, d^2z_h/a$ (6.4.4); $s_r^2 = a\,\Sigma\, d^2z_h/n^2$ and $v_v^2 = \Sigma\, d^2z_h/n$. This assumes that the factors $(1\text{–}f)$ are either negligible or properly included.

6.4B Paired Selection of Clusters

Paired selection frequently provides the best design for selecting clusters. It is *efficient* because it permits $H = a/2$ strata which, for fixed a, is more strata than if $a_h > 2$ were allowed. Paired selection is also *simple.* A specific application of these methods to area sampling of blocks and dwellings appears in 9.3.

1. *For sampling complete clusters,* form strata of $2F$ clusters each, and select 2 clusters at random from each stratum. The uniform sampling rate will be $2/2F = f_a = f$. To designate the sample clusters, draw two different random numbers from 1 to $2F$. For example, suppose a school population has 3000 classes of twelfth grade students. The classes vary in size from about 12 to 35, with an average of 24. A sample of *about* $n = 1440$ can be selected in $a = 60$ classes. Hence, form 30 strata of 100 classes each, and select at random 2 classes from every stratum. The sampling fraction is $2/100 = 1/50$.

2. *For subsampling with a uniform rate* f_b, two uniform selection rates will yield a uniform overall fraction: $f_af_b = f = 1/F = (1/F_a)(1/F_b)$. Form strata of size $2F_a$ each, and select 2 clusters at random, for a primary

selection fraction of $2/2F_a = f_a$. Then subsample with a uniform fraction f_b in each selected cluster. For example, the 3000 classes can be sorted into 150 strata of 20 classes each. From each stratum select 2 classes, hence $f_a = 2/20 = 1/10$. Then subsample with 1/5 in each selected class, for an overall rate of $1/10 \times 1/5 = 1/50$. The sample consists of 300 classes with an average of 4.8 students each.

3. *Control of subsample size* can be introduced if the cluster sizes vary greatly. Section 7.1 presents three simple methods for obtaining roughly equal subsamples, while maintaining a uniform overall sampling fraction. Sections 7.2 and 7.3 discuss formal selection with probabilities proportional to size.

4. *Unequal sampling fractions* can be introduced for different strata: $f_{ha} \cdot f_{hb} = f_h$. This may seem desirable because of difference in cost or variance factors between the strata. However, a uniform overall sampling fraction can be maintained, while introducing balanced changes into the two selection rates: $f_{ha} \cdot f_{hb} = f$.

After determining the number a of primary selections, establish $a/2$ strata, each of size $2F_a$. *For simple replicated subsampling*, select two clusters with two random numbers from 1 to $2F_a$. The same cluster can be selected twice, in which case two subsamples must be drawn from it. Subsampling with f_b is ordinarily done without replacement of the elements. It can utilize any convenient method; systematic selection with the interval F_b is often the easiest.

However, in practice we may prefer to select *two clusters without replacement*, thus insuring two different selections from each stratum. Typically this reduces the actual variance slightly, although the simple formulas below will not show this difference (see 5.3*B*, 7.4*A*, and 8.6). These two methods can be considered safe and simple.

Further reduction of the variance may be obtained by dividing each of the $a/2$ strata in half, establishing a strata, each of size F_a. Then *select a single cluster from each half stratum of size* F_a, with a random draw from 1 to F_a. Then subsample with f_b as before. The sample mean receives the benefits of the extra stratification, but the variance, computed with the simple formulas below, will not only fail to show the reduction but tends to show an increase (7.4*D*).

Systematic selection of clusters can be obtained with the interval F_a applied after a random start from 1 to F_a. This simulates a single selection from each stratum, if the ordering within the zones of size F_a can be assumed to be random. When it is safe, this method can be both efficient and extremely simple (see 4.1–4.2, 7.4*C*, and 9.3). For example, we can select every fiftieth classroom from a list of classrooms; the list may be ordered according to some stratifying variables.

The stratified ratio mean (6.4.1) can be written for paired selections as

$$r = \frac{y}{x} = \frac{\sum y_h}{\sum x_h} = \frac{\sum (y_{h1} + y_{h2})}{\sum (x_{h1} + x_{h2})}, \qquad (6.4.1')$$

where y_{h1} and y_{h2} represent the totals for the Y_i variable from the two primary selections in the hth stratum. Similarly, x_{h1} and x_{h2} represent the paired results for the base variable X_i, typically the number of elements in the two primary selections of the hth stratum. The variance is the special case for $a_h = 2$ of the general formula for stratified clusters (6.4.3). It becomes simply

$$\text{var}(r) = \frac{1-f}{x^2}\left[\sum D^2y_h + r^2 \sum D^2x_h - 2r \sum Dy_h\, Dx_h\right], \qquad (6.4.7)$$

where $Dy_h = (y_{h1} - y_{h2})$ and $Dx_h = (x_{h1} - x_{h2})$.

This may be expressed even more simply as

$$\text{var}(r) = \frac{1-f}{x^2} \sum D^2z_h. \qquad (6.4.8)$$

The computing units Dz_h may be obtained in two ways:

$$\begin{aligned} Dz_h &= (z_{h1} - z_{h2}) = (y_{h1} - rx_{h1}) - (y_{h2} - rx_{h2}) \\ &= Dy_h - r\,Dx_h = (y_{h1} - y_{h2}) - r(x_{h1} - x_{h2}). \end{aligned} \qquad (6.4.8')$$

If the sampling fractions vary between strata, the factors $(1-f_h)$ can be brought inside the summation, so that $\text{var}(r) = \Sigma\,(1-f_h)\,D^2z_h/x^2$. The procedures of computation are given in detail in 6.5*B*.

The formula can be derived readily from knowing (Remark 4.3.*I*) that: $\text{var}(y_h) = \text{var}(y_{h1} + y_{h2}) = (1-f_h)(y_{h1} - y_{h2})^2$. Then from (2.8.7) we have $\text{var}(\Sigma y_h) = \Sigma\, \text{var}(y_h)$. Substituting such terms in (6.6.3) yields (6.4.7).

For single clusters selected from half-strata, the pairs of clusters can be contrasted as if they were selected at random from the same full, *collapsed* stratum.

6.5 STATISTICS FOR STRATIFIED UNEQUAL CLUSTERS

6.5*A* *The Difference of Two Ratio Means*

The difference $(r - r') = (y/x - y'/x')$ of two ratio means can be denoted similarly to the single mean in (6.4.1):

$$r - r' = \frac{y}{x} - \frac{y'}{x'} = \frac{\sum_h^H y_h}{\sum_h^H x_h} - \frac{\sum_h^H y_h'}{\sum_h^H x_h'} = \frac{\sum_h^H \sum_\alpha^{a_h} y_{h\alpha}}{\sum_h^H \sum_\alpha^{a_h} x_{h\alpha}} - \frac{\sum_h^H \sum_\alpha^{a_h} y_{h\alpha}'}{\sum_h^H \sum_\alpha^{a_h} x_{h\alpha}'}. \qquad (6.5.1)$$

The variance for this difference can be expressed in the general terms for any two random variables:

$$\text{var}(r - r') = \text{var}(r) + \text{var}(r') - 2\,\text{cov}(r, r'). \qquad (6.5.2)$$

When the two means are obtained from samples independent of each other, the covariance term vanishes, and we simply add the two variances. This would also be true for two domains of the same survey, each domain based on a distinct and separate set of strata and primary selections. The comparisons from a national survey of two regions generally fits this pattern.

Another relatively simple case is the comparison of two variables based on the same sample, a problem treated in 12.10. For example, the prevalence of a disease (or attitude) can be observed as $r = y/x$ and $r' = y'/x'$ in two periods based on the same sample x. As another example, we could compare in the same sample x the proportions of sufferers from two strains of some disease. The difference in proportions would be

$$(r - r') = (y/x - y'/x') = (y - y')/x,$$

since the same $x = x'$ denotes the sample in which y and y' numbers were observed for the two strains. The two variates y and y' may be overlapping; that is, some people could have both strains, some only one, and some neither. The base x could be either the entire sample, or only some subclass. The variance of this difference can be computed as an ordinary ratio, using the primary variates $(y_{h\alpha} - y'_{h\alpha})$ in each primary selection. But it can also be computed with the general formulas developed below.

The difference of two means $(r - r')$ is computed frequently from the results of multistage samples. In most cases the two bases x and x' are neither independent nor identical. The lack of independence, hence the covariance term, results from using the same sampling units, especially the same primary units, to obtain the two compared means. Two types of comparisons are common: (1) *The comparison of two subclasses of the same sample* is a common way of presenting survey results; for example, the prevalence of a disease is compared between two age classes. (2) *Comparison of sample results from two periodic surveys* is often based on the same primary units, although the elements may be newly selected; for example, the prevalence of a disease (or attitude) may be compared for two yearly surveys. This may be done either for the entire sample or for a subclass, such as an age class. Typically, as in the above examples, both kinds of comparisons concern the same characteristic. It would be unusual to compare one disease in one period or age class with a different disease in another period or age class. But if this were desired, the variance formula would cover that too, because nothing need connect the two variables y and y'.

From (6.6.9) we obtain the general form of the covariance of r and r'; then, with the same methods as for (6.4.2), we obtain the second line:

$$\text{cov}(r, r') = \frac{1}{xx'}[\text{cov}(y, y') + rr' \text{cov}(x, x') - r \text{cov}(y', x) - r' \text{cov}(y, x')] \quad (6.6.9)$$

$$= \frac{1}{xx'}[\sum dy_h\, dy_h' + rr' \sum dx_h\, dx_h' - r \sum dy_h'\, dx_h - r' \sum dy_h\, dx_h']. \quad (6.5.3)$$

When computing the variance of a comparison, six other terms for the two variances must be added to the four terms for the covariance. That is difficult with pencil and paper, or even with desk computers. Here it is particularly helpful to introduce the computing form $z_{h\alpha} = y_{h\alpha} - rx_{h\alpha}$. Then (6.5.2) becomes

$$\text{var}(r - r') = \frac{1}{x^2}\sum d^2z_h + \frac{1}{x'^2}\sum d^2z_h' - \frac{2}{xx'}\sum dz_h\, dz_h'. \quad (6.5.4)$$

Now we merely need the different forms that $\Sigma\, dz_h\, dz_h'$ can take with different designs. These forms are similar to the computing forms for $\Sigma\, d^2z_h$ developed in (6.4.4) to (6.4.10). They can also serve as models for the four components of (6.5.3) if we need to compute them. The general form, corresponding to (6.4.4), is

$$\sum_h^H dz_h\, dz_h' = \sum_h^H \frac{(1-f_h)}{a_h - 1}\left[a_h \sum_\alpha^{a_h} z_{h\alpha} z'_{h\alpha} - z_h z_h'\right]. \quad (6.5.5)$$

For the special case of equal allocation $a_h = a_c$ and with proportional sampling $f_h = f$, the equivalent of (6.4.6) clearly follows. For paired selections, instead of (6.4.8) we have

$$\sum dz_h\, dz_h' = \sum (1 - f_h)\, Dz_h\, Dz_h' = \sum (1 - f_h)(z_{h1} - z_{h2})(z'_{h1} - z'_{h2}) \quad (6.5.6)$$

$$= \sum (1 - f_h)(Dy_h - r\, Dx_h)(Dy_h' - r'\, Dx_h'). \quad (6.5.6')$$

The equivalent for systematic sampling follows similarly in 6.5*C*.

6.5B Simple Variances for Complex Samples

With this presumptious title I mean to emphasize the efficacy of the computational methods presented here as a numerical example. Time and again, we shall refer back to this model for computing variances for a number of important designs.

(1)	(2)	(3)	(4)	(5)	(6)	(7)	(8)	(9)	(10)	(11)	(12)	(13)	(14)
h	α	y	x	y'	x'	Dy	Dx	Dy'	Dx'	z	z'	Dz	Dz'
1	1	11	19	5	12	2	3	−1 *	3	−0.102	−0.923	0.247 *	−2.481
1	2	9	16	6	9					−0.349	1.558		
2	1	8	10	1	1	2	0	−6	−12	2.157	0.506	2.000 *	−0.077
2	2	6	10	7	13					0.157	0.583		
3	1	6	13	2	10	−9	−7	−7	0	−1.596	−2.936	−4.910	−7.000
3	2	15	20	9	10					3.314	4.064		
4	1	13	23	7	12	8	15	3	6	−0.439	1.077	−0.765 *	0.039
4	2	5	8	4	6					0.326	1.038		
5	1	9	13	3	5	5	7	2 *	−1	1.404	0.532	0.910	2.494
5	2	4	6	1	6					0.494	−1.962		
6	1	4	10	6	13	−3	−3	4	9	−1.843	−0.417	−1.247	−0.443
6	2	7	13	2	4					−0.596	0.026		
7	1	5	7	3	6	−2	−3	0	2	0.910	0.038	−0.247	−0.988
7	2	7	10	3	4					1.157	1.026		
8	1	4	8	0	1	−1	−4	−4	−9	−0.674	−0.494	1.338	0.442
8	2	5	12	4	10					−2.012	−0.936		
9	1	9	12	2	13	0	−3	1	12	1.988	−4.417	1.752 *	−4.923
9	2	9	15	1	1					0.236	0.506		
10	1	9	20	10	18	5	10	9	16	−2.686	1.115	−0.843 *	1.102
10	2	4	10	1	2					−1.843	0.013		
Σu_1 or Σu^+		78	135	39	91	+22	+35	+19	+48	−0.881	−5.919	+6.247	+4.077
Σu_2 or Σu^-		71	120	38	65	−15	−20	−18	−22	+0.884	+5.916	−8.012	−15.912
Σu		149	255	77	156	+7	+15	+1	+26	+0.003	−0.003	−1.765	−11.835

TABLE 6.5.I For Computing Variances of Two Ratio Means and Their Differences

Paired selections are denoted with $\alpha = 1$ and 2, in 10 strata numbered with h from 1 to 10. The computations refer to two ratio means: $r = y/x$ and $r' = y'/x'$. These designate two subclasses based on $x = 255$ and $x' = 156$ sample elements All the information is contained in columns 3 through 6. The ratio means and their difference are found readily from the sums of these columns, and they are

$$r - r' = \frac{y}{x} - \frac{y'}{x'} = \frac{149}{255} - \frac{77}{156} = 0.5843 - 0.4936 = 0.0907.$$

Three alternative procedures for computing the two variances and the covariance are presented. The same procedures are simpler if the variance of only a single mean is needed. Then we need only half of the data columns and only one of the two sets of summed squares and not their cross products.

Procedure 6.5.I. With pencil and paper or desk computers, this is, I think, the easiest for computing variances of paired selections.

(*a*) Compute the quantities $Dy_h = (y_{h1} - y_{h2})$, $Dx_h = (x_{h1} - x_{h2})$, $Dy_h' = (y'_{h1} - y'_{h2})$, and $Dx_h' = (x'_{h1} - x'_{h2})$. These are in columns 7 to 10.

(*b*) Compute $Dz = (Dy - r\ Dx)$ and $Dz' = (Dy' - r'\ Dx')$ in columns 13 and 14.

(*c*) Compute $\Sigma\ D^2z$, $\Sigma\ D^2z'$, and $\Sigma\ Dz\ Dz'$.

(*d*) Compute the two variances and the covariance as

$$\begin{aligned} \text{var}(r - r') &= \frac{\sum D^2z}{x^2} + \frac{\sum D^2z'}{x'^2} - \frac{2\sum Dz\,Dz'}{xx'} \\ &= \frac{36.7724}{65{,}025} + \frac{88.208}{24{,}336} - \frac{55.3484}{39{,}780} \\ &= (5.66 + 36.25 - 13.91) \times 10^{-4} = 28.00 \times 10^{-4}. \end{aligned}$$

Computational Aspects. (*a*) Check that $\Sigma\ Dy = +22 - 15 = 7$ (obtained from the sums of positive and negative terms of Dy) equals the difference ($\Sigma\ y_{h1} - \Sigma\ y_{h2} = 78 - 71 = 7$) of first terms minus second terms. Similarly check the other three columns. (*b*) Check that $\Sigma\ Dz = +6.247 - 8.012 = -1.765$ equals $\Sigma\ Dy - r\,\Sigma\ Dx = 7 - (0.5843)15 = 7 - 8.765 = -1.765$. Also check that $\Sigma\ Dz' = +4.077 - 15.912 = -11.835$ equals $\Sigma\ Dy' - r'\,\Sigma\ Dx' = +1 - (0.4936)26 = -11.834$, within rounding errors. (*c*) Observe that the sum of products $\Sigma\ Dz\ Dz'$ requires separate computation of like signs (++ and −−) for the positive products and of unlike signs (+− and −+) for negative products. The unlike signs are marked

(*) in column 14. (*d*) Columns 11 and 12, each with $2H = 20$ computations of z values are not needed; but only the $H = 10$ computations of z values in columns 13 and 14.

Procedure 6.5.II. This is almost as simple as procedure *I* for paired selections, and for more selections per stratum it is probably the simplest.

(*a*) Compute the values $z_{h\alpha} = y_{h\alpha} - rx_{h\alpha}$ and $z'_{h\alpha} = y'_{h\alpha} - r'x'_{h\alpha}$ in columns 11 and 12.

(*b*) Compute $Dz_h = (z_{h1} - z_{h2})$ and $Dz_h' = (z'_{h1} - z'_{h2})$ in columns 13 and 14.

(*c*) Compute $\Sigma\, D^2z$, $\Sigma\, D^2z'$ and $\Sigma\, Dz\, Dz'$.

(*d*) Compute the two variances and the covariances as

$$\text{var}(r - r') = \frac{\sum D^2z}{x^2} + \frac{\sum D^2z'}{x'^2} - \frac{2\sum Dz\, Dz'}{xx'}$$

$$= \frac{36.7724}{65{,}025} + \frac{88.208}{24{,}336} - \frac{55.3484}{39{,}780}$$

$$= (5.66 + 36.25 - 13.91) \times 10^{-4} = 28.00 \times 10^{-4}.$$

Computational Aspects (*a*) Check that $\Sigma\, z_h = 0$ and $\Sigma\, z_h' = 0$. This can also be done by adding the first selections and second selections separately. Thus $\Sigma\, z_{h1} + \Sigma\, z_{h2} = -0.881 + 0.884 = 0.003 \doteq 0$. Similarly, $\Sigma\, z'_{h1} + \Sigma\, z'_{h2} = -5.919 + 5.916 = -0.003 \doteq 0$. (*b*) Check that $\Sigma\, Dz = -1.765 = \Sigma\, z_{h1} - \Sigma\, z_{h2} = -0.881 - 0.884$, the sum of first selections minus the sum of second selections. Also $\Sigma\, Dz' = -11.835 = \Sigma\, z'_{h1} - \Sigma\, z'_{h2} = -5.919 - 5.916 = -11.835$. (*c*) Note that columns 7 to 10 are not needed.

Procedure 6.5.III. This is the most direct procedure and well-suited to electronic computers. It yields separately all ten components for the variance of the difference. But it is difficult on desk computers.

(*a*) Compute Dy, Dx, Dy', Dx' in columns 7 to 10.

(*b*) Compute the four sums of squares and the six sums of products.

(*c*) Compute:

$$\text{var}(r) = \frac{1}{x^2}\left[\sum D^2y + r^2 \sum D^2x - 2r \sum Dy\, Dx\right]$$

$$= \frac{1}{255^2}\left[217 + 0.584^2(475) - 0.584(586)\right]$$

$$= 5.66 \times 10^{-4},$$

$$\text{var}(r') = \frac{1}{x'^2}[\sum D^2y' + r'^2 \sum D^2x' - 2r' \sum Dy'\, Dx']$$

$$= \frac{1}{156^2}[213 + 0.494^2(756) - 0.494(626)]$$

$$= 36.24 \times 10^{-4},$$

$$2\,\text{cov}(r, r') = \frac{2}{xx'}[\sum Dy\, Dy' + rr' \sum Dx\, Dx' - r \sum Dy'\, Dx - r' \sum Dy\, Dx']$$

$$= \frac{1}{255 \times 156}[240 + 0.584(0.494)(438) - 0.584(392) - 0.494(166)]$$

$$= 13.91 \times 10^{-4}.$$

Computational Aspects. (*a*) Check that $\Sigma\, Dy_h = \Sigma\, y_{h1} - \Sigma\, y_{h2}$, the sum of first selections minus the sum of second selections. Similarly, for $\Sigma\, Dx$, $\Sigma\, Dy'$, $\Sigma\, Dx'$. (*b*) There are six sums of products, and each must be computed in two parts for like signs (++ and −−) that yield positive products and for unlike signs (+− and −+) that yield negative products; the latter are marked with (*). (*c*) Columns 11 to 14 are not needed.

From the components of the variance we can obtain readily the coefficients of variation of the sample size x:

$$\frac{\sqrt{\text{var}(x)}}{x} = \frac{\sqrt{475}}{255} = 0.086, \quad \text{and} \quad \frac{\sqrt{\text{var}(x')}}{x'} = \frac{\sqrt{756}}{156} = 0.176.$$

These are high, reflecting the small size of the sample.

The factor $(1 - f)$ was omitted because f was negligible a common occurrence. If f is constant for all strata, the factor $(1 - f)$ can be easily applied at the end.

The *design effect* (8.2) is often worth computing both for what it reveals about the design, and as a check on gross computational errors. Because the means are proportions, the simple random variance can be easily computed as $p(1 - p)/(n - 1) = y(n - y)/n^2(n - 1)$. We compute these as

$$\frac{149 \times 106}{255^2 \times 254} = 9.53 \times 10^{-4} \text{ for } r, \quad \text{and} \quad \frac{77 \times 79}{156^2 \times 155} = 16.02 \text{ for } r'.$$

The design effects are then computed as

$$\frac{5.66}{9.53} = 0.59 \text{ for } r; \quad \frac{36.25}{16.02} = 2.26 \text{ for } r'; \quad \frac{28.00}{9.53 + 16.02} = 1.10 \text{ for } (r - r').$$

The first of these is surprisingly low, the second somewhat high, and the divergence between them somewhat dubious. We need not worry much about this case, because the wild numbers may be entirely due to sampling fluctuations in estimating the variances. These were based on only 10 differences; hence, they are subject to large sampling errors (8.6*D*).

6.5C *Variances for Systematic Primary Selections*

When the a primary selections have been selected systematically, the variance may usually be computed with the paired selection formulas, pairing them in the order of selection: first with the second, third with the fourth, etc. If the number of selections is odd, we may drop one of the selections (first or last or a random one); the variance of the sample total would then be corrected with the factor $a/(a-1)$. Or we could use a selection twice—for example, 1–2, 2–3, 4–5, 5–6, etc.; the correction factor would be $a/(a+1)$.

However, we may also utilize variance formulas specific for systematic sampling which utilize all the $(a-1)$ possible differences instead of only $a/2$ pairs. Hence, precision is greater with systematic than with paired formulas (8.6*D*). Both methods involve the approximations of collapsed strata (8.6*B*) and tend to overestimate somewhat the true variance. The systematic formulas are developed in 8.6*B*; there the symbol G is used instead of a to denote the number of primary selections.

The computing formulas resemble closely those for paired differences developed in 6.4*B*, and detailed in 6.5*B*. Because the systematic formulas utilize $(a-1)$ instead of $a/2$ differences, a factor of $a/2(a-1)$ is introduced into the variance of sample totals. Thus the variance of r may be expressed as

$$\begin{aligned}\text{var}(r) &= \frac{1-f}{x^2}\frac{a}{2(a-1)}\sum_g^{a-1} D^2 z_g \\ &= \frac{1-f}{x^2}\frac{a}{2(a-1)}\left[\sum_g^{a-1} D^2 y_g + r^2\sum_g^{a-1} D^2 x_g - 2r\sum_g^{a-1} Dy_g\, Dx_g\right], \qquad (6.5.7)\end{aligned}$$

where $Dy_g = (y_g - y_{g+1})$, $Dx_g = (x_g - x_{g+1})$, and $Dz_g = (z_g - z_{g+1})$.

There are similar expressions for var (r'). The variance of the difference $(r - r')$ is

$$\begin{aligned}\text{var}(r - r') = \frac{1-f}{x^2}\frac{a}{2(a-1)}\sum_g^{a-1} D^2 z_g + \frac{1-f}{x'^2}\frac{a}{2(a-1)}\sum_g^{a-1} D^2 z_g{}' \\ - \frac{(1-f)a}{xx'(a-1)}\sum_g^{a-1} Dz_g\, Dz_g{}', \qquad (6.5.8)\end{aligned}$$

where $$\sum_g^{a-1} Dz_g\, Dz_g' = \sum_g^{a-1} Dy_g\, Dy_g' + rr' \sum_g^{a-1} Dx_g\, Dx_g' - r \sum_g^{a-1} Dy_g'\, Dx_g - r' \sum_g^{a-1} Dy_g\, Dx_g'.$$

The computational procedures are similar to those of 6.5*B*. Consider the 20 primary selections in Table 6.5.I drawn in a systematic order. Procedure I would involve computing 9 additional sextuplets of values for columns 7, 8, 9, 10, 13, and 14. Procedure II requires 9 additional pairs of Dz_g values for columns 13 and 14, computed from the z values of columns 11 and 12. After computing these we have

$$\begin{aligned}\text{var}\,(r - r') &= \frac{1}{255^2}\frac{20}{38}(97.150) + \frac{1}{156^2}\frac{20}{38}(128.072) - \frac{1}{255(156)}\frac{20}{38}(50.580)\\ &= (7.86 + 27.70 - 6.69) \times 10^{-4}\\ &= 28.87 \times 10^{-4}.\end{aligned}$$

Procedure III utilizes directly the Dy and Dx values of columns 7, 8, 9, and 10, for which 9 additional quadruplets are required. After adding these we may compute

$$\text{var}\,(r) = \frac{1}{255^2}\frac{20}{38}[267 + 0.584^2(635) - 0.584(662)] = 7.85 \times 10^{-4},$$

$$\text{var}\,(r') = \frac{1}{156^2}\frac{20}{38}[388 + 0.494^2(1194) - 0.494(1116)] = 27.70 \times 10^{-4},$$

$$2\,\text{cov}\,(r, r') = \frac{1}{255(156)}\frac{20}{38}[124 + 0.584(0.494)366 - 0.584(265) - 0.494(100)] = 6.73 \times 10^{-4}.$$

Thus $\text{var}\,(r - r') = (7.85 + 27.70 - 6.73) \times 10^{-4} = 28.82 \times 10^{-4}$.

6.5D Expansions with Ratio Estimates

To estimate the population total Y we can compute the *simple unbiased estimate* $\hat{Y}_w = Fy$, where y is the sample total and $F = 1/f$ is a uniform expansion factor. The problem is essentially unchanged if $\Sigma F_h y_h$ is computed with diverse expansion factors. The variance of Fy is simply F^2 var (y).

However, we may have an alternative in the *ratio estimate* of the aggregate with the *auxiliary variable X:*

$$\hat{Y}_r = Xr = \frac{X}{x}y = \frac{X}{Fx}Fy. \qquad (6.5.9)$$

The statistical aspects are similar for some measurement problems which appear different in their substantive aspects. Generally, the population mean $\bar{Y}$ is estimated with the simple mean $\bar{y}$ of the sample. But suppose for every sample element, we observe not only the survey variables y_j but also the auxiliary variables x_j. Also suppose that the population mean $\bar{X}$ is available *for the very same variable*, Then the *ratio estimate* of the mean may be computed with the auxiliary variable $\bar{X}$ as

$$\bar{Y}_r = \bar{X}r = \bar{X}\frac{y}{x} = \bar{X}\frac{\bar{y}}{\bar{x}} = \frac{\bar{X}}{\bar{x}}\bar{y} = \frac{X}{Fx}\bar{y}. \qquad (6.5.9')$$

The factor X/Fx is used to adjust the sample mean $\bar{y}$. Similarly in Xr the expansion factor X/Fx is used to adjust the simple estimate Fy. The variances are

$$\text{var}(Xr) = X^2 \text{var}(r) \qquad \text{and} \qquad \text{var}(\bar{X}r) = \bar{X}^2 \text{var}(r). \qquad (6.5.10)$$

A few examples can illustrate some uses of ratio estimates. From Surveys of Consumer Finances [1950], mean liquid assets per household were obtained; these can be expanded with the Census Bureau's count of households to provide estimates of aggregate liquid assets. From the same surveys, estimates of assets per income were also obtained, and these can also be expanded with income data from outside sources.

Suppose, for the children of a school system, health data are available, based on a brief screening examination. If a sample of children is examined thoroughly, the ratio (r) of the results of the thorough examination (y) to screening (x) can be used to "calibrate" the results X of the screening.

The value and applicability of ratio estimates depend on several factors. First, the gains of ratio expansions over simple means can be stated compactly with the coefficients of variation C_y and C_x (from 6.6.19):

$$\frac{\text{Var}(\hat{Y}_r)}{\text{Var}(Fy)} = \frac{C_y^2 + C_x^2 - 2\rho_{yx}C_yC_x}{C_y^2} = 1 + \left[\frac{C_x^2}{C_y^2} - 2\rho_{yx}\frac{C_x}{C_y}\right]. \qquad (6.5.11)$$

Gains occur when the bracketed quantity is negative. For this, ρ_{yx} must be positive and preferably large; it must be greater than $C_x/2C_y$. For large gains, we need C_x/C_y between 0.5 and 1.3, and ρ_{yx} greater than 0.6 or 0.7 [see Hansen, Hurwitz, and Madow, 4.19]. The strength of the correlation ρ_{yx} between y and x is the key to the gains of the ratio estimate.

Second, the gains are small if C_y is much greater than C_x. If the y is a variable based on a small subclass of a much greater sample x, then C_y may be much greater than C_x. Hence, small subclasses may benefit little from ratio estimates.

Third, ratio expansions employ the factor X/Fx. Therefore, the X in the population and the x in the sample should be based on essentially identical measurements. A considerable difference between the two could introduce a large bias into the expanded estimate. On the other hand, the y measurement can differ freely. For example, if y represents a current medical examination for a sample and X represents a health record of several years ago for the population, the x must also be taken from the same old health records as X was.

Fourth, a survey's nonresponses are often not drastically different from the responses. In those cases the expansion Xr may be a great deal safer than the expansion Fy (see 13.4*B*)

Fifth, if X is not available for the entire population, it may be worth obtaining an estimate X' from a large sample. This may be based on a separate and independent sample (11.8*B*). But in *double sampling* (12.2*A*), x is based on a subsample of the larger sample for X'.

Sixth, an extension of the method to several auxiliary variates has been proposed by Olkin [1958]. The estimate would have the form $\hat{Y}_r = y(W_1X_1/x_1 + W_2X_2/x_2 + W_3X_3/x_3 + \cdots)$, with the W's so designated as to provide a good estimate.

Seventh, the relative efficacies of ratio estimates and regression estimates are discussed in 12.3*B*.

Eighth, the comparison $Kr - K'r'$ of two ratio means has the variance $K^2 \text{ var}(r) + K'^2 \text{ var}(r') - 2KK' \text{ cov}(r, r')$.

Example 6.5a In 6.3*a*, the proportion $r = y/x$ of rented dwellings was estimated as 58.62 percent with a standard error of 4.17. Because we know $X = 6786$, the total number of dwellings in the same population, we can estimate

$$\hat{Y}_r = Xr = 6786 \times 0.5862 = 3978 \text{ total number of rented dwellings.}$$

Its standard error is $X \text{ se}(r) = 6786 \times 0.0417 = 283$.

On the other hand, the simple direct estimate and its standard error are obtained from merely knowing that a sample of $a = 20$ blocks from A yielded 255 rented dwellings: $Y' = Fy = \dfrac{Ay}{a} = \dfrac{270}{20} 255 = 3442$. Its standard error is

$$F \text{ se}(y) = \frac{A}{a} \sqrt{(1-f)as_y^2} = A\sqrt{(1-f)s_y^2/a}$$

$$= 270\sqrt{0.926 \times 278.62/20} = 270(3.59) = 969.$$

The ratio of standard errors is $969/283 = 3.42$, and its square is the ratio $\text{var}(Fy)/\text{var}(Xr) = 3.42^2 = 11.7$. This measures the "statistical efficiency" of the ratio estimate against the simple unbiased expansion. In 6.3*a*, we also found that a simple random sample with the same number of elements would have a variance smaller by the factor 3.37; hence, the combination of the two effects of selection and estimation could alter the variance by the factor $11.7 \times 3.37 = 39.5$.

These computations concern a specific example and are subject to sampling errors. But many situations exist where substantial gains can be made with auxiliary variables in the estimate.

6.5E Separate Ratio Estimates

The ratio mean $r = y/x$ is known as the *combined* ratio estimate. The *separate* ratio estimate takes the form

$$r_w = \sum_h^H W_h r_h = \sum_h^H W_h \frac{y_h}{x_h} = \sum_h^H W_h \frac{\sum_\alpha^{a_h} y_{h\alpha}}{\sum_\alpha^{a_h} x_{h\alpha}}. \tag{6.5.12}$$

The stratum weights are constants known from other sources, and are applied to ratio means r_h computed separately in each stratum. The basic formulas of stratification (3.2) applied to ratio means (6.3) mean that

$$\text{var}(r_w) = \sum W_h^2 \,\text{var}(r_h) = \sum W_h^2 \frac{(1-f_h)}{x_h^2}[d^2y_h + r_h^2 d^2x_h - 2r_h\, dy_h\, dx_h]$$

$$= \sum W_h^2 \frac{1-f_h}{x_h^2} d^2z_h, \text{ where } d^2z_h = \frac{a_h}{a_h-1}\sum_\alpha^{a_h}(y_{h\alpha} - r_h x_{h\alpha})^2. \tag{6.5.13}$$

If the r_h vary considerably between strata, and if the W_h are known, the separate estimate may have a much lower variance than the combined estimate.

Thus the variance decreases for designs with many strata, each with only small numbers a_h of primary selections. However, the bias of the ratio means r_h may be large compared to their variances; and if they are all in the same direction, they will sum to a large bias for all strata. Hence, the separate ratio estimate may have a large ratio of bias to standard error. Because of this, and because good values of W_h are often unavailable, we omit a detailed treatment of this estimate. See the treatment by Cochran [1963, 6.10–6.12] and by Hansen, Hurwitz, and Madow [1953, 5.4].

*6.6 RATIO MEANS

6.6A Variance of the Ratio Mean

The sample value $r = y/x$ estimates the population value $R = Y/X$. Usually it is both proper and convenient to assume that the sampling design ensures that the sample totals y and x are unbiased estimates of Y and X. Thus, if a uniform sampling fraction f is needed, it can be assumed to be incorporated in the Y and X. Strict unbiasedness is needed only in a few places which will be noted.

Deviations of sample values from population values are denoted as $\delta_y = (y - Y)$ and $\delta_x = (x - X)$. Then $V_y^2 = E(\delta_y^2)$ and $V_x^2 = E(\delta_x^2)$ and $V_{yx} = E(\delta_y \delta_x)$ are the expected mean squared deviations of sample from population values. These are the variances, Var (y), Var (x), and the covariance, Cov (y, x), when y and x are unbiased.

We need to assume that the absolute value X is so large that (1) the occurrence of x near zero is a negligible rarity; and that (2) the relative error $|\delta_x/X|$ is small enough to permit dropping all but the leading terms of the needed Taylor expansions. These are not unreasonable demands on most survey results; especially in the common basic situation when x denotes the sample size. Hence a reasonable control of the coefficient of variation $C_x = V_x/X$ must be maintained.

First, we note:

$$(r - R) = \frac{y}{x} - \frac{Rx}{x} = \frac{\delta_y + Y - R(\delta_x + X)}{x} = \frac{\delta_y - R\delta_x}{x}$$

$$= \frac{\delta_y - R\delta_x}{X} \cdot \left(1 + \frac{\delta_x}{X}\right)^{-1}. \qquad (6.6.1)$$

Now we want to derive an expression for the *mean square error of r;* that is, for $E(r - R)^2$. The numerator alone can be easily obtained, since

$$E(\delta_y - R\delta_x)^2 = E(\delta_y^2) + R^2E(\delta_x^2) - 2\,RE\,(\delta_y\,\delta_x) = V_y^2 + R^2V_x^2 - 2RV_{yx}.$$

The standard practice at this point is to use an approximation, as if the denominator were constant:

$$\text{Var}\,(r) = E(r - R)^2 \doteq \frac{1}{X^2}[V_y^2 + R^2V_x^2 - 2RV_{yx}]. \qquad (6.6.2)$$

We shall denote this expression V_r^2. The sample estimate is computed by substituting sample values throughout, r for R, var (y) for $V_y^2 =$ Var (y), and so on:

$$\text{var}\,(r) \doteq \frac{1}{x^2}[\text{var}\,(y) + r^2\,\text{var}\,(x) - 2r\,\text{cov}\,(y, x)]. \qquad (6.6.3)$$

The justification for obtaining (6.6.2) from the expected square of (6.6.1) is difficult and imperfect. It amounts to obtaining the variance for the variate $(y - Rx)/X$ instead of the variate $r = (y - Rx)/x$.

One approach is to disregard fluctuations in the ratio $(x/X)^{-2}$ in (6.6.1) [Cochran, 1963, 2.9]. Practically the same effect is reached by expanding $(1 + \delta_x/X)^{-2}$, then disregarding all but the first term [Hansen, Hurwitz, and Madow II, 4.11; Sukhatme, 1954, 4a.5]. The validity of the approximation in large samples can also be shown with "propagation of error" methods [Rao, 1952, 5e.1].

The approximation (6.6.2) for the mean square error depends on having a reasonably small coefficient of variation. As a rule of thumb, aim for $C_x < 0.1$, but $C_x < 0.2$ is tolerable. Following Sukhatme [1954, p. 151], using terms up to X^{-4} and assuming y and x normal, the second approximation for $E(r - R)^2$ comes to $V_r^2[1 + 3C_x^2 + 6 \text{ Bias}^2 (r)/V_r^2]$. Here the bias ratio (developed in the next subsection) accounts for less than one-ninth of the error of approximation in V_r^2. Cochran [1963, 6.4] notes that the possible value of the bracketed term is less than $(1 + 9C_x^2)$, and that values between $(1 + 3C_x^2)$ and $(1 + 6C_x^2)$ are most likely. Hence, $C_x < 0.1$ would likely bring the accuracy within 3 or 6 percent. Even $C_x < 0.2$ would make the variance correct within 12 to 24 percent and the standard error within 6 to 12 percent. Following Fieller's [1940] approach, Hansen, Hurwitz, and Madow [1953, II, 4.12] come to similar conclusions. The conclusions of Hajek [1958] are also reassuring.

6.6B Bias of the Ratio Mean

To investigate the bias $E(r - R)$ of the ratio mean r, note a relationship of the covariance of r and x:

$$\begin{aligned}\text{Cov}(r, x) &= E[(r - Er)(x - Ex)] = E(rx - xEr - rEx + ErEx)\\ &= Ey - ExEr - ErEx + ErEx\\ &= Y - XEr = -XE(r - R) = -X \text{ Bias}(r).\end{aligned}$$

On the last line we assumed that y and x are unbiased estimates of Y and X. Observe that $\text{Cov}(r, x) = \rho_{rx}\sigma_r\sigma_x$, where ρ_{rx} is the correlation of the total sample values of r and x, and σ_r and σ_x are their standard errors. Then

$$\text{Bias}(r) = -\frac{\rho_{rx}\sigma_r\sigma_x}{X} = -\rho_{rx}\sigma_r C_x, \text{ and } \frac{\text{Bias}(r)}{R} = -\rho_{rx}C_r C_x. \quad (6.6.4)$$

This relationship was noted by Goodman and Hartley [1958] and Hansen, Hurwitz, and Madow [1953, II, 4.16]. Now

$$\frac{\text{Bias}(r)}{\sigma_r} = -\rho_{rx}C_x, \quad \text{and} \quad \frac{|\text{Bias}(r)|}{\sigma_r} \leq C_x. \quad (6.6.5)$$

The *bias ratio* presents the magnitude of the bias in r as a ratio of its standard error. Its absolute magnitude cannot exceed the coefficient of variation of x, commonly the sample size. Often it is much less because the correlation ρ_{rx} is weak. When r and x are uncorrelated, the bias vanishes.

These meaningful results were obtained without approximations, but they fail to provide a convenient method of computing the bias. For that we turn to (6.6.1) again. If in the expansion of $(1 + \delta_x/X)^{-1}$ we retain only the first two terms $(1 - \delta_x/X)$, we obtain the approximation for Bias (r):

$$E(r - R) \doteq E\left[\left(\frac{\delta_y - R\delta_x}{X}\right)\left(1 - \frac{\delta_x}{X}\right)\right]$$

$$= \frac{E(\delta_y - R\delta_x)}{X} + \frac{RE(\delta_x^{\,2}) - E(\delta_y\delta_x)}{X^2}$$

$$= \frac{RV_x^{\,2} - V_{yx}}{X^2}. \tag{6.6.6}$$

The first term vanishes, assuming that y and x are unbiased. With this expression of the bias it can also be shown [Cochran, 1963, 6.5; Kish, 1962] that the bias ratio cannot exceed the coefficient of variation of x; that is, $|(RV_x^{\,2} - V_{yx})X^{-2}|/V_r \leq C_x$. If y and x are not unbiased, the term neglected in (6.6.6) becomes [Bias (y) − R Bias (x)]/X. If the relative sizes of these biases are known to be small enough, they can be neglected. The value of the bias can be estimated as

$$\text{bias}\ (r) \doteq [r\ \text{var}\ (x) - \text{cov}\ (y, x)]x^{-2}. \tag{6.6.7}$$

Rather generally in practical survey work the bias of the ratio mean is negligible. Its relative importance expressed by the bias ratio is necessarily less than C_x. Moreover, it will often be much less than that because ρ_{rx} is often small. These conjectures have been the subject of several investigations; one of these, containing a large variety of reassuring data, was published by me [Kish, 1962].

Several practical considerations require control of the sample size C_x. The validity of the variance approximation requires that C_x be kept under 0.2 and preferably under 0.1. The effect of the bias ratio B on the mean square error can be stated as MSE $= \sigma_r^{\,2}(1 + B^2)$. Hence, with C_x less than 0.1 or 0.2, the effect of B on the mean square error is generally well under 1 to 4 percent.

6.6C The Difference of Two Ratio Means

The variance of the difference $(r - r')$ can be expressed as

$$E[(r - r') - (R - R')]^2 = E[(r - R)^2 + (r' - R')^2 - 2(r - R)(r' - R')]$$
$$= \text{Var}\ (r) + \text{Var}\ (r') - 2\ \text{Cov}\ (r, r').$$

For two independent samples we compute merely the sum of the two variances, but for samples that are not independent we need the covariance.

This is evaluated similarly to the variance, from the product of (6.6.1) for the two variates, disregarding all but the first term in the expansion of $(1 + \delta_x/X)^{-1}$:

$$\text{Cov}(r, r') = E\left[\left(\frac{\delta_y - R\delta_x}{X}\right)\left(\frac{\delta_{y'} - R'\delta_{x'}}{X'}\right)\left(1 + \frac{\delta_x}{X}\right)^{-1}\left(1 + \frac{\delta_{x'}}{X'}\right)^{-1}\right]$$

$$\doteq \frac{1}{XX'} E[(\delta_y\delta_{y'} + RR'\delta_x\delta_{x'} - R\delta_{y'}\delta_x - R'\delta_y\delta_{x'})]$$

$$= \frac{1}{XX'} [V_{yy'} + RR'V_{xx'} - RV_{y'x} - R'V_{yx'}]. \quad (6.6.8)$$

As with the variance (6.6.2), this can be estimated by computing

$$\text{cov}(r, r') \doteq \frac{1}{xx'} [\text{cov}(y, y') + rr' \text{cov}(x, x') - r \text{cov}(y', x) - r' \text{cov}(yx')]. \quad (6.6.9)$$

The bias of the difference is equal to the difference in the biases:

$$E[(r - r') - (R - R')] = E(r - R) - E(r' - R') = \text{Bias}(r) - \text{Bias}(r') \quad (6.6.10)$$

The magnitude of the bias is more meaningful when measured relative to the standard error. The *bias ratio* is

$$\frac{\text{Bias}(r - r')}{\text{SE}(r - r')} = -\frac{\rho_{rx}C_x\sigma_r - \rho_{r'x'}C_{x'}\sigma_{r'}}{(\sigma_r^2 + \sigma_{r'}^2 - 2\sigma_{rr'})^{1/2}}. \quad (6.6.11)$$

When the biases are important in comparisons, the correlations should usually have the same, rather than the opposite direction. If we want to know the worst that can happen, we must look at their absolute values:

$$\frac{|\text{Bias}(r - r')|}{\text{SE}(r - r')} \leq \frac{|\rho_{rx}|\, C_x\sigma_r + |\rho_{r'x'}|\, C_{x'}\sigma_{r'}}{(\sigma_r^2 + \sigma_{r'}^2 - 2\rho_{rr'}\sigma_r\sigma_{r'})^{1/2}}. \quad (6.6.12)$$

The limit is simple if the two means have the same denominator, because $(r - r') = (y - y')/x$ can be treated as an ordinary ratio mean; then the bias ratio cannot be greater than C_x. For two independent samples the covariance vanishes and an upper limit can be found. But the comparison of two samples that are not independent raises serious problems, if the denominator can approach zero (with $\rho_{rr'}$ approaching ± 1) without the numerator diminishing similarly. No necessary limit is known for this case. But it has been shown that the *larger of* C_x *and* $C_{x'}$ should serve as the plausible and simple "limit" in most practical situations; and a large number of survey results buttress this conjecture [Kish, 1962].

In situations where reasonable doubt remains, (6.6.7) and (6.6.10) may be utilized to compute the approximation:

$$\text{bias}\,(r - r') \doteq \frac{r\,\text{var}\,(x) - \text{cov}\,(y, x)}{x^2} - \frac{r'\,\text{var}\,(x') - \text{cov}\,(y'x')}{x'^2}. \quad (6.6.13)$$

After dividing it by se $(r - r')$, the bias ratio is obtained, and its effect on the results can be assessed.

6.6D The Product yx and Relvariances

To estimate YX, the product of two random variables yx may be treated more easily with the methods used for the ratio mean:

$$yx - YX = (Y + \delta_y)(X + \delta_x) - YX = X\delta_y + Y\delta_x + \delta_y\delta_x. \quad (6.6.14)$$

The bias of the product becomes

$$\text{Bias}\,(yx) = E(yx - YX) = V_{yx} = \rho_{yx}\sigma_y\sigma_x, \quad (6.6.15)$$

assuming that y and x are unbiased. [Otherwise the term X Bias (y) + Y Bias (x) also exists.] The bias may be computed as cov (y, x). But it should become negligible in most cases. The ratio of the bias to the mean square error developed below can be shown to be less than $[C_xC_y/(C_x + C_y)]$.

The *mean square error* of the product is

$$\begin{aligned} V_p^2 &= E(yx - YX)^2 = E(X\delta_y + Y\delta_x + \delta_y\delta_x)^2 \\ &\doteq X^2E(\delta_y^2) + Y^2E(\delta_x^2) + 2XYE(\delta_y\delta_x) \\ &= X^2V_y^2 + Y^2V_x^2 + 2XYV_{yx}. \end{aligned} \quad (6.6.16)$$

The approximation introduced in the second line disregards the terms $(2XV_{y^2x} + 2YV_{yx^2} + V_{y^2x^2})$ which are of the third and fourth orders, compared to the second-order terms retained. These higher-order terms can be computed, but they should be negligible in most situations. The mean square error can be computed by substituting sample values:

$$\text{var}\,(yx) \doteq x^2\,\text{var}\,(y) + y^2\,\text{var}\,(x) + 2yx\,\text{cov}\,(y, x). \quad (6.6.17)$$

If we can also assume that y does not vanish, then formulas of this section can be expressed in terms of *relvariances* and *relcovariances*. With these, some symmetries appear more clearly. Thus the relative mean square errors are

$$\frac{\text{var}(yx)}{(yx)^2} \doteq \frac{\text{var}(y)}{y^2} + \frac{\text{var}(x)}{x^2} + \frac{2\,\text{cov}(y, x)}{yx}; \tag{6.6.18}$$

$$\frac{\text{var}(r)}{r^2} \doteq \frac{\text{var}(y)}{y^2} + \frac{\text{var}(x)}{x^2} - \frac{2\,\text{cov}(y, x)}{yx}; \tag{6.6.19}$$

$$\frac{\text{bias}(yx)}{yx} = \frac{\text{cov}(y, x)}{yx}; \tag{6.6.20}$$

$$\frac{\text{bias}(r)}{r} \doteq \frac{\text{var}(x)}{x^2} - \frac{\text{cov}(y, x)}{yx}. \tag{6.6.21}$$

The expressions are stated in sample values for estimating population values, which can be stated by substituting capital letters. Note that all terms in the relative mean square error are in relvariances. These are squares of the coefficients of variation, which should have small values in large samples. The bias terms also contain relvariances. The bias terms must be squared when compared to the relative mean square errors; hence, they enter with fourth powers of the coefficients of variation.

PROBLEMS

6.1. In Problem 2.1 you selected 20 blocks at random. Consider these as $a = 20$ random clusters, in which the total dwellings are elements, and rented dwellings are a dichotomous variable.

(*a*) Compute the proportion of rented dwellings $r = y/x$ and its variance $\text{var}(r) = (1-f)(s_y^2 + r^2 s_x^2 - 2rs_{yx})a/x^2$, utilizing the terms computed in 2.1.

(*b*) Compute also $\text{var}(r) = (1-f)a \Sigma z_\alpha^2/(a-1)x^2$, where $z_\alpha = y_\alpha - rx_\alpha$.

(*c*) Compute the variance of a simple random sample of the same size as $(1-f)r(1-r)/x$. Compute the ratio of the actual variance to the simple random variance; note how large the *design effect* is, due to the homogeneity of this variable within blocks.

6.2. (*a*) For the ratio mean r in 6.1, and with the ancillary information that $X = 6786$ is the total number of dwellings in the population, compute the ratio estimate Xr of Y, the number of rented dwellings in the population. Compute also $\text{se}(Xr) = X\,\text{se}(r)$.

(*b*) Compute the simple expansion estimate Fy of Y, where $F = 270/20$, and its standard error $\text{se}(Ay/a) = A\,\text{se}(y/a) = F\,\text{se}(y)$.

(*c*) Compute the ratio of the variances of the simple expansion estimate and the ratio estimate: $\text{var}(Ay/a)/\text{var}(Xr)$.

(*d*) The actual value of Y is 4573; is Xr or Fy closer to Y? Do you expect this to occur always, frequently, seldom, or never? Why?

6.3.

y_α	24	34	11	35	17	25	24	23	40	20	253
x_α	96	103	92	109	101	104	92	100	108	95	1000
$\bar{y}_\alpha$	0.25	0.33	0.12	0.32	0.17	0.24	0.26	0.23	0.37	0.21	0.25

The percentages of "yes" answers in 5.7 correspond to the means $\bar{y}_\alpha = y_\alpha / x_\alpha$ here. The overall mean is $r = y/x = 253/1000 = 0.25$. The numbers x_α of elements are seen to vary between the 10 randomly selected clusters. (*a*) Take this variation into account in computing the variance of $r = y/x$ as $\text{var}(r) = [\text{var}(y) + r^2\text{var}(x) - 2r\text{cov}(y,x)]/x^2$, where $\text{var}(y) = (1-f)[a \Sigma y_\alpha^2 - (\Sigma y_\alpha)^2]/(a - 1)$. (*b*) Note that since the x_α vary only slightly, the variance of r is close to that obtained with the equal cluster formula $\text{var}(y) = (1-f) \Sigma (\bar{y}_\alpha - \bar{y})^2/a(a - 1)$. That the two should nearly be equal can be guessed from the small variation of the weights $x_\alpha/(x/a)$ in

$$\text{var}(r) = \frac{1-f}{a(a-1)} \sum \left[\frac{x_\alpha}{x/a} (\bar{y}_\alpha - r) \right]^2 .$$

(*c*) Show this to be equal to the results in (*a*) either by computing it, or by proving it equal to $(1-f)a \Sigma (y_\alpha - rx_\alpha)^2/(a - 1)x^2$, which is equal to (*a*).

6.4. Problem 4.7 presented results y_γ for a dichotomous variable based on 10 replications. These may be regarded as 10 clusters of unequal sizes x_γ.
(*a*) Compute the sample proportion r and its standard error.
(*b*) Compute the ratio of actual variance to simple random variance. This *design effect* should be close to 1, because the "clusters" actually represented scattered elements.
(*c*) Knowing that $X = 5988$, compute the estimate Xr of Y and its standard error.
(*d*) Compare the variance of Xr with the variance of the simple expansion Fy. Since the x_γ's varied little, the two should not differ much.

6.5. The numbers in 5.3 actually represent the sample sizes x_{ha} and x_{hb} of 10 paired selections. The corresponding values y_{ha} and y_{hb} are given in 5.1. (*a*) Compute the ratio mean $r = y/x = \Sigma (y_{ha} + y_{hb})/\Sigma (x_{ha} + x_{hb})$; compute its variance and standard error. (*b*) The ratio mean represents a proportion; estimate the variance of a simple random sample of the same size. Compute the design effect: the ratio of actual to simple random variance. (*c*) Make *rough* estimates of the variance we would obtain by doubling the sampling fraction: first, for a design of 20 paired selections; second, for a design of 10 paired selections, with subsamples that are twice as large.

6.6.

x_j	80	81	117	85	12	342	32	190	852	78	55	638	100	532	132	11	318
y_j	117	287	250	244	0	1806	64	328	4813	89	53	1368	316	1035	256	9	471
w_j	63	149	143	89	0	602	40	169	3327	16	22	521	212	284	157	5	280

These 17 complete clusters were selected at random from a population of 780. It illustrates how bad this design can be when the clusters vary greatly in size; the size measures X_i should have been used for stratification or selection probabilities. (*a*) Compute the simple expansion aggregate Fy and its standard error. (*b*) Compute the ratio mean $r = y/x$ and its standard error. (*c*) Compute the ratio estimate Xr and its standard error, if $X = 220{,}000$ and compare with (*a*). (*d*) Compute bias $(r) = [r \text{ var } (x) - \text{cov } (y, x)]x^{-2}$. Compute the bias ratio, *bias* (r)/se (r), and show that it is less than $c_x = \text{se } (x)/x$. Discuss the effect of the large value of c_x on the sample.

6.7. (*a*) In Problem 6.6 compute the difference of the ratio means $(w/x - y/x)$ and its standard error. (*b*) Compute the difference in the simple expansions $Fw - Fy$ and its standard error. (*c*) Compute the ratio estimate $Xy/x - Xw/x$ of the difference in aggregates, and its standard error, and compare with (*b*). (*d*) Compute the bias of the difference, bias (y/x) − bias (w/x). Compute the ratio of this difference to the standard error of the difference; show that this ratio is less than c_x.

6.8. The following data come from a multistage area sample. From each of $H = 5$ strata, two primary selections, *a* and *b*, were made; these were then subsampled to provide an epsem sample of elements. In the primary selections y_{ha} and y_{hb} are the results for the survey variable, and x_{ha} and x_{hb} for the count of elements; these are the sample totals for the *a* and *b* selections in the *h*th stratum. The data are fictitious and simple for easy hand computation.

h	1		2		3		4		5		
	a	*b*	*a*	*b*	*a*	*b*	*a*	*b*	*a*	*b*	Total
y_{ha}, y_{hb}	65	40	42	40	57	42	60	50	56	48	$y = 500$
x_{ha}, x_{hb}	120	90	96	98	100	80	110	120	88	98	$x = 1000$

(*a*) Compute the sample mean $r = \dfrac{y}{x} = \dfrac{\Sigma\,(y_{ha} + y_{hb})}{\Sigma\,(x_{ha} + x_{hb})}$.

(*b*) Compute var (r), the variance of r. The overall sampling rate f is 1: 10,000; hence $(1–f)$ may be neglected.

(*c*) Compute the simple random variance of r, knowing that r denotes the *proportion* of the element possessing the characteristic. Compute also the overall *design effect*: the actual variance divided by simple random variance.

(*d*) How much variance would you expect roughly from a similar design, applied to a similar but larger population, so that 2 selections would be made from each of 20 strata?

(*e*) How much variance would you expect roughly if the sampling fraction and the subsample size were increased by a factor of 4, in a sample confined to 5 paired selections?

6.9.

h		1		2		3		4		5		6		7		8		9		10		Total
y_{h1}	y_{h2}	5	2	1	3	0	1	2	3	2	3	4	2	1	3	5	2	0	3	4	4	50
x_{h1}	x_{h2}	14	12	8	10	7	12	7	10	8	10	10	7	8	8	11	10	9	11	14	7	193

These data represent paired selections from 10 strata. (*a*) Compute the ratio mean $r = y/x = \Sigma\,(y_{h1} + y_{h2})/\Sigma\,(x_{h1} + x_{h2})$ and its standard error. (*b*) The mean here represents the proportion of members who possess a dichotomous variable among x sample cases. Estimate the variance of a simple random sample of the same size, and then compute the ratio of the actual variance to it. (*c*) Estimate roughly the variance you could expect if the sample were increased fourfold by taking 80 primary selections, instead of 20, from the same 10 strata. (*d*) If the 80 selections were made as paired selections from 40 strata, how might that change the variance?

6.10.

h		1		2		3		4		5		6		7		8		9		10		Total
y'_{h1}	y'_{h2}	0	3	2	6	2	7	16	3	3	1	5	4	1	5	1	2	1	1	7	3	73
x'_{h1}	x'_{h2}	7	10	5	10	10	15	24	7	10	5	8	8	3	7	6	7	5	6	13	5	171

The data in 6.9 represent a subclass of an entire sample, and these data represent another subclass from the same primary selections. (*a*) Compute the ratio mean $r' = y'/x'$ and its standard error. (*b*) Compute the difference $(r' - r)$ and its standard error.

6.11. The mailing list of a magazine has a card for each of its 1,000,000 subscribers, filed in alphabetical sequence. A large sample is desired for a mailed questionnaire, but it is decided that the use of subscribers as sampling units would be too laborious. The following alternative procedures are proposed by various people:

(*a*) Take the names under the letter *H* because it has about the right number of cards, and no reason is known why it would be biased.

(*b*) Select one of the 26 letters with equal probabilities and take all the names under it.

(*c*) Divide the stack of cards into a great number of groups (say, 10,000) by measuring off equal lengths with a ruler. Then select the required number of groups (say, 100) with equal probabilities, taking all cards in the selected groups.

(*d*) Select five of the 26 letters with equal probabilities; then select with random choices the same number of cards, that is, one-fifth of the sample, from the cards of each of the five selected letters.

For each of these four procedures answer the following questions with yes or no and a very brief explanation. Add a brief characterization of the nature of each sample, and brief reasons for your answer to (5).

(1) Is it a *probability* sample?

(2) Is it a *measurable* sample?

(3) Is it an *epsem* sample?

(4) Is the simple mean an unbiased estimate of the population mean?

(4′) If not, can an unbiased estimate of the mean be constructed? Why not *or* how?

(5) Rank the four designs from most desirable (1) to least desirable (4).

6.12. A sample of $f = 1/20$ is wanted from about 20,000 students in a university. All students of selected classes will complete the questionnaires. From about 2000 classes 100 will be selected from 50 strata. (*a*) Describe the design in detail, inventing parameters, if necessary. (*b*) Describe some problems you anticipate and procedures for dealing with them. (*c*) Give formulas for computing a mean and its variance.

6.13. Problem 4.3 describes a selection with the interval 27 from the 270 blocks of Appendix E. Consider these as a sample of 10 clusters. (*a*) Compute the estimate $r = y/x = \Sigma\, y/\Sigma\, x$ of rented dwellings and its standard error. (*b*) Compute and compare the two estimates Fy and Xr, where $F = 270/20$ and $X = 6786$ dwellings. (*c*) Compute and compare the ratio of the variances var (Fy)/var (Xr). (*d*) Compute the ratio of the actual variance to the simple random variance, $(1-f)r(1 - r)/x$.

6.14. Problem 4.5 describes the paired selection of 10 blocks from 270 blocks of Appendix E. Consider them as a sample of clusters and compute (*a*)–(*d*) from 6.13.

6.15. Suppose $E(x) = X$ and $E(y) = Y$. Show that $r = y/x$ is an unbiased estimate of Y/X when the covariance of y/x with x is zero.

6.16. The ratio $w = \dfrac{y/N_y}{x/N_x}$ is used sometimes to compare disease rates in two populations. The values y and x are the number of cases found in a sample or census of each population. N_x and N_y denote the sizes of the two exposed populations, which are the bases of the y and x; they are considered known without appreciable error. Suppose that the disease is rare so that the prevalence rates y/N_y and x/N_x are small numbers. It has been stated that the variance of w is var $(w) = w^2\,(y + x)/yx$. Show that this is true under certain assumptions. State these assumptions. Describe circumstances when those assumptions may be unjustified, and likely consequences.

7

Selection with Probabilities Proportional to Size Measures (PPS)

7.1 CONTROL OF SAMPLE SIZE

The total sample size x can be subject to unduly large variation if it is based on a random selection of clusters that differ greatly in size (X_α). If we subsample the selected clusters at a fixed rate f_b, the expected size of the subsamples $x_\alpha = f_b X_\alpha$ are proportional to the unequal cluster sizes. The total sample size $x = \Sigma\, x_\alpha$ depends on which clusters happen to fall into the sample. Therefore, we try to avoid uncontrolled random sampling of clusters with large size variations; e.g., cities, blocks in big cities, and establishments.

Exact control of sample size is unnecessary and impossible in most situations. It may be too difficult to obtain either the information or the procedures for firmly controlling even the initial sample size. Moreover, nonresponses and subclasses introduce additional sources of variation. We should aim at an *approximate control* that is both feasible and desirable. The degree of control depends on the situation, and its appraisal is aided by examining briefly the reasons for controlling sample size.

First, the cost of data collection requires an upper limit on the overall sample size; and contractual limitation may also impose a lower limit. We may designate the sample size as $x(1 \pm t_p C_x)$, with C_x as its permitted coefficient of variation. For example, we could aim at a sample of no less than 1000 and no more than 1100. To meet both requirements with a probability of $P = 0.99$, we should aim at a sample of 1050 with C_x equal to about 0.02 to obtain about $1050(1 \pm 2.5 \times 0.02) \doteq 1050 \pm 50$. The total sample size is also affected by nonresponse and empty listings, but these problems are treated elsewhere (13.5*A*); here we concentrate on the effects of unequal cluster sizes.

Second, large differences and fluctuations in the sizes of clusters can cause administrative inefficiencies in the field work. For example, a sample of counties is easier and more efficient to administer if the workloads are roughly equal and predictable both between counties and between successive studies. If large differences exist, either some counties are too large or others too small to provide efficient workloads. We could, for example, require that, if $\bar{x} = x/a$ is the average size of primary selections (such as the workload per county), very few of these be either under $0.5\bar{x}$ or over $1.5\bar{x}$. This may mean a coefficient of variation of $V = 0.5/2.5 = 0.2$ for single selections, with $t = 2.5$ representing roughly $P = 0.99$ probability. Observe that $V = 0.2$ for single selections would also mean $C_x = 0.2/\sqrt{100} = 0.02$ for a sample of 100 selections.

Third, statistical efficiency tends to suffer from large inequalities of sample clusters. The technical aspects are discussed by Hansen, Hurwitz, and Madow [1953, pp. 267, 354]. But it is intuitively clear that it is undesirable to have one or two blocks happen to fall into the sample with very large clusters, if each contains a preponderance from one social class or ethnic group. This could have large effects on a moderate sized sample, and even more on the analysis of some subclasses.

Fourth, the valid use of the ratio mean requires control over C_x, the coefficient of variation of sample size. That control is needed both for holding the bias of the ratio mean below a negligible level, and especially for the approximations in deriving the variance of the ratio mean (6.6*A* and *B*). A value of $C_x < 0.1$ is sufficiently good, and even $C_x < 0.2$ can be tolerated.

For several reasons then, a reasonable control over the sample size is needed. Hence, we should maintain a check on the coefficient of variation of size C_x, and compute it from the sample results. Control of C_x is an important aspect of sample design. Essentially we can write $C_x = V_x/\sqrt{a}$, where a equals the number of primary selections and V_x is the coefficient of variation for single selections. Hence, C_x can be reduced by taking more primary selections. Since cost considerations limit a, the sampler looks for a procedure to reduce V_x. Therefore, we should avoid uncontrolled random selections from a population of clusters that are grossly unequal in size. There are several ways to control and decrease variations in sample size.

Stratification by size can reduce the variation in sample size. If the distribution of cluster sizes is rectangular, creating H strata can reduce the standard deviation of cluster sizes within strata by the factor H. This gain can be approximated in cases where extreme clusters are absent or are treated separately. Stratification obtains control over the sample size, although the sample contains complete clusters of unequal sizes. On the contrary, the other three methods depend on subsampling the clusters, and on controlling these subsamples.

Split and combine natural clusters to form artificial clusters less unequal in size. This method is most practical when a large majority of natural clusters can be accepted unchanged, within an acceptable range of variation. Unduly large clusters, probably a small proportion, are split into smaller units. Small clusters can be combined or attached to regular clusters.

We need a list of clusters with measures of their sizes. On this list, mark the numbers of the units assigned to large clusters. Also indicate combinations of clusters into single units. Then apply the selection procedure to the units thus formed. The labor of actual splitting can be confined to the selected large clusters. Diverse procedures for splitting can be used; the clusters can be split physically into compact units, or a subsampling procedure applied.

For example, suppose that in a U.S. city of moderate size, most residential blocks contain between 10 and 30 dwellings and that these can be accepted for clusters. Their range is 20, and their standard deviation 5 on a mean of 20 gives a coefficient of variation of 1/4. Large blocks can be split into units of roughly 20. Smaller blocks can be combined into units of 20. These numbers could also describe a sample of classes from a university, or work groups in a factory.

Size-stratified subsampling can be explained with a simple illustration. Suppose that we need to select elements with a uniform probability of $f = 1/60$. Suppose these exist in clusters of different sizes, but that we prefer subsamples of about 5 elements each. These aims can be achieved by sorting the clusters into nine size strata and varying the subsampling rates within them:

Range of cluster size	0–6	7–12	13–20	21–30	31–40	41–80	81–150	151–300	301+
$f_a \cdot f_b = \frac{1}{60}$	$\frac{1}{60} \cdot 1$	$\frac{1}{30} \cdot \frac{1}{2}$	$\frac{1}{20} \cdot \frac{1}{3}$	$\frac{1}{12} \cdot \frac{1}{5}$	$\frac{1}{10} \cdot \frac{1}{6}$	$\frac{1}{6} \cdot \frac{1}{10}$	$\frac{1}{3} \cdot \frac{1}{20}$	$\frac{1}{2} \cdot \frac{1}{30}$	$1 \cdot \frac{1}{60}$
Range of subsample size	0–6	3–6	4–7	4–6	5–7	4–8	4–8	5–10	5+

Observe that the sampling fractions are varied in pairs so as to maintain the uniform overall sampling fraction $f_{ha} \times f_{hb} = f = 1/60$. Often we can tolerate more variation, and four strata may suffice. We can also employ a less felicitous overall rate than 1/60. Furthermore, we can adjust the stratum boundaries to create strata containing numbers of clusters that are integral multiples of $F_{ha} = 1/f_{ha}$; then we can apply exactly the first stage sampling fraction.

Selection with probabilities proportional to size is discussed in detail below. But observe that the last two procedures also control subsample sizes with informal selections proportional to size.

Remark 7.1.1 The valid use of the ratio mean requires only that C_x be below 0.1 (or more loosely 0.2), and this ordinarily seems less stringent than the demands of cost and administrative convenience. However, the requirement that $C_x < 0.1$ can become critical for ratio means based on small subclasses. The sizes of subclasses within clusters may vary greatly, even with good control over the overall sizes of subsamples. This problem can become especially acute for subclasses which are unevenly concentrated in a small number of clusters in the sample. For example, we should be careful with subclasses, such as farmers and miners, based on occupations that appear unevenly concentrated in a few sample clusters; also with ethnic groups and with social classes, if they tend to be segregated in clusters.

If the coefficient of variation for individual primary selections is under 1, then $a \geq 100$ selections guarantees $C_x \leq 1/\sqrt{100} = 0.1$. If the individual variation is less than 0.5, then only $a \geq 25$ is needed to guarantee $C_x \leq 0.5/\sqrt{25} = 0.1$.

If the subclass is widely distributed, as are age and sex subclasses, the distribution may approximate the Poisson distribution. For that the coefficient of variation is $\sqrt{m}/m = 1/\sqrt{m}$, where m is the mean cluster size of the subclass. Hence, an average cluster of $m \geq 4$ would yield a coefficient ≤ 0.5 for individual clusters, and $C_x \leq 0.5/\sqrt{25} = 0.1$ for 25 or more clusters.

A quick check for C_x can be had by noting the *range* in the sizes of clusters. The ratio of *range* $(x)/x$ estimates C_x well for up to $a = 13$ random selections. For a greater number of random selections the estimate needs to be increased. On the other hand, stratified selection may reduce the variation. The range is $x_{\max} - x_{\min}$, and observing that $x_{\max}/x < 0.05$ can ordinarily yield the needed reassurance. The use of the range is discussed in Appendix C.

7.2 CONTROL OF SUBSAMPLE SIZE WITH PPS

Often the sampler wants to obtain equal subsamples (primary selections) from clusters that vary in size. Two illustrations of this problem follow:

1. From the 77,200 dwellings of a city, a sample of dwellings needs to be selected with probability of 1/200 for each dwelling. The two-stage sample will consist of subsamples of dwellings chosen from a sample of about 80 blocks, averaging about 5 dwellings per block. The number of

dwellings per block varies from 0 to perhaps 500. The selection could be made with constant probabilities f_a for blocks and f_b within blocks, with $f_a \cdot f_b = 1/200$. However, that design would permit the large variations in block size to be reflected undiminished in the sizes of the subsamples. We want to reduce this variation in the size of the subsamples, if we cannot entirely eliminate it.

2. A survey design calls for interviews with a sample of $n = 400$ employees of an organization to estimate their attitudes, with equal chance of selection (epsem) for each employee. An up-to-date payroll list contains the names of all the 77,200 employees, section by section. A work section has anywhere from 20 to 380 employees. Moreover, the sample is to be taken as a selection of 80 sections, with a subsample of 5 chosen from the employees of each of the selected sections. This clustering is judged to cut costs and administrative difficulties. It also provides a desired sample of sections for further analyses of section means and of relationships within sections.

The second example exhibits two reasons for equalizing the sizes of primary selections, in addition to reasons for controlling C_x in the overall sample. First, the means of the primary selections may be used as the elements for some kinds of analyses; for example, the subsample means for work sections may provide the basis of such an analysis. Therefore, equal sized subsamples may be preferred, because they tend to have equal variances (see 11.6). Second, for the analyses of the relationships that exist between the elements of the same primary selections (between members of the same work group), equal numbers may also be preferable.

At this point we should note the disadvantages of a design that seems to enjoy intuitive appeal, and frequent, though mistaken, use. Suppose that in the first stage someone selects a from A primary units with the equal probabilities $f_a = a/A$, and then selects a constant number b of elements from each of the a selected primaries. For example, after selecting either blocks or work sections with equal probabilities, he could select exactly b dwellings or employees from the a sample blocks or sections. The number of elements in the clusters can be designated as the variable N_α. Then the selection probability of any element becomes the variable $(a/A)(b/N_\alpha) = ab/AN_\alpha$; the selection probability of any element is inversely proportional to the size N_α of its cluster. Dwellings in large blocks and workers in big sections receive smaller probabilities than their counterparts. The selection bias can be compensated by weighting the sample elements proportionately to the N_α, if they are known. But this kind of weighting is ordinarily costly and inefficient (11.7C). On the other hand, when selection bias is permitted to remain unadjusted in the estimates, these are subject to large biases, if the N_α vary greatly *and* if the survey variable is correlated with the variable cluster size. Therefore, it is ordinarily preferable to avoid this source of variation in the selection probabilities of elements.

We concentrate on methods for selecting elements with equal probabilities denoted with the constant overall sampling fraction f. If to this we add the second condition, that a fixed number b of elements must be selected from the N_α elements in the selected primary units, then it follows that the first-stage units must be selected with probabilities proportional to the sizes N_α. That is,

$$\frac{N_\alpha}{Fb} \cdot \frac{b}{N_\alpha} = \frac{1}{F} = f \tag{7.2.1}$$

denotes the selection probabilities, where b and f are the fixed constants of selection. For a fixed overall rate $f = 1/F$, and for the variable fractions b/N_α in the second stage, the primary units must be selected with the probabilities N_α/Fb, proportional to the sizes N_α.

Measures of Size. Measures of size, denoted as Mos_α, must often be used instead of exact sizes N_α. The selection probabilities of PPS can generally be designated as

$$\frac{\text{Mos}_\alpha}{Fb^*} \cdot \frac{b^*}{\text{Mos}_\alpha} = \frac{b^*}{Fb^*} = f. \tag{7.2.2}$$

We shall refer time and again to this general formula, which permits great flexibility.

If the N_α are known, we can use $\text{Mos}_\alpha = N_\alpha$, and obtain a fixed and exact number of elements $b^* = b$ in each primary selection. In some situations the measure $\text{Mos}_\alpha = X_\alpha$ can be a measure of worth or influence other than size. The Mos_α can be any arbitrary measure assigned to the primary units to suit the aims of the sample design. However, *the same value of* $\textit{Mos}_\alpha$ *must be used for selection in both stages*, to cancel out its effect in the overall uniform selection probability for all elements.

Sampling rates vary within the selected primary units. The selection rate b^*/Mos_α must be applied to the contents of the units. However, we are free to apply it with any suitable method; and its application may involve several stages of selection. The concepts of *simple replicated subsampling* developed in the last two chapters are also pertinent here and will be illustrated with several examples.

Selection in the first stage is generally stratified, because it leads both to greater efficiency and simpler procedures. It also allows for easy recourse to different overall rates f_h for diverse strata, when needed. Furthermore, within any stratum a fixed overall rate f_h can be obtained with different combinations of primary and secondary selection rates.

The probability of selecting the αth unit must be made equal to Mos_α/Fb^*. If F and b^* are fixed constants and a is the number of primary selections, then $Fb^* = \sum_\alpha^A \text{Mos}_\alpha/a$. This sum of size measures represents a portion of the population that must generate a primary selection with an expected sample yield of b^* measures. Thus the sampling fraction can be visualized as $b^*/Fb^* = 1/F$.

To maintain a desired equal overall selection probability f, the selection probabilities (rates) b^*/Mos_α must be applied within the selected primary units. These rates must be based on the measures Mos_α regardless of the numbers of elements N_α actually found. To the degree that the ratios $N_\alpha/\text{Mos}_\alpha$ tend to be constant, we tend to obtain equal-sized subsamples. If those ratios are near 1, the subsamples will be near b^*; if the ratios are near any constant c, the subsamples will be near cb^*. Hence we search for measures which are proportional to the expected sample sizes. To the degree that the ratios $N_\alpha/\text{Mos}_\alpha$ tend to vary, the subsample sizes will also vary accordingly. Nevertheless, the overall probability f will be maintained.

Typically the measures of size Mos_α are not exactly equal to the number of elements N_α in the primary clusters. Sometimes they represent different units. For example, the Mos_α may denote the approximate units of 10 persons each, found by the last Census count in the cluster; whereas we want to sample the current numbers of dwellings. Then application of the rate b^*/Mos_α to the cluster obtains variable numbers of elements, with an average number different from b^*. Nevertheless, we shall keep the b^* as a fixed constant in the selection equation. These problems are discussed in detail in 7.5.

Selecting units with different probabilities entails some difficulties. Several practical procedures for doing it are presented with illustrations in the next section.

7.3 PAIRED SELECTION OF PRIMARIES

For several reasons, this is probably the single most important key to design in survey sampling. It facilitates stratified multistage selection with PPS from unequal clusters. By making two random primary selections per stratum, we can compute the variance without the need for assumptions about the listing order of primary units. The design carries stratification as far as possible for truly random selections with unbiased estimates of the variance. Moreover, computing the variance is somewhat simpler for pairs than for more selections per stratum.

The method, presented in two illustrations, can be applied widely to most situations involving unequal clusters. Consider it also as the basic method, of which the other methods presented in 7.4 are modifications. In Section 7.2 we introduced a problem of selecting $n = 400$ of the $N =$ 77,200 employees of a company; this requires $f = 400/77{,}200 = 1/193$. If we want subsamples of $b = 5$ from each selected work section, we need $a = n/b = 400/5 = 80$ sections in the sample. Furthermore, we can specify a design of paired selections of primaries; this requires two selections from each of 40 strata. The sampling rates are expressed with the selection equation,

$$\frac{2\ \text{Mos}_\alpha}{1930} \cdot \frac{5}{\text{Mos}_\alpha} = \frac{1}{193}.$$

Observe that each *zone* of 1930 employees (or measures) is expected to yield in two primary selections a sample of $2 \times 5 = 10$ employees (or measures), with an overall sampling fraction of $10/1930 = 1/193$.

In general, the selection equation is

$$\frac{2\ \text{Mos}_\alpha}{2Fb^*} \cdot \frac{b^*}{\text{Mos}_\alpha} = \frac{2b^*}{2Fb^*} = \frac{1}{F} = f. \tag{7.3.1}$$

Note that the zone size is $2Fb^*$. For the sample of the company, $F = 193$, and the zone size is $2 \times 193 \times 5 = 1930$. This is also $\text{Mos}_h = 2Fb^* = \text{Mos}_t/H = 2\ \text{Mos}_t/a$, when $a = 2H$ primary selections are drawn from H strata, and the sum of measures in the population is $\sum_\alpha^A \text{Mos}_\alpha = \text{Mos}_t$. Hence, the sampling fraction is $f = 1/F = ab^*/\text{Mos}_t$. For example, the total of 77,200 employees was divided into $H = 40$ zones of $\text{Mos}_h =$ 1930 each. Thus there will be $a = 80$ selections of $b^* = 5$ employees, for an overall sampling fraction of $80 \times 5/77{,}200 = 1/193$.

The aim of the first problem of Section 7.2 was stated as a selection with $f = 1/200$ from the 77,200 estimated dwellings of a city. We could expect a sample of about $77{,}200/200 = 386$ dwellings; but this is subject to nonresponse, to errors in the population estimate, and to fluctuations due to errors in the block measures. If we design $b^* = 5$ expected dwellings and 2 selections per stratum, the selection equation becomes

$$\frac{2\ \text{Mos}_\alpha}{2000} \cdot \frac{5}{\text{Mos}_\alpha} = \frac{1}{200}.$$

Each *zone* of 2000 dwellings (or measures) is expected to yield in two primary selections a sample of about $2 \times 5 = 10$ dwellings (or measures) with the overall sampling fraction of $10/2000 = 1/200$. A further problem arises because the sum of the size measures $\text{Mos}_t = 77{,}200$ is not an integral multiple of the zone size $2b^*F = 2 \times 5 \times 200 = 2000$. If we use this zone size, after 38 complete zones totaling 76,000 measures, we shall be left with a fractional zone of 1200. This zone can result in either 0, 1, or 2 selections; hence, the actual total number of selections becomes variable, with $a = 76$, 77, or 78. This can be avoided by adjusting the total measure to $\text{Mos}_t = 78{,}000$; or by using $f = 1/193$, and with the zone size $2Fb^* = 1930$, as in the first example; or by using $b^{*\prime} = 3$ and $2Fb^{*\prime} = 1200$ in the last stratum. The adjustment of size measures and sampling fractions is discussed in detail in 7.5*D*.

The design assumes two primary selections with replacement; but the elements are selected without replacement. Thus we have *simple subsampling* in the sense described in 6.2. Although selection with PPS is aimed at obtaining primary selections of about equal size, typically some variation in sample sizes remains. Hence, we need to express the mean and its variance in formulas for unequal subsamples. For paired selections this is

$$r = \frac{y}{x} = \frac{\sum (y_{ha} + y_{hb})}{\sum (x_{ha} + x_{hb})}. \tag{7.3.2}$$

Here y_{ha} and y_{hb} are the totals for the Y_i variable of the paired selections from the hth stratum. The x_{ha} and x_{hb} are similar paired totals for the X_i variable, which is often the count variable. Then the variance of the mean may be computed as

$$\text{var}(r) = \frac{1-f}{x^2}\left[\sum D^2 y_h + r^2 \sum D^2 x_h - 2r \sum Dy_h\, Dx_h\right], \tag{7.3.3}$$

with
$$D^2 y_h = (y_{ha} - y_{hb})^2,\ D^2 x_h = (x_{ha} - x_{hb})^2,$$

and
$$Dy_h Dx_h = (y_{ha} - y_{hb})(x_{ha} - x_{hb}).$$

This general and convenient formula for paired selections is illustrated in 6.5B. For equal sizes within pairs ($x_{ha} = x_{hb}$) the second two terms vanish. If we manage to achieve equal sizes for all primary selections ($x_{ha} = x_{hb} = x_{ia} = x_{ib}$), the variance may also be computed as

$$\text{var}\left(\frac{y}{x}\right) = \frac{1-f}{a^2} \sum (\bar{y}_{ha} - \bar{y}_{hb})^2, \tag{7.3.4}$$

where $\bar{y}_{ha} = y_{ha}/x_{ha}$ and $\bar{y}_{hb} = y_{hb}/x_{hb}$. This formula for the variance of a cluster means is similar to the variance formula (4.3.2) for n elements also selected in pairs. It should be a good approximation if the variation in the sizes of primary selections is small.

Paired selection with PPS can be applied to many situations with the following procedures:

For the cumulation procedure, two random selection numbers are needed for each zone of Mos_h *measures*, altogether $a = 2H$ selection numbers.

Select a random numbers, unrestricted and with replacement, from 1 to Mos_h, and write them down in the order drawn. Accept the first pair for the first stratum. Accept the second pair for the second stratum, after adding Mos_h to both. Accept the hth pair for the hth stratum, after adding $\text{Mos}_h(h - 1)$ to both. If the size measures Mos_h of the strata are unequal, the procedure may be easily modified to fit the situation. An alternative procedure is the *Lahiri procedure* described in 7.5*A*.

Subsampling within the selected primary units must be made with the selection rate b^*/Mos_α, to preserve the desired overall selection probability at f. But we have great freedom in choosing subsampling methods, and should use this freedom to achieve economy (7.5). If the actual size of the primary unit equals exactly the size measure Mos_α, we shall obtain equal subsample sizes. Otherwise, the subsample sizes will reflect variations in the ratio of actual size to Mos_α, discussed in subsection 7.5*C*.

Primary units are selected with replacement in the method here described, and any primary unit may be selected twice. Other procedures are described in 7.4. If a primary unit is selected twice, draw two subsamples (primary selections) from it, each with the selection rate b^*/Mos_α. To satisfy this, the size measure Mos_α of any primary unit must be at least $2b^*$; and we may need a procedure to establish a minimum "sufficient size" $2b^*$ for the measures Mos_α of all primary units (7.5*E*).

The selection zones are not completely independent strata if the size measures of some primaries "straddle" the zone boundaries, so that some primary units belong to two zones. Such primary units may become selected to represent both strata to which they belong. This problem can be avoided by creating completely independent strata without overlaps. These can be obtained with suitable small changes in assigning size measures to primary units. Flexibility in assigning size measures can also be used to change the stratum size, the subsample size, and the overall sampling rate (7.5*D*).

The design of the sample has two principal steps, although these can have many ramifications, involving many factors. *First, we decide on the overall sampling rate f*, after estimating the population size and the desired sample size. The relationship is simply $n = fN$, if the size measures represent elements. Otherwise, the sample size can be generally denoted as $m = f \times \text{Mos}_t$ (7.5*D*). *Second, we must choose either the number of clusters a or the cluster size b**; we cannot choose both, because $a \cdot b^* = n$ (or m). An approximate choice depends on economic factors (Chapter 8). Then we designate some convenient exact values for f, b^*, and a. The final specification of exact values for f, b^*, and a may involve some conflict with the size measures. These are discussed in 7.5*D*.

7.4 SELECTING PRIMARIES WITHOUT REPLACEMENT

The last section described a design of two primary selections with replacement from each stratum (or zone). Because of its statistical simplicity we consider that as a basic model and as a standard of comparison for the other designs which follow. Each of these has particular utility in specific situations.

Selecting primaries without replacement has two advantages over the procedure of 7.3 for selecting with replacement. First, it reduces the variance somewhat by reducing the component of the variance due to the primary units. Second, it eliminates an inconvenience that occurs occasionally in sampling with replacement: having to draw a double subsample from primary units that were selected twice. The possibility of a double selection in 7.3 led to the requirement that the minimum size of a primary unit be $2b^*$. But without the possibility of reselection the minimum size may be b^*.

Several procedures are presented for selecting primaries without replacement. In most situations, procedure C for systematic selection with PPS is the simplest.

7.4A Single Selections from Random Half-Strata

To draw a primary selections, we can first design $a/2$ strata, as for paired selections. Next, sort at random the primary units of each stratum into two *random halves* of equal sizes. Then draw, with a random number, a single primary selection from each of the a half-strata.

If all strata are of the same size, we first create $a/2$ strata, each of size $2Fb^*$. Second, we halve these to obtain a half-strata, each of size Fb^*. Then we draw a primary selection with the sampling fraction b^*/Fb^* from each half-stratum. This draw is in two (or more) stages, represented by the selection equation,

$$\frac{\text{Mos}_\alpha}{Fb^*} \cdot \frac{b^*}{\text{Mos}_\alpha} = \frac{b^*}{Fb^*} = \frac{1}{F} = f. \tag{7.4.1}$$

This design differs from paired selection with replacement, because it forces the selection of two different units from each full stratum of size $2Fb^*$. It is intuitively easy to consider it as a design for selecting *without* replacement two primaries per full stratum; and this conjecture [Goodman and Kish, 1950] was also demonstrated in detail [Rao, 1963]. Often it is a simpler procedure than other designs described in the next subsection B. The remarks there about variance computations also hold for this design A.

		Paired Selection		Systematic Selection	
Mos_α	Cum	Selection Number	$Mos_\alpha/5$	Selection Number	$Mos_\alpha/5$
(14)					
149	163				
10	173				
30	203			201	6.0
90	293				
56	349				
42	391				
113	504	486	22.6		
45	549	540	9.0	541	9.0
16	565				
67	632				
23	655				
18	673				
33	706				
89	795	718	17.8		
114	909			881	22.8
66	975				
61	1036				
25	1061				
46	1107				
58	1165	1165	11.6		
44	1209				
66	1275			1221	13.2
61	1336				
45	1381				

TABLE 7.4.I Selections with Probabilities Proportional to Size.

The $A = 270$ blocks of Appendix E are the primary sampling units of a two-stage selection. The size measures are the 1950 Census dwellings, which total 6786. The design calls for *about* $n = 100$ dwellings, each with equal probability $f = 1/68$, and with 5 dwellings from 20 blocks. For convenience, we establish $\Sigma\, Mos_\alpha = 6800$ with an arbitrary increase of 14 for the measure of the first block. Each *zone* of 680 measures yields two sample blocks.

Paired selection of blocks: $2\, Mos_\alpha/680 \times 5/Mos_\alpha = 1/68$. Pairs of random numbers are drawn from 1 to 680: 540, 486, 485, 38, etc. The first pair of selection numbers is 540 and 486. The second pair is $485 + 680 = 1165$ and $38 + 680 = 718$. And so on for 8 more pairs.

Systematic selection of blocks: $Mos_\alpha/340 \times 5/Mos_\alpha = 1/68$. Random start from 1 to 340 was 201. Successive addition of the interval 340 gives the selection numbers 201, 541, 881, 1221, . . . , 6661. (*Continued on p. 229.*)

If the procedure for separating the half-strata is not fully random, then some homogeneity may exist within them, although this effect may often be negligible. Subsection *D* deals with the creation of homogeneous half-strata.

*7.4B Selecting Two Primaries Without Replacement

Selecting two or more units without replacement and with unequal probabilities can be a troublesome task. It has been the subject of several investigations [Horvitz and Thompson, 1952; Yates and Grundy, 1953; Durbin, 1953; Rao, Hartley, and Cochran, 1962; Fellegi, 1963]. Most of these procedures are too complicated, I feel, for our purposes. We present three procedures (in addition to the random halves of procedure *A*) for selecting two primaries without replacement and with PPS, which are relatively simple to execute.

1. Often we can assume that the order of primary units within the stratum of size $2Fb^*$ is approximately random. Now select a random number from 1 to Fb^*; then add Fb^* to it. These two numbers are the two selection numbers from the stratum.
2. Often we can assume that there are in the stratum at least two primary units of the same size, for any particular size P_j. Make two selections with PPS *with* replacement. If a unit is selected the second time, substitute for it at random one of the other units of the same size. If exactly equal sizes do not exist, we can establish strict rules for matching primaries of the nearest size. When two sizes P_j and P_k are matched, the uniform overall sampling rate f may be retained by subselecting with $(P_j + P_k)/2$ whenever both units appear in the sample.
3. Select 6 units *with* replacement and with PPS. Then sort them into three pairs without allowing the same unit twice into a pair. Now select one of the pairs at random (with $P = 1/3$). The number 6 is suggested merely as a desirable compromise to avoid a double selection without undue extra effort. If the selection probability of a unit is 0.10, the chance of its appearing twice in two draws is 1/100; but its appearance more than three times in six draws is reduced to about 1/800.

With any of these procedures we can select two primary units with probabilities proportional to their sizes Mos_α from each stratum of size

---

Cumulate the size measure Mos_α. When a block measure reaches or passes a selection number, draw it into the sample.

Apply the selection rates $5/Mos_\alpha$ within blocks. This may often be done simply by applying the selection intervals $\text{Mos}_\alpha/5$. It may be applied either to listed dwellings, or to segments into which the block was previously divided.

$2Fb^*$. They are then subsampled with b^*/Mos_α so that the two stages can be represented with the selection equation:

$$\frac{2\,\text{Mos}_\alpha}{2\,Fb^*} \cdot \frac{b^*}{\text{Mos}_\alpha} = \frac{b^*}{Fb^*} = \frac{1}{F} = f. \tag{7.4.2}$$

The sample mean, based on an epsem sample, is self-weighting. For computing the variance, I suggest using the formula (7.3.3), based on sampling with replacement. This computation tends to overestimate slightly the true variance. It fails to account for the reduction achieved in the component between primary units by the selection without replacement. But computing the proper variance would be complicated and expensive. This problem is similar to that discussed in 5.6*A* and again in 8.6*B*.

7.4C Systematic Selection of Primaries with PPS

This is frequently the simplest procedure for applying selection with PPS and without replacement of primaries. First, determine the desired number a of primaries, the desired subsample size b^*, and the total size measure Mos_t in the population. From these, compute the zone size $\text{Mos}_t/a = Fb^*$. Now draw a random start from 1 to Fb^*, say, R_n. The selection numbers for the a primaries will be R_n, $R_n + Fb^*$, $R_n + 2Fb^*$, and so on, until you reach $R_n + (a - 1)Fb^*$. There will be a single selection number in each zone of size Fb^*. The selection probabilities in the two stages will be

$$\frac{\text{Mos}_\alpha}{Fb^*} \cdot \frac{b^*}{\text{Mos}_\alpha} = \frac{b^*}{Fb^*} = \frac{1}{F} = f. \tag{7.4.2}$$

This procedure is frequently applied to the selection of blocks with PPS in the first stage, where the Mos_α denote the census measures of block sizes. In the second stage, dwellings are subsampled with rates b^*/Mos_α from the selected blocks.

The model of paired selections yields $a/2$ contrasts for computing the variance. Its precision can often be increased by including all $(a - 1)$ contrasts in the computations, as presented in 6.5*C*.

Systematic selection would be the practical equivalent of single selection from random half-strata if: (*a*) strata of size $2Fb^*$ were first created; (*b*) the primary units within the strata were randomized; (*c*) and then systematic selection applied to this order. But the randomization of primary units generally would require more effort than could be justified. Nevertheless, the assumption that the units are randomly ordered within the strata is an acceptable approximation to actual situations. In such cases, we can use this simple selection method and apply the variance computations for paired selections (6.5*B*) with replacement, remembering the arguments as presented in 7.4*B*.

7.4D Single Selection from Each Stratum

In subsection *A*, each of the $a/2$ strata was subdivided into two random halves. But we can increase the precision of the sample mean if we subdivide each stratum into two relatively homogeneous strata, then draw a single primary selection from each. Often we have access to information that will allow us to divide each stratum into a "high" and a "low" half, according to some criteria we would like properly represented in the sample.

Essentially the situation is as follows. If we use design *A* for random halves, the variance of the sample mean would have a value $\text{Var}(\bar{y})$. Design *D* reduces the variance by some amount δ^2 (where 2δ is the difference between the means of the half-strata), so that the true variance of the sample is $\text{Var}(\bar{y}) - \delta^2$. However, in computing the variance, we must use the pairs of selections that represent two separate strata. Because of this stratification the computed variance tends to estimate a value of approximately $\text{Var}(\bar{y}) + \delta^2 = [\text{Var}(\bar{y}) - \delta^2] + 2\delta^2$. This term, $2\delta^2$, represents the overestimation of the achieved variance. To the extent that the stratification is effective, it reduces the true variance of the mean, but increases the computed variance. The symbolic statement of the overestimate is an approximation valid for a large number of primary units per stratum; see 8.6*B*.

One alternative is to use the overestimated variance, knowing that the sample results are actually more precise than we are able to estimate. This overestimation of the variance is always present in *collapsed stratum* procedures (8.6*B*). Its acceptance is often justified by the presence of the unestimated errors of response, to which survey results are subject. The gains δ^2 from the last bit of stratification tend to be small, so that the overestimate is small too. If the gains could be great, many practical people would still be loath to sacrifice it.

A second alternative may exist when the subdivision into half-strata is symmetrical, as described above. If $2\delta_1$ is the difference between "high" and "low" halves of the first stratum, and $2\delta_2$ is the similar difference in the second stratum, it may be a reasonable expectation that $2\delta_1 - 2\delta_2 = 0$, approximately, except for random variation. We assume, in other words, that the effect of the "high-low" separation is additive and equal for the first and second strata. Then the variance for the two strata may be computed by using $y_{1'a} = (y_{1a} + y_{2b})$ and $y_{1'b} = (y_{1b} + y_{2a})$, where the subscripts 1 and 2 denote strata, and a and b denote their high and low halves. This is the *grouped strata* procedure, discussed in 8.5*C*. An appropriate design for a primary selections needs $a/4$ pairs of strata, each with two full strata, and each of these further subdivided into two half-strata.

However, this symmetry is often lacking, and we have simply a strata, with a single primary selection from each. Then the first alternative of *collapsed strata* must be used. The variance is computed by forming pairs of strata on *a priori* grounds of expected similarity, using the two primary selections as if they were paired selections from the same stratum. This will overestimate the variance to the extent, often slight, that the primaries within the two paired strata differ. Collapsing triplets, instead of pairs of strata, would reduce the variability of the computed variance; but it would generally also tend to increase the overestimation (7.4*E*).

On the other hand, gross underestimation of the variance could result from two faulty procedures that must be avoided. First, to collapse pairs of strata on the basis of the similarity of sample results would yield a gross and meaningless underestimate of the variance. Second, if each primary selection is split into random halves, the variance computed by contrasting these halves estimates only the component of the variance *within* the primary units. By excluding the component between the primary units, it will grossly underestimate the true variance.

If the collapsed strata have great differences in size, the variance formulas need a modification (8.6*B*). But if the size inequalities are kept small, or moderate (say, under 10 or 20 percent), this complication can be avoided.

*7.4E Three or More Primaries per Stratum

Designs with more than two primaries per stratum are less common in everyday work. They permit less stratification, and require somewhat more difficult computational forms. But they have the advantage of yielding more precise estimates of variances. For a number of primaries fixed at a, paired selections yield only $0.5a$ contrasts; these, denoting approximate "degrees of freedom," determine the precision of the computed variance (8.6*D*). With three selections per stratum, the number of contrasts is increased to $0.67a$. After that the increase is slow: $0.75a$ for four, $0.8a$ for five, etc.

If the natural strata in the population are unequal, the sizes of primary selections can still be kept roughly equal if their numbers are varied proportionately to the stratum sizes. Strata with either two or three selections may be preferred when single selections are not wanted, and four or more selections can always be further subdivided.

The basic formula for the variance of a ratio mean, as shown in (7.3.3), still holds. Only the computing forms are different, as shown in 6.4A:

$$d^2y_h = \left(a_h \sum_\alpha^{a_h} y_{h\alpha}^2 - y_h^2\right) \Big/ (a_h - 1),$$

$$d^2x_h = \left(a_h \sum_\alpha^{a_h} x_{h\alpha}^2 - x_h^2\right) \Big/ (a_h - 1),$$

$$dy_h\, dx_h = \left(a_h \sum_\alpha^{a_h} y_{h\alpha}x_{h\alpha} - y_h x_h\right) \Big/ (a_h - 1). \tag{7.4.3}$$

Because three selections per stratum may become a preferred design, we note a convenient computing form:

$$d^2y_h = \tfrac{1}{2}[(y_{ha} - y_{hb})^2 + (y_{ha} - y_{hc})^2 + (y_{hb} - y_{hc})^2]$$
$$= \tfrac{1}{2}\left[3 \sum_\alpha^{3} y_{h\alpha}^2 - y_h^2\right], \tag{7.4.4}$$

with similar forms for d^2x_h and $dy_h dx_h$. This computation depends on the three contrasts between the three primary totals y_{ha}, y_{hb}, and y_{hc}. This holds because

$$(a - b)^2 + (a - c)^2 + (b - c)^2 = 2(a^2 + b^2 + c^2 - ab - ac - bc)$$
$$= [3(a^2 + b^2 + c^2) - (a + b + c)^2].$$

This is a specific illustration for $n = 3$ of the general rule that the mean of squared differences of all pairs equals $2s^2$, as discussed in Remark 4.3.*I*. The computing forms for paired selections ($n = 2$) use single contrasts. For $n > 3$ there are too many contrasts for convenient computing.

For three selections per stratum, the selection equation is

$$\frac{3\,\text{Mos}_\alpha}{3\,Fb^*} \cdot \frac{b^*}{\text{Mos}_\alpha} = \frac{b^*}{Fb^*} = \frac{1}{F} = f. \tag{7.4.5}$$

For k primary selections per stratum, use k instead of 3 in the first stage, and k_h when the number of primary selections varies from stratum to stratum.

Techniques for paired selections, developed in Sections 7.3 and 7.4, can be adapted for three or more selections. Those discussions regarding methods of selecting with or without replacement are still relevant. For three or more selections, the probabilities of obtaining double or multiple selections in primary units selected with replacement tend to be greater than for paired selections.

7.5 PROBABILITIES PROPORTIONAL TO SIZE: PROBLEMS AND PROCEDURES

7.5A Selection with PPS Measures

Selection with PPS is illustrated in tabular and in graphic forms. A population of 21 primary units is assigned the size measures Mos_α that are multiples of the planned subsample size $b^* = 5$. This allows the subselection rates b^*/Mos_α to be simple fractions. But this convenience is not essential to the method, and it is sometimes difficult to achieve. In either case the measures Mos_α are cumulated, and the selection numbers are

	(1) Mos_a	(2) Cum. Mos_a	(3) Zones	(4) Actual X_a	(5) Primary Selection Sizes x_a: *A*	(6) *B*	(7) *C*	(8) *D*
1	10	10		17				
2	10	20		36				
3	15	35		7			2	
4	20	55		29				
5	20	75		20				
6	40	115	100	44				6
7	25	140		26		5		
8	45	185		132	15, 14	14	14	14
9	15	200	200	14				
10	30	230		40				
11	10	240		34				
12	25	265		18			7	
13	50	315	300	12				2
14	10	325		0	0	0		
15	25	350		23				
16	5	355		0				
17	15	370		17				
18	55	425	400	47	4	4	4	5
					4	4	4	
19	50	475		59				
20	110	585	500	159	7	7		8
							7	7
21	15	600	600	0				

TABLE 7.5.I Selection with PPS with Four Alternative Methods

applied to this. In Table 7.5.I, column 1 contains the measures and column 2 their cumulation; this table corresponds to Figs. 7.5.I and 7.5.II.

The selection numbers are drawn to fit the design:

(*A*) For paired selection with replacement, draw 2 random numbers for each of the zones 1–200, 201–400, and 401–600.

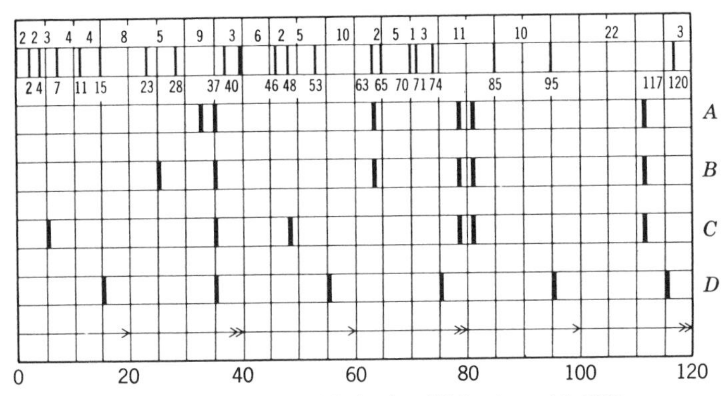

FIGURE 7.5.I Four Methods of Selecting with PPS

A represents paired selections *with* replacement of primary units, and the first two primary selections come from the same unit with size measure $\text{Mos}_\alpha = 9$. This is avoided in paired selections *without* replacement, represented by line *B*. For both *A* and *B* two primary selections are made with PPS from each *zone* of 40 cumulated measures, and the selection equation is $2\text{Mos}_\alpha/40 \times 1/\text{Mos}_\alpha = 1/20$.

C and *D* represent single primary selections from each half-zone of 20 cumulated units, with the selection equation $\text{Mos}_\alpha/20 \times 1/\text{Mos}_\alpha = 1/20$. Line *C* represents random selection, and note that it avoids the double selections from the same half-zone, that occur twice for methods *A* and *B*. Line *D* represents the systematic selection of primaries with the interval 20, after the random start 16. Note that the last two selections in *D* come from the same unit, with $\text{Mos}_\alpha = 22$ straddling the selection interval.

The zones are not explicit strata, and the unit with $\text{Mos}_\alpha = 11$ straddles the second and third zones. This permits its selection from both zones with methods *A*, *B*, or *C*.

The uniform overall sampling fraction of $f = 2/40 = 1/20$ is obtained from each zone with any method, regardless of the size Mos_α of the selected units. In each case two primary selections come from each zone of 40 measures. We can change the scale of measurement by noting the height of 5 elements for all our units. The areas of the primary units are all shown in multiples of 5 measures, and the unit showing a length of 9 contains $\text{Mos}_\alpha = 45$ measures. Then two primary selections of $b^* = 5$ are taken from each zone of $5 \times 40 = 200$ measures, and the selection equation becomes $2\text{Mos}_\alpha/200 \times 5/\text{Mos}_\alpha = 10/200 = 1/20$. This illustrates graphically how the inverse use of the measures Mos_α in two stages leads to the uniform overall selection probability f.

Because all measures Mos_α are integral multiples of the planned size $b^* = 5$, all the subselection factors b^*/Mos_α are simple fractions; but this is often inconvenient to achieve, and not essential to the method.

(*B*) For paired selections without replacement, the above must be modified according to one of the rules proposed in 7.4*B*.

(*C*) Single random selection from each half-zone (1–100, 101–200, 201–300, 301–400, 401–500, 501–600) is an easy procedure (7.4*A*) for approximating *B* above.

(*D*) For systematic selection, apply the selection interval 100 after a random start.

Probabilities-proportional-to-size selection is ordinarily obtained by *applying selection numbers to cumulated size measures.* Draw the selection numbers in accord with the design for selecting primaries. Cumulate the size measures in sequences; whenever the addition of a size measure

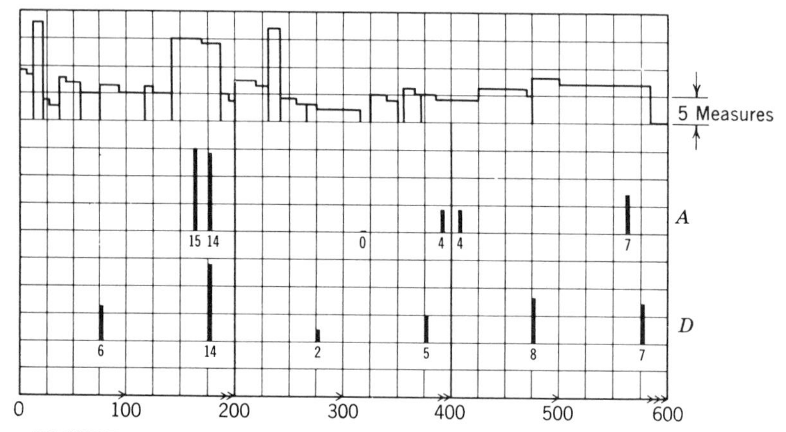

FIGURE 7.5.II The Variable Actual Sizes of Primary Selections

Two primary selections are taken from each zone of 200 measures. The primary units are selected with PPS, and their size measures Mos_α are represented by their $2\text{Mos}_\alpha/200 \times 5/\text{Mos}_\alpha = 10/200$.

The areas of the primary units represent their actual sizes X_α, and exhibit two sources of variation. First, their heights depart from the value 5 anticipated by their measures, because the actual sizes differ from the size measures; and three units are shown with zero actual sizes. Second, the heights of units are not even because the actual sizes are not even multiples of 5. Thus the eighth unit has a ratio of $X_\alpha/\text{Mos}_\alpha = 132/45$ of actual size to size measure. The expected sample size is $b^* X_\alpha/\text{Mos}_\alpha = 5(132/45) = 14.67$. The two primary selections drawn from it yield 15 and 14. The variation in the sizes of primary selections from any unit is presented as a single element; this is possible with a systematic subselection after a random start.

Line *A* shows the results of paired selections with replacement and the following sizes (x_{ha}, x_{hb}) for paired selections from three zones: (15, 14), (0, 4), (4, 7). Line *D* shows similar results of collapsed pairs (x_{ha}, x_{hb}) for systematic selection: (6, 14), (2, 5) (8, 7).

of a sampling unit causes the sum to reach or pass one of the selection numbers, that unit is thereby selected. We assume that the size measures are integers, although adaptation to fractions is not hard.

Cumulating measures is easy on an adding machine. The individual additions need not be written down; but costly errors can be prevented by writing down subtotals at reasonable intervals. For systematic selection, each interval can be cumulated separately.

If the strata are explicit, a separate cumulation begins with each stratum. But if zones are used, the cumulation of measures is continuous zone after zone. Within strata or within zones (or half-zones) the ordering is generally unimportant, and any convenient ordering present in the frame is acceptable.

Remark 7.5.I To illustrate another selection procedure, imagine a ledger of 9 pages, each with 10 lines, of which variable numbers $\text{Mos}_\alpha \leq 10$ are actually occupied. Suppose that the first Mos_α lines are occupied, and that the following $(10 - \text{Mos}_\alpha)$ lines are blanks. For example, the 9 units of the first zone in Fig. 7.5.II have $\text{Mos}_\alpha = 2, 2, 3, 4, 4, 8, 5, 9, 3$. Of the 9×10 lines, 40 are occupied and 50 are blanks. We can select a line with the equal probability 1/90, by drawing a random number 1 to 90. Or we can draw instead, first, a random page 1 to 9, and, second, a random number 1 to 10 to determine the selected line. Clearly, the Mos_α for any page determines whether an occupied line or a blank is drawn. Now, if we associate the drawing of an occupied line with the successful selection of its page, we have a procedure for selecting pages with PPS.

This exemplifies the *Lahiri procedure for PPS selection* [Lahiri, 1951]. Suppose there are A_h units in a stratum, each with its size measure Mos_α. Determine a number K_h not exceeded by any Mos_α ($\text{Mos}_\alpha \leq K_h$). First, select a unit with equal probability $1/A_h$. Then draw a random screening number R_1, 1 to K_h. If the random number does not exceed the unit's size measure (if $R_1 \leq \text{Mos}_\alpha$), accept the unit as selected. If the random number exceeds the size measure (if $R_1 > \text{Mos}_\alpha$), reject it; then repeat the process, drawing again with $1/A_h$ from all A_h units. Continue until a unit is accepted. For paired selection with replacement, a second selection can be similarly drawn from all A_h units. Three or more selections with replacement can also be made the same way. For selections without replacement refer back to 7.4*A*.

Clearly, the occupied lines on the pages need not be actually the first lines. Instead of occupied lines on pages, we can similarly deal with dwellings in blocks, or cards in files, or students in classrooms, provided a bound can be fixed for the largest size measure in the set of clusters. The *Lahiri procedure* is convenient if cumulation of the Mos_α is bothersome, or if the measures Mos_α are not easily available, though the units are listed and numbered. The stratum sum of measures $\sum_\alpha \text{Mos}_{h\alpha}$ must be available for computing the selection probability of the selected units.

7.5B Subselection within Primary Units

Within the selected primary units the subselection rate b^*/Mos_α must be applied. In 6.2 a fixed subselection rate $f_b = 1/F_b$ assured the uniform overall probability $f = f_a \cdot f_b$; but after PPS selection of primary units, the variable subselection rates $b^*/\mathrm{Mos}_\alpha = f_\alpha = 1/F_\alpha$ must be applied within the selected primaries. With this difference the principles of simple subsampling outlined in 6.2 hold also, and we present them only briefly here.

The subselection rates are simpler if the primary size measures Mos_α are made integral multiples of b^*, which will result in integral subselection intervals F_α. This is especially convenient when it has to be applied in the field. However, other difficulties often force us to accept the available measures Mos_α and then deal with fractional values of F_α.

We assume that the elements are selected without replacement, because reselecting them is typically neither opportune nor efficient. If the elements are selected in final compact clusters, these also are selected without replacement. The relative advantages of selecting elements or compact clusters are discussed in 9.4*A*. Both of these permit choices among a great variety of designs, and the principal types are briefly outlined below. The formulas of 7.3 and 7.4 for the sample means and their variances are generally valid for all of them.

Systematic subselection with the interval F_α can be applied directly to the elements or to the final clusters. An example of the former for selecting dwellings from block listings is detailed in 9.6, and of the latter for selecting compact segments in 9.5. Systematic selection is especially easy to apply in the field, when the actual number of elements or final clusters is unknown in advance.

Two segments from each primary unit can be obtained by dividing each into $2F_\alpha$ secondary units and selecting two of them either at random or systematically. More generally, create kF_α secondary units and select k of them, where k is 1, 2, 3, etc. This procedure avoids the fluctuation in sample size present with systematic selection of one or two segments; it is detailed in 9.7. It is easiest if F_α is an integer; decimal fractions can be converted to integers (7.5*F*), or they can be used directly. Suppose, for example, that $f_\alpha = b^*/\mathrm{Mos}_\alpha = 5/46 = 1/9.2$. Create 8 segments of measure 5 each plus one with measure 6, so that they sum to 46. Then select a random number 1 to 46 and add 23; these two numbers will select two segments with PPS; if the segment with measure 6 is selected, a subselection of 5/6 must be applied. *Multistage sampling* can be used to apply the subselection with f_α. Since this complicates the selection, it should be used only when needed (10.1).

Further varieties, such as *double sampling*, can be introduced as needed.

7.5C Variations in Actual Subsample Size

Probabilities-proportional-to-size procedures rarely attain complete control of the subsample size of individual clusters. In the PPS selection equations we denoted the planned subsample size as b^*; this was the uniform subsample based on the unequal measures Mos_α for the different primary clusters. But we emphasized that exact equality of subsamples is rarely attained. In the formulas for means and variances, the actual subsample size from the cluster was denoted x_α. The discrepancy between b^* and x_α can take a multitude of forms, which often cause misunderstanding. This can be avoided with a unified, systematic view of the reasons for the variations in the ratio (x_α/b^*) of actual to planned size. For this discussion, we utilize four successive measures of the subsample sizes:

b^*,	the *planned* subsample size based on the cluster measure Mos_α;
$E(x_\alpha'') = X_\alpha b^*/\text{Mos}_\alpha$,	the *expected* size, on the basis of the total cluster size X_α;
x_α'',	the *initially* designated subsample size;
x_α,	the *actually* attained subsample size.

The expected subsample size is obtained by applying the subsampling rate b^*/Mos_α to the total cluster size X_α. *The ratio of the expected to planned size is* $E(x_\alpha'')/b^* = X_\alpha/\text{Mos}_\alpha$; *it differs from unity as the actual total cluster sizes differ from their measures.* These differences often result in large variations in the subsample sizes, and they have several sources. (1) *Obsolescence of the size measures* Mos_α *can occur when there are rapid and unequal changes in the total actual sizes* X_α. For example, a city block where the 1960 Census measure is $\text{Mos}_\alpha = 20$ dwellings, may actually contain $X_\alpha = 108$ dwellings in 1965. The ratio 108/20 can be blamed either on Mos_α obsolescence or on X_α growth. (2) *Different units may have been used for measuring* Mos_α *and* X_α. For example, the Mos_α may represent Census dwellings, and the X_α the number of adults (or persons, or grocery stores, or dogs) found by the survey in the blocks. Even if the X_α represent survey dwellings, their operational definition may differ from the Census definitions of dwellings for the Mos_α. Furthermore, the X_α *may represent a subclass;* for example, dwellings containing an eligible respondent, such as a gainfully employed female. Use of *approximate and rough units* for size measures belongs here; for example, "block units" may be assigned for each multiple of 10 dwellings estimated for the block. (3) *Errors in the size measures* Mos_α *may occur* even in Census counts and processing, but especially when the Mos_α are based on rough estimates.

Furthermore, this may also cover the results of *arbitrary adjustments* inflicted on the Mos_α for other purposes. (4) *Errors can occur in the X_α*, the total cluster size found by the survey operation. For example, field workers may miss many elements, such as dwellings in sample blocks.

Differences between the designated subsample size x_α'' and its expected size $E(x_\alpha'') = X_\alpha b^*/\text{Mos}_\alpha$ can occur in three ways. (1) The expected size is often fractional, whereas the designated sample must be integral. For example, if the $X_\alpha = 108$ represents total survey dwellings listed in a block, and $4/20 = b^*/\text{Mos}_\alpha$ the sampling rate to be applied to the listing, then $108 \times 4/20 = 21.6$ is the expected size. Applying the selection interval $20/4 = 5$ systematically will obtain either 21 or 22 dwellings. Generally, the variation due to this source can be kept to a fraction of one element if the selection is applied directly to the list of elements. (2) If the selection with b^*/Mos_α involves clusters of elements, the discrepancy between the fractional expected and integral designated number of clusters causes a variation in elements that increases with the cluster size. This is one factor considered among the advantages of selecting listed dwellings rather than segments of dwellings (9.4*A*). Additional variation arises if the segments are unequal in size. (3) The sampling of subclasses gives rise to further discrepancy through the selection of variable numbers of "blanks," that is, nonmembers of the class. For example, suppose that among 108 dwellings in the block, a random $108/3 = 36$ contain a subclass member. Then the eligibles expected in the sample are $21.6/3 = 7.2$; but the designated number has a standard error of $\sqrt{21.6 \times 1/3 \times 2/3} = 2.2$, so that its designated numbers below 5 or above 9 can occur with about $P = 0.3$.

Differences between the designated size x_α'' and the finally attained actual sample size x_α are caused by nonresponse. The losses due to noncoverage could be included here also; but instead we considered them excluded from the operational definition of the cluster size X_α.

All three kinds of discrepancies give rise to random variability, which contributes to variations in the sizes of actual subsample clusters and to uncertainties in the achieved total sample size. Random variations tend to cancel, and the net effect on the entire sample diminishes with increasing numbers of primary selections. However, two of the discrepancies, symbolized by the ratios $E(x_\alpha'')/b^*$ and x_α/x_α'', have net average effects that do not cancel. These should be computed in designing sample size. They are related to the "eligibility rate" and the "response rate" we treat elsewhere (13.5*A*).

This entire discussion deals with methods for applying b^*/Mos_α as subselection rates to the selected primary units. This approach is needed to maintain the fixed overall sampling fraction f through multistage selection with PPS. Contrariwise, if a fixed number of elements b^* is selected, the sample selections must be weighted with $X_\alpha/\text{Mos}_\alpha$ to counteract the differences in selection probabilities; without these weights the estimates are affected by the selection bias that may or may not be serious.

7.5D Sampling Fractions, Strata, and Measure Adjustments

Probabilities-proportional-to-size designs result in k primary selections, each of b^* planned size measures, from a stratum of $M_h = kFb^*$ measures. Here M_h is the sum of the size measures of the primary units contained in the stratum. These selection factors may be varied between strata, and $M_h = k_h F_h b_h^*$ can denote the stratum factors in the selection equation:

$$\frac{k_h \text{Mos}_\alpha}{M_h} \cdot \frac{b_h^*}{\text{Mos}_\alpha} = \frac{k_h b_h^*}{M_h} = \frac{k_h b_h^*}{k_h F_h b_h^*} = \frac{1}{F_h} = f_h .$$

The sampling fraction is $f_h = k_h b_h^*/M_h$, and it can be varied by manipulating any one of the three factors, keeping the other two fixed. Thus the sampling fraction can be increased by increasing the planned subsample size b_h^*. More often we keep $b_h^* = b^*$ constant, and increase the sampling fraction by decreasing the stratum size M_h. Seldom would we vary k_h, the number of primary selections, to adjust the sampling fraction.

Similarly, we can keep a constant overall sampling fraction f_h, by varying two of the factors simultaneously and proportionately. By increasing and decreasing b_h^* and M_h together, we can maintain a constant overall sampling fraction f. For example, near the center of a city we can keep both the subsamples b_h^* and the stratum sizes M_h small. But in outlying suburbs, where the travel costs are greater, we may take larger subsample clusters b_h^* and keep the uniform f by increasing the stratum sizes proportionately.

Note also that the subsample size x_α varies with the ratio $X_\alpha/\text{Mos}_\alpha$ of the actual cluster size over its size measure. Hence, we can increase the actual subsample size of primary selections by assigning smaller size measures Mos_α to primary units. For example, in outlying suburbs we can deliberately assign smaller size measures than elsewhere.

In trying to keep a fixed stratum size M_h, we face a problem when the size measures Mos_α do not sum exactly to the desired stratum size. We can treat this problem with one of four alternative procedures.

1. Often the simplest procedure is to *create uniform zones* (*implicit strata*) *of size* M_h, permitting some primary size measures to straddle the stratum boundaries. Then *cumulative selection* can be applied to size measures cumulated in order, zone after zone. It is usually unimportant that the same primary unit may become selected from two strata. The creation of zones and cumulative selection from them is so simple that it is

the preferred procedure in many cases. However, if we desire to avoid zones in favor of explicit strata, three other procedures are available.

2. The size measures Mos_α may be adjusted arbitrarily, without disturbing the desired overall selection probability f. Increasing (or decreasing) the size measure results in decreasing (or increasing) the actual subsample size x_α, if the primary unit is selected. This may be only a slight addition to other sources of size variation. For example, obsolescense and inaccuracies of size measures and nonresponse may result in coefficients of size variation of 50 percent; arbitrary adjustments of 25 percent in a few units add little to the other sources of variability. One form of adjustment is adding fictional "blank" units, which would yield zero subsamples if selected.

3. The planned subsample sizes $b_h{}^*$ can be made proportional to the stratum sizes M_h, and thus maintain a uniform selection rate $f = k(b_h{}^*/M_h)$. This may involve some extra work in computing the intervals $\text{Mos}_\alpha/b_h{}^*$ within the selected primaries. These intervals become fractions, but this inconvenience often exists already.

4. We may permit a small variation in the sampling fractions, $f_h = 1/F_h = kb^*/M_h$, inversely proportional to the stratum sizes. These may be compensated in the results with weights proportional to F_h. If the variations in F_h are small and haphazard, the weights may be inconsequential and ignored.

Even when using zones, rather than explicit strata, the aggregate size M_t of the entire population must be reconciled with the zone sizes M_h. For designing equal-sized zones, we need $M_h = \text{Mos}_t/H$, where $H = a/k$ is the number of zones required for a primary selections, with k from each zone. For example, if we want $a = 80$ primaries, as paired selections from 40 zones for a population of 77,200 dwellings, we also need zones of $M_h = 77{,}200/40 = 1930$ each.

If the aggregate measure is $M_t = 77{,}260$, the zone size becomes $M_h = 77{,}260/40 = 1931.5$. We can use fractional zones; or avoid them by applying either the second or third procedure above to one stratum. For example, we could reduce the measures in the last stratum from 1990 to 1930, thus dissipating the excess 60 measures and creating 40 strata of 1930 measures each. If there were a deficit of 60 measures (that is, for $\text{Mos}_t = 77{,}140$), we could increase the measures by 60, or simply add a "blank" unit of 60 measures. Alternatively, using the third procedure, we can leave the last stratum at 1990 instead of 1960, and adjust proportionately the planned subsample size of $b^*(1990/1960)$. This spreads the increase of the subsample size over all units, whereas reducing the measures of a few units would concentrate it in those selections.

A measure Mos_α can be adjusted by adding or subtracting some quantity

d, whichever is more convenient or smaller. This adjustment changes the subsample size in proportion to $\text{Mos}_\alpha/(\text{Mos}_\alpha + d) = 1/(1 + d/\text{Mos}_\alpha)$, and the total adjustment and effect can be spread over several units. For example, increasing the measure by $d/\text{Mos}_\alpha = 0.2$ decreases the subsample by $1/1.2 = 5/6$, if the unit is selected; and decreasing the measure by $-d/\text{Mos}_\alpha = -0.2$ increases the subsample by $1/0.8 = 5/4$. The selection probability of the unit in the first stage changes in proportion to $(\text{Mos}_\alpha + d)/\text{Mos}_\alpha = 1 + d/\text{Mos}_\alpha$, so that the overall selection probability remains unchanged. In these two examples, the overall effects are $6/5 \times 5/6 = 1$ and $4/5 \times 5/4 = 1$.

The total measure Mos_t can also be adjusted by simply adding one or more nonexistent "dummy" clusters with the needed measure. If the dummy is selected, it results in a subsample of size zero; but this can also happen because of nonresponse or incorrect measures. Contrariwise, the total measure can be reduced by combining clusters and decreasing their combined size.

7.5E *Undersized and Oversized Clusters*

If a unit size measure is less than the planned subsample size (if its $\text{Mos}_\alpha < b^*$), the subsampling rate within the unit becomes $b^*/\text{Mos}_\alpha > 1$, which is troublesome. The problem can be solved formally with proper weighting or replication of the unit's elements. For example, if $b^*/\text{Mos}_\alpha = 1.2$, we can include all elements and duplicate a fifth of them. But this is inconvenient, and can cause appreciable inefficiency if large values of b^*/Mos_α need to be accommodated frequently. Hence, we may designate b^* as the minimum *sufficient size*. Moreover, in some situations we may designate cb^* as the minimum sufficient size; this permits a subsampling rate no greater than $1/c$, because $b^*/cb^* = 1/c$. With this device the maximum subsample size is kept to X_α/c, and we get some protection against units that have increased greatly in actual size X_α.

Units with small measures are often subject to sporadic, irregular, and relatively large growth; that is, great increase in the ratio $X_\alpha/\text{Mos}_\alpha$ of total actual unit size to size measure. This results in proportionately large subsamples. For example, a block with only 3 Census dwellings may contain 120 actual dwellings a year later. These "surprises" are difficult to handle (12.6*C*), and it is worth some effort to obtain protection against them. A higher value of c in the selection procedures provides greater protection. It also obtains c separate subsamples for future use. But it also costs more, because it requires listing more units. Therefore, choose the value of c as a compromise based on estimates of likely risks and added costs.

In the following procedures we assume that the elements of the undersized units are also selected with the common overall sampling fraction f. This results in a self-weighting sample, with its advantages. But if the cost of obtaining these elements is high, and if there are enough of them to merit special consideration, we may resort to disproportionate allocation. For example, if these elements cost 16 times as much as ordinary elements, we should sample them with $f/4$. However, large cost differentials seem to be rare.

There are four ways to handle undersized units:

1. Put the small clusters in a *separate stratum*, or strata, and select them with special procedures. Perhaps in this stratum we can just reduce the planned subsample size $b_h{}^*$. But often it is simplest to select directly with the desired overall sampling fraction f, and to accept all elements from them. Or we guard against surprises by selecting units with uniform c/f, then subsampling with $1/c$, for a two-stage selection of $c/f \times 1/c = 1/f$, where c may be 2, 3, or perhaps more. A separate stratum may be best when these units are numerous, and different from the ordinary units.
2. If the small units contain only a small portion of the population, separate strata may not be justified. Simply *assign the arbitrary sufficient measure* b^* to all units that have size measures less than b^*. Or, for extra protection, the minimum size cb^* may be assigned to all units with size measures less than cb^*; this results in the sampling rate of $b^*/cb^* = 1/c$. Then apply selection with PPS to all units collectively.
3. Field costs can sometimes be reduced with a procedure for *linking undersized units to other units before selection, creating units of sufficient size.* The linking creates a new unit with measure ($\text{Mos}_\alpha + \text{Mos}_{\alpha'}$), the sum of the size measures of the linked units. Both in selection and in subsampling, the linked units are treated as one. The minimum sufficient size may be b^* or cb^*. It may save field costs to link related or close units, such as two blocks near each other. It may be best to link an undersized unit to an ordinary larger unit; this creates desired heterogeneity within the unit created by linking. It also spreads the protection of the measure of the larger ordinary unit over the undersized unit linked with it.
4. If linking all undersized units is too laborious, it can be avoided with the following procedure; this achieves equivalent results, but involves only the selected units and their neighbors. *Yet it effects a strict linking system, equivalent to linking before selection.* (*a*) Select units with PPS, based on their size measures Mos_α. Note if the *selected unit and the next following unit* are *both* of sufficient size. This may be set at b^* or cb^*. For example, suppose that to obtain a planned size $b^* = 4$, we set a minimum sufficient size $cb^* = 3 \times 4 = 12$, to yield sampling rates no

greater than $4/12 = 1/3$. Imagine the following sequence of size measures:

$$\cdots 16, 25, \overline{37, 7, 0}, 13, 12, \overline{20, 6}, \overline{7, 3, 0, 1, 3}, 18, 20, \cdots$$

(*b*) If *both* the selected unit and the next following unit are of sufficient size, *no linking is made*. In many situations most selections are of this kind and need no further action. In the above sequence, the units with Mos_α of 16, 25, 13, 12, and 18 come into the sample alone. (*c*) If either the selected or the next following unit is less than sufficient, a linking is needed. Go forward on the list until the first unit of sufficient size is reached (not passed). Now go backward on the list, and form linkings of units of sufficient size: *cumulate the size measures backward* only until a sufficient size is just spanned. It may be necessary to go backward beyond the selected unit. That grouping of units which includes the selected unit is then taken into the sample as one linked unit. In the example, the following units come into the sample as joint units with combined measures, whenever any of the contained units get selected: (37, 7, 0), (20, 6), and (7, 3, 0, 1, 3).

It would be easier if we had to link only when the selected unit turned out to be smaller than sufficient; but this would lead to a selection bias against small units and in favor of large units. For example, the elements in the unit with $\text{Mos}_\alpha = 6$ would have their selection probability reduced by the factor $6/(20 + 6)$, because it would become $k6/M_h \times b^*/(20 + 6) = kb^*/M_h \times 6/(20 + 6)$, where $kb^*/M_h = f$ is the desired selection probability.

The last procedure treats the units with zero size measures as it does other undersized units, if care is taken to include them in the selection frame. The other procedures can also be made to include them. The zero units can pose special problems if they actually contain elements. First, an ordinary selection using simply the $\text{Mos}_\alpha = 0$ would exclude them. Second, if they do not appear on the list, a special search must be made for them. For example, the Census list of block measures (Appendix E) excludes the blocks without Census dwellings; a search of the map can be made to find and link them. On the other hand, if there are many zero units that are reasonably known to contain no survey elements, they may be excluded to save fruitless field work. For example, many city blocks are known to contain only parks, factories, offices, schools, and other buildings without any dwellings.

Oversized units may be conveniently denoted as those about as large as the stratum size M_h, or larger. If there are few, they may be left among the other units; then some can yield replicate selections for the sample if selected more than once. If there are more than a few, they can contain an appreciable portion of the population that merits special treatment in separate strata.

If a stratum contains only one unit, it becomes selected with certainty, and the overall sampling fraction f may be applied directly to the listing of elements. If the ordinary strata yield paired primary selections, we can also draw a paired selection from them to maintain simplicity in the variance computations. Or, if the size of the unit is about dM_h (where M_h is the size of an ordinary stratum), we can draw d paired selections from it.

If the listing costs of large units is much greater than for ordinary units, we place them in separate strata, to be treated with special procedures. We can increase the planned sample size $b_h{}^*$ and the stratum size M_h by the same factor; we maintain the uniform sampling rate $f = kb_h{}^*/M_h$, and the fixed number k selections per stratum. The increase in $b_h{}^*$ and M_h results in a similar decrease in the number of large primary units that must be treated in the sample.

7.5F Fractional Systematic Selection Intervals

In 4.1*B* we described a simple procedure for systematic selection with decimal fractions. It consists of designating provisional numbers with the fractional interval, by disregarding the decimal point; for example, using the interval 92 instead of 9.2. Then the provisional numbers are converted into integral selection numbers by *rounding down* to the nearest integers, just dropping the decimal fractions.

Using decimal fractions is often easy, especially in office work. It may even have an advantage because it tends to destroy the exact periodic character of integral intervals. But it may be inconvenient sometimes, as in selecting dwellings from block listings while in the field. Hence we may have a set of fractional intervals that we would like to convert to integrals intervals. Given an interval of $I + f$, where I is an integer and f is a decimal fraction, we want to convert it to I or $(I + 1)$; that is, a sampling rate of $1/(I + f)$ needs to be converted to $1/I$ or $1/(I + 1)$.

However, two "common sense" procedures for conversion lead to selection biases. The first of these is rounding to the nearest integer: taking I if $f < 0.5$, and $(I + 1)$ if $f \geq 0.5$. The second is taking $(I + 1)$ with probability f, and I with probability $1 - f$, according to a random number from 0 up to 1. For example, we may want to convert a sampling fraction of $1/1.5 = 2/3$. An epsem choice between $1/1$ and $1/2$ would mean an overall probability of $0.5(1/1) + 0.5(1/2) = 3/4$; this would increase the selection probability with a bias of $3/4 - 2/3 = 1/12$. It is true that this bias decreases rapidly with increasing I; it is within $1/660$ for $I = 5$ and within $1/4620$ for $I = 10$.

Remark 7.5.II Any of three simple procedures will convert a decimal interval to an integral interval.

(1) Take the interval $(I + 1)$ with probability $p = f\left(\frac{I + 1}{I + f}\right)$, and the interval I with probability $(1-p)$. For example, for $I + f = 1.5$, take interval 2 with probability $0.5(2/1.5) = 2/3$, and interval 1 with probability $1/3$. Thus the overall selection probability becomes $1/3(1/1) + 2/3(1/2) = 2/3$, as needed. Take a random number from 0 to 9; the result 0 through 5 designates the interval 2, whereas 6 to 8 designates the interval 1.

To convert the interval 3.4, we compute $p = 0.4\left(\frac{4}{3.4}\right) = 0.47$. Take a random number from 0 through 99; the result from 0 through 46 designates the interval 4, whereas the result from 47 through 99 designates the interval 3.

Since either an increase or decrease in the sample size is decided with a single random draw, this procedure is suitable only for deciding separately the subsampling intervals within the numerous units that comprise a larger sample. The solution for the equation

$$\frac{(1-p)}{I} + \frac{p}{(I + 1)} = \frac{1}{(I + f)} \quad \text{is} \quad p = \frac{f(I + 1)}{(I + f)} = f + \frac{f(1-f)}{(I + f)}.$$

We also note that the bias due to using f instead of p to decide between I and $(I + 1)$ is

$$(p - f)\left(\frac{1}{I} - \frac{1}{I + 1}\right) = \frac{f(1-f)}{(I + f)I(I + 1)},$$

and it decreases rapidly with I^3.

(2) Take the interval $(I + 1)$, and add a duplicate selection for one selection in a, where $a = (I + f)/(1-f)$. Here a is the solution for the equation $(1 + 1/a)/(I + 1) = 1/(I + f)$. For example, instead of the interval 1.5, use the interval 2, and add a duplicate for every third selection, because $a = 1.5/(1 - 0.5) = 3$. Or instead of the interval 3.4, use the interval 4, and add the a duplicate for one selection in 5.67, because $a = 3.4/(1 - 0.4) = 5.67$.

(3) Take the interval I, and delete one selection in b, where $b = (I + f)/f$. Here b is the solution for the equation $(1 - 1/b)/(I + 1) = 1/(I + f)$. For example, instead of the interval 1.5, use the interval 1, and delete every third selection, because $b = 1.5/0.5 = 3$. Or instead of the interval 3.4, use the interval 3, and delete one selection in 8.5, because $3.4/0.4 = 8.5$.

PROBLEMS

7.1. The payroll list of a company identifies all its 30,000 employees, section by section, each consisting of 25 to 100 employees. An epsem sample of 600 employees is needed for an interview survey of attitudes. The same sample will also serve for intensive studies of relationships within sections. A sample of 30 sections is chosen, with 20 employees selected from each of the selected sections. Use numbers where available, and define the other terms.

(*a*) Describe how you would select the sample sections and the persons within the sections, so that both of the requirements stated above will be satisfied.

(*b*) What is the formula for estimating the proportion of employees holding a certain attitude?

(*c*) What formula would you use to estimate the variance of that estimated proportion?

7.2. You want to select lines from a geographically ordered list of 768,000 lines, giving each line a probability 1/800 of being selected. After due investigation you decide on a systematic selection of clusters of eight lines.

(*a*) What interval will you use for selecting these clusters? How will you apply this interval? (A sentence or two.)

(*b*) Assume that 10 percent of the lines turn out to be "blanks" or not members of your target population; and that 5 percent turn out to be nonresponses. How do you propose to deal with these problems?

(*c*) How do you estimate the mean of a variable Y_i and the variance of the mean?

(*d*) Estimate also the total of the variable Y_i in the population and its standard error.

7.3. A city has about $N = 100{,}000$ dwellings in 2500 blocks averaging 40 dwellings per block. A sample of $f = 1/100$ is desired, with subsamples of about 10 dwellings from about 100 blocks. What combinations for selecting blocks (*a* through *c*) and dwellings within sample blocks (*d* through *k*) result in epsem selections of dwellings? State combinations of code numbers. For these epsem designs write the two-stage selection formulas with numbers to yield the overall $f = 1/100$.

The following plans are proposed for selecting blocks:

(*a*) Select blocks with equal probability 1/25.

(*b*) Select blocks with probabilities proportional to the number of dwellings M_α found by the census four years ago.

(*c*) Select blocks with probabilities proportional to measures Mos_α assigned as integers near $M_\alpha/5$, adjusted for any known or guessed changes in size.

The following plans are proposed for selecting dwellings:

(*d*) Select exactly 10 dwellings at random from the N_α dwellings actually found and listed by the survey interviewer.

(*e*) Select systematically exactly 10 dwellings from the N_α listings.

(*f*) Select dwellings systematically with the interval $M_\alpha/10$ applied to the N_α listings.

(*g*) Select systematically with the interval 4 applied to the N_α listings.

(*h*) Divide the block into B_α compact segments of roughly 5 dwellings each, and apply selection intervals 4 to these B_α segments.

(*i*) Select segments with the interval $M_\alpha/10$ applied to the B_α segments.
(*j*) Select exactly 2 from the B_α compact segments.
(*k*) Divide blocks into exactly Mos_α segments and select exactly 2 of them.

7.4. Among the epsem designs in 7.3 for selecting dwellings, choose two that you would prefer in two specific sampling situations. These need not refer to blocks and dwellings, but to any two-stage selection of clusters and elements. (*a*) Describe briefly the two situations and reasons for your preferences. (*b*) Give selection formulas both for paired selections and for systematic selection of clusters. (*c*) Show formulas for computing the means and standard errors, defining the terms.

7.5. You want an epsem sample from a population of about $N = 120{,}000$ students of a school system and made several decisions already: a sample of *about* $n = 6000$ students, hence $f = 1/20$ for all students; paired selections from 25 implicit strata (zones) of 4800 measures. For these measures of size you can obtain Mos_α, the number of students in each school last year; the total for all A schools was $\Sigma\,\text{Mos}_\alpha = 120{,}000$ students last year. If you obtain in each sample school a correct and up-to-date list of its N_α actual students, you can apply to it the selection interval of $\text{Mos}_\alpha/120$ and obtain approximately $120\,N_\alpha/\text{Mos}_\alpha$ students. The probability of selection of each student in the two stages, schools and within schools, will be as required:

$$\frac{2\text{Mos}_\alpha}{4800} \cdot \frac{120}{\text{Mos}_\alpha} = \frac{1}{20}.$$

In most schools students belong to specified *homerooms*. The number of students per homeroom varies moderately with an average of about 30. You would prefer to give questionnaires to entire homerooms, about 4 per sample school. The trouble is that the number of homerooms is unknown, and is bothersome to obtain for the entire school system.

State for each of the five proposed plans for selecting students within the schools whether the equal probability of 1/20 for every student is obtained (Yes) or not (No); if the answer is No, write down the overall selection probabilities for the students.

(*a*) To the list of B_α homeroom classes obtained in each sample school, apply the sampling rate $120/\text{Mos}_\alpha$, obtaining a sample of roughly $120/\text{Mos}_\alpha$ classes and including all students in the selected classes.
(*b*) Select exactly four homerooms with epsem from the list of B_α homerooms in the school.
(*c*) Apply the sampling rate $4/\text{Mos}_\alpha$ to the list of N_α students in the school, and take the selected student plus the next 29. What do you do if this count of 29 runs beyond the end of the school list?
(*d*) Apply the sampling rate $4/\text{Mos}_\alpha$ to the list of the N_α students, take each selected student plus all other students in his homeroom class.
(*e*) From the *corrected and up-to-date list* of N_α students in the school select exactly 120 with epsem.

7.6. Consider again the situation presented by Problem 7.5.

(*a*) Suppose a school gets selected into the sample twice, what should be done about selecting students within it?

(*b*) What is the "minimum sufficient size" school that will avoid the embarrassment of sampling rates greater than 1 in any school?

(*c*) Suppose a plan using unequal probabilities for students has been used. How could you correct for selection bias?

(*d*) Write down the formula for the variance of the mean $\bar{y} = y/6040$ for the 6040 students finally included in the sample. Define the terms you need; assume two schools selected at random with replacement from each implicit stratum.

(*e*) Suppose now that, having selected the sample schools with probabilities 2 $\text{Mos}_\alpha/4800$, you decide on a plan of selection different from any of the alternatives in 7.5. You create selection units (su's) that are roughly of typical homeroom size: about 30 students per su. Most homeroom classes will be single su's, but the larger homerooms will be made into 2 or 3 or k su's. You permit some variation in the numbers of students in the su's, but the variation will be less than in the homerooms. The design then calls for: selecting *exactly 4 su's* from each sample school; taking all students in the selected su, hence 1/2, or 1/3, or $1/k$ from a homeroom containing 2, or 3, or k su's maintaining the overall selection probability of 1/20. The selection formula for the two stages is 2 $\text{Mos}_\alpha/4800 \times 4/\text{su}_\alpha = 1/20$.

(e_1) How do you compute the number su_α of selection units for a sample school?

(e_2) Create selection units for a school that had a $\text{Mos}_\alpha = 228$ and that submitted the following list of homeroom classes, with the accompanying numbers of students: 23, 27, 25, 10, 30, 51, 62.

(e_3) Select exactly four units from the school in (e_2); describe briefly how you did it.

(e_4) To avoid fractional numbers of su_α, one can assign appropriate measures to the schools, instead of the Mos_α. How would you compute these measures? Show the corresponding selection formula for $f = 1/20$.

(*f*) Instead of any of the other designs, assume now that you *can* obtain the number of homerooms B_α in each of the A schools in the population. For $\Sigma B_\alpha = 4000$ in the entire population, write down the formula for the two stages of selection: first selecting 2 schools per stratum with probabilities proportional to numbers of homerooms; then exactly 4 homerooms in each sample school, taking all students in selected homerooms.

7.7. The entries for property tax assessments in a city are entered on pages with 30 lines. Each entry has a first *initial line*, but some occupy several lines; thus initial lines per page can be any integer up to 30. The pages are bound in nine volumes; a variable number, not over 1000, of pages per volume are numbered 1 to 1000. We want an epsem sample of about 1 percent of the estimated 75,000 entries. For six alternative sampling designs (*a* through *f*) state Yes or No to the questions: (1) Is it an epsem sample? (2) Is it a simple random sample (srs) of initial lines? If epsem, but not srs, identify the selection process with a few words. If not epsem describe departure in one sentence.

(*a*) Take every 100th page after a random start from 1 to 100, and include all initial lines found on the selected pages.

(*b*) Take an srs of 1 percent of the pages and include all initial lines found on the page. Draw four-digit random numbers; the first digit identifies the volume and the last three the page in the volume.

(*c*) Take a 1 percent srs of the pages and include any entry if any part of it falls on a selected page.

(*d*) Select pages by srs and take the first initial line on the page. Continue until 750 entries are selected.

(*e*) Select 250 pages by srs; then select three entries with srs from the initial lines on each page.

(*f*) Draw six-digit random numbers. The first number identifies the volume; the next three numbers identify the page in the volume; the last two numbers identify the ordered initial line on the page. If any of these denote a nonexistent volume, page, or initial line, the draw results in no selection. Continue until 750 different, valid, initial lines are selected.

7.8. Assume now that the entries (the initial lines) in 7.7 have actually been numbered consecutively from 1 to $N = 73{,}259$. Five new designs are suggested. For each of the five procedures below (*a* through *e*) answer Yes or No to questions (1) and (2) in 7.7.

(*a*) Select 750 different random numbers from 1 to N and take the corresponding entries.

(*b*) Select random numbers from 1 to N. Include in the sample not only the selected entry, but also all other initial lines found on the same page. Continue until 750 entries have been selected.

(*c*) Select 250 random numbers from 1 to N. The selected entry merely denotes the selection of the page on which it occurs. Now select with srs three of the initial lines on the same page.

(*d*) Select 250 random numbers from 1 to N. In addition to the selected entries, select with srs two more initial lines on the page. What would you do if a page contains only one or two initial lines?

(*e*) Select 250 random numbers from 1 to N. In addition to the selected entries, include also the two following entries, whether on the same page or not.

7.9. For the situation described in 7.7*b* an epsem design of entries is wanted. Give brief instructions for paired selections of pages from 100 strata. Give formulas for the mean per entry and its standard error.

7.10. For the situation described in 7.8*c* an epsem design of entries is wanted, selected as $a = 150$ subsamples of $b = 5$ entries each. Give brief instructions for paired selections from 75 strata. Give formulas for the mean per entry and its standard error.

7.11. Design an epsem selection of members of an organization with $f = 1/200$. First the branches are sorted into strata, so that $\sum_{\alpha} M_{h\alpha}$ denotes the sum of measures in the hth stratum. Despite attempts to keep them roughly equal, the stratum sizes $\Sigma M_{h\alpha}$ do differ some. In the first stage of selection, two branches are selected from each stratum with probabilities proportional to their sizes $M_{h\alpha}$; the selection probability is $2M_{h\alpha}/\Sigma M_{h\alpha}$. In the second stage, members are selected systematically from the selected branches. (*a*) Write the formula for computing the *intervals* for selecting members if the αth branch is selected in the hth stratum. (*b*) Compute these intervals for each branch of a stratum composed of these measures: 2000, 1200, 760, 40. (*c*) Compute the subsample sizes obtained if the actual sizes were equal to the measures. (*d*) Compute the subsample sizes if the actual sizes were 3000, 1500, 760, 30. (*e*) Suppose now that the stratum contained a fifth unit with 1000 members; recompute the intervals (*b*) and the subsample sizes (*c*). (*f*) Show the formula for computing the mean of an entire sample composed of 37 strata and its standard error.

7.12. A social security system contains a population of about $N = 12{,}000{,}000$ cards. These are kept in $A = 2400$ offices, from about 2000 to 30,000 in each. An epsem selection of cards with $f = 1/2000$ is to be selected in two stages. We aim to select 100 offices with *roughly* 60 cards from each selected office. Thus each group of about 120,000 cards should be represented by a sample office with roughly 60 selected cards. In the second stage d_α three-digit ending numbers will be selected, leading to a subsampling rate of $d_\alpha/1000$. We aim at roughly equal subsamples by selecting more ending digits in the smaller offices.

In the first stage, measures are assigned by first computing the ratio $N_\alpha/60{,}000$, where N_α is the number of cards in the αth office. Then the nearest value of $1/d_\alpha$ is assigned as the office's size measure. These measures are the reciprocals of the integers d_α, taken to three decimal places. The measures are then accumulated, and a paired selection is made with PPS from each stratum of 4.000 cumulated measures. The selection formula in two stages is:

$$\frac{2(1/d_\alpha)}{4.000} \cdot \frac{d_\alpha}{1000} = \frac{1}{2000}.$$

The number 60,000 represents 1000 times 60, the subsample size. The size 4.000 for strata leads to the overall rate $f = 1/2000$. It also yields paired selections from each stratum of about 240,000 cards. If the measures total exactly 200, there will be exactly 100 selections from 50 strata.

(*a*) Compute the measures for a stratum composed of offices with the following numbers of estimated thousands: 30, 30, 20, 20, 20, 15, 15, 15, 15, 12, 12, 12, 12, 6, 3.6, 2.4. Select two offices with the random numbers 0.020 and 3.980.

(*b*) Show how the selection of the offices with 30,000 and with 2400 both lead to $f = 1/2000$, and to subsamples of 60 expected cards.

(*c*) Show formulas for the sample mean and its standard error.

(*d*) Suppose that the population is $N = 6{,}000{,}000$, and that you want subsamples of roughly 200 cards each from 60 offices, for 30 pairs of selections with $f = 1/500$. Show the method for computing selection measures and the two-stage selection formula.

7.13. A two-stage epsem selection of adults in a city involves listing all adults found in sample blocks and selecting them systematically from the list. Adults average two per dwelling. The selection formula is $\text{Mos}_\alpha/1000 \times 5/\text{Mos}_\alpha = 1/200$. The measures of size Mos_α are numbers of dwellings counted in the blocks by the census four years ago. You are free to assume either a systematic selection of blocks or a paired selection from implicit strata. (*a*) In a few sentences give instructions for selecting blocks. (*b*) What would you do for blocks which have $\text{Mos}_\alpha > 1000$? (*c*) What would you do for blocks which have $\text{Mos}_\alpha < 5$? (*d*) Give reasons (5 to 7) why most subsample sizes will not equal either 5 or any other fixed number.

7.14. (*a*) Continue paired selections of Table 7.4.I, applied to the blocks of Appendix E. (*b*) Show formulas for computing a ratio mean and its variance. (*c*) Where necessary combine blocks so that each has at least a sufficient size of 10 measures.

7.15. Complete the systematic selections of 7.4.I as in (*a*)–(*c*) of 7.14.

8

The Economic Design of Surveys

8.1 PLANNED PRECISION BASED ON UNIT VARIANCE

Section 2.6 describes the planning of simple random samples to determine the expected variance for a fixed sample size, or the sample size necessary for a fixed desired variance. Now we generalize that method to a large variety of surveys. With proper interpretation of the terminology, it can be extended to many major designs. To begin, we state the variance of the mean in the generalized form:

$$\text{var}(\bar{y}) = \frac{1-f}{a} s_g^2, \tag{8.1.1}$$

where f is the overall uniform sampling fraction. Here a denotes *the sample size in numbers of primary selections*, whether elements or sample clusters. The s_g^2 represents the *unit variance* of a single primary selection; and it denotes the variation among the *means per element* of primary selections. Let us clarify these terms by stating their specific meaning for several major designs.

For all *element sampling* $a = n$, the number of elements in the sample. Specifically, for a simple random sample we had $\text{var}(\bar{y}_0) = (1-f)s^2/n$ (2.2.2). For proportionate sampling, the element variance is taken *within the strata*, so that $\text{var}(\bar{y}_{\text{prop}}) = (1-f)s_w^2/n$ (3.4.2). These led to the design of simple random samples in 2.6, and of proportionate element samples in 3.4B.

For *cluster sampling*, a represents the number of *primary selections*; s_g^2 represents the *unit variance of a single primary selection*, when this variance is computed from the *per element means* of the primary selections. For equal and compact clusters (5.2.3), the s_g^2 stands for $s_a^2 = \frac{1}{a-1} \Sigma (\bar{y}_a - \bar{y})^2$. In the subsampling of equal clusters, we have again $s_a^2 = \frac{1}{a-1} \Sigma (\bar{y}_a - \bar{\bar{y}})^2$. See (5.3.3) and Remark 5.3.V.

For clusters of unequal size, a similar form can be still maintained by introducing the weighted deviations of cluster means (6.3.6): $s_r^2 = \frac{1}{a-1} \Sigma \left[\frac{x_a}{x/a} (\bar{y}_a - r) \right]^2$, with s_r^2 representing s_g^2. We can utilize this form also for stratified sampling of unequal clusters, provided we understand that the variance of the unequal clusters is taken within strata established by the design. See Remark 6.3.I.

We can utilize this simple relationship between var $(\bar{y})$ and a and s_g^2, so long as we keep a fixed meaning for s_g^2 by keeping the sample design fixed. That means that the population and the strata, also the methods and rates of subsampling within primaries, are all assumed to remain fixed. This indicates that both the overall sampling fraction f and the number of elements in the sample n vary directly with the number of primary selections a. For example, comparing the sampling fractions, f and $\tilde{f}$, of two sample sizes in a three-stage design we should have, respectively,

$$\frac{a}{A} \cdot \frac{b}{B} \cdot \frac{c}{C} = \frac{n}{N} = f, \qquad \text{and} \qquad \frac{\tilde{a}}{A} \cdot \frac{b}{B} \cdot \frac{c}{C} = \frac{\tilde{n}}{N} = \tilde{f}. \tag{8.1.2}$$

Formulating the design problem this way permits the application of the simple relationship (8.1.1) to a large variety of designs that would be difficult to handle otherwise. Most cluster samples are stratified, and the variance formulas become complicated. A complete and explicit model for a three-stage sample, stratified in each stage, would require the estimation of six components of the total variance. See the complicated formulation of even a relatively simple design in Example 5.6*b*. Instead, we can often resort to some simplified procedures, which are illustrated below. Given the variance of a sample mean, based on a primary selections of a past survey, we can compute

$$s_g^2 = \frac{a}{(1-f)} \text{var}(\bar{y}), \tag{8.1.3}$$

and use it as the unit variance of a primary selection, to plan another survey with the same design.

Given an estimate of s_g^2, we can compute the sample size $\tilde{a}$ needed to achieve a desired variance V^2. From (8.1.1) we find that

$$\tilde{a} = \frac{(1-\tilde{f})s_g^2}{V^2}. \tag{8.1.4}$$

Since $\tilde{f}$ also depends on a, it is easier to solve this equation in two steps, as in (2.6.2): first find the preliminary $a' = \tilde{a}(1-\tilde{f})$, and then correct for the $(1-\tilde{f})$. Thus,

$$a' = \frac{s_g^2}{V^2}, \quad \text{and} \quad \tilde{a} = \frac{a'}{(1+f')} \quad \text{or} \quad \tilde{f} = \frac{f'}{(1+f')}. \tag{8.1.5}$$

From the preliminary number a' of primary selections we complete f', the preliminary sampling fraction. This overall fraction has specific meaning based on the details of a fixed design. It also implies that $\tilde{a}/a' = \tilde{f}/f'$, and the desired fraction $\tilde{f}$ is readily computed.

Formula (8.1.5) is proved by noting that $a' = \tilde{a}/(1-\tilde{f})$ implies $f' = \tilde{f}/(1-\tilde{f})$; hence $(1+f') = (1-\tilde{f}+\tilde{f})/(1-\tilde{f}) = 1/(1-\tilde{f})$; then $\tilde{a} = a'(1-\tilde{f}) = a'/(1+f')$.

Example 8.1a In Example 6.3*a* we found for the mean r of a random sample of $a = 20$, from $A = 270$ blocks, the value var $(r) = 0.001742$. We have, from (6.3.6),

$$\text{var}(r) = \frac{1-f}{a} s_r^2 = \frac{0.926}{20} s_r^2 = 0.00174.$$

We compute $s_r^2 = \dfrac{a}{1-f}\text{var}(r) = \dfrac{20}{0.926} 0.001742 = 0.03765$ for the variance of the mean of a single random block. From it, the variance of the mean of a random sample of 40 blocks should be predicted as

$$\text{var}(y) = \left(1 - \frac{40}{270}\right)\frac{1}{40}(0.03765) = 0.852 \times 9.41 \times 10^{-4} = 8.02 \times 10^{-4}.$$

Suppose we need a standard error of $V = 3.5$ percent; how many sample blocks are needed? Since the desired $V^2 = 0.035^2 = 12.25 \times 10^{-4}$, we find the preliminary

$$a' = \frac{s_r^2}{V^2} = \frac{376.5}{12.25} = 30.7 \quad \text{and} \quad f' = \frac{30.7}{270} = 0.114;$$

then $a = a'/(1+f') = 30.7/1.114 = 28$ blocks are needed.

Example 8.1b In Example 5.5*a*, for a sample of 40 clusters drawn (with $f = 0.01$) as paired selections from 20 strata, the variance of the sample mean was computed as var $(\bar{y}) = 13.30 \times 10^{-4}$. The single cluster variance is

$$s_a^2 = \frac{40}{0.99} 0.001330 = 0.05373.$$

For a sample of 80 clusters, we may compute

$$\text{var}(\bar{y}) = \frac{0.98}{80} 0.05373 = 0.000659.$$

This estimate would be entirely proper for four selections each from 20 strata. More likely, we would take paired selections from 40 strata. Additional stratification tends to yield some new gains, but in ordinary situations these could be ignored. Furthermore, we can compute rapidly that by doubling the number of primary selections from 20 to 40, we decrease the variance by a factor of 2, from 13.30×10^{-4} to 6.65×10^{-4}. This neglects the trivial change in $(1-f)$ from 0.99 to 0.98; hence, the corrected answer is 6.59×10^{-4}.

Example 8.1c In Section 6.5*B*, we computed $\text{var}(r) = 5.66 \times 10^{-4}$ for the mean of paired selections from 10 strata. The finite population correction is absent. Hence, we predict that 40 selections from the same 10 strata should yield a variance $5.66/2 \times 10^{-4} = 2.83 \times 10^{-4}$. Increasing the number of strata to 20, again for paired selections, should decrease the variance only slightly.

Alternatively, we can compute the unit variance of the primary selection as $s_r^2 = 20(5.66 \times 10^{-4}) = 113.2 \times 10^{-4}$. Then the variance of 40 units becomes $113.2 \times 10^{-4}/40 = 2.83 \times 10^{-4}$.

8.2 ESTIMATES OF UNIT VARIANCE; DESIGN EFFECT

How can we obtain the estimates of S_g^2 needed for the design formulas? Basically, this is a more complex form of the problem of estimating S^2, discussed in 2.6.

Past surveys of similar variables may provide data for computing s_g^2, as in the examples of 8.1. If the past variables differ from the planned data, the difference may be narrowed with judicious adjustments. Utilization of past surveys can be facilitated by the relative variance $C_g^2 = S_g^2/\bar{Y}^2$, when this is more comparable between surveys than S^2, and if estimates of $\bar{Y}^2$ are available.

Given $\text{var}(\bar{y})$ and a from a past survey, the planned variance $\widetilde{\text{var}}(\bar{y})$ or size $\tilde{a}$ of a future sample can be designed directly from

$$\frac{\tilde{a}}{a} = \frac{\text{var}(\bar{y})}{\widetilde{\text{var}}(\bar{y})} \cdot \frac{(1-\tilde{f})}{(1-f)}. \tag{8.2.1}$$

This supposes a fixed sample design, so that the unit variance is constant and $s_g^2 = a \text{ var}(\bar{y})/(1-f) = \tilde{a}\, \widetilde{\text{var}}(\bar{y})/(1-\tilde{f})$. It also assumes epsem sampling, and modifications are required for varied sampling fractions if these are not negligible. The factors $(1-\tilde{f})$ and $(1-f)$ are often negligible, and the sample sizes are inversely proportional to the variances in the two samples.

Except for the fpc's, the size of the population has no effect on the size of the needed sample. However, if we were to design for estimating an aggregate $N\bar{y}$ for a fixed absolute standard error, the population size would become important indeed (see 2.6). It is important and obvious—but so easy to forget!—that the size of the sample is inversely proportional to the *square* of the standard error. That is, forgetting about the "fpc":

$$\frac{\tilde{a}}{a} \doteq \left[\frac{\text{se}(\bar{y})}{\widetilde{\text{se}}(\bar{y})}\right]^2. \tag{8.2.2}$$

To decrease the standard error by a factor k, the sample size must be increased by the factor k^2.

Most studies are too small and too hurried to support a large enough *pilot sample* to yield useful estimates of S_y^2, of variance components, and of cost factors. If the pilot study is too small, its results may seem useless when they are much less dependable than expert guesses. On the other hand, a *sequential sample based on several samples* can sometimes be planned by building flexibility into the design (8.4D).

We need a technique for circumventing the difficulties of designing complex samples. The *design effect* or *Deff* is the ratio of the actual variance of a sample to the variance of a simple random sample of the same number of elements:

$$\text{Deff} = \frac{\text{Var}(\bar{y})}{(1-f)S^2/n} \quad \text{and} \quad \text{Var}(\bar{y}) = (1-f)\frac{S^2}{n}\cdot\text{Deff}. \tag{8.2.3}$$

This device is commonly used by practical samplers, and it deserves, I think, explicit recognition. In many situations where we cannot estimate directly either the variance components or S_y^2, we may be able to guess fairly well both the element variance S^2 and the Deff, from experience with similar past data. This comprehensive factor attempts to summarize the effects of various complexities in the sample design, especially those of clustering and stratification. The design effect may include even the effects of ratio or regression estimation, of double sampling, and of varied sampling fractions. Its ready applicability leads many samplers to include the ratio of the actual variance to $(1-f)s^2/n$ as a routine item in the variance computations.

We established (2.8.23) that s^2 computed from any large probability sample yields a good approximation of S^2. Actually $(n-1)s^2/n + \text{var}(\bar{y})$ would be a better approximation, and this equals $(n-1)s^2/n + s^2\,\text{deff}/n = s^2[1 + (\text{deff} - 1)/n]$. Hence s^2 is a good approximation when deff is near 1; in other cases it neglects a term of order $1/n$. I feel that the simplicity of using merely s^2 justifies ignoring the negligible improvement. (In the mean square error of s^2 the variance is of order $1/n$, whereas the neglected bias^2 is of order $1/n^2$.) Furthermore, when estimating Deff in (8.2.3) the precision of var $(\bar{y})$, based on a clusters, is much weaker than of s^2, based on n elements (8.6D).

We have already computed this ratio in examples for several designs. For stratified element samples Deff is typically less than one, and it expresses the variance reduction due to stratification, proportionate—which is S_w^2/S^2 (4.6.5)—or disproportionate (4.6.11). In cluster samples, the ratio is typically larger than one, expressing the losses due to clustering. In random selection of equal clusters (5.4.2), it expresses the effect of $[1 + \text{roh}(b - 1)]$ due to the homogeneity of the sample clusters. This can be extended as an approximation for $[1 + \text{roh}(n/a - 1)]$, where n/a is the average size of unequal clusters; this approximation holds well if the sizes of clusters are not extremely unequal, a caution observed in most good designs. In stratified clustered samples, the roh can be interpreted in terms of cluster homogeneity within the established strata.

We may use interchangeably the concepts of design effect and of the *element variance*, developed especially in Remarks 5.3.V and 6.4.I. The element variance is $V_v^2 = S^2$ Deff, estimated by $v_v^2 = s^2$ deff; v_v^2 is the variance of single elements incorporating all complexities of the sample design. Hence v_v^2 varies with every change in sample design, whereas s^2 refers to the same element variance in the population, disregarding the sample design. For $\bar{y} = y/n$ it would be computed as $v_v^2 = n \cdot \text{var}(\bar{y})/(1-f) = \text{var}(y)/n(1-f)$. For a ratio mean $r = y/x$, it is $x \text{ var}(r)/(1-f) = \text{var}(z)/x(1-f)$.

With this concept we may utilize the srs design formulas of 2.6 substituting v_v^2 for s^2. The concepts of element variance and element cost are combined in (8.3.2) for comparing the economies of sample designs.

We define two related concepts which are also helpful occasionally. Since $\text{var}(\bar{y}) = (1-f)s^2 \text{ deff}/n = (1-f)s^2/(n/\text{deff})$, we may define n/deff *as the estimate of the effective* n'. Thus when the effect of clustering is large, we can think of it as reducing the effectiveness of the sample size. If we know or guess a value of deff, we may use $n' = n/\text{deff}$ in srs design formulas.

Suppose that instead of $s\sqrt{\text{deff}}/\sqrt{n}$ somebody uses $s/\sqrt{n}$ for probability intervals such as $\bar{y} \pm t_p s/\sqrt{n}$. This is equivalent to using

$$\bar{y} \pm (t_p/\sqrt{\text{deff}})(s\sqrt{\text{deff}}/\sqrt{n}).$$

Hence we may consider that failure to adjust probability intervals for $\sqrt{\text{deff}}$ amounts to using *an effective* $t_p' = t_p/\sqrt{\text{deff}}$. For example, $\sqrt{\text{deff}} = 1.5$ causes a decrease from $t_p = 2.0$ to $t_p' = 2.0/1.5 = 1.34$, hence an increase of the error rate from 5 percent to 9 percent; or from $t_p = 3.0$ to $t_p' = 3.0/1.5 = 2.0$, hence an increase of the error rate from 0.2 percent to 5 percent.

When designing samples, we should not only consider the entire sample, but also pay attention to important subclasses for which estimates will be made in the analysis. Complete and separate designs for numerous subclasses would demand too much effort. However, we may assume as a reasonable rough approximation that S^2 is the same for subclasses as for the population. Then, if $\bar{M}_a$ is the fraction of a subclass in the population, $\bar{M}_a^{-1}S^2/n$ is the expected variance of its mean in an srs of size n. The difference of the means of two subclasses, which account for fractions $\bar{M}_a$ and $\bar{M}_b$ of the entire sample, would be $(\bar{M}_a^{-1} + \bar{M}_b^{-1})S^2/n$. Taking the design effect into account we should use S^2 Deff instead of S^2.

The design effect may be used to relate the expected variance of an actual design to simple random variance. This, in turn, can often be anticipated approximately, with some shrewd guesses about the population distribution. To aid these guesses is the aim of the following remarks about some simple distributions.

The binomial distribution has been the subject of many remarks, beginning with 2.4, because it is so common in survey work. For low values of P the variance $\sigma^2 = P(1-P)$ rises almost linearly with P, then it flattens out at $\sigma^2 = 0.25$ for $P = 0.5$. The standard deviation $\sigma = \sqrt{P(1-P)}$ is much less sensitive to changes in P. Whereas σ^2 and σ are symmetrical around $P = 0.5$, the relvariance $\sigma^2/\bar{Y}^2 = (1-P)/P$ decreases monotonically with increasing P. For low P values it is large, and nearly equal to $1/P$; it becomes 1.0 at $P = 0.5$; then it decreases to 0 for $P = 1$. The coefficient of variation $\sigma/\bar{Y} = \sqrt{(1-P)/P}$ is, of course, much less sensitive.

$\bar{Y} = P$	0.001	0.005	0.01	0.05	0.10	0.20	0.30	0.50	0.70	0.80	0.90
$\sigma^2 = P(1-P)$	0.001	0.005	0.010	0.048	0.09	0.16	0.21	0.25	0.21	0.16	0.09
$\sigma = \sqrt{P(1-P)}$	0.03	0.07	0.10	0.22	0.30	0.40	0.46	0.50	0.46	0.40	0.30
$\sigma^2/\bar{Y}^2 = (1-P)/P$	999	199	99	19	9.0	4.0	2.3	1.0	0.42	0.25	0.11
$\sigma/\bar{Y} = \sqrt{(1-P)/P}$	31.6	14.1	9.9	4.4	3.0	2.0	1.5	1.0	0.66	0.50	0.33

TABLE 8.2.I The Curves of Binomial Variation

Within clusters of size M, if the variation is binomial with the mean proportion P, the mean content of the units will be MP, the unit variance $MP(1-P)$, and the unit standard deviation $\sqrt{MP(1-P)}$. When the unit size M is large, the mean content MP may still remain small for rare characteristics, when P is very low and $(1-P)$ is nearly 1. The contents of different units ($y_i = 0, 1, 2, 3, \ldots$) may then have a *Poisson distribution* with mean MP, variance MP also, and the standard deviation $\sqrt{MP}$. The unit relvariance is $1/MP$, and the coefficient of variation $1/\sqrt{MP}$. However, inequalities of cluster size would tend to increase this to

$\sqrt{C^2 + 1/MP}$ (11.8C). For example, the distribution in area sampling units of persons with a congenital defect may approach this. But a contagious disease would tend to have larger relvariance, because both zero and large values would be more common.

The discrete rectangular distribution is worth describing, because it may be approximated by some actual distributions. First, take the simple distribution of $K + 1$ elements with values $0, 1, 2, 3, \ldots, K$; its mean is obviously $K/2$, and its variance $\sigma^2 = K^2[1/12 + 1/6K]$. This results from

$$\sigma^2 = \frac{1}{K+1}\sum_{i=0}^{K} i^2 - \frac{\left(\sum_i i\right)^2}{(K+1)^2} = \frac{K(K+1)(2K+1)}{6(K+1)} - \frac{K^2(K+1)^2}{4(K+1)^2}$$

$$= \frac{4K^2 + 2K - 3K^2}{12} = \frac{K^2}{12} + \frac{K}{6}.$$

The element variance is $\sigma^2 = K^2[1/12 + 1/6K]$, regardless of what uniform weights are placed at the $K + 1$ points, and we prefer to use relative frequencies with the constant weights $W_h = 1/(K + 1)$ at all $K + 1$ points. To make the model consistent with others, reduce the distance between the first and last points from K to the unit distance, with the distance $1/K$ separating neighboring pairs of the total of $K + 1$ points. This transformation reduces the mean to $1/2$ and the variance to $\sigma^2 = (1/12 + 1/6K)$. The variances for several values of K become

K	1	2	5	10	100	1000	10,000	∞
σ^2	0.25	0.167	0.1167	0.10	0.085	0.0835	0.08335	1/12

The variance for $K = 1$ coincides with the binomial for $P = 0.5$. For large values of K it approaches the $1/12$ shown for the continuous rectangular distribution in Table 8.2.II.

In that table, we present simple formulas for means and variances of combinations of regular geometric figures, with the weights W_h for its several discrete portions, where $\Sigma W_h = 1$ for the unit area. The mean is $\bar{Y} = \Sigma W_h \bar{Y}_h$, if the $\bar{Y}_h$ are the means of the discrete portions. The population variance is $\sigma^2 = \Sigma W_h[\sigma_h^2 + (\bar{Y}_h - \bar{Y})^2]$, where the σ_h^2 are the variances within the discrete portions, and $[\sigma_h^2 + (\bar{Y}_h - \bar{Y})^2]$ is the second moment of a discrete portion around $\bar{Y}$. For example, the binomial yields $\bar{Y} = Q \cdot 0 + P \cdot 1 = P$, and $\sigma^2 = Q(0 + P^2) + P(0 + Q^2) = QP^2 + PQ^2 = PQ$.

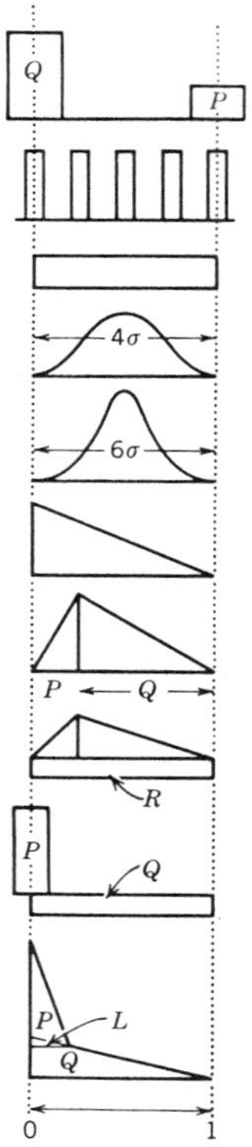

	Mean	Variance	Special Cases
A	P	PQ	$P = \frac{1}{2} \quad \sigma^2 = \frac{1}{4}$ max
B	$\frac{1}{2}$	$\frac{1}{12} + \frac{1}{6K}$	$K \to \infty \quad \sigma^2 \to \frac{1}{12}$ min
C	$\frac{1}{2}$	$\frac{1}{12}$	
D	$\frac{1}{2}$	$\frac{1}{16}$	
E	$\frac{1}{2}$	$\frac{1}{36}$	
F	$\frac{1}{3}$	$\frac{1}{18}$	
G	$(1 + P)/3$	$(1 - PQ)/18$	$P = \frac{1}{2} \quad \sigma^2 = \frac{1}{24}$ min
H	$\frac{R}{2} + \frac{(1 - R)(1 + P)}{3}$	$\frac{(1 - R)(1 - PQ)}{18} + \frac{R}{12} + \frac{R(1 - R)(1 - 2P)^2}{36}$	
I	$\frac{Q}{2}$	$\frac{Q}{12}(1 + 3P)$	$P = \frac{1}{3} \quad \sigma^2 = \frac{1}{9}$ max
J	for $L \neq 0$: $\frac{Q + PL}{3}$	$\frac{Q + PL^2 + 2PQ(1 - L^2)}{18}$	$L = \frac{2Q}{1 + 2Q} \quad \sigma^2 = \frac{3Q}{18(1 + 2Q)}$ min
	for $L = 0$: $Q/3$	$\frac{Q}{18}(1 + 2P)$	$P = \frac{1}{4} \quad \sigma^2 = \frac{1}{16}$ max

TABLE 8.2.II Variances of Several Finite Distributions

To facilitate comparisons, the distributions are presented with unit areas and unit width; if the width is changed to K, the variance is changed by K^2. Irregularities and discreteness of actual distributions would tend to increase the variances of smooth distributions. For example, $\sigma^2 = \frac{1}{12}$ of C is increased to $\sigma^2 = \frac{1}{12} + \frac{1}{6K}$, for the discrete rectangular of $K + 1$ points, shown with $K = 4$ in B. With stratification the variances may be decreased. Other distributions may be obtained by combining simpler forms. Thus, G, H, I, and J were obtained from C and F.

Note also that for any population of only *two portions* P and Q, we have $\bar{Y} = P\bar{Y}_p + Q\bar{Y}_q$, and $\sigma^2 = P[\sigma_p{}^2 + (\bar{Y}_p - \bar{Y})^2] + Q[\sigma_q{}^2 + (\bar{Y}_q - \bar{Y})^2] = P\sigma_p{}^2 + Q\sigma_q{}^2 + PQ(\bar{Y}_p - \bar{Y}_q)^2$. For example, the binomial variance is $\sigma^2 = P \cdot 0 + Q \cdot 0 + PQ(1-0)^2 = PQ$.

With the aid of these simple formulas and with the use of variances of the rectangular (C) and triangular (F) distributions, we can obtain the variances of composite geometrical figures, such as G, H, I, and J in Table 8.2.II. Some of these are reasonable approximations for actual curtailed distributions that arise in practice. For example, the two triangles in J can be made to fit approximately a curtailed skewed distribution.

8.3 MODELS OF COST FUNCTIONS

8.3A A General Model for Cost Factors

Cluster sampling ordinarily results in greater variance than a sample of the same number of elements selected individually. On the other hand, the cost per element is lower, sometimes much lower, for cluster sampling than for element sampling. Clustering should be used when its effect on lowering the cost per element is greater than the increase in the element variance. Furthermore, both the cost and the variance effects tend to increase with the degree of clustering, hence with the size of the clusters (5.5). Both must be determined for economic design.

Element sampling is preferable when it does not greatly increase the cost. Examples are: mailed questionnaires to a sample from a good list; a telephone survey in a city; interview survey of a sample drawn from a good city directory. On the contrary, large clusters may be used if these decrease the costs substantially without important increases in the variance. Perhaps the search for some rare and noninfectious diseases has these characteristics.

However, when clustering has substantial and opposing influences on cost and on variance, economic design depends on estimating and balancing those influences. It depends on making the right choices for sampling units. If clusters are chosen, then their kind, size, and number must also be decided. Our chief concern here is with clustering, the most important and common source of cost problems; but modifications of the model will facilitate working on other problems as well.

The aims of economic sample design may be stated simply in two alternative ways: minimum variance for fixed cost, or minimum cost for fixed variance. (If bias is also a factor, it should be included by using the mean square error instead of the variance, as in 13.1.) Being two expressions of the same principle, they generally lead to the choice of the same design. It would be desirable to have a single mathematical expression, the solution of which would lead to the optimum among all possible designs. This would have to incorporate at once, and in useful detail, the relative merits of all designs and all factors relevant to deciding among them. Although that is not possible, modern computers will offer new approaches.

The possible variations in sample design are numerous indeed. For example, in multistage sampling we must decide the nature of clusters for each stage; also the number of sampling units; hence, their size in numbers of sample elements; whether to select with random or systematic selection, the nature of stratification, the probabilities of selection, etc. All the cost factors would have to be guessed for each alternative design, and the appropriate variances estimated and compared. Therefore, in practice, the sampler must confine his comparisons to a few feasible designs, to which he is guided by his judgment, experience, reading, and consultants. Some classes of designs can be dismissed rapidly and without formal comparisons, for obvious reasons. The differences between the compared designs can be confined to simple changes, such as the size and number of clusters. Often we can take advantage of the monotonic nature of many cost and variance relationships; e.g., if b is found to be a more economical cluster size than $2b$, we can usually assume that it is also more economical than size $4b$.

Here we attempt to facilitate the comparison of designs with a general expression applicable to most situations. The great variety of expenditures of a survey can be summarized by four categories. This classification is more or less arbitrary, but it helps to simplify complex reality. Thus the total cost (T) of a survey is expressed as the sum of four component classes, each of which may contain several distinct factors:

$$T = K + C = K + K_v + nc + nc_v. \tag{8.3.1}$$

1. K is the class of *constant* cost factors, which do not change either with the number of sample elements nor with the type of sample design used. These would often include most of the costs of: (*a*) discussion and design of survey objectives and of the sample; (*b*) construction of the questionnaire or other instruments of observation; (*c*) analysis of results (not involving differential costs of estimates); (*d*) preparation, printing, and circulation of survey reports; (*e*) various overhead costs. Most of the analysis and presentation of the substance of the survey results belongs in this class.

2. K_v designates the class of cost factors which vary with changes in design, but not with the number of sample elements; the subscript stands

for the vth design. These might comprise the greater part of such factors as: (*a*) computation of sample estimates and their variances; (*b*) weighting of sample results; (*c*) some sampling office work, purchase of sampling materials, and preparation of sampling instructions; (*d*) training of field workers in sampling. For example, point (*a*) can refer to the cost difference of computing variances in cluster sampling versus the simple formulas (pq/n) of simple random sampling. Point (*b*) can refer to the costs of bookkeeping and machine weighting entailed in disproportionate sampling versus self-weighting samples.

3. nc denotes the total cost of the class of factors which are proportional to the number n of sample elements, but which are not affected by changes in sample design. These factors would tend to include: (*a*) making and recording observations, such as interviewing; (*b*) coding and machine-punching the observations; (*c*) some work in the sampling office. Note that the cost per element c need not be the same for all parts of the sample, but may be only an *average* cost per element.

4. nc_v denotes the total cost of the class of factors that are proportional to the number of sample elements, but that also vary with changes in design. Several factors would tend to fall here: (*a*) training and maintaining a force of field workers depends on the number of locations; (*b*) locating respondents depends on the travel cost among them; (*c*) the cost of an adequate frame can differ for different sample designs; (*d*) the costs of preparing sampling materials and selection procedures can differ also.

Note that c_v need be only the *average* cost per element for the design v, permitting variations within the entire sample. Since only a few aspects of the design can be specified, the others are taken simply on the average.

The costs proportional to the sample elements are $n(c + c_v)$, containing two factors, one constant and the other variable. The list of items appearing under the four classes of components above does not exhaust the possible kinds of expenditures; nor can the boundaries among cost factors be always drawn neatly. However, we can consider the nature of each important cost item, and either assign it to one of the components, or divide it into parts and assign these to the components.

Ordinarily the sampler has no precise data on cost factors, and must base his decisions on estimates or guesses. Often he can make good enough guesses to eliminate designs that would be obviously uneconomical. Moderate errors in estimating cost factors tend to result in only small departures from the optimal design. A good cost model helps to ask the right questions and to make good guesses.

The relative advantages of designs can be expressed in terms of the element cost and the element variance. Furthermore, using the cost model (8.3.1) and the design effect (8.2.3), a measure of the economy of two designs, v and v', has the ratio:

$$\frac{\text{Cost}_v \times \text{Var}_v}{\text{Cost}_{v'} \times \text{Var}_{v'}} = \frac{(\text{element cost} \times \text{element variance})_v}{(\text{element cost} \times \text{element variance})_{v'}} = \frac{(c + c_v) \times \text{Deff}_v}{(c + c_{v'}) \times \text{Deff}_{v'}}. \tag{8.3.2}$$

Design v' (or v) is preferable when the ratio is greater (or less) than one. This comparison of relative economy gives similar results whether we determine relative cost for fixed variance, or relative variance for fixed cost. Thus if "Cost" denotes the cost for a unit variance, and "Var" denotes the variance for a unit cost, the above ratio defines $\text{Cost}_v/\text{Cost}_{v'} \times \text{Var}_v/\text{Var}_{v'}$. A design is preferable if it has either a smaller cost per unit variance or a smaller variance per unit cost.

Observe that the larger the constant element cost c, the more it dominates the cost ratio, and the less important becomes the variable cost c_v. Thus for more expensive interviews we can afford greater travel costs, hence a more widespread sample.

The element cost $(c + c_v)$ in (8.3.2) disregards the factor K_v that may vary with the design, but not with n. When important, its effect should not be neglected; and it may be considered separately, as in (8.3.2). Or it may be included in the formula,

$$\frac{\text{Cost}_v \times \text{Var}_v}{\text{Cost}_{v'} \times \text{Var}_{v'}} = \frac{(c + c_v + K_v/n) \times \text{Deff}_v}{(c + c_{v'} + K_{v'}/n) \times \text{Deff}_{v'}}. \tag{8.3.3}$$

This has the disadvantage that K_v/n varies with n, hence is not a fixed unit cost. But within small relative variations in n, it may be considered a constant small addition to the other factors.

If the design effect is mainly due to clustering, it can be expressed as $[1 + \text{roh}(b - 1)]$, where b is the size of the subsamples; or it may be the average size, if the size variation is not great. Then the economies of using either b or b' for subsample size compare as

$$\frac{1 + \text{roh}(b - 1)}{1 + \text{roh}'(b' - 1)} \times \frac{(c + c_v)}{(c + c_{v'})}. \tag{8.3.4}$$

Often $\text{roh} = \text{roh}'$ approximately; also $1 + \text{roh}'(b' - 1) = 1$ when $b' = 1$, as in Example 8.3*a*.

Example 8.3a In Example 5.2*a*, 40 clusters of 10 elements each were selected. The mean of the sample was 46.25 percent, and the variance of the mean was computed as var $(\bar{y}) = 25.85 \times 10^{-4}$. What size srs would give the same variance as the actual cluster sample? An approximate answer can be found by using (2.6.2) and considering $S^2 \doteq PQ \doteq pq$:

$$n' = \frac{S^2}{V^2} = \frac{46.25 \times 53.75 \times 10^{-4}}{25.85 \times 10^{-4}}.$$

Then $f' = n'/N = 96.17/39{,}800 = 0.0025$, and $n = n'/(1 + 0.0025) = 96$.

Thus the number of elements in the sample had to be increased by the factor $400/96 = 4.17$ to compensate for the effects of clustering. This is the same design effect found in 5.4a, and expressed as $[1 + \text{roh}(b - 1)]$; the large cluster, $B = 10$, and the large roh $= 0.35$ bring about this great contrast in the necessary sample size. To choose the more economical design we must also consider (estimate, guess, or assume) some cost factors. For a brief interview, with some callbacks to locate respondents, assume the following factors:

Cost of conducting and coding an interview	$c = \$2.00$;
Cost of locating separate addresses	$c_1 = \$1.00$;
Cost of locating an address in clusters of ten	$c_{10} = \$0.20$.

The field costs of locating and interviewing 400 addresses in clusters of 10 each can be computed as $400\,(2.00 + 0.20) = \$880$. The same field costs for the srs of 96 individual addresses is $96(2.00 + 1.00) = \$288$. Thus on the basis of the field costs alone the srs is considerably more economical than the clustered sample. Though the cost per element is decreased in the cluster sample by the factor of $3.00/2.20 = 1.36$, this decrease is more than overtaken by the increase of 4.17 in the variance per element. The ratio of $880/288 = 3.06$ is also the ratio 4.17/1.36. Moreover, the variance computations of the srs (as pq/n) cost almost nothing; but the variance computations of the cluster sample need be done from the separate tabulation of the cluster means. Therefore, include \$30 for K_{10}, giving a total cost of $C = 30 + 800 + 80 = \$910$ for the clustered sample.

The clustered sample of 400 addresses costs \$880. For that, we could interview $880/(2.00 + 1.00) = 293$ individual addresses. The variance of an srs sample of 293 interviews would be about

$$\text{var}\,(p_0) = 0.99\,\frac{46.25 \times 53.75}{293} \times 10^{-4} = 8.40 \times 10^{-4}.$$

For the fixed cost \$880, the variance of the srs is lower than the variance of the clustered sample by the ratio $25.85/8.40 = 3.08$. This is essentially the same as the ratio of 3.06 obtained in 8.2*a* when comparing the relative costs for the same variance.

The element cost is less only by 1.36 for the cluster sample than for the element sample. Therefore, the ratio of element variances must be below 1.36 (instead of 4.17), before the cluster sample could be considered. That ratio should also be low enough to allow for the greater K_v of the cluster sample.

Note, however, that the element sample becomes costlier if the travel costs increase to $c_1 = \$10$ and $c_{10} = \$2$, because $(2 + 10)/(2 + 2) = 12/4 = 3.00$ for the ratio of element costs. Now the cluster sample would be more attractive if the increase of element variance were considerably lower than 3.

On the contrary, if the constant element cost c is increased to \$20, then the ratio of element costs becomes $(20 + 1)/(20 + 0.2) = 21/20.2 = 1.04$. The travel cost, although in the ratio $1/0.2 = 5$ fades to insignificance compared to the constant element cost. Under these conditions, cluster sampling should be avoided.

8.3B *Specific Cost Function for Cluster Samples*

The above approach is general, but it depends on comparisons made by constructing separate estimates for each likely design, and on separate estimates for the cost components of each design. Greater flexibility can be attained when we can assume and construct models of functional relationships between cost components and specific factors of the design. A model for cluster sampling uses this simple equation:

$$C = nc + aC_a, \tag{8.3.5}$$

where a is the number of primary clusters in the sample and C_a is the cost per cluster. The assumption is that cost factors related to clustering can be expressed with a term directly proportional to the number of sample clusters. This may approximate fairly well some sampling situations. For example, in a sample of n dwellings of a city taken in a sample blocks, the main cost factor, beyond the cost c of interviewing, could be the listing of dwellings; this would be C_a per sample block. Another example could be a sample of a cities, in each of which a sample is drawn from a list (of housing construction, or retail stores, or school teachers); the cost per city consists of obtaining a field worker, cooperation from the city government, access to the list, and ancillary information about the city.

We can relate this specific function (8.3.5) to our general model (8.3.1) by noting that $n = ab$, where b is the number of sample elements per cluster, and

$$C = nc + n\frac{C_a}{b} = n\left(c + \frac{C_a}{b}\right). \tag{8.3.5'}$$

Thus the specific function assumes that $K_v = 0$ and that $c_v = C_a/b$; this variable factor, being constant per cluster, is inversely proportional to the number of sample elements per cluster.

In the general model (8.3.1) the cost factors must be estimated separately for every value of b, which expresses the degree of clustering. On the contrary, when the specific function (8.3.5) can be assumed, it permits studying the effects of varying b, the subsample size. Moreover, this function can lead to neat results. The optimum (most economical) design can be reached when the size of the cluster is fixed at

$$\text{optimum } b = \left(\frac{C_a S_b^2}{c S_a^2}\right)^{1/2}. \tag{8.3.6}$$

Here the variance components refer to the variance of a two-stage sample, where, from (5.6.8), Var $(\bar{y}) = [S_a^2/a + S_b^2/n] = [S_b^2 + S_a^2 b]/n$, when sampling *with* replacement, or neglecting the factors $(1 - f_a)$ and $(1 - f_b)$ in sampling without replacement. Even better, by using

$$\text{Var}(\bar{y}) = (S^2/n)[1 + \text{roh}(b - 1)] = (S^2/n)[(1 - \text{roh}) + \text{roh} \cdot b],$$

the most economical subsample size can be expressed as

$$\text{optimum } b = \left(\frac{C_a(1 - \text{roh})}{c \text{ roh}}\right)^{1/2}. \tag{8.3.7}$$

These are specific illustrations of a more general rule about optima. Often the element variance may be expressed as a linear function $V^2 = (w + Wb)$ increasing with b, and the element cost as a linear function $C = (c + C_a/b)$ decreasing with b, where the other four symbols denote any positive constants. Then the product V^2C denotes variance for unit cost, and $V^2C = cw + WC_a + cWb + wC_a/b$. Differentiating, we get $d(V^2C)/db = cW - wC_a/b^2$, and find that $b = \sqrt{wC_a/Wc}$ yields the minimum for V^2C. This means minimum cost for unit variance, or minimum variance for unit cost. It leads to (8.3.6) and (8.3.7) if we use the element variances in square brackets [ ] above, and the element cost (8.3.5′).

This is also shown in (8.5.12), as well as the

$$\text{optimum } b = \left(\frac{C_a S_b^2}{c S_u^2}\right)^{1/2}, \qquad \text{where } S_u^2 = S_a^2 - \frac{S_b^2}{B} \tag{8.3.6'}$$

for sampling *without* replacement, and taking account of the finite population corrections.

A complicated function usually would approximate reality better than the simple function (8.3.5). However, it probably would not lead to such simple results, and it would require the difficult and hazardous estimation of more cost factors. Therefore, we may apply the simple function (8.3.5) as a useful though crude approximation, and force it on the available cost data. From a single sample design we cannot obtain the needed data for separating the cost factors, except by using detailed and difficult bookkeeping of the components of the field costs. Often this seems too laborious, and we must be satisfied with using educated guesses about the cost components.

If we have an experiment which contrasts two different degrees of clustering (with other variables kept constant), from the difference of the two values for the cost per element we may separate the two unknowns c and C_a. Comparing the cost per element in the two treatments, from $C/n = c + C_a/b$ and $C'/n' = c + C_a/b'$, we can get and solve for

$$C_a = \left(\frac{C}{n} - \frac{C'}{n'}\right)\left(\frac{b'b}{b' - b}\right), \tag{8.3.8}$$

and for $c = C/n - C_a/b = C'/n' - C_a/b'$.

The cost function for clustering can be extended to several stages as $C = nc + aC_a + abC_b + \cdots$, but this may be difficult to treat. Instead, we can solve the problem by using (8.3.5) successively stage by stage, each time with a single component for selection within the clusters.

b \ k	0.1	0.2	0.5	1	2	5	10	20	50	100
1	1.00	1.00	1.00	1.00	1.00	1.00	1.00	1.00	1.00	1.00
2	0.95	0.92	0.83	0.75	0.67	0.58	0.55	0.52	0.51	0.50
3	0.94	0.89	0.78	0.67	0.56	0.44	0.39	0.37	0.35	0.34
5	0.93	0.87	0.73	0.60	0.47	0.33	0.27	0.24	0.22	0.21
7	0.92	0.86	0.71	0.57	0.43	0.29	0.22	0.184	0.160	0.151
10	0.92	0.85	0.70	0.55	0.40	0.25	0.182	0.143	0.118	0.109
20	0.91	0.84	0.68	0.53	0.37	0.208	0.136	0.095	0.069	0.059
30	0.91	0.84	0.68	0.52	0.36	0.194	0.121	0.079	0.052	0.043
50	0.91	0.84	0.67	0.51	0.35	0.183	0.109	0.067	0.039	0.030
70	0.91	0.84	0.67	0.51	0.34	0.179	0.104	0.061	0.034	0.024
100	0.91	0.84	0.67	0.51	0.34	0.175	0.100	0.057	0.029	0.020

TABLE 8.3.I Variations in the Ratio of Element Costs

The entries denote the ratio $(C_a/b + c)/(C_a + c)$ of the element cost for samples in clusters of size b, compared to samples of individual elements. The columns denote the ratios $k = C_a/c$, and the rows the size b of the clusters. Within the columns note that when k, the relative cost of clusters, is small, the cluster size is unimportant, but as k becomes great, the large clusters result in a big reduction of element costs.

The entries can be generalized to stand for $(C_a/db + c)/(C_a/d + c)$, comparing the element costs of clusters of size db to clusters of size d. Then the columns denote $k = C_a/dc$, and the rows b. Finally, the entries can represent the generalized forms of the element costs $(c_v + c)/(c_{v'} + c)$; the columns denoting $k = c_{v'}/c$ and the rows $b = c_{v'}/c_v$.

In some situations, it may be argued that (8.3.5) does not approximate reality well enough. For example, if we vary a large number a of clusters over an area, the distance between pairs tends to be proportional to $1/\sqrt{a}$; therefore, the total travel distance among the a points tends to be proportional to $a/\sqrt{a} = \sqrt{a}$. Portions of the travel costs—overcoming "the friction of space"—may be proportional to the total travel distance, hence to $\sqrt{a}$. In that case, a close approximation to reality may be

$$C = nc + aC_a + \sqrt{a}C_a^*, \qquad (8.3.9)$$

with C_a^* denoting the cost of travel from one cluster to the next. But remember the advantage of a simple over a more complex model. Moreover, much of the travel costs are caused by starting to move rather than by keeping moving; hence, they tend to be proportional to a rather than to $\sqrt{a}$.

Remark 8.3.I The optimum allocation formula of Section 3.5 for stratified element sampling assumed a specific cost function, $J\Sigma n_h$ or $\Sigma J_h n_h$. In the terms of the general cost function (8.3.1), both specific functions disregard the factor K_v, yet this should also be considered. Particularly, the cost of weighting sample elements exists for all disproportionate samples, but not for proportionate samples.

The function $J\,\Sigma\, n_h$, adopted when minimizing the total number n of sample elements (3.5.1), amounts to assuming that $c_v = 0$, and that $J = c$. The function $\Sigma J_h n_h$, minimized in (3.5.2), can be handled with an average value of $(c + c_v) = \Sigma (n_h/n)J_h$. The variable cost factor is the weighted average cost per element; it varies according to the sample proportions assigned to different strata.

Example 8.3b In Example 8.3*a*, the cost per element was \$2.20 for clusters of $b = 10$, and \$3.00 for individual elements ($b = 1$). For the simple model (8.3.5) we can compute with (8.3.8) that

$$C_a = (3 - 2.2)\left(\frac{10 \times 1}{10 - 1}\right) = 0.8\,\frac{10}{9} = \frac{8}{9} = 0.9.$$

Note that this leads to the per element costs of $C_a = 0.9$ for $b = 1$, and $C_a/10 = 0.09$ for $b = 10$. (These values differ from the locating costs of $c_1 = 1.0$ and $c_{10} = 0.2$, which not being in the ratio 10:1, do not fit the simple model.)

From $C_a = \frac{8}{9}$, one obtains $c = 2\frac{1}{9}$, so that $2\frac{1}{9} + \frac{8}{9} = 3.0$ is the element cost for individual selections, and $2\frac{1}{9} + (\frac{8}{9})/10 = 2.2$ is the element cost in clusters of $b = 10$.

The optimum cluster size can be found with (8.3.7), when roh is known. For Deff $= 4.17 = [1 + \text{roh}(10 - 1)]$, we find roh $= 0.35$. Then element sampling ($b = 1$) is obtained as the

$$\text{optimum } b = \left[\left(\frac{8/9}{19/9}\right)\left(\frac{0.65}{0.35}\right)\right]^{1/2} \doteq 1.$$

But suppose that for another variable Deff = 1.45 = [1 + roh(10 − 1)], we have roh = 0.05. Then we get a different answer:

$$\text{optimum } b = \left[\left(\frac{8}{19}\right)\left(\frac{0.95}{0.05}\right)\right]^{1/2} = 2.83 \doteq 3.$$

8.4 PRACTICALITY

The emphasis throughout this book is on the practical, and it is difficult to write something special here without becoming banal. Yet I hope that this section may help some readers to shorten their apprenticeship in practical methods. These must be learned ultimately through experience. To shorten and ease that apprenticeship is an important function of practical theory. It is appropriate to discuss practicality in this chapter on economic designs, since economy is not abstract and general, but must be related to available human and material resources.

All components of the sample survey operation should be brought into a reasonable balance. It is dangerous to concentrate on one aspect—say, the statistical design or the office routine—and disregard others, such as the field work. This is often most difficult to control, and facilitating it with better and simpler procedures may deserve our best effort. Ordinarily we should resist the temptation to concentrate further refinements for the best parts of the operations at the expense of neglecting their weakest aspects.

However, I feel that the mere existence of poorly controlled portions does not justify neglecting other aspects. For example, the existence of large response errors in the field work does not justify neglecting sampling biases. The former may be inaccessible to available techniques or to the survey administration, whereas control of the latter may be feasible. The effort and cost available for controlling survey operations is best spent wherever it achieves the greatest reduction in the total errors of the survey.

8.4A Simple Designs

Available sampling materials differ greatly throughout the world, and they affect the sampling techniques actually employed or potentially feasible. In Great Britain, element sampling is practiced successfully on many populations, because they are highly concentrated and because good lists are available. Good population registers are also traditional in several other countries of Northern Europe. These circumstances are evident in descriptions of their sampling practices [Gray and Corlett, 1950; Moser, 1955; Dalenius, 1957]. They are also reflected in the development of their sampling theory.

On the contrary, for many populations in the world, element sampling is not feasible due to widespread or expensive travel, or for lack of good lists or other resources. Then cluster sampling is necessary in some form, and area sampling often provides the best solution. But even area sampling may be difficult where a good network of boundaries, such as interlacing roads, is lacking.

Ordinarily the complexities of cluster sampling should be avoided in favor of *element sampling*, unless strong cost considerations make this unreasonable. For example, in the United States, city directories with supplements (8.8) have often provided good samples that were essentially (about 95 percent) element samples. Field cost per element may be permitted to be higher than in cluster sampling, if it is saved in the cost of analysis, in office bookkeeping, and in variance computations.

Sometimes *complete clusters* can be selected simply in one stage and with equal selection probability, providing an epsem selection of the clustered elements. The inclusion of entire natural clusters facilitates the field work. For example, consider samples of students in classrooms, passengers in airplanes, or inhabitants in villages. If great inequality in cluster sizes is present, it can be reduced either with stratification, or with "split-and-combine" procedures (7.1). Even when the average size is somewhat greater than optimum, its inefficiency may be repaid with simpler field work.

The advantages of epsem selection have been often noted in easier office work and simpler analyses. Disproportionate sampling for optimum allocation can be justified only for large gains. Other departures should be generally avoided, because weighting elements can also result in increasing the element variance. But even if epsem selection results in some increase of the element variance, this may be compensated by the convenience of a self-weighting sample.

The advantages of two special simplified designs have been described elsewhere. *Paired selections* of clusters from strata is a basic design with many uses in survey practice. *Replicated samples* have broad applicability, especially in element sampling.

In choosing the seemingly best procedures, all important aspects of the sample survey operations should be considered. Just what are the best procedures for any situation depends on the degree of control that can be exercised over the operations. Less trained personnel requires simpler procedures, clearer and more definite instructions, and more immediate control. Think of three factors that should be kept in overall balance: (1) the degree of supervision, both technical and administrative; (2) the quality of personnel doing routine work, in both office and field; and (3) simplicity of procedures. Weakness in one of these factors may be balanced with strength in one or both of the other two. For example, for a fellow statistician I can design a complicated sample in an informal discussion of a few salient points. The instructions in Sections 9.5 to 9.8 have more detail, simplicity, and control, because they are for nonstatistical field workers. However, those instructions also assume an educated and motivated field staff. In some situations they would be too complicated; then greater simplicity or supervision, or both, would have to be imposed.

8.4B Practical Field Instructions

The field worker should be given simple, clear, adequate instructions and materials. These must be translated into practical procedures for specific survey situations, and I can state only a few guiding generalities. Examples of instructions appear in 9.5 to 9.8. Perhaps the best example of what *not* to do is to send an interviewer to a street corner with the injunction: "Get a random sample!" It violates all rules, being neither simple, nor clear, nor adequate. Another example is provided unwittingly by any statistician who says: "I wrote a perfect set of instructions, but they were widely misunderstood by the ignorant enumerators." Instructions are not good or bad in the abstract, but only in relation to the available material and human resources.

Perfectly foolproof field instructions are impossible, but we must always try to do better. We may not have the time or means to "optimize" our procedures in any absolute sense, but we should try to "satisfice." Nor should we ordinarily choose the "best procedure regardless of price"; the price of the procedure should be judged against its expected worth, along with other survey operations. We have some expectation, however vague and subjective, of the price and value of each operation, instruction, and procedure. Finally, the actual performance should be *measured* against expectations, to teach us how to improve our future instructions.

For a fixed cost, we want to maximize *the amount learned*, not necessarily the amount taught. The most complete written instructions do not necessarily give the best performance. Training the field worker in a complex and long set of instructions may cost more than it is worth. Reading fascinating rare exceptions may lead the confused interviewer to forget about the essential problems. We should aim our best efforts at teaching the interviewer to handle well the 98 percent of ordinary cases; he should know enough about the 2 percent of exceptions to notice them, and obtain procedures from either his manual or the central office.

The instructions should be more extensive and detailed and the training more thorough if the enumerator is far removed from control, as in widespread national samples. Where the enumerator can be kept under close control, for example, on a survey confined to a single city, we may economize on training, and omit complete instructions for the rare occurrences; the enumerator needs to know enough to recognize them as problems, and to call at the central office for clarification.

In writing instructions, use the imperative tense; it is clearer. Put the instructions in simple outline form, not in long narratives. Underline or capitalize the essential points. Provide headings and stubs that facilitate rapid reference to the needed sections. Use a reasonable structure and sequence in arranging the sections. Write for the field workers and not for yourself or for fellow statisticians. In providing ideas, arrangement, and language, put yourself in the field worker's place and see them with his eyes. Perhaps ask one of them to help rewrite the instructions. If practicable, pretest your instructions; or borrow them from good sources. Whenever feasible, give the purpose of each step in simple terms, to motivate the interviewer and to provide needed flexibility through understanding.

Often the field worker can better perform his various tasks separately, concentrating on each in turn. Hence, if the extra cost is not too great, it is better to divide complex field tasks into simpler units, to be performed in separated periods. In Sections 9.5 and 9.6 we discuss the desirability of separating the tasks of listing or segmenting from the final selection of the sample. Intensive interviewing may require so much concentration that the interviewer may not pay proper attention to simultaneous listing. Then this should be made a separate procedure. Sometimes listing and interviewing may be so divergent that they should be done by different workers engaged specifically for each task. Supplementary listing may be better executed when separated from both listing and interviewing.

Sometimes the statistician must sacrifice some of the theoretically available efficiency for the sake of practicablity. Goodman [1960] writes

"... However, an efficient sample which makes great demands on the field worker for its proper implementation may be carried through so badly that the undertaking can no longer be called scientific. In my work in underdeveloped countries I have found that an insistence upon practicability often requires a sacrifice of sampling efficiency. The emphasis on science highlights the need for adhering to a thoroughly practicable, even if inefficient, plan at all costs.

"As a final example, there is the case of the sample of villages which was selected in Pakistan for enumeration in the country's 1960 (first) Census of Agriculture. As selected and used this sample was one of the largest samples ever taken anywhere in the world but it could have been much smaller with a more efficient design. In planning the sample it was concluded that the only possible means of doing sub-sampling within selected villages would be to place complete reliance upon officials within the different districts (1) to see that accurate lists of households were obtained, and (2) to make random selections from these lists. At the same time it was well known that to the district officials the idea of sampling was relatively new and the necessity for really rigorous work in such matters was entirely unrecognized. The inescapable conclusion, therefore, was that if sub-sampling were to be attempted it could not be expected that the households finally to be enumerated would be a true probability sample of the rural households of Pakistan. Accordingly, the design that was decided upon was a stratified random sample of whole villages. The entire work of sample selection was done at a central point for each of the two Provinces, a total of about 12,000 villages being selected in all. Instructions were then issued that within each selected village all cultivators (farmers) were to be enumerated, which meant that although the Census was based on a sample the work of an enumerator within a village was in all major respects exactly the same as it would have been if a complete Census were being taken. The sample was, of course, a grossly inefficient one (however, a much more "efficient" operation than a complete Census would have been) but the rigorous adherence to a scientifically selected sample seemed a goal worthy of the price which was paid."

8.4C Random Numbers for Field Work

In many situations the field worker must apply a probability selection to certain lists. For example, selecting segments from blocks he has just segmented; selecting dwellings from block lists he has just prepared; selecting an adult from the list he has just prepared of adults in a household. To bring the lists into the office for the application of probability selection would require another trip back to the field, a trip we may want to avoid. However, entrusting the use of ordinary tables of random numbers to field workers removed from central supervision is often impractical. Though most may learn to apply them correctly, there may remain enough improper use to cause trouble, or at least uneasiness. Hence, we should resort to procedures which are simpler to apply, and over which it is easier to maintain control.

The random numbers should be selected in the office and sent to the field in a simple form, easy to supervise and check. They can be applied by using two practical methods: (*a*) a fixed order for the listing that leaves little room for doubt and is easy to check; (*b*) concealment of the selection numbers until after the list has been prepared. Only one of these conditions need be present in any single selection procedure. But for extra safety, we may use both of them when the extra cost is small, and when reasonable doubt remains about the order of listing. For example, the block listing (9.6) and the segmented listing (9.5*E*) contain both features. In the process of creating compact segments, the order is probably less well-defined and concealing the numbers becomes more important. When selecting a person from a household, the use of age and sex for ordering the listing leaves practically no latitude, and unconcealed tables of selection will suffice (11.3).

Selection numbers can be concealed behind opaque black tape. Or they can be sealed in an envelope, which the field worker is instructed to open only after he completes the listing.

Systematic selection often yields an easier field procedure than random choice. After a random start, the interval is applied a sufficient number of times; the field worker can be instructed to continue to apply the same interval as long as necessary. When the measures of size are good, but not perfect, we give the field worker selection numbers that allow for three or four times as many selections as we expect on the basis of the size measure. This safety limit avoids many unnecessary calls to the office, when, for example, the block has recently grown to two or three times its former size. It also helps the field worker in extreme situations, when the actual size turns out to be surprisingly large. This may mean that he has gone to the wrong unit, or that he encountered an unusual situation requiring special handling.

8.4D Sequential Control of Sample Size

It is difficult to fix the sample size and the sampling rate if their determining factors are subject to considerable uncertainties. (1) The *unit variance* S_g^2 of the design may be difficult to estimate. (2) The *unit cost* of collecting the data may be obscure. (3) If the population size N is unknown, the sampling rate, $f = n/N$, cannot be specified although the sample size may be decided. Even with good Census data about total persons and dwellings, the size of the survey population may be subject to wide uncertainty. For a sample of married women aged 18 to 40, their *eligibility rate* per dwelling may be obscure. (4) Vagueness about the rates of nonresponse and noncoverage introduce further uncertainties.

The sample size can be controlled by limiting the response rate. This is a bad approach, however, because it reduces the response rate below the feasible limit. The response rate and effort should be set as high as possible within the economic and practical limitations of survey resources, independent of the desired sample size.

I am also skeptical about any procedure for releasing a large sample in the field, then stopping the collection when the desired sample size is reached. Only if this were a proper random sample, guaranteed by strict instructions, could the method be justified. But a strict random selection can seldom be prescribed with economy, practicality, and the assurance that it will be well-carried out. A cut-off procedure can result in large selection biases in favor of ready respondents.

Separate *pilot studies* could be conducted just for dissolving the uncertainties. But this is seldom practiced, because we cannot afford large enough pilot studies to provide better estimates than we can guess with the aid of past experience and expert advice. The pilot studies I know about are mostly "feasibility studies," testing the survey operations, especially the questionnaire.

Flexibility in sampling rates can be attained with a well-designed supplemental sample. *The initial sampling rate should be a sensible minimum*, based on reasonable extreme expectations: lowest estimates of the unit variance S_y^2 and highest estimates of unit cost, population size, and response and coverage rates. Improved estimates of these factors are computed from the results of the initial sample. If the recomputed, desired sampling rate considerably exceeds the initial rate, we can release the needed supplement (13.5*A*).

However, interrupting the field work can be costly, and it may interfere with the survey's objectives. The disadvantages of interruptions of the field work are often so great that they prevent the sequential determination of the sample size. The interruption, if needed, should be minimized.

The initial sample should be large enough to set the field work into efficient motion. It can perhaps be divided into two parts. The part to be completed first can be the basis for computing the supplement and for releasing it. While this is done, the field force can be working on the second part.

The supplement should be planned in advance, and perhaps also partially prepared. Sometimes a large initial sample can be first selected, then separated into initial sample plus reserve for supplement. *The large sample should represent a reasonable maximum*, assuming highest estimates of unit variance and lowest estimates of unit cost, population size, and response and coverage rates. The minimum size is released for the initial sample and the rest held in readiness. After computing estimates from the initial sample, the needed supplement is determined and released to the field.

For simple random samples, the desired sample size can be reached exactly. A sample confined to a single factory or city may also be supplemented relatively easily. However, for complex samples we should be satisfied with obtaining an approximate desired size. For example, adding a supplement of 5 or 10 percent to a widespread national interview sample may be too troublesome; designing, distributing, and collecting the supplement may be too expensive. It may be better to tolerate small uncertainties in the sample size.

The method of supplementing should be carefully planned to facilitate the field work. It should also be designed to permit statistical analysis and variance computations without undue complications. For example, epsem selection should not be destroyed. It may be better to increase the subsample size than to add new primary units.

*8.5 OPTIMUM DESIGNS

The Cauchy inequality is a basic tool in mathematics. It states that for any two sets of real numbers x_i and y_i ($i = 1, 2, \ldots, I$), $(\sum x_i^2)(\sum y_i^2) \geq (\sum x_i y_i)^2$. The minimum for the product of the sums is at the equality, reached when the x_i's are proportional to the y_i's; that is, when $y_i^2/x_i^2 = K^2$ and $y_i/x_i = K$. We can obtain optimum conditions for a wide variety of sampling problems by applying the Cauchy inequality to variance and cost functions [Stuart, 1954]. These simple results are obtained because for most sample designs the total sampling variance and the total sampling cost can be expressed, exactly or approximately, as

$$\text{Var}(\bar{y}) = V^2 + W \qquad \text{where } V^2 = \sum \frac{V_i^2}{m_i}, \tag{8.5.1}$$

and

$$\text{Cost}(\bar{y}) = C + K_v \qquad \text{where } C = \sum C_i m_i.$$

The variance, except for a constant W, is the sum of several components, each of which is a *unit variance* V_i^2 *divided by the number* m_i of those units. The total cost, except for a constant K_v, is the sum of a similar number of components, each of which is a *cost per unit* C_i, *multiplied by the number* m_i of those units.

Optimum conditions are obtained by minimizing either total cost or total variance when the other is fixed. This condition is equivalent to minimizing the product V^2C when one of them is fixed. The product of the variance and cost functions, except for the irrelevant constant portions can be expressed as

$$V^2C = \left(\sum \frac{V_i^2}{m_i}\right)\left(\sum C_i m_i\right) \geq \left(\sum \sqrt{V_i^2 C_i}\right)^2. \tag{8.5.2}$$

The product V^2C is minimized (either V^2 or C is minimized with the other fixed) when

$$\frac{C_i m_i}{V_i^2/m_i} = m_i^2 \frac{C_i}{V_i^2} = K^2, \qquad \text{hence } m_i = \frac{KV_i}{\sqrt{C_i}}. \tag{8.5.3}$$

From this simple relationship we can find K for the optimal conditions in various situations. We can either (*a*) minimize variance for fixed cost, or (*b*) minimize cost for fixed variance.

(*a*) When the cost C is fixed at C_f, note that

$$C_f = \sum C_i m_i = K \sum V_i \sqrt{C_i} \qquad \text{and} \qquad K = \frac{C_f}{\sum V_i \sqrt{C_i}},$$

and

$$m_i = \frac{V_i}{\sqrt{C_i}} \cdot \frac{C_f}{\sum V_i \sqrt{C_i}}. \tag{8.5.4}$$

The optimal $V_{\min}$ is obtained as

$$V_{\min}^2 = \sum \frac{V_i^2}{m_i} = \frac{\sum V_i\sqrt{C_i}}{C_f} \sum \frac{V_i^2}{V_i/\sqrt{C_i}} = \frac{(\sum V_i\sqrt{C_i})^2}{C_f}. \tag{8.5.5}$$

(*b*) On the other hand, when V^2 is fixed at V_f^2, we obtain

$$V_f^2 = \sum \frac{V_i^2}{m_i} = \frac{1}{K} \sum V_i\sqrt{C_i} \qquad \text{and} \qquad K = \frac{\sum V_i\sqrt{C_i}}{V_f^2},$$

and

$$m_i = \frac{V_i}{\sqrt{C_i}} \cdot \frac{\sum V_i\sqrt{C_i}}{V_f^2}. \tag{8.5.6}$$

The optimal $C_{\min}$ is obtained as

$$C_{\min} = \sum C_i m_i = \frac{\sum V_i\sqrt{C_i}}{V_f^2} \sum \frac{C_i V_i}{\sqrt{C_i}} = \frac{(\sum V_i\sqrt{C_i})^2}{V_f^2}. \tag{8.5.7}$$

These optimal expressions are applied where specifically needed; we add here only two basic examples. One relates to stratified random sampling of elements:

$$\text{Var}(\bar{y}_w) = \sum \left(1 - \frac{n_h}{N_h}\right) W_h^2 \frac{S_h^2}{n_h} = \sum \frac{(W_h S_h)^2}{n_h} - \sum \frac{(W_h S_h)^2}{N_h}$$

and

$$\text{Cost} = \sum J_h n_h + K_v.$$

Application of (8.5.3) gives directly the optimal allocations:

$$n_h = \frac{K' W_h S_h}{\sqrt{J_h}}; \quad \text{and if } W_h = \frac{N_h}{N}, \quad \text{then } \frac{n_h}{N_h} = \frac{K'}{N} \frac{S_h}{\sqrt{J_h}}. \tag{8.5.8}$$

With fixed $C_f = \Sigma J_h n_h = \text{Cost} - K_v$, we obtain

$$K' = \frac{\sum J_h n_h}{\sum W_h S_h \sqrt{J_h}} \qquad \text{and} \qquad V_{\min}^2 = \frac{(\sum W_h S_h \sqrt{J_h})^2}{\sum J_h n_h}. \tag{8.5.8$'$}$$

With fixed $V_f^2 = \Sigma W_h^2 S_h^2 / n_h = \text{Var}(\bar{y}_w) + \Sigma W_h^2 S_h^2 / N_h$, we obtain

$$K' = \frac{\sum W_h S_h \sqrt{J_h}}{V_f^2} \qquad \text{and} \qquad C_{\min} = \frac{(\sum W_h S_h \sqrt{J_h})^2}{V_f^2}. \tag{8.5.8$''$}$$

When i takes on only two values, 1 and 2, the expressions for optimum allocations are especially simple and frequently used:

$$\text{optimum} \left(\frac{m_1}{m_2}\right) = \frac{K V_1/\sqrt{C_1}}{K V_2/\sqrt{C_2}} = \frac{V_1}{V_2}\left(\frac{C_2}{C_1}\right)^{1/2} = \frac{V_1}{V_2}\sqrt{k}. \tag{8.5.9}$$

Here $k = C_2/C_1$ in the cost function,

$$C = C_1m_1 + C_2m_2 = C_1m_1\left(1 + \frac{km_2}{m_1}\right).$$

Then with fixed $C_f = C_1m_1 + C_2m_2$, we obtain

$$V^2_{\text{min}} = \frac{(V_1\sqrt{C_1} + V_2\sqrt{C_2})^2}{C_1m_1 + C_2m_2} = \frac{C_1(V_1 + V_2\sqrt{k})^2}{C_f} = \frac{(V_1 + V_2\sqrt{k})^2}{m_1(1 + km_2/m_1)}. \tag{8.5.10}$$

On the other hand, for fixed V_f^2,

$$C_{\text{min}} = \frac{C_1(V_1 + V_2\sqrt{k})^2}{V_f^2}. \tag{8.5.11}$$

An important application occurs in two-stage sampling *with* replacement when

$$\text{Var}(\bar{y}) = \frac{S_a^2}{a} + \frac{S_b^2}{n},$$

with $n = ab$ and Cost $= aC_a + nc + K_v$.

Application of (8.5.9) and (8.5.10) yields

$$\text{optimum } b = \frac{n}{a} = \frac{S_b}{S_a}\sqrt{k} \quad \text{and} \quad V^2_{\text{min}} = \frac{(S_b + S_a\sqrt{k})^2}{n(1 + k/b)}, \text{ with } k = \frac{C_a}{c}. \tag{8.5.12}$$

This may be modified for random subsampling of equal clusters without replacement in both stages [from (5.6.5′) with $S_u^2 = S_a^2 - S^2/B$]:

$$\text{Var}(\bar{y}_{\text{es}}) = \left(1 - \frac{a}{A}\right)\frac{S_u^2}{a} + \left(1 - \frac{ab}{AB}\right)\frac{S_b^2}{ab} = \frac{S_u^2}{a} + \frac{S_b^2}{n} - \left(\frac{S_u^2}{A} + \frac{S_b^2}{AB}\right).$$

The third term in parentheses is constant and does not affect the optimum allocation; hence, application of (8.5.9) and (8.5.10) yields

$$\text{optimum } b = \frac{n}{a} = \frac{S_b}{S_u}\sqrt{k}, \quad \text{and} \quad V^2_{\text{min}} = \frac{(S_b + S_u\sqrt{k})^2}{n(1 + k/b)}. \tag{8.5.12′}$$

Remark 8.5.I *La Grange multipliers* are frequently utilized to derive formulas of optimum design. For example, suppose that the variance of the mean and the variable cost may each be denoted as the sum of two components:

$$V^2 = \frac{V_1^2}{m_1} + \frac{V_2^2}{m_2} \quad \text{and} \quad C = C_1m_1 + C_2m_2.$$

Then construct the function with the indeterminate La Grangian multiplier λ:

$$F = \frac{V_1^2}{m_1} + \frac{V_2^2}{m_2} + \lambda(C_1m_1 + C_2m_2).$$

Obtain the partial derivatives with regard to m_1 and m_2 and set them equal to zero:

$$\frac{\partial F}{\partial m_1} = -\frac{V_1^2}{m_1^2} + \lambda C_1 = 0, \qquad \frac{\partial F}{\partial m_2} = -\frac{V_2^2}{m_2} + \lambda C_2 = 0.$$

Hence $m_1^2\lambda = V_1^2/C_1$ and $m_2^2\lambda = V_2^2/C_2$,
and

$$\frac{m_1}{m_2} = \left(\frac{V_1^2}{V_2^2} \cdot \frac{C_2}{C_1}\right)^{1/2}.$$

If instead of two we have more components, we obtain $m_i = k\sqrt{V_i^2/C_i}$ as before (8.5). The constant of proportionality is obtained again by fixing either $V_f = \Sigma V_i^2/m_i$ or $C_f = \Sigma C_i m_i$. As long as the cost and variance function have the simple linear form of (8.5.1), the simple Cauchy inequality suffices. However, the La Grange multiplier can handle more complex functions. For example, if the cost function is $C = C_0\sqrt{m_1} + C_1 m_1 + C_2 m_2$, construct the function

$$F = \frac{V_1^2}{m_1} + \frac{V_2^2}{m_2} + \lambda(C_0\sqrt{m_1} + C_1 m_1 + C_2 m_2).$$

Then

$$\frac{\partial F}{\partial m_1} = -\frac{V_1^2}{m_1^2} + \lambda\frac{C_0}{2\sqrt{m_1}} + \lambda C_1, \qquad \frac{\partial F}{\partial m_2} = -\frac{V_2^2}{m_2^2} + \lambda C_2.$$

Hence

$$\frac{m_1}{m_2} = \left(\frac{V_1^2}{V_2^2} \cdot \frac{C_2}{C_0/2\sqrt{m_1} + C_1}\right)^{1/2}.$$

This is further developed by Hansen et al. [1953, Vol. I, 6.18 and II, 6.11].

*8.6 TECHNIQUES FOR COMPUTING VARIANCES

8.6A Replicated Subsampling

Imagine a population divided into clusters (sampling units) in several stages. Let $N = \sum_\alpha N_\alpha = \sum_\alpha\sum_\beta N_{\alpha\beta} = \sum_\alpha\sum_\beta\sum_\gamma N_{\alpha\beta\gamma}$ denote the population count; although more stages are possible, three stages possess the necessary generality. Suppose now that each of the last stage units is divided into *ultimate clusters* by applying a constant overall fraction $f/a = 1/aF$ through all stages. Assume now that this can be done without remainders. Here $f = 1/F$ is the intended overall sampling fraction and a the number of intended selections. For example, if $N = 20{,}000$, $F = 1000$, and $a = 2$, then 2000 ultimate clusters averaging 10 elements will be formed. In general the population consists of these aF imaginary clusters, averaging N/aF elements. The clusters may differ in size. Dividing the last stage sampling unit into ultimate clusters may be done by any method.

Now select a of the ultimate clusters with simple random sampling. The sample sum $y = \sum_{\alpha}^{a} y_\alpha$ is an unbiased estimate of the population value fY, where $f = a/aF$ represents the uniform probability of selection of each ultimate cluster; hence, also of each element. The variance of y is

$$\text{var}(y) = (1-f)\frac{a}{a-1}\left(\sum_{\alpha}^{a} y_\alpha^{\,2} - y^2/a\right). \tag{8.6.1}$$

It is unnecessary, of course, to divide the entire population into ultimate clusters. Instead we may employ any equivalent method for selecting a samples, each with the sampling fraction $1/aF$. Except for the last stage, each selection should be random, independent, and with replacement. Selection in the last stage is without replacement; if any unit is selected $a' \leq a$ times, a' similar samples should be selected from it. For the rare cases when the unit lacks enough elements, rules can be found for creating units of sufficient size (7.5E).

Remainders left beyond integral numbers of ultimate clusters pose merely an academic problem. This may be solved by considering the expected distribution of all possible clusters employing the same sampling design.

Independent replications can—and generally will—be taken in several strata. The sample total will be $y = \sum_{h}^{H} y_h = \sum_{h}^{H}\sum_{\alpha}^{a} y_{h\alpha}$; its variance is

$$\text{var}(y) = \sum_{h}^{H} \text{var}(y_h) = \sum_{h}^{H} dy_h^{\,2} = \sum_{h}^{H} (1-f_h)\frac{a_h}{a_h - 1}\left(\sum_{\alpha}^{a_h} y_{h\alpha}^2 - \frac{y_h^{\,2}}{a_h}\right). \tag{8.6.2}$$

Often $a_h = 2$, so the variance becomes

$$\text{var}\left[\sum_{h}^{H} (y_{h1} + y_{h2})\right] = \sum_{h}^{H} (1-f_h)(y_{h1} - y_{h2})^2. \tag{8.6.3}$$

8.6B Collapsed Strata

Suppose that we select two replicated subsamples from a stratum, each with the fraction $f/2$. Denote the population total of the stratum $2Y$, and the two sample estimates y_g and y_h, each having the expected value Y. The variance of *the paired selection* is

$$\text{Var}(y_h + y_g) = E(y_h - Y)^2 + E(y_g - Y)^2 = 2S_*^{\,2}, \tag{8.6.4}$$

where $S_*^{\,2}$ denotes the variance of a single selection and includes the factor $(1-f)$.

Now divide the stratum into two equal half-strata, and denote the two totals Y_h and Y_g. Select, with the fraction f, a single subsample from each half-stratum. The sample estimates, y_g and y_h, will have expected values Y_g and Y_h. The variance of the sum of the *two single selections* is

$$\begin{aligned} \text{Var}(y_h) + \text{Var}(y_g) &= E(y_h - Y_h)^2 + E(y_g - Y_g)^2 \\ &= E(y_h - Y)^2 - (Y_h - Y)^2 + E(y_g - Y)^2 - (Y_g - Y)^2 \\ &= 2S_*^2 - 2\delta^2, \end{aligned} \tag{8.6.5}$$

where $2\delta^2 = (Y_h - Y)^2 + (Y_g - Y)^2$. Thus the expected unit variance is reduced from S_*^2 to $S_*^2 - \delta^2$ when the selection method is changed from paired selections to a single selection per half-stratum.

However, the variance of single selections is computed with a collapsed stratum formula $(y_h - y_g)^2$. Its expected value is

$$\begin{aligned} E(y_h - y_g)^2 &= E[(y_h - Y_h) - (y_g - Y_g) + (Y_h - Y_g)]^2 \\ &= E(y_h - Y_h)^2 + E(y_g - Y_g)^2 + (Y_h - Y_g)^2 \\ &= 2S_*^2 - 2\delta^2 + 4\delta^2 \\ &= 2S_*^2 + 2\delta^2. \end{aligned} \tag{8.6.6}$$

Two covariances vanish because $E(y_h - Y_h) = E(y_g - Y_g) = 0$; and $E(y_h - Y_h)(y_g - Y_g)$ vanishes since selections are independent between half-strata. Thus, when the selection method is changed from paired selection to single selections from half-strata, the *computed unit variance is increased to* $S_*^2 + \delta^2$, *whereas the true unit variance of the sample is reduced to* $S_*^2 - \delta^2$. The overestimate in the computed unit variance is $2\delta^2$ [Cochran, 1963, 5A.11; Raj, 1964].

Ordinarily the gain δ^2 in the unit variance due to splitting the strata is not large. However, if careful stratification can obtain appreciable gains, it raises an important question for which no clear answer is available: Should we prefer a design that obtains an *actual gain* of δ^2, if it also causes an *apparent loss of* δ^2 in the computed variance? Not if we obey the many passages in statistical literature that stress the necessity for unbiased estimates of sampling errors. On the other hand, many practicing statisticians willingly reach for a modest and safe gain in efficiency, even at the cost of a slight overestimation of the variance. This overestimation, they argue, is probably less than the underestimation of total survey errors caused by undetected nonsampling errors. Perhaps in large or periodic surveys, we can take the precaution of building a measurement of δ^2 into the design, and adjusting the results with it.

We have emphasized the relatively secure basis of starting with $H = a/2$ full strata, then splitting these into half-strata for single selections from the

a half-strata. Suppose that precaution is omitted, and a strata are designed, then a single primary selection is drawn from each. Now pairs of strata must be *collapsed*, by matching *strata* which are nearly alike, to minimize δ^2. The strata should be nearly alike in size and in characteristics; strata that are "neighbors" in the selection process may approximately have such traits. Note that the entire strata must be compared and matched, not just the selected primary units, nor the sample results of primary selections. Matching the latter would probably cause severe underestimates of the true variance.

The two collapsed strata should be nearly the same size; if they are not, the variance formulas will need adjustment. For example, the variance of a ratio mean, $r = y/x$, for paired selections is

$$\text{var}(r) = \frac{1}{x^2}\left[\sum_h d^2y_h + r^2 \sum_h d^2x_h - 2r \sum_h dy_h\, dx_h\right], \qquad (6.4.2)$$

where $d^2y_h = (y_{h1} - y_{h2})^2 = 2[(y_{h1} - y_h/2)^2 + (y_{h2} - y_h/2)^2]$, with $y_h = (y_{h1} + y_{h2})$, and with similar forms for d^2x_h and $dy_h\, dx_h$. These forms also serve for pairs of collapsed strata of similar sizes. If the collapsed strata differ considerably in size, M_{h1} and M_{h2}, these should be introduced as weights:

$$W_{h1} = \frac{M_{h1}}{M_{h1} + M_{h2}} \quad \text{and} \quad W_{h2} = \frac{M_{h2}}{M_{h1} + M_{h2}}.$$

Then $d^2y_h = 2[(y_{h1} - y_hW_{h1})^2 + (y_{h2} - y_hW_{h2})^2]$ may be employed in the standard formula above, with similar adjustments for d^2x_h and $dy_h\, dx_h$. It may be more convenient to compute

$$\text{var}(r) = \frac{1}{x^2}\sum_h d^2z_h = \frac{1}{x^2}\sum_h 2[(z_{h1} - z_hW_{h1})^2 + (z_{h2} - z_hW_{h2})^2], \quad (8.6.7)$$

where $z_{h1} = y_{h1} - rx_{h1}$, and $z_h = y_h - rx_h$. All these terms have factors $(1-f_h)$ when necessary. These formula are developed by Hansen, Hurwitz, and Madow [1953, 9.15, 9.28, and Vol. II, 9.5]. For ratio means the adjustment is likely to have small effects for moderate differences in the sizes of collapsed strata. However, the lack of an adjustment can easily cause a gross overestimation of the variances of simple sample totals, such as $\text{var}(y) = \sum_h d^2y_h$.

Remark 8.6.I *When primary units are selected systematically, the variance depends on collapsing pairs of neighboring strata.* We must assume that the ordering of units within the half-strata is effectively random, ruling out any strong linear or periodic trends (4.2). With these assumptions, the preceding remarks apply directly when $G/2$ pairs are formed from G selections. The precision of the variance is enhanced by utilizing all $G - 1$ differences between successive pairs g and $g + 1$, with $g = 1, 2, \ldots, G - 1$:

$$E(y_g - y_{g+1})^2 = E[(y_g - Y_g) - (y_{g+1} - Y_{g+1}) + (Y_g - Y_{g+1})]^2$$
$$= \text{Var}(y_g) + \text{Var}(y_{g+1}) + (Y_g - Y_{g+1})^2.$$

The covariance terms vanish. That $E(y_g - Y_g)(y_{g+1} - Y_{g+1}) = 0$ assumes that deviations in neighboring strata are uncorrelated. Then

$$E\sum_g^{G-1}(y_g - y_{g+1})^2 = \sum_g^{G-1}\text{Var}(y_g) + \sum_g^{G-1}\text{Var}(y_{g+1}) + \sum_g^{G-1}(Y_g - Y_{g+1})^2$$
$$= \frac{2(G-1)}{G}\text{Var}(y) + \left[\frac{2}{G}\text{Var}(y) - \text{Var}(y_1) - \text{Var}(y_G)\right]$$
$$+ (G-1)4\delta^2,$$

where $\text{Var}(y) = \sum_g^G \text{Var}(y_g)$, and $4\delta^2 = \sum_g^{G-1}(Y_g - Y_{g+1})^2/(G-1)$. The variance of the sample total y is

$$\text{Var}(y) = E\left[\frac{G}{2(G-1)}\sum_g^{G-1}(y_g - y_{g+1})^2\right]$$
$$- \frac{G}{G-1}\left[\frac{\text{Var}(y)}{G} - \frac{\text{Var}(y_1) + \text{Var}(y_G)}{2}\right] - G2\delta^2. \quad (8.6.8)$$

The first term is employed to estimate the variance. The second term tends to vanish, because it is smaller by the factor $1/G$, and because the average variance of the first and last selection are probably not very different from the average variance of all selections. The last term does not generally vanish; it adds $2\delta^2$ to the unit variance. This causes an overestimation of the variance as in other forms of collapsed strata.

In the systematic formula for ratio means (6.5C) the two corresponding neglected terms can be shown [Kish and Hess, 1959] to be

$$-\frac{G}{G-1}\left[\frac{\text{Var}(y - Rx)}{G} - \frac{\text{Var}(y_1 - Rx_1) + \text{Var}(y_G - Rx_G)}{2}\right] - G2\delta_z^2,$$

where $2\delta_z^2 = \sum_g^{G-1}[(Y_g - RX_g) - (Y_{g+1} - RX_{g+1})]^2/2(G-1)$.

8.6C *Special Techniques for Variance Computations*

Order Statistics for Estimating σ. Utilizing the range for estimating the standard error is described in Appendix C. We can raise a general question of *estimating variances from cumulated distributions* when (*a*) a large number of random selections is available, and (*b*) the form of the distribution can be specified. We know, for example, that 2.5 percent of the normal distribution lies beyond each of the two limits $\pm 1.96\sigma$. Therefore, if $y_{2.5}$ and $y_{97.5}$ denote the two values in the cumulated distribution, then $0.255(y_{97.5} - y_{2.5})$ will estimate σ, since $0.255 = 1/(2 \times 1.96)$. The optimum choice for normal distributions is at the 7 percent point; $0.339(y_{93} - y_7)$ estimates σ, since $0.339 = 1/(2 \times 1.475)$ and 7 percent of the normal area lies beyond 1.475σ [Hansen, Hurwitz, and Madow, 1953, 10.17; Mosteller, 1946].

After estimating σ, the variance of the mean can be estimated as $\sigma/\sqrt{n}$ for simple random samples only; otherwise a *design effect* would be neglected. This method is rarely used for estimating the standard error. Nevertheless, it is valuable for estimating the element σ, when this is not obtained for the standard error. We can also compute estimates of σ from published data, for use in sample designs (8.2).

The distributions of large probability samples that are not simple random raise unsolved questions; still, their approach to population distributions may be conjectured (2.8C). Serious departures from normality may require another model for estimating the area fractions under the cumulated distribution curve. It may be convenient to note here the proportion P of the normal distribution that lies beyond the distance $t_p\sigma$ from the mean:

P	0.01	0.02	0.03	0.04	0.05	0.06	0.07	0.08	0.09	0.10	0.11	0.12	0.13	0.14	0.15	0.16
t_p	2.33	2.05	1.88	1.75	1.64	1.55	1.48	1.41	1.34	1.28	1.23	1.18	1.13	1.08	1.03	0.99

Computing s from Reduced Samples. Suppose we have a simple random sample of $n = 1200$. Computing labor can be reduced if we subselect a simple random sample of, say, $m = 60$ and compute the element variance s^2 from it. The variance of the mean will be $(1-f)s^2/1200$, or generally $(1-f)s^2/n$, where $s^2 = \left(\sum_j^m y_j^2 - \left(\sum_j^m y_j\right)^2/m\right)/(m-1)$. The precision of the variance is lower than if based on the full sample, but $m = 60$ may furnish a useful estimate. Complex samples can also be reduced in size for computing the unit variance. The number of primary selections (a_h) may be reduced with random selection for computing unit variances (s_{ah}^2) within strata; but the full values a_h must be used to estimate the variance (s_{ah}^2/a_h) of the stratum means.

Random Groups. Suppose again that we have a simple random sample of $n = 1200$, and that we divide it, with random subselection, into $g = 60$ groups of $k = n/g = 20$ elements each. If $y_\gamma = (y_{\gamma 1} + y_{\gamma 2} + \cdots + y_{\gamma k})$ denotes the total for the γth group, and $y = \sum_\gamma^g y_\gamma = \sum_j^n y_j$ for the entire sample, then

$$\text{var}(y) = \frac{(1-f)g}{g-1}\left(\sum_\gamma^g y_\gamma^2 - \frac{y^2}{g}\right). \tag{8.6.9}$$

This is readily proved by considering $g = n/k$ random replications from a population of N/k units, with a sampling fraction of $f = (n/k)/(N/k) = n/N$. If the selections are simple random, the variance of y is an unbiased estimate of $(1\text{-}f)nS^2$. Since $n = kg$, we note that

$$E(s_k^2) = S^2, \qquad \text{where } s_k^2 = \frac{1}{k(g-1)}\left(\sum_{\gamma}^{g} y_\gamma^2 - \frac{y^2}{g}\right). \tag{8.6.10}$$

This is reasonable, since we expect that the sum of k independent selections has variance kS^2.

However, note that in (8.6.9) we do not need to assume independence between the k selections that comprise a group. Nor must we have exact equality k for the group sizes. It is enough to assume g replications, each selected with the same method and the same sampling fraction $1/Fg$; the sum of the g samples will have the fraction $g/Fg = 1/F = f$. For example, g systematic samples may be selected, each with the interval Fg. If only a single systematic sample is selected with $1/F$, pairs of successive selections may be "collapsed" into two groups.

For independent selections in every stratum, the variance of the sample sum will be

$$\text{var}\left(\sum_h^H y_h\right) = \sum_h^H \frac{(1\text{-}f_h)g_h}{(g_h-1)}\left(\sum_{\gamma}^{g_h} y_{h\gamma}^2 - \frac{y_h^2}{g_h}\right). \tag{8.6.11}$$

With many strata, we may be satisfied with $g_h = 2$, yielding a single comparison per stratum. Then

$$\text{var}\left[\sum_h (y_{h1} + y_{h2})\right] = \sum_h^H (1\text{-}f_h)(y_{h1} - y_{h2})^2. \tag{8.6.12}$$

In addition to saving computing time, random groups better approximate normality, since they are based on sums of variables (8.6D) [Hansen, Hurwitz, and Madow, 1953, 10.16]. These are also the aims of the following method.

Combined Strata. Suppose that a variance computation consists of the sum of squared differences of paired selections from several strata: $\sum_h (y_{h1} - y_{h2})^2$. Now note that

$$E\left(\sum_h y_{h1} - \sum_h y_{h2}\right)^2 = E\left[\sum_h (y_{h1} - y_{h2})\right]^2$$
$$= E\sum_h (y_{h1} - y_{h2})^2 + E\sum_{h\neq g}(y_{h1} - y_{h2})(y_{g1} - y_{g2}). \tag{8.6.13}$$

If the paired selections (1 and 2) are independent between strata, the covariance of differences has zero expectation. Hence, we can add randomly selected halves of paired selections from several strata and combine them into computing units; such combinations will retain the expected value of the variance. This method is called "thickening the zones" by Deming [1960, 189–192].

The method can be applied to more than two selections per stratum. (The variance among several units may always be expressed as half of the mean of the squared divergences between pairs.) However, when several selections are available in each stratum, they may be first combined into pairs of grouped units. On the other hand, if g random selections are available in each stratum, they may also be combined across strata into g combined units (4.4).

8.6D Precision of Variance Estimates

The precision of variance estimates is a complex subject involving higher moments, particularly the standardized fourth moment $\beta = \Sigma\,(Y_i - \bar{Y})^4/N\sigma^4$. Our treatment is brief; for details and depth the reader must turn elsewhere [Hansen, Hurwitz, and Madow, 1953, 4.12, 4.21, 10.4–10.10; Cochran, 1963, 2.14]. The relvariance of s^2 for simple random samples of n is approximately

$$\frac{\text{Var}\,(s^2)}{\sigma^4} = \text{CV}^2(s^2) = \frac{1}{n}\left(\beta - \frac{n-3}{n-1}\right) = \frac{2}{n'} + \frac{\beta - 3}{n}$$

$$= \frac{1}{n'}\left(\beta - 1 - \frac{\beta - 3}{n}\right) \doteq \frac{\beta - 1}{n'}, \text{ where } n' = n - 1. \quad (8.6.14)$$

The relvariance and the coefficient of variation of s are approximately

$$\text{CV}^2(s) \doteq \frac{\text{CV}^2(s^2)}{4} \doteq \frac{\beta - 1}{4n'}, \quad \text{and} \quad \text{CV}(s) \doteq \frac{\text{CV}(s^2)}{2} \doteq \left(\frac{\beta - 1}{4n'}\right)^{1/2}. \quad (8.6.15)$$

The relvariance and coefficients of variation of s^2/n and $s/\sqrt{n}$ involve the same constants n^2, n, or $\sqrt{n}$ in both numerator and denominator; hence they remain the same as for s^2 and s, respectively.

For normal populations $\beta = 3$; then the coefficients of variation are approximately $\sqrt{2/n'}$ for s^2 and $\sqrt{1/2n'}$ for s. Nonnormality of the population can raise the value of $\sqrt{\beta - 1}$ considerably above $\sqrt{2}$, hence drastically increase the coefficients of variation. A few examples of the factor $\sqrt{(\beta - 1)/2}$ are given by Hansen, Hurwitz, and Madow [1953, 4.15]: inhabitants per dwelling 1.6; dwellings per block 3.5; size of farm 17; personal income 10; sales of retail stores 53. These are specific and local illustrations; other situations will produce widely divergent parameters. An investigation of U.S. incomes showed a factor of 200!

Values of β are subject to wild variations which may well elude detection in moderate size samples. We must be concerned with values which hardly ever even appear in the sample. Good sample design can sharply reduce the effects of extreme cases, often with stratification, or with a special investigation of the extreme values, or by excluding them from the population (11.4B).

Such methods can reduce the value of β to acceptable levels in element sampling. In stratified element sampling, the preceding formulas hold if the moments are computed within strata. The value of n' in the above formulas becomes approximately $(n - H)$, the number of degrees of freedom; this is further reduced if there are wide differences in sampling fractions, variances, or β's between strata [Hansen, Hurwitz, and Madow, 10.7–10.10]. If the β's are not too different (for example, if distributions are roughly normal within strata), the degrees of freedom are [Jones, 1956] about n' in

$$\frac{1}{n'} = \sum_h^H \frac{w_h}{n_h'}, \qquad \text{where } w_h = \left[\frac{W_h^2 s_h^2}{\sum_h^H W_h^2 s_h^2}\right]^2.$$

If s^2 is computed only from a subsample m, its precision is reduced accordingly: $\text{CV}(s^2) = \sqrt{(\beta - 1)/m'}$. It is interesting to note the effect of *computing s^2 from grouped data:* g groups, each comprised of $k = n/g$ elements. The relvariance of s_k^2 for g random groups is, substituting g for n and g' for n' in (8.6.14):

$$\text{CV}^2(s_k^2) = \frac{1}{g}\left(\beta_k - \frac{g-3}{g-1}\right) \doteq \frac{\beta_k - 1}{g'}. \tag{8.6.16}$$

When the k elements of each group are selected with simple random sampling,

$$\beta_k = \frac{\beta - 3}{k} + 3, \qquad \text{and} \qquad \text{CV}^2(s_k^2) = \frac{2}{g'} + \frac{\beta - 3}{n - k}. \tag{8.6.17}$$

For normal populations $\beta = 3$, and the relvariance of the variance is $2/g'$, which depends only on the number of groups regardless of the size of the sample. On the other hand, when we are concerned with large values of β, it is good to know that with large n the relvariance tends to approach the value $2/g'$ of normal populations. In other words, the value of β_k for individual groups approaches 3 with increasing group size (k), regardless of the element value β.

This is extremely important for cluster sampling. If the sample consists of g randomly selected clusters, (8.6.16) gives the relvariance for either the unit variance, or for the variance of the mean of g clusters. Here β_k is the value for cluster means. Suppose that the grouping of elements into clusters is nearly random, as manifested by roh $\doteq 0$. Then (8.6.17) also holds nearly; β_k for group means approaches the normal value 3 for large cluster size, and the relvariance of the variance approaches $2/g'$. On the contrary, when the clusters are very homogeneous (roh is high), the value of β_k for single clusters remains high, and $(\beta_k - 1)/g'$ remains considerably higher than $2/g'$.

Let us apply these ideas to a complex sample: a ratio mean based on $H = a/2$ paired selections. Suppose that control of subsample size eliminated extreme variations. We can then assume roughly that the H comparisons amount to that many degrees of freedom. Normality for cluster values may be approximated if the cluster sizes are fairly large, and the individual variable not very nonnormal. For ratio means, the variable being measured is essentially $z_{h\alpha} = y_{h\alpha} - rx_{h\alpha}$; this deviation may approach normality more readily than either of its components. When normality can be assumed, the coefficient of variation of the standard error becomes approximately $\sqrt{1/2H}$.

8.6E Components of the Variance

Analysis of the variance into its components becomes intricate in multistage samples of unequal size units. Full exploration of the topic for many designs is not feasible. I attempt to present the fundamental concepts that have widest applicability. The exposition will be facilitated by beginning with designs of equal clusters.

Suppose that $\bar{a}$ primary units are first selected from A at random, with the fraction $f_a = a/A$; then from the B secondary units within each selected primary unit, b are selected at random with the fraction f_b; the overall fraction is $f = f_a f_b$. The variance of the sample total $y = \bar{y}n = \bar{y}ab$ can be estimated (5.6A) with

$$\begin{aligned}\text{var}(y) &= (1 - f_a)ab[bs_a^2 - (1 - f_b)s_b^2] + (1 - f_b)abs_b^2 \\ &= (1 - f_a)[d^2y - (1 - f_b)b^2y] + (1 - f_b)b^2y. \qquad (8.6.18)\end{aligned}$$

Here $b^2y = abs_b^2 = \sum_\alpha^a \frac{1}{b-1}\left(b\sum_\beta^b y_{\alpha\beta}^2 - y_\alpha^2\right)$ estimates abS_b^2;

$$d^2y = ab^2s_a^2 = \frac{1}{a-1}\left(a\sum_\alpha^a y_\alpha^2 - y^2\right) \text{ estimates } ab[bS_a^2 + (1 - f_b)S_b^2];$$

$a^2y = d^2y - (1 - f_b)b^2y$ estimates $ab^2S_a^2$;

and

$$y = \sum_\alpha^a y_\alpha = \sum_\alpha^a \sum_\beta^b y_{\alpha\beta}.$$

The variance of the mean is estimated by var $(\bar{y}) = \text{var}(y)/n^2 = \text{var}(y)/a^2b^2$

The element variance is var $(y)/n$ = var $(y)/ab$. The unit variance of the mean of a single cluster is var $(\bar{y})a$ = var $(y)/ab^2$. These two quantities are more useful in design problems if divided by $(1-f)$. Thus, var $(y)/ab^2(1-f)$ is the estimate of the unit variance s_g^2 in the design formula (8.1.1).

These ideas can be readily carried into stratified sampling; for the sample total $(\Sigma\, y_h)$, the variance is

$$\text{var}\,(\sum y_h) = \sum (1 - f_{ha})(d^2y_h - \overline{1 - f_{hb}}b^2y_h) + \sum (1 - f_{hb})b^2y_h. \quad (8.6.19)$$

To achieve this simplicity, we assume that any necessary stratum weights k_h are introduced into the $y_h = \sum_{\alpha}^{a_h} y_{h\alpha} = \sum_{\alpha}^{a_h}\sum_{\beta}^{b_h} y_{h\alpha\beta}$, as well as k_h^2 into the d^2y_h and the b^2y_h. The sample size is $x = \sum_h k_h n_h = \sum_h k_h a_h b_h$, and the mean is $\bar{y} = y/x$; here x is not a random variable. If the weights are not constant within strata, they must be introduced earlier into the $y_{h\alpha}$ and the b_h to adjust all terms. The variance of the mean is var $(\bar{y})$ = var $(y)/x^2$. We may also utilize the concepts of an element variance $n$var $(\bar{y})$ and a unit cluster variance a var $(\bar{y})$, where $n = \Sigma\, n_h$ and $a = \Sigma\, a_h$. Both of these are more meaningful when divided by $(1-f)$, if the overall sampling fraction was uniform. If it was not, the factors $(1-f_h)$ may be disregarded when the f_h are small. If they are not, it may be better to compute unit variances and element variances separately within strata.

The above forms can be extended to cover ratio means, arising from unequal clusters:

$$r = \frac{y}{x}, \qquad \text{where } y = \sum_h^H y_h = \sum_h^H \sum_\alpha^{a_h} y_{h\alpha} = \sum_h^H \sum_\alpha^{a_h} \sum_\beta^{b_{h\alpha}} y_{h\alpha\beta}, \quad (8.6.20)$$

with a similar expression for x. Primary units are selected at random with probability $f_{ha} = a_h/A_h$; secondary units are selected at random with probability $f_{hb} = b_{h\alpha}/B_{h\alpha}$; the overall probability of secondary units (and elements) being chosen is $f_{ha}f_{hb} = f_h$. The variance of the mean is var (r); then

$$x^2\,\text{var}\,(r) = \sum (1 - f_{ha})[d^2z_h - \overline{1 - f_{hb}}b^2z_h] + \sum (1 - f_{hb})b^2z_h. \quad (8.6.21)$$

The definitions of these terms are analogous to those in (8.6.19), with the variable $z_{h\alpha\beta} = (y_{h\alpha\beta} - rx_{h\alpha\beta})$ taking the place of $y_{h\alpha\beta}$; also $z_h = \sum_\alpha z_{h\alpha} = \sum_\alpha \sum_\beta z_{h\alpha\beta}$. We assume that any necessary proportionate weighting has been applied to the basic computing terms $y_{h\alpha\beta}$ and $x_{h\alpha\beta}$. Along with the general terms, we present also their equivalents for paired selections.

Thus,

$$b^2 z_h = \sum_{\alpha}^{a_h} \frac{1}{b_{h\alpha} - 1}\left(b_{h\alpha} \sum_{\beta}^{b_{h\alpha}} z^2_{h\alpha\beta} - z^2_{h\alpha}\right) = \sum_{\alpha}^{a_h} (z_{h\alpha 1} - z_{h\alpha 2})^2, \text{ when } b_{h\alpha} = 2;$$

and

$$d^2 z_h = \frac{1}{a_h - 1}\left(a_h \sum_{\alpha}^{a_h} z^2_{h\alpha} - z_h^2\right) = (z_{h1} - z_{h2})^2, \text{ when } a_h = 2. \quad (8.6.21)$$

The variance of the ratio mean is estimated dividing by x^2. The components for the two stages are available either for the entire sample or within strata. We may also estimate the element variance with x var (r), and the unit variance with a var (r), where $a = \sum_h a_h$ is the total number of primary units. Both of these are more meaningful if divided by $(1-f)$, if f is the uniform overall fraction. If the f_h vary and cannot be disregarded, it may be preferable to compute separate estimates of the element and unit variances within the strata.

These formulas may be extended to three stages, and further. If the sampling fraction in the third stage is $f_{hc} = c_{h\alpha\beta}/C_{h\alpha\beta}$, then

$$x^2 \text{ var } (r) = \sum (1 - f_{ha})[d^2 z_h - \overline{1 - f_{hb}} b^2 z_h - \overline{1 - f_{hc}} f_{hb} c^2 z_h] \quad (8.6.22)$$
$$+ \sum (1 - f_{hb})(b^2 z_h - \overline{1 - f_{hc}} c^2 z_h) + \sum (1 - f_{hc}) c^2 z_h,$$

where
$$c^2 z_h = \sum_{\alpha}^{a_h} \sum_{\beta}^{b_{h\alpha}} \frac{1}{c_{h\alpha\beta} - 1}\left(c_{h\alpha\beta} \sum_{\gamma}^{c_{h\alpha\beta}} z^2_{h\alpha\beta\gamma} - z^2_{h\alpha\beta}\right),$$

and
$$z_{h\alpha\beta\gamma} = y_{h\alpha\beta\gamma} - r x_{h\alpha\beta\gamma}, \quad \text{and} \quad z_{h\alpha\beta} = \sum_{\gamma}^{c_{h\alpha\beta}} z_{h\alpha\beta\gamma}.$$

We assume again uniform overall rates $f_h = f_{ha} \cdot f_{hb} \cdot f_{hbc}$ within strata; also, that any necessary weights were applied to the basic terms $y_{h\alpha\beta\gamma}$ and $x_{h\alpha\beta\gamma}$. An analogous formula for equal-sized clusters can be readily written with y's in place of z's.

The sampling fractions f_{ha}, f_{hb}, and f_{hc} denote random selections of units with equal probabilities at each stage. However, probabilities proportional to size measures are frequently employed instead. A simple way to adjust for that situation is to imagine arbitrary sampling units, whose sizes are transformed in accord with the size measures. This is a realistic view of the sample design, since the selection is made from the transformed units.

The above formulas are in accord with the detailed treatment in Hansen, Hurwitz, and Madow [1953, Chapters 7, 8, 9]. Simple formulas are available for "nested" or "hierarchal" models of the analysis of variance, which assume infinite populations (no fpc's), and random selection at each stage [Anderson and Bancroft, 1952, 22.4].

PROBLEMS

8.1. An epsem selection of an occupational class in a city is needed with $f = 1/90$. In an epsem of dwellings, short screening questionnaires will locate the sample individuals from whom long interviews will be taken. Two alternative plans are made for the epsem of dwellings with $f = 1/90$.

(*A*) A sample of 1/90 of the blocks, taking all dwellings within the sample blocks.

(*B*) A sample of 1/15 of the blocks, followed by subsampling 1/6 of dwellings listed within them.

To decide the economy of the subsampling plan *B* over the one-stage cluster plan *A*, the effect of six factors in the design should be estimated or guessed.

State separately for each factor, holding the other five constant, whether it is true or false that: *A high value for the factor tends to make the subsampling design B more advantageous.* Explain your reasons briefly.

(*a*) The measure of homogeneity (intraclass correlation) of dwellings in the blocks for survey characteristics: the tendency for dwellings within blocks to be alike.

(*b*) The length (cost) of the interview.

(*c*) The costs of traveling to and locating the blocks.

(*d*) The cost of listing the dwellings.

(*e*) The proportion of dwellings that fall into the survey population.

(*f*) The average number of calls per dwelling necessary to complete the screening questionnaire.

8.2. Answer True (T) or False (F) for each of the eight statements below. Assume that the finite population correction is negligible.

Doubling the number of elements selected from each stratum in the sample will cut the sampling variance in half *whenever* the sample is:

1. A probability sample.
2. An epsem (equal probability of selection method).
3. A simple random sample.
4. A multistage area probability sample of dwelling units in the United States.
5. A proportionate stratified sample of elements.
6. A sample of entire clusters, each consisting of exactly *B* elements, where *B* is the size of the entire actual cluster for both designs.
7. A stratified sample of elements allocated to produce minimum variance.
8. A two-stage sample, holding the number of primary sampling units constant.

8.3. A sampler wants to select a sample of families from a city. An expert can define three kinds of areas according to densities of the survey population. It is roughly estimated that, due to different densities, costs per family will be respectively \$5, \$25, and \$50 in the three kinds of areas. Large differences may exist when covering a sample of families in a large province or a country; they may also occur in a city if screening questionnaires must identify a subclass (socio-economic or ethnic) that is very unevenly distributed.

Describe briefly a two-stage area sample of blocks and dwellings. Assume equal sizes for the three strata; also, that the design effect is close to 1, hence the element variance is close to S^2.

8.4. Suppose that for a two-stage sample $C = nc + aC_a = nc(1 + C_a/cb)$ approximates well the variable cost components; n denotes elements and a clusters in the sample. Assuming that $b = n/a$ is constant yields good approximations when variations in subsample sizes are moderate. The lowest variance per dollar is given by the

$$\text{optimum } b^* = \left(\frac{C_a}{c} \cdot \frac{1 - \text{roh}}{\text{roh}}\right)^{1/2},$$

where roh measures average homogeneity within clusters. Compute optimum b^* for the following nine situations, showing one decimal place:

roh \ C_a/c	0.2	2	20
0.01			
0.1			
0.4			

8.5. Table 8.3.I gives values of the ratio $(C_a/b + c)/(C_a + c)$ of the element cost for samples in clusters of size b, compared to samples of individual elements. The ratio of element variances is $[1 + \text{roh}\,(b - 1)]$. The product of the two ratios is the ratio of the products V^2C, which compares the economy of clusters of size b to element samples of the same size n. (*a*) Suppose that $C_a/c = 20$ and roh $= 0.4$. Compute the economy ratio $[(C_a/c + c)/(C_a + c)][1 + \text{roh}\,(b - 1)]$ for several values of b, including its optimum value computed in 8.4. (*b*) Repeat (*a*) for roh $= 0.1$. (*c*) Repeat (*a*) for roh $= 0.01$. (*d*) Plot the three curves.

8.6. (*a*) Repeat 8.5*a* for $C_a/c = 2$ and roh $= 0.1$.
(*b*) Repeat 8.5*a* for $C_a/c = 2$ and roh $= 0.01$.
(*c*) Plot the two curves.

8.7. (*a*) Repeat 8.5*a* for $C_a/c = 0.2$ and roh $= 0.1$.
(*b*) Repeat 8.5*a* for $C_a/c = 0.2$ and roh $= 0.01$.
(*c*) Plot the two curves.

8.8. Outline the main features of a sample design for *one* of the surveys below. Describe the various survey units: elements, sampling units, observational units, listing units. Mention serious problems you anticipate and procedures for dealing with them. Note in symbols the probabilities for selecting sampling units. Give estimates of means and their variances. Stick to essentials only and be brief. *Outlines* will suffice, without sentences or grammar.

(*a*) The career plans of senior students in the high schools of a state are obtained from questionnaires administered to students of a subsample of classes from a sample of listed high schools.

(*b*) The population composition (age, sex, etc.) in the rural areas of an African state is to be estimated from all households in a sample of listed villages.

(*c*) The distribution of purchases of clothing on charge accounts is to be measured from subsamples of accounts in a sample of stores in a large city.

(*d*) The fluctuation of inventories of specified goods during a three-year period is to be estimated from the records and shelves of a sample of stores in a city.

(*e*) A sample of a city of 400,000 and its metropolitan area must serve both as a sample of household expenditures and as a sample of the adults for interviews on their social and political attitudes.

(*f*) A survey investigates the characteristics of children who have or have not been vaccinated during the special campaign of the previous week. In a sample of blocks of a city all households are covered by interviewing the mother or another responsible adult.

8.9. For a survey chosen above, invent criteria and the necessary parameters; describe completely the design and size of the sample, the expected costs and standard errors.

8.10. Suppose that you have been asked to design the sample for *one* of the surveys listed below. Invent criteria and the necessary parameters; describe completely the design and size of the sample, the expected costs and standard errors.

(*a*) A survey of a city's voters about attitudes toward proposals to raise school taxes by 5, 8, or 10 percent.

(*b*) A survey to determine the prevalence of trachoma (an eye disease) among the village population of an underdeveloped country.

(*c*) A survey to estimate the yearly production of a major agricultural product, say rice, of an underdeveloped country.

8.11. From a simple random sample of $n = 400$ it was estimated that, of the $N = 500{,}000$ adults of a city, $p = 20$ percent or 100,000 were "heavy smokers." The standard error is $N\sqrt{pq/n} = 10{,}000$ adults. A researcher accepts a grant for making a similar estimate for a state with $N = 5{,}000{,}000$ adults, also with a standard error of about 10,000 adults. For designing the sample, it can be assumed that the percentage of heavy smokers in the state is also about 20 percent. He is horrified to hear from his statistician that he needs a very large sample. How large? Do you think that requiring the same *absolute precision* for the state aggregate is wrong?

Why? What sample size is needed to provide about the same *coefficient of variation* $0.02/0.20 = 0.10$ for the state sample?

8.12. Write down the *number* of the proper formula for the variance, or estimated variance, of the mean for each of the designs (*a*) to (*i*). The factors $(1-f)$ are neglected. (*a*) Any probability sample, (*b*) random selection of equal clusters, (*c*) stratified random sample of elements, (*d*) simple random sample, (*e*) sum of two unrestricted random selections, (*f*) any stratified sample, (*g*) random selection of unequal clusters, (*h*) difference of two unrestricted random selections, (*i*) paired selections of clusters from strata. Include also (*j*) the mean square error of any probability sample, also (*e*) and (*h*).

(1) s^2, (2) s^2/n, (3) $\sum_h^H \frac{W_h^2 s_h^2}{n_h}$, (4) $E[\bar{y}_c - E(\bar{y}_c)]^2$,

(5) $\frac{1}{x^2}\sum_h^H [(y_{ha} - y_{hb}) - \frac{y}{x}(x_{ha} - x_{hb})]^2$, (6) $\sum_h^H W_h^2 \text{ var } (\bar{y}_h)$,

(7) $E(\bar{y}_c - \bar{Y})^2$, (8) $\frac{a}{x^2(a-1)}\sum_\alpha^a \left(y_\alpha - \frac{y}{x}x_\alpha\right)^2$,

(9) $(y_1 - y_2)^2$, (10) $\frac{1}{a(a-1)b^2}\left(\sum_\alpha^a y_\alpha^2 - y^2/a\right)$.

8.13. Estimate the element variance σ^2 of a population in which a portion $P/2$ has the value $Y_i = 1$; another $P/2$ has the value 0; and the remaining $(1-P)$ is uniformly distributed between 0 and 1. Compare its σ^2 with those for the binomial with $P = 0.5$, and with a uniform distribution.

8.14. Estimate the element variance of a distribution composed of two equal triangles, two units high at $Y_i = 0$ and 1, and of zero height at $Y_i = \frac{1}{2}$.

PART II
SPECIAL PROBLEMS AND TECHNIQUES

9
Area Sampling

9.1 AREA FRAMES FOR DWELLINGS

Area sampling is a practical listing procedure that accommodates good frames for selecting dwellings. The dwellings serve as sampling units for persons, families, or for other populations that can be associated with dwellings, such as dogs, refrigerators, or home gardens. Area sampling can also be used for selecting many kinds of samples other than dwellings: stores, buildings, farms, crops, and other flora; fauna too, but their mobility causes other complications.

Avoiding abstract terms, I prefer to discuss a specific problem: selection of dwellings from the blocks of a city. The procedures will later be generalized to cover a county, then a state, and a national sample of counties. Discussing specific procedures for one set of problems, within the framework of the 1960 culture of the United States, has a parochial flavor. However, methods for treating these problems also have general relevance to other situations.

Area sampling provides a convenient and effective frame for dwellings and people for several related reasons. (1) By office mapping procedures the entire population of dwellings (elements) can be clearly identified with a defined list of blocks and segments (clusters). (2) These identifications possess permanence from the time of the listing through the survey period. (3) The field worker can readily and clearly identify block and segment boundaries, and dwellings within them. (4) The dwelling serves as a convenient medium for sampling persons, because it is readily identifiable, relatively stable, usually contains few persons, and every person can be identified uniquely with one and only one dwelling. Hence, the dwelling serves as a unique and identifiable sampling unit for a small cluster of persons; and for other populations that we can associate similarly with dwellings. The problems of identifying and selecting persons from dwellings are treated in 11.3; this chapter deals with selecting dwellings.

These desirable characteristics of the model are subject to troublesome imperfections and exceptions that can lead to selection biases. Good instructions are necessary to reduce these to manageable proportions, and the sampler must seek the best practical solutions within available resources. Perfection eludes us, but the methods in this chapter have yielded satisfactory results when applied carefully to appropriate situations. These results justify the necessary procedures: a listing of all the blocks comprising the entire city, a listing of dwellings or segments comprising each selected block, probability selection from the lists in both stages. Several short-cuts are sometimes employed, but these lack the same degree of reliability in actual field work. Some of these are: (*a*) simply "counting off" without listing, and then selecting every *k*th dwelling within sample blocks; (*b*) walking along the length of selected streets and selecting every *k*th dwelling; (*c*) selecting at random one of the four sides from the sample blocks. These procedures may look fine in the abstract, but they typically develop complications and sources of bias, such as dwellings removed from streets "inside" the blocks, the uncertain location of corner houses, the many difficulties of counting dwellings in large houses, etc.

Remark 9.1.I A *dwelling* can generally be defined as a group of rooms, or a single room occupied or intended for occupancy as separate living quarters by a family or some other group of persons living together, or by a person living alone. Ordinarily, a dwelling is a one-family house, or half of a "duplex house," or an apartment, or a flat. But some unusual dwellings must be sought behind stores, above garages, in basements, in trailers, tents, boats, shacks, in caretakers' and watchmen's quarters, etc. When in doubt about how to divide a building into dwellings, the possession of separate cooking facilities may be used to distinguish separate dwellings; but it should not be used to exclude members of the survey population. Though lacking cooking facilities, an apartment or a rented room with a separate entrance is generally defined as a separate dwelling.

While the above covers a large proportion of the U.S. population, marginal problems exist that must be decided according to the survey resources, the direct aims of the survey, and the desire for comparability with outside check sources, especially the U.S. Census Bureau. (1) A single room rented by a person not eating with the family, and having entrance from the hall is a "housing unit" in the Census Bureau since 1960, although it was not a "dwelling unit" before then. (2) The transient population, whose only dwelling is a room in a hotel or trailer or tent, must either be neglected or treated with special instructions. (3) Similarly, populations in institutions—hospitals, monasteries, military installations, prisons—can be included only with special procedures.

Remark 9.1.II The two-stage sample of a city, described in this chapter, requires several decisions. Furthermore, these decisions are not independent, and the final plans involve mutual adjustments.

(*a*) The size n of the sample in numbers of dwellings (elements) and the overall sampling fraction $f = n/N$ require: (1) some knowledge of the population size N, (2) consideration of allowed costs and required variances, and (3) allowance for the "design effects" of the complexities of the design on the variances. (Sections 8.1 to 8.3.) If different sampling fractions f_h are used for separate strata, these must be determined.

(*b*) The number of blocks a and the average number of sample dwellings b per block are determined jointly, since $n = ab$, and in accord with economic considerations (8.1 to 8.3).

(*c*) If, after inspecting maps and other materials, the desired average size b is found to be close to the average size $\bar{B}$ of most blocks in the city, subsampling can be avoided, and we can select entire blocks in a single stage. Then the assigned probability of selecting dwellings f is also the probability of selecting the blocks. Hence, we can select blocks with the interval $F = 1/f$, or select two blocks at random from every stratum of $2F$ blocks. Furthermore, other methods for one-stage sampling of clusters are possible (6.4). A design for selecting entire natural clusters, like blocks, often has practical advantages; and we may adopt it if the actual average size $\bar{B}$ is within perhaps a factor of 2 of the desired size b. Then after splitting the largest blocks into block units and retaining most blocks as single units, a direct selection of block units may provide a simple cluster selection (7.1). Nevertheless, we often must resort to subsampling, and most of this chapter is devoted to it.

(*d*) We must choose among alternative selection procedures. For selecting blocks, a systematic selection procedure is described in Section 9.3, with an alternative of two random selections per stratum. For subsampling dwellings within the sample blocks, 9.5 describes procedures for creating compact segments, and 9.6 an alternative for subsampling from a block listing of dwellings.

(*e*) To obtain a reasonable expectation of the final sample size, the number of planned sample dwellings must be adjusted for three factors: (1) the expected *response rate* to adjust for nonresponse; (2) the expected *coverage rate* to adjust for noncoverage, especially missed dwellings; and (3) the expected *eligibility rate*, when the number of eligible elements is not necessarily one in each dwelling. Nonresponse and noncoverage are treated in 13.3 to 13.6, and persons per dwelling in 11.3. Because it is subject to many guesses, the working sampling fraction may be rounded to a convenient number.

(*f*) Decisions must also be made about miscellaneous problems, some of which are treated in 9.4: the nature of maps and other materials; measures of size and their sources; supplementation, if any, for new or missed dwellings and for unusual places of residence; and possibly for other populations that may have to be added to the basic sample of dwellings.

Example 9.1a Suppose that the important characteristics for a projected study are proportions and are guessed to be in the neighborhood of 25 percent, so that $S^2 = PQ = 1875 \times 10^{-4}$. A standard error of 3 percent is desired, hence a variance of 9×10^{-4}. From other studies the "design effect" on the variance is estimated to be 1.24 so that the required sample size is $1.24 \times 1875/9 = 259$ (see Section 8.1). Suppose also that the response and coverage rates are expected, on the basis of past results, to be 90 and 95 percent, respectively; also, that the eligible population (females 21–55 years) average 0.80 per dwelling. Hence, to obtain the working sampling fraction, divide the initial fraction (or the sample size) by the factor $0.90 \times 0.95 \times 0.80 = 0.684$. Thus we aim at $259/0.684 = 379$ "Census dwellings" to obtain the final sample of 259. The city is estimated to contain 77,200 dwellings, so that the initial overall sampling fraction is $259/77{,}200 = 0.00335$. The adjusted sampling fraction is $0.00335/0.684 = 0.00492 = 1/204$. For convenience, the final sampling fraction of $0.005 = 1/200$ was adopted. This results in $77{,}200/200 = 386 = m$ expected sample dwellings. It should yield a final sample of about $386 \times 0.684 = 264$ elements.

9.2 PREPARING MAPS

Several necessary tasks are described below. Sometimes we can obtain maps and materials with some of these tasks already performed. The map may show all blocks with clear and acceptable block boundaries, and we need only stratify and number them, and assign measures of size. When selecting from the block statistics of the U.S. Census Bureau, we find the blocks of the city already numbered, ordered geographically, and showing the numbers of Census dwellings. Then perhaps we merely need a "linking" procedure to create blocks of "sufficient size" (7.5E), and new measures of size for blocks that have grown greatly since the Census.

Defining Population Boundaries. The boundaries for the area of the population covered by the survey must be defined explicitly and in accord with the survey objectives. The boundaries may be limited to the official city limits, or to only a defined portion, or they may extend to the surrounding built-up area. Within these limits we may exclude areas definitely known to contain no dwellings; such as parks, stadiums, airports, schools, etc. Furthermore, some areas may contain only populations explicitly excluded from the survey: e.g., military reservations, college dormitories, hospitals, prisons, etc. But places of this kind sometimes contain the dwellings of caretakers, staff members, and others who may belong to the survey population.

The survey population may be more restricted; for example, to some nationality, to multi-dwelling houses, boarding houses, or grocery stores. Areas *known* to be empty of the survey population can be left out of the surveyed area to save effort. We may even attempt to exclude areas which, although not completely empty of the survey population, are known to contain so few of them that a reliable limit can be placed on the bias due to their exclusion. Alternatively, these sparsely populated areas can be sampled with a reduced sampling fraction.

In numbering the blocks on a map, four important tasks must be accomplished: (1) defining the boundaries of the blocks; (2) assigning them measures of size; (3) sorting them into strata; (4) preparing the frame by assigning listing numbers. Although distinct tasks, they can usually be prepared simultaneously.

Dividing the Entire Area into Blocks. The block serves as a convenient sampling unit because it is not too large either in area or in numbers of dwelling units, because it has convenient and identifiable boundaries, and because it permits clear and unique association of the dwellings. Boundaries should be lines rather than areas, in the sense that they must contain no dwellings; streets, roads, railroad tracks, rivers and lakes, usually make good boundaries. If the river contains houseboats which should be included, one side may serve as a boundary. In some cities of the world, even the streets are the "homes" for some of the very poor. On the other hand, the entire open country between settlements can be regarded as empty boundaries in some regions where people are entirely concentrated in cities and villages.

For boundaries, choose existing physical landmarks, obvious and permanent features, that can be readily and clearly identified by the field worker. He, without engineering skills and tools, cannot identify a long, arbitrary, and imaginary line drawn on the map.

If the block is completely surrounded by streets or other natural and clear boundaries, then simply writing a number inside its sketch will identify it. However, ambiguous boundaries should be clarified by drawing them in distinctly, along boundaries that can be identified by the field worker. Any dwelling should have a chance to come into the sample from one and only one block. Hence, every dwelling should belong to a predesignated side of the boundary, regardless of which side gets selected into the sample. City limits and other political boundaries often result in block boundaries without visible physical landmarks. These may be tolerated if the occupants generally know in which city, township, county, etc., their dwelling is located. However, be wary of planned but non-existent streets which sometimes appear on maps. Checking map boundaries against an aerial photograph will help with this problem.

Stratifying Blocks. Ordinarily stratification involves simply making up groups, or strata, of relatively similar blocks. People in the same neighborhoods tend to be relatively similar—by reasons of selection, common influences, or mutual interaction. Geographical stratification involves simply outlining as strata the diverse neighborhoods in the city.

It may be worthwhile to discover and to place into separate strata some special populations such as blocks containing very large numbers of dwellings, large housing developments, large trailer courts, hotels, and institutions. Moderate effort will take care of the largest problems; stratification need not be perfect. Public officials can usually provide most of the information; doubtful areas may be treated with a brief field check. Some strata may have distinct problems, requiring special selection procedures. For example, the dwellings of a large housing development, treated as a separate stratum, can be selected in a single stage. Special populations, such as dormitories and institutions, may be sampled with separate procedures suited to the nature of the frame.

In some strata the design may call for increased sampling fractions to yield the precision desired for separate estimates for these domains (11.4). We may delineate strata for distinct areas in order to apply different sampling rates. For example, the sample for a population that is highly concentrated in some areas may be improved with higher overall selection rates for areas of concentration.

We can use different degrees of clustering in different strata, while maintaining a uniform overall selection rate. In strata with high costs per sample block, we can select fewer blocks, with larger subsamples per sample block. This can also be done by assigning smaller measures of size to the costly blocks. This procedure reduces the number of clusters and increases the subsample sizes in the same proportion, yet maintains the fixed overall probability. The high block cost may be due to listing, as in some congested central areas with hundreds of dwellings per block; or due to traveling costs, as in remote suburban areas.

Numbering Blocks. By writing serial numbers within the block boundaries we can accomplish three tasks simultaneously: identify them; establish their list; and define them, if the map boundaries appear unambiguous; otherwise, proper boundaries must also be drawn. The numbers can simply be the integers from one on; or they can be code numbers explicitly designating separate strata.

To keep count of the blocks, leaving none unnumbered; it helps to number them in small groups of contiguous blocks, proceeding from one group to another. A serpentine order within the group may be easy, and may yield some stratification. Numbering the blocks to conform with identified strata can introduce stratification into the numbered list. A systematic selection of blocks then brings this *implicit* stratification into the sample. Alternatively, we may designate *explicit* strata and select two or more blocks from each separately. In either case, the numbering of blocks should anticipate the stratified selection. If two blocks get selected from each stratum of $2F$ block units, numbering the blocks in sets of $2F$ units can establish these strata.

Assigning Size Measures to Blocks. Selecting dwellings within blocks with a constant sampling rate f_h would yield from each sample block a number of sample dwellings in proportion to the total dwellings contained in the block. But sometimes these totals vary a great deal and, for reasons of efficiency and convenience, we should prefer to keep the numbers of sample dwellings more nearly constant. This can be done by selecting with probabilities proportional to size. First, we increase the probability of selection of any block in proportion to its size, then we decrease the probability of selection within the block in the same proportion. The overall probability of selection of any dwelling in any block is kept constant, being the product of the probabilities in the two stages of selection: first blocks, then dwellings within sample blocks.

Assigning measures of size should decrease, sometimes drastically, the variation in the sizes of subsamples, but usually does not completely eliminate it. Some variation can remain, arising from three sources: (*a*) the measure is not an exact multiple of the selection interval in the block; (*b*) the measure of size is subject to error, and the ratio of actual size to measure of size varies; (*c*) nonresponse and using subclasses introduce further variations. All these possible sources affect the actual sizes of subsamples obtained from different blocks. But none of them are permitted to change the uniform probability of selection f maintained through the use of carefully controlled sampling rates. (See Section 7.5 for details.)

We use the design's flexibility to get what we need, by yielding where necessary and possible without undue sacrifice. Our approach stresses the importance of maintaining the uniform sampling rate f for all dwellings—either in the entire population or within specified strata. Maintaining the uniform sampling rate f and obtaining *roughly equal* subsamples from the blocks are usually both desirable and feasible. But trying to obtain *exactly equal* subsample sizes is often neither feasible nor necessary. This approach suffices for most situations; others may call for accepting unequal probabilities and weighting inversely to them.

Formal measures of size in numbers of dwellings can be justified if good information is available separately for blocks. Some sources of these materials are: (*a*) Block Statistics from the decennial U.S. Census; (*b*) block data from the City Engineering or Planning Office; (*c*) public utilities; (*d*) the Chamber of Commerce, Real Estate Board, local newspapers, or banks. Different sources are available in different places and times. If none are available, it may be worthwhile to resort to cheap, rough estimates, perhaps by a moving observer, and perhaps only for a sample of the blocks.

But measures of sizes need not always be the numbers of elements. It is enough if they are numbers approximately proportional to the elements. For example, a *block unit* may denote the "typical block size" in the city, say, 12 dwellings in some hypothetical city. We may then assign the uniform measure of one block unit to most of the blocks, and *two or more block units to the larger blocks*. The measures can either be tabulated separately or simply incorporated into the numbering system; thus a block with two (or Mos_α) measures gets two (Mos_α) block numbers.

Rough but useful estimates of dwellings in the blocks can be made from aerial photographs—but only where vertical buildings are not prevalent. There is much less information on ordinary city maps; but we may note that the numbers of houses tend to be proportional to the perimeters of blocks rather than to their areas.

Most data, materials, and maps are partly obsolete, and it is useful to discover the blocks with very large rates of increase—say, fivefold or more. Officials of the city or public utilities may help locate areas of extraordinary growth, usually a small portion of all blocks. We can assign measures of size without having complete and objective data. We should use judgment, even suspicions, about where heavy building exists or is taking place. The larger blocks on the outskirts of cities are particularly subject to heavy, spotty, and unexpected growth. To guard against these effects, we may assign larger measures than obsolescent maps indicate.

9.3 SELECTION RATES FOR BLOCKS AND DWELLINGS

This section presents some applications of multistage cluster methods to area sampling. It deals especially with systematic selection (7.4C) and paired selection (7.3) of primary units. In the two stages of selecting and subsampling blocks, a uniform overall selection probability f for dwellings is obtained. The methods can be modified to include unequal probabilities for dwellings.

Two methods will be described for selecting blocks, the primary units. Systematic selection is easier and is adequate in most cases. Paired selections of blocks require more care, but the interpretation of the variance is simpler, as discussed below. Each method can be applied either with equal probabilities or with probabilities proportional to size (PPS).

The subsampling of dwellings from selected blocks is treated in Sections 9.4 to 9.7. The easiest procedure may be a systematic selection from dwelling lists, prepared by enumerators for the sample blocks. Its alternatives are listing buildings or compact segments. The subsampling may be done in two stages, when large blocks make this economical. Systematic subsampling is more convenient for field applications, but random subsampling can be resorted to, if necessary.

The simplest procedure is a *systematic selection* of sample blocks, with the selection interval F_a applied to a list of block numbers. Since the blocks receive the equal probabilities $f_a = 1/F_a$, the selected blocks are also subsampled with the uniform probability $f_b = 1/F_b = f/f_a$. Thus the uniform overall probability f is the result of two uniform rates:

$$f = f_a \cdot f_b = \frac{1}{F_a} \cdot \frac{1}{F_b} = \frac{1}{F}. \tag{9.3.1}$$

The uniform overall rate f is determined to fit survey needs; in Example 9.1*a* it was fixed at $f = 1/F = 1/200$. Two numbers, F_a and F_b, are now chosen to attain a desired spread of sample dwellings. We aim at some compromise between the better representation in a greater spread and the lower cost of a more concentrated sample. One extreme would be an element sample of 1/200 from a list of the city's dwellings. Another extreme would be a 1/200 selection of entire blocks; if the blocks average 25 Census dwellings, the entire sample of about 380 Census dwellings would come from only about 15 blocks. A subsampling design of $1/40 \times 1/5$ would increase the number of sample blocks to about $5 \times 15 = 75$ blocks. This means selecting every 40th block (after a random start from 1 to 40), and subsampling with 1/5 in selected blocks.

Consider the sample spread in terms of the average subsample size. The blocks average $25 \times 0.684 = 17$ dwellings for the survey's population. A subsampling rate of 1/5 obtains an average subsample of $17/5 = 3.4$ dwellings. Around that average the subsample sizes will vary in proportion to variations of block sizes.

Considerable control over subsample size can be obtained with *rough measures* of block size. We can assign numbers of *block units* Mos_α (1, 2, . . .) to blocks in rough proportion to the number of dwellings.

Selecting block units with equal probabilities selects blocks with probabilities proportional to the Mos_α; then the dwellings must be subsampled with probabilities inversely proportional to the same Mos_α. The same measure Mos_α must enter in opposite and balancing ways in the two selection stages:

$$\frac{\text{Mos}_\alpha}{F_a} \cdot \frac{1}{\text{Mos}_\alpha F_b} = \frac{1}{F} = f. \tag{9.3.2}$$

For example, with $F_a \cdot F_b = 40 \times 5$, a block with $\text{Mos}_\alpha = 3$ receives a selection probability of 3/40; if selected, it is subsampled at the rate of $1/3 \times 1/5 = 1/15$. Despite the variation of the Mos_α between blocks, we can regard this as a selection of block units with $1/F_a$, and subsampling block units with $1/F_b$, for a uniform overall probability of $1/F_a \cdot F_b = f$ for the city's dwellings. It is convenient to find a basic, usual size of the city's blocks, so that $\text{Mos}_\alpha = 1$ can be assigned to most of the blocks. Larger blocks will then receive $\text{Mos}_\alpha = 2, 3$, etc. Blocks smaller than the basic size can be attached to others. This direct method may be preferred for rough measures of size; for example, when information comes from looking at an aerial photograph or a map of the city, or from eye estimates from a cruising automobile. When F_b and the Mos_α are all integers, the within-block intervals $\text{Mos}_\alpha F_b$ are integers, easier to apply than fractions.

Note that if the rates are $1/10 \times 1/20$, the subsample size averages $17/20 = 0.85$. With fairly good block measures, most sample elements would come from different blocks. Such a sample would have about the same characteristics as an element sample, with its low variances and its simpler statistics. Yet it can be had from listing one-tenth of the dwellings rather than the entire city. Nevertheless, listing 20 dwellings for every sample dwelling might still be too expensive.

If *precise measures* of size are available, they can be transformed into rough measures and utilized with procedure (9.3.2). But they can also be applied directly. Suppose the measures Mos_α represent numbers of dwellings found in the blocks by the last Census, and printed in the *Block Statistics* for the entire city. The uniform probability f for dwellings can be attained by selecting blocks with probabilities proportional to the Mos_α, then selecting dwellings from the sample blocks with probabilities inversely proportional to the Mos_α:

$$\frac{\text{Mos}_\alpha}{F_a} \cdot \frac{b^*}{\text{Mos}_\alpha} = \frac{b^*}{F_a} = \frac{1}{F} = f. \tag{9.3.3}$$

After a random start from 1 to F_a, the interval F_a is applied to the cumulated list of measures, Mos_α. The two numbers b^* and F_a are chosen to yield the desired sample spread. For example, $b^* = 4$ Census dwellings would yield an average of $4 \times 0.684 = 2.7$ final sample dwellings. Then

$F_a = Fb^* = 200 \times 4 = 800$ Census dwellings would be represented by each sample block, yielding altogether $M_t/F_a = 77{,}200/800 = 96.5$, hence either 96 or 97 sample blocks. The selection rates are $(\text{Mos}_\alpha/800)\cdot(4/\text{Mos}_\alpha) = 1/200$.

Within the sample blocks the diverse sampling rates b^*/Mos_α are applied. If b^* is an integer, and if the measures Mos_α denote exactly the actual numbers of dwellings (elements) in the blocks, each sample block should yield a subsample of exactly b^* elements. But there are frequently divergences between the measures and the actual sizes, because of changes, or mistakes, or different definitions (7.5C). Applying b^*/Mos_α maintains the desired probability f, and permits variations in subsample size.

If b^* is an integer and the Mos_α are made integral multiples of it, the intervals Mos_α/b^* are maintained as simple integers, facilitating selections within the blocks. On the other hand, systematic selection with decimal fractions can be done readily on an office computer (4.1B); other methods of dealing with fractional selection rates are described in (7.5F). If the choice of an integer for b^* would yield neither of the above advantages, a fractional value for b may be chosen.

The ratio M_t/F_a of the sum of block measures $(M_t = \sum_\alpha \text{Mos}_\alpha)$ to the zone size denotes the number of block selections. This can be made a desired integer either by manipulating F_a, hence also either b^* or f, or both; or by arbitrary changes in some of the block measures, so that they sum to the desired M_t. For example, if we change M_t from 77,200 to 76,800, the number of sample blocks will be exactly $76{,}800/800 = 96$. This can be done by subtracting a total of 400 measures from several large blocks.

Blocks with large measures can get selected more than once; a block can get selected either k or $(k + 1)$ times $(k = 0, 1, 2, \ldots)$, if its measure is $kF_a < \text{Mos}_\alpha < (k + 1)F_a$. If selected k times, k subsamples should be selected from it, each with the fraction b^*/Mos_α; or the selection rate can be changed to kb^*/Mos_α. This sampling fluctuation between k and $(k + 1)$ samples can be avoided by assigning the measure $\text{Mos}_\alpha = F_a$ to all blocks whose Census measure is larger than F_a; this automatically makes it a separate implicit stratum within which the overall rate f is applied.

On the other hand, blocks so small that $\text{Mos}_\alpha < b^*$, would result in sampling fractions greater than 1, since $b^*/\text{Mos}_\alpha > 1$. This would cause either a selection bias or weighting the sample elements with b^*/Mos_α. However, small block sizes can be avoided: the minimum measure $\text{Mos}_\alpha = b^*$ can be assigned to all blocks with smaller Census measures; or smaller blocks can be combined with larger blocks to maintain a *sufficient size* $\text{Mos}_\alpha \geq b^*$ for all primary units (7.5E).

Paired selection from zones (or strata) is an alternative method to systematic selection for choosing sample blocks. Aside from this difference in choosing block selection numbers, the same procedures are available for both methods. For selecting blocks with equal probabilities, the method of paired selection can be represented with the selection formula,

$$\frac{2}{2F_a} \cdot \frac{1}{F_b} = \frac{1}{F} = f. \qquad (9.3.4)$$

From each stratum of $2F_a$ blocks, two are selected at random. Larger blocks can be assigned 2 or more block units; denoting again with Mos_α the number of block units assigned to the αth block, the selection formula becomes

$$\frac{2\ \text{Mos}_\alpha}{2F_a} \cdot \frac{1}{\text{Mos}_\alpha F_b} = \frac{1}{F} = f. \qquad (9.3.5)$$

Finally, the Mos_α may represent formal measures of size, and again denoting the designed subsample size as b^*, the selection formula is

$$\frac{2\ \text{Mos}_\alpha}{2F_a} \cdot \frac{b^*}{\text{Mos}_\alpha} = \frac{b^*}{F_a} = f. \qquad (9.3.6)$$

Two random selections from each zone of $2F_a$ measures, instead of one systematic selection from each zone of F_a measures, distinguishes the paired selection procedures (9.3.4 to 9.3.6) from the corresponding systematic procedures (9.3.1) to (9.3.3). The following is perhaps the most convenient way to draw selection numbers for paired selections. Select $H = a/2$ pairs of random numbers, each from 1 to $2F_a$, where the number a of primary selections equals twice the number H of zones. The first pair is for the first zone; add $2F_a$ to each of the next pair for the second zone; add $2F_a(h - 1)$ to the hth pair for the hth zone, $(h = 1, 2, \ldots, H)$. The blocks are selected with replacement, and the same block may be selected twice from a zone.

Alternatively, the same a random numbers can be utilized for *a single selections* from a zones. Simply add $F_a(\alpha - 1)$ to the αth selection where $\alpha = 1, 2, \ldots, a$. This chooses one selection number and one block from each zone of F_a measures (7.4A). Selection formulas (9.3.1) to (9.3.3) are then more appropriate.

On the contrary, the method can be modified for k selections from zones of size kF_a, where k is an integer greater than two. Furthermore, different k_h, with k_h selections from a zone of k_hF_a measures in the hth stratum, is a further possible modification.

The implicit strata of size $2F_a$ are called zones when their boundaries are permitted to cut across some block measures. Such a block can be selected from both strata to which it belongs. If this seems inconvenient, place the stratum boundaries at the end of block measures, thus creating explicit strata. The exact stratum size $2F_a$ can be restored by altering the measures of some of the blocks. Or the stratum size $2F_{ha}$ may be allowed to vary between strata; then either the subsample size $b_h{}^*$ is varied proportionately, as in $(2\,\text{Mos}_\alpha/2F_{ha})\cdot(b_h{}^*/\text{Mos}_\alpha) = b_h{}^*/F_{ha} = f$; or the sampling fractions are permitted to vary between strata, as in

$$(2\,\text{Mos}_\alpha/2F_{ha})\cdot(b^*/\text{Mos}_\alpha) = b^*/F_{ha}.$$

As a rule, these complications are unnecessary, and we shall continue below to assume constant zone size $2F_a$.

Paired selections provide a simple model for computing the variance of the mean r:

$$\text{var}(r) = \frac{1-f}{x^2}\left(\sum_h D^2y_h + r^2\sum_h D^2x_h - 2r\sum_h Dy_h\,Dx_h\right) = \frac{1-f}{x^2}\sum_h D^2z_h.$$

The sample mean is a ratio mean; its variance is easier to understand when it is represented as a set of H paired selections denoted with the subscripts 1 and 2: $r = \sum_h (y_{h1} + y_{h2})/\sum_h (x_{h1} + x_{h2})$. The computing forms for the variance are: $Dy_h = (y_{h1} - y_{h2})$, $Dx_h = (x_{h1} - x_{h2})$. Or utilize the alternative form Dz_h, which can be computed either as $Dz_h = (Dy_h - r\,Dx_h)$ or as $Dz_h = (z_{h1} - z_{h2}) = (y_{h1} - rx_{h1}) - (y_{h2} - rx_{h2})$. Details of variance computations for r and for the difference $(r - r')$ of two means are given in (6.5B).

When single selections are made from each of a strata, these must be *collapsed* into pairs to permit variance computations. Similarly, for systematic selections the variance is computed as if neighboring pairs were selected at random from strata of size $2F_a$. When this assumption is doubtful (4.2), resort to one of the other methods. Another computing form utilizes all $(a - 1)$ successive pairs instead of only $a/2$ of them (6.5B).

Stratification should be considered jointly with the variance formulas. When sorting or numbering blocks, consider that each set of successive blocks with combined size measure $2F_a$ constitutes a stratum. Within these sets, homogeneity of blocks reduces the variance.

9.4 SELECTION PROCEDURES

9.4A Compact Segments Versus Listed Dwellings

Two methods compete as alternatives for subsampling dwellings within the blocks selected in the first stage. Section 9.6 presents a method for selecting individual dwellings after listing them separately, and Section 9.5 describes a method for dividing the block area into compact segments and selecting these segments as clusters of dwellings. The discussion in this section is relevant to subsampling clusters other than blocks; the essential contrast is between selection from a list of individual elements or other small sampling units, and selection of compact subclusters identified with boundaries within the primary cluster. The choice between the two methods depends on eight factors; the first three factors tend to favor the compact segments, the next three favor dwelling lists, and the last two may favor one or the other. The conflicting factors do not lead to a unique choice for all occasions; but I believe that practical samplers often prefer compact segments to individually listed dwellings.

1. *Coverage* seems to be more complete with compact segments. Practical samplers believe that block lists tend to be hastily prepared and may miss many dwellings, and that it is difficult to recover these during interviews (13.3). Compact segments can be covered more completely by the field workers, and their completeness can also be checked more readily.

2. *Stability* is another advantage of compact segments; as long as their boundaries remain identifiable, they can continue to serve as valid sampling units, despite moderate changes in their contents. On the contrary, blocks lists of dwellings become obsolete sooner.

3. *Simplicity* may also tend to favor the compact segments; it is often easier to train the interviewer to cover completely the defined segments than to identify individual dwellings. Though the creation of compact segments may require more skill than preparing dwellings lists, this preliminary work can sometimes be accomplished earlier and separately, perhaps by a specially trained crew.

4. The greater *homogeneity* of dwellings within the compact segments tends to increase the variance per dwelling. Subsamples of dwellings selected individually from their lists are spread around the entire block. This has the advantage of creating subsamples with less homogeneity than compact segments. This decrease in homogeneity (roh) is useful because the increase of the variance in segments is approximately proportional to roh$(b - 1)$, where b is the average subsample size and roh is the ratio of homogeneity within the segment clusters (5.5). For subclasses this factor decreases in importance in proportion to decrease in the average b. It tends to be even less for comparisons of subclasses. Hence, large compact clusters have less disadvantage when subclasses become more important.

5. *Variation in size* of the entire sample is increased by selecting segments. The application of intervals causes the size of the subsample to become variable. When selecting segments, the units of that variation

become entire segments, rather than dwellings. If the average size of the segment is b dwellings, the variance of the entire sample size increases by about b and the coefficient of variation by $\sqrt{b}$. Thus, using segments that average 4 dwellings tends to double the coefficient of variation due to this source. The variation in size between clusters also adds to this effect. However, both those sources are often smaller than the variation due to the ratio of actual block size to its measure of size; and the use of segments has little effect on this source of variation.

6. *Dwelling ratings* may be assigned by the enumerator during listing, for later use as auxiliary sampling variables. For example, economic ratings based on hasty appraisals of the exterior have been utilized for disproportionate sampling (11.4*C*). We can go further and obtain, from any responsible household member, a few characteristics—such as age, sex, occupation—to serve other designs; for example, double sampling.

7. *Costs* per element may be less for segment sampling. Segmenting the block may cost less than preparing a complete list of dwellings. The costs of locating and identifying the selected segment, and of making calls and recalls, may be less than those for an equivalent number of individual dwellings spread over large blocks. However, cost factors vary in different situations and for different procedures.

8. *Social interaction* (contagion) among neighboring dwellings may conceivably be induced over the survey period. Some surveyors fear greater refusal rate, and others "contamination" during the interviewing period, but evidence for this conjecture is lacking. Some types of interviewing may be facilitated by interaction. For example, neighbors can give information about finding people at home. Information from neighbors can be particularly valuable in screening operations for finding subsamples.

These conflicting considerations can lead to an average segment size smaller than the desired average size of the subsample. For example, an average of $b = 8$ dwellings per block can lead to an average of 2 segments per block, with an average of 4 dwellings per segment; or an average of $b = 10$ dwellings per block with an average of 2.5 segments per block. Often we can capture most of the advantages of factors 1, 2, 3, and 6, yet gain appreciably on 4 and 5, by aiming at several segments per block. Thus the average number and size of segments per block should be determined with economic and practical considerations in mind. The number of actual segments from the different blocks can vary to the degree that the size measures lack precision (unless this is avoided, as in 7.5B and 9.7). If we aim at an average of one segment per block, we can also get either none, or two, or more segments. Since zero segments are a costly nuisance, we should usually aim at more than one.

9.4B *Problems of Block Size*

Some problems are caused by the existence of blocks that are either too large or too small. Inaccuracies in the size measures of blocks raise other problems. Their descriptions and solutions appear in 7.5, and here we apply them specifically to blocks.

Blocks are too small if their size measure Mos_α is less than the designed subsample size b^*; accepting them would result in a subsampling rate b^*/Mos_α greater than 1. We may even set an arbitrary maximum subsampling rate $1/k$, say, 1/3or 1/6, then set kb^* as the minimum "sufficient size." For formal PPS selection, blocks below sufficient size are attached to others with a linking procedure that creates units of sufficient size.

This can be achieved simply with map numbering when block units are assigned. A block containing very few dwellings can be attached to a neighboring block of ordinary or larger size; this procedure also applies to blocks that, although seemingly empty of dwellings, may actually contain some. These dwellings then get the proper probability of selection; yet the procedure obviates separate visits to blocks that yield no—or too few—sample dwellings.

Very large blocks may need attention too. If there are many, and dwelling lists are too costly, we can put them in separate strata. We can then select these blocks with a reduced rate and proportionally increase the sampling rate within them, thus maintaining the uniform overall selection rate.

If the block is as large as the stratum size, it can be made a self-representing stratum. For example, if paired selections are taken from every stratum of $2Fb^*$ dwellings, each block which reaches that size can be made a separate stratum.

9.4C *Supplements for the New, the Missed, and the Unusual*

Procedures may sometimes perform well enough to be useful, yet not be fully satisfactory; if they performed worse, we would not use them, and if better, we would not have a problem. In these cases we add, to the regular procedures that cover most dwellings, some supplementary procedure(s) for covering some missed portions of the population. The supplementary frame should cover *all* the elements excluded from the regular frame, but *none* of those included in it. Rather than cover the entire population with the supplementary frame alone, we also retain the regular frame, whenever the combined use of the two frames is better or cheaper than either alone.

For adding the units missing from a frame, we may employ either a supplementary stratum or a linking procedure. Linking procedures are included for adding, to listed dwellings (in 9.6) and to listed segments (9.5D), the dwellings missed by the original lister, and those constructed after the original listing. The compact segments of 9.5, if properly designed, should be self-correcting to cover the entire area of the block without any "holes."

Instead of linking procedures, a supplementary stratum with separate procedures can be used to add the missing dwellings. Field workers must search some of the sample blocks and find all dwellings which do not appear on the listings. This search is confined to a calculated and designated portion of the sample blocks. For example, if the regular sampling rates were $1/40 \times 1/5 = 1/200$ for blocks and dwellings, select a supplement in 1/5 of the sample blocks and take all missed dwellings. This will yield the same sampling fraction $1/200 \times 1 = 1/200$ for the missed dwellings, but the search is confined to a fifth of the regular sample blocks. Or, because of its greater cost per interview, we could aim at a smaller sampling rate, say, 1/400, for the missed dwellings; this can be achieved by taking all missed dwellings found in only 1/10 of the sample blocks. The procedures must clarify what a missed dwelling is, so that *all those, and only those, not covered in the regular stratum receive their proper probabilities in the supplementary stratum.*

The separate stratum has the advantage that the field worker can concentrate on this special effort or, if desirable, specially trained personnel can be employed. The linking procedure also has advantages: it does not require a separate trip to the block; it can utilize the closer contact offered by the interview situation within segments; and it brings in the missed dwellings in smaller clusters. But a separate stratum may be needed to cover special populations, whose characteristics or distribution preclude their successful coverage with the regular procedures. For example, the populations of military installations, institutions, large commercial hotels, boarding houses, and trailer courts present various problems of coverage. Such populations either should be excluded from the survey, or included in a supplemental stratum with special instructions; the decision between these two "evils" should depend on balancing the damage done by their exclusion against the costs of special measures for their inclusion. The former tends to weigh more heavily with larger samples and the latter with smaller samples.

A separate stratum can also prove useful for covering spotty, concentrated, large-scale new construction, particularly large apartment buildings, common in the larger cities. Ordinary procedures of linking or even supplementary blocks cannot deal with them adequately. For example, in a block which had a measure of six dwellings, finding a newly built apartment house with 120 dwellings, which should be included in the sample according to regular procedures, presents a "surprise" (12.6*C*) and dilemma for those procedures. These unpleasant surprises can often be avoided by learning, perhaps at the City Hall, about all blocks with concentrated new construction of dwellings; for example, all blocks with more than k new dwellings built since the date of the measures of block size. Such blocks are put into a separate stratum, to which a desired sampling rate is applied.

Thus the sample of a city may come from four distinct strata: (1) the regular block sample; (2) individual and small-scale new construction and scattered missed dwellings, added by linking or with supplementary blocks; (3) a separate stratum for large-scale, concentrated, new construction; and, if desired, (4) a separate stratum for institutions, military, and other populations not in private dwellings.

We have been dealing with problems of associating households and persons as population elements to the frame units: the dwellings present in the frame. A frame unit containing zero households causes no trouble. When a frame unit is found to contain several households, they are all included with regular procedures. Special procedures, such as the supplemental stratum above, prevent the appearance of many households in one frame unit. Supplementary procedures exist for including households without frame units (i.e., missed dwellings), as well as for populations not in dwellings (e.g., institutions). Thus, three of the four kinds of frame problems mentioned in 2.7 are covered by the procedures we discussed. The fourth type of frame problem arises from households or persons occupying two or more dwellings. If these households are included in the sample with the selection of either of their dwellings, they get duplicate probabilities of selection. This problem is infrequent, except for the vacation homes of a growing portion of the population. Questioning respondents can reveal the existence of other dwellings; then rules for identifying the principal home can be applied, though annoying minor problems may persist.

9.4D Repeated Selection from a Listing

A survey organization may be in position to draw, within a reasonably short period, several samples from the same set of block listings, thus spreading the listing costs over several studies. If we can plan ahead, we can also increase the size, spread, and quality of listings. Since listings are "perishable," they become less valuable with time and eventually deteriorate beyond usefulness. Their useful life is limited to some reasonable period, which depends on the rate and the nature of the changes in the population. Within that period, the need for using supplementary procedures, to record additions and other changes, becomes increasingly important with the passage of time. Compact segments, denoting actual areas, automatically reflect changes within their boundaries, and handle these problems better than block listings. Eventually, boundaries and measures of size for blocks and segments undergo such drastic changes that trying to live with the old lists and maps becomes more troublesome and costly than preparing new ones.

Procedures for repeated selection can take advantage of the fact that when a sample is selected properly out of a frame, the remainder will reflect the sample itself. If the units are selected with equal probability, with epsem, the remainder in the frame will also be an epsem sample. Furthermore, if the sample is selected with random choice, the remainder will also constitute a random sample. If the units are drawn with systematic selection, the remainder will likewise constitute a systematic sample. The situation is easily perceived with the example of a systematic sample drawn with the interval k from a block list. There will be k equivalent samples, depending on which of the k random starts is used to begin the selection. Hence, if one of these random starts has already been used, the next selection is confined to the other $(k - 1)$ possible random starts. If two of those samples have been used, later samples will be confined to the other $(k - 2)$ random starts, and so on.

However, we may wish to select from a set of listings several samples with different intervals, or even different procedures. Whatever epsem procedure of selection is used, a simple rule will suffice: If the selection hits a unit (lands on a listing line) that has already been used, simply go on to the next line; and if that has been used, go on to the next line, with the understanding that the first line of the listing follows the last line in a circular fashion. Hence, whenever we select a sample from a listing, it is a good precaution to select an epsem sample from it. If we need to use different sampling rates for different strata, we should "eliminate" from the listings a corresponding fraction of the less used strata; so that, by having removed an epsem sample, we leave a remainder that is also epsem, facilitating further selection from the same listings.

9.4E Modification to Three or More Stages

Obtaining and utilizing auxiliary information—for defining boundaries, for stratification, and especially for size measures—for all the blocks in the city may appear too costly, especially if it involves field work. But we are not forced to choose between covering either all of the blocks or none; a compromise to obtain the information for only a sample of blocks may be more economical than either extreme.

We can introduce another *stage* before selecting the blocks: a selection of "chunks," or areas containing several blocks. These chunks should have clear boundaries, easy to define on the map. In many cases we can exploit existing defined units such as Census tracts and Census Enumeration Districts. These form particularly accessible preliminary units, with boundaries, size measures, and other information available. Such a preliminary stage, preceding the selection of blocks, seems desirable for samples of large cities and other large, complex populations (10.1 to 10.4). In the first stage, the primary sampling units, the chunks, are selected. The detailed work on boundaries and measures necessary in the second stage is confined to the blocks within the selected chunks. From these chunks the actual sample blocks are then selected in the second stage. Subsampling the blocks for listed dwellings or compact segments leads to the third stage of sampling.

Selection within the blocks may also require two stages or more. For example, if the blocks are large, we can create and select large segments, list them, and subsample dwellings. The selection of persons within dwellings can also add another stage. Two stages within the block will make a block selection design into a three-stage sample; and four stages result if the selection of chunks precedes the block selections.

We may take a formal view of the several stages of selection. The first stage selection results in a sample of *primary selections*, or PS's; these can be either chunks, or blocks, or any other units. For this first stage selection we may choose between alternative methods: systematic selection of PS's with intervals, or selection of two PS's per stratum, or other possible methods. The subsampling can be accomplished in a single second stage, or it may require two or more stages, resulting in an overall design of three stages or more.

In some situations, instead of introducing another stage of selection we can introduce a preliminary *phase* of selecting blocks; the blocks can be selected in *two phases* with "double sampling" (12.1). In the first phase a large initial sample of blocks is selected, perhaps with equal probability because of ignorance of their characteristics. After information, such as size measures, is obtained for this large sample of blocks, in the second phase the smaller actual sample of blocks is selected from the initial larger sample, perhaps with PPS.

9.4F Maps, Photos, and Similar Materials

Preparation of the map as a frame of blocks involves four tasks that require auxiliary information: defining the city's limits, dividing the city into strata, defining block boundaries, and assigning size measures. Where can we obtain this information? The answers vary considerably with current conditions in diverse places in the United States, to which our brief presentation must be confined.

For most cities of moderate or large size, we can obtain detailed maps which can be used to define and identify blocks. They vary in quality and amount of useful information. Some also contain information useful for stratification; for example, zoning maps and maps of market areas. Among the local sources of maps are the City Engineer, the city or county Planning Office, the Real Estate Board, the Chamber of Commerce, banks, newspapers, and public utilities. Perhaps the best single source of national scope is the Rand McNally Company, with offices in major cities. Three others are the Hagstrom Map Company and the Hearne Brothers, both of New York, and for cities on the West coast, Thomas Brothers of California. The Sanborn Map Company of New York sells detailed maps of individual blocks showing the location of individual structures and dwellings in the blocks, for most U.S. cities. These maps are rather expensive; their use is described in detail by Hansen, Hurwitz, and Madow [1953, 6.4].

In most state capitols of the United States, county highway maps are sold by an agency called the State Highway (or Road) Department (or Division or Commission), or something similar. These maps, generally $\frac{1}{2}$ inch or 1 inch to the mile, contain a great amount of detail, including dwellings as well as all roads for the thinly settled "open country" areas; but they are much less useful inside cities and densely settled areas.

Aerial photographs can be purchased for most of the United States from the Production and Marketing Administration, U.S. Department of Agriculture, Washington, D.C. They can be utilized for checking the validity of the block boundaries found on maps; for estimating the numbers of dwellings in blocks, in the absence of a great deal of vertical building; for dividing the city into strata; and, if needed, as substitutes for maps in small towns and in the open country.

For every city of over 50,000 population the Census Bureau publishes a volume of "Block Statistics." It contains a map of the city with all blocks numbered, defined, and identified; also tabular data for each block, including number of dwellings; this may serve as measure of size. Of course, this information becomes less reliable with the passing of each year after the Decennial Census. Most of these cities are centers of "tracted" Standard Metropolitan Statistical Areas, for which the Census Bureau publishes "Tract Statistics." The tracts vary in size, mostly from 1000 to 5000 population. Hence, they are too large for sampling units for most situations, but can serve for stratification.

Among the most helpful materials are the data by "Enumeration Districts" sold cheaply by the U.S. Census Bureau. These "ED's," averaging roughly 300 households or 1000 persons, are the areas covered by individual enumerators in the Decennial Census. The data include a few socio-economic indicators, as well as numbers of persons and dwellings, and descriptions of boundaries.

Available maps are often inadequate and obsolete in part. A special mapping effort for areas of difficulties and rapid growth may be worthwhile. Moderate variations in the actual block size from the expected can be neglected, and efforts directed to discover blocks that may contain, say, five to fifty times the expected numbers. Appropriate public or business officials may either supply the needed information or direct the sampler to potentially troublesome areas to be checked in the field.

If satisfactory maps are not available at all, field workers can prepare a crude "map," dividing the city into blocks. If the city is large, the first stage can consist of outlining some larger areas, that we called "chunks"; then a sample of these can be divided into the actual blocks. The field worker can outline the block boundaries, and assign size measures simultaneously. He can also assign some social classification, if needed for stratification.

9.5 COMPACT SEGMENTS

9.5A Procedures for Compact Segments

To divide the block into useful segments, we should be guided by three criteria. Because these criteria often conflict, compromises among them must be made—some in the office procedures, others by the field worker.

1. The desired average size of the segment must be determined. Considerations of economy lead to some reasonable compromise between the increased variance and the decreased cost of larger segments (8.1 to 8.3). However, for practical reasons we may sometimes favor segments larger than the theoretical economic optimum, because smaller segments may be too hard to define and lead to greater errors of coverage (13.3). In city blocks in the United States, average sizes of segments commonly used by different survey organizations range from 3 to 10 dwellings. In some cases, "natural" segment sizes tend to exist; for example, in many farm areas of midwestern United States about 4 to 8 farms per "section" is fairly common, with good road boundaries around that mile-square area.

2. The segments should preferably be roughly equal in size, measured in numbers of dwellings (or other units for other populations). Exact

equality cannot be expected and, in practice, segments from half to double size are often tolerated. The need for recognizable area boundaries often results in oversized segments; here we may have to resort to listing the dwellings, then dividing the listing into segments of about desired size.

3. The segment should have clear, identifiable boundaries, with enough permanence to remain distinguishable through its period of use. Roads, streets, tree lines, rivers, and utility lines are preferred. Yet sometimes arbitrary lines must be used, such as a straight line between two well-fixed points; or a straight line of sight perpendicular from a well-fixed point on a road. In thinly settled areas the arbitrary line can be distinguished well enough to locate any dwelling without doubt on the proper side. However, in thickly settled areas, especially with multi-dwelling and multi-storied houses, the segmentation into pieces of land sometimes becomes impossible, and we must resort to listing dwellings. Even here we may prefer (for better coverage) to use an entire house of several dwellings as one segment; or in multi-storied houses, to take each floor as a separate segment; or the floor can be divided into halves or quarters as compact segments. Alternatively, the listing can be divided into artificial "segments" of separated dwellings by using intervals for selection.

After creating the segments, the field worker is usually instructed to bring the block sketches into the office for checking and for selecting segments to be included in the sample. Irregularities and mistakes are corrected; the oversized segments, and especially listings of entire blocks, are divided into segments near the desired size; segments that are too small may be combined.

If all the segments in the block are numbered consecutively, this numbering comprises their listing. From this list the sample segments are selected with the within-block interval. If the design calls for an average of one segment per block, most blocks should yield a single segment per block; but variations occur, resulting in some blocks with zero sample segments and some with two or more. If the design calls for an average of b segments of size c each (where $bc = b^*$, the desired average subsample of dwellings), the variation in numbers of segments in the different blocks should be around the average of b. In any case, the yield, both in actual numbers and in actual sizes of sample segments in the block, is usually a random variable; this is due to the usual inaccuracies in the measures of size, and to the variations in the sizes of segments obtained.

After the sample segments are selected and identified in the office, they are sent out again for interviewing. Sending field workers out on the first trip solely for segmenting without interviewing may seem desirable under certain conditions. (1) The segmenting is done by a different crew, perhaps a few specialists. (2) It is preferred to separate the specialized tasks and instructions of segmenting from those of interviewing. (3) It is preferred to separate the *period* of segmenting from the interviewing period. (4) The segmenting will be utilized and its cost spread over a series of surveys. (5) The field worker's area is so small that separate travel for segmenting is inexpensive.

On the other hand, if the travel to the block is relatively expensive, we may want to combine the three distinct tasks of segmenting, selecting, and interviewing (or at least the first call) into one step. This requires proper training of field workers for the three tasks, with emphasis on preventing unconscious biases in selection. The interviewer must assign a specified order when numbering the segments he creates; or segment numbers selected in the office must be hidden—either in an envelope or behind black tape—and revealed only after the interviewer has assigned his numbers to the segments. The selection numbers must run comfortably beyond the expected average number to allow for reasonable variation in numbers of segments found in the block. Extraordinarily large blocks, however, are best reported to the office for checking and handling.

These methods can be readily generalized to include further subsampling. In the preliminary stage the field worker may be asked to create areas (blocks) that are too large for compact segments; then a selected sample of these areas is segmented and subsampled. In these situations the field worker need not concentrate on obtaining equal sizes for the larger areas; he creates areas with good boundaries and assigns reasonably good measures. For example, the first stage may consist of the primary selection with PPS of two Census Enumeration Districts (ED's) per strata in the population (of a county or a state). These primary units average about 250 dwellings. Then the field workers divide the sample ED's into identifiable blocks and assign size measures of approximate numbers of dwellings. These blocks may average 20 to 40 dwellings. Selecting sample blocks, probably with PPS, forms the second stage of selection. Finally, these blocks are subsampled either by creating compact segments, or by listing and selecting individual dwellings.

In thickly settled places such as apartment houses, segment listings of dwellings may be necessary. Then we may prefer to increase the spread of sample interviews. For example, in the illustration in 9.5B.IV of *Segment Listing* for a house, the 23 dwellings are divided into 5 compact segments: each is half a floor. Instead, we can introduce the interval 5 to obtain 5 segments; segment 1: basement, 105, 204, 301, 306; segment 2: 101, 106, 205, 302, 307; segment 3: 102, 201, 206, 303, 308; segment 4: 103, 202, 207, 304; segment 5: 104, 203, 208, 305.

9.5B Instructions for Segmenting

I. For your trip to the block you will need: (1) the *map* showing the general location of the sample blocks, outlined in red on the map of your work area; (2) the *Sketch Sheet* containing a detailed drawing on an enlargement of the sample block, in the Segment Folder; (3) these *Instructions.*

II. On the Block Sketch Sheet we have outlined the block to be subdivided, including the physical features shown on the map. You are to do the following.

(*a*) *Check the block boundaries to make sure you have correctly identified it.* The block boundaries are usually streets, roads, highways, railroad tracks, streams, or other recognizable physical features. However, some may be state or county lines, corporate limits, or other civil boundaries. If there are no street names, highway numbers, etc., shown on the sketch, or if the local names differ from those which we have used, *please enter the correct names or numbers on the sketch.*

(*b*) *Your main job is to divide the block into segments of about 4 dwellings each.* The segments will be sent out for interviewing on future surveys. Your sketches of segments need not be works of art, but they should be reasonably neat and accurate, so that at some future date an interviewer will be able to locate the sample segments you have outlined. The segments should contain a *rough average* of about 4 dwellings. However, in the interest of good segment dividers, often you will have to allow some variation in the size of the segment. Furthermore, you should not knock on doors to determine the number of dwellings at each address, just make a *rapid guess.* Three or 5 dwellings are as satisfactory as 4; 2 or 6 dwellings will do well; and in some situations we expect to see as many as 7 dwellings in the segment.

(*c*) We have drawn in on the sketch whatever seemed to be good subdivisions of the blocks as we found them on our maps, but you must revise these subdivisions to show the current facts as you find them. If the street, road, or stream which we have drawn does not really exist in the block, draw a wavy (∿∿∿) line through it. *Add any boundaries you find and need.*

Begin the canvass of the block at any convenient location. As you identify good segment boundaries, draw them on the block sketch and indicate what each is (irrigation ditch, alley, county road, etc.). If some road or other physical feature should not be used as a segment boundary because it forms a subdivision with no dwellings, or only 1 dwelling, note that fact on the sketch.

Preferred segment boundaries are easily recognizable and are usually relatively permanent features such as roads, railroads, streets, alleys, creeks, rivers, drainage or irrigation ditches, and power transmission lines. Fence lines and tree lines may be used. Civil boundaries and section lines are acceptable if they are well known locally. We discuss below ways of handling problems where good boundaries are hard to find.

(*d*) *On the sketch indicate the location of each residential structure.* Use an X for one-family dwellings. For structures containing more than one dwelling, indicate the approximate number by writing the figure over or in a rectangle locating the structure, thus: $\overset{3}{\square}$. If for some reason it is not feasible to draw in each dwelling, record only the dwelling count (approximate number) within the segment boundaries and prepare a *Segment Listing.* These cases occur in densely settled areas and for multi-dwelling and multi-storied apartment houses.

(*e*) *Report any new construction of dwellings going on in the block*, and look for and report signs of any new construction planned for the near future. Also tell us about any advertising on billboards, in newspapers, radio, or TV, of lot sales or new houses to be built in the block. This will permit us to include them, if they become occupied during the use of the block.

(*f*) For each trailer court and each motel, identify dwellings (rooms, suites, or other accommodations) or trailer spaces used for permanent guests; also, separate those provided for transient guests.

III. When you cannot find physical features to subdivide the block, choose the appropriate method among those described in the following paragraphs.

When the density of buildings is not too great, you can draw on the sketch a broken line (– – – –) to show imaginary lines and property lines as boundaries. Describe what the broken line represents: extension of streets, line of trees, property line, imaginary line between landmarks, etc.

(*a*) Clear property lines may be used to subdivide the block in the absence of natural boundaries. In this case, show the location of each structure in the segment and obtain one or more of the following facts about the structures adjacent to the property line.

(1) House numbers of adjacent structures.

(2) Names, obtained from mail boxes, of occupants of each adjacent dwelling.

(3) A description of the structures.

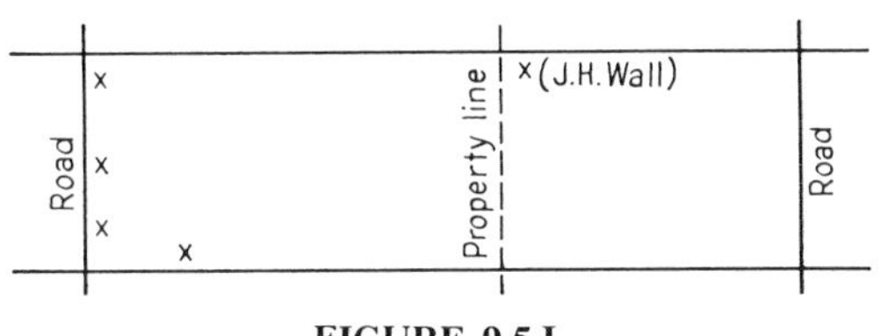

FIGURE 9.5.I

(*b*) A street (road) which is not cut through to another street may be extended by an imaginary straight line to some landmark or to another street (road). An imaginary straight line connecting two landmarks (nonresidential structures, bridges, trees, etc.) may be used as a segment boundary.

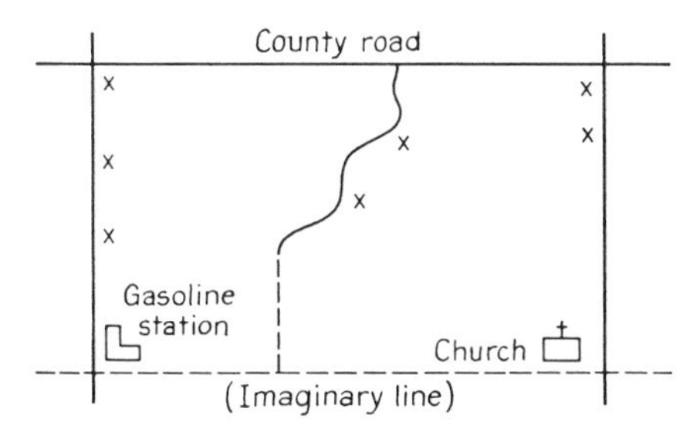

FIGURE 9.5.II

(*c*) Segment boundaries may also be defined by mileage along a road in the open country, where there are no dwellings that could be misplaced near the imaginary line.

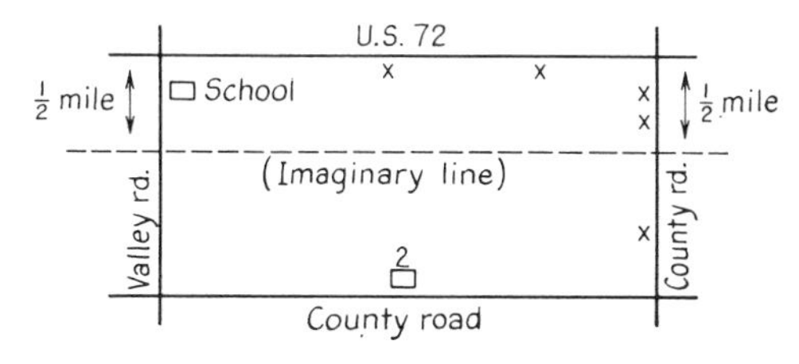

FIGURE 9.5.III

(*d*) House numbers may also be used to divide a block, provided the following procedure is observed. (1) As in the illustration below, the boundaries of the segment must be drawn in on the sketch. (2) The *right-hand boundary*, as you face the segment from the street, must be drawn *next to the last structure* included in the segment. (3) Where there are no alleys, common *back property lines* may be used to subdivide a block. (4) If a multi-unit structure contains 3, 4, or 5 dwellings, it may be regarded as a single segment.

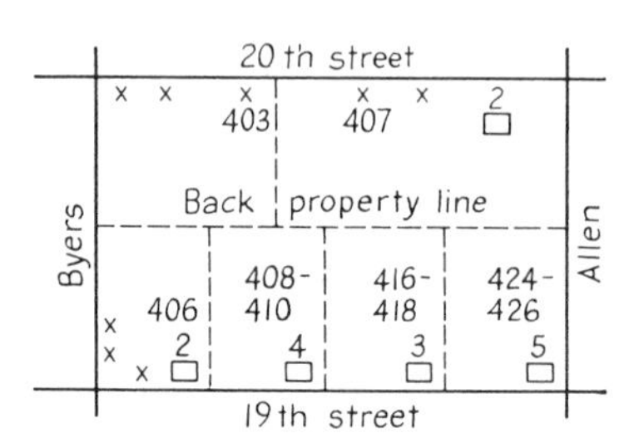

FIGURE 9.5.IV

(*e*) In multi-unit structures containing 6 or more dwellings, each floor, or part of a floor, may be considered as a segment. Indicate the location of the structure on the sketch and dash (- - - - -) an imaginary boundary around the structure. *Prepare a Segment Listing* and be sure to list all dwellings including the basement and any other unusual attached dwellings.

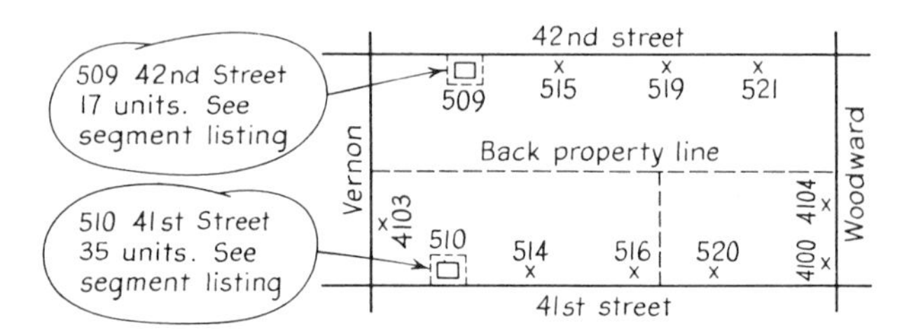

Note that 4103 Vernon plus 514 and 516 41st Street, form one segment.

FIGURE 9.5.V

IV. *Segment Listing* is our procedure for forming segments when the density of dwellings prevents their clear location and identification on the Block Sketch. Make out a separate Segment Listing for each such problem you encounter within the block. This block is shown in Figure V; three regular segments have been formed (called *a*, *b*, and *c*) and four segments (*d*, *e*, *f*, and *g*) on the Segment Listing for 509 42nd Street; hence, here at 510 41st Street, the five segments are called *h*, *i*, *j*, *k*, and *l*. An example of a Segment Listing follows.

Area: La Place City SEGMENT LISTING
Block: 356
1. Identify here clearly the house(s) included on this Segment Listing:
House with 23 dwellings at 510 41st Street

2. List each dwelling on a separate line. Take them in order, floor after floor. Indicate floor number and the apartment number.
3. If apartments have no numbers use some other identification, such as location on the floor. In case of any possible doubt include a floor sketch.

Dwelling	Dwelling address or floor	Apartment No. or description	Segment number	Project
1.	Basement	Superintendant		
2.	First floor	101		
3.	" "	102	H	
4.	" "	103		
5.	" "	104		

FIGURE 9.5.VI

9.5C *Instructions for Interviewing in Segments*

I. A detailed sketch of each block, outlined in red, is included in the *Segment Folder*. Segments are outlined in blue and assigned letter designations. The record on the cover of the Segment Folder identifies each segment and provides the following information.

(*a*) The segment number consists of three digits for the block number and a letter for the segment designation. These letters are listed in column 1 of the folder in alphabetical order.

SEGMENT FOLDER

Area Pearson County
Place Fisher Township
Block Number 517

Segmenting Date Nov. 2, 1961
Segmenting done by Smith

Segment Control Record

Segment Number	Project Number	Comments
A	412	One new dwelling built in 1961, plus a new trailer
B		Western boundary arbitrary extension of Hill pond
C		
D		
⋮		
⋮		

FIGURE 9.5.VII

(*b*) The project number in column 2 designates the project for which interviews are to be taken. This is crossed out for previous projects so that only the current project appears uncancelled.
(*c*) Brief remarks about a segment are entered in column 3.

II. The Segment Folder contains a listing sheet, *The Final Listing of Segment*, with heading items completed for the *sample segment* assigned on the current project. Prepare the listing of all the dwellings in the sample segment, proceeding as follows.

(*a*) Begin at a convenient location and proceed clockwise around the sample segment; record the description or address of each dwelling in the sample segment. List first the dwellings you identify with the marks (X and ☐) on the sketch and draw a line under these. Then list any additional dwellings you may find.
(*b*) Indicate the location of *any new or additional dwelling* you find by drawing a small circle (○) on the sketch. Postpone action and describe the situation to the office if you find more than four new dwellings in a segment, or if you find dwellings you cannot *clearly* place in a specific segment.
(*c*) Identify each dwelling on the sketch by entering the line number (col. 1 of the Final Listing) next to its X, ☐, or ○ on the map, or on its proper Segment Listing. For multi-unit structures, indicate the range of line numbers; e.g., ☐ 5–6 denotes two dwellings listed on lines 5 and 6 of the listing sheet.

FINAL LISTING OF SEGMENT

Area Pearson County
Place Fisher Township
Block Number 517A

Date of Listing Nov. 2, 1961
Interviewer Smith
Project 412

Line Number	Address or description of dwelling	Interview Result	Comments
1.	Big white frame, porch across front, green trim	Int.	
2.	Buff brick, brown roof, carport on N.E.	Ref.	
3.	White frame, green trim, carport on N.W.	N.A.H.	
4.	White asbestos siding, two doors on front, green roof	Int.	
5.	Pink brick, white roof, double carport on N.W.	Int.	
6.	Trailer on property of line 1	Int.	Here for full year
7.	White frame, grey roof, 2 picture windows, between lines 2 & 3	Int.	Built March, 1961
8.			
9.			

FIGURE 9.5.VIII

III. Assign a separate Cover Sheet to each dwelling *whether vacant or occupied* (unless specifically instructed otherwise for the project). *For every dwelling* in the sample segment, complete either an Interview Cover Sheet or a Noninterview Form (on the Cover Sheet).

IV. Use your presence in the block to indicate needed improvements you notice. If you note any new dwellings, or dwellings under construction, or other changes, or previously missed dwellings, or any reason for correcting the Block Sketch, please make appropriate notes on the inside of the Segment Folder. Indicate the date of any changes on the sketch.

9.5D Segmented Listing

A modification of segmenting involves listing all the dwellings in consecutive order and then breaking the listing into "segments"—of, say, 4, 6, or 10 dwellings. This procedure was introduced in 9.5*B*.IV for the Segment Listings of oversized divisions in densely settled portions and for apartment houses; but its use can be extended to the entire block. First, the procedures of Section 9.6 are used to list all dwellings in the block. Second, these are divided into segments; here again some compromises can be made between equal sizes and good physical boundaries for the segments. Third, the within-block interval is applied, after a random start, to select segments. Fourth, the sample segments thus selected are assigned for interviewing. As with the other procedures, steps 2, 3, and 4 are best done in the office. But if we prefer to save the extra field trip, the four steps can be combined into one trip; in this case the selected numbers should be "hidden" in an envelope or behind black tape.

A key part of this procedure is to make the interviewer understand that every dwelling within the block boundaries must belong clearly to one and only one segment—whether it was listed or not. He must understand that the segment must include *every dwelling lying from the first listed dwelling in that segment up to, but not including, the first listed dwelling of the next segment.* This procedure, using a "half-open interval," is an example of "linking" missing units with specified units in the frame (2.7*B*).

I believe that this compromise between segmenting and listing should usually obtain results somewhere between those of compact segmenting and listing. The problems of homogeneity and interaction are the same as for compact segments, but variation in size may be somewhat less because finding "boundaries" may be easier. In simplicity it may excel and in cost it may often do well, if the listing is not expensive. The key question is: Will this modified procedure prove as good as compact segments in coverage and in stability? The answer depends on the specific situation. I believe that often it will do well enough, particularly for smaller and localized surveys; but in many circumstances the problems of linking may prove troublesome. Imperfections can arise when the order of location of new dwellings is not clear in relation to the listed dwellings; also when the interviewer is so busy with the main job of interviewing that he does not notice some missed or new dwelling. A "theoretically sound procedure" may work poorly in practice. When trying to link, with half-open intervals, missing dwellings to single listed dwellings, our experiences were negative [Kish and Hess, 1958].

Deming [1960, p. 228] writes that these procedures "go smoothly in the field, with simplicity, economy, and statistical efficiency." His procedures actually should be even less costly, because the interviewer needs to enter only the first dwelling of each segment. But it should be used only when the ordering of dwellings around the block leaves little room for doubt in locating dwellings into their proper segments. The essence of his instructions (from pages 232, 235, and 237) follows to the end of this section.

1. Go to the corner marked "Start here." Start walking around the block counterclockwise. As you proceed around the block you will count only the dwelling units on your left.

2. Record on the Segment Sheet the address of the first unit on Pearl River Street. This address begins segment No. 1. The numeral 1 appears in the left-hand column of the Segment Sheet. Count 4 more units, and the segment is complete. Record in the proper column the number of dwellings in Segment No. 1 (5 dwellings in this example).

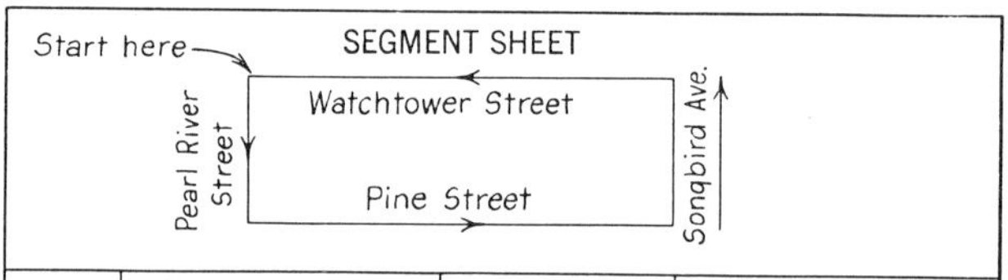

SEGMENT SHEET

Start here

Watchtower Street

Pearl River Street

Pine Street

Songbird Ave.

Segment number	Beginning address: Street	Beginning address: Number	Visual estimate of the number of dwellings	Remarks
1	Pearl River St.	135	5	
2	" " "	145	5	
3	Pine St.	34	4	Apts. 1, 2, 3, 4
4	" "	38	6	Apts. 1 - 6
5	" "	42	5	Basement and apts. 1, 2, 3, 4

FIGURE 9.5.IX An interviewer's segment-sheet. Each address begins a segment, which includes this address and every dwelling up to but not including the next address. The map was drawn in the office before the interviewer commenced work [from Deming, 1960, p. 232].

3. Segment No. 2 will begin with the next dwelling. Record this address. Count 4 more units, and Segment No. 2 is complete. Record the number of dwellings in this segment (5 again).
4. Follow this procedure throughout the entire block.

In rural areas and in towns where there are no streets and addresses you will describe the first house in a segment, not by number, but by the name of the owner, by location, by size and description, by distance from main roads. The color of a house is risky, as people sometimes paint their houses, and an interviewer coming at a later date might be confused.

Note that you are recording the beginning-addresses. These beginning-addresses define or create the segments, as we said earlier.

As you might suspect, it will not always be so simple to break a block into segments. Here are some of the possibilities that will require deviation from the foregoing ideal situation. As noted before, the beginning point of each segment must be definite and unmistakable. Consequently, you should never begin a new segment in the middle of a small structure that has multiple entrances (apartment houses are different and will come later). It is better to let the whole structure be 1 segment, even if it contains 6 or 8 dwellings. Break it into 2 segments only if you can do so unmistakably. Remember that there must be no subsequent doubt about which is the beginning-address of any segment.

When you have finished creating the segments, open the sealed envelope and find listed there the numbers of segments that have been chosen for you to interview. IT IS OF THE UTMOST IMPORTANCE THAT YOU FINISH YOUR JOB OF CREATING SEGMENTS BEFORE YOU OPEN THIS ENVELOPE. TO OPEN THE ENVELOPE PREMATURELY WOULD PROFIT YOU NOT IN THE LEAST, BUT IT WOULD SERIOUSLY IMPAIR YOUR WORK.

The envelope will contain 7 numbers. You will interview in every segment that bears one of these numbers. Do not worry if there are numbers for which you have no segments. For example, your envelope may contain the numbers 3, 18, 19, 35, 45, 51, whereas you had created only 19 segments in your area. You would then disregard the numbers 35, 45, and 51. The supply of numbers is purposely more than you can usually use.

If, however, you have more segments than random numbers, the explanation can only be one of three possibilities:

(*a*) This area has grown a great deal since the map that we used was made.

(*b*) Your segments are mostly smaller than the intended size (choice of definite boundaries may have forced you into this).

(*c*) You are working in the wrong area, in spite of the precaution that you took.

If you are in the wrong area, discard the work that you have done; go to the right area and proceed in the regular manner. If this is not the explanation, however, please halt your work; send to the home office your revision of the map, your list of segments, and any possible explanation in regard to growth. Await further instructions.

9.6 LISTING DWELLINGS

9.6A Instructions for Listing Dwellings

I. Proceeding Around the Block. Begin listing at the place on the block shown by the X in the upper-right corner of the Sketch Sheet. This is usually the North side next to the Northwest corner of the block. *Proceed in a clockwise direction around the block, i.e., so that the buildings of the sample block are on your right as you walk along.* This clockwise direction is shown by the arrow (→) next to the X on the Sketch Sheet. You must cover the *entire area inside the boundaries* and nothing outside it.

If there are any small streets or alleys on the block, be sure to check them all for possible dwellings. Go in and out of such intersecting streets and list them as you come to them. List first that side of the street or alley which you come to first. If the alley cuts through your block, you may find it easier to consider separately the two halves of the block listing, first one half, then the other. When you do this, explain what you did by including a little sketch on the Listing Sheet.

II. Rules for Entering Addresses or Descriptions. Proceeding around the block, you encounter dwellings in a specified order. In that same order, list the address or description of *each dwelling on a separate line* of the Listing Sheet. Beginning with line 1 on the Listing Sheet, continue listing the addresses of all dwellings that you find around and inside the block as you proceed clockwise. If there are more dwellings on the block than lines on the listing sheet, use one or more Continuation Listing Sheets as necessary.

Watch for Obscure Dwellings. Often a building has an obscured dwelling in the rear. This is particularly true of stores, shops, etc., where someone may live at the rear of the structure, or over the store. A shack or lean-to alongside a building may be a dwelling. Be sure to list these dwellings too. Remember that your job is to list every place where someone lives, or might live, within the area of the boundaries of the block. Be on the lookout for obscure dwellings in the middle of your block—away from the street.

How do we tell from the outside, without interviewing, how many dwellings there are in a building? Houses that seem clearly to be single family homes should be listed as such. It is true that some will turn out to contain two dwellings, but our selection procedures will help you to deal with them. If you suspect that the house may contain two or more dwellings, search for clues: doorbells, mailboxes, separate entrances, and sometimes inquire briefly.

In apartment houses note the doorbells or mailboxes, or ask the superintendent (and remember his apartment). Watch for separate entrances. You will not need to inquire inside special dwelling places such as apartment hotels, commercial hotels, rooming houses, trailer camps, and motels; make a detailed description of such situations so that these dwellings can be distinguished. Try to find out and specify those units which are or can be used for regular, permanent occupants—as distinguished from transient quarters.

III. Order of Listing Unnumbered Dwellings Within a Building. A systematic order of listing dwellings that have no numbers is important. Many buildings have their own numbering system which you can use readily. However if you find unnumbered dwellings in a building, use the following rules for the order of listing:

(*a*) List the bottom floor first (basement, if any) and work up.
(*b*) List the dwellings on the right first and then the left.
(*c*) List the front dwellings first and then the rear.
(*d*) The ground or street floor is called the "first floor" (whether or not there are dwellings there); the first floor above the ground floor should be called second floor and so on up. For example, a dwelling unit above a one-story store would be "second floor above store."

In duplex houses attached side to side, list the right side first and then the left side, because that is the order you will encounter when going clockwise around the block. If the two entrances of the duplex are front and rear, take the front first and then the rear. *If you come across some complicated arrangement*, make a sensible system in accord with the above; *include a sketch and explain the layout* on the listing sheet or on an attachment to it.

IV. Writing the Description or Address. You will need to identify each dwelling with sufficient clarity so that anyone who has never seen it can locate it on the block by using the listing you made. If there is only one dwelling in the house, you usually need only to identify the house. However, if there is more than one dwelling in the building, you need to *identify* both the building and each dwelling in it.

DWELLING ADDRESS LISTING SHEET

Survey Research Center July 1954
PSU Clark, Arkansas
Place Arkadelphia
Block No. 3
Interviewer L. C. Johnson

X→ Adams St.
Madison St.
See attached sketch
Washington Blvd.
Newton Ave.

List on this sheet the address or description of *every dwelling* located on the area *within the boundaries* of the block outlined above. Start listing at the corner marked X (see above diagram) and proceed systematically around the block in a *clockwise* direction as shown by the arrow (→). (Clockwise is to your left as you face toward the block.) If there are streets or alleys that cut into the block, go in and out of each as you come to it. Mark on the sketch such streets and alleys, and any other useful changes or additions. For dwellings without street numbers, give good descriptions. Each dwelling address should be on a separate line. Do not use a line for anything but an address or description of a dwelling. Write all other relevant information about this block around the edges or on an attached sheet. As you list, fill out the rating column and the zone number or P. O. address in the box below. (For further instructions, see Chapter on Listing in your manual.)

BE SURE THAT EVERY PLACE OF RESIDENCE WITHIN THE BLOCK IS LISTED AND THAT ALL APARTMENTS ARE NUMBERED OR OTHERWISE IDENTIFIED.

LINE NO.	ADDRESS OR DESCRIPTION OF DWELLING (AND APT. NO. OR LOCATION)	RATING	PROJECT NO.
1	201 Adams St. duplex right	M	
2	201 Adams St. duplex left	M	
3	207 Adams St. first floor	ML	
4	207 Adams St. second floor	ML	
5	211 Adams St. White, 2 story, number not clear, large bay window	M	
6	1842 Washington Blvd.	MH	
7	1836 Washington Blvd. front	ML	
8	1836 Washington Blvd. rear	L	
9	1822 Washington Blvd.	M	
10	Dwelling above garage at 1822 Washington Blvd.	ML	

(OVER)

The Post Office address of this block is:
Arkadelphia, Ark.

Note: There are about 10 vacant lots between lines 5 & 6

FIGURE 9.6.I

(*a*) Where the dwelling (and the building in which it is located) has a clearly discernible number, just use the number and the street name as on lines 6 and 9 of the illustration (Fig. 9.6.I). When the building has more than one dwelling, each dwelling is designated, as on lines 12, 13, 14, and 15.

(*b*) In multiple-family houses which contain unnumbered (and unlettered) dwellings, describe each dwelling by its location. Examples

Page 2

LINE NO.	ADDRESS OR DESCRIPTION OF DWELLING (AND APT. NO. OR LOCATION)	RATING	PROJECT NO.
11	1820 Washington Blvd. Second floor above a grocery store	L	
12	1816 Washington Blvd., basement apt.	ML	
13	1816 Washington Blvd., apt. # 1	L	
14	1816 Washington Blvd., apt. #2	L	
15	1816 Washington Blvd., apt. #3	L	
16	Alley near 1816 Washington Blvd., one story white bungalow, green trimmings, 3 red steps to front door	ML	
17	? Washington Blvd. Grey 1½ story, blue roof, large oak tree in front yard	ML	
18	208 Madison St. basement apt.	ML	
19	208 Madison St., first floor right, front apt.	M	
20	208 Madison St., first floor right, rear apt.	M	
21	208 Madison St., first floor left, front apt.	M	
22	208 Madison St., first floor left, rear apt.	M	
23	208 Madison St., second floor, front apt.	ML	
24	208 Madison St., second floor, rear apt.	ML	
25	212 Madison St., home behind tailor shop	ML	
26	Note: No dwellings on Newton Ave. Strictly		
27	commercial and there are no dwellings above or behind stores.		
28			
29			
30			

If you run out of lines on this sheet and there are more dwellings on the block, attach a "continuation sheet," and continue listing.

FIGURE 9.6.I (Continued)

are on lines 1 and 2, lines 3 and 4, lines 7 and 8. See also the seven apartments on lines 18–24.

(*c*) In the cases of the dwellings on lines 5, 10, 11, and 25 there was only one dwelling in each building, but the additional description might be needed to locate it.

(*d*) When the dwelling or the building in which it is located (or both) is unnumbered, use a *description*, as on lines 16 and 17. Use the more *permanent* and more obvious features such as the material (brick, frame) and color of the building, the type of roof it has, the location of the porch, the location of the building in relation to a numbered building on the street, etc.

V. Listing Vacant Dwellings and Dwellings Under Construction. So that possible occupants during the survey period may receive their proper chance of selection, include on your listings dwellings that are under construction and vacant, but habitable later—excluding only houses clearly and finally abandoned. If selected and found still empty, no harm results: the interviewer simply notes "vacant" or "no dwelling."

Similarly, enter stores which may have a dwelling behind or above them. But do not list those stores, shops, factories, offices, vacant areas, and public property which clearly cannot contain dwellings.

Little harm is done to the sample if a doubtful case turns out not to be a dwelling at the time of interview. However, if the doubtful cases actually contain dwellings, much harm can be done to the sample if those dwellings have been left out of the listing. When you remain in doubt as to whether a building or part of a building should be listed, *list it while you are at the location, and write up the situation* in a note attached to the Listing Sheet. Describe the layout of the building, the arrangement of possible dwellings, and the system you used in listing. When in doubt, list. However, you can eliminate most doubtful cases by reasonable investigation.

VI. What to Do If a Block Contains No Dwellings. If you go to a sample block and find there are no dwellings on it, briefly describe on the Listing Sheet what there is on the block. For example: "The entire block is occupied by commercial buildings. These were inspected and no dwellings were found." If a sample block is only partially occupied by dwellings, explain what occupies the rest of the block.

Feel free to enter notations on the margins of the Listing Sheets, or to write explanatory notes on an attached sheet of paper. Rough sketches of the arrangements of dwellings in a difficult building should be attached to the Listing Sheet whenever you think it will help explain the situation. Add special remarks about unusual conditions and cases found in the block; also about unusual living quarters, such as institutions, that have been excluded from the survey population.

9.6B Selecting Listed Dwellings

The lines of the Block Listing Sheet constitute listing units from which a sample must be selected. Taking a systematic selection is usually the easiest: the within-block interval is applied in each block after selecting a random start. The interval is determined by the procedures of 9.3, or some other method, for selecting first blocks, then dwellings within the sample blocks.

By applying the intervals to the listing sheets we select the sample lines; these designate the *sample addresses* to be included in the sample for the project. These should be designated clearly in the project column of the Listing Sheet.

The number of sample dwellings in each sample block will not be exactly the planned average b; it will vary from block to block. That number is zero if the actual number of dwellings is less than the random start in the block. The extent of the variation depends mainly on the ratios of the actual to the expected sizes of the blocks, second, on the application of intervals after random starts. The use of proper intervals assures desired selection probabilities for the listed addresses, and will reflect the actual situation in the blocks as found by the field worker.

These variations are explained in detail in 7.5*C*. Procedures for applying intervals, including fractional intervals, are given in 7.5*B* and *F*. Instead of systematic, a truly random selection of dwellings can be selected, if desired (7.5*B*). Instead of individual dwellings, the sampling unit within the block can be designated as a segment of adjacent dwellings (9.5*D*).

The same listing may be used for several surveys. Where F_b is the interval (uniform or minimum), up to F_b distinct samples, each using the same interval F_b in the block, can be taken without using the same dwellings twice; or the interval can be changed to suit the requirements of different surveys (12.6). The costs of listing the blocks can thus be spread over several surveys.

This procedure assumes that, after listing the blocks, the field worker returns the Listing Sheets to the office where the sample addresses are selected, and these designated lines are then returned to the field for interviewing. An alternative plan calls for doing the interviewing (or the first call) at the time of the listing. This plan saves the enumerator an extra trip to the block and may be useful when travel cost is an important consideration; but there are some practical disadvantages in giving the interviewer the two tasks simultaneously. If this plan is used, the numbers of the sample lines, preselected in the office, should be hidden in an envelope or behind opaque black tape until the block listing is completed—so that knowledge of the numbers will not influence the order of listing (8.4*C*).

9.6C Identifying Dwellings at Sample Addresses

The sampling interval applied to the listing sheet selects only lines on that sheet and the associated listing units, the "sample addresses." The interviewer should be given clear, simple, adequate, and workable instructions; also, materials to locate the sample addresses, and to identify the dwellings properly associated with them. A block address card containing the sample addresses and describing the block boundaries may be sufficient. But the Block Listing Sheet, with the selected lines clearly shown, gives more information and allows for better judgment on what should or should not be included at the sample address. If the original sheets are kept in the office, cheap copies can be made. Outlining the sample blocks on a city map can be helpful for locating them and for planning routes.

Most of the time the sample address will be either a single family house or a clearly labeled apartment, and the interviewer will easily find the dwelling uniquely associated with it. But problems can arise either because the lister was misled by outside appearances or because of changes since the time of listing. The lack of one-to-one correspondence of sample addresses to dwellings (of listing units to sampling units in 2.7), if not negligible, calls for special attention—especially since the interviewer tends to concentrate on his main task of interviewing. The *instructions for identifying dwellings at sample addresses* that follow are designed to deal with these problems.

I. Checking for "Extra Dwellings" at the Sample Address. Sometimes a sample address which was listed as a single dwelling actually contains more than one dwelling. These "Extra Dwellings" (for example, a basement apartment, servant's quarters over a garage, etc.), not seen at the address when the block was listed, must be included in the survey if we are to have an unbiased sample.

When you go to a sample address, ask yourself the question, or inquire if necessary: "*Are there any other dwellings at this specific address?*" (in addition to the one the lister obviously had in mind). This information can be obtained by inspection, or if necessary, from the residents of the sample address. In most cases the answer to this question will be "No" since the greater part of the listing is usually accurate. If the answer to this question is "Yes," *interviews should be taken at all extra dwellings discovered at the specific sample address.* At the top of the Cover Sheet for each of these "extra" interviews, write "Extra Dwelling." This procedure is designed to pick up 1, 2, 3, or 4 dwellings in one spot missed by the lister; if you find 5 or more, do not interview them, but send us quickly a description of the situation and wait for further instructions.

However, do not take interviews at other dwellings that have been listed separately and are not in the sample. If you happen to notice actual dwellings missing from the listings (dwellings that should be associated with specific listed addresses not in the present sample), describe their location on a note attached to the listing sheet. If you find missed dwellings, and you cannot decide clearly whether they should be associated with the sample address or with other listed addresses, send in quickly the description of the situation and wait for instructions.

II. Combined Addresses. You may find, occasionally, that the reverse situation exists: 2 dwellings have been combined since the time of listing; or you find that *what was listed as* 2 *dwellings actually is only* 1. *Interview at the existing dwelling only if the first of the* 2 *listed addresses was designated for the sample.* Describe the situation briefly on the Cover Sheet. If the sample address is the second of the 2 listed dwellings, consider that the sample dwelling is now nonexistent; take no interview, but complete the noninterview form on the Cover Sheet and explain the reason for the noninterview. Similarly, if 1 actual dwelling has been listed as 3 or more addresses, it comes into the sample only when the first is selected.

III. Addresses Without Dwellings. *The sample address can be a vacant dwelling, or a store, shop, or office containing no dwellings.* These seemed doubtful to the lister who entered them "just in case." If you find it still vacant, take no interview and make no substitutions. Merely describe briefly the facts on the Cover Sheet for the sample address and send it in quickly.

IV. Checking for Dwellings Between the Sample Address and the "Next Listed" Address. When you approach a sample address, you should ask yourself another question, and inquire if necessary: "*Are there any dwellings between the sample address and the next listed address?*" In most cases, the answer to this question will be "No." In other words, most of the time you will find that the address given as the "next listed" address is actually the next dwelling adjoining the sample address. The "next listed address" is the address immediately under the sample address on the Listing Sheet. If the sample address is the last address on the sheet, the next address is the first address on the sheet. If there are some commercial structures without dwellings between the two addresses, your answer will be still "No," because we are interested only in dwellings; but in case of doubt, you should take a quick look at such places to see whether there are any dwellings in them (e.g., stores where the proprietor may live in the back, or upstairs). But do not include in the sample a dwelling that has been merely listed out of order and appears unselected elsewhere on the Listing Sheet; check the Listing Sheet to be sure. If you find dwellings between the sample address and the next listed address, handle the situation as follows:

(*a*) *If there are less than* 5 dwellings (either newly constructed or missed in the original listing) between the sample address and the "next listed" address, *interview all of them*. For each interview taken, write "New Construction" or "Missed Dwelling" (as the case may be) in large letters across the top of the Cover Sheet for each interview. Also write a brief note on the back of the Cover Sheet *describing where you found the new construction or missed dwelling*.

(*b*) *If there are* 5 *or more* newly constructed or missed dwellings between the sample address and the "next listed" address, *write the office quickly* about the situation, list the addresses of all such dwellings and wait for instructions.

(*c*) If the sample address is the last listed address in a multi-unit structure, check to see if there is a basement apartment which was missed in listing. *Unlisted basement dwellings* are to be associated with the last listed unit in the same structure, and are to be included in the sample whenever the last listed address is selected for interviewing.

(*d*) Be sure to check for extra, missed, and new construction dwellings even though the specific assigned address turns out to be a non-interview.

V. Changes in Sample Blocks. If you notice any change in the sample block which is not indicated on the Listing Sheets, let us know about it. For example, if some of the house numbers have been changed or if there is any large-scale demolition of dwellings, attach a note to the Listing Sheet explaining the situation and its location.

We are especially interested in large-scale new construction in the block. While you are in the vicinity of a sample block, take a quick look at the entire block. Report to us, if you notice *any unlisted large-scale new construction of dwellings* (5 or more dwellings) in the block. This would include things like newly constructed apartment buildings, a new housing development, a row of new houses, a newly developed trailer camp, etc. The purpose of this check is to allow us to correct our listing sheets, to protect us from having these large groups of new dwellings come into a later sample in one big cluster. If any large new construction areas are found, make a notation on the Listing Sheet or Block Address Record *telling us the location, type, and size* of the construction, and the date of completion and occupation, if you can find out. We will then be able to take account of the information for future surveys. Of course, if it is occupied during the current survey, you would need to let us know about it immediately so we can decide whether it should be in the sample on the current survey.

9.7 CREATING SEGMENTS FROM LISTED BUILDINGS

9.7A Special Features

This procedure should prove economical for creating segments from blocks that contain many multiple-dwelling houses. For selecting segments in a large city, it may be preferred to procedures described in the last two sections. Unlike the listing procedure of 9.6, it does not require prelisting individual dwellings for entire blocks. Listing entire buildings is easier than identifying and listing individual dwellings.

The segmenting procedure of 9.5 asks the enumerator to create segments of about equal size with clear boundaries, and to draw these boundaries on the block sketch. These tasks are avoided with the present procedure. The enumerator identifies and lists entire buildings, or the floors of large apartment buildings, and assigns to each a rough guess of the number of dwellings.

The block lists of buildings are sent to the office for creating segments. In most situations this step is needed, because dividing buildings into segments of about equal size is a complicated task for enumerators. Hence, other procedures should be employed when segmenting and interviewing must be done simultaneously.

We can take advantage of the presence of the lists in the office to assign convenient size measures. The segmenting procedure in subsection 9.7*D* requires creating exactly Mos_α segments, where Mos_α is the size measure of the block in number of segments. Then the predetermined number of segments is selected from the Mos_α segments. In subsection 9.7*D* exactly 2 segments are selected, but generally any integer b^* can be specified.

Creating segments in the office permits modifications based on information about buildings or dwellings obtained by enumerators. We may resort to stratified selection of segments; or even to forming heterogeneous segments.

Selecting 2 (or b^*) from Mos_α segments is easy if the Mos_α are integers. The instructions in subsection 9.7*A* assign integers of 2 or more. It would be possible to utilize directly the numbers of dwellings as size measures, but this would result in fractional values of Mos_α. These could be treated with the procedures of 7.5*F*.

The instructions of subsection 9.7*A* yield on the average more than 5 Census dwellings per size measure. New growth tends to increase this, but nonresponse decreases the actual yield.

The procedure is flexible. For example, instead of 2 segments of about 5 dwellings, we could easily design for 1 segment of 10 dwellings. In one city we selected 2 segments of 10 dwellings each; then from the average of 20+ dwellings we obtained an average of 3 members of the survey population, persons over 65 years old.

9.7B *Selecting Blocks*

1. Assign size measures Mos_α to all blocks in the Block Statistics volume as follows:

Census Dwellings	10–14	15–19	20–24	25–29	30–34	35–39	40–44
Mos_α	2	3	4	5	6	7	8

Rule: Multiply by 2 and discard the last digit.

2. Blocks of 1–9 are not sufficient. Attach them with a bracket to a neighboring block—given a choice, attach to a larger block.

3. For some blocks special measures have been previously assigned, according to official information about large-scale new construction.

4. After blocks are selected, search for unlisted blocks with zero dwellings. If a block numbered X is selected, the next listed block should be numbered $X + 1$. If any block numbers are missing, find them on the city maps. Unless obviously without dwellings (parks, museums, schools, monuments, etc.), attach these blocks to the selected block X.

5. Select blocks with the sampling fraction $1/2F$ applied to the size measures. This fits the two-stage selection equation

$$\frac{\text{Mos}_\alpha}{2F} \cdot \frac{2}{\text{Mos}_\alpha} = \frac{1}{F} = f \qquad \text{or} \qquad \frac{2\,\text{Mos}_\alpha}{4F} \cdot \frac{2}{\text{Mos}_\alpha} = \frac{1}{F} = f$$

for 2 segments per block. The measures can be selected systematically with the interval $2F$; or 2 selections can be made from each *zone* of $4F$ measures (7.4).

Instead of 2 segments, another digit b can be substituted above. If explicit strata are needed, some block measures can be altered to make complete strata with $4F$ measures (7.5D).

9.7C *Instructions for Listing Buildings*

The information you collect will be used to construct segments of roughly 5 dwellings each, for detailed canvassing and interviewing. We say "roughly 5 dwellings," because we expect most of them to vary from 4 to 6 and some even from 3 to 8 dwellings. Your work consists of listing on the Building Listing Sheet every building located within the boundaries of the sketch, with its address, identification, and a rough guess about the number of dwellings in it.

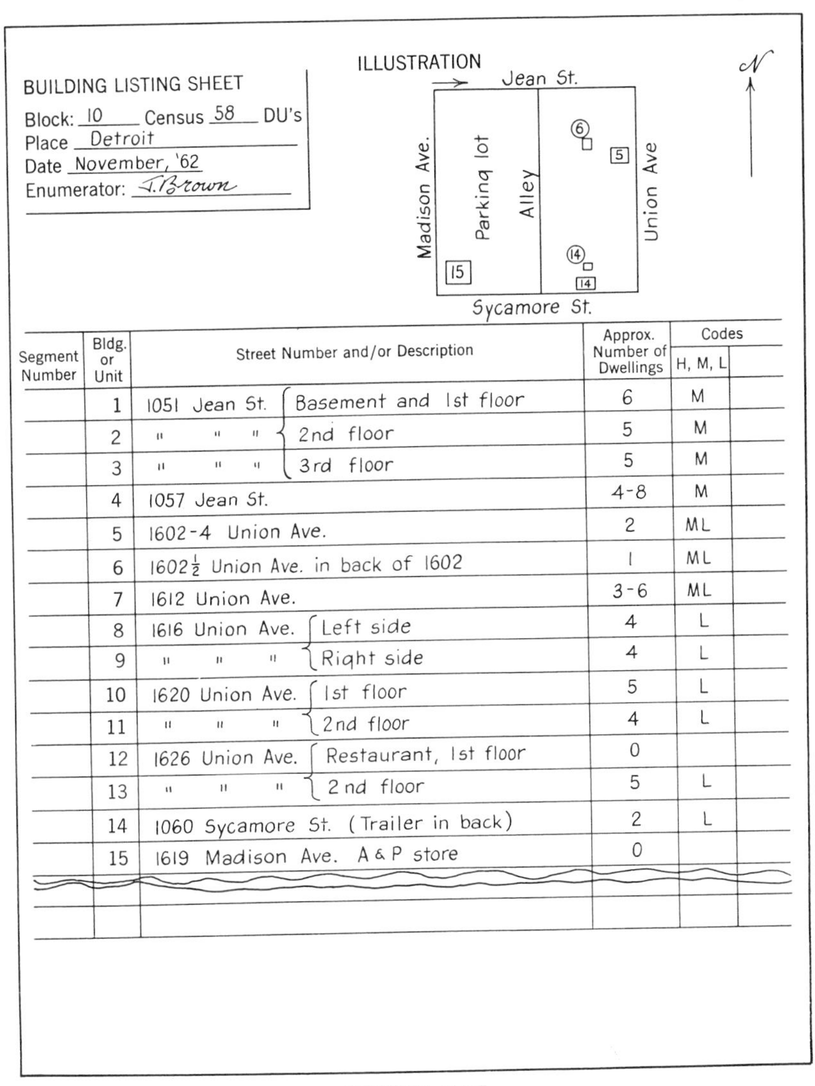

BUILDING LISTING SHEET

Block: 10 Census 58 DU's
Place Detroit
Date November, '62
Enumerator: J. Brown

Segment Number	Bldg. or Unit	Street Number and/or Description		Approx. Number of Dwellings	Codes H, M, L	
	1	1051 Jean St.	Basement and 1st floor	6	M	
	2	" " "	2nd floor	5	M	
	3	" " "	3rd floor	5	M	
	4	1057 Jean St.		4-8	M	
	5	1602-4 Union Ave.		2	ML	
	6	1602½ Union Ave. in back of 1602		1	ML	
	7	1612 Union Ave.		3-6	ML	
	8	1616 Union Ave.	Left side	4	L	
	9	" " "	Right side	4	L	
	10	1620 Union Ave.	1st floor	5	L	
	11	" " "	2nd floor	4	L	
	12	1626 Union Ave.	Restaurant, 1st floor	0		
	13	" " "	2 nd floor	5	L	
	14	1060 Sycamore St. (Trailer in back)		2	L	
	15	1619 Madison Ave. A & P store		0		

FIGURE 9.7.1

1. *Check the block boundaries to make sure you have correctly identified the block.* These boundaries are usually streets, roads, highways, railroad tracks, streams, or other recognizable physical features. However, some others may be civil boundaries, corporate limits, township or county lines.

2. We have drawn on the sketch the boundaries and street names from our maps, but *you must revise them* to show the current facts as you find

them. *Add any missing streets, alleys,* and other boundaries as you find them. Eliminate any street or boundary, or any part of one, that appears on the sketch but does not currently exist in the block, by drawing a wavy line (∿∿∿) through it. We drew our boundaries in red so that your corrections with pencil or pen will show clearly.

You can write any remarks about the block in the space next to the sketch. We have written the number of dwellings the Census found in this block, in the upper left-hand box on the sketch sheet. This may help you decide if you are in the right block. But this figure may be out-of-date, and sometimes it can be just plain wrong.

3. Begin at the corner marked with an arrow (→) and proceed systematically around the block in a clockwise direction. As you walk around it, your block should be on your right side. If there are streets or alleys that cut into the block, go in and out of each as you come to it. You may change the starting point if you need to, and change the arrow on the sketch accordingly.

4. *List every building* located in the area *within the boundaries* outlined above. List them in clockwise order as you cover the block quickly. If a house is inside the block, away from the boundaries, spot it on the sketch with its listing number (thus, ⑧), and note this fact in its description. Enter any remarks you need on the back of the sheet, identifying the *building number*.

5. Many buildings will seem to be single-family homes. You can list them easily, writing 1 for the number of dwellings. If it seems to be a two-family house, list it on a single line and write 2 for its number of dwellings, etc.

List also commercial and other buildings without dwellings; include a description (gas station, tailor, barber shop, fire station, etc.) and enter 0 for dwellings, if you think nobody lives there. Thus we shall have a complete list of all separate buildings within the block. The only buildings you need not list are private garages and shacks that obviously belong to specific houses. If a trailer is parked behind a house, you should note that fact when listing the house.

If a building contains anywhere from 3 to 8 dwellings, list it and give your estimate of the number of dwellings.

6. If a house seems to contain more than 8 dwellings, break it up into smaller divisions and list each division on a separate line, writing the appropriate number of dwellings for each division. The divisions can be of any size up to 8, but divisions between 3 and 5 dwellings are best. Use whatever system seems easiest for forming the divisions. We hope you can form the divisions and assign approximate dwellings quickly, without walking to the upper floors. Usually each floor can be used as a division.

The floors should be listed, beginning with the first (ground) floor; this should include a single basement apartment (the janitor's). Then the second floor, one flight up, and so on up. If there are several apartments on the basement floor, they should be listed first. If the ground floor is occupied exclusively by stores and shops, you should note this fact on the listing sheet. Then proceed with the floors above the stores. In some buildings other divisions may seem easier and more natural; use them instead of floors. If you cannot use natural divisions (floors, etc.), indicate the apartment numbers assigned to each. If you cannot divide the building into divisions, note the approximate total number of dwellings in the building and we shall subsample it.

7. You may need to look only into larger apartment houses to find the number of dwellings and a system of division. But usually we expect you to estimate the number of dwellings *quickly* from the outside, without any detailed investigation. We need only your quick guess, not exact figures. If in doubt, you can write either the range of your guess (such as 4–8 dwellings), or its midpoint (6 dwellings), or your best guess (5 dwellings); write whichever seems easiest to you.

Range	1–3	2–4	3–5	4–6	3–7	4–8	5–9
Midpoint	2	3	4	5	5	6	7

8. Write the street name and number; and both numbers for a duplex (e.g., 218–220) or an apartment house (410–420), if they are so posted. For houses without good numbers, give brief but good descriptions of permanent features. List each building on a separate line. If you divide the building into divisions, list each division on a separate line; but note with ditto marks (″) or a brace ({) all the lines that belong to a single building.

9. A dwelling is generally a group of rooms (or a single room) occupied (or intended for occupancy) as separate living quarters by a family or some other persons living together, or by a person living alone. Ordinarily a dwelling is a one-family house, or half of a duplex, or an apartment, or a flat. But some unusual dwellings are behind stores, above garages, in basements; or in trailers, tents, shacks, or boats; or in caretakers' and watchmen's quarters. Usually dwellings have separate cooking facilities; but some small apartments and rooms with separate entrances do not; yet they are dwellings where people live.

10. We have also shown the number of dwellings found in the block by the last Census (or other sources). For many reasons these will differ from what you listed. If the difference is great and you see its likely cause

(demolitions, large-scale new construction, etc.), write it for us in a few words on the back of the sheet.

9.7D *Segmenting Blocks and Selecting Segments*

The entire block is taken and no segmenting is needed if $\text{Mos}_\alpha = 2$; the block need not be sent out for listing. Segmenting is needed if $\text{Mos}_\alpha > 2$. (Mos_α of 1 or 0 is not used.)

The five criteria for making segments are given in decreasing order of importance:

1. Make the number of segments equal to the measure Mos_α used to select the block; or an integral multiple or fraction as described in 4.
2. Form compact segments, consisting of contiguous buildings or floors. You can join the beginning of the list with its end, but try to observe large physical gaps. If the sampling block consists of two or more physically separated actual blocks or areas, observe these boundaries. In a few cases, double (or multiple) segments can be created for later subsampling.
3. Form segments of *about* equal numbers of dwellings.
4. Try for roughly 5 dwellings per segment. The measure Mos_α represents roughly the multiples of 5 Census dwellings, hence in most cases the ratio of listed dwellings to measure will be near 5. If the dwellings are more numerous than expected and this ratio is closer to 10 (between 8 and 13), make 2 Mos_α segments. If the ratio is closer to 15 (between 13 and 18), make 3 Mos_α segments. (This procedure can also be used to create segments smaller than 5.) If the dwellings are fewer than expected and the ratio is closer to 3, then $\text{Mos}_\alpha/2$ segments can be formed if Mos_α is divisible by 2.
5. Evidence of differences between dwellings can be used for creating heterogeneous segments
6. *Selecting Segments.*

Plan A. Select two segments at random from the Mos_α created segments; but if $k\ \text{Mos}_\alpha$ segments were created, select $2k$ segments.

Plan B. If Mos_α is odd, eliminate 1 of the Mos_α segments; but 1 of the segments if $k\ \text{Mos}_\alpha$ is odd. Then make up pairs of *unlike* segments. Select one of these pairs; but if $k\ \text{Mos}_\alpha$ segments were created, select k pairs.

9.7E *Instructions for Sampling the Segments*

We assigned a separate *Segment Listing Sheet* for each selected segment. There are two of these for most sample blocks in this sample. They are identified at the top of each *Segment Listing Sheet*—as for example, "Block 60, Segment 1"—and attached to the *Building Listing Sheet* for each sample block. *Do not get your Segment Listing Sheets mixed up.* One segment in the block may be a Take-All segment and the other a Take-Part segment, and you would cause much trouble if you used the wrong *Segment Listing Sheet* for the selected segments.

SEGMENT LISTING SHEET

a. Place ____________________

b. Interviewer ____________________

g. SELECTED LINE NUMBERS ____________________

c. Sheet ____ of ____ sheets

d. Block Number ________

e. Segment No. ________

f. TYPE OF SEGMENT:

Take-All ▱

Take-Part ▱

Line No. (1)	Description (or address) of Dwelling (2)	Project No. (3)
1		
2		

FIGURE 9.7.II

Your work in sampling each segment has five parts:

1. *Identify the Selected Segment* on the Building Listing Sheet against the actual buildings you see in the block.
2. *Find* any actual buildings within the selected segment that are *missing* from the *Building Listing Sheet* and list them there.
3. *List all dwellings* within the selected segment on the *Segment Listing Sheet*.
4. Identify the sample dwellings.
5. In all *sample dwellings identify the respondents* and population members.

The Study Instructions have a specific part to take care of the fifth step. The other four steps are described below.

Steps 1 and 2. *Identify Sample Segments and Check for Missed Dwellings.* Selected segments are identified with a purple bracket to the left of the group of addresses on the Building Listing Sheet with the project number 711 in the same color next to the brackets. Identify and check the actual addresses against those listed on the Building Listing Sheet and look for missing dwellings.

Add to the end of the Building Listing Sheet, below the pencil line, any addresses you find *within the selected segments* that do not appear on the Building Listing Sheet. Write the estimated number of dwellings at the address. Also note, on the left, the line numbers on the Building Listing Sheet between which it should be located.

Include in the segment all addresses located between the first and last addresses, plus any missing addresses you find between the *last address listed in the segment and the first address listed for the next segment.* Thus the selected segment contains all actual buildings between the first building listed in the selected segment and *up to the first building of the next segment,* but not including it. (If the selected segment is the last one on the Building Listing Sheet, its "next segment" is the very first segment on the next Building Listing Sheet.)

For example, the last listed address in the selected segment may be 444 Jones Street. On the Building Listing Sheet, the address succeeding it is 327 Smith Street. On checking the actual addresses in your segment you find that the lister missed 446 Jones Street. Include 446 Jones St. in the selected segment and add it to the end of the Building Listing Sheet, as well as on the Segment Listing Sheet.

But missed addresses in the rest of the block belong to other segments. Thus an actual address before the first listed address of the selected segment belongs to the segment preceding it. None of these addresses belong to the selected segments. But if you happen to notice them, add them to the Building Listing Sheet.

If you cannot decide whether a building belongs in the selected segment or not, write to us about it.

If the selected segment includes the entire block, identifying the segment becomes simple.

Step 3. *List All Dwellings Located Within the Selected Segments on the Segment Listing Sheet.* Keep in mind when listing your addresses that an address may contain 1 dwelling, or there may be 2, 3, 4, 5, etc., dwellings within the structure that has one address number on the outside. Use a separate line for the listing of each *dwelling.* For example, you have 444 Jones St. in your selected segment on the Building Listing Sheet. In listing this address on the Segment Listing Sheet you may find 4 dwellings within 444 Jones St. You then list on one line 444 Jones St., apartment, left downstairs; on the next line list 444 Jones St., apartment, right downstairs; and so on until you have listed each *dwelling* within the address 444 Jones St. List your segment beginning with the first address of the segment, and follow in order through the last address of the segment.

While you work in the segment, especially during your interviewing, you may discover dwellings you missed during your listing. Add these dwellings to the end of the Building Listing Sheet and to the Segment Listing Sheet with any notes that may help to locate it.

Step 4. *Identify the Sample Dwellings on the Segment Listing Sheet.* For *Take-All Segments* this means simply including as *Sample Dwellings all* the dwellings within the selected segment, and placing the project number in col. 3 of the Segment Listing Sheet, on each line of which you have entered a dwelling. In a Take-All Segment, if you find you have recorded more than 20 dwellings on the Segment Listing Sheet, contact the office for further instructions before interviewing at those dwellings beyond the 20th line. These additional dwellings will have to be dealt with separately.

For *Take-Part Segments* you must first identify the *Sample Dwellings.* These are segments that were judged in advance to be too large for our sample. You will find on the Segment Listing Sheet a place for "*Selected Line Numbers*" (item *g*) under a black tape. *Leave the tape in place until after* you have listed the dwellings in the segment. Then lift the tape and find which lines have been marked as selected line numbers. Write the project number in the right-hand column (col. 3) for these lines. These are the *Sample Dwellings* for this project—the sample dwellings you will contact to carry out your interviewing instructions.

You will notice that we checked every second, or every third line for sample dwellings. In most blocks you will find that we wrote down enough or more than enough numbers. But we did not know exactly how many you would find. If the sample dwellings run past our expectations, please write us about the situation in the segment, and you will receive further instructions.

You will try to write down only dwellings on the Segment Listing Sheet, omitting commercial, etc., buildings after you investigate and find no dwellings in them. But write down vacant dwellings, and write "Vac" next to the project number on that line. Try to list each dwelling on a single line. But you may find, after entering the dwelling and questioning, that what you listed as a single dwelling when judging from the outside, contains 2 or 3 or, perhaps, no dwellings. Do not change the sample dwelling designation. Just write on the margin or the back of the Segment Listing Sheet what the situation is.

9.8 SAMPLING CITY DIRECTORIES

9.8A City Directories as Frames for Dwellings

City Directories provide lists of addresses from which samples of dwellings can be developed. They are available for most U.S. cities in the range of 10,000 to 1 million population. They are products of private and often local enterprise, and their exact nature and quality vary. But they do have valuable features in common. They can be purchased from the R. L. Polk Company of Detroit or its several branches; and they can often be found at libraries and Chambers of Commerce. Providing dwelling lists is not the chief aim of directories, and their utilization requires careful modification and supplementation.

The first section contains an alphabetical list of residents. This does not provide a good frame because too many people move in a year or two, the typical age of the latest directory lists. The sample is selected from the "street guide" section, which contains streets in alphabetical order, and addresses on them in numerical order. With proper procedures, we can economically obtain good samples from them. Furthermore, these procedures illustrate the utilization of imperfect frames with supplements.

Our first concern is with dwellings missing from the directory for one of three reasons. From SRC experience, I judge that directories *omit* about 5 percent or less on the average, subject to local differences. *New construction* averages less than 2 percent annually; since most directories are renewed typically every 2 or 3 years, this source averages perhaps about 3 percent. These are not proper estimates but mere guesses; and local differences are obviously great. Parts of the city may be *not covered by the directory*. The areas of coverage of the directory appear subject to irregularities and uncertainties; the sampler should approximately determine this at the outset. These three sources of missing dwellings differ, but all can be compensated for with a block supplement.

The second kind of imperfection is the reverse of the first: directory listings containing no population dwellings. The listing may be a *commercial* or other nonresidential address, a *vacant* residence, or *an additional family* in the same dwelling. Furthermore, the directory may list many dwellings *outside the defined area* of the survey population. We should try to exclude most of these, and treat the others as blanks.

The third imperfection is the existence of 2, seldom more, dwellings at one listing. The fourth is its opposite, the existence of 1 dwelling where 2 are listed. The instructions below attempt to deal with all of them.

The supplement of sample blocks is designed for selecting a sample of dwellings missing from the directory. But it is not the best procedure for large new apartment houses; another supplement may be constructed for them, based on information from some reliable official source.

A third supplement, to represent hotels, motels, trailer courts, and other special dwelling places, may also be preferred to their direct selection from the directory. Any building appearing in one supplement must be excluded from the population of the directory, and from the other supplements. Thus the sample can consist of four strata, all mutually exclusive: (1) directory listings, (2) block supplement, (3) large new buildings, (4) hotels and other special places.

Different sampling fractions may be employed for the several strata, but for the sake of a simple self-weighting sample, a proportionate sample is often best. If $f = 1/F$ is the sampling fraction for selecting directory addresses, the block supplement may consist of all unlisted dwellings found in a fraction f of the blocks. However, for blocks where we expect many unlisted addresses, we could sample blocks with, say, $3f$ (or kf), and then subsample them with the fraction 1/3 (or $1/k$). This is the reason for the Take-Part Blocks in 9.8D. Such blocks could be assigned in areas of rapid new growth, and in areas not well covered by the directory.

Although we may utilize up to three supplements, still 90 or 95 percent of the sample may come from the directory. The directory can often provide a listing that is both more economical and satisfactory than any alternative. The combination of directory plus block supplement probably provides a better coverage than a block sample alone, because the directory coverage utilizes rather different procedures.

For small samples of, say, 100 to 300 cases, the coverage of a carefully compiled recent directory may be good enough without supplements. It may be better and cheaper than an improvised block sample.

Often the simplest and best procedure is selecting single addresses. If 90 or 95 percent, or more, of a sample represents individual dwellings from the directory, it will probably have properties similar to a simple random sample. This makes statistical analysis simpler, including variance computations. However, cluster sampling may be more economical for small samples in large cities; perhaps the lowered cost can justify the added complexity of the variance computations. The addresses can be selected readily in clusters. Then some appropriate design of clusters can be chosen, perhaps paired selections or a replicated sample.

If the directory is large, labor-saving devices may be utilized. A first stage can be introduced for selecting the pages or the half-page columns of directories. Instead of counting addresses, we may designate the printed lines as elements, and select them with appropriate rules.

9.8B *Selecting Lines from Directories*

1. Assign measures m to the addresses as follows:

0	For addresses fairly certain to contain no dwellings.
1	For 0 to 3 dwellings at the address. This includes vacant homes and stores that show no dwelling, but may have one.
2	For 4 to 6 dwellings at the address.
M	For 6 to 60 dwellings at address, divide by 3 and discard remainder. Thus 17 dwellings have $M = 5$.
M'	For over 60 dwellings at address, divide by 6 and discard remainder.
A	For apartment houses not indicating separate dwellings.
H	For hotels, motels, trailer courts, and other specified places

In apartment houses, the separate apartments are generally not separately identified; then the entire building must be taken as a multiple-dwelling address. But where clear identifying apartment numbers exist, the apartments can be sampled as individual addresses. Write the measures on the margin next to the addresses. Write the measures 0, H, and A in a different spot (or color).

2. Apply selection procedures to assigned measures 1, 2, M, or M'.

3. Copy each address with the measure 1 on a Regular Take-all Address Card. Also write down the next listed address and the number of apparent directory dwellings. For guessing these, use clues such as different family names and telephone numbers.

4. Copy each sample address with measures of 2 or M on a Take-Part Subsampling Address Sheet. Also write down the next listed address and the number of apparent directory dwellings. On the bottom of the sheet write Sample Dwelling numbers by applying interval 2 or M after a random start. Write down three more numbers than expected.

The treatment for the M' is similar to the M, but add remarks such as "large building, expected 6 (or 7) dwellings." Note that the selection probability of these dwellings is $mf \cdot (1/m) = f$, as required, whether the measure m is 2, M, or M'.

Special Strata. (1) Hotels are best sampled as a separate stratum, if at all. The search for permanent guests should be confined to an economical number. (2) In rare cases, the addresses of entire apartment houses appear without indication of the number of dwellings. If enough of these appear, they should be treated separately. (3) Perhaps investigating a few dozen addresses with zero dwellings indicated will provide desired reassurance. (4) If the directory contains many streets excluded from the survey population, it may be best to identify and exclude them before selection. If this purification is imperfect, simply omit any that happen to fall into the sample.

Possible Modifications. (1) The above size measures avoid the problems of sampling individual dwellings from houses containing only 2 or 3 of them; it is often difficult to distinguish these in the directories. But if we prefer to avoid selecting 2 or 3 dwellings from one house, we can employ a different set of measures. For example, $M = 1$ for 0–1 dwellings; $M = 2$ for 2 dwellings; $M = 3$ for 3 dwellings; for 4 to 60 dwellings, divide by 2 and discard remainder; for over 60 dwellings, divide by 4 and discard remainder. (2) The stratum M' was created to avoid having to list over 60 dwellings for only 3 or 4 sample dwellings. The measure M' is designed to obtain 6 or 7 sample dwellings. (3) The above selection requires only one pass through the list of addresses. However, the addresses with measures $1, 2, M$, and M' can be divided into strata with separate selection from each. (4) The directory can be divided into geographic strata before selection.

Selection Procedures can be chosen to fit the situation and these are described in appropriate sections. (1) A *systematic* selection with the interval $F = 1/f$ is the simplest and often good enough. (2) Random *paired selections* from each *zone* of $2F$ measures is possible. (3) *Replicated* sampling is possible. (4) *Clusters* of consecutive addresses can be selected. (5) *Double sampling* may be employed for stratification.

9.8C Instructions for Sample Dwellings at Selected Addresses

1. *Regular Take-All* Addresses are so marked on the *Address Card*, together with the number of dwellings indicated in the directory. These may be 1, 2, or 3; and 0 is marked at addresses where the directory showed no dwellings, but needed checking. These numbers are based on uncertain clues in the directory. You must find the actual number of dwellings, write it on the Address Card, and list each with its identification. Then make out a face sheet for each dwelling.

Regular TAKE-ALL ADDRESS CARD

Sample Address 816 Madison St.

The next listed address is 824

Apparent directory dwellings 2 Dwellings ACTUALLY FOUND 3

REMARKS Including grocery store at rear address

DWELLING LIST 1. Second floor, front 4.
2. " " back 5.
3. Attic apt. 6.

FIGURE 9.8.I

To clarify which address is meant, in some cases we also made *Remarks* about addresses which seemed to be associated, but should not be included at the sample address because they had separate chances of selection; for example, "not rear" or "not 722½." If a rear apartment does not have such a designation or a separate number, it should be included at the address. For further clarification you will also find on most cards *the next higher listed number* that seemed to follow the sample address in the directory's street list.

2. *Take-Part Subsampling Address Sheets* pertain to addresses where 4 or more dwellings appeared to exist in the directory. First, list all dwellings on a separate line in proper order. If necessary, continue on the back side and on the next sheet. *Then* lift the black tape and check the designated sample dwellings. Include these in the sample and make out a face sheet for each.

TAKE-PART SUBSAMPLING ADDRESS SHEET

Sample Address 78 Hamilton Place

The next listed address is 86

Apparent directory dwellings 62 Dwellings ACTUALLY FOUND 63

Remarks Large building expected 7 dwellings

Line No.	Dwelling Number or Description	Sample Check
1		
2		✓
15		

SAMPLE DWELLINGS 2, 12, 22, 32, 42, 52, 62, 72, 82, 92

FIGURE 9.8.II

In most cases we expect 3 or 4 sample dwellings, but we designate about 6. In some Large Buildings, so noted under Remarks, we designate a greater number of expected sample dwellings.

9.8D Instructions for Block Supplement for Missed Dwellings

1. You have Supplement Sheets for Missed Dwellings for several blocks. Check the entire area *within* the block boundaries to pick up dwellings that were missed by the directory. You are given either (*a*) the entire directory, or (*b*) a Listing of Directory Addresses attached to each Supplement Sheet.

The directory typically lists both sides of the street, both even and odd numbers. But you are concerned only with the addresses located *inside* the block boundaries. If an alley or street cuts into or through the block, look also for those addresses in the directory.

2. Begin at the corner marked with the arrow, and proceed around the block in the indicated direction. As you reach them, go in and out of alleys or streets that cut into the block.

3. Compare the dwelling addresses you actually find in the block with those listed in the directory. When you find an unlisted address, ascertain the number of dwellings and enter each dwelling on the next empty line. Enter each missed dwelling on a separate line and skip no lines. If you need more lines, continue on back of the sheet. Write in column 2 the street address or full description of each dwelling. For explaining doubtful cases and problems use bottom or back of sheet.

4. Do not list any of the following:

(*a*) Another dwelling at a listed address, if that dwelling, although unlisted, would be linked and selected with the listed dwelling at the same address.

(*b*) Unlisted commercial addresses that clearly contain no dwellings. But investigate an unlisted commercial address if it may contain a dwelling. If in doubt, include it.

(*c*) Special Exempted Buildings listed on the bottom of the supplement sheet. These belong to other strata, because they are either (1) hotels and other special places, or (2) large new buildings. They are listed and sampled separately.

5. Follow one of three procedures, according to instructions on Supplement Sheet. *Note:* In some cases one of the procedures (*a*), (*b*), or (*c*) may be chosen for all blocks.

(*a*) *Take-All Blocks.* Include all dwellings listed on the Supplement Sheet in the sample. Fill out separate Face Sheets for each dwelling and proceed to interview. But if more than 5 (or x?) dwellings are thus designated, bring list to office before interviewing.

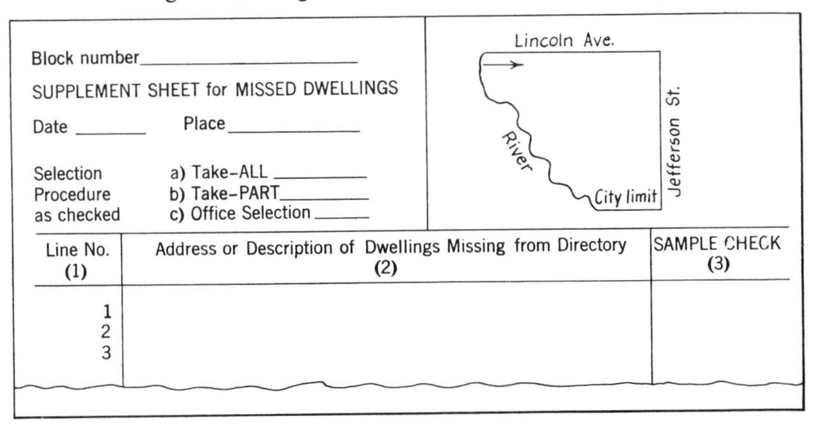

Block number________

SUPPLEMENT SHEET for MISSED DWELLINGS

Date ______ Place ________

Selection Procedure as checked
a) Take-ALL ______
b) Take-PART ______
c) Office Selection ______

Line No. (1)	Address or Description of Dwellings Missing from Directory (2)	SAMPLE CHECK (3)
1		
2		
3		

FIGURE 9.8.III

(*b*) *Take-Part Blocks. After* you have completed listing all missed dwellings on the Supplement Sheet, lift the black tape below (or open sealed envelope). Enter a check in column 3 for the designated lines. Proceed to interview at designated lines that contain a dwelling. But if more than 3 (or x?) dwellings are thus designated, bring list to office before interviewing.

(*c*) *Office Selection Blocks.* Bring list into office, where it will be checked and sample dwellings selected and designated.

PROBLEMS

9.1. We need a sample of a subclass (nationality, income, or occupation) of a city's population. About 80 percent of the subclass is highly concentrated in a clearly delineated area, where it comprises 60 percent of the residents. That area accounts for about one-quarter of the city's blocks and population. In other parts of the city they constitute only about five percent of the population and are fairly well scattered throughout the blocks. The average block size is about 40 dwellings throughout the city. Discuss a design involving selection of blocks and dwellings. Give formulas for selection rates, and for the mean and its variance.

9.2. You want to select a sample of about $n = 600$ from about $N = 60{,}000$ dwellings in a city with a sampling fraction of $f = 1/100$. Data from past surveys point to an economic design of an area sample of 100 blocks with subsamples of about 6 dwellings per block. There are no exact size measures for blocks, but preliminary work on aerial photographs and field checks shows that most of the blocks average about 15 dwellings. This average comes from a skewed distribution of block size: about 90 percent of the blocks contain 0 to 20 dwellings, about 10 percent contain 20 to 60 dwellings, and a few dozen blocks contain large apartment dwellings. Several reasonable sampling plans are possible; describe one briefly, including sampling fractions. Give formulas for the mean and its variance.

9.3. A national epsem of students is selected in 3 stages; students from schools and schools from districts; sample districts are defined as those located within selected counties. Most districts are contained entirely within single counties, but some lie partly in a sample county and partly outside. (*a*) For each of the following five procedures state whether it can (yes) or cannot (no) lead to the desired epsem selection of students. (*b*) Discuss briefly the advantages and disadvantages of several. (1) Accept schools located within sample counties. (2) Accept students living within sample counties. Accept (or reject) entire district: (3) if its unique headquarters (central office) is in (or out of) a sample county; (4) if the larger part is in (or out of) a sample county; (5) according to a random choice, with probabilities either equal or proportional to the portions.

10
Multistage Sampling

10.1 THREE-STAGE AREA SAMPLES

A two-stage design of selecting sample blocks, and then subsampling segments or listed dwellings from them, often serves well for moderate-sized cities. It may also function satisfactorily for large cities if good data are available for selecting blocks. It may be spread over a county or metropolitan area by applying it to all strata into which the county was divided. We could continue this process and apply the two-stage design to a whole province or state, and even to an entire country divided into strata, provided the sample was increased proportionately as new strata were added to the population. But if the sample is small, trying to stretch it over a large population dispersed over a wide area—a state, region, or nation—can result in serious field problems. We want to discuss these problems and ways to overcome them by introducing more stages of selection.

The problems concern two cost factors: listing and interviewers' travel. We can readily select a sample of blocks for the largest city if a good list is available, with information about block boundaries, number of dwellings, and perhaps stratification variables. If the information exists for a county or state, then selecting a sample from a long list is still not too expensive, provided the list is orderly. However, if we must divide a map into blocks and also attach size measures and stratifying variables, the task can become formidable for a large city. The situation is even more serious if we must cover a county or state; also, if the block information must be gathered in the field.

Even if good lists of blocks were available, a direct selection of blocks for a national sample of field interviews would be too expensive, because too widely spread. For mailed questionnaires this limitation on spreading the sample does not apply. It can also be greatly relaxed for telephone interviews in the United States.

It may be possible to extend the coverage of a two-stage sample by using much larger units in the first stage. For example, in the first stage we can select a national sample of counties, of which there are about 3000 in the continental U.S.; or towns and townships, which are ten or twenty times more numerous. But having to divide entire counties or towns directly into small segments in the second stage may be too expensive. We should probably use an intermediate stage, say, blocks, thus creating a three-stage design.

Let us put some numerical values on these constraints. The sample segments commonly used contain between 2 and 12 dwellings. This limit of 12 sample dwellings per block can also be reached alternatively by selecting 3 segments of 4 dwellings each. For sampling fractions of segments from blocks, perhaps 1/4 and 1/15 are reasonable limits. Fractions larger than 1/4 are seldom worth using, because one would rather avoid the complications of another stage of sampling. On the contrary, fractions smaller than 1/15 are usually avoided, because the cost of field segmenting becomes too great. (This limit can be raised for cheaper segmenting methods or for more frequent use of the blocks.) Thus we have a lower limit of $2 \times 4 = 8$ dwellings, and an upper limit of $12 \times 15 = 180$ dwellings, in sample blocks ordinarily used for field segmenting and interviewing.

The limits are vague in regard to tolerable ratios for the number of population blocks to sample blocks. If the blocks are already created and neatly listed, a sampling fraction of several hundred can be practical, and with machine sorting it can run to thousands. But if we ourselves must create and map the blocks (boundaries, measures, stratification), perhaps the limits of 5 to 50 population blocks to sample blocks will seem reasonable. Then we get the upper and lower limits of $8 \times 5 = 40$ to $180 \times 50 = 9000$ dwellings in the population for every sample block, each with a segment selected in two stages. Hence a sample, let us say, of 1200 dwellings in 100 sample blocks can be spread, with the two extreme fractions of $1/50 \times 1/15 = 1/750$, at the utmost over a city of $750 \times 1200 = 900{,}000$ dwellings, a metropolis of 3,000,000 persons. Even before reaching this limit, another stage will seem desirable if we want segments smaller than 12, or segmenting intervals less than 1/15, or block selection intervals under 50. Hence, circumstances can force us to introduce—reluctantly because we wish to avoid its complications—another stage of selection.

How do we choose the proper sampling units for the first stage, from which to subselect blocks in the second? First, we want the *number of primary units in the population to be large;* at least five times greater than the number of selections; otherwise it may not seem worthwhile to introduce this stage. We want that ratio to be much higher—perhaps 100 or 1000 or more—when selecting the primary units is inexpensive, but dividing them into blocks is expensive.

Second, we need *units with good boundaries*, measures of size, and preferably with information useful for stratification. If necessary, we create these units; but we prefer to obtain them cheaply, hence ready-made. One looks for important statistical units, especially those used for tabulating and publishing Census results.

Third, we prefer some *uniformity in the size* of the sampling units. However, we must often accept great variation and then try to control it with stratification or with probabilities proportional to size. Fourth, we prefer well-known and recognized *administrative units.* If these are commonly known by the respondents, the identification of the sampling units is facilitated. Fifth, we prefer units that remain *stable* from the time of obtaining the unit information to the survey period. Beyond that, we often prefer units that remain comparable to past and future data. Sixth, we may desire *comparability* to other specialized sources of data related to the survey.

States are too large and too few, but counties are widely used as primary units for national samples. Counties are stable, recognized administrative units; they are selected with PPS, because they vary greatly in size. The Census Tracts in metropolitan areas are useful, fairly stable units. For tracts and, of course, for counties, a great deal of Census data is available in published form. The Enumeration Districts (ED's) of the U.S. Decennial Census are not stable, recognized units; but they are useful for current work because they are small and uniform in size. The average size of ED's is about 250 dwellings, or about 800 persons, and there were over 200,000 of them in 1960. Data for ED's can be purchased for moderate cost from the U.S. Census Bureau. ED's containing only institutions and special dwelling places can be separated from others. Their inhabitants may also appear in the size measures of other EDs'.

For special studies we may consider school districts, voting precincts, health districts, etc. Sometimes the population being surveyed may have its own organizational subdivisions that can be used for sampling units. We may select the primary units with the techniques described in 6.4, 7.4, or those in 9.3 for selecting blocks. We need only generalize the methods of Sections 9.5 to 9.7 to separate the former second stage into two stages: one for blocks and another for segments of dwellings (see also 9.4*E*).

For example, suppose that for the $\text{Mos}_t = 2{,}400{,}000$ dwellings of a state we decide on a sample of about 2000 dwellings; and that these are to be selected in 200 ED's with about 10 dwellings from each ED. Suppose there are 9600 ED's in the state, and that 2 ED's are to be selected per stratum, with equal probability from each of 100 strata containing 9600/100 = 96 ED's. We have

$$\frac{2}{2F_a} \cdot \frac{10}{10F_b} = \frac{2}{96} \cdot \frac{10}{250} = \frac{1}{1200} = f = \frac{2000}{2{,}400{,}000}.$$

The sampling fraction 2/96 for ED's was determined to obtain exactly 2 sample ED's from each of 100 strata; this fraction, together with the desired overall f, fixed the second fraction 10/250. This needs to be broken (if we do not want to segment entire ED's) into two fractions: one for selecting blocks, another for selecting segments or dwellings from sample blocks. We can choose two constant sampling fractions that multiply to 1/25; say, $1/5 \times 1/5$ or $1/10 \times 2/5$. Formally, then, for three stages we may have fixed selection rates:

$$\frac{a}{aF_a} \cdot \frac{1}{F_b} \cdot \frac{1}{F_c} = \frac{a}{aF_aF_bF_c} = \frac{1}{F} = f. \tag{10.1.1}$$

Alternatively, we can take blocks with probabilities proportional to their sizes $\text{Mos}_{\alpha\beta}$; thus $\text{Mos}_{\alpha\beta}/75 \times 3/\text{Mos}_{\alpha\beta} = 1/25$ yields about three dwellings per block. Or one can select with PPS a fixed number of blocks, say, 2, from the total measure of size Mos_α of the blocks in the ED, and let this determine the sampling rates within the two selected blocks:

$$\frac{2\,\text{Mos}_{\alpha\beta}}{\text{Mos}_\alpha} \cdot \frac{\text{Mos}_\alpha}{50\,\text{Mos}_{\alpha\beta}} = \frac{1}{25}.$$

More formally and generally, we write for the three stages:

$$\frac{a}{aF_a} \cdot \frac{b\,\text{Mos}_{\alpha\beta}}{\text{Mos}_\alpha} \cdot \frac{c\,\text{Mos}_\alpha}{\text{Mos}_{\alpha\beta}(bcF_b)} = \frac{1}{F_a} \cdot \frac{1}{F_b} = \frac{1}{F} = f. \tag{10.1.2}$$

Contrary to the foregoing, suppose now that we select, with probabilities proportional to their measures of size Mos_α, 2 ED's from each of 100 strata. Suppose that the total measure of all ED's in the state is 8,000,000 persons. Thus the primary stratum size is 8,000,000/100 = 80,000, and

$$\frac{2\,\text{Mos}_\alpha}{2F_a} \cdot \frac{b^*}{\text{Mos}_\alpha} = \frac{2\,\text{Mos}_\alpha}{80{,}000} \cdot \frac{100/3}{\text{Mos}_\alpha} = \frac{1}{1200},$$

denotes the two sampling fractions, where $100/3 = (80{,}000/2)/1200$ is the constant needed to balance the equation. It should get about $2000/100 = 20$ dwellings per stratum, or 10 dwellings per ED, but these numbers are subject to variation (7.5*C*). The second fraction, b^*/Mos_α, must be divided into two sampling fractions. We can select the blocks and segments with two rates that multiply to $100/3\ \text{Mos}_\alpha$.

Alternatively, we may draw with PPS exactly 2 blocks from each ED. In the third stage we should get the average of about 10 dwellings per ED and 5 per block, but the sampling fraction is used simply to balance the equation. Thus,

$$\frac{2\ \text{Mos}_{\alpha\beta}}{\text{Mos}_\alpha} \cdot \frac{50/3}{\text{Mos}_{\alpha\beta}} = \frac{100/3}{\text{Mos}_\alpha}.$$

We assume that the sum of block measures $\sum_\beta^{B_\alpha} \text{Mos}_{\alpha\beta} = \text{Mos}'_\alpha$ is equal to the measure Mos_α of the ED. If that were not true, the last equation (which is both the most complex and most general) becomes, for the three stages,

$$\frac{2\ \text{Mos}_\alpha}{80{,}000} \cdot \frac{2\ \text{Mos}_{\alpha\beta}}{\text{Mos}'_\alpha} \cdot \frac{50\ \text{Mos}'_\alpha}{3\ \text{Mos}_{\alpha\beta}\ \text{Mos}_\alpha} = \frac{1}{1200}.$$

Generally, we have in symbols

$$\frac{a\ \text{Mos}_\alpha}{aF_a} \cdot \frac{b\ \text{Mos}_{\alpha\beta}}{\text{Mos}'_\alpha} \cdot \frac{c\ \text{Mos}'_\alpha}{\text{Mos}_{\alpha\beta}\ \text{Mos}_\alpha(bcF_b)} = \frac{a}{aF_a} \cdot \frac{bc}{bcF_b} = \frac{1}{F} = f. \quad (10.1.3)$$

Although this looks complex in symbols, it can—and often has been—set up as a practical procedure that can be made routine, with adequate checks. It is just an elaboration of the techniques described in detail in Sections 7.2 to 7.5, 9.3, and 9.4.

10.2 A NATIONAL SAMPLE OF PRIMARY AREAS (COUNTIES)

By a *primary area* we mean an area designed to be covered by a small group of interviewers (usually 1 to 4). We explore the design of *primary selections* that can serve as efficient primary areas. But first let it be understood that the two units need not be the same; we may prefer to have many more primary selections (to decrease the variance) than primary areas (to decrease the interviewer training costs). For a sample of a city, county, or metropolitan area we could individually select dwellings or segments or blocks, then assign them in convenient groups to interviewers. A moderate-sized sample, even of most states of the United States, can be selected in individual segments; for example, a selection of 2000 dwellings in 500 segments could be assigned to 50 interviewers; each would need to cover only about 1/50 of the state near his home, and obtain an average of 40 dwellings in 10 segments. A widespread sample of many primary selections can be covered either by a staff traveling from a common center, or by a staff located at key points from which they cover the nearest primary selections.

We are reluctant to confine the primary selections to the primary areas, which are relatively sparse and few, because they are expensive. Then, under what conditions are confined samples preferable? (1) When hiring and training of interviewers is expensive. (2) When continuing operations permit the high cost of training local interviewers to be spread over many samples. For a single survey a smaller group of traveling interviewers may be cheaper. (3) When training separate interviewers for each primary area is less expensive than sending them to cover several areas. Hence, the sizes of efficient primary areas depend on the mobility of the interviewer. (4) When the primary areas are fairly heterogeneous units, the design may have high sampling efficiency. (5) When call-backs are necessary to obtain enough responses, they require a longer survey period. This tends to argue against small clusters and against plans based on traveling among small primary selections.

The discussion has centered around field problems, but we should mention two others. (6) Listing, mapping, and sampling costs are sometimes high per primary area—if they are not overcome by adopting a different sample design. (7) Research plans may require obtaining information from the primary areas, beyond the survey data. The auxiliary information may be analyzed jointly with survey variables. When the information is expensive per area, this becomes a factor against greatly increasing the number of areas.

For national interview samples in the United States the county is a favored choice for primary areas and primary selections. It provides an economic compromise between reducing the travel cost per sample dwelling and spreading the sample over diverse classes of the population. Discussing the reasons for this choice can also illustrate the general problems of choosing primary areas. Although the national samples of dwellings of the Survey Research Center are most salient for me, their principal features also appear in other well-designed samples.

Counties are recognized administrative and statistical areas. The Census Bureau and other agencies publish substantial data about them. We use these for size measures and for stratification. Other data can be used for improving the statistics by correlations and other statistical devices. Counties are the subjects of detailed maps, most important of which are the maps (1 inch to 1 mile) available in most states from the road (or highway) departments, and aerial photographs (9.4*F*). County boundaries are stable and known to the people who live near them.

The larger units, the states, are too big and too few to be good sampling units. The next lower units, usually called townships, towns, or cities, are too small and too homogeneous. Most sample counties contain a moderate or large city near the center; several smaller towns or cities; unincorporated congested areas around the cities; and surrounding all these, the open country where farmers and many nonfarmers live. The sample from each county tends to contain most of these types and a variety of occupations, economic classes, and other social groups. This heterogeneity of sampling units is statistically efficient for samples of the entire population; it also provides widespread bases for subclasses and for surveys limited to urban or rural portions of the population. Disregard of this principle is illustrated by the impractical sample of a large firm; each interviewer's entire workload for several weeks was contained within single Census Tracts or Enumeration Districts, comprising only several hundred dwellings. Imagine the interviewer traveling each day through different kinds of neighborhoods just to reach the rather uniform population of one small area! There probably was some saving in cost per interview; but surely that was dearly bought by reducing drastically the number of areas representing each type (urban, rural, etc.) of dwelling in the sample.

The central city in the county is the most likely place to find a suitable interviewer. The interviewer travels to the other areas—smaller cities, towns, villages, and open country segments—mostly on roads which radiate from his city. Most counties are about 20 or 30 miles wide, and most interviews are within a radius of 15 miles; but the average travel distance is much less, because more than half the dwellings are usually in or adjacent to the central city. With his automobile the interviewer can go to any sample segment, complete one or more interviews, and return home the same day.

Exceptions to this idealized picture of the county can be handled by dividing very large counties and combining small ones. We try to reduce travel cost per interview, but also to create heterogeneous primary areas from diverse and contrasting units. Major departures from single counties involve using as primary areas the Standard Metropolitan Statistical Areas (SMSA's) consisting of counties around larger cities. These often contain 2 to 4 counties, seldom more; the increased travel from the central city is usually feasible, and the combination is considerably more heterogeneous than its components.

Thus the general type of the primary area, and the primary selection with it, is chosen to suit the field interviewers. We extend its boundaries, taking advantage of our ability to hire and train educated people with automobiles, for relatively low hourly pay, and employed only when needed. Having most segments within brief driving time from home, we save the expense and trouble of their eating and sleeping away from home. Actually most of our interviewers are women, who manage to cook and eat dinner at home, doing most of the interviewing before and after dinner, and during week ends—when respondents tend to be most available. There are many educated (high school or college) women, married and not working, but with previous work experience, who are willing and able to do this work on a part-time and irregular basis, preferably and usually for several years. Interview surveys in the United States are based currently on the availability of automobiles, and an educated but not regularly employed portion of the labor force.

At the SRC any single survey may last from 3 to 10 weeks; the number of surveys is irregular and unpredictable. The average interviewer works fewer than half of the weeks during the year, and usually less than 15 hours during those weeks. The hiring and training is done by 5 skilled supervisors, who must travel extensively for this purpose. This process, and the interviewers' acquired skills, constitute an expensive and important asset. This places a premium on the continued employment of the interviewers and the sample counties for several years at least. After the initial training, most of the further training and instructions for specific surveys are handled with mailed materials.

Obviously, these conditions vary with the needs and resources of survey organizations. The Census Bureau's Surveys (10.4) last 1 week each month. Such a short period is difficult to maintain if several call-backs are needed, and if a larger workload per interviewer is preferable. Sometimes enumerators are hired specifically for a single survey. On the contrary, some surveys are conducted by interviewers working full-time and continuously [Kish, Lovejoy, and Rackow, 1960].

Choice of the county for the primary area is specific to some situations, depending on needs and resources. Other situations require different solutions, conforming to their populations, problems, and resources. I think that if an interviewer must be confined to primary areas, these should generally be as large as he can cover economically. The primary areas should be as numerous as economy permits, in order to spread the sample widely. To spread the sample most widely, each primary area should have only a single interviewer. Most of our counties, however, have two, or even three, which permits discussion of problems and mutual help. Furthermore, they provide insurance against the uncertain availability of each interviewer for any specific project. If the probability of a single interviewer not working on some particular project is 0.2, having to temporarily staff 20 percent of the counties might be expensive; but with 2 interviewers this risk becomes more like $0.2^2 = 4$ percent, and can be more easily handled with traveling interviewers.

With 20 hours per week available per area (2 interviewers $\times$ 10 hours), and with generous allowances for irregularities and delays, it seems that a range of 20 to 60 interviews per area (10 to 30 per interviewer) is suitable for interviewing periods ranging from 3 to 10 weeks. These factors seem to me the major criteria that determine roughly, but actually, the range of sample sizes for our primary clusters. Most of the SRC national surveys range from 1400 to 4000 interviews. Since $70 \times 20 = 1400$ and $70 \times 60 = 4200$, the range of 20 to 60 desirable interviews per area, and the range of 2400 to 4200 total desired interviews, point to about 70 for the desired number of primary areas, or 75 to allow for some growth.

The number 75 of primary areas, divided into the 1960 population of 180 millions, gives 2.4 millions as the average number of persons each primary area should represent (or about $50{,}000{,}000/75 \doteq 670{,}000$, dwellings). Having first identified all primary sampling units (PSU's), counties and SMSA's, each is assigned a measure of size; we use persons, but dwellings or occupied dwellings would also be suitable. The PSU's are then sorted into strata of about 2.4 million persons each, followed by selection proportional to size (PPS) of a single primary area from each stratum.

The 12 largest SMSA's in the United States range from about 1.5 million to 15 million, and account altogether for 51 million or 35 percent of the population. Each of the 12 areas became a "self-representing" primary area, covered by a team of interviewers. Within these areas the primary selections are mostly segments in the central cities, and cities and towns in the surrounding suburban area.

The other primary sampling units, the smaller SMSA's plus about 3000 counties, account for 130 million, which were sorted into 62 strata, of approximately 2.2 million each. In each stratum were grouped PSU's that appeared similar on several characteristics, principally geography and size of central city. We used any information we could for this purpose, and there was no attempt at uniformity. On the contrary, we used percent nonwhite in the South, and percent Republican in the North, each in the portion where it had more discriminatory value. We can use any simple system for putting into one stratum only PSU's that resemble each other on important survey variables (3.5). The average number of PSU's per stratum is about 50; but at one extreme we have a stratum containing only 3 SMSA's, each having more than half a million persons;

while at the other extreme are strata containing more than 200 farming counties with less than 10,000 persons in each.

This last extreme demonstrates forcibly the need for concentrating the sample into primary areas. Instead of allowing a sample of 20 to 60 cases to be scattered into 200 counties, spread over hundreds of thousands of square miles, we concentrate it into a single county. Interviews from the sample county represent the entire scattered stratum. In any one stratum this representation is far from perfect because the counties within strata do actually differ, despite our stratification; but the procedure applied to all 62 strata produces good overall results.

The selected primary sampling units become the primary areas and are used repeatedly year after year, preserving the continuity and investment of a core of trained interviewers, and of sampling materials. Furthermore, repeated use of the same PSU's reduces the variance of the measures of change (12.4*B* and 12.5*C*). On the contrary, segments and dwellings are freshly chosen for each survey, except for deliberately planned panel studies. Blocks are in the sample usually from 1 to 3 years, until all or most segments and dwellings have been exploited on some survey (12.6).

We shall now illustrate how a primary selection is obtained from a stratum of PSU's. One stratum contains 22 PSU's in the Midwest, each containing a moderate-sized city; the size measures of the 22 PSU's add to 2,112,808. A random number from 1 to 2,112,808 is drawn, and then the measures are cumulated in order until the random number is reached or surpassed. Any of the 134,606 numbers from 1,072,227 to 1,206,832 would select Washtenaw County, Michigan; thus the probability of selecting Washtenaw is

$$\frac{134{,}606}{2{,}112{,}808} = 0.0637 = \frac{1}{15.7}.$$

Suppose now that Washtenaw County is part of a national sample in which every dwelling in the United States is given a selection probability of 1/10,000. Instead of selecting 1 dwelling in 10,000 directly from all dwellings in the entire stratum, we confine the selection to Washtenaw County, and therein select dwellings with this sampling fraction:

$$\frac{15.7}{10{,}000} = \frac{1/10{,}000}{0.0637} = \frac{1}{637}.$$

The overall selection probability of 1/10,000 for selecting dwellings in the stratum is achieved in two stages of selection, with the two sampling fractions

$$\frac{1}{15.7} \cdot \frac{1}{637} = \frac{1}{10{,}000}.$$

We could conceivably list all the 40,000+ dwellings in Washtenaw County; or divide the entire county into segments; say, about 10,000 segments of about 4 dwellings each. Then we could apply the sampling fraction 1/637 to the list of dwellings or segments. But listing all dwellings would be too expensive in any county; and segmenting the entire county may be economical only in smaller counties. Instead, the sampling fraction of 1 in 637 will be applied to Washtenaw County in two or more stages, as described in 10.3*C*. The numbers just given for Washtenaw and its stratum were for 1950, and they will be corrected to the 1960 data used in the next section.

State	Counties	1950 Pop.	Pop. in 1000's of Largest City	State	Counties	1950 Pop.	Pop. in 1000's of Largest City
Ill.	Champaign	120,070	40	Mich.	Washtenaw	134,606	48
Ill.	Vermillion	92,104	38	Ohio	Allen	113,431	50
Ind.	Elkhart	84,512	36	Ohio	Ashtabula	78,695	24
Ind.	Grant	76,182	30	Ohio	Columbiana	98,920	24
Ind.	Howard	70,064	39	Ohio	Erie	71,921	29
Ind.	LaPorte (pt)	90,411	28	Ohio	Jefferson	96,495	36
Ind.	Madison	124,243	47	Ohio	Licking	76,062	34
Ind.	Tippecanoe	83,008	36	Ohio	Marion	69,744	34
Ind.	Wayne	74,978	40	Ohio	Muskingum	98,117	41
Mich.	Calhoun	143,097	49	Ohio	Richland	108,473	44
Mich.	St. Clair	113,557	36	Ohio	Scioto	94,118	37
	Subtotal	1,072,226			Total	2,112,808	

TABLE 10.2.I Stratum 7 in North Central Region

How would we obtain an overall sampling fraction of 1/10,000? Suppose that the design calls for a national sample of about 4800 households (occupied dwellings), after considering the number of eligible respondents per household. Suppose also that we obtain estimates of 0.98 for the coverage rate, 0.89 for the response rate, and 55.9 million for the current number of United States households. Thus the effective population covered by the survey is estimated at $55{,}900{,}000 \times 0.98 \times 0.89 = 48{,}210{,}000$. The overall sampling fraction should be $4{,}800/48{,}210{,}000 = 1/10{,}044$; and 1/10,000 is acceptable. This sampling fraction is applied to each of the primary areas, after taking into account their probabilities of selection. In the 12 largest metropolitan areas, selected with certainty, the sampling fraction of 1/10,000 is applied to the entire area, in two or three stages. However, in the other primary areas (smaller SMSA's and counties), each selected from strata containing several similar primary sampling units, the selection rate within the county is adjusted to form the product 10,000 together with the selection probability of the primary area; within Washtenaw County we have the sampling fraction 1/637, so that $0.0637 \times 1/637 = 1/10{,}000$.

10.3 SAMPLING A COUNTY

10.3A Measures for Major Strata

We begin with the distribution of the population in the county. Table 10.3.I summarizes some relevant data from the 1960 Census about Washtenaw County, Michigan, used for illustration. Ann Arbor is the central and largest city, and Ypsilanti is the second largest. Eastlawn is a populous unincorporated area, recognized and delimited by the Census.

	Persons		Dwellings	
	Total	In Dwellings	Occupied	Total
Ann Arbor	67,340	56,370	19,726	20,752
Ypsilanti	20,957	18,913	6,077	6,563
Eastlawn (U.)	17,652	17,636	4,476	4,708
Chelsea	3,355	3,173	965	1,062
Milan (pt.)	2,847	2,843	868	916
(Subtotal)	(6,202)	(6,016)	(1,833)	(1,978)
Dexter	1,702	1,702	503	532
Manchester	1,568	1,568	484	524
Saline	2,334	2,322	708	764
(Subtotal)	(5,604)	(5,592)	(1,695)	(1,820)
Remainder Rural	45,352	39,884	11,083	13,129
Remainder Urbanized	9,333	9,278	2,382	2,458
Remainder (Subtotal)	(54,685)	(49,162)	(13,465)	(15,587)
County Total	172,440	153,689	47,272	51,408

TABLE 10.3.I 1960 Census Count of Washtenaw County, Michigan

The counts of total persons came from the U.S. Census Population Vol. I, Part A, "Number of Inhabitants," Table 7; except the "remainder rural," obtained by subtraction from total rural persons in Table 6. Data for total and occupied dwellings and persons in dwellings were taken from Housing Volume HC(1), Tables 14, 15, 23, 24, 26, and 27. The Census term for dwellings was "housing units."

The separate figures for rural and urban remainders were obtained from total urban and rural counts, by subtracting the counts for the five urban and the three rural places, respectively. The county boundary passes through Milan, and of its 3616 people, 2847 were counted in Washtenaw County. The proportion of 2847/3616 was used to obtain the three other estimates for Milan (2843, 868, and 916), because they were not given separately. The researcher can decide whether he wants to maintain strictly the county boundaries, as was done here, or to include the entire town.

Chelsea and Milan are two other urban places (over 2500 total population); whereas Dexter, Manchester, and Saline are rural. The remainder of the county surrounds these eight places, with 54,685 persons living in the "open country." Another 9333 were counted in several small, settled areas, outside city limits, but delimited as *urbanized areas* by the Census.

The unusually large (10,970) difference between total persons and persons in dwellings in Ann Arbor is due mostly to students living in University dormitories and fraternity houses. The difference (5468) in the rural remainder is found mostly in two large hospitals and a prison. Most of the large difference (2046) between total and occupied dwellings in the rural remainder can probably be explained by summer vacation homes. These differences demonstrate that the researcher may improve the bases for his sampling design and fractions by choosing the most suitable population estimates he can find with reasonable effort. For example, a survey of household heads is preferably based on occupied dwelling counts, when these are available. Because in time Census data become less accurate, the researcher should seek recent estimates from local authorities, or adjustments of the old Census counts.

We also search for maps and other materials to construct a frame for the county's dwellings or other target populations. Since blocks and segments will serve as sampling units, we require good boundaries and size measures for small areas; information for stratification may also be useful. A detailed and recent map of the county, showing streets, roads, and other boundaries is valuable; the county highway map may also indicate the approximate location of dwellings in rural areas. Aerial photos can be very helpful, providing actual and recent information. Separate maps for cities and towns show block boundaries, and perhaps dwelling density. For cities over 50,000 the Census publishes *Block Statistics*. Census data for Tracts and Enumeration Districts may be used; or one may decide to use City Directories or special lists. Finally, from local sources we may obtain information about blocks with many new dwellings, to adjust the data from Census and maps. All these sources are described in Section 9.4*F*, and all of Chapter 9 is relevant.

With information from maps and tables, we divide the county into several major strata. Most cities and towns will be placed, probably by using city and town boundaries, into distinct strata. However, city and town limits are often arbitrary lines without a physical boundary to mark them, and this makes them difficult to identify in the field. We may find it more convenient to extend the corresponding stratum boundaries to the nearest identifiable, natural boundary. This causes no confusion if the analysis of domains (urban, rural, etc.) is not affected.

10.3B Uniform Design for All Strata

Suppose first that you have decided on a sampling fraction of 1/40 for the county's dwellings, because of the following assumptions. Occupied dwellings have increased by 1.12 in three years, and a coverage rate of 0.98 and response rate of 0.94 are expected on this survey. We then expect an effective population of $47{,}272 \times 1.12 \times 0.98 \times 0.94 = 48{,}773$ occupied dwellings or households. The number of sample households desired also depends on the number of eligible elements per household. Suppose that a sample of $n = 1200$ sample dwellings is wanted, leading to the sampling fraction $1200/48{,}773 = 1/40.64$, and the convenient fraction 1/40 is adopted.

Suppose also that the design calls for selecting about four dwellings from each of about 300 sample blocks. This and the sample size of 1200 dwellings were decided together, considering factors of cost, variance, and the organization's resources and past experience, all with due regard for survey aims. Assume that the principles of economic design were considered in choosing the sample size of 1200 dwellings and the sample cluster size of 4 dwellings. Also that the two-stage sample of blocks and dwellings was adopted to avoid listing either about 50,000 dwellings or 13,000 segments for the entire county.

We must choose between feasible methods of selecting dwellings in two stages, several of which are described in detail in Chapter 9. We must also decide whether to use one method throughout the county for the sake of simplicity; or, on the contrary, to use for each stratum—the several cities, towns, and rural areas—that specific method which appears best suited to it. In this subsection, suppose that we intend to treat the entire county with essentially one method; let us see, in turn, how each method would be applied to this problem.

1. If we decide on a uniform within-block fraction of 1/5, we should create blocks of average size of about $5 \times 4 = 20$, to yield the expected sample clusters of 4 dwellings. The sampling fraction for the block would be $5/40 = 1/8$. For selecting 2 blocks from every stratum of 16 blocks, the two sampling fractions are $2/16 \times 1/5 = 1/40$.

Creating blocks of about 20 dwellings each is easy if we have some dwelling information, and if we permit considerable variation in actual sizes (say, 10 to 40 dwellings). For larger blocks without feasible divisions, assign integral Mos_α measures of block units, so that the selection fractions become $2\ \text{Mos}_\alpha/16 \times 1/5\ \text{Mos}_\alpha = 1/40$. The consecutive numbering of blocks can be employed to create strata of 16 block units. From each of these strata of 16 blocks (hence, of about $16 \times 20 = 320$ dwellings) we expect about two segments, about 4 dwellings each, to come into the sample. In the first stage we may make two random selections from each set of 16 units. Alternatively, we may select blocks systematically with the interval 8.

2. We may decide to accept physical blocks mostly as they are, assigning formal size measures Mos_α to them, and select them with probabilities proportional to size. For two blocks per stratum, the sampling fractions are

$$\frac{2\ \text{Mos}_\alpha}{320} \cdot \frac{4}{\text{Mos}_\alpha} = \frac{1}{40},$$

if the Mos_α are defined directly in estimated dwellings. On the other hand, if we define the Mos_α in count units of 4 dwellings, to the nearest convenient integer, we get to use only integral intervals $1/\text{Mos}_\alpha$ within the blocks:

$$\frac{2\ \text{Mos}_\alpha}{80} \cdot \frac{1}{\text{Mos}_\alpha} = \frac{1}{40}.$$

When numbering blocks, create strata of about 320 dwellings, or 80 count units of 4 dwellings each. We may make two random selections of blocks per stratum, or decide to select blocks systematically with the interval of 160 measures, or 40 count units. For sampling with PPS, some minimum sufficient measure is needed for the blocks (7.5*E*).

3. Dwellings can be subsampled from a list of dwellings within sample blocks (9.6). Alternatively, we may divide the sample blocks into compact segments of about 4 dwellings each, and then select sample segments (9.5). The design aims at an average of about one segment per block, but this will be subject to sampling variation, both because the actual/measure ratio of size varies and because of variation in segment sizes. For example, if we are taking every fourth segment, three segments can yield either 0 or 1 sample segment; five segments yield either 1 or 2, and so on. Procedures for controlling this are described in 7.5*B*.

4. When segments are selected with a constant interval, we may introduce a systematic selection for all the segments created in all sample blocks. For example, if the sampling fraction within blocks is 1/5, we can select with the interval 5 right through all the segments created in all the sample blocks. There must be reasonable reassurance that the dangers of systematic selection are avoided by the natural irregularities of the material. This procedure will largely eliminate the variation in sample size that arises when a separate random number (from 1 to 5) is applied to the number of segments created in different sample blocks. In sample blocks that had two (or Mos_α) measures, the interval should be 2 (Mos_α) times greater. In these blocks elimination with the fraction 1/2 (or $1/\text{Mos}_\alpha$) may be applied first; then the interval 5 can be applied to the adjusted set of segments.

10.3C Separate Procedures for Strata

Although a uniform design for the entire county has the virtue of simplicity, we may abandon it in favor of separate procedures within strata, for several reasons. First, in various strata we may need distinct sampling fractions, and these may lead to dissimilar preferred designs. Second, optimum cluster sizes may differ in some strata, and this may cause variations in design. Third, distinct procedures in some strata may suit specific field conditions, or take advantage of special frames only locally available.

An example of the last point would be sampling the "open country" directly in a single stage, selecting small segments (of 4 to 8 dwellings) outlined on a good, recent county map showing rural dwellings. In smaller towns and villages, small blocks could perhaps also be outlined directly on their maps or aerial photographs, and selected directly.

In cities—such as Ann Arbor and Ypsilanti in Washtenaw County—dwellings could be selected from city directories, probably adding a block supplement (9.8). Special procedures may be economical for certain special groups of dwellings. Ann Arbor has several large apartments and university housing developments; these could be put into separate strata, their dwelling lists obtained, and the sampling fraction of 1/40 applied directly.

Some strata, if they are to be covered, require special procedures. We may want to search for permanent residents in hotels, trailer camps, and other special dwelling places, which are large and mostly occupied by transients. University dormitories and fraternity houses in Ann Arbor contain over 10,000 students. There are several hospitals and a large prison in the county. After obtaining information about their nature and size, it can be decided whether they should be included or excluded.

10.3D The County as Part of a National Sample

In an abstract sense, the present problem only repeats the last two sections. Yet we may recognize several practical differences when the county sample is only a small part of a larger sample, which is regional or national in scope. First, the county sample will usually be smaller when it is only one of many primary selections. Second, because control of field work is remote, the interviewer must become more self-reliant; hence, instructions—although as simple as feasible—must be relatively complete. Third, because uniform instructions for many areas are preferable, tailoring methods and procedures to the peculiarities of individual counties is inhibited; nevertheless some flexibility may be applied for major types of counties—for example, in metropolitan *versus* rural areas—and for major problems.

One possible method of selection introduces three stages applied uniformly to the entire county: first, some large units, perhaps Census Enumeration Districts or Tracts are selected; second, the selected ED's or Tracts are divided into blocks, either in the field or on the map; third, in the selected blocks either segments are created or dwellings are listed. These represent the second, third, and fourth stages of selection, because the county was the primary selection.

In Washtenaw County about 70 dwellings will be obtained by applying the fraction 1/637 to the county's dwellings; this is about the same number that the rate 1/10,000 would have yielded if applied to the entire primary stratum. The sample might be obtained in 10 ED's (or Tracts), with an average of 7 dwellings from each; for example, the 7 dwellings could come from two segments in one block, or in two blocks.

If we decide on exactly 10 selections for the second stage, the sampling fraction for it is Mos_β/M_b, where Mos_β denotes the measure of the ED, and $M_b = \Sigma\, \text{Mos}_\beta/10$, the sum of ED measures for each of 10 implicit strata. We may select them systematically with the interval M_b, or as two random selections from each of the five implicit strata, each of size $\Sigma\, \text{Mos}_\beta/5$. If we use occupied dwellings for measures, then $M_b = 47{,}272/10 = 4{,}727.2$. We should place the ED's in some meaningful strata (perhaps urban to rural) before selection.

We could use a simple two-stage selection, eliminating the block selections, if we can justify dividing the entire ED into segments. This would require creating about 60 or 80 segments of four dwellings in ED's of about 250 to 300 dwellings. The two sampling fractions would be

$$\frac{\text{Mos}_\beta}{M_b} \cdot \frac{M_b}{637\ \text{Mos}_\beta} = \frac{1}{637}.$$

Since that would often be uneconomical, we prefer selecting blocks and segments in two stages. The possible variations are many, and we can mention only a few of them. In all cases, we force the overall product of sampling fractions in the three stages to be 1/637. Suppose, first, that we use the simple fraction 1/4 for selecting segments or dwellings within sample blocks. This immediately determines that the selection fraction for blocks is the second fraction in

$$\frac{\text{Mos}_\beta}{M_b} \cdot \frac{4M_b}{637\ \text{Mos}_\beta} \cdot \frac{1}{4} = \frac{1}{637}.$$

If we create blocks of about 32 dwellings, and segments of about 4 dwellings, we shall tend to select single blocks and two segments. If we create blocks of 16 dwellings, we shall tend to select two blocks, each with single sample segments. But there will be variations in both; variations in numbers of sample blocks is more severe, since we may have difficulty in creating equal blocks. Variations in sample size can be reduced if we obtain and utilize measures of block size. With information obtained from maps, or in the field by enumerators, we may divide the entire area of the ED into blocks, and assign measures of size to them. Then we can force the ED sample to yield exactly two sample blocks, and permit variation only in the last stage:

$$\frac{\text{Mos}_\beta}{M_b} \cdot \frac{2\ \text{Mos}_{\beta\gamma}}{M_c} \cdot \frac{M_b M_c}{2 \times 637 \times \text{Mos}_\beta \times \text{Mos}_{\beta\gamma}} = \frac{1}{637}.$$

Here $M_c = \Sigma\ \text{Mos}_{\beta\gamma}$, the sum of the measures of blocks in the ED. This can be made into a careful routine. Instead of two blocks, we can select one block or C_β blocks in the ED.

Perhaps we can largely avoid the selection of ED's, and select blocks and segments directly. To do this, we need detailed maps and other information for the cities and towns in the county. With this knowledge, we design procedures best suited to each stratum within the county. Within each stratum we select with the rate 1/637. To form strata we inspect the county's data (Table 10.3.I), keeping in mind that it takes $637 \times 8 \doteq 5000$ dwellings to provide about 8 sample dwellings; we should consider about 4000 to 5000 as the minimum size for strata. Suppose that in Ann Arbor we select blocks and segments, using probabilities proportional to measures of block size (Mos_β), expressed in dwellings. If we select the blocks systematically, and aim at four dwellings per block, we have

$$\frac{\text{Mos}_\beta}{637 \times 4} \cdot \frac{4}{\text{Mos}_\beta} = \frac{1}{637}.$$

Suppose that in Ypsilanti and in Eastlawn we assign to blocks count units C_β for each estimated 4 dwellings. Then the sampling fraction becomes

$$\frac{C_\beta}{637} \cdot \frac{1}{C_\beta} = \frac{1}{637}.$$

This same sampling fraction can also be applied to the remaining 13,465 dwellings, which are mostly in the "open country." Whereas in the cities we form mostly blocks of 12 to 20 dwellings ($C_\beta = 3$ to 5), in most of the open country, if our map and aerial photo show dwellings and roads, we can form segments of 4 dwellings ($C_\beta = 1$) directly, then select the segments in a single stage. This saves the trip and effort needed to divide blocks into segments.

The five towns are too small to be sampled separately. We can select one of them with probabilities proportional to size, then select segments from it. The measure of the town is Mos_β, and in terms of occupied dwellings the measures for the five towns add to 1833 + 1695 = 3528. Thus,

$$\frac{\text{Mos}_\beta}{3528} \cdot \frac{C_\gamma}{637 \times \text{Mos}_\beta/3528} \cdot \frac{1}{C_\gamma} = \frac{1}{637}.$$

For example, if Dexter were selected with probability 503/3528, the selection rate within Dexter should be 637 × 503/3528 = 90.8. This fraction 1/90.8 applied to 503 dwellings comes to 503/90.8 = 5.33 expected dwellings, the same as if the 1/637 were applied to the 3528 directly. This can be used to check the arithmetic. If Dexter grew faster or slower than the stratum, the sample size will be subject to that source of variation. But inequalities of block size and the application of intervals are even greater sources of variation. For example, in Dexter we expect either one or two segments. To reduce this source of variation we should guard against forming too many separate strata with separate random starts.

10.4 A NATIONAL SAMPLE OF DWELLINGS

This famous series of survey samples has been well described in Technical Paper No. 7 [U.S. Census Bureau, 1963], the source of this brief description. Three functions performed by the survey should be distinguished:

1. "The Current Population Survey (CPS) is a sample survey conducted monthly by the Bureau of the Census to obtain estimates of employment, unemployment and other characteristics of the general labor force and of the population as a whole or of various subgroups of the population. The survey provides national estimates, with measurable reliability, of the size and composition, and changes in composition, of the labor force such as the number of unemployed and the number of people employed in agricultural and nonagricultural pursuits. It also provides data on hours worked, duration of unemployment, and the like.

2. "In addition, periodic studies are made of personal and family income, migration, educational attainment, and other demographic, social and economic topics.

3. "It has been recognized that the monthly interviews in sample households included in the CPS provide a valuable resource for the collection of additional information and so the survey is also used for many other special data collection functions."

To these three direct functions, three other effects should be added. First, the large field organization for these surveys provides assistance and cooperation for other surveys, such as the National Health Survey. Second, the samples served as catalyst for a group of statisticians led by M. H. Hansen and W. N. Hurwitz, with profound effect on the statistical activities of the Census Bureau and other government agencies. It is an outstanding example of the benefits of scientific teamwork. Third, this group has become a center of research and training, with hundreds of its trainees performing statistical sampling throughout the world.

The several functions of the survey have evolved historically and are still changing. We describe briefly its primary function in 1963, which consisted of monthly samples of about 35,000 dwellings. Information about the household and all its members was elicited from a "responsible adult member." The "ultimate sampling units" were area *segments* containing about six dwellings on the average. All dwellings located within the boundaries of *take-all segments* were included in the sample. Where lack of clear boundaries prevented defining segments with roughly 6 dwellings, larger segments were defined and their dwellings were listed. These *take-part segments* were assigned two or more units, and an appropriate subsample selected systematically from the list of dwellings. The ultimate clusters of 6 dwellings contained about 8 members of the labor force, and about 20 persons. A uniform overall sampling fraction was obtained for these ultimate sampling units, hence for the associated dwellings, and then for their occupants. That uniform fraction was attained in two stages; by first selecting primary sampling units with probabilities proportional to size, and then segments within the units with probabilities inversely proportional to the same size.

In 1963 the sample consisted of 357 primary sampling units, each a county or group of counties. Except for the largest metropolitan areas, a crew of four enumerators covered the average PSU: about 16 segments of 6 dwellings each, during one week of each month. An average of eight PSU's comprised the territory of each district supervisor, who was in charge of recruiting, training, and supervision.

About 56 percent of the population and sample belonged to "self-representing areas," populous enough to comprise singly an entire stratum. There were 112 such areas, each with over 242,000 population in the 1960 Census; and almost all were Standard Metropolitan Statistical Areas (SMSA's), some consisting of a single county, others of several counties.

The other 44 percent of the population, of which only about a sixth were in SMSA's, resided in 1800 primary units, most containing one or two counties, and altogether containing most of the country's area in about 3000 counties. These were sorted into 245 strata, averaging 320,000 persons and roughly equal in population size. Each of these strata contained two or more primary units, from which one was selected into the sample with probabilities proportional to size.

"Stratification proceeded with primary attention being given to rate of population increase, degree of urbanization, color, principal industry, and type of farming. In defining the strata, these characteristics were considered jointly to produce what appeared to be the best possible grouping of PSU's into strata.

"The tentatively formed strata were also reviewed to see whether they were homogeneous with respect to a number of housing characteristics. The strata were thus defined on the basis of available objective measures, supplemented by expert judgment, in an effort to maximize the heterogeneity between and homogeneity within strata.

"A great many professional man-hours were spent in the stratification process. However, it is questionable whether the amount of time devoted to reviews and refinements paid off in appreciable reductions in sampling variances. Intuitive notions about gains from stratification can be misleading. Methods of stratification that appear to be very different often lead to about the same variances."

Two interesting features of the PSU selections, described in the report, can merely be mentioned here. First, a controlled selection was employed chiefly to insure good geographical distribution; it utilized a method described in 12.8 and by Goodman and Kish [1952]. Second, a sample of 230 PSU's selected in 1954 was expanded to 330 PSU's in 1956, and to 357 PSU's in 1963. The changes utilized new information about size measures and stratifying variables. The necessary changes in probability were accomplished, yet the number of new PSU's among the 357 was confined to 53. The methods for reducing the required changes are similar to those mentioned in 12.7.

Within the sample PSU's the selection rates W_{hi} had to balance the selection probability P_{hi} of the PSU, to bring about a uniform overall selection fraction f: $P_{hi}W_{hi} = f$. For example, a county was selected with $P_{hi} = 11{,}829/440{,}830 = 1/37.267$; to achieve $f = 1/2244.671$, the within rate must be set at $37.267/2244.671 = 1/60.232$; then $37.267 \times 60.232 = 2244.671$. This example, from pages 8–19 of Technical Paper No. 7, describes a sample in 1954, when stratum sizes were larger because then there were 230, not 357, of them; and the sampling fraction was smaller because the sample contained 25,000 dwellings.

The fraction 1/60.232 was employed essentially for selecting separate segments. However, to reduce the labor of segmenting the entire county, two auxiliary units were introduced. (Since the average county was larger by a factor of about four, the within county fractions were on the average more like 1/240 than 1/60.) One was the ED, the Enumeration District of a single enumerator in the decenniel census, an area with well defined boundaries. This contained an average of 250 dwellings, but with a great deal of variation. Each was assigned a measure q (properly q_{hij}) of the number of segments it was estimated to contain, an integer nearly equal to its number of suspected dwellings divided by 6. To the cumulated measures the interval 60.232 was applied. Since the ED had a selection probability of $q/60.232$, the next step could have been selecting segments with $1/q$ within the ED's.

However, since the ED's averaged 40 segments, another stage of selection was introduced by dividing ED's into *blocks* on the basis of information from available maps and similar sources. These were areas with well-defined boundaries and with estimated integral measures r (properly r_{hijk}) of contained segments; that is, dwellings divided by 6. However, these measures r had to be adjusted to sum to q, the measure assigned to the ED. One of these blocks was selected with probability r/q, with a random number from 1 to q.

Now the block had to be divided into r segments. It was sent to the field to be divided into *take-all* area segments with 6 dwellings and good boundaries. Where this could not be accomplished, larger *take-part* segments were created, a measure s of two or more was assigned, its dwellings were listed, and a "segment" of $1/s$ selected.

The probability of a segment being selected within the county (or PSU) was $q/60.232 \times r/q \times 1/r = 1/60.232$. The ED's and blocks were used to reduce the extent of field work required to identify segments. However, ED's and blocks were merely auxiliary to selecting individual segments, each one in a separate ED and block, spread throughout the county.

A separate stratum and procedure was designed for a supplement to cover blocks and segments with unexpectedly large numbers of dwellings. These were areas on which the information from Census ED's and maps was deficient, due to mistakes or obsolescence, especially in areas of large-scale new construction.

Another important supplement covered the population living in *institutions* and large *special dwelling* places. These places were enumerated in separate ED's which were distinguished and excluded from the selection of the regular dwelling sample.

"These EDs because of their special characteristics, required separate treatment in sampling, as described below.

"The EDs sampled separately belonged to two groups. One was the group of *institutional EDs*, made up of places such as prisons and penitentiaries, mental institutions, nursing homes, homes for the aged, children's homes, etc. The resident population of institutional EDs was divided, for sampling purposes, into *staff* (including family members, if living in the ED) and *inmates*. A sample of resident staff members and their families in institutional EDs is included in the Survey each month. A sample of inmates is included only one month in each year, usually March or April, when data are collected for the entire civilian population.

"The second group of EDs sampled separately consisted *of special dwelling place EDs, made up of large hotels, general hospitals, armed forces installations, convents*, etc., whose residents have common living arrangements of some kind, but are not considered to be part of the institutional population. A sample of persons (excluding military personnel) living in these EDs is included in the Survey each month. Also in this group are some EDs consisting of large apartment buildings, which were included because they could not be distinguished easily from other types of special dwelling place EDs at the time the list of these EDs was prepared from 1950 Population Census materials."

These ED's were selected from the same primary areas as the dwelling sample, with the sampling fraction set at 1/2 of the fraction for dwellings. Thus corresponding to the previous example from 1954, the sampling fractions would be

$$P_{hi}\left(\frac{\text{ED pop.}}{4489.342 \times 15P_{hi}}\right)\left(\frac{15}{\text{ED pop.}}\right) = \frac{1}{4489.342}.$$

The interval $4489.342 \times 15P_{hi}$ was employed to select ED's with probabilities proportional to the Census ED population. Within the ED's that were institutions, the third fraction was applied to obtain a sample with the expected size 15.

A complex rotation design was employed to improve the survey statistics (Sections 12.4 and 12.5), especially the comparisons with neighboring months, and those exactly a year apart. Each monthly sample comprised eight equal portions, only one brand new. The segments of each portion were in the sample four consecutive months, then out eight months, then in again for four months.

The rotating sample can be represented for a 16-month period as shown on page 382. In July, 1963, the new portion is represented by d_4, which also appears in August, September, and October, 1963. After a lapse of eight months, it reappears in July, 1964, for four more months. The first four appearances are represented with lower case letters, and the four reappearances with upper case. In July, 1963, $A_1B_1C_1D_1$ represent the four portions overlapping with July, 1962; and $a_4b_4c_4d_4$ represent four overlaps with July, 1964, when they reappear as $A_4B_4C_4D_4$. The six overlaps with the neighboring August are shown as $b_4c_4d_4B_1C_1D_1$, and the six overlaps with the neighboring June are $a_4b_4c_4A_1B_1C_1$. There are four overlapping portions with samples two months removed, and two overlapping portions with samples three months removed.

July, 1963	$a_4b_4c_4d_4$	$A_1B_1C_1D_1$
August, 1963	$a_5b_4c_4d_4$	$A_2B_1C_1D_1$
September, 1963	$a_5b_5c_4d_4$	$A_2B_2C_1D_1$
October, 1963	$a_5b_5c_5d_4$	$A_2B_2C_2D_1$
November, 1963	$a_5b_5c_5d_5$	$A_2B_2C_2D_2$
December, 1963	$a_6b_5c_5d_5$	$A_3B_2C_2D_2$
January, 1964	$a_6b_6c_5d_5$	$A_3B_3C_2D_2$
February, 1964	$a_6b_6c_6d_5$	$A_3B_3C_3D_2$
March, 1964	$a_6b_6c_6d_6$	$A_3B_3C_3D_3$
April, 1964	$a_7b_6c_6d_6$	$A_4B_3C_3D_3$
May, 1964	$a_7b_7c_6d_6$	$A_4B_4C_3D_3$
June, 1964	$a_7b_7c_7d_6$	$A_4B_4C_4D_3$
July, 1964	$a_7b_7c_7d_7$	$A_4B_4C_4D_4$
August, 1964	$a_8b_7c_7d_7$	$A_5B_4C_4D_4$
September, 1964	$a_8b_8c_7d_7$	$A_5B_5C_4D_4$
October, 1964	$a_8b_8c_8d_7$	$A_5B_5C_5D_4$

The estimates prepared from these samples were complex, and only three principal features are noted here. (1) Two adjustments were made for six "color-residence" groups; white and nonwhite each for urban, rural-farm, and rural-nonfarm residences. First, the results were multiplied by the inverse of the response rates of households within the six groups. Second, a ratio adjustment of stratum to PSU total was introduced to reduce the contribution to the variance arising from selecting PSU's. (2) A ratio correction of 56 age-color-sex groups was made: 14 age groups, for males and females, whites and nonwhites. This ratio estimate (see 6.5*C* and 12.3*B*) utilized information on the population size of the 56 groups, based on vital statistics (births and deaths) projections from the last Census. This ratio adjustment not only reduced the sampling variation, but also adjusted for differences in nonresponse and noncoverage. (3) The composite estimate utilized not only information from the current survey, but also data from past months, by means of the correlation obtained from the overlapping portions (12.4*C*).

Computing variances for the complex estimates becomes an intricate task. Yet methods similar to those in Section 12.11 could be developed. However, the large electronic computers of the Census Bureau are well adapted to applying methods of replication (14.2). Reliance can also be placed on relative stability of variances in the periodic surveys. Tables of approximate standard errors were computed and published regularly for monthly estimates and for month-to-month changes.

Finally, a few words about field and office procedures. Since the respondents were any "responsible adult" from the household, as much as 70 to 80 percent of households yielded information on the first call. Altogether the response rates ran from 94 to 97 percent. The net coverage rate (13.3) was 98 to 99 percent of the decennial Census coverage. A widespread field quality check aided supervision, measured overall error rates, and contributed to research for future improvements. Similar objectives were served by constant checks on the auditing and tabulating procedures.

Remark 10.4.1 *A List of Case Studies.* During the last two decades many descriptions of sample surveys have been published. Readers will have their own lists of favorites, but the following references may be helpful. It is an informal list of some of the best and most available I could find from a variety of fields.

Chapter 12 of Hansen, Hurwitz, and Madow [1953] contains five excellent case studies: *A.* The Sample Survey of Retail Stores; *B.* The Current Population Survey; *C.* Estimation Procedure for the Annual Survey of Manufactures; *D.* Some Variances and Covariances for Cluster Samples of Persons and Dwelling Units; *E.* Sample Verification and Quality Control Methods in the 1950 Census.

From the work of the Survey Research Center, note the "Methods of the Surveys of Consumer Finances," by Katona, Kish, Lansing, and Dent [1950]; and "A Nation-wide Sample of Girls from School Lists," by Bergsten [1958].

Chapter 11 of Deming [1960] includes several case studies. "A Population Sample for Greece" appears as Chapter 12 in Deming [1950], essentially reproduced from an article by Jessen, Blythe, Kempthorne, and Deming [1947].

Inventory problems are treated with replicated sampling designs in several chapters of Deming [1960]: "A Sample of Employees in Several Factories" (Ch. 6); "A Survey of Business Establishments with Correction for Nonresponse" (Ch. 7); "Quick Study of Accounts in Ledgers," and "A Large-Scale Study of Accounts of Freight Shipments in Multiple Locations" (Ch. 8). Chapter 11 of Deming [1950] also deals with "Inventories by Sampling." "Sampling Techniques in Accounting," by Trueblood and Cyert [1957], includes several case studies.

Concerning agricultural samples, the Statistical Laboratory of Iowa State University (Ames) has issued many research bulletins [Jessen 1942; Jessen and Houseman 1944]. Of the many reports from the Indian Statistical Institute, two are especially famous [Mahalanobis, 1944; Lahiri 1954]. Dalenius [1957, Ch. 3] deals with Swedish samples. Shaul [1952] wrote several articles about samples in Central Africa.

Each Bulletin in *Series C* of the Statistical Papers of the United Nations describes several "Sample Surveys of Current Interest."

"History of the Uses of Modern Sampling Procedures," by Stephan [1948], is interesting. We add here "Sampling for the Social Survey," by Gray and Corlett, and "Recent Developments in the Sampling of Human Populations in Great Britain," by Moser [1955]; also You [1951] and Zarkovich [1956]. The "Bibliography on Sampling Theory and Methods," by Murthy et al. [1962], seems to be remarkably complete.

11

Sampling from Imperfect Frames

11.1 EMPTY LISTINGS AND FOREIGN ELEMENTS; VARIABLE FRAME DENSITIES

Problems caused by imperfect frames, discussed in 2.7, are treated in greater detail in several following sections. In each section a common, basic approach is applied to several related imperfections.

This section treats problems that occur when the target population is a subclass M of the entire listed frame population N. Listings are *empty* if they contain no elements of the target population: either if they are blanks containing no elements of any kind, or if they contain only foreign elements. When listings are either blanks or contain only single elements, the formulas for subclasses in element sampling can be applied. Section 4.5 deals with subclasses of stratified element samples. For simple random samples the mean can be treated with the ordinary formulas of 2.2, and estimates of aggregates are treated in 11.8.

The presence of blanks introduces variation in the size of the sample. This variation is a function of the proportion $\bar{M} = M/N$. For a sample of n listings, the average sample size m is $n\bar{M}$ with a standard error of $\sqrt{n\bar{M}(1 - \bar{M})}$. This neglects the factor $(1 - f)$, and assumes that either the selection was simple random, or the distribution of the M elements among the listings was random. The standard error of $m = n\bar{m}$ can be estimated with $\sqrt{m(1 - \bar{m})}$, and the coefficient of variation with $\sqrt{(1 - \bar{m})/m}$; this is small if the proportion of blanks is small, but approaches $\sqrt{1/m}$ when sampling rare elements.

The variation in sample size due to blanks increases the variance of some statistics. True, the simple random mean is not affected, because its variance equals essentially the variance of m selections (Remark 2.2.II). But for means of stratified samples, the presence of blanks increases the element variance by $(1 - \bar{m}_h)(\bar{y}_h - \bar{y}_w)^2$ within each stratum. For small proportions $(1 - \bar{m}_h)$ of foreign elements this will be small; but when searching for rare elements, this tends to destroy the gains of proportionate stratification (4.5*B*). For simple estimates Fy of the aggregate Y, the increase of the element variance amounts to $(1 - \bar{m})\bar{y}^2$ for simple random sampling (11.8), and to $(1 - \bar{m}_h)\bar{y}_h^2$ in all strata of stratified random sampling (4.5*A*).

The problem is generally trivial if the blanks can be readily identified before making expensive measurements, and if their proportion is known. Then the blanks can be generally disregarded. A simple random sample of size m can be selected; or a stratified sample with size m_h if the M_h are known. A systematic sample of elements can usually be adjusted also. The problem becomes more serious if the blanks cannot be identified in advance, and their proportion is unknown.

Variations due to empty listings have two effects on the variance: it becomes less subject to control, and it is increased for some statistics. Both effects can be avoided by purifying the frame of empty listings. When this is too expensive, empty listings must be tolerated. Then we may utilize one of the following remedies, depending on the specific situation.

1. If reliable values of the population size M and of stratum sizes M_h are available, they can be used to improve the sample values. These adjustments tend to eliminate the increase in the variance. The value M can be used to produce a better aggregate $M\bar{y}$ than the simple estimate Fy (11.8). The values of M_h can be used in post-stratification (3.4*C*). The values may be complete Census counts, or good estimates from outside sources or from double sampling procedures.
2. If reliable values of M and M_h are not available, increased variances for aggregates and stratified means must be tolerated. Even rough advance estimates will help to adjust the selection rates and thus obtain some control over the sample size and over the variance. If estimates are not available, perhaps a pilot study can provide them.
3. If advance estimates are not available, we may draw a smaller first sample and use its results for obtaining approximately the needed final sample size. Suppose, first, a sample of size m' is obtained with a sampling fraction of $f' = n'/N'$. For a final sample of size m, one needs $m'' = m - m'$ more elements. Next, draw the second sample with the fraction

$f'' = f'm''/m'$, and the two efforts together will yield about m elements, with a combined fraction of $f = f' + f'' = f'(m' + m'')/m'$.

This procedure can also yield approximately the desired sample sizes m_h within each of several strata. After obtaining m_h', the second selection fraction should be $f_h'' = f_h' m_h''/m_h'$.

In the first effort, try to have a sample large enough to set the field work into efficient motion, but not so large that it yields a larger preliminary sample $m' > m$ than required. Thus the first effort f' should be set according to some reasonable *minimum* expectation of the population proportion M/N in the frame (see 8.4*D*).

Avoid any system of individual substitutes for blank listings. Such substitution is a common mistake that can lead to serious selection biases. Substituting a neighboring element for a selected empty listing transfers an extra chance of selection to that element. Different parts of the frame can have different densities of elements; here lower density means fewer elements per listing in parts with more empty listings. Elements in the parts with lower densities receive greater chances of selection than elements in denser parts. For example, substitution of the next element for each empty "0" in (1, 1, 1, 1, 1, 1, 0, 2, 0, 0, 3, 0, 0, 0, 4) gives two chances to the 2, three to the 3, and four to the 4. The selection of one element with substitution has a mean of $[(6 \times 1) + (2 \times 2) + (3 \times 3) + (4 \times 4)]/15 = 35/15$, whereas the population mean is 15/9. Selection bias generally results if the variable density of elements is correlated with the mean content of elements.

A list of names, or a file of cards, is a unidimensional frame in which the names or cards are the listings. The presence of blanks introduces variation in the density of elements in the frame. This concept can be readily extended to different dimensions. Case 5 below illustrates time as the dimension, with seconds serving as listings. Case 6 extends the concept to two dimensions with random points on a map as the listing.

Selection bias also occurs if, after drawing elements with epsem, clusters of *equal numbers of elements* are selected, disregarding any intervening blanks. Epsem can be maintained by specifying *equal numbers of listings* for sample clusters, accepting the unequal clusters of elements contained in each. The epsem selection of listings needs equal probabilities of selection to equal clusters for listings, not of elements. However, equal clusters of elements would have unequal numbers of listings, and selection probabilities proportional to those unequal numbers (7.2).

Example 11.1a A few examples should help to present the general problem. In each of these, the proposed substitution can lead to selection biases, because elements near empty listings receive greater probabilities of selection. This would destroy an initial epsem selection by substituting nearby elements for selected empty listings.

1. A store's file of charge accounts is numbered and ordered according to when the account was opened. Accounts are selected with epsem; but if a closed (canceled) account is selected, it is proposed to substitute the next active account for it. Since the older parts contain more closed (empty) accounts, substitution would give active accounts a greater chance in older than in newer parts.

2. From a hospital's card file of patients a systematic sample of every 50th card is selected; these are regarded as paired selections for every hundred. Some are blank because they fail to meet certain qualifications for the study. For these blank selections, substitutions are proposed: random selections to restore the paired selections from each hundred. But parts of the file may differ seasonally, secularly, or due to epidemics. The presence of blanks may be associated both with the proportion of blanks and the characteristics of the qualified cards.

3. Cards in a file are selected with epsem. Then it is proposed to proceed forward until three active cases are selected, disregarding the cards of inactive cases. This would give greater chances to those active cards which follow inactive cards.

4. Persons arriving at a store's entrance are counted mechanically and an epsem is selected for interviewing. If any are blanks (because they are children, for example), the next eligible person is proposed as a substitute. Adults following children and other blanks receive increased selection probabilities.

5. Automobiles passing a point are to be sampled by selecting the seconds of time at random. If no automobile passes the point at the exact second, the next automobile is proposed for substitution. Automobiles receive lower probabilities during busy than during quiet hours.

6. Random points are selected on a map by laying down a grid with selection points equally spaced in both dimensions. It is proposed to select the three farms nearest each selection point. But the larger (or more widely spaced) the farm, the greater its chance of being nearest the points selected with equal probability. However, equal chances are maintained if we select all farms in equal areas specified around each point. This can be a specified circle around the point, or a specified equal area of any shape. If the areas are of unequal sizes, the grid selects areas with probabilities proportional to their sizes. But irregular and unequal segments can be selected with epsem, after listing them.

Remark 11.1.I When the size M of a subclass is known, it can be used to estimate the unknown size N of a population. In an epsem sample of size n, observe the size m and the proportion $\bar{m} = m/n$ of the subclass. An estimate of the population size is $N' = M/\bar{m} = Mn/m$. The coefficient of variation of this estimate is about $\sqrt{(1 - \bar{m})/m} = \sqrt{(n - m)/nm}$. (Because the cv of $M/\bar{m}$ equals that of $1/\bar{m}$; and this, if small, is about equal to the cv of the proportion $\bar{m}$). The method is applicable to social research, but the available literature deals with fisheries and wild life. It is used to estimate the size N of *mobile population* (fish in a lake or deer in a state) with the *capture-tag-recapture* method [Craig, 1953; Chapman, 1954; Hammersley, 1954; Darroch, 1958].

A number M of elements are captured, tagged, and released; then another sample of size n is caught, and the number m of tagged animals is counted. There are problems about the assumptions of independence and the effects of capture and tagging, and others regarding births, deaths, and migration in the population. These lead to models for the specific populations, and the structures and parameters of the model influence the estimates. Therefore, the definition of probability sampling (known probabilities of selection) does not apply.

11.2 DUPLICATE LISTINGS; OVERLAPPING FRAMES

11.2A How to Avoid the Problem

Selection problems arise when some of the elements appear two or more times on the list, so that several listings represent the same element. An epsem selection of listings results in selection probabilities proportional to the number of listings for any element. The problem is similar for clusters that appear repeatedly on the list.

To clean up the entire list may be too expensive. However, sometimes we can change to another frame with the desired unique listings. For example, a sample of classes from a U.S. high school would give each student a listing in each of his classes; but each student is listed uniquely in one "homeroom" in most schools. A second example: The file of address cards of a particular insurance company would give duplicate chances to policyholders who have two or more policies; but a special metal clip identifies uniquely the oldest insurance for each policyholder, and the clipped cards provide the desired unique listing of the population of policyholders.

Sometimes we can accept the population defined by the listings in the frame, which includes the replication of some units. Consider three examples. (1) An epsem of insurance policies would provide a self-weighting estimate in which each policyholder was weighted by his number of policies; and this could be an acceptable measure of his importance to the company. (2) If we take all the students from an epsem of all classes of a university, the selection probability of each student is proportional to the number of classes he takes; this may be a desirable weighting of his importance for a survey of student needs. (3) The users of a facility (library, clinic, store, rail terminal, etc.) can perhaps be contacted easily at the time of their appearance at an entrance. The probability of selection of each *user* is proportional to his number of *uses*, his appearances during the survey period. This may be an acceptable weight of his importance, and the researcher may even prefer the self-weighting sample of *uses* he obtains automatically. In all these instances, the same person may appear two or more times in the sample selection; these duplicate appearances should be accepted in the sample estimates; if duplicate interviews are unnecessary, then the single interview should be replicated. The mean of the sample $\bar{y} = \frac{1}{n} \sum_{j}^{n} y_j$ estimates $\frac{1}{M} \sum_{i}^{M} p_i Y_i$. In this population mean, each element is weighted by its replications p_i. This population consists of the $\sum_{i}^{M} p_i$ listings, which also serve as elements. For epsem selections from this population the usual formulas will hold.

11.2B Creating Unique Listings

In many situations the element i appears on p_i listings in the frame, and we need to estimate $\bar{Y} = \Sigma Y_i/M$. A *unique identification* of one of the p_i listings as the only selecting unit can often provide the remedy. An element is selected into the sample, if and only if the unique listing is selected, but not with the selection of any of its other $(p_i - 1)$ listings. The other $(p_i - 1)$ listings become empty listings; thus a frame of $N = \sum_{i}^{M} p_i$ total listings is converted into M elements and $(N - M)$ blanks. This changes the problem from replicate listings to one of empty listings, with procedures and consequences already described (11.1). In the formulas for variances of totals and stratified means, the elements appear as a subclass of all listings. The ratio $\bar{m} = m/n$ of valid selections estimates the proportion $\bar{M}$ of elements per listing.

On many lists, one of the p_i replicate listings can be designated uniquely, readily, and safely. Practical procedures must be found to determine for each selected listing whether it is the unique selector or a blank. The choice of a good procedure depends on the nature of the list and on the clerks' ability. The unique selector may be the first, or the oldest listing, or the largest entry (or last, youngest, smallest). (1) If all listings of an element appear together on the list, the first becomes the unique selector and the other $(p_i - 1)$ listings become blanks. (2) If the listings are not together, they may contain the information (cross references and characteristics) needed for deciding the selection; when ties are present, they can be broken with several criteria used in sequence. Or the unique selector can be decided with a random choice of $1/p_i$. (3) If the listings do not contain the needed information, a search for possible replicates must be made throughout the frame; this search may be easy if the list has a good alphabetical or numerical order. In the case of clustered selections, special care may be necessary to identify replicate listings in different clusters and to specify the unique selector. For example, a person and a family may have two dwellings, and a farmer may have farms in different counties.

Note that eliminating only the duplicate selections actually found in the sample would not correct the selection bias. An element with one listing has a selection probability f, an element with two listings has an expected appearance in the sample of $2f$, and those with k listings have kf. But if f is small, only about f^2 duplications actually appear in the sample from duplicate listings; and only about $3f^2$ duplications and f^3 triplications from triplicate listings. Eliminating only these would leave most of the selection bias uncorrected. To correct the situation we have to search and purify the entire *population list* of replicates of sample selections; this is the purpose of unique identification.

In some situations it is difficult to find the replicates and to designate the unique selectors; yet we can discover readily, perhaps on the listing itself, the number p_i of replicate listings of each selection. Then we may choose, at random and with probability $1/p_i$, one of the p_i listings to be the selection listing. For each selected listing we determine a secondary probability $1/p_i$ to remain in the sample, and $(p_i - 1)/p_i$ to be eliminated. If an element appears twice in the sample, it gets $2/p_i$ probability of remaining. (If $p_i = 2$, one selection is retained; if $p_i = 4$, one selection is retained with probability 1/2. Triplicate appearances are negligibly few if f is small.) This reestablishes a self-weighting sample; the initial probabilities p_i have been reduced by the secondary reselection with $1/p_i$. The subclass formulas for selection from a list with blanks hold here also. In practical situations this procedure is almost as good as one with unique identification.

This last method may sometimes be improved slightly with a small modification. Separate the selections into groups according to their respective selection probabilities: $1, 2, 3, \ldots, p_i$. Then select at random the proper fraction $(1/2, 1/3, \ldots, 1/p_i)$ of elements in each group, retaining all those with $p_i = 1$. This stratification of the sample reduces the variation of the sample sizes between the groups with different sizes.

11.2C Weighting the Selections with $1/p_i$

In some situations we can discover neither a unique listing nor the number of replications until the selected units have been interviewed, or generally, the cost of observations has been incurred. Thus we have a sample of n elements, and we know the value of the characteristic y_i for the elements, and the numbers p_i of their replicate listings, which denote their probabilities of selection. In the simple sample mean the elements would appear weighted by these probabilities. Often we cannot ignore these differential weightings, and the possibly large bias this would cause if a considerable correlation exists between the y_i and the p_i. We are then forced to some form of reweighting of the sample observations. We do this reluctantly because weighting complicates the analysis and, usually, increases the variance. But these penalties are often less than the bias that would be risked when using unweighted data with a selection bias. A statistical investigation may promote a better choice.

When the observations are inexpensive, the simplest procedure is to eliminate the proportion $(1 - 1/p_i)$ of selections, and to retain $1/p_i$ of the selections with p_i replications. The retained observations constitute an epsem selection. However, discarding completed observations would in most cases unduly increase the variance, and weighting the observations with $1/p_i$ will produce a more precise estimate (see 11.7*B* and *C*). Note that if an element is selected twice, it should either appear twice with the weight $1/p_i$, or receive the weight $2/p_i$ once.

To derive the formulas for weighted means, first think of an imaginary population of $N = \sum_i^M p_i$ listings, from which we select n listings with $n/N = f = 1/F$. Imagine also that in this population each element is divided evenly among its p_i listings, thus creating p_i identical units, each having the value of the variable $U_i = Y_i/p_i$. Note that the total of this population is

$$U = \sum_i^N U_i = \sum_i^M p_i U_i = \sum_i^M Y_i = Y,$$

equal to the total of the target characteristic. Thus, with the sample total $u = \sum_j^n u_j = \sum_j^n y_j/p_j$, we can construct the unbiased estimate of Y as $Fu = (N/n)u$. This follows because $E(\Sigma\, u_j) = nE(u_j) = (n/N)Y$. The usual formulas will estimate the variance; for srs, the var $(u) = (1-f)n$ $(\Sigma\, u_j^2 - u^2/n)/(n - 1)$. However, if M is known, the ratio estimate $M\bar{y}$ of Y will have considerably lower standard error, equal to $M[\text{se}\,(\bar{y})]$. If M is known, we can estimate $\bar{Y}$ with $Fu/M = u(N/nM)$. But, because of its lower variance, we generally prefer to use the ratio mean:

$$\bar{y} = \frac{u}{v} = \frac{\sum u_j}{\sum v_j} = \frac{\sum y_j/p_j}{\sum 1/p_j}. \tag{11.2.1}$$

The count variable $V_i = 1/p_i$ (of the U_i) distributes evenly the count variable 1 (of the Y_i) over the p_i appearances, and

$$V = \sum_i^N V_i = \sum_i^M p_i V_i = \sum_i^M 1 = M.$$

Note that $E(\sum^n V_i) = nE(1/p_i) = (n/N)M$; thus $Fv = (N/n)v$ is an unbiased estimate of M. This is of interest when M must be estimated.

The mean $\bar{y}$ can be treated with the usual variance formulas for ratio means. For a simple random selection of n listings, apply (6.3.4) to the mean $\bar{y} = u/v$ and obtain

$$\text{var}(\bar{y}) = \text{var}\left(\frac{u}{v}\right) = \frac{1-f}{v^2}\sum z_j^2, \tag{11.2.2}$$

where $z_j = u_j - \bar{y}v_j = (y_j - \bar{y})/p_j$. From the term $\Sigma(z_j - \bar{z})^2$ we can delete $\bar{z} = 0$. For a stratified element sample of Σn_h listings, we have $\bar{y} = \frac{u}{v} = \sum_h^H u_h \Big/ \sum_h^H v_h = \sum_h^H \sum_i^{n_h} u_{hi} \Big/ \sum_h^H \sum_i^{n_h} v_{hi}$, and its variance (from 6.4.4),

$$\text{var}\left(\sum_h^H u_h \Big/ \sum_h^H v_h\right) = \frac{1-f}{v^2}\sum_h^H \frac{n_h}{n_h - 1}\sum_i^{n_h}(z_{hi} - \bar{z}_h)^2, \tag{11.2.3}$$

where $z_{hi} = u_{hi} - \bar{y}v_{hi} = \dfrac{y_{hi} - \bar{y}}{p_{hi}}$, and $\bar{z}_h = \dfrac{1}{n_h}\sum z_{hi}$. For selections from unequal clustered designs, the usual variance formulas apply also to the variables U_i and V_i, and no special formulas are needed. We can state generally that the mean of the weighted observations $U_i = Y_i/p_i$ can be treated as a ratio mean, with the count variable $V_i = 1/p_i$ in the denominator.

If the values of p_i cannot be ascertained clearly from the listing, they should be obtained together with the values y_i, perhaps in the interviews. If these obtained values of p_i differ considerably from the actual selection probabilities, they may bias the estimates. If M is known, the check of Fv against M gives some indication of the possible presence of consistent biases in the p_i; but this would not check biases that tend to cancel. Perhaps a sample of the obtained values can be checked against their values on the list. Or we may base a check on the proportion of duplicate listings ($p_i = 2$), because the number of single to duplicated appearances in the sample should approximate $(1-f)/f$, as shown below.

Remark 11.2.1 If the values of the p_i of the sample selections are unknown, we face a predicament. Nevertheless, the prevalence of duplicate listings can be estimated from the number of duplicate selections appearing in the sample, under the following conditions: (1) The selection of $n = fN$ listings is epsem. (2) Either the selection is random, or the distribution of duplicates on the list is random relative to the selection procedure. (3) The proportion of triplicates and higher replicates is negligible, but the proportion of duplicates in the population is large enough to yield a fair number of duplicate sample selections.

If the number of elements with single listings is M_1, and the number of elements with duplicate listings is M_2, the total number of elements is $M_1 + M_2 = M$

among $M_1 + 2M_2 = N$ listings. For a sample of $n = Nf$, the expected yield is $M_1f + 2M_2f = Nf = n$, and the duplicate listings yield an expected number of $2M_2f$ selections. But among these selections only M_2f^2 pairs of duplicate selections are expected to appear. These account for $2M_2f^2$ selections, hence only for the fraction f of $2M_2f$ total selections from duplicate listings. If the sample actually contains d_2 pairs of duplicates, $d_2/f^2 = d_2F^2$ is an estimate of M_2. Therefore, $nF - d_2F^2$ estimates $M_1 + M_2 = M = N - M_2$, the number of distinct elements in the population.

The number of pairs of selections d_2 can be considered a Poisson variable, with a standard error of $\sqrt{d_2}$, and a coefficient of variation of $1/\sqrt{d_2}$. The estimate F^2d_2 has a standard error of $F^2\sqrt{d_2}$. Often this error can be intolerably large, and we should try for a better method of estimating the extent of duplication on the list.

If instead of the simple count M, the aggregate $Y = Y_1 + Y_2$ of the variable Y_i is needed, we note that $yF - y_2F^2$ can estimate Y. Furthermore, the mean Y/M can be estimated with $(yF - y_2F^2)/(nF - d_2F^2)$. To my knowledge this matter has not yet received attention. To be precise, the probability of selecting a single member from a pair is $2(N - n)n/N(N - 1)$, and of selecting both of them, $n(n - 1)/N(N - 1)$. In practice, we may use the approximations $2f(1 - f)$ and f^2. The proportions of single plus double selections are $2f(1 - f) + 2f^2 = 2f$, and the ratio of single to double selections arising from duplicate listings is $(1 - f)/f$.

Now consider a frame of N listings, with $N = M_1 + 2M_2 + 3M_3 + 4M_4 +$ etc., where iM_i denotes the number of listings of M_i elements with i replications. The expected selections for duplicates ($i = 2$) is the same as before. From triplicate listings we expect $3M_3f$ selections. But among these, only $3M_3f^2(1 - f)$ pairs of duplicate selections appear. The expected number of triplets M_3f^3 is typically negligible. The proportions of single, double, and triple selections are $3f(1-f)^2 + 6f^2(1-f) + 3f^3 = 3f$; and the ratio of expected single to double selections is $3f(1-f)^2/6f^2(1 - f) = (1 - f)/2f$.

Duplicate and triplicate listings can be generalized to i replicate listings. Then the probabilities of j appearances in the sample can be shown to equal $\binom{i}{j}\binom{N-i}{n-j}\Big/\binom{N}{n}$, for successive values of ($j = 0, 1, 2, \ldots, i$). In practical cases, these values can be well approximated with the expansion of $[(1-f) + f]^i$. Thus single appearances have the probability of $(1-f)^{i-1}$; and pairs of duplicate selections have the probability $i(i - 1)f^2(1-f)^{i-2}/2$.

From the number d_i of each kind ($i = 2, 3, 4$, etc.) of duplicate selections appearing in the sample, it would be possible to estimate each M_i with the factors $d_i/[i(i - 1)f^2(1-f)^{i-2}/2]$. However, it is necessary to distinguish the numbers (d_2, d_3, d_4, etc.) of pairs of duplicate selections arising from different sources. Perhaps survey methods can obtain this information for the duplicates, which comprise only a small portion of the total sample. If not, perhaps a mathematical model can be constructed, based on knowledge of the population distribution [Goodman, 1952].

11.2D Matching from Overlapping Lists

Suppose we need to sample a population defined by the sum (union) of H lists. If every element appears on only one list, we take an ordinary sample from H strata. But if elements appear on two or more lists, we must deal with duplicate listings. This situation is called "matching lists" or "overlapping lists" in the literature. For example, a population of professionals may be specified as the members of any of several associations. The replications can be ignored if they are known to be a small enough proportion, or if the target population is redefined to include them. If it is not too costly, we can eliminate replications from all lists. These can be represented in the following scheme for $H = 4$:

List	1	2	3	4
Uniques	M_1	M_2	M_3	M_4
Replicates	0	M_{12}	$M_{23} + M_{13.2}$	$M_{34} + M_{24.3} + M_{14.23}$
	N_1	N_2	N_3	N_4

From the first listing, we accept all N_1 listings as the M_1 unique elements of the first list. From the N_2 listings of the second list, we eliminate the M_{12} listings which also appear on the first list, and obtain the M_2 unique elements of the second list. From the third list of N_3, we eliminate the M_{23} which appear on the second list, and the $M_{13.2}$ which appear on the first, though not on the second. From the fourth list, we eliminate those found on the third, those on the second but not on the third, and those on the first, but neither on the third nor on the second. And so on. Thus the unique identification is defined as the appearance on the primary list of a specified order.

To save effort, this procedure of unique identification can also be carried out on a sample basis. From the first list, all $n_1 = N_1 f_1$ selections are accepted. The $n_2 = N_2 f_2$ selections from the second list are searched for duplications among the population N_1 of the first list, and any found are eliminated; but this number m_{12} will be a variate around its expected value $M_{12} f_2$. Hence, $m_2 = n_2 - m_{12}$ is also a variate which can be regarded as the subclass that appears as the proportion $\bar{m}_2 = m_2/n_2$ from the second list. Similarly from the $n_3 = N_3 f_3$ selections of the third list, a search of the second list eliminates m_{23}, and a search of the first list a further $m_{13.2}$; this leaves the random subclass proportion of $\bar{m}_3 = m_3/n_3$ from the third list. And so on to the fourth and later lists. The unique listings $m_1 + m_2 + m_3 + \cdots = \Sigma\, m_h$ denote a stratified sample of subclasses, except for the first stratum in which $\bar{m}_1 = m_1/n_1 = 1$. Now the formulas of stratified subclasses (4.5) apply; e.g., the element variance within the stratum is $[v_{yh}^2 + (1 - \bar{m}_h)(\bar{y}_h - \bar{y}_w)^2]$ for the stratified mean. The second term denotes an increase in the element variance due to the presence of blanks. This can be eliminated if the population values M_h are known and used in computing the sample mean.

The search for duplicate listings may be too expensive if they lack a systematic ordering scheme, or if the number of lists is large. In either case, it may be better to accept all selected listings, and to determine the number of lists on which each appears. Each selected element, including actual double appearances in the sample, is weighted by its $1/p_{hi}$, and is then treated with the procedures of 11.2*C*. The probability of selection of any element is $p_{hi} = \Sigma\, f_h$, the sum of sampling fractions applied to lists where the element appears. These joint sampling fractions may be subjected to optimum allocation [Hartley, 1962].

Remark 11.2.II In what order should the lists be sampled to decrease the cost of searching for duplicates? If searching a particular list is expensive, because it is disorganized or complicated, it should be placed at the end, because there it needs to be searched less often than at the beginning. But if all lists are in good condition, the costs are chiefly proportional to the numbers n_h for which the search must be made. Then the last place should be assigned to the smallest n_h, which designates the shortest list when the sampling fractions are uniform $(f_h = f)$.

Remark 11.2.III If separate estimates are needed for each list, replacements must be found for the $N_h - M_h$ excluded duplicate listings. Yet each sample case can be made to serve each list to which it belongs. This can be done by reversing the direction of the search. Among the n_1 selections from the first list we can find the m_{21} duplicate listings by searching the N_2 listings of the second list. These m_{21} duplicated listings plus the m_2 single listings represent the N_2 listings of the second list. The m_{21} duplicates should be weighted with f_2/f_1 if the two fractions differ. Similarly, the N_3 listings of the third list can be represented with the m_3 selections from it, plus the m_{32} duplicates from the second list (weighted with f_3/f_2), plus the $m_{31.2}$ duplicates that appear on the first and third, but not on the second list (weighted with f_3/f_1). And so on.

Remark 11.2.IV If we can neither eliminate duplicate listings nor find the values of p_{hi} for weighting the selections, we face a predicament similar to that described in Remark 11.2.I. Estimates can be produced by noting the number d_{ij} of duplicate selections that actually appear in the sample; but these estimates have large variances. If two lists comprise the population, when using sampling fractions f_1 and f_2 we expect $f_1 f_2 M_{12}$ duplicates in the sample. After finding d_{12}

duplicates in the sample, we can estimate M_{12} with $d_{12}F_1F_2$. Then $N_1 + N_2 - d_{12}F_1F_2$ estimates $N_1 + N_2 - M_{12} = M_1 + M_2 = M$, the number of elements on the two lists. To be useful, the d_{12} must be reasonably large, because the standard error for d_{12} is $\sqrt{d_{12}}$; and for $d_{12}F_1F_2$ it is $\sqrt{d_{12}}F_1F_2$. We assume small sampling fractions.

With three lists the estimate would become $N_1 + N_2 + N_3 - d_{12}F_1F_2 - d_{23}F_2F_3 - d_{13}F_1F_3 + d_{123}F_1F_2F_3$. The last term attempts to account for elements that appear on all three lists, because they have been subtracted among the duplicates three times instead of twice, as desired. But triplicate appearances are rare, and estimates based on them are unreliable because they involve large weights (11.7). They are given in detail by Goodman [1949, 1952], and by Deming and Glasser [1959].

11.3 SELECTING PERSONS FROM DWELLINGS; SMALL UNEQUAL CLUSTERS

11.3A Dwellings as Clusters of Persons

Selecting a sample of persons from an epsem selection of dwellings raises questions of procedure. Similar problems occur generally when sampling units, selected with epsem, contain small unequal clusters of elements. Selecting one element out of the p_i at the ith unit produces a sample of elements with unequal probabilities $1/p_i$. It would be epsem if the units were selected with the probabilities p_i. But this information is generally not obtained if it is expensive and the variation in p_i is small.

There are other examples of epsem selection of small, unequal clusters of elements: finding 2 to, say, 5 dwellings at listed addresses where we ordinarily expect single dwellings; finding two or more families in a dwelling; finding an average of 1.2 "spending units" per household. The preferred solutions vary with conditions and resources. For example, all persons per dwelling present a different problem than adults per dwelling, with a higher mean (3.3 versus 2.0 in the United States in 1960) and a much higher skewness. Persons per building or dwellings per block are other matters altogether, and do not belong in the category of small clusters.

Within the dwelling we should distinguish the element, the unit of analysis, from the observational unit, called respondent in interview surveys. Generally, the two are identical in attitudinal surveys, and when the observation is made directly by the enumerator. But the two need not be the same; for example, in many surveys of the Census Bureau one respondent ("any responsible adult") gives information about all elements in the household. Furthermore, an adult respondent is preferred when the elements are animals or inanimate objects, and often also for questions dealing with children. Ordinarily the respondent is either uniquely specified or is any one of several persons of a specified type. Hence, we concentrate on the problems of selecting elements. No selection problems arise if the element is uniquely designated, whether a person (e.g., the household head or homemaker) or not (e.g., the household, or dwelling, or furnace). But if households contain more than one member of the population, they become clusters of population elements. The sample of dwellings must be translated into a sample of persons, or of adults, or other elements; and the procedure must meet the demands of validity, efficiency, and practicality.

One can include in the sample every population element found in the sample households. This may be desirable under any of these conditions: (1) There is seldom more than one member in the household. For example, consider these average (approximate in the U.S. of 1960) populations per household: families and unrelated individuals, 1.02; spending units, 1.2; members of the labor force, 1.2; women of child-bearing age, 1.01. (2) The information about all members can be obtained simultaneously and cheaply. The Census Bureau's surveys obtain satisfactory labor force information from "any responsible adult." (3) The intraclass correlation within the household for the survey variables is either negligible or negative. This is true for some items strongly related to both age and sex, because not many dwellings have more than one adult of the same age and sex; e.g., labor force participation, and some apparel-buying habits.

When none of the above conditions hold, the situation is unfavorable for including the entire cluster. The cluster of all persons or all adults though small, is not of negligible size. Furthermore, their characteristics, behaviors, or attitudes often have considerable positive correlations, and these correlations increase the element variance with a factor of about $[1 + \text{roh}(\bar{n} - 1)]$, where $\bar{n}$ is the average of elements per dwelling and roh is the intraclass correlation. The larger roh is, the less new information is provided by the $(\bar{n} - 1)$ elements beyond the first. Including these extra $(\bar{n} - 1)$ elements may still be economical if the information can be provided cheaply and adequately; for example, when one respondent can give it about all persons in the dwelling, and the main cost comes from getting that one interview. None of the favorable conditions hold for many surveys of behavior and attitudes of the adult population; the clusters are not negligible, the correlations are high, and the element cost (interviewing, coding, and processing) is high. The information must be obtained from the adult himself and he must be selected. The selection procedure should be chosen to suit objectives, costs, and resources.

11.3B Selecting One Adult from the Household

For some interview surveys of attitudes of the adult population, selecting a single adult in each dwelling is preferred. (1) Multiple interviews would lead to undesirable interview situations, because the second respondent could overhear or discuss the questions; they would also be inefficient because of high correlations. (2) An interview in every household avoids futile calls on dwellings without interviews.

To be applied and checked easily, the interviewer needs a simple procedure for ordering the members of the household. A clear and rigid proce-

List ALL persons age 21 and over in the dwelling.

Relationship to Head (1)	Sex (2)	Age (3)	Adult No. (4)	Check R (5)
HEAD	M		2	
Wife	F	40	5	
Head's father	M		1	
Son	M	22	3	
Daughter	F	20	X	
Wife's aunt	F	44	4	✓

Number persons 21 or over in the following order–oldest male, next oldest male, etc.; followed by oldest female, next oldest female, etc. Then use selection table below to choose R (Respondent)

SELECTION TABLE D

If the number of adults in the dwelling is:	Interview the adult numbered:
1	1
2	2
3	2
4	3
5	4
6 or more	4

TABLE 11.3.I Part of "Cover Sheet" Showing Listing of Adults and a Selection Table (one of eight possible types).

dure eliminates the need for hiding the selection numbers. The following procedure has been applied successfully for years at the SRC and elsewhere, since its first report [Kish, 1949].

A *cover sheet* is assigned to each sample dwelling; it contains the dwelling's address, a form for listing the adult occupants, and a table of selection (such as Table 11.3.I). At the time of the first contact with the household, the interviewer lists each adult separately on one of the six lines of the form; each is identified in the first column by his relationship to the head of the household (wife, son, brother, roomer, etc.). In the next two columns the interviewer records the sex and, if needed, the age of each adult. Following this the interviewer assigns a serial number to each adult; first the males are numbered in order of decreasing age, followed by the females in the same order. To assign these serial numbers it is necessary to obtain the actual ages only in that small portion of households in which there are two adults of the same sex, not connected by parent-child relationship. For example, the answer, "I, my wife, her mother, and our son, who is 25" identifies the adults of a household as numbers 1, 4, 3, and 2. Then the interviewer consults the table of selection; this table tells him the number of the adults to be interviewed. One of the eight tables (*A* to *F*) is printed on each cover sheet.

We prepare cover sheets containing the eight types of selection tables in the correct proportions. Where sample addresses are available in advance, the office assigns specifically to each address its own cover sheet with a selection table. However, advance listings are not available for sampling unlisted segments, nor for newly discovered or newly constructed dwellings. For these, packs of cover sheets are stapled together, and the interviewer is instructed to assign the *next* cover sheet at the time of the

Proportions of Assigned Tables	Table Number	If the number of adults in household is:					
		1	2	3	4	5	6 or more
		Select adult numbered:					
1/6	*A*	1	1	1	1	1	1
1/12	*B*1	1	1	1	1	2	2
1/12	*B*2	1	1	1	2	2	2
1/6	*C*	1	1	2	2	3	3
1/6	*D*	1	2	2	3	4	4
1/12	*E*1	1	2	3	3	3*	5
1/12	*E*2	1	2	3	4	5*	5
1/6	*F*	1	2	3	4	5	6

TABLE 11.3.II Summary of Eight Tables Used for Selecting One Adult in Each Dwelling

The proper fractional representation of each adult is approximated closely with only the eight tables. They are exact for adults in households with 1, 2, 3, 4, and 6 adults. Because numbers above six are disallowed, a small proportion of adults are unrepresented; perhaps one in a thousand (see Table 11.3.III), generally young females. There is compensation for them in the overrepresentation of number 5 in households with 5 adults, where number 3 is also overrepresented arbitrarily. These differences are typically trivial compared to nonresponse problems. If greater precision is needed, simple changes will obtain it; in 5-member households, rotate the *E* table among them; in households with more than 6 members, accept some duplicate interviews by selecting number 7 with number 1, 8 with 2, etc.

first contact with the address, in a strictly defined order. Each of these, whether they result in interviews or not-at-homes or refusals, must be returned to the office for checking.

When dwellings are selected with equal probabilities, the chance of selection of a single adult becomes inversely proportional to the number of adults p_i in the dwelling. This selection bias can be corrected by weighting each response with p_i. This weighting results in the ratio mean $\bar{y} = \Sigma p_i y_i / \Sigma p_i$ and in more complex computations, as discussed in 11.7*A*.

The weighting also tends to make the variance greater than for a self-weighting sample of the same size. But this increase is not great for selecting a single adult, because of the high concentration within a small range of sizes: over 70 percent of dwellings contain two adults, and almost all of the rest have either one, three, or four adults. The increase of the variance can be computed to be about 10 percent by applying (11.7.6) to Table 11.3.III.

Number of Adults in Dwelling	1	2	3	4	5	6 or more
Proportion of Dwellings	0.146	0.730	0.090	0.028	0.004	0.002

TABLE 11.3.III Proportion of Dwellings with Different Numbers of Adults (From an SRC Survey of 2000 U.S. Adults in 1957)

Another result of the high concentration is that, unless the variable has a very high correlation with the number of adults in the households, the difference will be small between the weighted estimate and a biased estimate in which each respondent has equal weight. In small samples, that difference may be negligible compared to the sampling error. Fairly large differences in a characteristic may exist between households of different sizes, and still produce only small differences between weighted and unweighted means. Checks of weighted versus unweighted means have been made at the SRC for many studies over the years; all differences were found to be imperceptible or negligible. The routine procedure at SRC is to make such checks for a few dozen variates, and after that reassurance, to continue using the unweighted estimates.

Remark 11.3.I The low numbers of selection are concentrated in tables *A*, *B*, and *C*; the addresses to which these tables have been assigned will yield mostly male respondents. Since evening calls are necessary to find most male respondents at home, the interviewer may concentrate his evening calls at those addresses. Conversely, the interviewer may better utilize his time during the day by calling at addresses to which tables D, E, and F are assigned. We aid this process by assigning (at random) either A, B, and C, or D, E, and F cover sheets to all (2 to 5) sample addresses within a block or within a segment cluster.

Remark 11.3.II The selection table can also be hidden behind black tape or in an envelope, to be opened by the field worker only after listing the elements. This is desirable if reasonable doubts remain about the objectivity of the ordering actually obtained for the elements. When feasible, objective ordering has the advantage of easier checking than hidden selection numbers.

Selection Table Number	Number p_i of elements in cluster					
	1	2	3	4	5	6
1	✓				✓	
2	✓					
3	✓					
4	✓			✓		
5	✓		✓			
6	✓		✓			
7	✓	✓			✓	
8	✓	✓				
9	✓	✓				
10	✓	✓		✓		
11	✓	✓	✓		*	✓
12	✓	✓	✓	✓	✓*	✓

TABLE 11.3.IV Twelve Alternative Tables for Selecting the Element Checked Last.

Deming [1960, p. 240] describes a procedure in which the interviewer, after he lifts a tape, selects only the last element checked. The checks are so arranged as to provide equal probabilities of selection to each element, regardless of their number p_i in the cluster. By assigning one of the above 12 tables with probability 1/12, we assign the probability $1/p_i$ to each of the p_i elements. This is true because in each column each number appears checked $12/p_i$ times. The exception is the triple appearance of numbers 3 and 5 in the column for $p_i = 5$, marked with asterisks, and these can be rotated.

11.3C Epsem Selection of Persons

The procedures and tables of the last subsection can be modified to select a constant portion, say, half or third, of all persons found in sample dwellings. Applied to an epsem selection of dwellings, it yields an epsem selection of persons, regardless of the unequal household sizes. The disadvantages are that some households yield more than one interview, and some none at all. With large fractions, say half, the former predominates; with small fractions, say every fourth person, the latter. Selecting half of the elements can be done with two selection tables: *A* selects elements numbered 1, 3, 5, etc., and *B* selects 2, 4, 6, etc. Similarly, one-third of the elements can be selected with three tables: *A* selects 1, 4, 7, etc., *B* selects 2, 5, 8, etc., and *C* selects 3, 6, 9, etc. More complicated procedures can be adopted, if needed. Selection bias is prevented either with a strict scheme of ordering persons in the dwelling (e.g., according to sex and age), or with hidden random starts.

Sometimes a feasible alternative procedure consists of listing all persons in all sample dwellings before the interviewing. Then the selection fraction (1/2 or 1/3) is applied to the list either in the office, or in the field with hidden selection numbers.

We can also design a sample without duplicate interviews from households with a reduced amount of weighting. For example, we can select half of the single-adult households, and select adults with $1/p_i$ in all other households; then weight with $p_i/2$ the interviews from households with $p_i > 2$.

If information is inexpensively available about the numbers p_i of persons in sample dwellings, then PPS selection can achieve the advantages both of epsem selection and of single interviews in dwellings. First, obtain the numbers ($p_i = 1$, 2, 3, 4, 5, or 6) of persons in a sample of dwellings. Second, select dwellings with $p_i/6$; taking 1/6 of single-person dwellings, 2/6 of two-person dwellings, and all of the six-person dwellings. Third, select one person with $1/p_i$, perhaps with the procedures of 11.3B. The selection probability is $p_i/6 \times 1/p_i = 1/6$ for all persons in the sample dwellings. Hence, the number of sample dwellings must be $6n/\bar{p}$, where $\bar{p}$ is the mean number of persons per dwelling, and n is the desired number of persons plus allowance for nonresponse. For example, with $\bar{p} = 2.0$ adults per dwelling, to get 1000 responses from 1100 adults, we need $6 \times 1100/2 = 3300$ dwellings. Note that only a third of the sample dwellings is approached for an interview. In dwellings with $p_i > 6$, we can either weight with $p_i/6$ or permit duplicate selections from persons numbered higher than six. By choosing $p_i = 4$ instead of $p_i = 6$ as the upper limit, we can reduce the proportion of wasted dwellings, but increase the proportion of duplicate interviews from households.

Sometimes a procedure for selecting a lower fraction of not-at-homes may be practical. The interviewer can list the persons at home (*H*) and persons not-at-home (*NAH*) in each sample dwelling, then apply separate selection tables to the two types. For example, we can use fractions of 1:1 for the *H*, and 1:4 for the *NAH*. However, disproportionate rates are seldom justified economically, because the element costs are not sufficiently different.

We have discussed five alternative procedures for selecting elements from sample dwellings: (1) Taking all in the dwellings. (2) Selecting 1/2. (3) Selecting 1/3. (4) Selecting one with $1/p_i$. (5) Using PPS with $p_i/6 \times 1/p_i = 1/6$. We can summarize by discussing the chief economic disadvantage of each, and illustrate with numbers from the distribution of adults in 1957 U.S. households, from Table 11.3.III.

Increase in element variance due to clustering can be denoted with the factor roh$(\bar{n}_s - 1)$, where $\bar{n}_s$ is the average of selected persons per sample dwelling and roh is the intraclass correlation in the dwelling for any specified characteristic (see 5.5). Here sample dwellings include only those that contribute at least one person to the sample. The variance is increased by about roh when all adults of the household are taken, and $(\bar{n}_s - 1) = 1.0$. The increase is only 0.09 roh for selecting 1/2, 0.03 roh for selecting 1/3, and absent for the other procedures. These numbers, 1.0 for the first procedure and 0.09 for the second procedure, represent the ratio of second to first interviews from dwellings.

Increase in interview problems and bias due to second interviews is hard to measure. Their magnitudes depend on the proportions of second to first interviews; these are 1.0, 0.09, and 0.03 for procedures 1, 2, and 3, and absent in the other procedures.

Increase in element variance due to unequal weights can be denoted with the factor $(\sum W_h k_h)(\sum W_h/k_h)$, where W_h is the proportion of sample elements that receive the weight k_h (see 11.7.6). Only the fourth procedure for selecting with $1/p_i$ incurs an increase of about 0.10, as well as the *increased costs of weighting* in computing and tabulating, which distinguish weighted estimates from self-weighting estimates. Unweighted estimates escape these increases in variance and cost, but are subject to *increases in the mean square error*, represented by the fraction Bias²/Variance. Fortunately, this has been found negligible in many studies of adults from U.S. households.

Element cost for the different procedures can be compared, utilizing three component factors. First, denote with c the basic cost per element (person) constant for all procedures, including interviewing, coding, and processing; this amounts to nc for n elements with any procedure. Second, denote with dc the cost of obtaining entrance, cooperation, and household information from each of the m contacted sample dwellings. The total cost is $mdc = ndc(\bar{m})$, where d is the ratio of this factor to c, and $\bar{m} = m/n$ the ratio of dwellings to elements. Third, denote with bc the cost of sampling each of the m_b "blank" dwellings that must be discarded without interviews in some procedures. The total cost is $m_b bc = nc(\bar{m}bp_b/q_b)$, where b is the ratio of this factor to c, and $p_b/q_b = m_b/m$ represents the ratio of blank dwellings to sample dwellings. The three components result in the total cost:

$$nc' = nc + m(dc) + m_b(bc) = nc(1 + \bar{m}d + \bar{m}bp_b/q_b). \quad (11.3.1)$$

Numerical values of $\bar{m}$ and p_b/q_b can be obtained for each procedure from distributions of elements per dwelling, such as in Table 11.3.III. The second term is smallest, $0.5d$, for taking all adults. It is $0.91d$ for selecting 1/2, $0.98d$ for selecting 1/3, and d for the fourth and fifth procedures.

The third term is zero for taking all elements and for selecting a single element. It is $0.07b$ for selecting 1/2, $0.5b$ for selecting 1/3, and $2.0b$ for selecting with PPS.

11.4 RARE ELEMENTS; HIGH SKEWNESS

11.4A Selection Techniques for Rare Traits

A small proportion $\bar{M} = M/N$ of a population of N elements possesses a trait denoted with the variable $Y_i \neq 0$, whereas for the vast majority $(N - M)$ the variable $Y_i = 0$. The search for rare and important elements can have any of three aims; selection problems are similar for all three. (*a*) We may need to know only M or $\bar{M}$, the number or proportion of the members of the class with the trait; for example, the prevalence of a rare disease, or purchasers of an expensive item. Estimating M or $\bar{M}$ represents the same problem, if the population base N is known accurately, relative to M. (*b*) We may need $\bar{X} = Y/M = \bar{Y}/\bar{M}$, the mean of the Y_i variable within the class, disregarding the majority $(N - M)$ for which $Y_i = 0$. Here Y and $\bar{Y}$ denote the sum and the mean of the Y_i in the entire population. For example, Y_i may denote the individual's medical costs caused by the disease and $\bar{X}$ the mean cost in the class. (*c*) We may need $\bar{Y} = Y/N = \bar{M}\bar{X} + (1 - \bar{M})0$, the mean in the entire population; for example, the mean cost of a disease for the population. Of course, $\bar{M}$ is only the special case of $\bar{Y}$ when $Y_i = 1$, and denotes the mere presence of the variable for the M class members. Note also that estimating $\bar{Y}$ from $\bar{X}$ is easy, if $\bar{M}$ is known well enough relative to $\bar{X}$.

Degree of rarity and cost of measuring the trait are the two main factors that determine the relative merit of different selection methods. Sample surveys are used widely and successfully to obtain inexpensive measurements of traits that are not too rare. Screening the population to find a class may be practical if $\bar{M}$ comprises 10 or 20 percent, but far less so if $\bar{M}$ is one in 100 or 10,000. Very rare traits are easy to sample only if finding them is inexpensive; an extreme example would be a centralized and complete list of persons with a well-recognized rare disease.

Rare traits that are expensive to find create a predicament. Adroit applications of some of the following techniques bring about a reasonable approach. When choosing between them, consider the trait's rarity, as well as the available survey resources, including knowledge about large concentrations of the trait. This could mean lists, or known geographical concentrations, or high correlations with other clear indicators.

1. *Multipurpose samples* can discover several rare traits in one sample and divide the costs among the cooperating investigations. This approach can also yield data about relationships between the traits. Many social and economic surveys are based on such cooperation. In market research, samples commonly cover several products, each relatively rare.

2. *Cumulation* of a rare population can be obtained as a by-product of gathering other variables in continuing surveys with changing samples that cover large populations in time. The results must be interpreted as average over the period covered by the surveys, but this concept is often reasonable (12.5*D*). First, past surveys may already contain the data in an acceptable uniform manner; though too rare in any single sample, the cumulation may provide an adequate base for analysis. Second, the measurement can be added to future surveys if the enumerators can perform it. Such additions increase slightly the average interview length, if applied only to a small portion; but special emphasis is necessary to keep enumerators alert to rare events. Third, the surveys may perform only a screening operation to identify the rare individuals; the difficult measurements on them are done separately. If the identification is imperfect, it can still provide the first phase of a two-phase sample.

3. *Large clusters* can decrease drastically the costs of locating and screening population elements. For example, a sample of complete blocks of a city, or of entire villages of a country, can be searched (screened) for persons with the trait. Ordinarily we avoid large clusters, because of their adverse effects on the variance. But even large clusters of the entire population will yield only small clusters of a rare trait, if this is widely spread. For example, entire blocks may be sampled for persons over 65 years of age; entire villages may be searched for persons with an identifiable disease. If, on the contrary, the trait is concentrated in small areas, these areas often can be recognized and stratified accordingly.

4. *Controlled Selection*. To cut expenses the search must sometimes be confined to only a few dozen primary clusters, such as hospitals, schools, or cities, for example. We may then carefully try to control the selection, so as to make it representative of the population with respect to several

relevant control variables. The control cells may far outnumber the permitted selections; we can then resort to controlled selection (12.8). This may be an expedient approach to sampling problems that are now too often solved by restricting the sample to one or two clusters purposively chosen (hospitals, schools, or cities).

5. *Disproportionate Stratified Sampling.* If we can find small strata that contain large portions of the rare trait, it is efficient to increase their sampling fractions. In other words, if most of the rare trait can be located in small pockets of the total population, it should be sampled heavily. Large gains can accrue if over 90 percent of the rare trait can be located within 10 percent of the total population. Geographical strata seldom achieve the high separations of class densities necessary for efficient use of disproportionate sampling. Instead, we must look for special lists of the class with the trait, or with a highly correlated variable.

However, only small gains can be derived from the presence of only one of the two desirable conditions: high concentrations of only a moderate portion, or mild concentrations of a large portion of the rare trait. For example, U.S. cities have exclusive neighborhoods that are full of rich people; but oversampling them will not bring large gains, because most rich people live scattered elsewhere. Furthermore, rich people who live in exclusive neighborhoods differ in many ways from those living elsewhere. Similarly, visitors to a set of clinics may contain a high concentration of the class with a trait, yet most of that class may never reach any clinic. Visitors to clinics may well represent the entire class, including nonvisitors; on the contrary, the two kinds of class members can differ drastically. An exotic example is found in interstellar space: although almost empty compared to stellar density, it contains nevertheless a large portion of the total mass of matter, and its composition differs from stellar matter.

To achieve large gains from disproportionate allocation, we need a high degree of separation of the rare trait. Optimum allocation requires that the sizes n_h of samples selected from strata be proportional to $W_h\sqrt{\bar{M}_h/J_h}$, where $\bar{M}_h$ is the proportion in the rare class and J_h is the element cost. The element costs often differ little, and then the sampling rates should be proportional to $\sqrt{\bar{M}_h}$, approximately. Large differences in $\bar{M}_h$ are necessary to obtain even moderate differences in $\sqrt{\bar{M}_h}$. Here we use $\sqrt{\bar{M}_h}$ as a common rough approximation for the standard deviations of elements within strata. These are actually $\sqrt{\bar{M}_h(1-\bar{M}_h)}$ for estimating $\bar{M}$, the proportion of rare elements; and they are about $\sqrt{\bar{M}_h T_h^2}$ for estimating the mean $\bar{X}$ of the rare elements, where the T_h^2 are element variances (4.5*C*).

At one extreme we may have some *complete strata* with $f_h = 1$. At the other, we may omit some *cut-off strata* with $f_h = 0$, because they are known to contain only a negligible portion of the rare trait or none at all. The meager results of searching those strata may not be worth its cost, which can be better used to improve the samples from the other strata.

6. *Two-phase Sampling* (Double Sampling). The first phase is a preliminary screening to distinguish two or more strata in which proportions of the rare elements differ greatly. Disproportionate sampling can then be applied in the second phase. The allocation of effort between the first-phase screening and the second-phase depends on the utility of the screening, and on the relative costs of the two phases (12.1).

Screening operations can assist several of the proposed techniques, and may involve brief interviews, cheap tests or observations, or scanning data on filed cards. Applied to either the population or a large sample, the screening operation produces a set of screened inclusions, the potential members of the rare class. These are then measured with more precise, standard processes in order to discard the false positives.

What about the false negatives, members of the rare class among the screened exclusions? Even good screening can miss some, and poor screening can easily miss 10 percent. Without strong assurance of safety, at least a sample check of screened exclusions is advisable on large surveys. If this locates no rare elements, it provides an upper limit for the error. If the false exclusions rise beyond a negligible proportion, the screened exclusions become a stratum to be sampled at a reduced rate.

Screening is useful when we find a cheap test that has *very few false negatives* (erroneous exclusions), and *not many false positives* (erroneous inclusions). Generally, the screening should err more often on the side of false positives than of false negatives. Correcting the former requires some extra work, but its unit cost is less than for false negatives, which can cause either bias or reduced efficiency. The costs and the proportions of errors of both kinds should enter into the design of the screening process.

7. *Batch testing* can help us find rare elements, if part of the material, when tested in batches, can reveal the presence of even a single rare element. For example, a search for positive cases of some rare disease can be made in batches of n cases each, if a positive test result follows the presence of even a single positive individual in the batch. One method constructs batches of n and directs that when a positive batch is found, individuals in it are tested separately until a positive case is found; then the remainder is tested and if that is positive, individuals in it are tested until a positive case is found; and so on. The optimum size (n_{opt}) for the batch is near $3/2\sqrt{p}$, where p is the percentage of positives in the population. This method requires only 51 tests per 100 individuals when

$p = 0.10$ (with $n_{opt} = 5$), only 14 tests when $p = 0.01$ (with $n_{opt} = 16$), and 4 tests when $p = 0.001$ (with $n_{opt} = 47$) [Sterrett, 1957]. A modification calls for dividing any batch that tests positive into halves and testing both; any positive half is halved again and both quarters are tested; and so on. This method is somewhat better when $p < 0.01$ and when p is unknown [Roy, 1961].

8. *Special lists*, when available, provide the best means for locating a class of rare elements. If these are very rare, widely scattered, and difficult to identify, a list may be the only way to find them. The ideal list contains the entire target class (or a good sample of it) clearly identified, together with necessary information, such as a current address.

Such lists can sometimes be produced with ingenuity, perhaps with tolerable redefinitions of the class, and occasionally by combining several lists. *Snowball sampling* is the colorful name [Goodman, 1961] for techniques of building up a list or a sample of a special population by using an initial set of its members as informants. For example, consider asking an initial group of deaf people to supply the names of other deaf persons they know [Deming, 1963]. Members of some populations may well know about each other, as the deaf do in cities or areas that are not too large; perhaps also the owners of pigeons, or of racing cars. However, in other cases it may be difficult to assign reliable measures of the probability of inclusion.

Extra effort is needed if the list contains impurities, foreign elements, and replicate listings. The worst problems are caused by incompleteness, the *noncoverage* of elements missing from the list. When the list is good but imperfect, we must decide whether to search for the missing elements or to omit them. If the missing portion is small and extremely difficult to find, we may be forced to treat it as a cut-off stratum, and accept the bias. But if needed and feasible, it may be included in a special stratum with a supplemental procedure. The decision depends on several considerations.

First, the element cost is high when the missing units are rare, widely scattered, and hard to locate. Hence, their inclusion should be designed with a considerably lower sampling fraction. Second, the bias of their exclusion should be judged against the other sources of errors, hence also against the size of the sample. The relative effect of a fixed bias increases with the precision and size of the sample. The size of the bias should be balanced against the standard error, and the cost of reducing the bias against the reduction in the standard error it can buy. Third, the effect of exclusion is different for estimating frequency totals and for mean values of a trait. The exclusion of missing values has little effect on the mean, but great effect on the total, if the missing values closely resemble the values found on the special list. Contrariwise, if the missing values are very small, they affect the mean more than the aggregate (13.4*B*).

Supplements are useful for sampling a portion missing from the special list, when that portion, although small, is not negligible. Consider the rare elements as coming from two strata: the special list to which the sampling fraction f_a is applied, and the supplement for the rest of the population, sampled with the fraction f_b. Because the element cost is typically much higher in the supplement, the sampling rate f_b should be considerably lower, according to optimum allocation of resources. Sometimes a *linking* procedure can provide the supplement (11.5).

In summary, a *special list* that includes the rare class and nothing else is best, especially for very rare traits. If the list is good but imperfect, a *supplemental sample* may find the missing units. Lacking a good list, we try to separate a small *special stratum* containing most of the rare class, from a large stratum containing much less. The special list or stratum is sampled with *disproportionately high rates.* If we cannot find such special strata, we try to create them with a cheap *screening* operation. The efficiency of the screening may be increased with *double sampling* and *large clusters.* The costs of screening may be shared in *multipurpose* surveys, or the rare class may be *cumulated* in a periodic survey operation.

Example 11.4a The existence of concentrated areas of rare elements does not ordinarily guarantee their usefulness for disproportionate sampling. An outstanding economist wanted to design a sample of high incomes based on oversampling special high-income tracts (U.S. 1950 Census) in Standard Metropolitan Areas. There are many such tracts with large proportions of high incomes, and it seemed reasonable that increased sampling rates in those tracts would yield many high-income families. He found that the High's, although only 2.64 percent in the total population, accounted for 20 percent in the special tracts. These special tracts contained only 1.2 percent of the U.S. population, but 9 percent of the High's. The divisions were in percentages:

	In Special Tracts	In Remainder	In U.S.
High Incomes	0.24	2.40	2.64
Others	0.96	96.40	97.36
All Incomes	1.20	98.80	100.00

Actually, any oversampling of the special tracts would be of little value because it affects only 9 percent of the High's; even their complete census would miss 91 percent of the variance. On the other hand, increasing the area of special tracts to include more High's would quickly increase the screening loss from the 4 to 1 (0.96/0.24) to 10 or 20 to 1, and this would probably become inefficient. This situation is typical of many others: the existence of areas of concentration can mislead us into overlooking the dispersion of most of the target population among the residual.

Oversampling areas of concentration can be efficient for computing a biased estimate; this demands the bold assumption that rare elements from special areas and those from the residual do not differ substantially. It would be unrealistic to accept an assumption of no difference between high-income people from wealthy tracts of metropolitan areas, and those from the rest of the country.

11.4B Highly Skewed Populations

Sampling from highly skewed populations strains the assumptions about the normality of the distribution of estimates. Very large elements, though rare, can have important effects on the results of even moderately large samples. If the extreme cases are a mere handful, the researcher should know enough to isolate them from his cross-sectional selection. For example, suppose we are sampling farms in a state, and that all farms are below 500 acres, except for two giants of 100,000 acres each; the researcher should know enough to separate them before he selects his cross-sectional sample. When large elements, though a small proportion, are neither few nor easily segregated, problems arise. Examples are frequent in sampling establishments such as business firms, farms, organizations, clubs, and schools. But some individual characteristics are also highly skewed. Income, assets, and other financial characteristics are obvious examples; but the number of books read per year and, perhaps, many other voluntary activities, as well as some health data, such as number of days spent in a hospital, are also highly skewed variables. Several alternative sampling methods, particularly suited to specific problems, are available.

Disproportionate allocation utilizes high sampling fractions for strata containing high proportions of the elements with large values. It requires a list containing information about one or more ancillary variables highly correlated with the survey variables.

Procedures for finding rare elements, discussed above, may be used to find the large elements. Sometimes a special list can be found or prepared, to be sampled at a high rate, perhaps 100 percent. The remainder of the population becomes the supplemental stratum for selecting small elements with a lower rate. Both strata can be divided into several strata to be sampled separately, with different rates. See the case study of "The Sample Survey of Retail Stores" [12.A in Hansen, Hurwitz, and Madow, 1953].

Truncation of the large values, $Y_i > Y_{max}$, may be necessary: (*a*) if the sampling rates cannot select enough of the large elements; and (*b*) if this "upper tail," though small, accounts for a large enough portion of the aggregate Y to have a sufficient influence on the estimates. We may accept a *truncated* distribution, discarding from the sample the few values that appear with Y_i greater than some fixed $Y_{\max}$. Sometimes the truncated variable is accepted and presented as such, with an extreme category of "$Y_{\max}$ and over." If the values of the large elements ($Y_i > Y_{\max}$) are needed for producing estimates of the entire population, they must be estimated from other sources. We may use some model of the distribution, together with data from other sources, and perhaps a cumulation from several surveys.

Transformation of the variable Y_i can sometimes produce a distribution with much less skewness than the original. The logarithmic transformation of Y_i to $\log(Y_i)$ has been proposed for transforming some common skewed distributions into fairly normal distributions. This is a complex topic; the method has not been used often in surveys, though it has been tried with income [Aitchison and Brown, 1957].

Remark 11.4.1 *The effect of skewness* of the population distribution on sample estimates has been treated in statistical literature. These treatments are generally theoretical and assume simple random sampling; hence they do not directly apply to common survey practice. This purpose is better served by several sections in the books of Hansen, Hurwitz, and Madow [1953, 10.5–10.11] and Cochran [1963, 2.13–2.14]. We must confine ourselves to a few simple generalizations. (1) The skewness of the population distribution is reflected in the distribution of sample means, but it is reduced in proportion to the sample size. Hence, the means of even moderate size surveys have much less skewness, and are much more nearly normal, than the distribution of population elements. This "size" in complex samples is not simply the number of elements n; perhaps *effective size* $=$ *n/design effect* is a better rough measure (8.2). Hence, strong clustering tends to destroy the normalizing effects of a large n. On the other hand, good stratification and truncation can reduce considerably the skewness of sample means and bring it closer to normal. (2) The distribution of the standard error of the mean $s_{\bar{y}}$ seems to be more affected by skewness than the mean $\bar{y}$ itself. But the distribution of $t_{\bar{y}} = (\bar{y} - \bar{Y})/s_{\bar{y}}$, the standardized distribution of sample means, appears to be less affected [West, 1952]. (3) The sampling variability of the computed value for the standard error $s_{\bar{y}}$ is also higher in skewed distributions. The increase may be seen in the approximate expression for the coefficient of variation of $s_{\bar{y}}$: $\text{SE}(s_{\bar{y}})/S_{\bar{y}} \doteq \sqrt{(\beta - 1)/n}$ (8.6*D*). Here $\beta = \Sigma(Y_i - \bar{Y})^4/N\sigma^4$; it equals 3 for normal distributions, but it can have very large values for highly skewed distributions. The formula strictly applies only to simple random sampling; and in cluster sampling all symbols must refer to clusters as sampling units. (4) The skewness of the distribution of the mean increases the probability of error beyond that expected from normal intervals of the type $\bar{y} \pm t_p s_{\bar{y}}$. Positive skewness decreases the probability of error on the negative side ($\bar{y} - t_p s_{\bar{y}}$), but increases it on the positive side ($\bar{y} + t_p s_{\bar{y}}$). Cochran notes that this effect is kept within reasonable limits by making $n > 25\ G_1^2$ for simple random samples, where $G_1 = \Sigma(Y_i - \bar{Y})^3/N\sigma^3$.

Example 11.4b Skewness has been a problem in surveys of annual personal incomes conducted by the Survey Research Center. In samples of about $n = 2500$, the standard error was about $120 for a mean of $5000. The proportion of incomes over $25,000 was about 1/200; over $100,000 about 1/2500; and over $1,000,000 about 1/250,000. The numbers have changed some since 1953, but not the essential problem. These small groups of rare individuals can have large effects on the variance of the mean. Inclusion of an income of $1,000,000 would occur only once in 100 surveys, and would raise the mean by more than $400. The expected number of incomes over $100,000 is 1, but 0, 2, or 3 can occur easily (the probabilities are 0.37, 0.37, 0.18, 0.06, 0.02 for 0, 1, 2, 3, and 4 or more), and each income adds more than $40 to the mean. The expected number of incomes over $25,000 is over 12, and each adds over $10 to the mean. Where should we place the cut-off point? The answer depends on the situation and resources. I believe that the estimate for the class over $100,000, with a coefficient of variation near 1, could be done better with an expert's judgment based on outside figures. Perhaps this is true even down to the $25,000 limit (14.3).

11.4C Screening with Field Ratings

Sometimes sampling units can be sorted cheaply into distinct strata by field workers. This screening procedure may consist of brief interviews or rapid eye estimates. It should be inexpensive; also nondestructive and inoffensive to the respondent, so that it does not hinder the main measurement. For example, a screening interview about income may increase the refusals in the subsequent main interviews.

The aim is to divide the population into strata to be sampled with different rates. The screening succeeds when it establishes strata internally homogeneous and as divergent as possible. To estimate population means, large differences of standard deviations are especially helpful. This may be achieved by creating strata with very divergent proportions of the large elements, which comprise a rare class.

It may be economical to set the screening rate so that the stratum with the highest sampling fraction will be completely included in the final sample. At the other extreme, the lowest stratum can be omitted altogether if it contains no rare elements, and the target population consists only of rare elements. Hence, good screening aims to produce a large stratum ideally containing none of the rare elements.

For example, to find individuals over 65 years of age, ask any household member for ages of all members over 50; if nobody is at home, ask two neighbors, and if both say that nobody over 50 lives in the dwelling, it may be omitted. The leeway of 15 years produces some false inclusions to be eliminated, but reduces considerably the false exclusions. Similarly, the screening test for a rare disease should err on the side of including some false positives.

Selection from field ratings is most simply done in the office; this is feasible when sampling a city, for example. However, for a sample spread over a country, separate trips for rating could incur a heavy cost, which can be avoided by rating, selecting, and interviewing on one trip. Selection numbers are hidden in envelopes or behind tapes (8.4*C*). They can be used after rating with either of two procedures: (*a*) The enumerator enters his rating (for a block, or neighborhood) on a separate sheet for each stratum; then lifts a tape that reveals the hidden numbers for selecting sample elements. (*b*) Without separate sheets, one set of hidden selection numbers can yield separate instructions for different strata. For example, with selection numbers 1, 2, 3, 4, all four numbers select in the "High" stratum, only 2 and 4 in the "Medium" stratum, and only 4 in the "Low" stratum, yielding rates of 4/4, 2/4, and 1/4.

Example 11.4c We can examine the socio-economic ratings assigned by Survey Research Center interviewers at the time of listing dwellings located within sample blocks (9.6). The interviewer wrote one of the letters *L*, *M*, or *H* next to each dwelling on the block listing sheet, indicating a hasty personal judgment about the income of the dwelling's occupants, by merely glancing at the building. The proportions placed in the three classes were $W_L = 0.38$, $W_M = 0.53$ and $W_H = 0.09$. The success of this field rating procedure can be judged from Table 11.4.I, which summarizes results from a fuller report, based on a 1954 survey [Kish, 1961*b*].

Note that for items 1, 2, 3, involving very skewed population distributions, the means in the three strata differ considerably; for item 1, income, they are $2800, $4200, and $8000 for the Low, Medium, and High strata. The difference in standard deviations, $1700, $2800, and $8100 for income, are even greater and more important for optimum allocation. That optimum is achieved when the sampling fractions are in proportion to the standard deviations; these proportions for the Medium and High Strata appear in the next two columns as $k_M = S_M/S_L$ and $k_H = S_H/S_L$. Optimum allocation for income calls for sampling the Highs 4.7 times, and the Mediums 1.6 times as heavily as the Lows. Optimum allocation results in the optimum efficiency 1.36 for income, in 1.45 for assets, and 1.42 for debts. The efficiency is computed, in contrast to that of proportionate sampling, (1:1:1 for the three strata) as the inverse of the variances for equal numbers of elements. Note that the simple allocation of 1:2:4 for Lows, Mediums, and Highs achieves for the three items almost as high an efficiency as the optimum.

In item 4, note similar gains for estimating the proportion of the rare individuals who have incomes over $10,000. Item 5 refers to the skewed distribution

	Means for Strata (in $100 units)			Standard Deviation (in $100 units)			Optimum Loading		"Efficiency" Compared to Proportionate	
	L	M	H	L	M	H	k_M	k_H	opt	1:2:4
1. Mean income	28	42	80	17	28	81	1.6	4.7	1.36	1.32
2. Mean liquid assets	7	18	58	17	34	101	2.0	5.9	1.45	1.41
3. Mean personal debt	3.1	3.8	6.6	5.4	8.3	26.9	1.5	5.0	1.42	1.37
4. Percentage with income over $10,000	0.4	5	31	6.4	21	46	3.3	7.2	1.43	1.37
5. Mean liquid assets of managers	11	33	73	x	x	x	2.4	9.1	1.76	1.60
6. Percentage below median income	70	44	32	46	50	47	1.1	1.0	1.00	0.85
7. Standard of S_h constant	...	...	...	S	S	S	1.0	1.0	1.0	0.83

TABLE 11.4.I Data for Optimum Allocation of Five Skewed Variables

of assets for the rare subclass of managers; note the large gains from the combined effect of a skewed variable for a rare subclass.

However, most of the survey items showed losses from the 1:2:4 allocation. Even the estimate of the median income suffered a decrease to 0.85 of its efficiency. Its error depends on the proportions with less than median income in the three strata (12.9). These proportions were 70, 44, and 32 percent; the standard deviations ($\sqrt{PQ}$) were so alike that the optimum is near the proportionate. Most of the survey items were proportions of the "standard" type, where the standard deviations for the three strata are alike. For this type of item, proportionate sampling is best; 1:2:4 allocation results in an efficiency decreased to 0.83, as shown on the last line.

The following is a modified version of the field instructions for the 1954 Survey. One change increases the strata limits from $3000 and $6000 to $5000 and $10,000. The other encourages the double ratings *LM* and *MH* for borderline cases, giving five strata that can be combined as needed.

Instructions for Rating Dwellings

A. In addition to listing the addresses of dwellings in the sample blocks assigned to you, we ask you to rate each of the dwellings. From observation, assign an economic rating to each listed dwelling whether occupied or vacant; for a vacant dwelling, assign a rating comparable with the ratings for similar dwellings in the neighborhood. This rating is a simple judgment made from outside observation. Precise ratings are neither necessary nor possible. From outward appearance you cannot guess accurately the income level of the occupants of the dwelling. For these ratings to be useful to us it is not necessary that you be right in every case; it is enough if your judgment is correct in a majority of cases.

B. The ratings which you are to use are defined as follows:
   1. Your guess of H means that the dwelling probably is occupied by a family with a yearly income of $10,000 or more, or a weekly income of about $200.
   2. Your guess of M means that the dwelling probably is occupied by a family whose yearly income is from $5000 up to about $10,000, or whose weekly income is from $100 to $200.
   3. L means that your guess is that the dwelling probably is occupied by a family with income under $5000 a year; that is, less than about $100 per week.

C. There may be many cases where you dislike assigning a single rating. If it makes your job easier and quicker, use double letter ratings to indicate that your guess is on the borderline.

D. Record your rating to the right of the address on the listing sheet; you will notice that a special column with the heading "Rating" has been provided for this purpose.

11.5 SUPPLEMENTS FOR INCOMPLETE FRAMES

General problems of incomplete frames are treated in 2.7*A* and 13.3; other sections contain specific examples and practical procedures. Here we discuss methods for adding a supplement to include a sample of the elements missing from the principal frame. Suppose that most of the population can be found cheaply on a list, or with a procedure that establishes an efficient principal frame, yet the extent of noncoverage is sufficient to justify a search with a separate supplement. For example, in 9.8 a sample of city directory addresses is supplemented with an area block sample of dwellings missed by the directory.

Economic conditions typically require that the principal frame contain a good majority of the population; otherwise the supplemental procedure may simply be used alone. A relatively poor principal frame (say, with 50 percent coverage) may be used if it saves enough money. On the other extreme, if a frame is nearly perfect, adding the supplement may be a pedantic exercise. Before designing a probability sample for an extremely small proportion of the population, we should consider its cost, its relation to sampling errors and to other errors of response and nonresponse. In a sample of 1000 elements, a 2 percent noncoverage amounts to 20 elements; we may consider whether to omit them, add them with some *ad hoc* procedure, or to make an expert guess about them.

If the elements on the principal list have much lower sampling costs or higher standard deviations than those in the supplement, a higher sampling fraction may be indicated by optimum allocation.

The procedures and uses of supplements are varied. Several lists and several supplements can be combined to cover the target population adequately. Different sampling procedures can be applied in the several frames. For example, in 9.4*C* separate procedures are mentioned for newly constructed homes, for large apartment buildings, for the institutional population, and for the suburban areas; four possible supplements to the principal frame for a city's dwellings. The procedures of observation may also differ, such as a mailed questionnaire to an element sample from a special list, supplemented with personal interviews from an area block sample.

A special list can provide the principal frame for a class of important elements, with a supplement for the rest of the population. The supplement may be an area sample and may have a lower sampling fraction. For example, a list of large stores plus an area sample of small stores comprise the surveys of business of the U.S. Census Bureau [Hansen, Hurwitz, and Madow, 1953, 12.A].

A supplement's adequacy to produce missing units should not be merely assumed, but investigated. For example, dwellings missed by enumerators would often be missed again in a supplement conducted with similar procedures. Generally, the more dissimilar the two methods, the better their joint coverage should be. This may explain the apparently successful coverage of dwellings achieved by combining city directory lists with enumerators' block supplements [Kish and Hess, 1958].

The supplement may contain elements that also appear on the principal list; these will have duplicate chances of selection from overlapping frames (11.2*D*). The most direct way to deal with duplications is to prevent their appearance in the supplement. For example, the instructions for block supplements (9.8*D*) call for listing only dwellings that do not appear in the city directory; the area sample of small stores above included only stores not appearing on the list of large stores. On the supplemental list, special elements from the principal list become foreign elements treated as blanks.

However, it may not be feasible to keep the supplement "clean" of extraneous, special elements. One remedy consists of checking the supplemental sample and eliminating from it any element that appears on the full principal list; it is not enough to eliminate only those which appear in the sample from that list. If the entire list is included in the survey, eliminating duplicate inclusion becomes easier.

When identification of elements on the principal list A, selected with the rate f_a, is not feasible, they also receive the selection probability f_b in the supplement B. Hence, if the two samples are independent, the probability of selection of an A element is $f_{a'} = f_a + f_b$; if duplicate appearances in the sample are excluded, then $f_{a'} = f_a + f_b - f_a f_b$. Duplicate chances of selection can be tolerated, if all the A elements in the entire sample, whether obtained in the A sample of the special list or the B sample of the supplement, as given, are weighted with $1/f_{a'}$ rather than with $1/f_a$. It is, of course, necessary to identify all A elements in the B sample, so they may receive the weight $1/f_{a'}$, rather than $1/f_b$. The reweighting may be designed into the sampling rates f_a and f_b (11.2D).

A diagram can clarify the combined selection of large elements from a principal A list, plus a supplement B for small elements. The two strata are represented by the two rows. The situation is simple when only the cells AL and Bs have cases, with both As and BL empty; but we are often troubled by the appearance of small elements on the special list (As), and even more by large elements in the supplement (BL). To avoid sampling bias, both large and small elements (both AL and As) receive the weights $1/f_a$. Also, both small (Bs) and large (BL) elements from the supplement

	Rates	Large	Small	Both
Special list A	f_a	AL	As	$A.$
Supplement B	f_b	BL	Bs	$B.$
		$.L$	$.s$	Total

should receive the weight $1/f_b$. For example, to sample the farms of a state, we established and sampled with $f = 1/12$ a special list of large farms, with 1000 plus acres at the last Census; then added an area supplement of all other farms at 1/120. We find that some farms from the list are now small (As) yet receive only the weight 12. But we are especially troubled by the appearance of large farms from the area sample (BL) that must receive the weight 120.

The appearance of small elements (As) from the special list is the lesser problem. If few or cheap, they can be disregarded. If many and costly, a procedure can be devised to eliminate them from the special list. More troublesome are the large elements picked up in the supplement (BL), because the special list omitted them (false exclusion of a screening). Selected with the small rate f_b, they represent others that did not appear in the sample. Given the large weight $1/f_b$, they greatly increase the variance (as well as the variance of the variance). This is the penalty for false exclusions in screening, and often we must put up with it (but see 12.6C).

11.6 OBSERVATIONAL UNITS OF VARYING SIZES

One of the frightening statements made about American education, around 1957, was that half of the American high schools offered no physics, a quarter no chemistry, and a quarter no geometry. It was later noted that, although those backward schools were numerous indeed, they accounted for only 2 percent of high school students. There were many more small schools than large ones, but the small proportion of large schools accounted for a large proportion of students. Moreover, the curricula and facilities of large and small schools can differ greatly. Presenting average school characteristics gives a misleading picture if mistaken for conditions facing the average student.

This issue arises whenever groups of elements of greatly differing sizes serve not only as sampling units, but also as observational units. The group characteristic of each unit is observed and assigned a single score, then the results are reassigned to the elements which comprise the parent population. The researcher should be really interested in the effect of the unit characteristic on the individual element; but often the observational unit is used, automatically and mistakenly, as the unit of analysis also. In my experience, the researcher suffering from this confusion invariably converts to the proper element for his unit of analysis after being shown the difference. The reader can learn several aspects of this problem by making his choice between the *group mean* $\bar{Y}_g$ and the *element mean* $\bar{Y}_e$ in the following problems.

(*a*) We want to estimate the prevalence of swimming pools in the high schools of a state. $\bar{Y}_g$ percent of the schools have swimming pools, but $\bar{Y}_e$ percent of the students go to schools with pools. $\bar{Y}_e$ is considerably larger than $\bar{Y}_g$, because larger schools have pools much more often than small schools.

(*b*) In a voluntary civic organization, $\bar{Y}_g$ percent of the branches are in large metropolitan areas, but $\bar{Y}_e$ percent of the members come from these areas. $\bar{Y}_e$ is much larger than $\bar{Y}_g$ because the branches are larger in those areas. To what extent is this organization metropolitan? Being metropolitan is also correlated with other variables to be estimated.

(*c*) To forecast industrial employment and mobility in a state, the heads in a sample of manufacturing plants are interviewed regarding their plans to expand, to stay in the state, or move out of it. The results can be presented in terms of percentages of employees $\bar{Y}_e$ or as percentages of plants $\bar{Y}_g$, and the two means differ, because large and small plants differ.

(*d*) In a certain industry $\bar{Y}_g$ percent of firms operate with a specified type of organization (or leadership, or safety measures), and $\bar{Y}_e$ percent of employees are subject to it. The two figures differ greatly, because small and large firms are dissimilar. Which measures better the prevalence of the several types?

(*e*) To estimate the prevalence of museums in the cities of a country, we can choose between $\bar{Y}_g$, the proportion of cities with museums, and $\bar{Y}_e$, the proportion of people living in cities with museums.

From the values $\bar{Y}_\alpha$ for all A units in the population, we can compute either the unweighted mean of the units:
the *group mean*,

$$\bar{Y}_g = \frac{1}{A}\sum_\alpha^A \bar{Y}_\alpha, \tag{11.6.1}$$

or the *element mean*,

$$\bar{Y}_e = \frac{Y}{N} = \frac{\sum N_\alpha \bar{Y}_\alpha}{\sum N_\alpha} = \frac{\sum Y_\alpha}{\sum N_\alpha}, \tag{11.6.2}$$

using the numbers of elements in the units as weights. Note that $N_\alpha \bar{Y}_\alpha = Y_\alpha = \sum_\beta Y_{\alpha\beta}$ is the aggregate for the unit of the variable $Y_{\alpha\beta}$, which is the same for every element in the unit. As an element variable, this represents the extreme of complete uniformity and homogeneity within the unit.

The two means, $\bar{Y}_e$ and $\bar{Y}_g$, can differ emphatically when the distribution of the sizes N_α is very skewed, so that a small proportion of all units accounts for a large proportion of all elements, and the variable $\bar{Y}_\alpha$ is strongly correlated with the sizes N_α. The element mean $\bar{Y}_e$ is greater or less than the group mean $\bar{Y}_g$, according to a positive or negative correlation $(R_{N_\alpha \bar{Y}_\alpha})$.

Choice of the proper mean precedes the sample design, since the design depends on which mean is more appropriate. If we had data for the entire population (all the A values of N_α and $\bar{Y}_\alpha$), which mean would we choose? Since the element mean will most often be chosen, we turn to designing samples for it. As in many other situations, recognizing the chief issues is crucial.

The sampling problem resembles the selection of unequal size clusters, but with some special features. First, only a single observation is made on each cluster; replicate measurements of the same variable $\bar{Y}_\alpha$ would be wasted. Second, the homogeneity is extreme, with roh $= 1$, because the variable $\bar{Y}_\alpha$ is the same for all elements in the unit. Third, variations in the unit sizes N_α must be accepted as given, because the units cannot be divided.

Two other assumptions are implicit in the following brief statistical treatment. First, that the variables $\bar{Y}_\alpha$ have similar variances, regardless of unit size. Second, that the unit cost of obtaining $\bar{Y}_\alpha$ is similar for large and small units. Compared to variations in the sizes N_α, these assumptions are often reasonable. When they are not, disproportionate allocation may be introduced.

If a units are selected with equal probability a/A, the simple mean of the sample $\bar{y}_{gs} = \sum_\alpha^a \bar{y}_\alpha/a$ is an unbiased estimate of $\bar{Y}_g$. To estimate the element mean $\bar{Y}_\alpha$ we must introduce the weights n_α and use the ratio mean

$$\bar{y}_{es} = \sum_\alpha^a n_\alpha \bar{y}_\alpha / \sum_\alpha^a n_\alpha. \tag{11.6.3}$$

Its variance merely applies the proper formula for a ratio mean from Chapter 6. If the n_α are grossly unequal, those weights will make this estimate inefficient; a few large selections will tend to dominate the estimate and its variance (see 11.7). In this extreme situation, selection with probabilities proportional to size (PPS) seems particularly appropriate.

If the units are selected with probabilities proportional to the n_α, a good (and unbiased) estimate of the element mean is the simple mean

$$\bar{y}_{ep} = \frac{1}{a} \sum_\alpha^a \bar{y}_\alpha. \tag{11.6.4}$$

This demonstrates the chief advantage of selecting units with PPS for estimating the element mean. The ordinary self-weighting mean of sample observations is simple to compute, and is efficient under the usual conditions described above. If a selections have been drawn with replacement from the entire population, the variance of the sample mean may be estimated simply as

$$\text{var}(\bar{y}_{ep}) = \frac{1}{a(a-1)} \sum_\alpha^a (\bar{y}_\alpha - \bar{y}_{ep})^2. \tag{11.6.5}$$

This can be readily perceived by considering a selections drawn with replacement from a population of N elements. In this population, N_1 elements have the value $\bar{Y}_1$, N_2 have $\bar{Y}_2$, N_α have $\bar{Y}_\alpha$, etc. See also Cochran [1963, p. 254]. To retain a simple presentation, we assume that if a unit is selected two or more times, it remains in the sample and the estimate as often as it was selected.

We typically use stratification in these situations, resorting to formulas appropriate for stratified samples. Suppose that the units have been sorted into H strata, and that a_h units have been selected with PPS from the hth stratum. The weight of the stratum is W_h, which typically denotes N_h/N, the proportion of elements it contains. The element mean is estimated by

$$\bar{y}_{ep} = \sum W_h \bar{y}_h, \qquad \text{where } \bar{y}_h = \frac{1}{a_h} \sum_{\alpha}^{a_h} \bar{y}_{h\alpha}. \tag{11.6.6}$$

Assuming that the a_h selections in the strata were made with replacement, its variance is

$$\text{var}\,(\bar{y}_{ep}) = \sum_h^H \frac{W_h^2}{a_h(a_h - 1)} \sum_{\alpha}^{a_h} (\bar{y}_{h\alpha} - \bar{y}_h)^2. \tag{11.6.7}$$

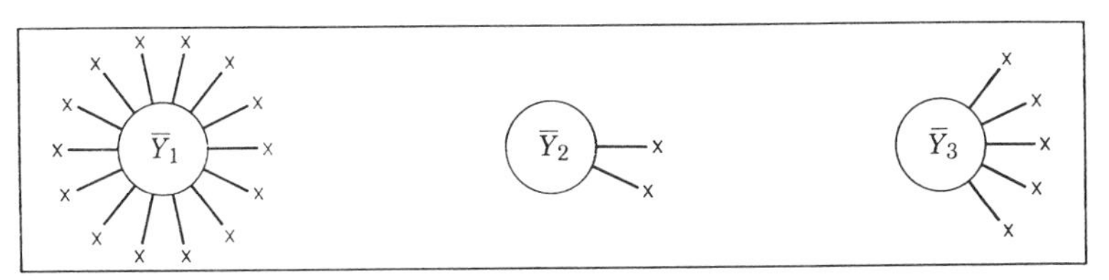

FIGURE 11.6.I Representation of Group ($\bar{Y}_\alpha$) and Individual (x) Variables.

A population of $A = 3$ units is represented with circles. The lines represent the $N = \Sigma N_\alpha$ elements in the population. Each element possesses both unit variables ($\bar{Y}_\alpha$) that are equal for all N_α elements in the unit, and element variables (x) that are distinct for each element.

The sample can be made self-weighting if the number of selections is made proportional to the stratum sizes ($a_h = kW_h$). Then for a total sample size of $a = \sum_h a_h$ we have

$$\bar{y}_{ep} = \frac{1}{a} \sum_h^H \sum_{\alpha}^{a_h} \bar{y}_{h\alpha}, \tag{11.6.6$'$}$$

and

$$\text{var}\,(\bar{y}_{ep}) = \frac{1}{a^2} \sum_h^H \frac{a_h}{a_h - 1} \sum_{\alpha}^{a_h} (\bar{y}_{h\alpha} - \bar{y}_h)^2. \tag{11.6.7$'$}$$

If strata of equal sizes are formed, and paired selections (1 and 2) are drawn from each, the last formulas become

$$\bar{y}_{ep} = \frac{1}{a} \sum_h (\bar{y}_{h1} + \bar{y}_{h2}), \quad \text{and} \quad \text{var}\,(\bar{y}_{ep}) = \frac{1}{a^2} \sum_h (\bar{y}_{h1} - \bar{y}_{h2})^2. \tag{11.6.7$''$}$$

Some large units may be larger than N/a, the designed population size per selection, and some almost as large. These should be taken into the sample with certainty, and will not contribute to the variance. At the other extreme, we may have strata filled with many units, each much smaller than the strata; for these it will matter little whether we select with or without replacement; we may disregard the factor $(1-f)$ in computing the variance. In between, we may have strata with units somewhat, but not much, smaller than the strata, where the factor $(1-f)$ should be considered (7.4).

Suppose that after selecting with probabilities proportional to the size measures n_α, we find different desired "true" sizes n_α' for the selected units. The differences may be due to changes in size, discrepancies in the units of measurement, etc., (7.5*C*). Although the n_α' and n_α should be highly correlated, the differences may not be negligible; for example, we could underrepresent the fast-growing units. Then weight each $\bar{y}_\alpha$ in the sample with $\bar{x}_\alpha = n_\alpha'/n_\alpha$ and, using $\bar{y}_\alpha' = \bar{x}_\alpha \bar{y}_\alpha$, obtain the ratio mean

$$\bar{y}_{ep} = \frac{\sum \bar{y}_\alpha'}{\sum \bar{x}_\alpha}. \tag{11.6.8}$$

For a stratified selection, and with the selections proportional to stratum sizes, the ratio mean becomes

$$\bar{y}_{ep} = \sum_h \sum_\alpha \bar{y}_{h\alpha}' / \sum_h \sum_\alpha \bar{x}_{h\alpha}. \tag{11.6.9}$$

Here $\bar{y}_{h\alpha}' = \bar{y}_{h\alpha}\bar{x}_{h\alpha} = \bar{y}_{h\alpha}n_{h\alpha}'/n_{h\alpha}$, the variable weighted to correct for the change in desired size from the measure of size. The variances of the ratio mean may be found in Sections 6.3 and 6.4.

With proper weighting, we can also obtain an estimate of the group mean $\bar{Y}_g$, when required. If the units were selected with probabilities proportional to n_α, each selected mean should be weighted with $1/n_\alpha$. The ratio mean for a proportionate stratified selection would be

$$\bar{y}_{gp} = \frac{\sum_h \sum_\alpha \bar{y}_{h\alpha}/n_{h\alpha}}{\sum_h \sum_\alpha 1/n_{h\alpha}}, \tag{11.6.10}$$

and its variance is that of a stratified ratio mean (6.4).

Let us recapitulate the estimation of the group mean $\bar{Y}_g$ and element mean $\bar{Y}_e$ from observations made on a sample of group values $\bar{y}_\alpha$. If the selection of the groups is epsem, the simple mean estimates $\bar{Y}_g$; but to estimate $\bar{Y}_e$, the sample values $\bar{y}_\alpha$ need to be weighted by their sizes n_α. On the contrary, if the groups are selected with PPS, the simple mean estimates $\bar{Y}_e$; but to estimate $\bar{Y}_g$, the values $\bar{y}_\alpha$ need to be weighted with $1/n_\alpha$.

Remark 11.6.1 Selecting groups with PPS is particularly well suited to studies in which both individuals and the groups to which they belong are used as units in separate analyses. For example, in a study of a large organization some results concern individual members, and others deal with the groups (units, branches) of the organization. By selecting groups with probability proportional to their size measures, and then subsampling elements within groups with probabilities inversely proportional to the same measure, we obtain an epsem selection of elements with about equal numbers of elements per unit. The simple mean of individual values estimates their population mean $\bar{Y}$; and the simple mean of the sample group values $\bar{y}_\alpha$ estimates the element mean $\bar{Y}_e$ of the group values.

Note, however, that selecting all N_α elements (or any constant fraction) from groups selected with PPS would not yield an epsem selection of individuals nor a self-weighting mean for estimating $\bar{Y}$.

Subsampling with PPS is also generally efficient when the group values are computed as means of the individuals selected from the group. Because the group values are based on about equal numbers of elements, they are subject to about equal variances. This typically enhances the efficiency of statistical analysis of group values.

Joint analysis of group variables $\bar{Y}_\alpha$ and individual variables $X_{\alpha i}$ is possible. Each population element (represented with the lines in Fig. 11.6.I) possesses both kinds of variables. A sample of elements permits their joint analysis.

Remark 11.6.II The proper representation of group values is a primary issue in many areas of social and economic research. Sources of groups values are varied. First, they can represent simply the means of individual values; for example, the mean income or the proportion of home owners as characteristics of cities. Second, they may be group values arising from individuals; consider, for example, the population sizes and densities of cities. Third, the values may belong specifically to the group; for example, the climate, altitude, or age of a city, its form of government, or the presence of museums. Whatever the origin of the group values, we may prefer either the element mean $\bar{Y}_e$ or the group mean $\bar{Y}_g$. The choice between them should depend primarily on which conveys a more meaningful summary value.

Remark 11.6.III Results are often tabulated for domains defined by size classes of units. Researchers tend to define domains roughly equal in numbers of elements, a measure of their importance. The domains of large units contain fewer units than the domains of numerous small units. Under these conditions PPS selection has another advantage over selection with equal probabilities for all units. If the latter is used, the domains of large units, though important in numbers of elements, contain few units and receive very few selections. With PPS selection, the domains receive equal numbers of selections to the degree that they contain equal numbers of elements.

These domains can serve as strata too. If the number of selections in the domains based on size are proportional to their element sizes, and if within these strata variations in size are small, efficiencies will be roughly equal for selecting with either PPS or with equal probabilities within strata.

Remark 11.6.IV Suppose that we have obtained an epsem ($f = n/N$) selection of n directly from a population of N individuals, without using groups in the selection. Now, suppose that the group variables $\bar{y}_\alpha$ have been obtained from the sample individuals, and become the basis for estimating the population mean. For example, for an epsem selection of persons we can obtain characteristics of the family, or county, or university to which they belong. For a group of size N_α, the expected representation in the sample is fN_α. Therefore, the simple mean of the values of $\bar{y}_\alpha$, taken for all the n sample elements, will be an unbiased estimate of the element mean $\bar{Y}_e$. To estimate the group mean $\bar{Y}_g$ we should use the weights $1/n_\alpha$ in the estimate $\bar{y}_g = \Sigma(\bar{y}_\alpha/n_\alpha)/\Sigma(1/n_\alpha)$, summed over the n sample cases. (But it would be a mistake to take simply all the group values that happen to appear in the sample, once or several times, for an unbiased estimate of $\bar{Y}_g$. Why?)

Remark 11.6.V A curious special case of the unit variable $\bar{Y}_\alpha$ occurs when it represents the unit size N_α. Suppose, for example, we were to ascertain for each element in the population the size of his family, or of his city, or of his organization. For the variable N_α the element mean is

$$\bar{Y}_e = \frac{\sum N_\alpha N_\alpha}{\sum N_\alpha} = \frac{\sum N_\alpha^{\,2}/A}{\bar{N}} = \frac{\sigma^2_{N\alpha}}{\bar{N}} + \bar{N} = \bar{N}(C^2_{N\alpha} + 1). \qquad (11.6.10)$$

The variable N_α obtained from an epsem selection of elements will estimate $\bar{Y}_e$. Clearly, when the variance $\sigma^2_{N\alpha}$ of unit sizes is great, this mean, called the "contraharmonic mean" of the unit sizes N_α, is much greater than the simple arithmetic mean $\bar{N}$. In other words, the variable representing the size of the unit to which elements belong has a larger mean than the mean size of the units. Although the mean number of adults per household (11.3.III) is only 2.02, the mean number of household adults comes to 2.24 for the average adult. The large range of social organizations produce more striking effects. For 130 American cities having over 100,000 inhabitants the average size was 390,000; but the size of the city in which the average person lived was 2,000,000.

If we ask people, "How many siblings do you have?" the answer will be $N_\alpha - 1$; and the mean of these is one less than the mean of N_α, or $\bar{N}C^2_{N\alpha} + \bar{N} - 1$.

The variances, can be computed, as for any ratio mean, for the variables $N_\alpha^{\,2}$ and N_α.

11.7 WEIGHTING PROBLEMS

11.7A Aims and Procedures of Weighting

In the simplest self-weighting situation, each element has the weight $1/n$ in the mean, 1 in the sample total, and $F = 1/f$ in the population total, where f is the uniform sampling rate for all population elements. The statistical treatment remains simple if, instead of F, we use any uniform F' for all elements. This can be an attempt to compensate for such imperfections as nonresponse and noncoverage when these are assumed uniform; thus $F' = F(1 + d)$, where d is a fraction computed to compensate for an actual sampling fraction $f/(1 + d)$, different from the designed f. Similarly, a ratio estimator yX/x can introduce the factor $F' = X/x$ instead of F. It may produce benefits and creates no disturbance if its effect is uniform for all sample cases. We are concerned here only with disturbances caused by unequal weighting of different sample elements. Despite the advantages of simplicity that favor self-weighting samples, unequal weights may be introduced into parts of the sample for various reasons:

(*a*) *Disproportionate sampling* may have been introduced deliberately into the sample, either to decrease the variance by optimum allocation among strata, or to produce large enough samples for separate domains. Usually these differences in sampling rates are great (factors of 2 or 5 or even 100 are possible) and are compensated by inverse weights in the statistics.

(*b*) *Inequalities in the selection frame and procedures* may create unequal selection probabilities. These may be serious if they affect a large portion of the sample; then they usually are corrected with weights inverse to those probabilities.

(*c*) *Differences in nonresponse and noncoverage* between parts of the sample can be balanced with unequal weights. These corrections are—or should be—rather small, and can perhaps be justified only in large samples. They require knowledge about the size of the nonresponse within the defined parts and some assumption about randomness; subjective judgment is involved in this effort to compensate for inadequacies of field operations.

(*d*) *Statistical adjustment of estimates may decrease the variance* by introducing unequal weights. For example, post-stratification adjusts the data to known stratum sizes. Ratio estimates may utilize diverse weights for separate parts of the sample. In multiple classification problems, complex (least-square) adjustments of cells can be made to agree with marginal totals. This is also related to adjustments for producing "standardized populations" used by demographers and economists. These are only four examples of the many uses of statistical weighting of data. They usually concern large masses of data, and are not strictly sampling problems, in the sense that they arise in the same form in data from complete censuses. [See Deming, 1943; El-Badry and Stephan, 1955.] *Adjustment of data from a judgment sample* or a convenient source, although sometimes confused with the above, is a distinct topic (13.7).

Before introducing unequal weights we should consider the several factors that it may involve: (1) reduction of some biases; (2) possible introduction of other biases; (3) increase of the variance; (4) complication of computations. Unfortunately, it does not seem feasible to put all this vague advice into a general yet still useful formula. Except for self-weighting samples, we must choose between accepting a potential bias, or increasing variance and the cost of the sample. On the one hand, large, or potentially large, biases should be avoided. But the elimination of a small bias should not be bought at the cost of a greater increase in the variance (13.1). Small differences for accidental and uncontrollable factors may often be neglected. Disproportional allocation for a few strata can be treated with a few divergent and simple weights.

Weighting can be achieved with diverse procedures, each with its own disadvantages. (1) Statistics for each weight class can be computed separately with equal weights, then *weighted as they are combined* into the overall statistics. This is economical if there are few weight classes and few statistics to compute. (2) Separate weights, punched on data cards or tape, can be *applied to each element* in each computation. This procedure is becoming increasingly feasible with modern machines, especially electronic computers.

For small scale computations, a self-weighting deck of cards is convenient; it can be produced with either random duplication or elimination of elements. It can be produced easily if there are only a few classes of weights, but it causes an increase in the variance, as shown in the next subsection. (3) *Duplicating sample elements* can produce a self-weighting sample. The maximum increase in variance from this procedure is 12.5 percent, and usually less than that. If used for large differences between weights, it creates an unwieldy, large deck. Furthermore, it introduces complications into variance computations, unless the sample was already complex. (4) *Eliminating sample elements* can effect a self-weighting sample, in which each element represents a different element. This can simplify variance computations. On the other hand, it increases the variance at the rate $e/(1 - e)$, where e is the proportion eliminated; that increase can be too costly if e is not small. (5) *A combination of elimination and duplication* may increase the variance less than duplication alone. A small proportion of the cases are eliminated from strata with small weights, and a larger portion are duplicated in the remaining strata, which have larger weights.

These procedures can be illustrated with a case, where weights are used to correct for different response rates r_h in several strata of a sample, which was initially self-weighting. The weights yield an adjusted sample total, which leads to estimates of either the mean or the aggregate in the population. (1) The factors r_h may be applied to y_h, the stratum totals in the sample, to produce the weighted statistic $\Sigma\, y_h/r_h$. (2) Each element y_{hj} may be weighted separately with r_h to yield $\sum_h (\sum_j y_{hj}/r_h)$. (3) We can increase the actual response r_h in each stratum to a synthetic response at a fixed constant level K, by duplicating the proportion $(K - r_h)/r_h$ of the elements in the stratum, because $[1 + (K - r_h)/r_h] = K/r_h$. The K could be 1, the theoretical top, but need be only the highest response r_{max} among the strata. (4) We can decrease the actual r_h to the fixed K by eliminating the proportion $(r_h - K)/r_h$, because $[1 - (r_h - K)/r_h] = K/r_h$. Here K would be the lowest response r_{min} among the strata. (5) With a proper choice of K, closer to the top response than the bottom response, we can combine duplication and elimination, and do better than with either alone. For one survey involving adjustments for response rates, ranging from 100 to 60 percent in subclasses, duplication plus elimination increased the variance only 6 percent, whereas duplication alone would have increased it by 9 percent.

It typically does not pay to use many digits for the separate weighting of elements. Two-digit weights are good enough, if the ratio of largest to smallest weight is not over 10. The weights run from 10 to 99; the error introduced by rounding is at most $1/20^2 = 1/4$ percent for the lowest and $1/200^2 = 1/400$ percent for the highest weight. Thus the increase in the variance, compared to more precise weights, will be safely under 1/4 percent. [Tukey, 1948; Murthy and Sethi, 1961.]

11.7B Losses from Random Duplication and Elimination

Suppose that we take an unrestricted random sample of n elements from a population and then give different weights to randomly selected subsets of elements, so that $n_h = nW_h$ have weights k_h. The variance of the weighted mean of the $n = \Sigma n_h$ elements will be

$$\operatorname{Var}\left[\frac{\sum k_h y_h}{\sum k_h n_h}\right] = \frac{\sum n_h k_h^2 S^2}{(\sum k_h n_h)^2} = \frac{nS^2 \sum W_h k_h^2}{n^2(\sum W_h k_h)^2} = \frac{S^2}{n} \cdot \frac{\sum W_h k_h^2}{(\sum W_h k_h)^2}. \quad (11.7.1)$$

When each element has equal weight, the variance is S^2/n; the factor following it represents the effect of the unequal weighting of randomly selected subsets. The larger the discrepancy between the weights, the greater this factor. Therefore, weighting with random replication of elements should be confined to two successive integers, with k applied to $(1 - W)$, and $(k + 1)$ to W of the elements, the sum of which is $(1 - W)k + W(k + 1) = k + W$. Its variance is

$$\frac{S^2}{n} \cdot \frac{(1 - W)k^2 + W(k + 1)^2}{(k + W)^2} = \frac{S^2}{n}\left[\frac{k^2 + 2kW + W}{k^2 + 2kW + W^2}\right]$$
$$= \frac{S^2}{n}\left[1 + \frac{W - W^2}{(k + W)^2}\right] = \frac{S^2}{n}\left[1 + \frac{W(1 - W)}{(k + W)^2}\right]. \quad (11.7.2)$$

Thus the unit variance increases by $W(1 - W)/(k + W)^2$, *when the portion* W *receives* $(k + 1)$ *replications* and the portion $(1 - W)$ receives k replications. For any fixed k, we find by differentiating that the maximum possible increase amounts to $1/4k(1 + k)$; this occurs when $W = k/(1 + 2k)$. When some cards are duplicated and others are not, $k = 1$ and the loss is $W(1 - W)/(1 + W)^2$; the maximum loss is 0.125, occurring when $W = 1/3$. The losses decrease rapidly for larger k:

k	1	2	3	5
Maximum increase	1/8	1/24	1/48	1/120
Occurs at W	1/3	2/5	3/7	5/11

Suppose that weighting with k_h must be done separately in each of H strata. Suppose the smallest among the k_h equals K, and we decide to reduce the k_h of the other strata to it by duplicating the portion $W_h = (k_h/K - 1)$. The result will be a self-weighting sample; the total increase in variance will be the sum of the increases in the variances of the separate stratum means. Noting that $(1 - W_h) = (2 - k_h/K)$ and $(1 + W_h) = k_h/K$, we see that the increase of the unit variance in each stratum can be represented by $W_h(1 - W_h)/(1 + W_h)^2 = (k_h - K)(2K - k_h)/k_h^2$. This is based on the assumption that only duplications are involved, and this is true when $1 \leq k_h/K \leq 2$. If some of the $k_h/K > 2$, then triplications, etc. are also involved. The portion $W_h = (k_h/K - c_h)$ is given $(c_h + 1)$ replications, and the portion $(1 - W_h)$ is replicated c_h times. The unit variance in each stratum is increased by $(k_h - Kc_h)(K + Kc_h - k_h)/k_h^2$.

Random elimination of the proportion e from an unrestricted random sample of n increases the variance from S^2/n to $S^2/n(1 - e)$; hence, the increase in the variance due to the elimination is

$$\left(\frac{S^2}{n(1 - e)} - \frac{S^2}{n}\right) \div \frac{S^2}{n} = \frac{1}{1 - e} - 1 = \frac{e}{1 - e}. \tag{11.7.3}$$

This increase, due to discarding a portion of cases, is greater than the increase due to duplication. The maximum for the latter occurs when $W = 1/3$, and it amounts to 0.125; but eliminating $e = 1/3$ would increase the unit variance by $(1/3)/(2/3) = 0.500$, and eliminating $e = 2/3$ would increase it by 2.00. Hence, elimination should be used, if at all, only for small portions of the sample. When $e = 0.05$, the increase is 0.0526 for elimination; the increase for duplicating $W = 0.95$ is still less, 0.0125.

Suppose that there are several strata to be adjusted with factors k_h, among which K *is the largest*, so that the estimate of the sample mean can be represented as $\Sigma W_{h'}\bar{y}_h = K\Sigma W_h\bar{y}_h(k_h/K)$. The weighting can be accomplished by eliminating the proportions $e_h = (1 - k_h/K)$, so that the retained portions, $(1 - e_h) = k_h/K$, will accomplish the desired weights. The increase of the unit variance in each stratum is

$$\frac{e_h}{1 - e_h} = \frac{1 - k_h/K}{k_h/K} = \frac{K - k_h}{k_h}. \tag{11.7.4}$$

11.7C Losses from Oversampling Strata

If the sampling fractions of a stratified sample diverge from optimum allocation, the variance is increased. We express the increase in variance as a function of the divergence of sampling rates. We emphasize the common situation when optimum allocation would be proportionate: How great is the increase in the variance due to unequal sampling rates? It can be large if the rates vary considerably; for example, if in two halves the rates are in the proportion 1:4, the increase in variance is 1.56. For an unrestricted random sample the increase is similar.

The weights k_h in a stratified sample are inversely proportional to the sampling fractions $f_h = f/k_h$; hence $n_h = f_h N_h = fN_h/k_h$. The stratum sizes are assumed proportional to the numbers of elements, hence $W_h = N_h/N$ and $n_h = mW_h/k_h$, where $m = fN = \Sigma fN_h = \Sigma n_h k_h$ denotes the properly weighted total sample size. The number of elements in the sample is $n = \Sigma n_h = m\Sigma W_h/k_h$.

The element variance for the mean of a stratified sample may be denoted

$$\begin{aligned} nV^2 = n \operatorname{Var}(\sum W_h\bar{y}_h) &= n\sum W_h^2\frac{(1 - f_h)}{n_h} S_h^2 \\ &= \frac{n}{m}\sum W_h k_h S_h^2\left[1 - \frac{f}{k_h}\right] \\ &= \left[\sum W_h k_h S_h^2\left(1 - \frac{f}{k_h}\right)\right][\sum W_h/k_h]. \end{aligned} \tag{11.7.5}$$

Instead of fixing n, the variances may be compared for a fixed cost $C = \Sigma J_h n_h = m\Sigma J_h W_h/k_h$. Therefore, the variance obtained for a unit cost C/m is

$$V^2C = \left[\sum W_h k_h S_h^2\left(1 - \frac{f}{k_h}\right)\right]\left(\sum\frac{J_h W_h}{k_h}\right). \tag{11.7.5'}$$

This form is useful for investigating the efficiency of various allocations, including the proportionate: $(1 - f)(\Sigma W_h S_h^2)(\Sigma J_h n_h)$. But now we want to investigate the results of different allocations when the S_h^2 and J_h are equal. This case is approximated in many situations where the cost J_h

per element varies little between the strata, and where the S_h^2 refers to a multitude of different items, many of them percentages. If the sampling fractions f_h are small, we can also disregard the $(1 - f/k_h)$ and have

$$nV^2 = \sum W_h k_h \left(\sum \frac{W_h}{k_h}\right). \tag{11.7.6}$$

This compares the actual variance to the variance of proportionate sampling, when this is optimum (S_h^2 and J_h constant).

The same result is obtained if different sampling rates are applied to parts of an unrestricted random sample. The initial size is $m = \Sigma\, n_h k_h = Nf$, then $n_h = mW_h/k_h$ are selected in the hth part, altogether $n = \Sigma\, n_h = m\Sigma\, W_h/k_h$. If the mean is computed as $\Sigma\, k_h y_h/\Sigma\, k_h$, the variance can be denoted as $(\Sigma\, n_h k_h^2 S^2)/(\Sigma\, n_h k_h)^2$; the ratio of this to S^2/n is

$$\frac{(\sum n_h)(\sum n_h k_h^2)}{(\sum n_h k_h)^2} = \left(\sum \frac{W_h}{k_h}\right)\sum W_h k_h. \tag{11.7.6'}$$

Expanding the above, we obtain

$$\begin{aligned} nV^2 &= \sum_h^H W_h^2 + \sum_{h<i}^H \frac{W_h W_i}{k_h k_i}(k_h^2 + k_i^2) \\ &= \left(\sum_h^H W_h^2 + 2\sum_{h<i}^H W_h W_i\right) + \sum_{h<i}^H \frac{W_h W_i}{k_h k_i}(k_h^2 + k_i^2 - 2k_h k_i) \\ &= 1 + \sum_{h<i}^H W_h W_i \frac{(k_h - k_i)^2}{k_h k_i}. \end{aligned} \tag{11.7.7}$$

The second term expresses the relative increase in variance due to departures from proportionate sampling. For any fixed spread from the lowest to highest weights (from $k_{\min} = 1$ to $k_{\max} = K$), and for any fixed weight $W_{\min}$ for the lowest group, the maximum loss is attained when all of the remaining weight is subject to K. *Thus, for any fixed spread among the weights, the greatest loss is incurred when there are only two strata.* Hence, the case of two strata ($H = 2$) is interesting because it is an extreme as well as a common case. The ratio of the variance when $H = 2$, $k_1 = 1$, and $k_2 = K$ is

$$nV^2 = 1 + W(1 - W)\frac{(K-1)^2}{K}. \tag{11.7.8}$$

The increase is the same whether the larger or smaller stratum has the greater weight, and reaches its maximum when both strata are equal. For large K it approaches and is a little greater than $W(1 - W)(K - 2)$.

For the general case of many strata we must confine ourselves to a few helpful illustrations. For comparability we suppose that the range of weights is from 1 to H, the same as the number of strata. We use (11.7.6)

W \ K	1.5	2	3	4	5	10	20	50
0.5	1.042	1.125	1.333	1.562	1.800	3.025	5.512	13.005
0.2	1.027	1.080	1.213	1.360	1.512	2.296	3.888	8.683
0.1	1.015	1.045	1.120	1.202	1.288	1.729	2.624	5.322

TABLE 11.7.I Effect of Two Divergent Selection Rates.

The sampling rates and the weights are in the proportion $1:K$ in two strata, which have the sizes W and $(1 - W)$. The variance of oversampling is presented relative to proportionate sampling, when the latter is optimum (11.7.8). Note the large increases for large weight differences, especially when W is near 0.5.

to find the ratio of actual variance to proportionate variance, assuming that the latter is optimum.

1. For a rectangular distribution of stratum size, that is, for *uniform size strata*, where $W_h = 1/H$, *with the* k_h *increasing linearly from* 1 *to* H, the variance is

$$\sum W_h k_h \left(\sum \frac{W_h}{k_h}\right) = \frac{1}{H^2}\left(\sum h\right)\left(\sum \frac{1}{h}\right) = \frac{H(H+1)}{2H^2}\left(\sum \frac{1}{h}\right)$$

$$= \frac{H+1}{2H}\left(\sum \frac{1}{h}\right). \tag{11.7.9}$$

2. *If the strata are of uniform size,* $W_h = 1/H$, *and the* $k_h = 1$ *for all but one stratum in which* $k_h = H$, the variance is

$$\sum W_h k_h \left(\sum \frac{W_h}{k_h}\right) = \frac{1}{H^2}\left(\sum k_h\right)\left(\sum \frac{1}{k_h}\right)$$

$$= \frac{1}{H^2}(H - 1 + H)\left(H - 1 + \frac{1}{H}\right)$$

$$= 1 + \frac{(H-1)^3}{H^3}. \tag{11.7.10}$$

3. *For a triangular distribution of stratum size* we must have $W_h = h/\Sigma h = h/[H(H+1)/2]$, because $\Sigma W_h = 1$. Note that we get the same results, whether the weights are $k_h = h$, increasing linearly with stratum size from 1 to H, or $k_h = H/h$, decreasing linearly with stratum size; in either case we get

$$\sum W_h k_h \left(\sum \frac{W_h}{k_h}\right) = \frac{(\sum 1)(\sum h^2)}{[H(H+1)/2]^2} = \frac{H[H(H+1)(2H+1)/6]}{H^2(H+1)^2/4}$$

$$= 1 + \frac{H-1}{3(H+1)}. \tag{11.7.11}$$

H	3	4	5	10	20	50
$(\Sigma\, 1/h)(H+1)/2H$	1.222	1.302	1.370	1.611	1.808	2.295
$1+(H-1)^3/H^3$	1.296	1.422	1.512	1.729	1.857	1.941
$1+(H-1)/3(H+1)$	1.167	1.200	1.222	1.273	1.301	1.320

TABLE 11.7.II Effect of Unequal Rates in Three Models.

The variances are compared, using formulas (11.7.9-11), to proportionate sampling, when the latter is optimum. Three distributions of stratum sizes and weights are presented. In each case the range of weights, 1 to H, equals the number of strata. Note that for large ranges the variances are not increased nearly as much as in the extreme case of two strata.

Remark 11.7.I The finite population correction is frequently negligible because the divergence of $(1-f)$ from 1 is trivial compared to other sources of error. In social surveys the sampling fractions may range, say, from 1/50 to 1/50,000 and it would be pedantic to worry about $(1-f)$. When the sampling fraction is large, the researcher may regard the population itself as a sample from an infinite population; then using $(1-f)$ would be inappropriate (2.3 and Problem 2.13).

When $(1-f)$ is not negligible, it can often be easily included in the computations if the selection is epsem either in the entire sample or within strata. The variance may be denoted briefly and generally as $(1-f)V^2$ or $\Sigma\,(1-f_h)V_h^2$.

However, there are rare situations in which the finite population correction is neither negligible nor simple, when it could interfere with an otherwise simple computation of the variance. (1) Suppose that the primary sampling units are counties; then farms are selected within them with disproportionate sampling rates, the large farms receiving high rates. Or dwellings can be selected with unequal rates from a primary selection of blocks. The unequal selection rates will be compensated with unequal weights. The simple variance formulas based on primary selections (6.5*B*) can be used, except for the irregularities in the factor $(1-f')$. (2) Suppose that replicated sampling is used for an inventory, and that within each replication disproportionate rates are used for selecting items of contrasting types. (See part *c* of Problem 4.6.) The simple variance formulas of replicated samples (4.4) need to accomodate the factor $(1-f')$. (3) Suppose that selection is made from a list of elements; that duplicate listing of elements is common; and that accepting and weighting for unequal selection probabilities is used (2.7 and 11.2*C*). The variance formula may be essentially simple except for the complications in the factor $(1-f')$.

In such cases, one may look for reasonable approximations of an average value f' in order to substitute for the separate values f_g that actually exist within the sample. Denoting with V_g^2 the several components of the total variance, we may consider that we need a value of f' such that $(1-f')\Sigma\, V_g^2 = \Sigma\,(1-f_g)V_g^2$. If this model is reasonable, then $f' = \Sigma f_g V_g^2/\Sigma\, V_g^2$; hence, f' is a weighted mean of the f_g values, where the weights are the respective variances V_g^2.

We need only approximate weights, because even moderate errors in the weights will ordinarily have small effects on the value of $(1-f')$. Instead of laborious computations of precise estimates of their separate values, we may utilize reasonable guesses about the relative values of the V_g^2. For similar reasons, we may often be able to ignore the separate effects of covariance terms V_{gh}. Their effects on the factor $(1-f')$ are likely to be proportional to $\sqrt{(1-f_g)(1-f_h)}V_{gh}$; hence, their effects are near the average of those of V_g^2 and V_h^2.

These demonstrations are heuristic rather than rigorous. Yet I believe the methods will provide reasonable approximations of these factors when they are needed—which is seldom.

11.8 ESTIMATING TOTALS

11.8A The Advantages of $N\bar{y}$ over Fy

Here we deal with special problems involved in estimating aggregates, having elsewhere emphasized the mean. If the population size N is known, the mean $\bar{y}$ leads directly to the estimate $N\bar{y}$ of the aggregate Y and to its standard error $N\text{se}(\bar{y})$. Problems arise when N is unknown.

The sampling fraction f is known for probability samples, and $F = 1/f$ can be used to estimate aggregates as Fy and its standard error as $F\text{se}(y)$. This estimate is equivalent to $N\bar{y}$ when F is equivalent to N/n, hence generally for designs with fixed sample size n; this holds for simple random and stratified random element sampling, and for samples and subsamples of equal clusters. But in many designs f is fixed without our either knowing N or fixing n; then the two estimates, $N\bar{y}$ and Fy, can differ drastically. Sampling of unequal clusters, involving the ratio mean, are of this kind. Then the sample size n is a random variable, and questions arise about estimating the sample total $Y = N\bar{Y}$. Even when the size of the entire sample is fixed, the sizes of subclasses become random variables. This section deals with subclasses of srs; subclasses in stratified random samples were handled in 4.5; subclasses of equal clusters become unequal clusters, and the theory for unequal clusters handles subclasses readily.

Thus, when n is random, though Fy can provide an unbiased estimate of the population total Y, we usually prefer $N\bar{y}$ because of its lower variance. (In Example 6.5*a* the variance of Fy is 11.7 times greater.) If we cannot obtain N without error, perhaps we can obtain a good estimate $\tilde{N}$ from an outside source or from double sampling. ($\tilde{N} = Fn$ from the same sample can yield no extra gain because it merely restates $Fny/n = Fy$.) Sometimes, instead of N, another auxiliary variable X can be used with ratio or regression estimation. If neither N nor any suitable expansion factor can be found, Fy may need to be used.

The estimate $N\bar{y}$ is also more useful than Fy for dealing with the bias from nonobservation (noncoverage and nonresponse). The bias in $N\bar{y}$ is proportional to the *difference* between the observed and the target population means. Contrariwise, the estimate Fy has the negative bias of the nonobserved total (see 13.4B). Generally, $N\bar{y}$ is less dependent than Fy on knowing f exactly; this can be of practical significance when control over f is not complete. For example, suppose that a crew of enumerators is sampling one in F of every vehicle on a road for 40 minutes per hour, and then takes a 20 minute rest; a mechanical count of all vehicles on that road during the hour would guard against both fluctuations in traffic volume and variations from the exactly 40 minutes of work. The lack of such variations is a less reasonable assumption than the one needed when using a mechanical count of total traffic: that the *average* characteristic of the traffic did not change much. Thus the simple expansion Fy depends entirely on the sample total y, and is subject both to variable errors and biases. When these are present, and if N is available, $N\bar{y}$ is often a better estimate. Sometimes y is corrected, for example, by adjusting it for nonresponse; in that case the two estimates may be equivalent.

11.8B Totals from Subclasses of SRS

Suppose that, having drawn a simple random sample of n from a population of N, we use a specified subclass m of the sample to infer the mean or total of some characteristic of the subclass M of the population. The subclass contains the proportion $\bar{M} = M/N$ of the population and the proportion $\bar{m} = m/n$ of the sample, and this is a random variable. The subclass may be created for survey analysis, or it may result from the presence of foreign elements in the frame. The subclass mean $\bar{Y} = Y/M$ is estimated by the sample mean $\bar{y} = y/m$ with variance $(1\text{-}f)s_y^2/m$, because the analysis is *conditional* on obtaining a sample of size m. The total $Y = M\bar{Y}$ can also be estimated easily as $M\bar{y} = (M/m)y$ if the subclass size M in the population is known. The variance is $\text{var}\,(M\bar{y}) = M^2\,\text{var}\,(\bar{y}) = (1\text{-}f)M^2s_y^2/m = (1\text{-}f)(M/m)^2ms_y^2$. Frequently M is not known and the estimate must be

$$\tilde{Y} = \frac{N}{n}\sum y_i = \frac{N}{n}\,y = N\bar{m}\bar{y}. \qquad (11.8.1)$$

The sample total $y = \Sigma\, y_i$ can be viewed as the total for the variable y_i, which takes the value $y_i = x_i$ for the m members of the subclass, and the value $y_i = 0$ for the $(n - m)$ nonmembers. This is merely the special case for one stratum of the stratified development expressed in formulas (4.5.7) and (4.5.8). The variance can then be computed for the sum of n variables:

$$\text{var}(\tilde{Y}) = (1\text{-}f)\frac{N^2}{n}s^2, \qquad \text{where } s^2 = \frac{1}{n-1}\left(\sum y_i^2 - \frac{y^2}{n}\right). \quad (11.8.2)$$

Note that in $\Sigma\, y_i^2$, as in $\Sigma\, y_i$, the $(n - m)$ zero values are absent. But an equivalent formula gives more insight into the nature of the variation:

$$\text{var}(\tilde{Y}) = (1\text{-}f)\frac{N^2}{n-1}\bar{m}[v_y^2 + (1-\bar{m})\bar{y}^2],$$

where

$$v_y^2 = \frac{1}{m}\left(\sum y_i^2 - \frac{y^2}{m}\right). \quad (11.8.2')$$

This follows from (11.8.2) because

$$(n-1)s^2 = \sum y_i^2 - \frac{y^2}{n} = \left(\sum y_i^2 - \frac{y^2}{m}\right) + \left(\frac{y^2}{m} - \frac{y^2}{n}\right)$$
$$= n\bar{m}[v_y^2 + (1-\bar{m})\bar{y}^2].$$

Thus, when $Fy = (N/n)y$ must be used instead of $(M/m)y$, the unit variance is increased from s_y^2 to $[v_y^2 + (1 - \bar{m})\bar{y}^2]$. This uses the approximation that $N^2/n(n - 1) \doteq (M/m)^2$. Similarly, in population formulas the unit variance is increased from S_y^2 to $[\sigma_y^2 + (1 - \bar{M})\bar{Y}^2]$.

Because the number m of subclass members found in the sample is a random variable, the unit variance of Fy is increased by about $(1 - \bar{m})\bar{y}^2$. In most cases the mean $\bar{y}$ is safely positive, and the variance (11.8.2) may be expressed in terms of the relvariances to a good approximation $[(n - 1) \doteq n]$:

$$\text{var}(\tilde{Y}) = \tilde{Y}^2(1\text{-}f)\frac{c_y^2 + (1-\bar{m})}{m} = \tilde{Y}^2[cv^2(\bar{y}) + cv^2(\bar{m})]. \quad (11.8.3)$$

Here $\qquad cv^2(\bar{y}) = (1\text{-}f)(v_y^2/\bar{y}^2)/m = (1\text{-}f)c_y^2/m,$

and $\qquad cv^2(\bar{m}) = (1\text{-}f)\bar{m}(1 - \bar{m})/\bar{m}^2 n = (1\text{-}f)(1 - \bar{m})/m.$

Note that the unit relvariance is increased by $(1 - \bar{m})$, which approaches unity for small subclasses. The relvariance of $\tilde{Y} = Fy$ is also seen as the sum of relvariances for $\bar{m}$ and $\bar{y}$, whose product is the basis for the estimate. A numerical example is given by Hansen, Hurwitz, and Madow [1953, 4.15].

The above also sheds light on a related problem. Suppose that the population mean of a variable is $\bar{X} = X/N$; that $X_i > 0$ for M elements, and for the other $(N - M)$ elements $X_i = 0$. How much can one reduce the element variance by confining the sample to the positive portion ($\bar{M}$) of the population? In other words, we want to compare the variances for the first estimate $M\left(\sum_i^{m'} y_i/m'\right)$ and the second estimate $N\left(\sum_i^{n} x_i/n\right)$; for this comparison, the m' in the first design is increased to be as large as n would be in the second design. The element variances are σ_y^2 and σ_x^2; for the same number of elements, the variances of the two designs will be in the ratio $M^2\sigma_y^2/N^2\sigma_x^2 = \bar{M}^2\sigma_y^2/\sigma_x^2$.

We have seen above that $\sigma_x^2 = \bar{M}[\sigma_y^2 + (1 - \bar{M})\bar{Y}^2]$. Hence $\sigma_y^2 = \sigma_x^2/\bar{M} - (1 - \bar{M})\bar{Y}^2$. Thus, using $\bar{X} = X/N = Y/N = \bar{Y}\bar{M}$ and $C_x^2 = \sigma_x^2/\bar{X}^2$, we have for the ratio of the variances for the two designs:

$$\frac{M^2\sigma_y^2}{N^2\sigma_x^2} = \bar{M} - \frac{(1 - \bar{M})\bar{X}^2}{\sigma_x^2} = 1 - \left[(1 - \bar{M}) + \frac{(1 - \bar{M})}{C_x^2}\right]$$

$$= 1 - \frac{(1 - \bar{M})(C_x^2 + 1)}{C_x^2}. \qquad (11.8.4)$$

Hence $(1 - \bar{M})(C_x^2 + 1)/C_x^2$ denotes the decrease in the unit variance obtainable by confining the survey to the positive ("active") portion of the population. We should balance this gain against the cost of the screening operation necessary to eliminate the zero cases from the survey.

11.8C Product Estimate $\tilde{N}\bar{y}$ of Aggregates

We may treat any estimated aggregate as the product $\tilde{N}\bar{y}$, consisting of the estimates $\tilde{N}$ for the population size and $\bar{y}$ for the sample mean. For any product of two random variables, the variance is approximately (6.6.17)

$$\text{var}(\tilde{N}\bar{y}) = \tilde{N}^2 \text{var}(\bar{y}) + \bar{y}^2 \text{var}(\tilde{N}) + 2\tilde{N}\bar{y} \text{cov}(\tilde{N}, \bar{y}). \qquad (11.8.5)$$

When N is a constant known without error, the variance is expressed by the first term alone; the other two vanish. At the other extreme, if $\tilde{N} = Fn$ is obtained merely from the same sample that yields $\bar{y} = y/n$, the variance of the aggregate is $F^2 \text{var}(y)$, where $F = 1/f$ is the inverse of the fixed sampling fraction.

When $\tilde{N}$ is obtained from an outside source, from a sample independent of $\bar{y}$, the covariance term vanishes, and

$$\text{var}(\tilde{N}\bar{y}) = \tilde{N}^2 \text{var}(\bar{y}) + \bar{y}^2 \text{var}(\tilde{N}). \qquad (11.8.6)$$

If we denote by d the ratio of coefficients of variation of $\tilde{N}$ to $\bar{y}$, so that $\text{var}(\tilde{N})/\tilde{N}^2 = d^2 \text{var}(\bar{y})/\bar{y}^2$, we have

$$\text{var}(\tilde{N}\bar{y}) = \tilde{N}^2 \text{var}(\bar{y})[1 + d^2]. \qquad (11.8.6')$$

The contribution of the variance of $\tilde{N}$ to the variance of $\tilde{N}\bar{y}$ is proportional to the ratio d^2 of the relvariances of $\tilde{N}$ to $\bar{y}$. If N is a known constant, var $(\tilde{N}) = d = 0$, and the variance reduces to N^2 var $(\bar{y})$. If $\tilde{N}$ is a variable based on a large sample much more precise than $\bar{y}$ (say, $d = 0.1$), we can treat it as a constant without error.

The expression (11.8.6) may be a good approximation even when $\tilde{N}$ and $\bar{y}$ come from related samples, because the correlation between the sample mean and the sample size is often small. If $\tilde{N} = Fn$ because it comes from the same sample as $\bar{y} = y/n$, it is a constant if n is fixed; this is true in some designs either exactly or approximately. But in other designs n is not fixed, and the variance of $\tilde{N}$ and d^2 may be large. This can happen when sampling unequal sized clusters; or when the $\tilde{N}$ refers to the estimated size of a small subclass of the entire sample, elsewhere denoted as $\tilde{M}$. Then we should search for sources of known N, or for better estimates than the sample itself can provide. If we can get estimates of $\tilde{N}$ with a coefficient of variation 0.2 as large as that of $\bar{y}$, so that $d = 0.2$, then its effect on the variance is confined within $d^2 = 0.04$. If good estimates of $\tilde{N}$ are not available from outside sources, one may resort to *double sampling* with ratio or regression estimates (12.2); bearing in mind that increasing the size of the preliminary sample will decrease d^2 proportionately.

The estimate $\tilde{M}(y/m)$ of aggregates for small subclasses can be greatly improved by obtaining precise estimates for $\tilde{M}$, because the ratio d^2 of the relvariance of $\tilde{M}$ to the relvariance of (y/m) may be great. Therefore, we should search for a good estimate of the subclass size $\tilde{M}$.

If an estimate $\tilde{M}$ of the subclass size cannot be found, $\tilde{N}(y/n)$ may have to be used where $\tilde{N}$ represents the entire population, or a large subclass. However, the relvariance for $\tilde{N}$ may be much smaller than for (y/n); and if d^2 is small, eliminating the variance of $\tilde{N}$ does little good. Therefore, introducing a precise value for $\tilde{N}$ has negligible benefit for small subclasses with large relvariances. For example, in the estimate $\tilde{N}\bar{y} = \tilde{N}(y/n) = \tilde{N}p$ for a binomial variate p, the variance formula (11.8.6) becomes

$$\text{var}\,(\tilde{N}p) = \tilde{N}^2p^2\left[\left(\frac{1-p}{np}\right) + \frac{\text{var}\,(\tilde{N})}{\tilde{N}^2}\right]. \tag{11.8.6''}$$

When p is close to 1, the first term may be small compared to the second; if we eliminate this second term, we reduce the total variance considerably. But if p is close to 0, the first term may be large compared to the second, and eliminating the latter yields only a slight gain. For example, in a traffic survey y may refer to a small subclass denoting a specified Origin-Destination pair for vehicles, of which we need the total $\tilde{N}\bar{y}$. Ascertaining the constant N by means of automatic traffic counters can eliminate the second term of the variance, but this will decrease the variance of this estimate only slightly. Most of the variation comes from happening to find in the sample a variable number y, and a variable small proportion p, of the specified kind of vehicles. (Nevertheless, introducing the correct N from an automatic traffic counter can eliminate a possible bias arising from an incorrect value of F, due to field errors in its application.)

The expansion of the sample total y to an estimate $\tilde{Y}$ of the population total can utilize an auxiliary variable X_i, different from the element count. Of the three variables involved in $\tilde{X}y/x$, the total $\tilde{X}$ is the base for inference to the target population, and we want assurance that it does refer to that. Since x is the link between y and $\tilde{X}$, it must be attuned to both; it must refer to the same characteristic and the same population as $\tilde{X}$, both in extent and in content; this correspondence should be operational, not only intentional. For example, if y/x is a mean per family or per dwelling, the definition of family or dwelling for $\tilde{X}$, the outside (Census) total for families or dwellings, should be the same as for x. A large difference in the meaning of x and $\tilde{X}$ can result in a large bias in $\tilde{X}y/x$, that more than wipes out the gain in the standard error.

The gain in the relvariance of the ratio total Xy/x over the simple total Fy has been seen in the variance of the ratio total (6.5C):

$$\text{var}\left(\frac{y}{x}X\right) = \frac{X^2}{x^2}\left[\text{var}(y) + \left(\frac{y}{x}\right)^2 \text{var}(x) - 2\left(\frac{y}{x}\right)\text{cov}(y, x)\right].$$

The gain is the sum of the second and third terms in the brackets, compared to the first. In Example 6.3a, these were $279 + 231 - 492 = 18$, so that the ratio variance is estimated only $18/279 = 0.065$ as large as the simple mean's. The first term of (11.8.5) is the variance of the ratio estimate; if the total $\tilde{X}$ is a variate, its contribution $(y/x)^2\,\text{var}(\tilde{X})$ should also be considered; also the covariance, if present and not negligible.

PROBLEMS

11.1. A researcher wants to obtain an epsem sample of the families in a city who have one or more children in grammar or high school. He obtains a list of schools, and an alphabetical list of the names and addresses of all children in each school. He now proposes to take an epsem sample of all names on the combined school lists, then visit at each of the selected addresses and interview the mother. (*a*) Evaluate briefly this sampling plan. (*b*) Suggest briefly one or two alternative ways of obtaining the desired sample from the school lists. (*c*) Suggest also a method for selecting children without the use of school lists.

11.2. To select an epsem sample of the farms of a county, a researcher constructed a grid with equidistant points. He put this grid down on the map after selecting a random start in two directions; thus every point on a detailed map of the county had the same probability of being selected. Now he wants to give the following instruction to the enumerator: "At each sample point shown on your map, take the three farms which are located nearest to the point."

× × ×

× × ×

(*a*) Evaluate briefly this sampling plan.

(*b*) Suggest briefly an alternative sampling procedure that would utilize the same selection of sample points.

(*c*) Suggest briefly an alternative sample design not using the grid.

11.3. In 1964 a researcher took an epsem sample of the private dwellings in a city. The respondent was "any responsible adult" found at home. He compared the sample of his respondents with 1960 Census data on adults in the city. Give four basic reasons why this comparison makes no sense.

11.4. A company making expensive watches wants to take a sample of its potential customers. A sample of telephone directory listings is proposed, with equal probabilities for all listings in the United States, clustered in sampling units. What problems do you see in the differences between the frame and the target population?

11.5. From the list of all $N = 100{,}000$ school children (all 13 grades) in a city, a sample of $n = 1000$ children was selected with equal probability. In interviews with their mothers, one of the questions was: "Last year did your family take a family vacation with all your children?" The proportion of yes answers was strongly correlated with the number of school children in the family.

Number of school children in family	1	2	3	4
Number of cases in sample	210	280	270	240
Proportion of "Yes" answers	0.80	0.60	0.40	0.20

In this imaginary situation no family has more than four children in school. Estimate: (*a*) the proportion of families with school children who had a family vacation; and (*b*) the proportion of school children who went on a family vacation. (*c*) Is $pq/1000$ the proper estimate of the variance for (*a*) or for (*b*)? Why?

12
Special Selection Techniques

*12.1 TWO-PHASE SAMPLING IN GENERAL AND IN STRATIFICATION

12.1A The Economics of Two-Phase Sampling

"It is sometimes convenient and economical to collect certain items of information on the whole of the units of a sample, and other items of information on only some of these units, these latter units being so chosen as to constitute a sub-sample of the units of the original sample. This may be termed two-phase sampling. Information collected at the second or sub-sampling phase may be collected at a later time, and in this event, information obtained on all the units of the first-phase sample may be utilized, if this appears advantageous, in the selection of the second-phase sample. Further phases may be added as required.

"It may be noted that in multi-phase sampling, the different phases of observation relate to sample units of the same type, while in multi-stage sampling, the sample units are of different types at different stages.

"An important application of multi-phase sampling is the use of the information obtained at the first-phase as supplementary information to provide more accurate estimates (by the method of regression or ratios), of the means, totals, etc., of variates obtained only in the second phase." [UN, 1950.]

This UN definition of *two-phase* sampling is generally accepted in survey sampling, where it is also called double sampling; elsewhere in statistics these words may refer to sequential sampling. The idea may be extended to more than two phases in *multiphase* sampling, but a description of two-phase sampling will suffice.

The central concept involves selecting the basic sample of n elements not directly from the population of N elements, but from a preselection of a larger sample of n_L elements. Ancillary information from the larger preliminary sample of n_L elements is used to improve the final sample of n elements. Two-phase sampling is a compromise solution, typical in sampling, for a dilemma posed by undesirable extremes. The statistics based on the sample of n can be improved by using ancillary information from a wider base; but this is too costly to obtain from the entire population of N elements. Instead, information is obtained from a large preliminary sample n_L, which includes the final sample n.

Recourse to ancillary information for improving the sample is the central motif common to several sampling methods. The application of each method can be combined with two-phase sampling by reducing the base of the ancillary information from the population to a preliminary large sample. The principles are applied to stratification in this section, and to regression and ratio estimates in the next, plus brief examples of a few other possible applications.

In the course of practical sampling, two-phase sampling is typically introduced into designs already complicated with stratification or clustering. However, we must examine the effects of two-phase sampling in the naive context of simple random selection, for which derivations are simpler and available. Comparisons of efficiency can be viewed as applying to the simple random component of the entire variance, or to the numbers of elements in the sample size. Then with the design effect (8.2) we can adjust for the effects of other complexities.

Typically, two-phase sampling is practical only if the per element cost of the ancillary information (c_L) is less by a large factor (say, 10 to 100) than the per element cost (c) of the principal measurements. Hence, the ancillary information must be obtained with entirely different and much cheaper methods. For example, the second phase of a crop measurement may involve cutting, preparing, and weighing yields from a sample of small plots, but the first phase requires only eye estimates of the standing crops from a moving automobile. In another example, the sample measurement may involve a lengthy interview or a medical examination, but the ancillary information can be read from cards in a larger file of cases.

On the other hand, if the ancillary information can be had practically free of charge (say, cheaper by a factor of 10,000 or 100,000), then we may take it from the entire population, rather than develop a preliminary sample. Perhaps we can justify and clarify these thoughts by a closer look at a simple cost function:

$$\text{Total variable cost} = cn + c_L n_L. \tag{12.1.1}$$

Here c is the cost of the regular measurement on each of the n elements in the second phase, and c_L is the cost of the ancillary information for each of the n_L elements in the first-phase sample. Denoting the ratios of the two cost factors as $k = c_L/c$, the combined cost is seen as $c(n + kn_L)$. This would buy in a single-phase sample $n_0 = n(1 + kn_L/n)$ elements at a cost of cn_0. But in two-phase sampling the sample must be reduced by the factor $n_0/n = (1 + kn_L/n)$ to pay for the ancillary information in the initial sample.

The gains from ancillary information will seldom be great enough to permit drastic reductions of the final sample size. On the other hand, trivial reductions of final sample size should not concern us. Thus let us assume that to buy the benefits of a two-phase sample, we are willing to reduce the final sample by the factor $n_0/n = 2$, hence make $kn_L/n = 1$. In most situations the ancillary information is helpful only if its base is larger by a substantial factor; if this factor is $n_L/n = 10$, the ratio k of per element cost should be under 0.1 to repay its cost. On the other hand, if the ancillary information is so cheap that k is near the sampling fraction f, with $n_L \doteq n/f = N$, the preliminary sample n_L may become the population N. Hence, a reduction of the final sample by a factor of $n_0/n = 2$, due to two-phase sampling, is likely only if the cost per element c_L of the ancillary information, though not trivial, is much less than the cost c per element of the basic measurement. For example, note in Table 12.2.I that in two-phase sampling with regression and $r_{yx} = 0.8$, a cost factor of $k = c_L/c = 1/7$ is needed to reduce the sampling cost to $E = 0.8$ of its value under single-phase sampling; $k = 1/55$ is necessary to reduce the cost to $E = 0.5$.

The reduction of the sampling cost with two-phase sampling should be sufficient to pay for the increased cost of the more complex statistical analysis and presentation. We can introduce this factor in the cost functions as

$$G_0 + cn_0 \qquad \text{for single phase sampling,}$$

$$G_d + cn\left(1 + \frac{kn_L}{n}\right) \qquad \text{for two phase sampling.} \tag{12.1.2}$$

The difference in these factors can be written as $G_d - G_0 = D = (D/cn_0)cn_0$; hence the two cost functions are equal when $(1 - D/cn_0)n_0/n = (1 + kn_L/n)$. The extra cost of analysis may be considered as if bought for the cost of a proportion D/cn_0 of the original interviews. For example, if two-phase sampling reduces the sampling cost by only 0.8, the saving of 0.2 of the original cost of cn_0 may be needed merely to pay for D, the increased cost of analysis. Since this extra expense for the increased complexity is a fixed charge, it looms as a larger proportion D/cn_0 in less costly than in more expensive samples, where cn_0 is large. Therefore, only large decreases in the element cost justify double sampling in small or moderate sized samples. D/cn_0 will seem small in large, complex samples, where the variance computation is already complex (6.3). Hence, double sampling can be freely introduced into the later stages of multi-stage samples (12.6).

12.1B Stratification in Two-Phase Sampling

Whereas in ordinary stratification we can use population values for the stratum weights W_h, in two-phase sampling we must use their estimates w_h, obtained in the preliminary sample of size n_L. Thus the estimate of the mean is

$$\bar{y}_{wd} = \sum w_h \bar{y}_h. \tag{12.1.3}$$

Note four points pertinent to this design: (1) Generally, the expected value of the estimated mean is $\bar{Y}$, exactly or nearly, since we are dealing with estimates with zero or little bias. (2) The estimate is unbiased if a simple random sample of size n_L is selected first from the population of N elements, then sorted into H strata, for selecting the final sample of n_h from the n_{Lh} elements in the hth stratum. (3) The variance formulas are simplified by assuming a simple random sample in both phases and disregarding the finite population corrections $(1 - n_L/N)$ and $(1 - n_h/N_h)$. The initial sampling fraction n_L/N is often small, then so is n_h/N_h; on the other hand, if n_L is very large compared to n_h, the variance (12.1.4) approaches that of an ordinary stratified sample. (4) When the weights refer to the population proportions of elements in the strata, then $W_h = N_h/N$; estimated weights are the proportions of elements in the preliminary sample: $w_h = n_{Lh}/n_L$. A good estimate of the variance is

$$\text{var}(\bar{y}_{wd}) = \sum w_h^2 \frac{s_h^2}{n_h} + \frac{1}{n_L} \sum w_h (\bar{y}_h - \bar{y}_{wd})^2. \tag{12.1.4}$$

The first term equals approximately the variance of an ordinary stratified sample. The second term measures approximately the increase of two-phase over ordinary stratified sampling; it becomes small if n_L is large.

For an ordinary *proportionate stratified sample* of the same size $n = \Sigma n_h$, the variance is the first term in $\text{var}(\bar{y}_0) = \Sigma W_h^2 s_h^2/n_h + \Sigma W_h(\bar{y}_h - \bar{y}_w)^2/n$; its gain over a simple random sample is the second term (4.6.2). The first term is approximately the same as in (12.1.4); hence a two-phase proportionate sample has smaller variance than a simple random sample of the same size n, by approximately

$$(1/n - 1/n_L) \sum W_h(\bar{y}_h - \bar{y}_w)^2 = (1 - n/n_L)(1/n) \sum W_h(\bar{y}_h - \bar{y}_w)^2.$$

Thus the gain of two-phase proportionate sampling resembles that of ordinary proportionate sampling, but is less by the factor $(1 - n/n_L)$. The typically modest gains of proportionate sampling may justify using the ancillary information if it is very cheap; but not if its cost requires a substantial reduction in sample size from n_0 to n (12.1.2). We must balance an increase of the first term by $(1/n - 1/n_0) \Sigma W_h S_h^2$ against a decrease of $(1/n_0 - 1/n_L) \Sigma W_h(\bar{y}_h - \bar{y}_w)^2$ in the second; also, remember to allow for the cost of increased complexity.

For *optimum allocation* of the sample sizes n_h, two-phase sampling may give large gains in favorable situations. The allocation of $n_h = kW_h S_h/\sqrt{J_h}$ can be applied here as an approximation. If the weights are the relative stratum sizes $w_h = n_{Lh}/n_L$ obtained from the sample, the sampling rates should be made approximately in the ratios $n_h/n_{Lh} = kS_h/\sqrt{J_h}\, n_L$, probably rounded to some convenient integer. Furthermore, the highest of these rates can often be fixed at $n_h/n_{Lh} = 1$, by taking in the most critical stratum all cases that the preliminary sample can supply. The size of n_L can be designed accordingly; a greater n_L would yield only the gains of proportionate two-phase sampling, which are usually small. The other sampling rates should diverge considerably; only large differences produce worthwhile gains from disproportionate sampling.

Remark 12.1.I The population value of the variance is

$$\text{Var}(\bar{y}_{wd}) = \sum W_h^2 \frac{S_h^2}{n_h} + \frac{1}{n_L} \sum W_h(\bar{Y}_h - \bar{Y})^2 + \frac{1}{n_L} \sum W_h(1 - W_h) \frac{S_h^2}{n_h}. \tag{12.1.5}$$

The first term is the variance of a stratified sample, with sample sizes n_h in the strata. The second term will often be small compared to the first, because n_L is large; it equals the reduction in variance that would be gained by a proportionate sample (but not by the double sample) of size n_L. The third term is caused by sampling variations of weights in the first term; it will be negligible compared to the first, since it is smaller by the factor $(1 - W_h)/W_h n_L$. The unbiased estimate of the variance (12.1.5) is

$$\text{var}(\bar{y}_{wd}) = \frac{n_L}{n_L - 1} \sum \left(w_h^2 - \frac{w_h}{n_L}\right) \frac{s_h^2}{n_h} + \sum \frac{w_h(\bar{y}_h - \bar{y}_{wd})^2}{n_L - 1}. \tag{12.1.6}$$

In the simple approximation (12.1.4), the first term is seen to involve the factor $[1 + (1 - 1/w_h)/(n_L - 1)]$ and the second term $[1 + 1/(n_L - 1)]$. In most cases these can be neglected. Cochran [1963, 12.1–12.4] gives detailed derivations of (12.1.5) and (12.1.6), with formulas for the gains from optimum allocation, and the choice of n_L and n for economic allocation of total costs.

12.1C Complex Two-Phase Selections

Two-phase sampling can often be effective in multistage sampling and in other designs more complex than those assumed for simple stratification in 12.1*B* and the estimators in 12.2. An exhaustive treatment is not possible here, but brief sketches of some likely prospects may help. (However, before selecting any complex sample, remember to write down a valid computing formula for its variance.)

1. Suppose a population of N can be sorted into G strata according to some variables; certain other variables, desirable but expensive, can be obtained only for a first-phase sample of n_L cases. First, stratify the population into the G strata, and obtain a first-phase selection of n_g/N_g, such that $n_L = \Sigma n_g$. For these n_L obtain the ancillary information and sort them into substrata, with n_{gLh} in the hth substratum of the gth stratum; then select a fraction n_{gh}/n_{gLh}. Assuming that both selections were random within strata, estimates of the mean and its variance can be obtained by applying further stratification to (12.1.3) and (12.1.4):

$$\sum_g^G W_g \bar{y}_g = \sum_g^G W_g \sum_h^{H_g} w_{gh} \bar{y}_{gh}, \tag{12.1.7}$$

and

$$\sum_g^G W_g^2 \operatorname{var}(\bar{y}_g) = \sum_g^G W_g^2 \left[\sum_h^{H_g} w_{gh}^2 \frac{s_{gh}^2}{n_{gh}} + \frac{1}{n_g} \sum_h^{H_g} w_{gh} (\bar{y}_{gh} - \bar{y}_g)^2 \right]. \tag{12.1.8}$$

2. In multistage sampling, the first-stage units can be selected in two phases. For example, suppose that the first-stage units are hospitals, available in a good list, and that stratification according to type of internal organization is deemed valuable. Obtaining the information to classify the entire population would be too expensive, but this could be done for a first-phase sample of n_L. Two-phase sampling permits statistical inference to the population and uses most of the information (if n_L is large) that the classifying variable can yield. On the contrary, too often in these cases the "sample" is confined to the few units (hospitals) about which the information happens to be available. These may be atypical for many reasons and lead to false inferences.

3. Two-phase sampling may prove valuable in the last stage of a multistage sample to select elements with disproportionate allocation. Suppose, for example, that a brief examination or interview yields information for stratifying dwellings into classes; and that this classification leads to disproportionate allocation due to divergent standard deviations on critical variables. Dwellings are rated High, Medium, and Low; then selected with different fractions (11.4C).

4. A first-phase sample can serve to introduce clustering in order to reduce travel costs, if enough can be saved to justify the increases in complexity and the variance. For example, the list may contain automobile registrations, or fishing licenses, or social security numbers of hospital discharges; from this a preliminary sample of n_L cases is selected.

The listings show clear and current addresses, which can be sorted cheaply into small clusters. The sorting may be done according to post office address, or with rough and cheap spotting on maps; a small portion of hard-to-locate cases are put into a miscellaneous stratum. After sorting or spotting, make up clusters of neighboring elements containing roughly equal numbers of elements. Suppose that the preliminary sample of n_L was k times larger than the final desired sample of n, travel within the clusters can be reduced by the factor k, say, 5 or 10 or 20. Distinguish strata of $2k$ clusters each, and select at random two clusters from each stratum. The variance is computed by contrasting the results from the two random selections from each stratum (6.5*B*).

*12.2 TWO-PHASE RATIO AND REGRESSION MEANS

12.2A Two-Phase Ratio Means

Both ratio and regression means depend on a first-phase sample of n_L elements, in which only the ancillary variable x is observed, and whose mean is $\bar{x}_L$. The final sample consists of n elements, on which both the ancillary variable x_i and the survey variable y_i are measured. The final sample may either be a subsample of the first-phase sample of size n_L, or be independent of it; the small differences in the variance formulas will be indicated. The data for the $\bar{x}_L$ may come from an outside source, provided that the means, $\bar{x}$ and $\bar{x}_L$, measure essentially the same variable. Violation of the last condition can cause a serious bias.

Double sampling may be viewed as improving the survey mean $\bar{y}$ by using the correlation between $\bar{y}$ and $\bar{x}$ in the final sample, together with the value $\bar{x}_L$ of the ancillary variable from the larger first-phase sample. Contrariwise, one may regard the estimate $\bar{x}_L$ as the main statistic, which is then improved with a more accurate measurement $\bar{y}$ on the smaller sample and the correlation between $\bar{y}$ and $\bar{x}$. For example, eye estimates of crop yields from a large sample may be improved by accurately weighing a subsample, then using the correlation of the two measurements in the small sample to correct the mean of eye estimates from the large sample.

In standard sampling theory, it has been assumed that both samples n_L and n are simple random; the variance formulas (12.2.2) and (12.2.3) have been derived with that assumption. More general assumptions lead to complications, though less for ratio than regression estimates.

For two-phase ratio means, the mean $\bar{y}$ is adjusted by the ratio for the ancillary variable of the first-phase mean $\bar{x}_L$ to the final sample mean $\bar{x}$:

$$\bar{y}_{rd} = \bar{y}\,\frac{\bar{x}_L}{\bar{x}} = \frac{\bar{y}}{\bar{x}}\,\bar{x}_L = r\bar{x}_L. \qquad (12.2.1)$$

Here $\bar{y}$ and $\bar{x}$ are the means from the final sample of size n, and $\bar{x}_L$ the mean in the first-phase sample of n_L. Or we can regard $\bar{x}_L$ as the main variate corrected by the ratio $r = \bar{y}/\bar{x}$ from the final sample.

Cochran [1963, 12.8] derives variances for situations where both n_L and n are simple random. These are approximately

$$\text{var}(\bar{y}_{rd}) = \frac{1}{n}[s_y^2 + r^2 s_x^2 - 2rs_{yx}] + \frac{2rs_{yx} - r^2 s_x^2}{n_L}, \tag{12.2.2}$$

when the n is subsample of the n_L; and

$$\text{var}(\bar{y}_{rd}) = \frac{1}{n}[s_y^2 + r^2 s_x^2 - 2rs_{yx}] + \frac{r^2 s_x^2}{n_L}, \tag{12.2.2'}$$

when the samples n_L and n are independent.

The difference between the two kinds of estimates is small when n_L is much larger than n. The variance tends to be smaller when the n is drawn as a subsample of the n_L (12.2.2) than when the two samples are independent (12.2.2′). First, for fixed cost the n_L may be larger in the first equation than in the second, because the former includes and the latter excludes the subsample n. Second, in the numerator we can typically expect that $(2rs_{yx} - r^2 s_x^2) \leq r^2 s_x^2$; that is, $r_{yx} s_y \leq r s_x$, where $r_{yx} = s_{yx}/s_x s_y$. Even if the samples n_L and n were drawn independently, we may combine the two to compute the estimate $\bar{x}_L$.

The samples n_L and n can be obtained with a selection method other than simple random. If the variables $r = \bar{y}/\bar{x}$ and $\bar{x}_L$ come from two independent samples, we can show with a derivation similar to the above, that generally and approximately

$$\text{var}(r\bar{x}_L) = [\text{var}(\bar{y}) + r^2 \text{var}(\bar{x}) - 2r \text{cov}(\bar{y}, \bar{x})] + r^2 \text{var}(\bar{x}_L). \tag{12.2.3}$$

When both samples are simple random, the special case of (12.2.2′) results. This formula may also be regarded as an example of the variance for the product of two independent variables; from (6.6.17) we obtain $\text{var}(r\bar{x}_L) = \bar{x}_L^2 \text{var}(r) + r^2 \text{var}(\bar{x}_L)$. The factor $\bar{x}_L^2/\bar{x}^2$ is omitted from (12.2.3); its variation around 1 is relatively small, and I am not certain whether its inclusion would provide a more precise estimate of the variance. If the two samples are not independent, the term $+2r\bar{x}_L \text{cov}(r, \bar{x}_L)$ belongs here; it accounts for the difference between the two forms of (12.2.2). Ordinarily this covariance between r and $\bar{x}_L$ would be small, because $\bar{x}$ is a small part of $\bar{x}_L$, and because the correlation between r and $\bar{x}_L$ would also be small.

12.2B Two-Phase Regression Estimates

The mean of the ancillary variable from the first-phase sample is $\bar{x}_L$. We must compute from the final sample of n not only $\bar{y}$ and $\bar{x}$, but also the regression coefficient of y on x. Thus the regression estimate is

$$(\bar{y}_{rdg}) = \bar{y} + b(\bar{x}_L - \bar{x}) = (\bar{y} - b\bar{x}) + b\bar{x}_L, \tag{12.2.4}$$

where
$$b = \frac{\sum (y_i - \bar{y})(x_i - \bar{x})}{\sum (x_i - \bar{x})^2} = \frac{s_{yx}}{s_x^2} = r_{yx}\frac{s_y}{s_x}.$$

If both n_L and n have been selected with simple random, an adequate and simple approximation for the variance is

$$\text{var}(\bar{y}_{rgd}) = \frac{s_y^2(1 - r_{yx}^2)}{n'} + \frac{r_{yx}^2 s_y^2}{n_L} \tag{12.2.5}$$

$$= \frac{s_y^2}{n'} - \left(\frac{1}{n'} - \frac{1}{n_L}\right)\frac{(s_{yx})^2}{s_x^2}. \tag{12.2.5'}$$

The best approximation is ordinarily obtained if we use $(n - 2)$ for n'; this can be derived from Cochran [1963, 12.5–12.7]. The second expression comes readily by noting that $r_{yx}^2 s_y^2 = b^2 s_x^2 = (s_{yx})^2/s_x^2$. The approximation of the variance is the same whether n is a subsample of the n_L or is independent of it. We can utilize $\bar{x}_L$ from an outside source, provided we feel confident that its measurement is the same as for $\bar{x}$.

Two-phase regression estimates may be useful in complex samples, including cluster selection. Another valuable generalization would involve the use of a set of several predictor variables, which we may denote with the vector $\{\mathbf{x}\}$. Unfortunately this theory has not been worked out. With plausible reasoning we can note that the regression mean,

$$(\bar{y}_{rgd}) = \bar{y} - b\bar{x} + b\bar{x}_L = \frac{y - bx}{n} + \frac{bx_L}{n_L},$$

can be treated as the sum of two random variables, with the approximate variance

$$\text{var}(\bar{y}_{rgd}) = \frac{1}{n^2}\text{var}(y - bx) + \frac{1}{n_L^2}\text{var}(bx_L) + \frac{2\,\text{cov}(y - bx, bx_L)}{nn_L}.$$

The residual variable $(y - bx)$ can be computed for the sample of n; if more than one predictor is involved, bx stands for $\Sigma b_i x_i$. The same values are used to compute the predicted value bx_L in the second term; this becomes $n_L^{-2} b^2 \text{var}(x_L)$ for a single predictor x. The last term tends to be small, since the summation in the numerator has n elements, whereas the denominator is nn_L. Theoretical uncertainties regarding the regression mean in complex samples may make us prefer the ratio mean.

If we apply the simple cost function $C = cn + c_L n_L = c(n + kn_L)$ to the variance (12.2.5), we can show (8.5.9) that the optimum allocation of the total cost C is reached when

$$\frac{n}{n_L} = \left(\frac{1 - r_{yx}^2}{r_{yx}^2} k\right)^{1/2}. \tag{12.2.6}$$

If we compare the variance of double sampling at this optimum (8.5.10) with a simple random sample obtainable for the same cost C, we get

$$E = \frac{\text{var}(\bar{y}_{rgd}) \text{ opt}}{\text{var}(\bar{y}_0)} = (\sqrt{1 - r_{yx}^2} + \sqrt{r_{yx}^2 k})^2. \tag{12.2.7}$$

We want to know the combination of r_{yx} and k from which we can expect some gain; that is, when $E < 1$. Better still, we need enough gain to overcome the cost of extra computations and complications; that is when $E < (1 - D/cn_0)$, in accord with (12.1.2). Errors in estimating the parameters in (12.2.5) will also prevent us from reaching the optimum. We then find that

$$(\sqrt{1 - r_{yx}^2} + \sqrt{r_{yx}^2 k})^2 < E, \quad \text{when } \sqrt{k} < \sqrt{E/r_{yx}^2} - \sqrt{(1 - r_{yx}^2)/r_{yx}^2}.$$

E \ r_{yx}	0.95	0.9	0.8	0.7	0.6	0.5
1.0	1.9	2.5	4	6	9	14
0.8	2.8	3.8	7	14	40	400
0.5	6	10	55	...	...	...
0.2	55	2800	...	...	...	...
0.1	...	...	...	...	...	...

TABLE 12.2.I Values of c/c_L that Reduce Optimum Variance of Regression Means by E, for Given r_{yx}

For example, with $r_{yx} = 0.9$, a cost ratio of $c/c_L = 10$ will reduce the variance to $E = 0.5$; the regression mean at its optimum has a variance 0.5 of the variance that a simple random sample would obtain for the same total cost. To justify the extra cost of computations of regressions means would require $E \leq 0.8$. This can only be obtained with very high r_{yx} or with high values of $c/c_L = 1/k$, that is, low first-phase cost c_L. Moderate gains of $E = 0.5$ arise only with a combination of high r_{yx} and low c_L. Spectacular gains of $E = 0.2$ are rare.

When k (and c_L) approach zero, E approaches $(1 - r^2_{yx})$, the variance of ordinary regression means, because we can take enormously large n_L/n. If r were extremely high (r_{yx} near 1), we could reduce E to the neighborhood of $k = c_L/c$; but it is unlikely that we can buy cheaply any variable X_i with the precision needed to give those extremely high correlations.

12.2C Comparisons of Regression, Ratio, and Stratified Means in Two-Phase Sampling

The ancillary variable X_i, if related to the survey variable Y_i, can be utilized in alternative ways to reduce the variance. The following comparison of the advantages of the alternatives is relevant not only to two-phase sampling, but also to ordinary cases where the first phase includes the entire population. Moreover, instead of a single ancillary variable, one can generalize to a vector of several ancillary variables $(X_{1i}, X_{2i}, \ldots)$. Some can be used with either a regression or ratio mean, others with stratification. For example, a regression mean using a linear relationship can also be stratified according to some qualitative variable, such as geography or occupation.

1. The relative gains of regression means are limited to R^2_{yx}, since the element variance is not less than $(1 - R^2_{yx})S_y^2$. Actual gains are less, due to the second term in (12.2.5), added expense, and missing the precise optimum. For appreciable gains, we need very high R_{yx} and low cost for the ancillary measurements; that combination is rare. Regression means must also bear the cost of complex computations, whereas ratio and stratified means are easier to compute. With complex samples there also remain theoretical gaps in variance formulas.
2. The variance for the ratio mean cannot be lower than for the regression mean, and it may be considerably higher. The reasons are the same as for single-phase estimates (12.3*B*).
3. The relative gains from proportionate stratification are also about R^2_{yx}, if the correlation with the stratifying variable is linear. The choice between the two methods may depend on whether the computations of regression means or the sorting for stratification are more expensive.
4. Stratification often has special advantages over regression and ratio means. First, if the relationship of Y_i and X_i is not linear, proportionate stratification can yield larger gains. Second, if the stratifying variable is not quantitative, but a multinomial classification, stratification is still possible, but the others are not. Third, disproportionate allocation can yield much greater gains when large differences in $S_h/\sqrt{J_h}$ can be discovered for low $k = c_L/c$.

5. Regression and ratio means can be used on samples of n selected independently of the ancillary variate $\bar{x}_L$, if this is already available. However, in stratification the n must be a subsample of the first phase n_L, unless post-stratification is used.

6. In all cases, estimates of the aggregate $N\bar{y}$ and its standard error can be readily obtained from the mean $\bar{y}$, if the N is known.

*12.3 REGRESSION MEANS; COMPARISONS WITH RATIO AND DIFFERENCE MEANS

12.3A Regression Means

Throughout the book we have been using information from ancillary variables to improve selection procedures, especially with stratified and clustered selections. Sometimes they can be better utilized for improving the estimation procedures, as shown with post-stratification (3.4*C*) and ratio estimates (6.5*C*). Other estimation techniques are also possible; some complex estimates are developed in the Census Case Studies [Hansen et al., 1953, Ch. 12]. Clearly, it would be impossible to describe many estimates here. We confine the discussion to regression and difference means, the two principal estimates developed in survey sampling literature.

The regression mean is defined as

$$(\bar{y}_{rg}) = \bar{y} + b(\bar{X} - \bar{x}) = (\bar{y} - b\bar{x}) + b\bar{X}, \tag{12.3.1}$$

where the regression coefficient b is

$$b = \frac{s_{yx}}{s_x{}^2} = \frac{\sum (y_j - \bar{y})(x_j - \bar{x})}{\sum (x_j - \bar{x})^2} = r_{yx}\frac{s_y}{s_x}.$$

It attempts to improve the sample mean $\bar{y}$ of the variable Y_i by utilizing an ancillary variable X_i, its population and sample means $\bar{X}$ and $\bar{x}$, and the linear correlation of the two variables in the sample. If the sample is assumed to be simple random of size n, its variance is well approximated with

$$\begin{aligned}\text{var}\,(\bar{y}_{rg}) &= \frac{1-f}{n'}s_y{}^2[1 - r_{yx}^2] = \frac{1-f}{n'}s_e{}^2 = \frac{1-f}{n'}\left[s_y{}^2 - \frac{(s_{yx})^2}{s_x{}^2}\right] \qquad (12.3.2)\\ &= \frac{1-f}{n'}\frac{1}{n-1}\left[\sum (y_j - \bar{y})^2 - \frac{\{\sum (y_j - \bar{y})(x_j - \bar{x})\}^2}{\sum (x_j - \bar{x})^2}\right]\\ &= \frac{1-f}{n'}\frac{1}{n-1}\sum \{(y_j - \bar{y}) - b(x_j - \bar{x})\}^2.\end{aligned}$$

Regression totals $N\bar{y}_{rg}$ and its standard error can be obtained by multiplying estimates for $\bar{y}_{rg}$ by the size N of the population.

The variance of a regression mean is $\{(1-f)S_y^2/n\}(1-R_{yx}^2)$, hence less than the variance of a simple random sample of the same size n by the factor $(1-R_{yx}^2)$. But the computations and complications of regression estimates cost something, say, D. Furthermore, obtaining the information on the X_i variable may increase the element cost to c, greater than c_0 for simple random. If we equate the total costs of a regression mean and a simple random mean as $c_0 n_0 = D + cn$, we obtain $n_0/n = (D/c_0 n + c/c_0)$ as the ratio of the sample sizes of a regression mean to a simple mean for the same cost. Then see that the regression mean reduces the variance to the extent that

$$\left(\frac{D}{nc_0} + \frac{c}{c_0}\right)(1 - R_{yx}^2) < 1. \qquad (12.3.3)$$

The regression mean is subject to bias, but its effect on the mean square error is of order $1/n$. Since the bias is of order $1/n$ compared to the standard error of order $1/\sqrt{n}$, the orders of variance and bias² have the relationship $(1/n + 1/n^2) = (1 + 1/n)/n$ in the mean square error.

The value of n' that yields the best approximation is $n' = n - 2$. In any case, this is a correction of n for the error in b, which is of order $1/n$ compared to s_y^2. This knowledge encourages shortcuts for computing b in large samples: using only a subsample; plotting a simple graph from a few points, especially near the ends; using a simplified regression method based on three class means [Jowett, 1955; Gibson and Jowett, 1957]; or guessing b on the basis of other evidence, then using a difference mean (12.3.5).

A detailed theory of regression means is readily available only for simple and stratified random selection of elements. However, large samples are often complex and clustered. Flexibility and simplicity can be gained in large samples by disregarding the error in b, since this is of order $1/n$ compared to the main error s_y^2. If we disregard the errors in b, a regression mean may be viewed as subject only to the *residual error* $e_j = (y_j - bx_j)$, instead of the error in the variable y_j. If we consider only the errors in $\bar{y}$ and $\bar{x}$ in $\bar{y}_{rg} = (\bar{y} - b\bar{x}) + b\bar{X}$, its variance becomes

$$\text{var}(\bar{y}_{rg}) = \text{var}(\bar{y}) + \text{b}^2 \text{var}(\bar{x}) - 2b \text{ cov}(\bar{y}, \bar{x}). \qquad (12.3.4)$$

This permits great flexibility. For example, the variable X_i may be measured either on the elements as Y_i is, or on other units, or perhaps on the clusters. If the means have a variate u as their base, so that $\bar{y} = y/u$ and $\bar{x} = x/u$, then

$$\text{var}(\bar{y}_{rg}) = \text{var}\left(\frac{y - bx}{u}\right), \qquad (12.3.4')$$

which can be computed directly from the sample residuals $(y_j - bx_j)$. Moreover, the bx need not refer to a single variate, but can represent a vector $\{\mathbf{b}_p \mathbf{x}_{pj}\}$ of multivariate regression.

Remark 12.3.I Cochran [1963, 7.1–7.5] derives the estimated variance of $\bar{y}_{rg}$. First, disregarding the errors in b,

$$\text{var}(\bar{y}_{rg}) \doteq \frac{1-f}{n} s_e^2,$$

where
$$s_e^2 = \frac{1}{n-1} \sum [(y_i - \bar{y})^2 - b^2(x_i - \bar{x})^2] = s_y^2 - \frac{(s_{yx})^2}{s_x^2}.$$

Next consider the classical regression model $y_{ij} = \bar{Y} + B(x_i - \bar{X}) + e_{ij}$. Assume: (1) linearity in the regression of Y_i on X_i; (2) homoscedasticity, that is, constant $E(e_{ij}^2)$ for different magnitudes of X_i; and (3) an infinite universe, so that $(1 - f) = 1$; we then can use this approximation to terms of order $1/n^2$:

$$\text{Var}(\bar{y}_{rg}) \doteq \frac{S_e^2}{n}\left(1 + \frac{1}{n}\right) \doteq \frac{S_e^2}{n-1}.$$

Then with $s_{y.x}^2 = \dfrac{1}{n-2} \sum [(y_i - \bar{y}) - b(x_i - \bar{x})]^2$, and since $E(s_{y.x}^2) = S_e^2$, we obtain $E\left(\dfrac{s_e^2}{n-2}\right) = E\left(\dfrac{s_{y.x}^2}{n-1}\right) = \dfrac{S_e^2}{n-1} \doteq \text{Var}(\bar{y}_{rg})$.

Example 12.3a In Examples 2.2a and 6.3a we used a sample of $n = 20$ blocks selected with srs from a population of $N = 270$ blocks. Now compute the *regression mean* of the number of renter occupied dwellings per blocks, using total dwellings as an auxiliary variable:

$$\bar{y}_{rg} = \frac{y}{n} + b\left(\frac{X}{N} - \frac{x}{n}\right) = \frac{255}{20} + 0.6329\left(\frac{6786}{270} - \frac{435}{20}\right)$$
$$= 12.75 + 0.6239(25.13 - 21.75) = 14.86,$$

where $b = \dfrac{s_{yx}}{s_x^2} = \dfrac{419.57}{672.51} = 0.6239.$

The estimated variance is

$$\text{var}(\bar{y}_{rg}) = \frac{1-f}{n-2} s_y^2[1 - r_{yx}^2] = \frac{0.926}{18} 278.62(0.06050) = 0.8674,$$

because

$$r_{yx}^2 = \frac{(s_{yx})^2}{s_x^2 s_y^2} = \frac{(419.57)^2}{672.51 \times 278.62} = 0.93950.$$

The ratio mean in 6.3a was $r\dfrac{X}{N} = (0.5862)25.13 = 14.73$, with $\text{var}(r\bar{X}) = \bar{X}^2 \text{var}(r) = 25.13^2 \times 0.001741 = 1.0995$. The simple mean in 2.2$a$ was $\bar{y} = 12.75$ with a variance of 12.90. The large decreases in the variances for both the ratio and regression means were due to the large correlation $r_{yx} = \sqrt{0.93950} = 0.969$. The variance for the regression mean is a little lower than for the ratio mean.

12.3B Comparisons with Ratio and Difference Means

If k is any constant, we can construct a *difference mean:*

$$(\bar{y}_{\text{diff}}) = \bar{y} + k(\bar{X} - \bar{x}). \tag{12.3.5}$$

If $\text{E}(\bar{x}) = \bar{X}$, then $\text{E}(\bar{y}_{\text{diff}}) = \bar{Y}$. Hence the introduction of the ancillary variable as a difference with *any constant* multiplier does not result in any bias. Its effect on the variance equals the last two terms in

$$\begin{aligned} \text{Var}(\bar{y}_{\text{diff}}) &= \text{Var}(\bar{y}) + k^2 \text{Var}(\bar{x}) - 2kR_{yx}\sqrt{\text{Var}(\bar{y})\,\text{Var}(\bar{x})} \\ &= \text{Var}(\bar{y})(1 - R_{yx}^2) + \text{Var}(\bar{x})(k - B)^2. \end{aligned} \tag{12.3.6}$$

The first line is obtained simply as the variance of $(\bar{y} - k\bar{x})$, since the constant $k\bar{X}$ has zero variance. To obtain the second line, first define R_{yx} and B in terms of the variances and covariances of $\bar{y}$ and $\bar{x}$:

$$[\text{Cov}(\bar{y}, \bar{x})]^2 = R_{yx}^2 \text{Var}(\bar{x})\,\text{Var}(\bar{y}) = B^2[\text{Var}(\bar{x})]^2. \tag{12.3.7}$$

Then note that

$$\text{Var}(\bar{x})(k - B)^2 = \text{Var}(\bar{x})\,k^2 - 2kR_{yx}\sqrt{\text{Var}(\bar{x})\,\text{Var}(\bar{y})} + R_{yx}^2\,\text{Var}(\bar{y}).$$

If they result from a simple random selection of size n, then $\text{Var}(\bar{y}) = (1 - f)S_y^2/n$; $\text{Cov}(\bar{y}, \bar{x}) = (1 - f)S_{yx}/n$; and $\text{Var}(\bar{x}) = (1 - f)S_x^2/n$.

The minimum value of the variance for the difference mean is reached when $k = B$: when we use a constant k equal to the regression coefficient. The minimum is approached if we can find a constant k with small error, so that $(k - B)$ is small compared to the variances. The sample estimate b usually differs little from B; its effect is accounted for by using $n' = n - 2$ instead of n in (12.3.2). The regression mean may be regarded as the special case of the difference mean with $k = b$.

We can rarely find a constant k so close to B that it can be used with confidence. Perhaps the regularities of some periodic surveys allow us to guess confidently a k near the unknown true B; then we can get the gains of regression means without computing b. Perhaps the simplest case is a common-sense adjustment with $k = 1$ and $\bar{y}_{\text{diff}} = \bar{y} + (\bar{X} - \bar{x})$. For example, an inaccurate or obsolete Census datum $\bar{X}$ can be adjusted with the difference $(\bar{y} - \bar{x})$ from a sample survey, producing $\bar{y}_{\text{diff}} = \bar{X} + (\bar{y} - \bar{x})$. Population Census data can be adjusted with the differences $(\bar{y} - \bar{x})$ found in quality checks or in other surveys. "For example, X and Y may represent the same characteristic for two different dates. In this event, if the nature of the characteristic has not changed tremendously between the two dates, B will be reasonably close to 1." [Hansen, Hurwitz, and Madow, 1953, 11.2.]

The ratio mean also can be regarded as the special case of a difference mean with $k = \bar{y}/\bar{x}$, so that $\bar{y}_r = \bar{y} + (\bar{y}/\bar{x})(\bar{X} - \bar{x}) = \bar{X}\bar{y}/\bar{x}$. Its variance is

$$\text{Var}(\bar{y}_r) = \text{Var}(\bar{y})\left[1 + \left(\frac{\bar{Y}}{\bar{X}}\right)^2 \frac{\text{Var}(\bar{x})}{\text{Var}(\bar{y})} - 2R_{yx}\left(\frac{\bar{Y}}{\bar{X}}\right)\sqrt{\text{Var}(\bar{x})/\text{Var}(\bar{y})}\right].$$

Hence, writing $C^2(\bar{x}) = \text{Var}(\bar{x})/\bar{X}^2$ and $C^2(\bar{y}) = \text{Var}(\bar{y})/\bar{Y}^2$, we obtain

$$\begin{aligned}\frac{\text{Var}(\bar{y}_r)}{\text{Var}(\bar{y}_{rg})} &= \frac{\text{Var}(\bar{y})[1 + C^2(\bar{x})/C^2(\bar{y}) - 2R_{yx}C(\bar{x})/C(\bar{y})]}{\text{Var}(\bar{y})(1 - R_{yx}^2)} \\ &= \frac{(1 - R_{yx}^2) + C^2(\bar{x})/C^2(\bar{y}) - 2R_{yx}C(\bar{x})/C(\bar{y}) + R_{yx}^2}{(1 - R_{yx}^2)} \\ &= 1 + \frac{[C(\bar{x})/C(\bar{y}) - R_{yx}]^2}{(1 - R_{yx}^2)}. \end{aligned} \tag{12.3.8}$$

The last term, always positive, is the relative increase in the variance of the ratio mean over the regression mean. The variance of the ratio mean is always greater than the variance of the regression mean. Often this gain of the regression mean may be too small to overcome its greater computational costs. The gain is small if R_{yx} is close to $C(\bar{x})/C(\bar{y})$. This relationship holds approximately for any population for which the least squares line of regression of Y_i on X_i is approximately through the origin; that is, when B is approximately equal to Y/X [Hansen, Hurwitz, and Madow, 1953, 11.2].

12.3C Stratified Regression Means

If we select a stratified sample of n_h elements from N_h elements, and we compute a separate regression mean $\bar{y}_{rgh}$ in each of the H strata, we can compute the *separate stratified regression mean* as

$$\bar{y}_{rgs} = \sum W_h \bar{y}_{rgh} = \sum W_h[\bar{y}_h + b_h(\bar{X}_h - \bar{x}_h)], \tag{12.3.9}$$

and the separate stratified regression total as

$$\tilde{Y}_{rgs} = \sum N_h \bar{y}_{rgh} = \sum N_h[\bar{y}_h + b_h(\bar{X}_h - \bar{x}_h)], \tag{12.3.9'}$$

where
$$b_h = \frac{\sum y_{hi}x_{hi} - n_h\bar{y}_h\bar{x}_h}{\sum x_{hi}^2 - n_h\bar{x}_h^2} = \frac{s_{yxh}}{s_{xh}^2} = r_{yxh}\frac{s_{yh}}{s_{xh}}.$$

Their variances are

$$\text{var}(\bar{y}_{rgs}) = \sum W_h^2 \text{var}(\bar{y}_{rgh}) \quad \text{and} \quad \text{var}(\tilde{Y}_{rgs}) = \sum N_h^2 \text{var}(\bar{y}_{rgh}), \tag{12.3.10}$$

where

$$\text{var}(\bar{y}_{rgh}) = \frac{1-f_h}{n_h'} s_{yh}^2[1 - r_{yxh}^2] = \frac{1-f_h}{n_h'}\left[s_{yh}^2 - \frac{(s_{yxh})^2}{s_{xh}^2}\right].$$

This is essentially the variance of the stratified residual variable $(y_{hj} - bx_{hj})$; applying (12.3.2) to each of the strata with $n_h' = n_h - 2$, and with the general stratified formulas (3.3) applied to the regression mean and its variance. But note the conditions of their usefulness: (1) We must know not only the stratum sizes W_h or N_h (with $\Sigma W_h = 1$ and $\Sigma N_h = N$), but also the population value of the auxiliary means $\bar{X}_h$. (2) We must compute for each stratum the sample means $\bar{y}_h$ and $\bar{x}_h$ and the regression coefficients. (3) The samples of n_h elements within the strata must be simple random, either by actual selection or by reasonable assumption. The n_h must be large enough to ensure that the bias is small enough within each stratum. This is necessary to guard against the possibility that the biases in all strata will have the same sign, and will grow to a considerable amount compared to the variance, which may become small when many strata are used. Hence, avoid too many strata and any strata with small n_h. Numerous strata are usually unjustified anyhow, because computation costs rise, and because regression coefficients do not differ enough. (4) The separate regression mean performs best when large differences exist among the regression coefficients b_h for the different strata, and when the bias is small, either because the regression of the Y_{hi} on the X_{hi} is linear or the n_h are large enough to reduce it.

When the regression coefficients b_h are suspected of being similar, a *combined stratified regression mean* may be used:

$$\bar{y}_{rgc} = (\bar{y}_w - b\bar{x}_w) + b\bar{X} = \sum W_h(\bar{y}_h - b\bar{x}_h) + b\bar{X} \qquad (12.3.11)$$

$$= \frac{1}{n}\sum_h^H (y_h - bx_h) + b\bar{X} = \frac{1}{n}\sum_h^H\sum_i^{n_h}(y_{hi} - bx_{hi}) + b\bar{X},$$

where $$b = \frac{\sum\sum(y_{hi} - \bar{y}_h)(x_{hi} - \bar{x}_h)}{\sum\sum(x_{hi} - \bar{x}_h)^2} \quad \text{and} \quad \bar{y}_h = \frac{1}{n_h}\sum^{n_h} y_{hi}.$$

This assumes either epsem or elements already weighted to compensate for unequal probabilities of selection. The variance of this mean can be approximated with a formula similar to (12.3.10):

$$\text{var}(\bar{y}_{rgc}) = \sum W_h^2 \frac{1-f_h}{n_h'} s_{eh}^2, \qquad (12.3.12)$$

where $$s_{eh}^2 = \left[s_{yh}^2 - \frac{s_{yxh}^2}{s_{xh}^2}\right] = \frac{1}{n_h - 1}\sum_i^{n_h}[(y_{hi} - bx_{hi}) - (\bar{y}_h - b\bar{x}_h)]^2,$$

the element variance of the residual variable. Here $n_h' \doteq n_h - 1 - 1/H \doteq n_h - 1$ can be used. This formula assumes that the overall sample size n is so large that, compared to the overall variance, the variation in b may be disregarded. Several considerations affect comparisons of the combined and the separate stratified regression mean: (1) The combined mean requires computing only one b instead of separate b_h for each stratum. (2) The combined mean has a larger variance to the degree that the regression coefficients B_h vary widely among the strata. (3) Under fairly common conditions the bias of the combined mean is of order $1/n$; whereas for the separate mean the bias of order $1/n_h$ in each stratum may mount up to larger proportions. Under extreme conditions the bias of the combined mean may amount to $1/n_h$ of the smallest stratum. The subject of stratified regression means is complex [Cochran, 1963, 7.7–7.9].

12.4 CORRELATIONS FROM OVERLAPS IN REPEATED SURVEYS

12.4A Variances of Differences between Two Overlapping Samples

Suppose that the two samples are taken; that n_x is the size of the first one, n_y the size of the second, and that n_c of these are overlaps, derived from the same elements. Our present interest is chiefly in the difference of the two means; they can be expressed as simple means per element of the sample totals x and y:

$$d = \bar{x} - \bar{y} = \frac{x}{n_x} - \frac{y}{n_y}. \tag{12.4.1}$$

This difference can represent a wide variety of research objectives. We may be interested in comparing two different characteristics, or in measuring the change in one, or perhaps in comparing two overlapping populations. The variance can be expressed as

$$\text{Var}(d) = \text{Var}\left(\frac{x}{n_x}\right) + \text{Var}\left(\frac{y}{n_y}\right) - 2\,\text{Cov}\left(\frac{x}{n_x}, \frac{y}{n_y}\right). \tag{12.4.2}$$

In many designs the sample sizes n_x and n_y are constants with no effect on the variances in the numerators. For all such designs the variance of the difference becomes

$$\text{Var}(d) = \frac{\text{Var}(x)}{n_x^2} + \frac{\text{Var}(y)}{n_y^2} - \frac{2\,\text{Cov}(x, y)}{n_x n_y}. \tag{12.4.3}$$

The effect of the covariance on the sum is the opposite of its effect on the difference:

$$\text{Var}(\bar{x} + \bar{y}) = \frac{\text{Var}(x)}{n_x^2} + \frac{\text{Var}(y)}{n_y^2} + \frac{2\,\text{Cov}(x, y)}{n_x n_y}. \tag{12.4.4}$$

To bring out the essential points in this section we assume the simplest design: simple random samples of elements from much larger populations. Thus we can disregard the factors $(1 - f)$ for each of the distinct simple random samples—of sizes n_c, $(n_x - n_c)$, and $(n_y - n_c)$, respectively—that jointly make up the two samples n_x and n_y. In these designs the variance of the difference becomes

$$\text{Var}(d) = \frac{n_x S_x^2}{n_x^2} + \frac{n_y S_y^2}{n_y^2} - \frac{2n_c S_{xy}}{n_x n_y} \tag{12.4.5}$$

$$= \frac{S_x^2}{n_x} + \frac{S_y^2}{n_y} - \frac{2R_{xy}S_xS_yP_xP_y}{n_c}. \tag{12.4.5'}$$

Here we denote the proportion of the overlap in the first sample as $P_x = n_c/n_x$, and in the second sample as $P_y = n_c/n_y$. We want to note how the variance of the difference changes as P_x and P_y vary between 0 and 1, because of the distinct meanings the difference takes in various situations.

Let us now apply this general formula to four special cases. We use population values denoted with capital letters, but we can compute the usual sample statistics s_x^2, s_y^2, and s_{yx} as estimates of S_x^2, S_y^2, and S_{xy}; and even s_{xy}/s_xs_y as an estimate of R_{xy}.

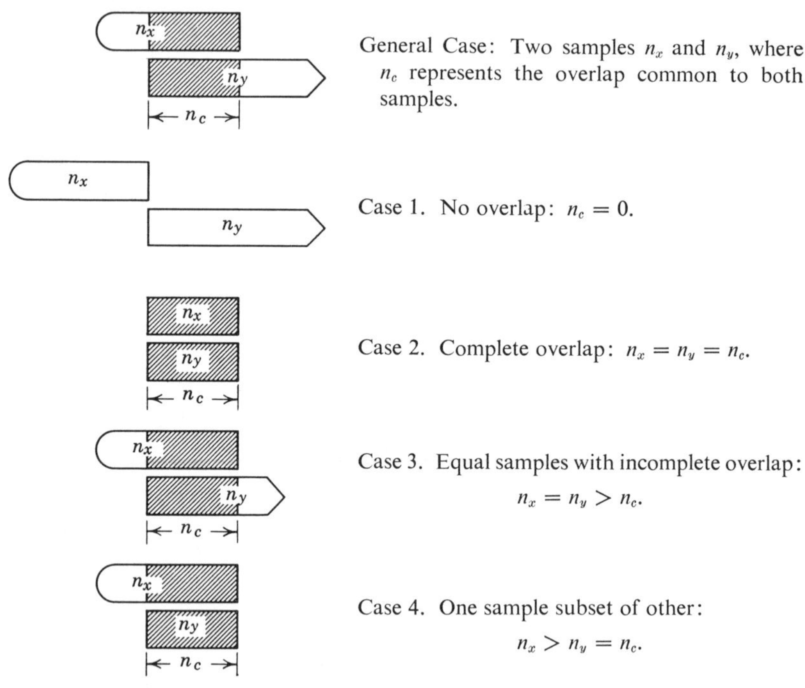

TABLE 12.4.I Types of Overlaps in Two Samples

Case 1. When the two samples do not overlap at all, $n_c = 0$ and $P_x = P_y = 0$. The covariance vanishes, and we find the common form for two independent samples:

$$\text{Var}(d) = \frac{S_x^2}{n_x} + \frac{S_y^2}{n_y}. \tag{12.4.6}$$

Furthermore, when

$$S_x^2 = S_y^2 = S^2, \qquad \text{then Var}(d) = S^2\left(\frac{1}{n_x} + \frac{1}{n_y}\right), \tag{12.4.6'}$$

and when

$$n_x = n_y = n, \qquad \text{then Var}(d) = \frac{2}{n} S^2. \tag{12.4.6''}$$

Case 2. When the overlap is complete, $n_x = n_y = n$ and $P_x = P_y = 1$. The variance of the difference of two identical samples can also be viewed as the differences of two measurements on a single sample:

$$\begin{aligned} \text{Var}(d) &= \text{Var}\left(\frac{x}{n} - \frac{y}{n}\right) = \frac{1}{n}(S_x^2 + S_y^2 - 2S_{xy}) \\ &= \frac{1}{n}(S_x^2 + S_y^2 - 2R_{xy}S_xS_y) && (12.4.7) \\ &= \text{Var}\left(\frac{x-y}{n}\right) = \frac{1}{n} S^2_{(x-y)}. && (12.4.7') \end{aligned}$$

Furthermore, when

$$S_x^2 = S_y^2 = S^2, \qquad \text{then Var}(d) = \frac{2}{n} S^2(1 - R_{xy}). \tag{12.4.7''}$$

The statistical advantage for measuring differences of a completely overlapping sample over two independent samples of n each can be denoted simply with the factor $(1 - R_{xy})$. The correlation coefficient measures the reduction in the variance when the cost can be stated simply in terms of the $2n$ measurements necessary in either case. When both X_i and Y_i measurements can be obtained for the price of one, the factor $(1 - R_{xy})/2$ represents the advantage of correlated over two independent measurements of $n/2$ each. The difference may represent the change over time of one characteristic. It can as well denote the difference between two characteristics, or responses to two stimuli.

Case 3. When the overlap is incomplete for two equal-sized samples, $(n_x = n_y = n > n_c)$, then $P_x = P_y = P < 1$ and

$$\text{Var}(d) = \frac{1}{n}[S_x^2 + S_y^2 - 2PS_{xy}] = \frac{1}{n}[S_x^2 + S_y^2 - 2PR_{xy}S_xS_y]. \tag{12.4.8}$$

Furthermore, when

$$S_x^2 = S_y^2 = S^2, \qquad \text{then Var}(d) = 2\frac{S^2}{n}(1 - PR_{xy}). \quad (12.4.8')$$

For periodic surveys of time series, constant-size samples with partial overlaps are a favored design. The gains due to correlation come only from the overlapping portion and are directly proportional to it.

Case 4. When one sample $n_y = n_c$ is a subset of the larger sample n_x, then $P_y = 1$, but $P_x < 1$ and

$$\text{Var}(d) = \frac{S_x^2}{n_x} + \frac{S_y^2}{n_c} - \frac{2S_{xy}}{n_x} = \frac{S_y^2}{n_c} + \frac{S_x^2 - 2R_{xy}S_xS_y}{n_x}. \quad (12.4.9)$$

Furthermore, when

$$S_y^2 = S_x^2 = S^2, \qquad \text{then Var}(d) = S^2\left[\frac{1}{n_c} + \frac{1 - 2R_{xy}}{n_x}\right]. \quad (12.4.9')$$

The practical consequences of this are important and not obvious. If $(S_x^2 - 2R_{xy}S_xS_y) < 0$, then the extra measurements $(n_x - n_c)$ can increase rather than decrease the variance of the comparison. When $S_x^2 = S_y^2$ and $R_{xy} > 0.5$, then increasing n_x beyond the overlap reduces the gain obtainable from the positive correlation in the overlap.

Case 4a. When the two variables X_i and Y_i are the same, $R_{xy} = 1$ and

$$\text{Var}(d) = S^2\left(\frac{1}{n_c} - \frac{1}{n_x}\right) = \frac{S^2}{n_c}\left(1 - \frac{n_c}{n_x}\right) = \frac{S^2}{n_c}(1 - P_x). \quad (12.4.10)$$

These situations, when the mean of a sample is compared to the mean of its subset, occur when comparisons are needed between a population and its subclass. They seem to confuse people, who compute the variance as if the two means were independent. For example, in an article, traffic accident data were compared for the United States, for its regions, and for separate states; the statement of sampling errors overlooked the reduction in the variance due to the overlap P_x. Other examples occur in the use of the "method of expected cases," which essentially compares the means of subsets with their parent group.

We note that the overlap portion n_c contributes nothing to the variance (12.4.10), and that the rest $(n_x - n_c)$ is its only source. This can be extended naturally to the situation where $n_x = N$, the entire finite population. The difference between the known mean of the sample of n_c and the unknown mean $\bar{Y}$ of the population of N is due to our ignorance about the $(N - n_c)$ cases not included in the sample. Thus we have $\text{Var}(\bar{Y} - \bar{y}) = (S^2/n_c)(1 - n_c/N) = \text{Var}(\bar{y})$, which may be considered another demonstration of the way the "finite population correction" arises due to the correlation of the sample mean $\bar{y}$ and the population mean $\bar{Y}$.

The variance of differences is usually reduced by the correlation of overlapping units, because the covariance term has a negative sign, and the correlations themselves are usually positive. This is especially true when measuring change in characteristics, because some stability of characteristics exists in most units from one period to another. Nevertheless, negative correlations do exist, and in those situations the variances of differences suffer increases from the overlapping units.

For the variance of sums the covariance term (12.4.3) has a positive sign, and positive correlations in the overlap commonly cause increases of the variance. For sums of overlapping samples the above formulas hold, after changing the minuses to plus signs. The only exception is (12.4.10), which becomes $S^2\left(\frac{1}{n_c} + \frac{3}{n_x}\right)$; this is an unlikely case, because it would mean adding a subset to its total.

When unrestricted sampling is used, computing formulas are obtained by substituting s_x^2, s_y^2, and s_{xy} directly into the formulas of this subsection. If the design is complex, we must revert back to the basic simple form of (12.4.3) whenever the nature of the samples permits it; otherwise, we resort to (12.4.2). An example of the latter would be the difference of two ratio estimates, $d = (y_1/x_1 - y_2/x_2)$. Its variance takes the form

$$\text{var}(d) = \frac{\text{var}(z_1)}{x_1^2} + \frac{\text{var}(z_2)}{x_2^2} - \frac{2\,\text{cov}(z_1, z_2)}{x_1 x_2}, \qquad (12.4.11)$$

where $z_1 = (y_1 - r_1 x_1)$ and $z_2 = (y_2 - r_2 x_2)$. This computing form (6.5*B*) takes care of various complex samples we have described. It also handles the different kinds of overlaps possible within the same primary selections in multistage sampling. For example, the overlap may refer not to the same elements (say persons), but only to the same dwellings, or to different dwellings in the same blocks, or to new blocks in the same counties. Under these conditions the correlations in the overlaps refer only to the similarities of different elements in the same clusters (12.5*C*).

Simple random formulas take a special form for binomial variables, and (12.4.5) becomes

$$\text{Var}(p_x - p_y) = \frac{p_x - p_x^2}{n_x} + \frac{p_y - p_y^2}{n_y} - \frac{2n_c(p_{xy} - p_x p_y)}{n_x n_y}. \qquad (12.4.12)$$

To avoid confusion with P for the proportion of overlap, we write $S_x^2 = p_x(1 - p_x) = p_x - p_x^2$ and $S_y^2 = p_y - p_y^2$. The proportions with the attributes are p_x among the n_x and p_y among the n_y. The covariance is $S_{xy} = p_{xy} - p_x p_y$, where p_{xy} denotes the proportion having the attribute in both samples. The differences between population values and their sample estimates are neglected here, as they can be in large samples.

In the case of complete overlaps, $n_c = n_x = n_y$ and we have

$$\begin{aligned} n_x \operatorname{Var}(p_x - p_y) &= p_x - p_x^2 + p_y - p_y^2 - 2p_{xy} + 2p_xp_y \\ &= (p_x - p_{xy}) + (p_y - p_{xy}) - (p_x - p_y)^2 \\ &= p_{10} + p_{01} - (p_{10} - p_{01})^2. \end{aligned} \tag{12.4.13}$$

Here p_{10} denotes the proportion with the attribute *only* in the first (or x) observation, and p_{01} the proportion only in the second (or y) observation. Also note that $(p_x - p_y) = (p_x - p_{xy} - p_y + p_{xy}) = (p_{10} - p_{01})$. This formula corresponds, as it should, to another development of two completely overlapping binomials in (12.10.2) and Fig. 12.10.I.

12.4B Gains from Overlaps in Measuring Change

For measuring change in the means of two samples, the variance can often be reduced by using the same sampling units in the two successive surveys. As compared to results from two independent samples, the variance of the difference between means is reduced to the degree that the means from overlapping units are positively correlated. These effects are presented simply for unrestricted random samples on the first line of Table 12.4.II. For example, from the same dwellings included in two yearly surveys (the 1951–52 Surveys of Consumer Finances), we found a correlation coefficient of $R_{xy} = 0.8$ in automobile ownership. Hence, in

Statistic	No Overlap	Complete Overlap	Partial Overlap
Difference: $(\bar{y}_2 - \bar{y}_1)$	$2S^2/n$	$(1 - R)(2S^2/n)$	$(1 - PR)(2S^2/n)$
Sum: $(\bar{y}_1 + \bar{y}_2)$	$2S^2/n$	$(1 + R)(2S^2/n)$	$(1 + PR)(2S^2/n)$

TABLE 12.4.II The Effect of Overlap on the Variances of the Difference and of the Sum of Two Means

The top line presents variances for the difference between the means of two unrestricted random samples, each of size n. The bottom line presents variances for the sum of the two means. The first column shows it for the case of two independent samples without overlapping. The second column represents complete overlap, when two samples consist of the same n elements. The third column shows an overlap of the proportion P: of the n elements in the first sample, Pn are also included in the second, but Qn are replaced by an equal number of new elements. $R = R_{xy}$ denotes the correlation coefficient between the variables X_i and Y_i.

measuring the change in car ownership with the difference $(y_2 - y_1)$, a complete overlap should reduce the variance from this source by a factor of $(1 - 0.8) = 0.2$; and an overlap of $P = 0.5$ would reduce it by a factor of $(1 - 0.5 \times 0.8) = 0.6$. Unlike car ownership, the purchase of new cars had a correlation of about -0.1; hence, a complete overlap would increase the variance of the change by the factor of $(1 + 0.1) = 1.1$; and an overlap of 0.5 by a factor of $(1 + 0.5 \times 0.1) = 1.05$.

Generally some stability over time exists for most characteristics. Hence positive correlations, small or great, generally arise from the use of overlapping units. Researchers often take advantage of those correlations for measuring differences. The effect on differences of complete overlaps is $(1 - R_{xy})$. The effect of a partial overlap $(1 - PR_{xy})$ is proportionate to the portion P of overlap. The effect on the sums of two means is exactly the opposite: $(1 + R_{xy})$ for complete overlap and $(1 + PR_{xy})$ for partial overlap. We may write R briefly for R_{xy}.

Overlapping samples are especially important for measuring change of a characteristic in a population, but they also have relevance for comparing the means of two different characteristics in a population. Furthermore, they can deal with problems of comparing two partially overlapping populations.

Circumstances may prevent a completely overlapping design, but permit a partial overlap. For partial overlaps statisticians went further and devised a better statistic for measuring change than the simple difference between means. Due to the correlation, each element in the overlap P contributes less to the variance by $(1 - R)$, than do elements in the non-overlap remainder Q. The minimum variance of the differences is obtained by decreasing the weights in the Q portion by the factor $(1 - R)$. This estimate of the difference with minimum variance is

$$\hat{D}(\bar{y}_2 - \bar{y}_1) = \frac{P}{1 - QR}(\bar{y}_{p2} - \bar{y}_{p1}) + \frac{Q(1 - R)}{1 - QR}(\bar{y}_{q2} - \bar{y}_{q1}). \quad (12.4.14)$$

Here $(\bar{y}_{p2} - \bar{y}_{p1})$ is the difference in means in the P portion, and $(\bar{y}_{q2} - \bar{y}_{q1})$ is the difference in means in the Q portion. The former gets the weight P and the latter $Q(1 - R)$; both are reduced to relative weights when divided by their sums, $P + Q(1 - R) = 1 - QR$. The variance of this weighted difference is

$$\text{Var}\,[\hat{D}(\bar{y}_2 - \bar{y}_1)] = \left(\frac{1 - R}{1 - QR}\right)\frac{2S^2}{n}; \quad (12.4.15)$$

the factor in parentheses shows the reduction due to overlap. The effect of the correlation on the variance is $(1 - R)/(1 - QR)$ on the weighted difference, whereas it is $(1 - PR)$ for the simple difference. These values are shown in Table 12.4.III for four degrees of overlap P and for several values of R. The greatest decrease of the variance occurs when there is complete overlap, with $P = 1$ and $Q = 0$. For this case the decrease is $(1 - R)$, the same as for the simple difference. But for partial overlaps with high correlations, the weighted difference will obtain much greater reductions than the simple difference. For example, to measure the change in car ownership with $R = 0.8$ and $P = 0.5$, the simple unweighted mean reduces the variance by $(1 - 0.5 \times 0.8) = 0.6$. The weighted estimate

P		Negative Values of R_{xy}					0	Positive Values of R_{xy}						
		−1.0	−0.8	−0.6	−0.4	−0.2		0.2	0.4	0.6	0.8	0.9	0.95	1.0
1/3	*a*	1.33	1.27	1.20	1.13	1.07	1.00	0.93	0.87	0.80	0.73	0.70	0.68	0.67
	b	1.20	1.17	1.14	1.11	1.06	1.00	0.92	0.82	0.67	0.43	0.25	0.14	0
1/2	*a*	1.50	1.40	1.30	1.20	1.10	1.00	0.90	0.80	0.70	0.60	0.55	0.52	0.50
	b	1.33	1.29	1.23	1.17	1.09	1.00	0.89	0.75	0.57	0.33	0.18	0.10	0
2/3	*a*	1.67	1.53	1.40	1.27	1.13	1.00	0.87	0.73	0.60	0.47	0.40	0.37	0.33
	b	1.50	1.42	1.33	1.24	1.12	1.00	0.86	0.69	0.50	0.27	0.14	0.07	0
1.0		2.00	1.80	1.60	1.40	1.20	1.00	0.80	0.60	0.40	0.20	0.10	0.05	0

TABLE 12.4.III Effects on the Variance of Differences of R_{xy} for Several Proportions of Overlap (P)

These effects are $a = (1 - PR)$ for the simple difference (12.4.8′), and $b = (1-R)/(1-QR)$ for the weighted difference (12.4.15). Two equal, unrestricted samples are assumed.

will reduce the variance by $(1 - 0.8)/(1 - 0.5 \times 0.8) = 0.2/0.6 = 0.33$. The weights would be $0.5/0.6 = 0.83$ for the overlap portion of $P = 0.5$, and $0.5(1 - 0.8)/0.6 = 0.1/0.6 = 0.17$ for the nonoverlap portion of $Q = 0.5$.

The variance of the weighted estimates compared to the simple estimates of the difference is $[(1 - R)/(1 - QR)]/(1 - PR) = (1 - R)/(1 - R + PQR^2)$; the extra gain becomes appreciable for large R and for moderate values of P and Q, when the term PQR^2 is substantial compared to $(1 - R)$. Again, for $R = 0.8$ with $P = 0.5$ we have $(1 - 0.8)/(1 - 0.8 + 0.25 \times 0.64) = 5/9$, which corresponds to the variances 0.33/0.60 we found before.

In practical situations the correlations R_{xy} are not known exactly, and their estimates must be used for designing the overlaps and computing the weights. Hence, the theoretical limits of efficiency are not actually achieved, but the variance is not sensitive to small changes in the weights, and approximate values of r_{xy} for computing weights will yield most of the available gains. Hence, approximate weights will suffice; using only part of the sample may also be sufficient. When dealing with periodic series, estimates can be computed from preceding surveys. The use of estimates r_{xy} from the sample also lead to a technical bias in the estimates; this effect is slight if the samples are not small.

To measure change we can sometimes choose between two independent samples of n elements each and a sample of n elements with duplicate measurements. When this choice is available, the sample of duplicate measurements is usually preferable because it has the variance $(1 - R)2S^2/n$, as against the variance $2S^2/n$ for the two independent samples. If the correlation R is positive, the duplicate measurements have a lower variance. Furthermore, sometimes the cost is proportional to the number of subjects rather than measurements, and the cost of two samples of n each equals the cost of $2n$ duplicate measurements. The latter has a variance of $(1 - R)S^2/n$, which has the ratio $(1 - R)/2$ compared to the two independent samples, and represents a very large reduction.

Often, however, duplicate measurements on the same elements are not possible. But it does not follow that one must use two entirely independent samples. Even if we cannot measure the same elements twice, we may take advantage of other correlations. This leads to the use of "matched samples" for measuring change, discussed in the next section (12.5).

Sometimes the correlation of duplicate measurements can be used to reduce the variance of the single estimate $\bar{y}$. For example, suppose we have n interviews with adults on how they expect to vote this year ($\bar{y}$), and on how they voted in a recent election ($\bar{x}$); also that both responses are without large bias and have relatively small error. Assume that the population has not changed essentially since the recent past election, in which the total vote is known to be $\bar{X}$. The variance of the simple mean ($\bar{y}$) is S_y^2/n; the variance of the estimate $\bar{X} + (\bar{y} - \bar{x})$ is $S^2_{(y-x)}/n = 2S^2(1-R)/n$. Hence, if the correlation is greater than 0.5—as is likely for voting behavior—the estimates will be more precise when based on the change applied to a known earlier value.

Other considerations often interfere and prevent us from choosing our design on statistical efficiency alone. A reliable standard mean, like $\bar{X}$ in the preceding illustration, is seldom available; even if it were obtained, we might not feel confident that we can draw a sample for measuring the new treatment mean from the same population. Even if the persons in the population are physically the same, the passage of time may have changed them or their response. This change in the population can be viewed as a bias that would confuse the measurement $(\bar{x} - \bar{y})$ of the difference of the two means.

The role of the correlation for reducing the variance of differences in overlapping units is not widely understood. Some fail to appreciate the large gains it can yield. On the contrary, others think that differences must always be measured with duplicate measurements, though this is not true. For example, suppose that from many measurements the effect of a standard treatment is well known to be $\bar{X}$ with variance S_x^2; and that we want to test a new treatment and to estimate the difference $(\bar{Y} - \bar{X})$ between the two treatments. Now suppose that the duplicate measurements, n old (X_i) plus n new (Y_i) measurements, cost as much as making only the new measurement on $2n$ cases. How large must the correlation be between the two measurements Y_i and X_i to make the variance of the n duplicates lower than the variance of $2n$ new cases? The answer is given by $(S_x^2 + S_y^2 - 2RS_xS_y)/n < S_y^2/2n$. The former is the variance Var $(\bar{y} - \bar{x})$ of the n duplicates with correlation R between them. The latter, Var $(\bar{y} - \bar{X})$, shows only the variance of the $2n$ new cases because it takes advantage of the known value of $\bar{X}$. The above yields $R > (S_x/2S_y + S_y/4S_x)$ as the minimum value that R must have for the duplicate measurement to yield a lower variance. When $S_x = S_y$, then $R > 0.75$ indicates the minimum correlation needed.

Now, for a second model, suppose that the number of subjects is the main cost factor, and the n duplicates cost only as much as n new measurements would. Then we must have $(S_x^2 + S_y^2 - 2RS_xS_y)/n < S_y^2/n$, which is satisfied by $R > S_x/2S_y$. When $S_x = S_y$, the duplicate measurements have lower variance to the degree that $R > 0.5$. The second model is proper when the X_i measurements cost nothing extra when taken along with the Y_i measurements on the new cases. The first model holds when the two measurements cost the same. If the X_i measurements cost more than the Y_i, the necessary minimum R will come to even more than 0.75.

12.4C Sums and Current Estimates from Overlapping Samples

When similar data are available from two or more successive surveys, we may want to utilize the results of their sum. For example, a series of monthly aggregates of the incidence of unemployment, or of illnesses, or consumer purchases, or store sales may be summed to yield estimates of yearly aggregates. From a series we may also compute a mean of their means. This estimates the mean condition prevailing over the period covered by the series. The means of attitudes and behavior from five yearly surveys could estimate the mean condition prevailing over the five years. This design is valuable when the distribution is reasonably stable, and if the results of a single sample would be subject to large errors. It seems particularly appropriate for measuring some rare characteristics (11.4). It has been used in studies of human fertility and life insurance ownership, among others.

Given two independent samples of the same size, each with variance S_j^2, the variance is $2S_j^2$ for the simple sum (or difference) and $S_j^2/2$ for their mean. For J independent and equal samples, the variances are JS_j^2 for their simple sum and S_j^2/J for their mean. When samples are overlapping, their sum is bound to show the effects of their correlation. Usually the correlations are positive and tend to increase the variances of sums. The effects on means is the same as on sums, since these two differ only by a constant factor.

Correlation in overlapping samples has an effect on the sum of two samples that is opposite to its effect on their difference. In the second row of Table 12.4.II this effect appears as $(1 + R)$ for a complete overlap, and $(1 + PR)$ for a proportion P of overlap of two samples. For car ownership, the correlation of $R = 0.8$ increases the variance of a complete overlap by 1.8; the mean for two samples would have a factor of 0.9, only slightly better than a single sample. If J samples overlap completely, the mean of their means has the variance:

$$\text{var}\,(\sum \bar{y}_j/J) = (\sum S_j^2 + \sum S_j S_k 2R_{jk})/J^2.$$

If all the S_j^2 and the R_{jk} may be considered uniform, or if one may consider an average value for them, the variance is

$$\text{var}\,(\sum \bar{y}_j/J) = (S_j^2/J)[1 + (J - 1)R].$$

If the samples overlap to the uniform fraction P, the place of R is taken by PR.

In symmetrical opposition to the difference, the sum (or average) of the two means can be improved by weighting the nonoverlap portion Q by the factor $(1 + R)$. This expresses the amount by which the elements in the Q portion have less variance than in the P portion. The variance can be proven to be lowest when the weights are in the proportions P and $Q(1 + R)$ for the overlap and nonoverlap portions, respectively. The formula for this weighted sum is

$$\hat{S}(\bar{y}_1 + \bar{y}_2) = \frac{P}{1 + QR}(\bar{y}_{p1} + \bar{y}_{p2}) + \frac{Q(1 + R)}{1 + QR}(\bar{y}_{q1} + \bar{y}_{q2}), \quad (12.4.16)$$

and its variance is

$$\text{Var}\,[\hat{S}(y_1 + y_2)] = \left(\frac{1 + R}{1 + QR}\right)\frac{2S^2}{n}. \quad (12.4.17)$$

This is minimum when $Q = 1$ and $P = 0$. Hence, in the presence of positive correlations for estimating sums or means of surveys, it is best to avoid any overlap.

If the overlap is unavoidable, the weighted sum increases the variance by a somewhat smaller factor $(1 + R)/(1 + QR)$ than the factor $(1 + PR)$ for the simple sum; but the improvement is not great. This may be seen in Table 12.4.III by comparing the bottom rows for the weighted sums with the top rows for unweighted sums. Because the table was prepared for differences, the sign of R must be reversed for the sum; the values of $(1 - R)/(1 - QR)$ for negative values of R give us symmetrically the values of $(1 + R)/(1 + QR)$ which we need for positive values of R. For example, the $R = 0.8$ for car ownership, in an overlap of $P = 1/2$, increases the variance of the simple sum by $(1 + R/2) = 1.40$; for the weighted sum this is improved modestly to an increase of $(1 + 0.8)/(1 + 0.8/2) = 1.29$.

The simple mean of one sample is not affected by the correlation of any overlap with a preceding or following sample; it remains simply S^2/n for an unrestricted sample. But estimates weighted for regression can improve the means of *partially* overlapping samples. Suppose we have a time series of unrestricted samples each of size n, of which Pn are overlaps with the preceding period and Qn are nonoverlaps. Then the current estimate is a weighted estimate of two parts: the mean $\bar{y}_q$ of the new Qn cases weighted with $Q(1 - QR^2)$, plus the mean $\bar{y}_p$ of the Pn cases improved with a regression estimate on the mean $(\bar{x} = P\bar{x}_p + Q\bar{x}_q)$ of the preceding sample. The weighted mean is

$$\bar{y}_{ri} = \frac{P}{1 - Q^2R^2}[\bar{y}_p + R(\bar{x} - \bar{x}_p)] + \frac{Q(1 - QR^2)}{1 - Q^2R^2}\bar{y}_q$$

$$= \frac{RPQ}{1 - Q^2R^2}(\bar{x}_q - \bar{x}_p) + \frac{P}{1 - Q^2R^2}\bar{y}_p + \frac{Q(1 - QR^2)}{1 - Q^2R^2}\bar{y}_q. \quad (12.4.18)$$

The brackets enclose the regression mean (12.3.1) of $\bar{y}_p$ on the mean $\bar{x}$ of a larger preceding sample; the correlation coefficient $R = BS_y/S_x$ is substituted for B on the assumption that $S_y = S_x$. The variance of the current mean can be reduced, with the correlation R found in the overlap Pn, by utilizing the portion Qn of the preceding sample. The reduction of the element variance due to correlation has the value $(1 - QR^2)$; the weight Q of the nonoverlap is decreased in proportion. The sum of the two weights is $P + Q(1 - QR^2) = 1 - Q^2R^2$. Then

$$\text{Var}(\bar{y}_{ri}) = \frac{S^2}{n}\left[\frac{1 - QR^2}{1 - Q^2R^2}\right], \quad (12.4.19)$$

because

$$W_pS_p^2 + W_qS_q^2 = P(1 - QR^2)S^2 + Q(1 - QR^2)S^2 = (1 - QR^2)S^2.$$

This holds equally for positive or negative R, which appears squared. Optimum results are obtained when the overlap is designed for $P = 1 - 1/(1 + \sqrt{1 - R^2})$. With these proportions, for any R we can produce the optimum variance:

$$\text{opt Var}(\bar{y}_{ri}) = \frac{S^2}{n}\left(\frac{1 + \sqrt{1 - R^2}}{2}\right). \tag{12.4.20}$$

Considerable gains are possible only for large R. In Table 12.4.IV, note both the optimum reductions of the variance and those available with $P = 1/3$. It is interesting that this overlap of $P = 1/3$ performs practically

R (or $-R$)	0.5	0.6	0.7	0.8	0.9	0.95	0.98	0.99
$\left(\frac{1 - (2/3)R^2}{1 - (2/3)^2R^2}\right)$	0.94	0.90	0.86	0.80	0.72	0.67	0.63	0.61
Optimum Variance $(1 + \sqrt{1 - R^2})/2$	0.94	0.90	0.86	0.80	0.72	0.66	0.60	0.57
at $P = 1 - (1 + \sqrt{1 - R^2})^{-1}$	0.46	0.44	0.42	0.38	0.30	0.24	0.17	0.12

TABLE 12.4.IV Improvement of Current Estimates by Regression on a Partially Overlapping Sample

The correlation coefficient (+ or −) is in the first row. The second row shows the reduction in the variance by means of regression on an overlap of $P = 1/3$. The third row shows the reduction of the variance for optimum overlap.

as well as the optimum P, which varies from 0.46 to 0.12 as we go from $R = 0.5$ to $R = 0.99$. An overlap of $P = 1/4$ performs slightly better than 1/3, when $R > 0.9$. Interesting elaborations of this technique can improve current estimates by utilizing auxiliary data from *several* previous rotating samples of a time series. Optimum schemes of rotation have been designed by several investigators [Hansen, Hurwitz, and Madow, 1953, 11.7; Cochran, 1963, 12.11; Yates, 1960, 7.19, 8.8; Patterson, 1950; Eckler, 1955].

12.5 PANEL STUDIES AND DESIGNS FOR MEASURING CHANGES

12.5A Panels versus New Samples

Panel denotes a sample in which the same elements are measured on two or more occasions. Some research problems dictate a clear choice between demanding or prohibiting the use of panels. Only panels can give information about the gross change behind a net change. For example, suppose that the prevalence of a behavior, attitude, or disease changes from 10 to 15 percent from one independent sample to another; we do not know whether the incidence of new cases is 5 or 15 percent—corresponding to 0 or 10 percent reverse change—or something in between. Sometimes theory and past research can supply a model; a disease may be chronic and the net change (5 percent) must equal the gross change; on the contrary, a disease or other event may occur only once, and the prevalence (15 percent) measures the new incidence. Sometimes changes can be traced merely from the memory or records of current cases, then applied to the earlier estimate; if 8 of the current 15 percent cases had the characteristic earlier, then 2 of the former 10 percent lost it, and 7 percent acquired it. Often both theory and memory are lacking or unreliable, and only panels can yield needed data on gross changes. Only panels permit studies of individual changes; these may be needed not only for counting the frequency of changes, but also for research on the dynamics of causation and relationships.

On the contrary, the use of panels may be disallowed by the "destructive" nature of the measurement. A measurement may be destructive because it destroys the subject; a treatment is "destructive" if it cures the subject of a condition that we are unwilling (if human) or unable (if animal) to inflict again. Testing the difference between two alternative questions given to the same subjects may be unrealistic or unreliable; the term *contamination* refers to tests and questions that destroy the innocent or unbiased nature of the subject. Measuring changes in persons from one year to another may be unreliable because too many may refuse to be interviewed twice, or because reluctant respondents may give poor answers. Furthermore, locating again a sample of respondents a year later is expensive, due to travel costs and the obstacles to finding those who have moved.

More often than not, neither the advantages nor the disadvantages of panels are absolute; they must be balanced against each other. Into this balance we place the relatively clear statistical arguments outlined in the last section, along with the broader but less clear problems of measurement. The choice is often complicated, because surveys have divergent aims that tend to lead to different optimum designs. Stability of unit characteristics over time result in strong correlations, which have conflicting results on the variances of different estimates. The variance of differences between samples is least for a completely overlapping panel, but this increases the variance of the sum (or average), which is least for zero overlap. The variance of simple current estimates is not affected by the overlap; but it tends to be least when the overlap is about one-third, if the estimates have been weighted by regression. Conflicts between optimum designs are severe for variates that possess strong panel correlations. Conflicts should be resolved in accord with guesses about the relative values of conflicting survey objectives. Conflicting aims may be resolved by compromises in partial overlaps.

12.5B Partial Overlaps in Rotation Designs

Partial overlaps can be understood most readily in the context of rotating designs of periodic surveys. As the simplest example, note a design for a periodic survey involving overlaps of $P = 1/2$ in successive samples, with two successive measurements on each new selection:

$$ab\text{–}bc\text{–}cd\text{–}de\text{–}ef\text{–etc.} \tag{12.5.1}$$

The b half-sample appears in the first and second samples; the c half-sample in the second and third samples, and so on. For a second example, take a design of a $P = 2/3$ overlap, with three successive measurements on each new selection:

$$abc\text{–}bcd\text{–}cde\text{–}def\text{–}efg\text{–etc.} \tag{12.5.2}$$

Each sample is composed of three equal portions. The c portion appears in the first, second, and third samples; the d portion in the second, third, and fourth samples, and so on. The third example changes the second into an overlap of $P = 1/3$:

$$abc\text{–}cde\text{–}efg\text{–}ghi\text{–}ijk\text{–etc.} \tag{12.5.3}$$

Portion c appears in the first and second samples, portion e in the second and third samples, and so on for g, i, etc. But d, f, h, etc., each appear only in a single sample. Thus half of the new selections receive two successive measurements, but the other half only one. If the samples are quarterly surveys, a yearly survey consists of the sum of four samples, in which nine of the 12 portions are distinct. If the overlap is made $P = 1/4$ (with *abcd–defg–ghij*–etc.), then only a third of each new selection is subject to two measurements. The fourth example is more complicated:

$$eaf\text{–}fbg\text{–}gch\text{–}hdi\text{–}|\text{–}iej\text{–}jfk\text{–}kgl\text{–}lhm\text{–}|\text{–}min\text{–}njo\text{–etc.} \tag{12.5.4}$$

This design could serve a series of four quarterly samples. Each portion occurs twice in succession and, after a lapse of three quarters, a third time a year after the second appearance. There is an overlap of $P = 1/3$ with the previous quarter, and another overlap with the previous year. These overlaps strengthen the two salient comparisons of quarterly and yearly surveys. Current estimates can also be strengthened with regression weights based on overlaps of 1/3, which was shown to be close to optimum, with support from 2/3 of two other samples. The yearly estimates contain four quarters and involve 12 parts, of which nine are different.

The complicated design of the monthly Current Population Surveys contain overlaps with neighboring months and with a year's span (10.4). Finally consider an asymmetrical overlap:

$$abbb\text{–}bccc\text{–}cddd\text{–}deee\text{–etc.} \qquad (12.5.5)$$

A quarter of each new sample is an overlap with three-quarters of the preceding sample. The overlap can take advantage of information gained in one survey to introduce disproportionate allocation in the next, for variables unevenly distributed in the population, such as income and savings. By using different sampling rates, the original sample is reduced to one-third its original size (*bbb* to *b*); for example, these proportions and rates in three strata could be $0.04 \times 1 + 0.20 \times 1/2 + 0.76 \times 1/4 = 0.33$. The three kinds of elements of the *b* sample would have weights of 1, 2, and 4, respectively, whereas in the new *c* portion the weights are all 1.

Instead of, or in addition to, overlaps, periodic surveys can benefit from recall of past information, if this can be obtained reliably from either records or memory. This information, correlated with past surveys, can be utilized, with proper precautions and estimating procedures, to reduce the variance; this is done in the Surveys of Retail Stores of the U.S. Census Bureau [Hansen, Hurwitz, and Madow, 1953, 12.A; Woodruff, 1963].

12.5C Choice of Overlapping Units

When planning a study involving overlaps, we naturally think of overlapping elements. These units yield the highest correlations for measuring net differences and are essential for measuring gross changes. When an overlap of elements is too difficult to achieve, before altogether abandoning the idea of an overlap, consider a compromise plan involving other units. Although such plans yield lower gains than an overlap of elements, they may be more feasible.

It may be too expensive to locate and identify the elements of a past sample, and to include them again in later samples. For example, in a sample of U.S. families, about 20 percent would have changed dwellings in a year's time. To follow and find them for interviewing may be costly, especially for those who have moved out of the areas (counties) covered by the field teams. This problem is simpler with mailed questionnaires and telephone interviews. Identification of the same family units can be complicated when births, deaths, marriages, divorces, additions, and removals change the family composition. Dwellings maintain more stability than families, both in location and identification. Panels of compact segments are usually even more stable than dwellings (9.4*A*, 10.4). It is true that dwellings and segments used as panel units yield smaller correlations than families; but the reduction affects only the dwellings of the 20 percent who moved. We can regard dwellings or segments as panel units with their own correlations. For many characteristics the in-and-out movers will tend to be correlated also, although less than the nonmoving families.

However, the loss from movers is more damaging for studies of gross changes; these should essentially be confined to the population of nonmoving and reidentified units. Moves would reduce the sample base of nonmovers to 80 percent in one year, and to 65 percent in two years. If nonresponses of two samples are added, panel losses can quickly reduce the sample base of identical families to about 50 percent [Sobol, 1959]. Under these conditions a supplement of movers may have to be added to the panel of segments, to provide a combined sample of identical families. The sampling rate of movers may be lower, corresponding to their higher element cost. The SRC has based reinterview studies on dwellings and segments designed so that the two samples represented correctly the entire populations for both periods and the net change between them. Furthermore, sometimes a supplement of movers is added to provide a broader base for studies of the gross change and individual changes.

Births and deaths in the population cause problems in procedures and in defining the population. *Migration* into and out of the population can be defined to include births and deaths, because of methodological similarities. Migration can refer not only to location, but to any defined entering or leaving the population. All these problems can be involved in samples of a city's total population, or its resident blue-collar employees, or the employees of a factory. For studying gross change and individual changes we may prefer a narrow definition of the population: elements that are members during the period covering both surveys. For studies of net changes and of separate periods, a broader definition is preferable; the population as it exists within similar limits at the time of each survey period; hence all migrations of the population will be included in measures of net change. The design for measuring net change must ensure the inclusion in the second sample of representation of new units "born" or created since the first one. For example, a supplement for new dwellings may be necessary; or the procedures may be self-correcting, such as the use of compact segments.

Problems of destructive measurement, poor responses, or refusals, may lead us to avoid any overlap of elements in samples used for comparisons.

Nevertheless, some correlation may be retained by basing comparisons on pairs of sampling units that are "neighbors" or "matching" in some meaningful sense. Neighboring pairs on the frame listing may be correlated, if the ordering is meaningful. For example, we selected triplets of neighboring addresses from a city directory list; one address from each triplet was then assigned to each of three successive yearly Detroit Area Studies [Sharp and Feldt, 1959; Sharp 1961]. Comparisons of yearly surveys benefited from the correlations of neighboring addresses. Or consider a sample of segments split into halves, with half of each segment assigned to one of two samples of a comparison. Halving the segment can also be done by the interviewer, who assigns face sheets from a bound pack in which A and B forms alternate. A sample of a county can be selected in meaningful pairs: one member of each pair assigned to each sample. Finally, primary units, such as counties, can also be selected in meaningful pairs. Generally, the utilization of the same sampling units for two samples reduces the variance of their comparison. The largest reduction comes from using the same elements, whose correlations are highest. The higher we go in the hierarchy of sampling units, and away from the elements, the lower the correlation available for reducing the comparison; however, some correlation remains even up to the primary units. In several studies we found that, for comparisons of national samples based on the same counties, the variance was reduced by correlations in the proportion $(\sigma^2 + \sigma^2 - 2R\sigma^2)/2\sigma^2 = 1 - R$. From values of 0.75 (Table 14.1.IV) and of 0.85 [Kish, 1963] we deduce rough correlations of 0.25 and 0.15 for these partial overlaps.

Sometimes a selected sample of individuals can be paired and matched, on the basis of available and meaningful data that induces some correlation in the pairs. One member of each pair is placed alternately into one of the two samples; the matching should reduce the variance for measuring differences and changes. The pairs should be selected with probability sampling; correlations are never strong enough to justify the naive disregard of sampling principles that characterizes some studies based on matched pairs.

12.5D Splitting a Large Survey into Repeated Samples; Time Sampling

In 12.4C we discussed the possibility that several periodic surveys, each designed chiefly for measuring one specified period, can be combined into an average result for the entire period covered by the surveys. Now we look at the other side of the same coin. Instead of representing a period with a single large survey, important advantages can be obtained from dividing it into several repeated smaller surveys. For example, some large surveys for measuring traffic volume, typically concentrated in a few weeks, could perhaps be better spread over a year. The budget of the triennial surveys of automobile traffic over New York's bridges was successfully spread over the three-year period [Kish et al., 1961*a*]. I think that some of the activity of the decennial Census, now focused on a single day every 10 years, could better be split into separate periodic samples, covering the country in 120 monthly surveys, or 40 quarterly surveys.

Obviously this is a considerable and controversial subject, without unique answers for all situations. Still, some brief and general remarks may be helpful. The possible advantages of repeated surveys are of three kinds:

(*a*) A design of repeated samples can improve quality or reduce costs, or some of both. This is in contrast to an abrupt, hasty, and isolated effort to conduct a large survey. Hiring a large force of enumerators and clerks for a single survey can be an expensive and risky adventure. The "input" cost of hiring and training the enumerators for a single survey looms large in relation to a short "output" time. On the other hand, minimizing that input cost can lead to poor quality.

(*b*) Separate samples of a repeated design can yield additional information about variations that occur between the periods, and can also estimate seasonal and secular trends. Such sampling may also detect the effects of irregular or sudden changes in the predictor variables. The effects of catastrophes—such as epidemics, stock market crash, bad foreign news, or the threats of war—may be detected in surveys measuring related variables. Repeated surveys can provide up-to-date information to facilitate and speed decisions that depend on knowledge of fluctuations in the survey variables.

For example, consider a survey period that straddles the date of an important event, such as an election. The sample can be split into two random parts, *A* and *B*, preferably about equal. The enumerators are instructed to finish as much as possible of part *A* before the critical date, and to begin *B* after the date. If the separation of the two samples, before and after the critical date, is good (even if not perfect), a satisfactory estimate of the event's effect can be computed. Without random separation, merely comparing responses before and after the event would confuse possible differences between early and late responses with the differences due to the event.

(*c*) The sum of repeated surveys over the entire period can lead to better statistical inference than a single, concentrated, *one-shot* survey. *Probability selection of time segments from an entire interval permits statistical inference from the sample to an average condition over the interval.* On the contrary, inference from a "typical" time segment on a *one-shot* survey to an entire interval demands judgment and assumptions about the nature of variation, or lack of it, over the entire interval. The choice of a single time segment is exposed to the risks of seasonal, secular, and catastrophic variations, known or unknown. The sum of repeated surveys relies on averaging out the variations over the repeated surveys.

The advantages of the *one-shot* survey over a continuous operation need only brief mention, since they are better known:

(*d*) A specific date may fit some requirement, seldom scientific, but perhaps legal, or perhaps imposed by relations to other sources of data.
(*e*) A complete census, or heavy coverage during a short period, may be needed to investigate relationships among the survey variables.
(*f*) It may be easier and cheaper to administer than a repeated sample spread over a long interval. This may be especially true of small samples and of geographically widespread samples.
(*g*) Most surveys use *one-shot* samples, to permit analysis and presentation of all its results rapidly.

Just as a target population varies in space, we can profitably consider time as another dimension of variation. The population varies from year to year and week to week, as it varies among regions and among counties. We emphasize here the central idea of variation, but not its distribution. For that purpose the ordered direction of time is not relevant; space can be made continuous by using latitude or altitude. The analogy of time and space is close when the elements are mobile; but the analogy is less close when elements have stability in space and time, because correlation is greater for a unit over time than for units neighboring in space. We use space here as the clearest illustration, but other analogies are relevant with dimensions based on social, physiological, and other variables.

Points (*b*), (*c*), (*d*), (*e*), and (*f*), regarding time have widely known counterparts in spatial dimensions, which can help us understand better their counterparts in time. The separate periods of a repeated sample can be regarded as strata and domains of the entire survey interval. The spread of samples over time yields comparisons of temporal domains, similar to comparisons of spatial domains (*b*). Inference from a "typical" area (or period) to an entire country (or time span) involves much personal and vague judgment. In contrast, a probability sample of space (or time) permits statistical inference about it (*c*). On the other hand, sometimes the sample of one small area, or of one short period, may be specifically demanded (*d*). This can permit the study of interrelationships within the area or period (*e*). Concentrating the sample in one area or a short period may be cheaper (*f*).

Our twin concerns about spreading a sample in space and in time are not always independent and may well conflict. For example, a sample of 1200 dwellings can be spread over 12 monthly samples, if confined to a small area. Or it may be spread over a *one-shot* national sample. But spreading it into 12 national samples of 100 dwellings each would raise many practical problems. We may have to decide whether spreading in space or in time is more relevant for our research. That is but one aspect of the constant necessity to choose an economic set of variables from a much greater number of potential candidates.

However, when we are able to spread the sample over both space and time, we can create designs that balance the two. For example, a design of 12 monthly samples may be too small to yield reliable regional data each month. But the monthly samples could be stratified over the regions, and the regional samples stratified over the separate survey periods. Then we can use either the separate periods or the separate regions as domains; in either case, the other dimension provides a stratified sample with correspondingly improved spread.

Among potential fields of application of these ideas are auditing, accounting, and inventories, where the common practice of the judgment choice of a "typical" week is being supplanted by the probability choice of a spread of dates. Another fruitful area is the measurement of traffic flows [Kish et al., 1961*a*].

12.6 CONTINUING SAMPLING OPERATIONS

The last two sections described techniques for the repeated collection of a series of similar data. *Here we deal with techniques generally applicable to any repeated use of the same selection frames with similar sampling methods.* These techniques can be used by organizations that carry on sampling operations on a continuing basis, whether the objectives of successive surveys are changed or not. We cannot dwell here on the advantages of adequate survey staffs in the office and in the field, accustomed to working as teams and experienced in their special problems. Nor shall we discuss the savings from merely distributing over many surveys the costs of expensive sampling materials, such as maps, instructions, forms, manuals. These advantages and savings are substantial; they may be properly allocated with cost accounting, similarly to investments in large machine tools for mass production.

If we plan for a continuing survey operation, we can introduce considerations beyond those of designs for single surveys. We present briefly several simple techniques especially suited to continuing operations. Each technique takes deliberate advantage of treating a specific sample as a subsample drawn from a group of similar samples. Thus both the costs and errors of a single sample can be reduced by spreading them over the group.

A clear example of this is the initial sample of segments divided into 12 monthly installments for the Current Population Surveys of the U.S. Census Bureau (10.4). Knowledge about the initial sample is used to improve each of the 12 subsamples. For example, segments with unusually large numbers of households are placed in a "large segment universe" and treated with special procedures (12.6*C*).

12.6A Master Frames

A large sample can be selected in an initial first phase, and later serve as a frame for selecting actual samples as needed. An extremely large-scale example is the "Master Sample of Agriculture" [King and Jessen, 1945], where rural areas on the maps of over 3000 U.S. counties were divided into segments of about four farms each. After selecting a systematic sample of 1/18 of the segments, the materials were duplicated and made available, with instructions, at low cost. As a modest contrast, the list of dwellings in a sample of a city's blocks can be utilized to provide several two-stage samples. The possible instances of this general method are too varied for a systematic treatment. However, we can illustrate the method with procedures of the Survey Research Center; we also enlarge their scope with occasional extensions. The actual sample for each new survey is not selected directly from the entire population, but from a frame of segments and dwellings that was selected earlier from the entire population. The preparation of this frame is usually accomplished separately from any survey, preferably during slack periods. We try to do the field work during months of good weather. Most of this listing is done by our field interviewers. In other situations this kind of work can be better accomplished separately by a specially trained staff.

We aim to obtain frames that can provide sample segments for periods of from one to three years. The length of the period is varied between cities and counties, with considerations of efficiency in mind. For example, we prepare smaller frames in cities and counties where we expect rapid growth and obsolescence. From this frame we subsample as needed for any specific survey. The sampling rates are so designed that the product of the two probabilities of selection—first into the frame, and second out of the frame—equals the probability desired for the particular survey. For example, an initial rate of 4/10,000 of the 50 million U.S. dwellings in 1960 yielded a frame of about 20,000 dwellings; selecting from it with $f_2 = 0.08$ resulted in a sample of about 1600 dwellings, with the overall $f = 0.08 \times 4/10{,}000 = 32/1{,}000{,}000$. The segments and dwellings are new for each study, unless a panel study is deliberately designed. The blocks contain an average of eight segments, and typically provide segments for two or three years, until all or most of the segments are used. The county is used for a longer, though indefinite, time; most of the present counties have been in the sample for about 15 years and represent a considerable investment in trained interviewers and sampling materials.

Because the frame must serve for a year or two, or even three, *obsolescence* becomes a special problem. Hence, segments are preferred to dwelling lists, otherwise supplements for new construction may be necessary. This problem of obsolescence is outweighed by the many advantages of master frames. First is the obvious *economy* of spreading the cost of listing operations over several surveys, especially when listing the master frame costs little more than listing for a single survey. Since the frame can be prepared in slack periods, *convenience* and *speed* result from the presence of the master frame, ready for use. Some of the economy is transferred into higher *quality* for the listing operations. When costs are spread over many surveys, we can spend more for creating good segment boundaries, for information on measures of size, and for stratification. For example, economic ratings can be assigned to dwelling listings, and later used for disproportionate allocation (11.4*C*). Four of the procedures involving a master frame for current samples at the Survey Research Center are:

1. Instead of sampling segments directly from a map, we ask the interviewer to create segments out of sample blocks. A rural block is an area that appears on the office maps to have good boundaries and to contain about 30 dwellings. We send the interviewer a *block sketch*, and ask him to revise it to reflect the current situation, to indicate the location of dwellings, and to add any internal boundaries that may be used to divide the block into segments of about four dwellings each. In urban areas the block may be all or part of a city block; it is subsampled after a preliminary visit by the interviewer, who reports dwelling counts and locations, numbers of floors and apartments in apartment buildings, and related information. This permits the office to create and send to the interviewer segments with recent measures of size and with recognizable boundaries. For successive studies we select segments without replacement from the block, and continue until most segments have been utilized for some survey.

2. For selecting tracts or blocks from Census listings we use double sampling; first we select a relatively large sample, stratify it, and then draw from it as needed. It costs little to increase the size of the initial sample from the Census lists. The cost of drawing the initial sample of tracts or blocks can be divided by the number of times it is used. Furthermore, for the actual sample we obtain the gains of stratifying the larger initial sample by geography and income indicators.

3. When using a city directory listing of addresses, we also select a larger frame, then subsample repeatedly. We first select clusters of 10 or 20 directory lines, then subdivide these into "segments" of about 4 dwellings.

4. We also use double sampling in area supplements for picking up new growth and missed dwellings. First we select an initial sample of blocks eight times larger than needed for a single survey, and obtain rough size estimates for them. Then for a single survey we can subselect either 1/8 of the blocks with little growth, or 1/8 of the dwellings from the blocks with much growth.

12.6B Techniques for Cumulating Information

Cumulating cases from several surveys can provide data either for analysis in a time series, or for combined analysis. When combined into one large sample, care must be taken to define the population viewed over the lapse of time, to standardize the survey instruments, and to account clearly for the sampling units used in the joint sample. *Cumulation of rare elements*, too costly to screen in a separate effort, can become the by-product of continuing operations.

Knowledge about sampling errors is cumulated with computations from similar surveys. Knowledge of the magnitudes of *cost factors* and of the *components of the total variance* can be cumulated; this is the best source of data for optimum designs and planned precision. Furthermore, we can increase the precision of estimated variances by averaging them over similar surveys. This is especially helpful when the computed variances, based on few primary selections, hence few degrees of freedom, are subject to large errors.

The cumulation of evidence on response and coverage rates permits better control of sample size. Planned cumulation of field results and analysis of factors associated with them can reveal the sources of nonresponse errors, and how to cope with them. Cumulated data can also yield knowledge about the sources and magnitudes of the *errors of response*.

Replacement procedures can reduce the effects of nonresponse by simulating an increased number of recalls (13.6*D*). The replacement consists of adding to the sample nonresponse addresses from a recent survey, conducted with similar methods over the same population. The effect is about equal to doubling the number of recalls, but without a corresponding increase in expense.

Transfers of large elements to special strata can be made as operations continue over a population; this increases the efficiency of disproportionate stratification. An example of this is reported by Woodruff [1963] for the Monthly Retail Trade Survey of the U.S. Census Bureau. A panel representing a 6 percent yearly area segment sample is divided into 12 equal parts, each providing a monthly sample of 1/2 percent. At the end of the year all stores that qualify for a "large sample cut-off" are transferred to a special stratum, to be sampled in the future monthly surveys at the rate of $12 \times 1/2 = 6$ percent. This stratum is additional to the "list sample of large stores"—which, identified beforehand, do not enter the area sample at all.

12.6C A Stratum for "Surprises"

The sample of a highly skewed distribution may be disturbed by the appearance of a few large units. If these appear with small selection probabilities, they receive large weights and have strong effects on the sample mean and its variance. We would prefer to have more of these units appear in the sample with larger selection probabilities and smaller weights. This can be accomplished with disproportionate allocation if the large units are identified before selection. But treating large units after selection is troublesome, as shown by the example below.

When trying to obtain good measures of size, area samplers are in constant competition with home builders; they are sometimes painfully surprised to find 20 or even 200 dwellings in a small segment where they expected to find 4. For the sake of equal probabilities, the sampler may accept all of the surprise dwellings, with the corresponding increase in variance and cost. Beyond some acceptable size he may decide to cut the sample take, thus accepting some bias, but probably with a lower mean square error than an unbiased procedure. This source of error can be reduced over the long run of continuing operations by *averaging in a surprise stratum the rare events* from single surveys over similar events from similar recent surveys.

The population expansion from a surprise of x_g dwellings (or other elements) from the gth survey, taken with the overall sampling rate of f_g, is x_g/f_g. The average of such expansion over G surveys is

$$\left[\sum_g^G w_g x_g / f_g\right] \Big/ \sum_g^G w_g, \tag{12.6.1}$$

where w_g is the weight given to the gth survey. In many situations, w_g should be made proportional to the f_g; then the average becomes

$$\sum x_g / \sum f_g. \qquad (12.6.2)$$

The last and current survey has the overall sampling rate of f_G, and the sample "take" from the surprise stratum for this survey should then be

$$f_G \sum x_g / \sum f_g. \qquad (12.6.2')$$

The summation can be over a period large enough (perhaps two years) to provide an average over a probability Σf_g. This should be considerably greater than f_G, but the length of the period should not cause a large bias as the result of obsolescence. The age and nature of the lists used in the summation should approximate the current sample.

The interviewer in the field must be given clear instructions to know when to recognize a "surprise segment" beyond a cut-off size; he sends to the office current information about the segment, and waits for instructions. The cut-off size should not be too large, because that would permit an undue increase of the variance, and yield too small a base for the stratum. But it should not be too small either, because that would cause too many interruptions of normal field work, and permit the growth of a potential source of bias.

For the national sample of dwellings of the Survey Research Center we have operated a surprise stratum for years [Kish, 1959*b*]. Currently, our cut-off size is 12 dwellings for segments, where we normally expect an average of 4 dwellings. This results in a surprise stratum that averages about 2 or 3 percent of the sample. For a national sample of 2000 dwellings this means a stratum of about 10 or 15 segments of about 4 dwellings each, coming from about 8 samples of the past year or two.

12.6D Inertias of Continuing Operations

One result of continuing operations is the presence of inertia in different parts of the design. For example, in designing some modest size studies for the Survey Research Center, we find it cheaper and easier to utilize our standard 74 primary areas (or perhaps half of them) than to design and staff 6 or 12 new counties just for one study. This arises from having trained interviewers and sampling frames and materials available in our standard primary areas. It contradicts the rule that it is always cheaper to use fewer sampling units.

Continuing operations also bring about conservatism in methods. Much of this is the justifiable result of having available certain good, economic, reliable methods that have been tested by long experience. But we suspect that there must also exist many less justifiable types of conservatism; we naturally think first of methods that seem to have worked well enough—at least without obvious catastrophes. It is difficult to view new problems with fresh, unbiased eyes. But we should always strive for that fresh point of view and question familiar methods, trying to separate seasoned timber from dead wood.

12.7 CHANGING SELECTION PROBABILITIES

12.7A A Procedure for Minimum Changes of Selections

Continued use of a set of sampling units has several advantages. First, they may represent important investments in office work, sampling maps, materials and computations, and especially in a selected and trained field staff. The interviewers in each county of a national sample may represent an investment of hundreds of dollars. Second, using the same sampling units can decrease considerably the standard errors of comparisons (12.5*C*) [Kish, 1962]. This factor looms especially large in panel samples.

However, to continue using an entire selection of sampling units would restrict the sample to the set of probabilities with which they were selected initially. These initial probabilities were based on initial size measures, which may differ considerably from new measures that would better suit the demands of current surveys. The difference between the initial and the desired new size measures may be due to differential changes of size among the sampling units, perhaps as revealed by the latest Census. It can also be due to differences in survey populations and variables. For example, the initial measures may have been based on all persons, but the current sample may call for a population of college students, or physicians, or farmers, or farm products. We know that variations in the ratios of actual size to size measures cause variations in subsample sizes, and that some variation in subsample size can be tolerated (7.5*C*). But if this variation becomes too large, we may want to change size measures. If measures of size for the new population are available, and if these differ considerably from those of the initial selection, then we should prefer to use them.

We present a method for changing from a set of initial probabilities p_j to a set of new probabilities P_j with a minimum number of actual changes of sampling units. There are $D + I$ units in the stratum. Of these, D receive a decrease in probability, and I receive an increase, which may also be zero. Subscripts distinguish the two sets of sampling units, so that

$$P_d < p_d \quad \text{and} \quad P_i \geq p_i, \tag{12.7.1}$$

where ($d = 1, 2, \ldots, D$) and ($i = D + 1, D + 2, \ldots, D + I$).

The initial probabilities used for the selection of sampling units from a stratum, and the new probabilities to which we want to switch, both sum to unity in the stratum:

$$\sum_d^D p_d + \sum_i^I p_i = 1 = \sum_d^D P_d + \sum_i^I P_i. \qquad (12.7.2)$$

The sum of the probability increases must equal the sum of the probability decreases:

$$\sum_i^I (P_i - p_i) = \sum_d^D (p_d - P_d). \qquad (12.7.3)$$

The rules for changing probabilities are:

(*a*) If the initially selected sampling unit shows an increase (or no change in probability), retain it in the sample, as if selected with the new probability P_i.

(*b*) If the initially selected unit shows a decrease, retain it in the sample with the probability P_d/p_d; that is, assign a probability of $1 - P_d/p_d$ for dropping it. Thus the compound probability of original selection plus retention is made $p_d(P_d/p_d) = P_d$.

(*c*) If a decreasing unit is dropped from the sample, select a replacement among the increasing units, with probabilities proportional to their increases. The probability of selection of the *i*th unit is $(P_i - p_i)/\sum_i^I (P_i - p_i)$. Thus the total selection probability of an increasing unit consists of p_i initially, but it is properly increased if any of the D decreasing units had been selected, then dropped:

$$p_i + \sum_d^D (p_d - P_d) \frac{P_i - p_i}{\sum_i^I (P_i - p_i)} = P_i. \qquad (12.7.4)$$

Often only one sampling unit is selected from each stratum. But the rules are equally valid for two or more fixed numbers of selections with replacement from the stratum. This method was first presented by Keyfitz [1951].

When selection probabilities are changed, the situation may also require changes in the boundaries of sampling units. Some initial units may be split, others combined. Parts of one unit may be transferred to others. The initial selection probabilities p_j of a unit can be divided among several portions; or several of the initial p_j values may be combined into one new unit. Rules can be specified for these changes, provided two vital precautions are observed: The presence of an initial selection in a sampling unit must have no influence either on the boundary changes or on the procedure for identification of initial with final units. The creation of entirely new units can be noted with $p_j = 0$, and the complete elimination of initial units with $P_j = 0$.

We assumed above that the sampling units are retained in their initial strata. A modest volume of shifting units can be handled by simply assigning $P_j = 0$ to initial units placed out of a stratum, and $p_j = 0$ to new units placed into it. If the number of shifts is large, other problems may arise. These are discussed in 12.7*D*.

12.7B Modifications of the Procedure

It is desirable to make the selection probabilities approximately proportional to the unit sizes, but exact proportionality, like exact size measures, are typically neither available nor necessary. To apply the "Keyfitz procedure," it is necessary and sufficient that the sum of changes in probabilities be zero, so that $\Sigma P_j = \Sigma p_j = 1$ be satisfied. Within that requirement we can adjust the selection probabilities to satisfy some criteria of change large enough to be recognized as important, and to deliberately neglect smaller changes.

For many sampling units the changes in probabilities are so small that they can be deliberately neglected. If these units are numerous, and if neglecting their changes adds little to the overall variance, we can reduce the number of changes significantly with little sacrifice in the variance. To these units we can reassign the initial probabilities; they constitute a set of units for which $p_i = P_i$ is arbitrarily assigned. This "flexible" procedure reduces the number of units we must switch.

We used the flexible procedure in changing from the 1940 to 1950 Census measures for the 54 counties, representing that many strata, in the national sample of the Survey Research Center. In choosing criteria we balanced the costs of changing counties against the increase in survey variances due to small distortions in the selection probabilities. These were considered relative to other sources of variation in sample size. We decided on the following procedure: (*a*) Define an important increase as 10 percent or more ($P_j/p_j \geq 1.10$). Compute the sum of the increases in the stratum. (*b*) Add enough of the largest decreases (smallest values of P_j/p_j) to balance *exactly* the sum of the increases; that is,

$$\sum (P_i - p_i) + \sum (P_d - p_d) = 0.$$

(*c*) Consider all other counties as not having changed probabilities, thus with $P_i = p_i$.

Details are given elsewhere [Kish, 1963], together with two other modifications. Instead of merely accepting a change from one Census period to another, we projected the growth of fast-growing areas

(California, Florida) forward into the middle of the service period of the sampling frame. Second, faced with rather small probabilities of change $(1 - P_d/p_d)$ in each of 16 strata, we did not draw independently within each stratum. Instead, we controlled the number of changes by cumulating the probabilities of change from one stratum to another, then applied an interval of one, after a random start. Thus the actual number of changes, which was three, was controlled within a fraction of the expected number of changes.

12.7C A Simple Procedure for Burgeoning Units

A simple procedure can handle a problem confined chiefly to large growth in a small proportion of the primary units. For example, suppose we have a sample of blocks selected with the initial probabilities p_j, proportional to the initial sizes N_j. Suppose also that for a new sample we are willing to retain the original probability p_j for all blocks, except for those that have at least doubled in size. That is, if the new size is $N_j' < 2N_j$, we accept the initial p_j; but for the *growth* blocks, with $N_j' \geq 2N_j$, we want new probabilities P_j proportional to N_j'.

Generally, place into new *growth strata* the portion $(N_j' - N_j)$, denoting the size increase of those primary units which are designated as *growth units*. Then select from these growth units a sample with probabilities proportional to the values $(N_j' - N_j)$. In most situations, the constant of proportionality will be the same (b/f) as for the initial selection, preserving the uniform overall sampling fraction f; but a different fraction can be introduced for the growth stratum. Thus the growth units are selected with the sum of two probabilities, one in the growth stratum and another in the initial stratum:

$$\frac{N_j' - N_j}{b/f} + \frac{N_j}{b/f}. \tag{12.7.5}$$

If selected, these primaries should be subsampled with the rates b/N_j' so that the overall selection probability becomes

$$\left(\frac{N_j' - N_j}{b/f} + \frac{N_j}{b/f}\right)\frac{b}{N_j'} = f. \tag{12.7.6}$$

Note that, as desired, the planned subsample is b when a unit becomes selected. This can occur for any unit both in the initial strata and in the growth strata if it is selected from both. The procedure amounts to splitting the growth units into two parts, consisting of an initial size N_j plus a growth size $(N_j' - N_j)$, and subjecting them to separate selections. The ordinary units, which do not qualify as growth units, remain subject to the initial selection rates $[N_j/(b/f)]b/N_j = f$.

12.7D Shifting Units between Strata

The same changes or differences in population distributions that lead to changing selection probabilities can also motivate changing the strata. The best strata for the new population may be so different from the initial strata, that shifting the units between strata can be justified, despite its difficulties.

To avoid selection bias, a keystone of any procedure must be a guarantee that the sorting of units into new strata is not affected by their having been initially selected. This can be guaranteed either with definitions that permit no latitude, or by a performer ignorant of the identity of the selected units, or both for extra safety. The guarantee is necesssary to ensure that selection probabilities before sorting are equivalent to selection probabilities after sorting into the new strata.

A simple procedure was mentioned at the end of 12.7A. All units in a new stratum that did not belong to it initially are treated as newly created, with $p_j = 0$; some new strata may consist entirely of such new units. On the contrary, units that belonged initially, but are absent from the new stratum, are treated as eliminated, with $P_j = 0$. Rules are needed to identify uniquely each of the initial strata with only one of the new strata. The rules must avoid selection bias, and should preferably maximize the retention of selected units.

Simple procedures are also available if the selection process has the benefit of strong symmetries, such as result from random choice with equal probabilities. The simplest example is an initial selection with simple random sampling of n units from the entire population of N units; that is, all $\binom{N}{n}$ combinations equally probable. These can be sorted into arbitrary strata with procedures that guarantee no effect on the sorting from the initial selections. The n selections can be accepted as equivalent to random selections within the strata. The stratum samples can then be increased or decreased at random to obtain the numbers of random selections needed from each stratum. The symmetries of simple random selections are sufficient; no more complicated proofs are needed.

If the initial selection was a proportionate stratified random sample, with the same equal selection probabilities in each stratum, the above procedure can still yield reasonable approximations to random results. The initial stratification would tend to yield somewhat improved results. Disproportionate sampling rates can also be accommodated by making appropriate increases or decreases in the selection probabilities.

However, the situation becomes complicated if the selection probabilities are unequal, and if we do not want to use the simple procedure proposed above because the volume of shifts is large. Complex procedures are described in two articles [Kish, 1963; U.S. Bureau of the Census, 1963].

*12.8 MULTIPLE STRATIFICATION, LATTICE SAMPLING, CONTROLLED SELECTION

When economic considerations restrict the number of primary selections to a, then $a/2$ is the upper limit on the number of ordinary strata that can be created for paired selections; and a strata is the limit for single selections per stratum. Yet sometimes other pressures exist for greater control over the selections than can be expressed with only a strata. For example, a national sample of interviews in the United States may need to be restricted to 50 or 100 counties, because of the cost of staffing counties. But to satisfy various requests for adequate control over the distribution of sample counties would require many more strata than 100.

First, data about counties are available for many variables that may be related to the survey variables. Population density, industry, education, income, ethnicity, politics, history are some of these; attaining a good geographical distribution can alone consume many strata. When faced with many good variables for stratification, instead of forming fine strata with only 1 or 2 variables, it is generally better to create crude divisions of only 2, 3, or 4 classes for several variables (3.6H and I). Nevertheless, even with only 3 classes for each, 6 variables require 3^6 or 729 strata altogether.

Second, choosing *a priori* the best few from many available variables is always risky. It is especially perplexing when the survey serves many objectives. Furthermore, when the sample must serve a variety of surveys, the objectives multiply in number and diversity. Economic behavior may be the prime concern of some surveys; socio-cultural variables for others; and political attitudes for still others. The stratifying variables that are best for some objectives may not function well for others. Thus pressures accumulate to control the sample with several variables.

Third, beyond the rational need for reducing the variance, further pressure for control can arise for public relations reasons. These become most evident in demands for good geographical distribution. Even a geographically stratified sample will exhibit apparent gaps in the distribution of counties on a map. By concentrating attention on geographical gaps and their possible effects, some people get overanxious about these effects on the sample.

Fourth, estimates for subclasses, especially when based on only portions of the primary selections, can place heavy demands on the sample, increasing the need for good control. For example, estimates for a region and for specified types of counties commonly cause severe control problems. Furthermore, these demands typically tend to increase along with the size of the sample. For example, in 1963 the U.S. Census Bureau used a sample having 357 areas instead of the former 68 (10.4), but was hard pressed for statistics from individual states which were represented in the sample by less than a dozen areas.

These problems for national samples of counties are discussed by Frankel and Stock [1942] and Goodman and Kish [1950]. Similar problems occur for many other samples where the number of selections must be kept low, while demands for control along many variables are high. A controlled selection of hospitals is described by Hess, Riedel, and Fitzpatrick [1961]; a controlled sample of cities of the U.S. was selected by the U.S. Bureau of Labor Statistics [1961] for its price index.

Even a generation ago, conflicting needs of control and randomization were widely thought to be irreconcilable; see the debate between purposive and random methods in the Bulletin of the International Statistical Institute [1926]. These opinions have led to balanced designs for experiments, including Latin squares, especially since Fisher's classic treatise [1935]. They also have given rise to the complex design of probability samples since Tschuprow's [1923] and Neyman's [1934] classic statement of the problem of stratification.

Demands for control often outrun the limitations of stratified random selection. Then researchers sometimes resort to purposive selection, abandoning probability sampling altogether. This could be illustrated with many case histories. In one case a random selection of counties was examined on a map, then adjusted by judgment for irregularities in its distribution. In another case random samples were drawn and inspected for good distribution, especially geographical; literally hundreds were discarded as unsatisfactory before the attempt was abandoned. (Then a controlled selection was drawn that satisfied all requirements.) Such methods are subject to the unknown biases of purposive selection.

Several methods have been proposed for imposing more control within the requirements of probability sampling. To present each of these with enough detail to teach their application would take more space than we can afford within the proper limits of this book. But brief descriptions with references for further details should be helpful.

The Latin squares familiar in experimental designs come to mind as a possible model for controls in selecting samples. For example, selecting any of the 6 patterns of controlled selection in Table 12.8.I will yield a sample of 3 cells; it represents the 2 variables of strata (rows) and controls (columns) at 3 levels each. Furthermore, the Latin letters can denote

Strata	Control Classes 1	2	3	Totals
I	$A\alpha$	$B\beta$	$C\gamma$	1.0
II	$C\beta$	$A\gamma$	$B\alpha$	1.0
III	$B\gamma$	$C\alpha$	$A\beta$	1.0
Totals	1.0	1.0	1.0	3.0

(*a*) 3 × 3 Graeco-Latin square

Prob.	1/6	1/6	1/6	1/6	1/6	1/6
Cum.	1/6	2/6	3/6	4/6	5/6	6/6
Patterns	A	B	C	α	β	γ
I	1	2	3	1	2	3
II	2	3	1	3	1	2
III	3	1	2	2	3	1

(*b*) Patterns of controlled selections

TABLE 12.8.I Latin Squares and Graeco-Latin Square. Represented as Controlled Selections

One cell must be selected from each *stratum*, I, II, and III. Instead of selecting a cell at random from each stratum, we can select it so that each of the *controls*, 1, 2, and 3, appears exactly once in the sample. This condition is satisfied by each of the $3 \times 2 \times 1 = 6$ preferred *patterns of controlled selection*: A, B, C, α, β, and γ. Of the total of 3^3 possible selections, the other 21 possible patterns are undesirable, and receive zero probabilities of selection. Each of these would contain a duplicate or triplicate cell from one of the control columns; for example, of the type (1, 1, 2) or (1, 1, 1).

Every controlled pattern is of the type (1, 2, 3), permitting a single cell of each control column. Each pattern receives the same selection probability 1/6. Every cell appears in two patterns, for a combined probability of $1/6 + 1/6 = 1/3$; for example, the cell (I 1) appears in patterns A and α. A random number from 1 to 6 can choose a controlled selection.

The *3 × 3 Latin square* is familiar in experimental designs for comparing three treatments. First, either the A, B, C, or the α, β, γ set is chosen at random, then each treatment is assigned to one of the 3 patterns of the set. Thus each treatment appears balanced in one control and one stratum. For example, A appears in I 1, II 2, and III 3.

The *3 × 3 Graeco-Latin square* carries the controls farther. The patterns represented by the Latin letters A, B, C, may denote further controls. Then each of 3 treatments is assigned at random to one of the Greek letters α, β, γ. Thus each treatment appears balanced in one stratum, one column control, and one Latin letter control. For example, α appears in I1A, II3B, and III2C.

another control, and selecting a Greek letter at random yields a sample that represents three variables (rows, columns, and Latin letters) at 3 levels each.

The number of possible combinations increases rapidly to $6! = 720$ in a 6×6 Latin square and to $8! = 40{,}320$ in an 8×8 Latin square. In an 8×8 design, we can select a sample of 8 cells, balanced at 8 levels each for 2 variables in a Latin square, and for 3 variables in a Graeco-Latin square. For applications of these methods to selecting a sample and computing its errors, consult Hansen, Hurwitz, and Madow [1953, 11.4], or the discussions of *lattice sampling* by Yates [1960, 10.14] and Patterson [1954].

We can go even further. Instead of 3 selections, we can draw 9 selections from a 3 × 3 square. The Latin letters can represent a third control variable that will be balanced in the entire sample. Utilizing both Greek and Latin letters permits balancing 4 control variables at 3 levels each for a sample of 9 selections [Frankel and Stock, 1942].

Each cell in a Latin square is composed of primary sampling units with the requisite characteristics. After selecting a set of cells, one or more units are drawn to represent the sample. The balanced design of the square requires cells of equal size. This can be attained if the stratifying variables are continuous or sufficiently flexible to permit equal divisions that are equally balanced among several variables. If the symmetry can be forced on the data, this method yields relatively easy computations. Latin squares are a convenient special class, having $p = q$, from a broader class of $p \times q$ balanced designs.

However, the inequality of stratum sizes can be rather rigid, and these situations necessitate methods that require less symmetry than lattice sampling. Yates [1960, 3.4, 8.4] presents briefly a scheme of multiple stratification. A method of two-way stratification by Bryant, Hartley, and Jessen [1960] is also reviewed briefly by Cochran [1963, 5A.5]. In these methods the sample is balanced along the "marginals" that represent single stratifying variables, but not in individual cells. Yet the actual size of individual cells can differ from the size expected on the basis of proportionality with the marginals. If these differences between actual and expected size are large, the method presents problems: if ignored, they may cause bias.

A method of *component indexes* [Hagood and Bernert, 1945] combines many stratifying variables into a few stratifying indexes, each a weighted composite of several correlated variables. The method requires numerical variables (not geography, for example), which are separable into highly correlated groups. It requires much computation, and its value is still to be demonstrated empirically.

A method for balancing samples on asymmetrical controls is *controlled selection*, presented by Goodman and Kish [1950], and reviewed by Hess *et al.* [1961], and the U.S. Census Bureau [1963]. First, ordinary stratification is achieved, probably to the extreme of a single selection from each stratum. Then, instead of independent selection within the strata, further *controls beyond stratification* are introduced to obtain better balance for other stratifying variables. Table 12.8.II is designed to facilitate understanding of an example of controlled selection, with the aid of the simpler

Strata	Control Classes 1	2	3	Totals
I	0.1	0.5	0.4	1.0
II	0.5	0.3	0.2	1.0
III	0.1	0.7	0.2	1.0
Totals	0.7	1.5	0.8	3.0

(*a*) Distribution of controls within strata

Prob. Cum.	0.1 0.1	0.1 0.2	0.1 0.3	0.1 0.4	0.2 0.6	0.2 0.8	0.2 1.0
Pattern	1	2	3	4	5	6	7
I	1	3	3	2	2	2	3
II	2	2	2	1	3	1	1
III	3	1	2	3	2	2	2

(*b*) Patterns of controlled selections satisfy controls

Strata	Control	Cell Probs.	0.1 0.1	0.1 0.2	0.1 0.3	0.1 0.4	0.2 0.6	0.2 0.8	0.2 1.0	Prob. Cum.
			1	2	3	4	5	6	7	Pattern
I	1	0.1	0							
	2	0.5				0.4	0.2	0		
	3	0.4		0.3	0.2				0	
II	1	0.5				0.4		0.2	0	
	2	0.3	0.2	0.1	0					
	3	0.2					0			
III	1	0.1		0						
	2	0.7			0.6		0.4	0.2	0	
	3	0.2	0.1			0				

(*c*) Accounting for remaining probabilities

TABLE 12.8.II Illustration of Controlled Selection

As in the Latin square, the strata sum to 1.0 each, because a single selection is drawn from each. However, the symmetry of the Latin square is lacking, because the cells are not all equal (1/3), and the class totals are not equal.

Therefore, the combinations (1, 2, 3) cannot be had with probability 1, but only with probability 0.5; see patterns 1, 2, 4, and 7. Because class 2 totals 1.5, it must appear twice in 0.5 of the patterns. Since class 1 is absent for 0.3, that portion of the patterns will have the combinations (2, 2, 3); see patterns 3 and 5. Since class 3 is absent for 0.2, that portion of the patterns have the combination (1, 2, 2); see pattern 6.

The combinations are controlled so that each satisfies the marginal totals (0.7, 1.5, 0.8) within the fraction of one selection. That is, the control allows only combinations with 0 or 1 selection from class 1; 1 or 2 selections from class 2; and 0 or 1 selection from class 3. Those restrictions are obeyed by the preferred situation of the comparable Latin square in Table 12.8.I. The 3 strata may represent counties with high, medium, and low population densities. One county is to be selected from each stratum, which have about the same total population. Instead of drawing them independently from the strata, the selections are controlled in 3 classes which may represent 3 regions. The cells are unequal, because population densities differ between regions. One of the patterns of controlled selection from Table 12.8.II(*b*) is drawn with a random number from 1 to 10. For example, 7 would select the cells I 2, II 1, and III 2. Then from each selected cell a county (sampling unit) can be selected with probability proportional to its size. (If the county contains the proportion P_{hi} of the measures in the *h*th stratum, its selection in two stages can be represented as $P_{hg}(P_{hi}/P_{hg})$, where P_{hg} is the measure of the *g*th cell in the *h*th stratum.)

Any departure from simple random sampling can be regarded as a control that enhances the probabilities of preferred combinations by eliminating or reducing undesirable combinations. Stratification is such a control: with a single selection per stratum, it eliminates all combinations containing two or more selections from any stratum. When the possibilities of stratification have been exhausted, *controls beyond stratification* further enhance the probabilities of preferred combinations, by eliminating or reducing them for other combinations which are undesirable because they violate defined control classes. The control classes can denote other variables not represented in the stratification.

The administration of controls can take diverse forms. I present here a method we have often applied at SRC, consisting of three basic rules. First, it is applied after exhausting stratification with a single selection per

---

patterns of controlled selection appearing in (*b*). Only preferred combinations appear: (1, 2, 3) for 0.5; (1, 2, 2) for 0.2; and (2, 2, 3) for 0.3. Excluded are the less desirable combinations such as (1, 1, 1), or (1, 1, 2), or (1, 1, 3), or (2, 2, 2), etc.

The Latin square in Table 12.8.I only permits combinations of (1, 2, 3); but here others are allowed, because the control class totals are not equal. Another dissimilarity is that the patterns are not equally probable, because the cell totals are not equal. Third, the permitted patterns represent a subjective subset of all preferred patterns. For example, whereas the pattern (1, 2, 3) appears, the pattern (1, 3, 2) does not. In this sense Table 12.8.II(*b*) resembles not the entire Table 12.8.I(*b*), but only its *ABC* half, thus excluding arbitrarily the $a\ \beta\ \gamma$ half. Note that the cell probabilities shown in (*a*) are obeyed by the pattern probabilities of (*b*). For example, cell I 1 appears for 0.1 in pattern 1; cell I 2 for 0.5 in patterns 4, 5, 6; cell III 2 for 0.7 in patterns 3, 5, 6, 7, etc. We need a worksheet to account for the probabilities remaining in the cells as they get used up by pattern after pattern; Table 12.8.I(*c*) is a possible form for such a worksheet. Note that each cell becomes 0 when it is completely used up.

stratum. Second, the probability of each cell, created by the combination of strata and controls, is maintained. Third, control classes are maintained with preferred patterns in which either k or $k + 1$ integral selections are permitted from any class which has a total size between those two integers.

Our specific illustration has other simplifications not often present. More decimal places in cell sizes also require creating more patterns. The number of control classes need not equal the number of strata. Difficulties increase rapidly with the number of strata, and especially with the number of control classes. For example, if the totals for 5 control classes are (1.2, 0.8, 1.3, 0.4, 0.3), should we prefer the combination (1, 1, 2, 3), or (1, 1, 3, 4), or (1, 1, 3, 5)? This kind of freedom can be utilized for more restrictions, which then incur more complications and subjective choices. More dimensions of control can be introduced, but they also complicate the computations. Although practical problems can be more complicated, the illustration contains the method's essentials.

The feasibility of controlled selection has been demonstrated with many samples, including the national sample of counties for the Survey Research Center and for the Current Population Surveys of the U.S. Census Bureau [1963], in addition to the two quoted above. We can expect it to reduce variances below those obtained with ordinary stratification, for reasons similar to the gains of stratification and Latin square designs. Goodman and Kish [1950] presented evidence of 20 percent reductions of variances between counties, and more evidence would be welcome.

However, the method has several serious drawbacks. First, its application is difficult and requires some skill and much care. Second, choosing (into a selection table) some patterns to be admitted from all possible preferred patterns depends on subjective judgment. (For Latin squares, all preferred patterns can be admitted.) A probability selection for all cells and sampling units is maintained with the random selection of a pattern for the sample. But an uncomfortable arbitrariness exists in the choice of patterns admitted into the selection table. Work is in progress to overcome these shortcomings by modifying the available procedures, and by enlisting the aid of electronic computers.

Available methods have another weakness: the reductions of the variance actually obtained are not reflected in the computed variances. Since the variance is computed like that of an ordinary stratified sample, it overestimates the sampling fluctuations to which the statistic is actually subject. Thus the sample statistic tends to be more precise than is indicated by its computed variance. Whether or not such methods are justified is a debatable question on which statistical philosophy has no clear, single answer.

It may be safer for nonspecialists to avoid controlled selection in favor of simpler and more standard probability methods. On the other hand, it is better to resort to it than to abandon probability sampling for a subjective selection.

12.9 STANDARD ERRORS FOR MEDIANS AND QUANTILES

Median designates a value Y_M of a variable such that $Y_i < Y_M$ for half of the population elements. It is the most frequently desired *quantile* or *percentile*—which are general terms for position measures. Other familiar position measures are *deciles* and *quartiles.* The median is the second quartile and the fifth decile. Although this discussion centers on the median, it is applicable to other quantiles as well.

We may want to estimate, for example, the yearly income which is not attained by half of the families. Interest in the median is most common for highly skewed distributions, when the median diverges considerably from the mean. In such distributions the variance may be less for the median than for the mean, because the latter is greatly influenced by large values on the far tail of the distribution.

The variance of a median can be computed conveniently and approximately by an indirect method, based on computing variances for proportions for complex samples. It consists of several steps, which are justified elsewhere [Hansen, Hurwitz, and Madow, 1953, 10.18, II, 10.7; Woodruff, 1952]. It extends to complex samples a method designed for simple random samples [Wilks, 1948; Mood, 1950, 388; Kendall and Stuart, 1958, Vol I, 10.10]. The method requires that in the neighborhood of the median the distribution of the sample be smooth and reasonably similar to the population distribution.

1. Obtain the cumulated distribution of the sample values y_j, if the sample is self-weighting; or of the weighted values y_j/p_j, otherwise. Actually, only the neighborhood of the median is needed in detail; elsewhere large classes can be cumulated. Compute the sample median value y_m, the boundary for half of the cumulated distribution in the sample. The sample value $p_m = 0.5$ estimates the proportion of Y_i values less than the unknown Y_m.

2. Compute the standard error of the proportion $p_m{}^*$ that have values $y_j < y_m{}^*$. Here $p_m{}^*$ is a value near 0.5; it corresponds to a value of $y_m{}^*$ that facilitates computations, probably a nearby class boundary. It is assumed that the standard error of $p_m{}^*$ is insensitive to small changes in $p_m{}^*$, hence that the computed se $(p_m{}^*)$ equals the desired se (p_m). If the

sample is simple random, the standard error is simply $\sqrt{(1-f)P_M(1-P_M)/n} = 0.5\sqrt{(1-f)/n}$. However, for complex samples, computations with appropriate formulas are necessary.

3. Find the desired lower limit $p_l = 0.5 - t_p \text{ se } (p_m)$, or the upper limit $p_u = 0.5 + t_p \text{ se } (p_m)$, or both. These are limits (with probabilities corresponding to t_p) for estimating the proportion P_m of population elements which have $Y_i < y_m$.

4. On the cumulated distribution of the sample find the values y_l and y_u, which correspond to the limits p_l and p_u. These are limits for the population median Y_M, with probabilities corresponding approximately to t_p.

We have extended this method for medians to obtain limits for the difference $dy_m = y_m - y_m'$ of two medians. This can be used to measure the change of the median between two periods, or the difference between two subclasses from one sample. For example, we may want to estimate the change in the median family income from one year to another, or the difference between two occupations or two age groups. Either lower or upper limits, or both, may be required for differences of medians.

1. On the cumulated distributions of the two samples, compute the sample medians y_m and y_m', corresponding to $p_m = p_m' = 0.5$.

2. Compute the standard error, se (dp_m), of the difference $dp_m = p_m^* - p_m^{*\prime}$, denoting the proportions in the two samples that are under a convenient value y_m^*, which is near both sample medians.

3. Find $dp_l = dp_m - t_p \text{ se } (dp_m)$ and $dp_u = dp_m + t_p \text{ se } (dp_m)$. These are lower and upper limits for the difference $dP_m = P_m - P_m'$ between the two population proportions: P_m with values $Y_i < y_m^*$, and P_m' with values $Y_i' < y_m^*$.

4. The corresponding limits for the difference of the medians are $dy_l = dp_l(Dy/Dp)$ and $dy_u = dp_u(Dy/Dp)$.

The ratio Dy/Dp denotes an estimate of the ratio, or slope, of difference in medians to a corresponding difference in proportions of the cumulated curves. If the two curves are not reasonably parallel, problems arise in estimating the slope of the difference. One problem concerns the sampling fluctuations in the two cumulated curves, hence in their difference. Another concerns the exact place where the difference in the two curves is to be measured. Research may yield optimum and exact solutions for these problems, but these are lacking at present. In our experience, these were not crucial for practical applications. We employed $dy_m = y_m - y_m'$ for Dy, and for Dp the mean of the vertical differences at those same two points.

12.10 TRINOMIALS AND MATCHED BINOMIALS

By *trinomial* variable, I mean a trichotomous variable D_i that is assigned the values -1, 0, or $+1$. This term is consistent with the term binomial for dichotomous variables that are assigned values of 0 or 1. Any three consecutive integers would yield the same standard error and the same distribution, except for adding a constant to the mean. The values 0, 1, and 2 for the variable $Y_i = D_i + 1$ are more convenient for subscripts than -1, 0, and $+1$. Thus the numbers and proportions in the three classes of a self-weighting sample of n are denoted as

$$p_0 = n_0/n \text{ with } \begin{cases} Y_i = 0; \\ D_i = -1 \end{cases} \quad p_1 = n_1/n \text{ with } \begin{cases} Y_i = 1; \\ D_i = 0 \end{cases}$$

$$p_2 = n_2/n \text{ with } \begin{cases} Y_i = 2 \\ D_i = +1. \end{cases}$$

The sample total for the variable D_i $(-1, 0, +1)$ is $y = \sum_j y_j = n_2 - n_0$, and the sample mean is

$$\bar{y} = \frac{y}{n} = \frac{n_2 - n_0}{n_2 + n_1 + n_0} = \frac{n_2 - n_0}{n} = \frac{n_2}{n} - \frac{n_0}{n} = p_2 - p_0 \,. \quad (12.10.1)$$

The variable $Y_i(0, 1, 2)$ has the mean

$$\frac{2n_2 + n_1}{n_2 + n_1 + n_0} = \frac{n_2 - n_0}{n_2 + n_1 + n_0} + 1.$$

The trinomial variable with values k, $k + 1$, $k + 2$ has the mean

$$\frac{(k + 2)n_2 + (k + 1)n_1 + kn_0}{n_2 + n_1 + n_0} = \frac{n_2 - n_0}{n_2 + n_1 + n_0} + (k + 1).$$

Both of the above have the same mean $(p_2 - p_0)$, except for a constant 1 or $(k + 1)$; they also have the same standard error.

Casting them into the relatively simple form of the mean for this variable D_i provides a ready solution to several problems of survey research. It is interesting and profitable to secure a single solution to several seemingly disparate problems.

1. *Scale values* of 0, 1, 2 (or k, $k + 1$, $k + 2$) are often assigned to three values of a variable that can be ranked, but not measured with a metric. Then $(p_2 - p_0 + 1)$ is the mean of the variable. The same methods also hold for any sequence k, $k + c$, and $k + 2c$ of real numbers, except that the standard errors acquire the factor c.

2. *The difference between two categories of a multinomial* is sometimes required. For example, a multinomial may express preferences for several presidential candidates, and $(p_2 - p_0) = (n_2 - n_0)/n$ measures the difference between two candidates. The categories not involved in the comparison jointly make up the proportion $p_1 = n_1/n$. (However, instead of $p_2 - p_0$, in some situations one may want the proportion $n_2/(n_2 + n_0)$, disregarding the n_1 cases that do not fall into the compared categories.)

3. *Test-retest* and *before-after* are terms employed for designs in which the same subjects undergo two observations in succession. For example, the proportions of positive answers can be measured before and after an educational campaign. Then $n_2 = n_{10}$ denotes the number of positive changes; $n_0 = n_{01}$ denotes the number of negative changes; and $n_1 = n_{11} + n_{00}$ denotes the number remaining unchanged, whether positive or negative. The difference $(n_{11} + n_{10}) - (n_{11} + n_{01}) = n_{10} - n_{01} = n_2 - n_0$ measures the number of changes from first to second observation, and $(p_2 - p_0)$ measures the change in proportions.

4. Similarly we can measure the difference between dichotomous *answers to two different questions*, probably but not necessarily in the same survey. For example, we can estimate the difference between positive preference for candidate A on one question and the positive views on another question. The difference is $n_{10} - n_{01} = n_2 - n_0$, because n_{11} and n_{00} do not contribute to the difference. *Duplicate observations* may be made on a sample, especially for obtaining a *quality check*. For a dichotomous variable, the mean difference is measured with $p_{10} - p_{01}$.

5. *Matched pairs of binomial variables* can also be readily treated. Consider, for example, that n pairs of subjects have been matched for comparing the positive (or success) rates in two classes (experimental versus control, or boys versus girls). The $n_2 = n_{10}$ denotes the number of pairs where only the first class was positive; $n_0 = n_{01}$ denotes pairs with success only for the second class; and $n_1 = n_{11} + n_{00}$ denotes pairs with success for both or neither. Again $(p_2 - p_0)$ measures the difference in success rates.

For a simple random sample, the element variance can be estimated from the sample as s^2, where

$$(n-1)s^2 = \sum y_j^2 - \frac{(\sum y_j)^2}{n} = (n_2 + n_0) - \frac{(n_2 - n_0)^2}{n}.$$

The variance of the mean is var $(\bar{y}) = (1\text{-}f)s^2/n$, hence for simple random sampling

$$\text{var}\,(p_2 - p_0) = \frac{1-f}{n-1}[p_2 + p_0 - (p_2 - p_0)^2]. \qquad (12.10.2)$$

		Second Variable		
		+	0	Total
First Variable	+	n_{11}	$n_{10} = n_2$	$n_{1.}$
	0	$n_{01} = n_0$	n_{00}	$n_{0.}$
	Total	$n_{.1}$	$n_{.0}$	$n_{..} = n$

TABLE 12.10.I Two Dichotomous Variables Observed on a Sample

Note the *net difference* $n_{1.} - n_{.1} = (n_{11} + n_{10}) - (n_{11} + n_{01}) = n_{10} - n_{01} = n_2 - n_0$. This difference, expressed in proportions, is $p_{1.} - p_{.1} = p_2 - p_0 = n_2/n - n_0/n = (n_2 - n_0)/n$. The variance of this difference can be computed with formulas appropriate to specific sample designs. For simple random samples the variance is $\text{var}\,(p_2 - p_0) = (1\text{–}f)[(p_2 + p_0) - (p_2 - p_0)^2]/(n - 1)$ approximately.

The above should not be confused with another statistic commonly presented in fourfold tables. If the two rows (or columns) represent two distinct samples, interest would be in the difference $n_{11}/n_{1.} - n_{01}/n_{0.}$ (or in $n_{11}/n_{.1} - n_{10}/n_{.0}$). Such differences between proportions based on two *separate* samples have been dealt with in several sections devoted to distinct sample designs.

This kind of fourfold table is the common subject for Chi-square tests of the difference of proportions found in two separate samples. However, the correlated net difference $(n_2 - n_0)/n$ has also been subjected to Chi-square tests. In both tests the independence of simple random sampling must be assumed. On the contrary, the standard error of differences may be computed for a wide variety of complex sample designs.

The factors $(1\text{–}f)$ and $n/(n\text{–}1)$ ordinarily can be neglected, and in many comparisons $(p_2 - p_0)^2$ is small compared to $p_2 + p_0$. For example, the difference between 20 and 30 percent would yield var $(0.3 - 0.2) = 0.3 + 0.2 - 0.1^2 = 0.50 - 0.01 = 0.49$. It is easy to remember that *for simple random sampling the mean* $(p_2 - p_0)$ *has the standard error* $\sqrt{(p_2 + p_0)/n}$ approximately. In a few words, in matched simple random samples the *net change* $(p_2 - p_0)$ *has an element variance equal to the gross change* $(p_2 + p_0)$. Or, multiplying through by n, we note that $(n_2 - n_0)$ has the approximate variance $(n_2 + n_0)$, and the standard error $\sqrt{n_2 + n_0}$. More precisely, including the term neglected above, the variance of the difference $(p_2 - p_0)$ from a simple random sample is, neglecting $(1\text{–}f)$,

$$\frac{(p_2 + p_0) - (p_2 - p_0)^2}{n - 1} = \frac{(n_2 + n_0) - (n_2 - n_0)^2/n}{n(n - 1)}.$$

In complex samples appropriate formulas hold for the mean of the variable D_i, with values $-1, 0, +1$. Alternatively, $(p_2 - p_0)$ can be treated as the difference of two means, correlated because of a common base. For stratified cluster samples, the methods of 6.5 can be applied directly to the difference $(p_2 - p_0) = (n_2/n - n_0/n)$. Alternatively, it can be treated as the ratio mean $(n_2 - n_0)/n$, and the variance becomes

$$\begin{aligned}\operatorname{var}\left(\frac{n_2 - n_0}{n}\right) &= \frac{1-f}{n^2}\left[\operatorname{var}(n_2 - n_0) + \left(\frac{n_2 - n_0}{n}\right)^2 \operatorname{var}(n)\right.\\ &\qquad \left. - 2\left(\frac{n_2 - n_0}{n}\right)\operatorname{cov}(n_2 - n_0, n)\right]\\ &= \frac{1-f}{n^2}\left[\operatorname{var}(n_2) + \operatorname{var}(n_0) + \left(\frac{n_2 - n_0}{n}\right)^2 \operatorname{var}(n)\right.\\ &\qquad \left. - 2\,\frac{n_2 - n_0}{n}\{\operatorname{cov}(n_2, n) - \operatorname{cov}(n_0, n)\}\right]. \end{aligned} \quad (12.10.3)$$

The Chi-square test has been applied to some of these problems, treated separately [Cochran, 1950; Mosteller, 1952; McNemar, 1962, p. 225]. This is essentially $(n_2 - n_0)^2/(n_2 + n_0)$, the square of the difference divided by its variance, under the null hypothesis $n_2 = n_0$. It applies the exact theories available for tests of null hypotheses in small samples, including the "Yates correction," all based on the assumption of simple random sampling. However, there are great advantages in treating these problems in large samples as estimated means with proper standard errors. First, instead of being confined to testing null hypotheses, we can make inferences with the probability intervals $(p_2 - p_0) \pm t_p \operatorname{se}(p_2 - p_0)$. Second, the formulas for standard errors of complex samples can be applied directly to the mean $(p_2 - p_0)$. Third, the logical structure of this statistic $(p_2 - p_0)$ can be seen more clearly in its application to several distinct problems.

Note that the element variance is

$$\begin{aligned} p_2 + p_0 - (p_2 - p_0)^2 &= p_2 - p_2^2 + p_0 - p_0^2 + 2p_2p_0 \\ &= p_2q_2 + p_0q_0 + 2p_2p_0 \end{aligned} \quad (12.10.4)$$

These terms correspond to $\operatorname{var}(p_2 - p_0) = \operatorname{var}(p_2) + \operatorname{var}(p_0) - 2\operatorname{cov}(p_2, p_0)$. Hence, the covariance is $\operatorname{cov}(p_2, p_0) = -p_2p_0$, due to the restriction that p_2 and p_0 are competitive parts of the same sample, rather than the means of two independent samples. At one extreme is the complete dependence of a dichotomy, when $p_2 + p_0 = 1$, and $p_2 - p_0 = 2(p_2 - 0.5)$; then $\operatorname{se}(p_2 - p_0) = 2\operatorname{se}(p_2) = 2\operatorname{se}(p_0)$. Even when $p_2 + p_0 = 0.3 + 0.2 = 0.5$, the covariance has a considerable effect:

$$p_2q_2 + p_0q_0 + 2p_0p_2 = 0.21 + 0.16 + 0.12 = 0.37 + 0.12 = 0.49.$$

At the other extreme is the case of two independent samples, when the covariance is zero.

Remark 12.10.I The results of the multiple classification of an attribute may be denoted with the proportions $p_0, p_1, p_2, \ldots$. Suppose we want the variance of the sum $(p_2 + p_0)$ of proportions in two categories. For simple random sampling the element variance is

$$(p_2 + p_0)[1 - (p_2 + p_0)] = p_2 + p_0 - p_2^2 - p_0^2 - 2p_2p_0 = p_2q_2 + p_0q_0 - 2p_2p_0.$$

Since $\quad \text{var}(p_2 + p_0) = \text{var}(p_2) + \text{var}(p_0) + 2\,\text{cov}(p_2, p_0),$

we see again that the covariance of two categories of a multinomial is $-p_2p_0$. Hence, the element variance of $(p_2 - p_0)$ is

$$p_2q_2 + p_0q_0 + 2p_2p_0 = [(p_2 + p_0) - (p_2 - p_0)^2].$$

*12.11 STANDARD ERRORS FOR COMBINATIONS OF RATIO MEANS

12.11A Linear Combinations of Ratio Means

The principles of linear combinations of random variables are straightforward and comprehensive (2.8*D*). To apply these principles, we need convenient computing forms. Here we formulate computing units which can be combined readily by simple summation. The ratio mean can be stated in the general form (see 6.5)

$$r_j = \frac{y_j}{x_j} = \frac{\sum_h y_{jh}}{\sum_h x_{jh}},$$

where the subscript j refers to a specific ratio mean, and h to strata containing independent selections. If W_j is a constant, the variance of W_jr_j can be computed as

$$\text{var}(W_jr_j) = \frac{W_j^2}{x_j^2}\sum_h d^2z_{jh} = \sum_h\left(\frac{W_j\,dz_{jh}}{x_j}\right)^2. \qquad (12.11.1)$$

The basic computing units dy_h, dx_h, and dz_h have been defined and illustrated in 6.5, both for the general case of a_h primary selections in the hth stratum, and for the important special case of $a_h = 2$. The desirability of simple and flexible computing units are also advanced by Keyfitz [1957], but only for the special case of paired selections. However, those methods can be utilized generally with the computing units below.

The covariance of two ratio means, r_j and r_k, has a similar form. Thus, for a linear combination, the variance is

$$\text{var}\left(\sum_j^J W_j r_j\right) = \sum_j^J W_j^2 \text{ var}(r_j) + 2\sum_{j<k}^J W_j W_k \text{ cov}(r_j, r_k)$$

$$= \sum_j \sum_h \left(\frac{W_j\, dz_{jh}}{x_j}\right)^2 + 2\sum_{j<k}\sum_h \left(\frac{W_j\, dz_{jh}}{x_j}\right)\left(\frac{W_k\, dz_{kh}}{x_k}\right) \quad (12.11.2)$$

$$= \sum_h \left[\sum_j \left(\frac{W_j\, dz_{jh}}{x_j}\right)^2 + 2\sum_{j<k}\left(\frac{W_j\, dz_{jh}}{x_j}\right)\left(\frac{W_k\, dz_{kh}}{x_k}\right)\right]$$

$$= \sum_h \left[\sum_j \frac{W_j\, dz_{hj}}{x_j}\right]^2 = \sum_h \left[\sum_j W_j\, de_{hj}\right]^2. \quad (12.11.2')$$

The first computing form (12.11.2) gives more information about each of the several components, if these are needed. However, it would be time-consuming to obtain $J(J-1)/2$ covariances if J is even moderately large. The second computing form involves simple additions of the J components within the strata. Both forms depend on the following basic computing unit:

$$W_j\, de_{hj} = \frac{W_j\, dz_{hj}}{x_j} = \frac{W_j}{x_j}(dy_{hj} - r_j\, dx_{hj}) = W_j r_j\left(\frac{dy_{hj}}{y_j} - \frac{dx_{hj}}{x_j}\right), \quad (12.11.3)$$

which has been defined in 6.5 for several kinds of computations. Thus the most convenient computing form may be sought for summation within strata. Then these forms can be squared and summed.

A simple example is the difference $(r_1 - r_0)$ of two ratio means, when $J = 2$, with $W_0 = -1$ and $W_1 = 1$:

$$\text{var}(r_1 - r_0) = \sum_h (de_{1h} - de_{0h})^2$$

$$= \sum_h de_{1h}^2 + \sum_h de_{0h}^2 - 2\sum_h de_{1h}\, de_{0h}$$

$$= \sum_h \left(\frac{dz_{h1}}{x_1} - \frac{dz_{h0}}{x_0}\right)^2$$

$$= \frac{1}{x_1^2}\sum_h d^2z_{h1} + \frac{1}{x_0^2}\sum_h d^2z_{h0} - \frac{2}{x_1 x_0}\sum_h dz_{h1}\, dz_{h0}.$$

Another example is a simple summed score based on one sample. Here $r = \sum_j r_j = \Sigma\, y_j/x$, when all $W_j = 1$, and all $x_j = x$. Then

$$\text{var}\left(\sum_j^J \frac{y_j}{x}\right) = \frac{1}{x^2}\sum_h \left(\sum_j dz_{hj}\right)^2 = \frac{1}{x^2}\sum_h \left[\sum_j (dy_{jh} - r_j\, dx_h)\right]^2$$

$$= \frac{1}{x^2}\sum_h (dy_h - r\, dx_h)^2,$$

where $dy_h = \sum_j dy_{hj}$. The Y_{ji} variable may be summed either for each element or for the value of the primary selection $y_{h\alpha} = \sum_j y_{h\alpha j}$.

12.11B Double Ratios; Comparisons; Indexes

A *double ratio* is a ratio of two ratio means:

$$R_1 = \frac{r_1}{r_0} = \frac{y_1/x_1}{y_0/x_0}. \tag{12.11.4}$$

The ratio mean of one sample divided by the ratio mean of another sample can be a useful statistic in survey research. Economists sometimes designate a mean score for the current period divided by the mean score for the base period as a *relative*. The base score can be set arbitrarily at 100 (or 1); then r_0 and y_0 are safely removed from zero. The values of x in both denominators are assumed to be safely nonzero. (However, small values of x_0 can be tolerated if $R_1 = y_1x_0/y_0x_1$ is accepted directly.)

If we consider $R_1 = r_1/r_0$ as the ratio of two random variables, its variance can be approximated with

$$\text{var}(R_1) \doteq \frac{1}{r_0^2}[\text{var}(r_1) + R_1^2 \text{var}(r_0) - 2R_1 \text{cov}(r_1, r_0)] \tag{12.11.5}$$

$$= R_1^2\left[\frac{\text{var}(r_1)}{r_1^2} + \frac{\text{var}(r_0)}{r_0^2} - \frac{2\,\text{cov}(r_1, r_0)}{r_1r_0}\right]. \tag{12.11.5'}$$

This involves theoretical problems and we must be satisfied with large sample approximations. These are discussed in a paper [Kish, 1962] on which this subsection is based. This reference contains a large volume of empirical work, including an investigation of the bias of R_1, which was found to be real but negligibly small. A detailed derivation is given by J. N. K. Rao [1957], with applications to forest surveys. An earlier application was made by Keyfitz to periodic labor force surveys [Yates, 1960, 10.5].

We must search for convenient computing forms, preferably those that lend themselves to simple summations. Note that

$$\frac{\text{var}(r)}{r^2} = \frac{\sum d^2z_h}{r^2x^2} = \sum_h\left(\frac{dy_h - r\,dx_h}{y}\right)^2 = \sum_h\left(\frac{dy}{y} - \frac{dx}{x}\right)_h^2, \tag{12.11.6}$$

with a similar expression for the covariance of r_1 and r_0. Hence the variance of (12.11.5′) can be computed as

$$\text{var}(R_1) \doteq R_1^2 \sum_h\left[\left(\frac{dy_1}{y_1} - \frac{dx_1}{x_1}\right) - \left(\frac{dy_0}{y_0} - \frac{dx_0}{x_0}\right)\right]_h^2. \tag{12.11.7}$$

If we were estimating the product r_1r_0, the above formula would have a plus sign in the middle. This form has pleasing symmetry, but fewer divisions are required by the following equivalent of (12.11.5):

$$\text{var}(R_1) \doteq \sum_h E_{1h}^2 = \sum_h (e_{1h} - R_1e_{0h})^2; \tag{12.11.8}$$

where

$$e_{1h} = \frac{de_{1h}}{r_0} = \frac{dz_{1h}}{r_0 x_1} = \frac{dy_{1h} - r_1\, dx_{1h}}{r_0 x_1},$$

and

$$R_1 e_{0h} = \frac{r_1}{r_0}\frac{de_{0h}}{r_0} = \frac{r_1}{r_0}\frac{dz_{0h}}{r_0 x_0} = r_1 \frac{dz_{0h}}{r_0 y_0} = r_1 \frac{dy_{0h} - r_0 dx_{0h}}{r_0 y_0}.$$

The simplicity of the e_h units appears particularly suitable for variances of linear combinations of the R_j. The difference of two double ratios is needed to measure the change in relatives from period 1 to period 2. Then

$$\operatorname{var}(R_2 - R_1) = \operatorname{var}(R_2) + \operatorname{var}(R_1) - 2\operatorname{cov}(R_2, R_1)$$
$$\doteq \sum_h E_{2h}^2 + \sum_h E_{1h}^2 - 2\sum_h E_{2h}E_{1h} = \sum_h (E_{2h} - E_{1h})^2 \tag{12.11.9}$$
$$= \sum_h [(e_{2h} - e_{1h}) - (R_2 - R_1)e_{0h}]^2 \tag{12.11.9'}$$
$$= \sum_h (e_{2h} - e_{1h})^2 + (R_2 - R_1)^2 \sum_h e_{0h}^2 - 2(R_2 - R_1)\sum_h e_{0h}(e_{2h} - e_{1h})$$
$$= r_0^{-2}\operatorname{var}(r_2 - r_1) + (R_2 - R_1)^2 r_0^{-2}\operatorname{var}(r_0) - 2(R_2 - R_1)r_0^{-2}[\operatorname{cov}(r_0, r_2) - \operatorname{cov}(r_0, r_1)]. \tag{12.11.9''}$$

The first term can be computed with simpler methods, and only with data from current surveys. It should often be a good approximation to the entire expression. The other two terms will often be relatively small. Then the easily computed variance for a change of scores can be used to estimate the more difficult variance of a change of relatives. This conjecture was upheld in a set of empirical investigations [Kish, 1962], in which approximations proved to be excellent.

The computing units can be employed for other linear combinations of the relatives. An *index* may be formed from the weighted sum $\sum_j W_j R_{1j}$ of several relatives. Its variance will be

$$\operatorname{var}\left(\sum_j W_j R_{1j}\right) = \sum_j \operatorname{var}(W_j R_{1j}) + 2\sum_{j<k} \operatorname{cov}(W_j R_{1j}, W_k R_{1k})$$
$$= \sum_j W_j^2 \sum_h E_{1jh}^2 + 2\sum_{j<k} W_j W_k \sum_h E_{1jh}E_{1kh}$$
$$= \sum_h \left[\sum_j W_j E_{1jh}\right]^2 = \sum_h \left[\sum_j W_j(e_{1jh} - R_{1j}e_{0jh})\right]^2. \tag{12.11.10}$$

The change from period 1 to period 2 of such an index will have the variance

$$\operatorname{var}\left(\sum_j W_j R_{2j} - \sum_j W_j R_{1j}\right) = \sum_h \left[\sum_j W_j(E_{2jh} - E_{1jh})\right]^2 \qquad (12.11.11)$$

As in (12.11.9″), here again a much simpler computational formula

$$\sum_j (W_j/r_{0j})^2 \operatorname{var}(r_{2j} - r_{1j})$$

has been found to be an excellent approximation [Kish, 1962].

PROBLEMS

12.1. The variance of ratio means for two-phase samples is given in (12.2.2′), assuming simple random selection in both phases. Assume the cost function is $C = cn + c_L n_L$ for the two phases. Assume also that you can make a good guess about $K = c_y/c_x = (s_y/\bar{y})/(s_x/\bar{x}) = (s_y/rs_x)$, the ratio of coefficients of variation; also about r_{yx}, the correlation between the two variables.

(*a*) Show that the optimum ratio of the two sample sizes is

$$\text{optimum } \frac{n}{n_L} = \sqrt{(K-1)^2 + 2K(1 - r_{yx})}\sqrt{c_L/c}.$$

(*b*) Suppose that you guess $K = 0.8$, $r_{yx} = 0.9$, $c = \$20$, $c_L = \$1$. Compute the numerical value for n/n_L.

(*c*) Suppose you also guess $s_x^2 = 1.1$ and $r^2 = 0.91$. For $C = \$40{,}000$, what standard error se $(\bar{y}_d)$ should you expect?

(*d*) Show similarly for (12.2.2) that the optimum ratio of the sample sizes is

$$\text{optimum } \frac{n}{n_L} = \left[\frac{(K-1)^2 + 2K(1 - r_{yx})}{2r_{yx}K - 1}\right]^{1/2}\left(\frac{c_L}{c}\right)^{1/2}.$$

12.2. A sample of $n = 1000$ persons was selected individually, and it may be treated as a simple random selection. Attitude toward a proposal was asked twice, and the positive (Y and y) and negative (N and n) answers were tabulated to the two questions, as follows: $Yy = 380$; $Yn = 80$; $Ny = 40$; $Nn = 500$. The difference in the positive answers, therefore, is $46 - 42 = 4$ percent. (*a*) Compute the standard error of the difference. (*b*) Estimate the standard error of two independent samples obtained for a similar cost. (*c*) Compute the coefficient of correlation r_{yx}.

12.3. Suppose that a population is composed of two subclasses with known weights, so that $W_1\bar{y}_1 + (1 - W_1)\bar{y}_2 = \bar{y}$ is the estimate of the population mean, based on the estimates of the subclass means. It may happen that we need both $\bar{y}$ and the difference $\bar{y}_1 - \bar{y}_2$. If W_2 is small, it may be cheaper not to have to compute $\bar{y}_1$ separately. (*a*) Show that $\bar{y}_2 - \bar{y}_1 = (\bar{y}_2 - \bar{y})/W_1$.

(*b*) Show the variance formula for the general case. (*c*) Show this variance in terms of var ($\bar{y}_2$) and var ($\bar{y}$), when the entire sample is simple random of size n. (*d*) Show variance in (*c*) when $S_1^2 = S_2^2$.

12.4. Construct a table of selection patterns, controlled so that:

(*a*) Each pattern has one selection from every stratum.

(*b*) In any pattern the representation of any control class is within 1 of its marginal probability.

(*c*) The sum of all patterns preserves the indicated cell probabilities.

	Control Classes			
Strata	*a*	*b*	*c*	Totals
I	0.2	0.4	0.4	1.0
II	0.2	0	0.8	1.0
III	0.2	0.4	0.4	1.0
Totals	0.6	0.8	1.6	3.0

12.5. The Recreation Department of a state wants sample estimates about cars with out-of-state licenses leaving the state: length of and reasons for visit, money spent in state, etc. Estimates of yearly means and aggregates are needed for the entire population and for important subclasses. A sample of $f = 1/400$ of out-of-state cars leaving the state is stopped briefly and its occupants interviewed. From automatic traffic counters, estimates are available of total traffic for each road point intersecting the state boundary, for different hours, days of week, and month of year. It is assumed that out-of-state traffic is roughly proportional to total traffic: about 0.6 of total traffic.

Investigations show a two-hour interviewing period at a road to be a convenient sampling unit. They also indicate that about 12 interviews of 3 minutes each is a reasonable, average workload for a two-man crew, one of whom is usually interviewing and the other counting off the out-of-state cars.

Describe salient features of the sampling plan, the probabilities of selection in different stages; also, formulas for estimating means, aggregates, and their variances.

12.6. In Problem 6.16, state and discuss a variance formula for the case when N_y and N_x are also subject to appreciable sampling variations.

PART III
RELATED CONCEPTS

13

Biases and Nonsampling Errors

13.1 RELATION OF BIAS TO VARIABLE ERROR

We have explored the sources and formulas of sampling variance for many kinds of sample designs, but we have largely deferred the problems of bias and nonsampling errors to this chapter. Now, the statistician might take a rigid attitude of complete and total opposition to any bias, no matter how small—as toward sin. He then could try always to use only mathematically unbiased estimators, and insist that the unavoidable biases of measurement, not being strictly sampling problems, are not his concern. However, the rational attitude is relativistic: it considers variable errors and biases jointly as parts of the total survey error, which should be reduced and (hopefully) minimized within available resources.

Bias refers to systematic errors that affect any sample taken under a specified survey design with the same constant error. Systematic biases of measurement are generally distinguished conceptually from variable errors, which often are vaguely assumed to be random.

We may begin by merely contrasting the variable errors of sampling with the nonsampling biases of measurement. That familiar distinction delineates roughly the most influential types of survey errors. Ordinarily, sampling errors account for most of the variable errors of a survey, and biases arise chiefly from nonsampling sources. A more complete and complex model in the next section also includes sampling biases and variable nonsampling errors.

In sampling theory, a widely accepted model combines the variable error and the bias into the *Total Error*. This, often called the *root mean square error* (RMSE), replaces the standard error (SE), and the mean square error (MSE) replaces the variance. This relationship between the variable sampling error and the bias is

$$E[\bar{y}_c - \bar{Y}_{\text{true}}]^2 = E[\bar{y}_c - E(\bar{y}_c)]^2 + [E(\bar{y}_c) - \bar{Y}_{\text{true}}]^2. \qquad (13.1.1)$$

The expectation is taken over the distribution of all possible values of the estimator $\bar{y}_c$ (see 1.3). The mean square deviations of the possible sample results from the *true value* are analyzed into two components: the mean square deviation of the variable errors around the average value $E(\bar{y}_c)$ of the survey design; plus the square of the deviation of that average from the true value. Thus the Total Error, or the root mean square error is

$$\text{Total Error} = \sqrt{\text{VE}^2 + \text{Bias}^2}. \qquad (13.1.1')$$

When the variable errors, VE, are caused only by sampling errors, VE^2 equals the sampling variance; other components will be discussed in the next section. The deviation of the average survey value from the true population value is the bias; this is mostly caused by measurement biases.

The terms *accuracy* and *precision* are widely used to separate the effects of bias. Precision generally refers to small variable errors; sometimes it denotes only the inverse of the sampling variance; in any case it excludes the effects of biases. Accuracy refers to small total errors, and includes the effects of biases. A precise design must have small variable errors; an accurate design must be precise *and* have zero or small bias. A survey design with a large bias is still precise if its variable errors are small, but not accurate. Thus both designs C and D of Fig. 13.1.I are precise (compared to A and B), but only D is accurate. In a rough way we can relate these concepts to those of *reliability* and *validity* in some sciences, particularly psychology. Design C has reliability but not validity; design B has validity but not reliability; A has neither; and D has both. However, these concepts are used more for the errors of several measurements on single individuals, rather than the errors of sample means; reliability refers chiefly to precision of measurements; and validity to lack of bias in the measurements.

Total errors, including variable errors, refer only to average results. We concern ourselves with the usual and expected results, not with the possible result of an individual sample. For example, the results of some samples from D, the best design, can be seen as further from $\bar{Y}_{\text{true}}$ than some samples from A, the worst. Curiously, although A is worse, it has more samples close to $\bar{Y}_{\text{true}}$ than the better design C; but A also has many more samples far away from $\bar{Y}_{\text{true}}$ than C has.

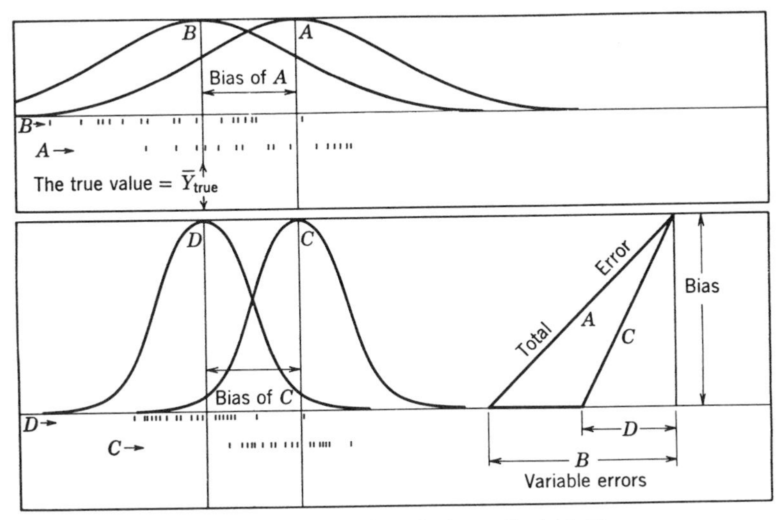

FIGURE 13.1.I Relation of Bias to Variable Error

The normal curves represent the sampling distributions of estimates for each of four survey designs. The heights of the curves measure the probability of varying values of estimates. Variable errors correspond to the standard deviations of the curves, and biases to the distance from the curves' centers to the true values, $\bar{Y}_{\text{true}}$. Designs A and C exhibit large biases, whereas B and D appear unbiased. Designs C and D are more *precise*, because they have smaller standard errors, than designs A and B. Of the four, only design D is *accurate*, because both its variable error and its bias are small. On the contrary, curve A exhibits the greatest average distance from the true value.

Each of the four distributions is also represented by a set of 20 dotted values (normal deviates), resembling shots aimed at the target of true value. The shots of A and C are off to one side, whereas B and D center on target. The shots of C and D are less scattered than those of A and B.

On the right triangles, the heights represent bias and the bases represent variable errors. The hypotenuses A and C measure total errors, combining a similar bias with large and small variable errors, respectively. The effects of eliminating bias appear in the total errors of B and D, equal to their variable errors, the horizontal sides. The total error for C remains large despite its smaller variable error. Only D has a small total error, combining low values for both variable error and bias.

These and many similar statements depend on accepting the model (13.1.1′) for the total error that should be minimized. Justification may be found in some quadratic loss function for errors. It is conditional on accepting the variance to measure all variable errors of normally distributed statistics from large samples; see also 13.8. More general views also involve more complex statements; see Schlaifer [1959, Ch. 31].

An example of a C-type design might be a mail questionnaire with high precision from a large sample, but with low accuracy because of the large bias due to weak response methods, deficient frame, large nonresponse, or a combination of these factors. Design A can represent the same mail questionnaire reduced in size. Design B shows the unbiased results of good methods, but the imprecision of small sample size; this could be due either to few interviews, or to many interviews in a few clusters. Design D shows unbiased methods combined with the high precision of large sample size.

Right triangles illustrate well the root of squared magnitudes. With the two sides representing the variable errors and the bias, the hypotenuse measures the total error. Thus eliminating the bias from C in the triangle of Fig. 13.1.I, reduces the total error drastically to D, a small variable error. For the larger variable error B, the increase to the total error A is relatively less. If the bias is small compared to the variable error, its effect on the total error is also small.

Biases can have important effects on the total errors of a survey; a precise design may be highly inaccurate if it has a large bias. Statements of the standard error, excluding the effects of biases (also often of some nonsampling variable errors), underestimate the total errors of the survey. Thus, as statements about the "*true population value*" $\bar{Y}_{\text{true}}$, intervals based on standard errors can result in much greater errors than suspected (13.8).

Hence, instead of the elusive true value, the population value $\bar{Y}$ is proposed as a simpler target for statistical inference based on standard errors. Inferences from the sample value $\bar{y}$ to the population value $\bar{Y}$, and comparisons between them, are essentially free of bias; by definition $\bar{y}$ is conceived as taken under "the same essential survey conditions" [Hansen, Hurwitz, and Madow, 1953], and $\bar{Y}$ as the result of "equal complete coverage" [Deming, 1960]. Thus the problems of estimating the effect of biases are separated from estimates of sampling errors, which generally can be measured with greater accuracy and objectivity. That separation does not eliminate the need for estimating the bias ($\bar{Y} - \bar{Y}_{\text{true}}$). If this is included with all variable errors, the resulting mean square error yields the total survey error (13.8). When biases are objectively unknown, subjective estimates may be combined with the sampling error [Schlaifer, 1959, Ch. 31].

Bias of size B increases the mean square error of a simple random sample from variance σ^2/n to the quantity $(\sigma^2/n + B^2)$. The relative effects of variance and bias are shown by the two terms $(1/n + B^2/\sigma^2)$. Whereas B^2/σ^2 remains constant, $1/n$ decreases directly with sample size. This can be generalized by using "effective sample size" in complex samples (8.2).

We must avoid the common mistake of fixing our sights exclusively on the entire sample, as if it represented the sole objective. In many surveys, results relating to subclasses are equally important. For a subclass of size n/k, the relative effects can be expressed as $(k/n + B^2/\sigma^2)$; for a small subclass (large k), the variance is much larger, and the relative effect of the bias is proportionately less important. The assumptions of uniform σ^2 and B^2 for the subclasses is reasonable for average conditions.

Comparisons of subclass means $(\bar{y}_1 - \bar{y}_2)$ are often the primary survey objectives. The relative effect of bias in comparisons of two simple random samples can be illustrated with $[(k_1 + k_2)/n + (B_1 - B_2)^2/\sigma^2]$. The biases B_1 and B_2 will often be similar, in which case their combined effect will tend to vanish. Although this situation is probably approached frequently, we should not assume that biases will always cancel out from comparisons.

Example 13.1a From interviews with a sample of home owners, a mean value of $9200 per home was obtained. The standard deviation (s_y) of individual home values was $5700 [Kish and Lansing, 1954]. Later a sample of the homes was evaluated by professional appraisers. The duplicate observations, although simple, yielded estimates of the home owners' bias. This was defined as the difference between the home owners' and the appraisers' estimates, accepting the latter as the "true value" of the home. A positive mean bias in the home owners' estimates was expected; it was comforting that this was only $320, about 3.5 percent of the mean home value.

But *is* a 3.5 percent bias small? To perceive its effect on the total error, it should be considered jointly with the standard error. In thousands of dollars, the total error of a simple random sample of size n would be $\sqrt{5.7^2/n}$ without bias; the bias increases it to $\sqrt{5.7^2/n + 0.32^2} = \sqrt{5.7^2/n + 0.1024}$. Note in Fig. 13.1.II that the small bias has moderate effects on a sample of $n = 100$, but becomes dominant for $n = 1000$, and overwhelming in a sample of $n = 10{,}000$. The last line shows values of $n' = 5.7^2/(5.7^2/n + 0.32^2)$, the number of unbiased observations that would yield the same total error as n observations with a mean bias of 0.32; this is a measure of the effect of the bias on the total error for diverse sample sizes.

Thus a relatively small bias can be the predominant source of error for the mean of a sample of 1000. However, before reaching a hasty conclusion, we must inquire about the diverse objectives of the survey. First, separate estimates for subclasses may be as important as the overall estimate. A subclass of 10 percent has only 100 cases, so the relative effect of the bias becomes merely moderate. This conclusion is buttressed by the statement in the above report that investigation of important subclasses disclosed little variation in the mean bias. Second, this finding is even more crucial for comparisons of subclasses, because these then become essentially free of bias.

Two approximations have been implicitly adopted in these comparisons, but neither disturbs seriously their realism. First, although the individual home values have a skewed distribution far from normal, their means are roughly normal, even for a few hundred cases. Second, although samples of homes would be complex, they can be compared on the common basis of simple random sampling, with appropriate factors for the "design effects" (8.2).

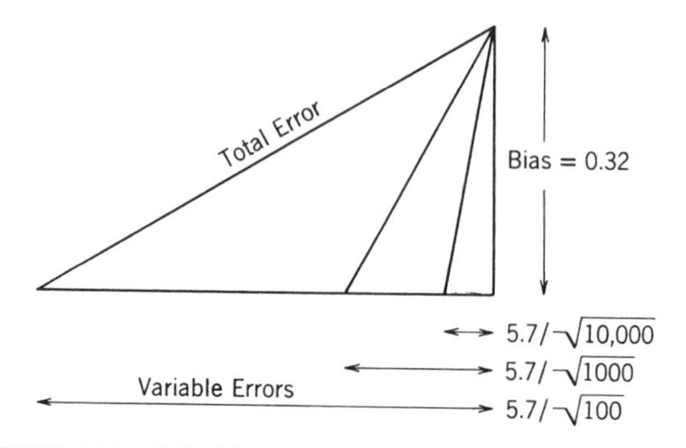

	$n = 100$	$n = 1000$	$n = 10{,}000$
$\sqrt{5.7^2/n}$	0.57	0.18	0.06
$\sqrt{5.7^2/n + 0.32^2}$	0.65	0.37	0.32
$5.7^2/(5.7^2/n + 0.32^2)$	76	240	308

FIGURE 13.1.II Illustration of Effect of Constant Bias on Diverse Sample Sizes. (Bias = 0.32 and $\sigma = 5.7$)

However, can we reduce the effect of the bias with information about its magnitude? In this instance the bias was measured to have a mean of \$320, with a standard error of \$170. Therefore, instead of the mean $\bar{y} = 9.20$, we can more sensibly accept the mean $(\bar{y} - \text{bias}) = 9.20 - 0.32 = 8.88$. The variance of this estimate is $\sqrt{5.7^2/n + 0.17^2} = \sqrt{5.7^2/n + 0.0289}$, and this effect of the bias is reduced in the ratio 0.1024/0.0289. The variance of the two quantities is approximately the sum of the two variances, because the covariance of $\bar{y}$ and bias was small.

However, the biases of most survey data are not measured. Often the users and readers of survey data will introduce their own subjective estimates for the bias. In the *personalistic* (Bayesian) view (14.3) the line between known and unknown values is often blurred, and they propose a formal statistical introduction of a "prior distribution" on the bias. This can take a form similar to the above computation [Schlaifer, 1959, Ch. 31].

13.2 SOURCES OF SURVEY ERRORS

13.2A Biases and Variable Errors

The concept of *individual true values* of a variable for each population element is a good point of departure. This concept is developed by Hansen,

Hurwitz, and Madow [1953, II, 12.2] to satisfy three criteria:

"(1) The true value must be *uniquely* defined.
"(2) The true value must be defined in such a manner that the purposes of the survey are met.
"(3) Where it is possible to do so consistently with the first two criteria, the true value should be defined in terms of operations which can actually be carried through (even though it might be difficult or expensive to perform the operations)."

In actual situations, the criteria may conflict and require a choice or compromise among the three. Neglecting the criterion of practical operations can increase the nonsampling biases unduly. On the contrary, we can theoretically eliminate the nonsampling biases by defining the true values solely in operational terms. But this approach can lead far from the purposes of the research, and also fail to produce a practical unique definition.

For some items, the true values can be obtained relatively easily, although even age and sex data have many errors. For others, such as cash, income and assets, or voting behavior, the true values can be clearly defined but difficult to obtain. For still others, even the definition of true values is obscure; for attitudes and opinions, for example. These lead to concepts of a "sample" from a universe of possible measurements.

In any situation the researcher must choose a set of *essential survey conditions* which can—hopefully—yield results close to the true values. Some of these operations refer to the selection and estimation procedures, and others to the processes of observation. These include hiring and training interviewers and coders, the procedures of interviewing, coding, and tabulating, etc. The essential survey conditions define theoretically the expected (average) results of the sample observations, hence also the biases of the survey. Therefore, it is important to state these essential conditions as completely as possible. Yet the description must always remain incomplete. Furthermore, these conditions only determine the bias in the observations; effects of variable errors remain in the sample results.

Surveys involve many operations and every one of them is subject to error. They may be formulated simply as

$$y_i = Y_i + \sum_r D_{ir}, \tag{13.2.1}$$

where Y_i is the ith individual true value in the population, and D_{ir} is the deviation, or error of its measurement due to the rth source. Hence y_i is a value with measurement errors of the ith population element. This model is too simple to be useful. First, it is wise to separate the constant biases from variable errors; hence,

$$y_i = Y_i + \sum_r B_r + \sum_r V_{ir}. \tag{13.2.2}$$

Here $D_{ir} = B_r + V_{ir}$, with B_r constant for all elements, and V_{ir} variable with zero mean. This arbitrary separation is the first modification toward a serviceable model; it is still too general to be an adequate frame for the concepts and measurements of empirical work. The variable V_{ir}, specific for each element and each error source, may be simple conceptually but is empirically not feasible. Further specifications are needed, but then we cannot retain full generality without becoming too complex. For example, the formula

$$y_i = Y_i + \sum_r B_r + \sum_r e_r \tag{13.2.3}$$

would be pleasingly direct, with the e_r's denoting independent random variables, each normal with mean 0 and variance σ_r^2, but it will not suffice because it makes three assumptions that are often unrealistic. First, covariances between elements (σ_{ijr}) may exist for some measurements. Second, covariances between measurements (σ_{irs}) may exist, as well as covariance (σ_{iyr}) between a measurement and the survey variable. We need not mention other possible covariances to be convinced of the problem's complexity. Third, the uniform and normal variances (σ_r^2) assumed for all elements may need to be modified. It could be defined for a single measurement on each element in a finite (N) population. But I think that it is more realistic, as well as more flexible, to specify "random" rather than "fixed" models, defining the variance over a hypothetical infinite universe.

To emphasize a few salient aspects, a simple model must sacrifice many others. Survey conditions vary, and much also depends on personal judgment; the reader is strongly advised to consult a variety of other presentations, especially Cochran [1963, Ch. 13], Hansen, Hurwitz, and Madow [1953, II, Ch. 12], Sukhatme [1954, Ch. 10], and Zarkovich [1963]. My present choice emphasizes simplicity, denoting the squared deviation of the sample value from the true value as

$$(\bar{y} - \bar{Y}_{\text{true}})^2 = \left(\sum_g B_g\right)^2 + \sum_v \frac{S_v^2}{m_v}. \tag{13.2.4}$$

The first term is the square of the combined bias, which is the algebraic sum of all bias terms. The second term represents the sum of all variance terms, representing diverse sources, each expressed as a unit variance divided by the number m_v of these units. A factor $(1 - f_v)$ may be inserted where needed for a "fixed" model, and other complications like stratification can be introduced. In place of the common subscript r, the separate subscripts, g for biases and v for variable error, emphasize that the two sets often refer to different sources of error; for any single source one may be important and the other negligible, or defined as zero.

The model presents errors for the sample mean jointly, rather than for individual elements, as (13.2.1) does. It corresponds more closely both to empirical measurements, and to the utilization of error magnitudes. It especially affords a more reasonable base for desirable assumptions, such as stability and normality of error variables. On the other hand, the model needs modification for other statistics. It must be modified for aggregates and for comparisons of means, as discussed in several places. For regressions and other analytical statistics, the model must be drastically different.

The model contrasts biases to variances. It can accommodate covariances also, often implicitly. Covariances between elements due to clustering is denoted by a component S_v^2/m_v, where S_v^2 is a unit variance and m_v the number of units. Similarly, covariances between elements due to measurement errors can be represented by S_v^2/m_v also; thus, S_v^2 may represent the component due to interviewers (or coders), and m_v their number in the sample (13.2*B*). All of these bias and variance components are measured on a per element basis.

Covariances between measurements should refer to joint influences on the sample mean. Furthermore, here it should refer to deviations from their expected (average) value; effects on the expected value are included among the biases. Such covariance, often small, may be represented by $2S_{uv}/m_v$. Similarly, we may denote as $2S_{yv}/m_v$ the covariance between element values and a measurement error. These may have negative values in the sample, whereas variance terms are always positive. The joint effects of two error sources may be represented by $(S_u^2 + S_v^2 + 2S_{uv})/m_v = (S_u^2 + S_v^2 + 2R_{uv}S_uS_v)/m_v$, which is always positive and less than $(S_u^2 + 2S_uS_v + S_v^2)/m_v$.

Several broad differences distinguish biases from variable errors. First, the variable errors of sampling and the nonsampling biases ordinarily cause more concern and have greater effects on survey results than their counterparts, sampling biases and variable nonsampling errors. Nevertheless, both variable errors and biases can arise either from sampling or nonsampling operations. This double dichotomy gives rise to a fourfold classification of errors: variable errors of sampling or nonsampling, and biases of sampling or nonsampling. Many potential sources of errors can be found in each of these classes, since every operation is a potential source of variable errors and biases.

Second, the different biases can be considered as a set of constants, determined by the essential survey conditions, although their values usually remain largely unknown. Each bias can be considered positive or negative, according to whether it increases or decreases the value of the estimator. A bias has the same effect B_g on one sample value as on another, whether of the same or different size, and the same as on the expected (average) sample value $E(\bar{y})$. The effect is either exactly or almost the same $(B)_g$ on the population value $\bar{Y}$, and the difference $E(\bar{y}) - \bar{Y}$ is generally small, as discussed later. Biases represent the difference between expected sample value and the "true value," so that the *total survey bias* $= \Sigma B_g = E(\bar{y}) - \bar{Y}_{\text{true}}$. Contrariwise, variable errors measure the sources of the difference $\bar{y} - E(\bar{y})$ between the estimate and its expected value. The variable errors would fluctuate if we were to select different samples with the same design; they would be larger or smaller, plus or minus, for different samples. The specific value of the actual deviation of the sample cannot be known; but it can be conceived and measured in long-run terms of the distribution for all possible values of the estimator. This fluctuation is measured with the standard error of the estimator $\sqrt{E[\bar{y} - E(\bar{y})]^2}$, the root mean square deviation around their zero mean. This is defined here to include all variable errors, both sampling and nonsampling.

Third, the total bias is the algebraic sum of all biases ΣB_g. Some may distort in the positive direction, and others in the negative, thus partially cancelling each other. Hence the reduction of one source may even increase the total bias. Variable errors take the positive form $\Sigma S_v^2/m_v$ of variances.

Fourth, most biases cannot be reduced by increasing the size of the sample, but only by improving the quality of some operation—by doing something *better*. Contrariwise, the reduction of variable errors depends on increasing the number m_v of units of some kind—by taking *more* of something—either sampling units, or observations, or observers.

Fifth, variable errors can be measured by noting the deviations between *internal replications* of units within the sample. Measurement requires that replications of units—whether sampling units, or observations, or observers—be properly designed to separate the sources of variations. Measurement of biases essentially depends on a different method, *external* to the survey proper. This may be done in either of two ways. A good *quality check* sample of two methods, which compares an accepted standard with the survey's method, can measure individual biases, their variations, and the separate sources of bias as well. A *comparison with an outside source* can ascertain only a *net bias*, the sum of several or all biases of a type.

Sixth, the effects of biases and variable errors differ for various statistics. For example, biases with large effects on the mean of the entire sample can have negligible effects on comparisons of means or on regression statistics. On the contrary, variable errors with negligible effects on the means of large samples can have large effects on comparisons of small subclasses and on regression statistics.

13.2B Descriptions of Diverse Errors

Frame biases can distort the values of large samples as much as of small ones. Their sources can be located in the selection procedures, but in probability sampling they can usually be corrected with proper estimation procedures. Frame biases tend to follow failure to adjust the estimate for unequal selection probabilities. For example, selecting one of a variable number of adults from each of an epsem sample of dwellings results in unequal selection probabilities from the population of adults; weighting can eliminate this bias (11.3*B*). Another example is provided by an epsem selection from a list that contains replicate listings of elements; the bias can be removed by reweighting. The elimination of these sources of bias, either in selection or estimation, is a serious task for the sampler, often

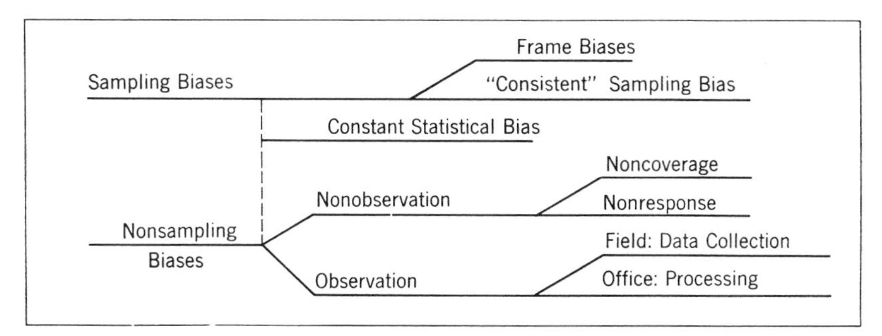

FIGURE 13.2.I Classification of Sources of Survey Biases.

treated in this book, especially in Section 2.7 and Chapter 11. In a complete population coverage the same biases can occur, although their treatment may be simpler, except for missing units or noncoverage.

Consistent sampling biases, caused by biased but consistent estimators, decrease with increasing sample size and vanish entirely from the population value in a 100 percent sample. I use "consistent" in that statistical sense; actually, this bias is "inconsistent" in that, unlike other biases, its value varies with the sample size. For any fixed sample size it, like other biases, has a constant effect on the estimator. Good sample design can ordinarily reduce these consistent biases to negligible proportions compared to other sources of errors. An example is the ratio $r = y/x$ as an estimator of $R = Y/X$ (6.6); another would be $\Sigma (y_j - \bar{y})^2/n$ as an estimator of σ^2.

By *constant statistical bias*, I mean a bias in statistical estimation which affects equally (exactly or approximately) samples of any size, as well as the population value based on a complete coverage. An example would be the use of the mean ratio $(1/a) \sum_{\alpha}^{a} y_{\alpha}/x_{\alpha}$ as an estimator of $Y/X = \sum_{\alpha}^{A} Y_{\alpha}/\sum_{\alpha}^{A} X_{\alpha}$. Another example of a persistent bias would be the use of the median to estimate the mean of a skewed distribution. If sample design includes estimation procedures, these biases should also be considered part of the sampling bias; they can be avoided through use of proper statistical estimation procedures. These situations, usual in sampling, lead statisticians to speak, not of biased selection, but only of biased sample design or biased estimators. In an unbiased sample design the combined total of selection and estimation biases is zero. These biases resemble nonsampling biases in that they affect equally the population value based on complete coverage. Furthermore, it seems to me, that the sampler cannot cover the estimation procedure completely without taking over the entire statistical analysis—an unrealistic assignment.

Nonsampling biases pose profound problems for scientific measurement. Their sources lie in the broad fields of measurement, hence in the substantive fields of the various sciences. These biases affect the population value as much as the sample value and (together with any constant statistical bias) account for the difference between the population value and the true value. This difference delineates the gap between the defined survey objectives and the entire set of essential survey conditions. Good practice demands that as many of these as practicable be described explicitly in the research report, although a complete description remains impossible.

Among the nonsampling biases of most surveys we can distinguish the biases of observation from those due to nonobservation. The latter arise from failure to obtain observations on some segments of the population, due either to noncoverage (incomplete frame) or nonresponse. Biases of observations are caused by obtaining and recording observations incorrectly. We may again distinguish two classes of biases. One arises in the field performance of observations, which may consist of interviewing, enumerating, counting, or measuring; these are response biases. Others, processing biases, are produced during the office processing, and in coding, tabulating, and computing.

Variable errors include both sampling and nonsampling errors. *Sampling errors* may consist of any number of components, depending on the design. For example, the variance of a mean based on a design involving three stages of random selection with replacement from equal clusters (5.6) would be represented with three components:

$$\text{Var}(\bar{y}) = \frac{S_a^2}{a} + \frac{S_b^2}{ab} + \frac{S_c^2}{abc}.$$

The denominators for the three stages show a primary sampling units, ab second stage units, and $n = abc$ elements selected in the third stage. In Fig. 13.2.II the three components appear combined into the hypothenuse $\sqrt{A^2 + B^2 + C^2}$ for sampling errors. The model needs modifications to accommodate complex sample designs. For example, selection without replacement introduces finite population corrections; stratification reduces the unit variance; unequal clusters and weighting introduce further

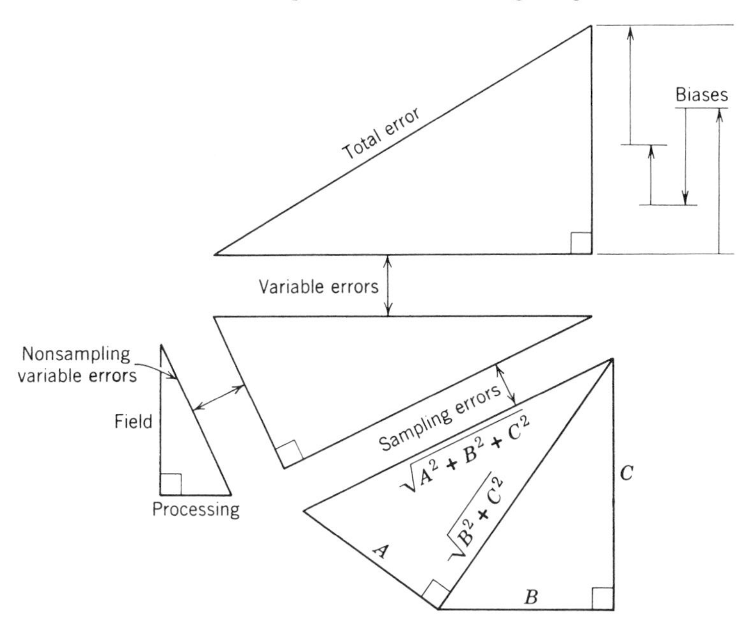

FIGURE 13.2.II Classification of Sources of Survey Errors

Sampling errors are shown arbitrarily with three components. Variable errors, sampling and nonsampling, combine with their summed squares. The total bias is the algebraic sum of all biases, sampling and nonsampling.

complications. Nevertheless, there is heuristic value in the basic formula of variance components, each a unit variance divided by the number of such units.

Nonsampling variable errors may also be represented as sums of squared components. This similarity to the more familiar properties of variable sampling errors is a virtue of the model. Figure 13.2.II shows components for field and processing errors. These can be subdivided in many ways; for example, processing errors into coding and tabulating components, and field errors into components for interviewers and for supervisors. Nonresponse errors can be included, but they tend to affect biases more than variable errors.

The graphical representation of variances needs modification if covariances are present. This may be done with a separate covariance component, which must be subtracted if it is negative; or by changing the right angle to γ, with the correlation coefficient $\rho_{ab} = \cos \gamma$ in the relation $C^2 = A^2 + B^2 - 2\rho_{ab}AB$.

In the model $\Sigma S_v^2/m_v$, the number m_v of units must be determined for each error component. In the error of response to an interview, a component may be assigned to the respondent, and this error may be uncorrelated between respondents. Each gives ordinarily only one response, and the number of units is equal to the number of respondents, hence $m_v = n$. If the sample of respondents is clustered, their response errors may be correlated within clusters, as their characteristics are. This random respondent error is ordinarily included in the computed variance of the mean.

The mean effect of a single interviewer on all n_i responses he obtains has been termed his own interviewer bias.

"Each interviewer has an individual average 'interviewer bias' on the responses in his workload, and we consider the effect of a random sample of these biases on the variance of sample means. This effect is expressed as an *interviewer variance* which decreases in proportion to the numbers of interviewers. Its

	Range of roh
Kish [1962]	
46 Variables in first study (a = 20)	0 to 0.07
48 Variables in second study (a = 9)	0 to 0.05
Percy G. Gray* [1956] (a = 20)	
8 "factual" items	0 to 0.02
Perception of and attitudes about neighbors' noises	0 to 0.08
8 items about illnesses	0 to 0.11
Gales and Kendall* (1957] (a = 48)	
Semi-factual and attitudinal items about TV habits	0 to 0.05
Hanson and Marks [1958] on 1950 U.S. Census Data (a = 705)	
31 "Age and Sex" items	0 to 0.005
19 simple items	0 to 0.02
35 "difficult" items	0.005 to 0.05
11 "not answered" entries	0.01 to 0.07

TABLE 13.2.I Results from Several Investigations on Interviewer Variances Computed as roh $= s_a^2/(s_a^2 + s_b^2)$.

* For these investigations we computed the above values from the published test statistics and this translation involves some uncertainties [Kish, 1962].

contribution to the variance of sample means (s_a^2/a) resembles other variance terms: it is proportional, directly to the variance per interviewer and inversely to the number of interviewers. This contribution to the total variance may be substantial; by failing to take it into account (as when estimating the variance simply with s^2/n), one may be neglecting an important source of variation actually present in the design, having been introduced by the sampling of interviewers' biases. The interviewer variance s_a^2 should be viewed as a component of the total variance, denoted as $s^2 = s_b^2 + s_a^2$, where s_b^2 is the variance without any interviewer effect, all three terms being measured per element.

"We studied the effects of the variable biases of interviewers on responses to a variety of items, mostly attitudinal, included in two surveys. Among these we included subclass means and comparisons among pairs of subclasses. The data yielded important results: (1) We can obtain responses with low or moderate interviewer variance on highly 'ambiguous' and 'critical' attitudinal questions. The range of roh's was mostly 0 to 0.07 in the first study, 0 to 0.05 in the second, and their average is about 0.01 or 0.02; these effects are not generally higher than for most items in a good census. (2) Nevertheless, these seemingly small roh's, combined with moderate workloads, can often increase the variance by factors as high as two or three, because the ratio of increase is $[1 + \text{roh}(\bar{n} - 1)]$ where $\bar{n}$ is the average number of interviews per interviewer. (3) The interviewer effects on subclass means were shown to be smaller in accord with $[1 + \text{roh}(\bar{n}^* - 1)]$ where $\bar{n}^*$ is the average number of subclass members per interviewer. (4) In the comparisons of subclass means, the interviewer effects tend to zero, in accord with an additive model of the effects." [Kish, 1962.]

0.000	0.001	0.002	0.005	0.01	0.02	0.03	0.04	0.05	0.06	0.07
All 96 Variables										
2	18	34	55	68	83	89	94	95	97	100
			31 "Age and Sex" Items							
6	45	81	100							
			19 "Simple" Items							
11	16	68	84	100						
			35 "Difficult" Items							
	3	14	26	51	77	86	94	97	97	100
			11 "Not Answered" Entries							
					27	36	64	64	82	100

TABLE 13.2.II Cumulative Percentages of Interviewer Variances (roh) on 96 Items of the 1950 Census

Computed by Kish [1962] from data by Hanson and Marks [1958]. Four categories separated by *a priori* judgment.

Even low interviewer variance can have a large effect on the entire sample; for example, with roh = 0.01, an interviewer's workload of 51 cases increases the variance of the entire sample by the factor $[1 + 0.01(51 - 1)] = 1.5$. In a variance computed for element sampling, this source of variable errors remains omitted from the estimate. It can be included by considering each workload as a cluster. For cluster samples it becomes part of the computed variance if each interviewer or group of interviewers is confined to a single primary cluster. Thus, when counties are the primary units and interviewers are confined to counties, the computed variance of the mean includes this source of variable errors. Other sources, such as coders and interviewer supervisors, may still be omitted. It is rarely possible to design samples so that the variance includes all sources of variable errors. That interpenetrating replicated samples (4.4) can accomplish this is a strong argument for their use in large scale surveys [Deming, 1960; Lahiri, 1957; Mahalanobis, 1946].

The measurement of nonsampling variable errors in surveys is a complex subject with a large literature. The reader can find further discussions and references in Cochran [1963, 13.8–13.17], Deming [1960, Ch. 13], Hansen, Hurwitz, and Madow [1953, II, Ch. 12], Sukhatme and Seth [1952], Hyman *et al.* [1954, Ch. 6], Zarkovich [1963, Ch. 13–14].

13.2C Some Effects of Errors

For practical purposes, we may often resort to a simple model for the deviation of a sample mean from its true value:

$$(\bar{y} - \bar{Y}_{\text{true}}) = (\bar{y} - \bar{Y}) + (\bar{Y} - \bar{Y}_{\text{true}}). \tag{13.2.5}$$

The first term incorporates all variable errors, both sampling and nonsampling; the second term contains all nonsampling biases. This model serves adequately in practice: it assumes that the sampling and statistical biases are negligible, and that the effects of variable sampling errors on the population value are also negligible. Thus $E(\bar{y}) = \bar{Y}_p = \bar{Y}$, or the difference between them may be neglected. The differences between these three—the expected sample value, the population value, and the estimand—are developed below.

Now we re-examine, in more detail than in 1.3, the relationship of the different sources of error to sample values, population values, and true values. The total error of a sample value, its departure from the true value, may be analyzed into four components:

$$\bar{y} - \bar{Y}_{\text{true}} = [\bar{y} - E(\bar{y})] + [E(\bar{y}) - \bar{Y}_p] + [\bar{Y}_p - \bar{Y}] + [\bar{Y} - \bar{Y}_{\text{true}}]. \tag{13.2.6}$$

The first component, $[\bar{y} - E(\bar{y})]$, denotes the deviation of the specific sample value from *the average value* of the sampling distribution; that is, the distribution of all possible sample values, given the same essential survey design, including both its sampling and nonsampling aspects. It is convenient to think of this difference as accounting for all the variable errors of the survey, sampling and nonsampling.

The difference $[E(\bar{y}) - \bar{Y}_{\text{true}}]$ between the average of the sampling distribution and the true value incorporates the various sources of biases, and it can be subdivided into the last three components. $\bar{Y}_p$ denotes the *population value* that would be obtained if all the elements in the population were included in the sample, subject to the same essential survey conditions. This concept serves to separate the errors due strictly to the sampling process from the other errors inherent in the survey design; sampling errors are represented by $[\bar{y} - \bar{Y}_p] = [\bar{y} - E(\bar{y})] + [E(\bar{y}) - \bar{Y}_p]$; and nonsampling errors by $[\bar{Y}_p - \bar{Y}_{\text{true}}] = [\bar{Y}_p - \bar{Y}] + [\bar{Y} - \bar{Y}_{\text{true}}]$.

As a rule, we can assume that $\bar{Y}_p$ is approximately a constant on which variable errors have no effects. This model involves assuming an indefinitely large population of observations, observers, coders, and other units whose variability could otherwise affect $\bar{Y}_p$. In these terms, the assumptions for constant $E(\bar{y})$ and constant $\bar{Y}_p$ coincide. Actual complete censuses are subject to variable errors, and we can conceive of the population value as a random variable with a sampling distribution. Usually, this variation of $\bar{Y}_p$ may be disregarded, either because the population is much larger than the sample, or because the actual population is in turn conceived as a sample from a larger population, or a hypothetical infinite universe. But situations arise where the $\bar{Y}_p$ obtained is subject to nonsampling variability. Census data for small towns, covered by a few interviewers, have been shown to be subject to large effects of interviewer variability [Hansen, Hurwitz, and Bershad, 1961].

Clearly the difference $[E(\bar{y}) - \bar{Y}_p]$ includes the "consistent sampling bias," which disappears when the sample embraces the entire population. The proper allocation of the frame biases seems less clear. Theoretically, the population value obtained under the "same essential survey conditions" should be equally subject to them. But, actually, treating frame problems tends to be simpler in complete censuses; it becomes easier to handle the presence of clusters or foreign elements, and duplicate listings can be checked more readily. Still, the problem of missed units remains equally troublesome (2.7). It is practical to treat frame biases as part of the sampling process; their elimination or economical reduction forms an important aspect of the sampler's craft.

In summary, we contrast the sample value with the concept of the population value that would be yielded by an "equivalent complete coverage," and consider that the difference $[\bar{y} - \bar{Y}_p]$ arises from errors peculiar to the sampling process: variable errors, the consistent sampling bias, and sampling frame bias. On the contrary, the constant statistical biases affect the population value as much as the sample value. To accommodate them is the function of the component $[\bar{Y}_p - \bar{Y}]$; here $\bar{Y}$ is the *estimand*, the population value free of biases of statistical estimation. This value is the target of the sample design, and $[E(\bar{y}) - \bar{Y}] = 0$ defines an unbiased sample design, including both selection and estimation. Probably this is the most frequent meaning of a statistical *parameter*; however, its definitions concentrate on estimation, neglecting selection and observation. Because *parameter* lacks an accepted definition covering these problems, *estimand* is used here.

The estimand $\bar{Y}$ serves to define an unbiased sample design, disregarding the nonsampling biases of observation and nonobservation. These biases account for the difference $[\bar{Y} - \bar{Y}_{\text{true}}]$ of the estimand from the *true value*, this being the population value free of any errors of any kind, the target of the entire survey design. This concept separates the sampling and statistical biases in $[E(\bar{y}) - \bar{Y}]$ from the nonsampling biases of observation and nonobservation in $[\bar{Y} - \bar{Y}_{\text{true}}]$. This separation, however, amounts to a redefinition of the population to exclude the portion encompassed by the biases of nonobservation. Alternatively, these biases may be included either in $[E(\bar{y}) - \bar{Y}_p]$ or in $[\bar{Y}_p - \bar{Y}]$, leaving $\bar{Y}_p$, or at least $\bar{Y}$, free of the biases of nonobservation.

From the complex model outlined above, we can retreat to the usual simple model when the sample design achieves its aim of eliminating or reducing the sampling and statistical biases to negligible proportions; that is, when $[E(\bar{y}) - \bar{Y}]$ is negligible. Then the total errors of the survey consist only of $[\bar{y} - E(\bar{y})]$ and $[\bar{Y} - \bar{Y}_{\text{true}}]$, the former incorporating the variable errors, and the latter the nonsampling biases of observation and nonobservation. The standard error formulas are designed to measure the variable sampling errors.

I cannot even attempt separate discussions of the effects of various biases and variable errors on diverse statistics. Differences in effects do exist, which should not be forgotten, despite the great concentration of sampling literature on means and aggregates. We shall illustrate them briefly, mostly with examples described elsewhere. If the mean of nonresponses is the same as the mean of responses, then nonresponse has no effect on the sample mean $\bar{y}$ (nor on the estimate $N\bar{y}$), but would bias the simple expansion estimate Fy of the aggregate. On the contrary, noncoverage of a portion of a sample which is void of the Y_i variable does not bias the estimate Fy of the aggregate, but will bias the sample mean (13.4*B*). The effects of biases on ratio and regressions means can be entirely different from the effects on simple means. If the bias is the same for the mean of a small subclass as for the entire mean, its effect is less important relative to the larger variance of the subclass (13.1).

Serious biases of the individual means may have much less effect on the difference of two means if both are similarly affected by the bias. Thus the comparison of two subclasses, say, two age groups of a sample, can be nearly unbiased even for variables which are poorly measured. Similarly, the comparison of statistics from a series of periodic surveys can possess great validity if the survey processes have been kept under control. The home values reported (13.1) by the respondents were found, on the average, to be $320 higher than the appraisers' estimates, but investigations of socio-economic and city size subclasses disclosed no sizeable differences in this bias. Thus the comparisons of these subclasses would be largely free from the bias. Generally, subclass comparisons should be unaffected by biases which are *additive*; that is, constant for the subclass means.

Nevertheless, it is reckless to assume that all comparisons are automatically free of large biases. Indeed it is possible for comparisons to have larger biases than the means; this would be caused by a large *interaction* effect between the variables and the biases in the subclasses. The interviewer variance effects that I reported [1962] appeared to be similarly additive; subclass means showed proportionately reduced effects, and subclass comparisons were free of them.

The effects on regression coefficients may be entirely different from the effects on means. For example, a constant bias has no effect on them; but random variable errors in the measurements reduce the correlations. The diverse effects of bias on different statistics can also be illustrated with the results of a hypothetical monthly sample. Assume that an estimate $\bar{y}$ is produced each month with a standard error of 1 and a bias B; that the monthly samples are independent; and that the survey operations are under good control so that the bias is the same each month. Then the monthly change is measured with a standard error of $\sqrt{2}$ and no bias. On the other hand, a yearly average based on the entire year's sample would have its standard error reduced to $1/\sqrt{12}$ while its bias remained B. Hence, the *relative* effect of the bias increases greatly to $B\sqrt{12}$.

13.3 NONCOVERAGE; INCOMPLETE FRAMES, MISSING UNITS

Errors of nonobservation result from failure to obtain data from parts of the survey population, and we may distinguish two sources: noncoverage and nonresponse. Nonresponse refers to failure to obtain observations on some elements selected and designated for the sample; it may be due to refusals, not-at-homes, unreturned or lost questionnaires, etc. Unlike nonresponse, noncoverage denotes failure to include some units, or entire sections, of the defined survey population in the actual operational sampling frame. Because of the actual (though unplanned and usually unknown) zero probability of selection for these units, they are in effect excluded from the survey result. We do *not* refer here to any *deliberate* and *explicit* exclusion of sections of a larger population from the survey population. Survey objectives and practical difficulties determine such deliberate exclusions. For example, many surveys of attitudes are confined to adults, deliberately excluding persons under 21 years; residents of institutions are often excluded because of practical survey difficulties. These explicit exclusions differ both in intent and in result from noncoverage caused by failures in the procedures. When computing noncoverage rates, members of the deliberately and explicitly excluded sections should not be counted either in the survey population or in the noncoverage. Defining the survey population should be part of the stated *essential survey conditions*.

For some surveys clear, convenient, and complete frames exist; they permit the sampler to avoid noncoverage, or to measure it readily. It may be easy to cover the employees of a firm, telephone subscribers of a city, or students of a school. The national registers of countries in northern Europe provide fairly easy and complete frames of their populations. However, for many frames, including area samples of dwellings, reducing and measuring noncoverage requires formidable labor. Although few studies and estimates of noncoverage have been published, we should not conclude that it is not an important source of survey error.

Noncoverage includes the problems of "incomplete frames," a term that seems to imply omissions in preparing the frame. But it also refers to "missed units," omissions due to faulty execution of survey procedures. Missing elements have been discussed before (2.7*B*), but the importance of noncoverage and the difficulties of measuring it justify this separate treatment.

Instead of the yes-or-no (1 or 0) dichotomy of coverage versus noncoverage, degrees of coverage can be treated as a variable and measured for identified subclasses. For example, we would find a higher coverage of home-owning families, than of renters, and ordinary renters would be higher than migrants. For brevity, we avoid that treatment here, but the description given later for nonresponse is also applicable to noncoverage. Although in its formal aspects it resembles nonresponse, we separate noncoverage because it is much more difficult to measure in most survey situations.

Noncoverage refers to the negative errors of failure to include elements that would properly belong in the sample. There occur also positive errors of *overcoverage*, due to the inclusion in the sample of elements that do not belong there. The term *gross coverage error* refers to the sum of the absolute values of noncoverage and overcoverage error rates. The *net noncoverage* refers to the excess of noncoverage over overcoverage, and it is their algebraic sum. The net noncoverage measures the gross coverage errors only if overcoverage is absent. Net noncoverage is also acceptable if the effects of a small overcoverage are cancelled by part of the noncoverage; this is true if the erroneously included elements are like those erroneously excluded, except that there are fewer of them.

Practitioners believe that in most social surveys noncoverage is a much more common problem than overcoverage, so that net noncoverage is an acceptable measure of coverage problems. But this is not true for all surveys. On the contrary, when small grids are used to mark plots for selecting stalks of small grains, overcoverage typically occurs. Many stalks appear on the borderline, and the field workers, when in doubt, tend to include them. The area of borderline is proportional to the perimeter; for a given plot shape, this is proportional to the linear measurement, whereas the area is proportional to its square. Therefore, a large plot has a smaller proportion of borderline area than a small plot. Because of this, Sukhatme [1947] and others advocate using larger plots than would be designed when trying to minimize only the sampling variance [Zarkovich, 1963, Ch. 17]. The contribution to the mean square error of a large bias may overcome the precision obtained from a widespread sample of small plots.

The advantages of larger segments are also pertinent to social surveys, where the main problem is noncoverage. We may prefer to use larger compact segments, even when smaller segments or individually selected dwellings would reduce sampling variances. Boundary problems are proportionately less in larger segments, and better boundaries can be found more often for larger segments than for small ones. This argument dominates specific recommendations for selecting compact segments (9.4*A*) or entire villages (8.4*A*). It applies not only to area samples, but also to other frames and lists. For example, selecting larger segments from a list of names can also reduce the problems at the pairs of segment end-points.

Whereas nonresponse can be measured from sample results, the extent of noncoverage can be estimated only against some check obtained *outside the survey procedure itself*. This resembles the problem of measuring other kinds of bias, and two difficult alternatives exist. One involves an auxiliary investigation, a *quality check* with improved procedures. This can reveal cases of overcoverage and of noncoverage, as well as their sum, the *net* noncoverage. Since it permits analysis of individual cases and subclasses, it may reveal causes and lead to correction. However, unless the quality check procedure is very good, the error estimates can be badly biased. Moreover, on good surveys errors of noncoverage tend to be rare and spotty. Hence, trying to find and measure them with reasonable accuracy becomes an extensive and expensive operation, only available for some of the larger surveys.

The alternative calls for a comparison of the internal sample estimate of the population total *with an external estimate* from reliable sources. This involves *expanding* the sample count x (including nonresponse) by F, the inverse of the sampling fraction; or computing $\Sigma F_h x_h$, if sampling probabilities vary among strata. The internal sample estimate Fx is compared to the outside estimate $\tilde{X}$ of the same total. Then $Fx/\tilde{X}$ and $1 - Fx/\tilde{X}$ are the estimates of the proportion of coverage and noncoverage. This reveals only the net noncoverage, hiding whatever overcoverage the sample may contain. But it may suffice for many social surveys, where noncoverage is a much more substantial problem than overcoverage.

For example, if the number of dwellings in a city is estimated at 500,000 and a sample of 1 in 1000 obtains 460 dwellings (including nonresponse), then the estimate of noncoverage is $1 - (1000)460/500{,}000 = 1 - 0.92 =$ 8 percent. If the coefficient of variation is 0.03 for the sample size x, and negligible for $\tilde{X}$, we can estimate that, within one standard error, the noncoverage lies between 5 and 11 percent. With two standard errors, we can say that the noncoverage is less than 14 per cent.

Acceptable estimates of a population's comparable elements can frequently be found or constructed from a census count or another large sample. This must be reliable, current, and comparable. Furthermore, the basic units must be similar, not only by formal definition, but also operationally. Any differences in the two operational definitions must be small enough to permit meaningful comparisons between the two sources. Hence the choice of the base may be crucial. For example, we may have to decide whether to compare dwellings, or all persons, or only all adults, or some other population. The choice will depend on appraisals of the two data sources and the survey aims.

What measures can we adopt to *decrease the effect of noncoverage*? Assume that the sampler, using his own and others' cumulated experience, has done his best to design a set of good procedures. What other improvements are possible? (1) *A subsample* of sampling units may be sent out for a special *quality check*, either with similar, but preferably with improved or drastically different procedures (11.5 and 9.8*D*). The sample estimates can be adjusted accordingly. For example, a sample of block listings may be reassigned for a similar but independent ("blind") listing, or for a relisting by a special crew, or for a check from independent sources (such as city directories). (2) A *linking procedure* may be attached either for the entire sample or for a subsample. For example, with an ordered list of dwellings on a block, we can design a "half-open interval" for including any missed dwelling following the sample dwelling and up to, but not including, the next nonsample dwelling (9.6*C*.IV). (3) With knowledge about diverse rates of noncoverage in different parts of the sample, weighting *may* reduce its effect on estimates of the means. (4) Increasing the segment size can often decrease the coverage errors.

Which method will work best in any situation? The answer requires more than *a priori* reasoning; often only experience will reveal the practical efficacy of a procedure for the specific situation and resources.

Published reports of noncoverage rates are rare, and they hardly contain an unbiased selection of survey situations. Furthermore, because they depend so much on specific conditions, they are hard to evaluate, to relate to each other, or to our own survey situation. Investigators who report about difficult coverage situations, such as area samples of a country's dwellings, are also likely to have better procedures. Reports about selections from ready population lists, such as a firm's employees, would not be very informative. It is of limited value to know whether such a list is 98 percent or 99.9 percent complete. We may merely be skeptical and consider 100 percent as likely to be an uninvestigated assumption or an approximation.

It is generally easier to attain a good coverage rate for a sample of a single city or county than for an entire country. Immediate and direct supervision of the field work permits tighter control than for a field staff spread across a large country and directed by mail. Coverage rates over 95 percent, perhaps over 97 percent, should really be expected from better samples of single cities [Sharp and Feldt, 1959]. But the coverage of over 97 percent of census figures reported by the Census Bureau samples (10.4) and by SRC samples [Kish and Hess, 1958] represent probably the best for area samples of U.S. dwellings. This last reports how an improvement from 90 to 97 percent was attained; I suspect that 10 percent noncoverage for national samples is often exceeded. Experienced samplers do not expect high coverage rates in difficult situations without extraordinary efforts. A skilled colleague was not surprised, although not pleased, to find (with a quality check) 20 percent noncoverage on one of his samples in an underdeveloped area.

The decennial U.S. Census is able to attain a coverage of about 97 or 98 percent of the total population [Eckler, 1953; Coale, 1955]. The coverage is 95 percent for corn production, much less for some minor and scattered crops (berries, for example), and more for some concentrated, highly commercial crops. The percentage covered is often much better for production volume than for numbers of producing farms, because farms with small volume are more often missed.

The effect of noncoverage is different—and less in most cases—on sample means, than on simple totals of the Fy kind. The effect is likely to be still less on most differences between means (13.4*B*).

13.4 NONRESPONSE

13.4A Sources of Nonresponse

Nonresponse refers to many sources of failure to obtain observations (responses, measurements) on some elements selected and designated for the sample. The nonresponse rate can be measured well if accurate accounts are kept of all eligible elements that fall into the sample. These are necessary for understanding the sources of nonresponse, for its control and reduction, for predicting it in future surveys, and for estimating its possible effects on the surveys. Furthermore, reporting the extent of nonresponse has become an accepted responsibility for better surveys.

These aims can be better served if the many possible sources of nonresponse are sorted into a few meaningful classes. A good classification of nonresponse depends on the survey situation; the classification given below may be considered suggestive, rather than definitive. The terminology centers around problems of interview surveys; but the general treatment is applicable to other sampling situations, and some of the necessary translation is indicated. Identification of sample elements may be also necessary when the classifications refer to the respondent (or unit of observation) and not necessarily the element. For example, when "any responsible adult" can give the information about the entire household, only one adult need be at home. Or when the elements are children, but the housewife is the respondent, her presence and cooperation are needed, not theirs.

1. *Not-at-homes* (NAH's) have diverse sources and effects in different survey situations. First, the *nature of the respondent* makes a difference. Norwegian farmers spend more time at home than urban Americans, especially New Yorkers; housewives more than male employees; and it is much easier to find a "responsible adult" in the household than a specified respondent. Second, the *time of calls* is important. Daytime is particularly bad for finding employed (or student) members of households; evenings and weekends have favorable interviewing hours; seasonal variation occurs, especially during vacations. Third, *the interview situation* can be altered sometimes, perhaps by making advance appointments by

telephone, or by mailed notice; and even more easily by taking advantage of a group situation in a school, or factory, or office. Fourth, *call-backs* to find the not-at-homes increase the overall response rate. We may distinguish between "*nobody-at-home*" and "*respondent absent*" situations: the latter permits eliciting information about the household and about the respondent's hours of availability.

The term *temporarily unavailable* would be a useful generalization for this category, denoting a *deferment* rather than a denial of the interview, or other observation. A respondent may be too busy, tired, or ill at the time, but will be cooperative on another call. This can also refer to a busy signal in telephone interviews; to delay and hesitation in answering mailed questionnaires; also to the postponement of medical or meteorological observations due to temporary obstacles.

2. *Refusals* depend on several factors. First, *the nature of respondents* may differ, and their dispositions may vary from cheerfully cooperative to hostile. Differences may occur between cultures, social classes and demographic categories. Second, the *auspices* of research and the *technique* of the interviewer are important. This involves motivating the respondent to cooperate through some expected reward, usually intangible but sometimes financial [Kahn and Cannell, 1957]. Third, the nature of the *question* may be important; also, the sequence of questions. In diverse cultures income, religion, sex, and politics may be either difficult or easy to discuss. The effects of these several factors on the refusal rates do not appear independently, because they interact. For example, the members of a profession (say, statisticians) may answer questions about the profession readily, when asked by their own professional organization, but less readily to other sponsors or about other matters.

Several of these factors are temporary and changeable. A person may refuse because he is ill-disposed or approached at the wrong hour; another try, or perhaps another approach, may find him cooperative. Hence, all refusals need not be considered an "all or nothing" matter. Deming [1953] proposes a model with partial and temporary refusals, which can be reduced, like not-at-homes, with further effort, although a "hard core" of refusals remains.

Since most refusals can be considered permanent, a general term for this category is *unobtainable*, denoting a *denial* rather than a deferment of the observation, whether by interview, telephone, mailed or distributed questionnaire, or instrumented observation. Repeating the attempt will not bring success; this will come either not at all, or only through a drastic change in procedures. From this view, respondents known to be away during the entire survey period belong in this category, rather than among the not-at-homes.

3. *Incapacity or inability* may refer to illness, physical or mental, which prevents a response during the entire survey period. For some surveys, a language barrier may play this role; for others, complete or even partial illiteracy. (One respondent was reported completely drunk by the time he reached home from work every evening during the entire survey period.) If generalized, this group could fit into the previous unobtainable category. However, it may help to distinguish unwillingness or refusal, to "perform the role" of respondent, from inability or incapacity to do so. In the rare situation where this group is large, some one or two causes probably predominate; this group should be identified, and perhaps obtained with different methods.

4. *Not found* is a category that may be large in mailed surveys, or for movers in panel studies. Such respondents are either not identified or not followed, because this would be too expensive. Cases of *not attempted* interviews belong to the same general category. They could be caused by inaccessibility (lighthouse keeper, desert hermit, forest ranger), including dangerous surroundings; also, floods or storms preventing access for the entire survey period. It can include cases of *missing field staff* that may occur in a distant sampling area, inaccessible within the survey period and resources.

5. *Lost schedules* includes information lost after a field attempt. They can be lost in the mail or destroyed in the office. Some may be found *unusable* because of poor quality or cheating. Others may remain *unfilled* because they were lost or forgotten before the interview, either before or after they reached the interviewers. Similar in effect is the *rejection* of interviews, because they were obtained by mistake from wrong elements, improperly substituted for sample elements.

The above categories refer to nonresponse involving the entire interview or questionnaire. But in otherwise completed schedules, some *questions may remain not ascertained.* The reasons may resemble any of the above; the most common are refusals or incapacity on the part of respondents; or it may be due to omissions or recording unusable or invalid answers by the enumerator.

All categories of nonresponse refer to eligible respondents, and should exclude the ineligibles. These could be vacant dwellings, or households without the specified kinds of population elements. The ineligibles must be clearly noted to complete the accounts, but they do not belong among the nonresponses. The nonresponse rate should be computed from responses and nonresponses among the eligibles only.

Ambiguity and confusion can arise when the specification of the respondent lacks precision either in definition or in the survey operation;

for example, when contacting a household by telephone or mail questionnaire. Although these errors belong to the broader province of response errors, they have some bearing on errors of nonresponse.

Usually one, two, or three specified sources of nonresponse predominate; these should be tabulated and reported separately, leaving a joint category for "others." The dominant sources are not the same for all surveys. For interview surveys in dwellings, "not-at-homes" and "refusals" may be reported, plus a grouping of "others." In panel studies a group of "movers not found" may become large. In addition to complete nonresponses, small percentages of "not ascertained" appear separately for many questions.

13.4B Effects of Nonresponse

To show the diverse effects of nonresponse on several survey statistics, denote by

$$\bar{Y} = Y/N = W_1\bar{Y}_1 + W_2\bar{Y}_2 \tag{13.4.1}$$

the mean of some characteristic, where W_1 and W_2 denote the proportions of response and nonresponse ($W_1 + W_2 = 1$), and $\bar{Y}_1$ and $\bar{Y}_2$ denote the means of the characteristics in the two segments. These are the average values expected under the "essential survey conditions." For simplicity we shall discuss nonresponse, but the model is equally relevant for noncoverage. A joint treatment of nonobservation results if W_2 represents the sum of the proportions of nonresponse and noncoverage.

The use of mean response $\bar{Y}_1$ to estimate the mean $\bar{Y}$ causes a bias $(\bar{Y}_1 - \bar{Y})$. The *relative bias* (RB) of the sample mean is

$$\text{RB}(\bar{Y}_1) = \frac{(\bar{Y}_1 - \bar{Y})}{\bar{Y}} = \frac{(\bar{Y}_1 - W_1\bar{Y}_1 - W_2\bar{Y}_2)}{\bar{Y}} = W_2\frac{(\bar{Y}_1 - \bar{Y}_2)}{\bar{Y}}. \tag{13.4.2}$$

If the nonresponse mean $\bar{Y}_2$ differs little from the response mean $\bar{Y}_1$, the relative bias will remain small, even for moderate values of W_2. If both W_2 and $(\bar{Y}_2 - \bar{Y}_1)$ are small, the bias should be negligible. For the bias to be important, a large nonresponse must coincide with large differences between the means of the two segments.

The population aggregate $Y = N\bar{Y}$ can be estimated with the simple expansion Fy, whose expected value is $N_1\bar{Y}_1 = NW_1\bar{Y}_1$. The relative bias is

$$\text{RB}(N_1\bar{Y}_1) = \frac{NW_1\bar{Y}_1 - N\bar{Y}}{N\bar{Y}} = -\frac{NW_2\bar{Y}_2}{N\bar{Y}} = -W_2\frac{\bar{Y}_2}{\bar{Y}}. \tag{13.4.3}$$

Hence, if a subclass, which is difficult to include in the frame, is known to contain negligible amounts of the characteristic (because $\bar{Y}_2 \doteq 0$), excluding it by some "cut-off" method will not substantially bias the simple expansion estimate $N_1 \bar{Y}_1$. However, when $\bar{Y}_2 \doteq \bar{Y}_1$, the relative bias of the simple expansion is about the size of the noncoverage W_2; if the total population size N is known, it can be used to compute W_1, hence to estimate Y with $(F/W_1)y_1$, whose expected value is $N\bar{Y}$. Thus the expansion factor can be increased for the nonresponse, assuming that the means of the responses and of nonresponses are equal. In this adjusted expansion the relative bias of $N\bar{Y}_1$ is the same as that of $\bar{Y}_1$; that is, $\text{RB}(N\bar{Y}_1) = \text{RB}(\bar{Y}_1)$.

When comparing two means, the difference $(\bar{Y}_a - \bar{Y}_b)$ has the bias $[(\bar{Y}_1 - \bar{Y})_a - (\bar{Y}_1 - \bar{Y})_b] = [W_2(\bar{Y}_1 - \bar{Y}_2)]_a - [W_2(\bar{Y}_1 - \bar{Y}_2)]_b$. Hence, the relative bias is

$$\frac{[W_2(\bar{Y}_1 - \bar{Y}_2)]_a - [W_2(\bar{Y}_1 - \bar{Y}_2)]_b}{\bar{Y}_a - \bar{Y}_b} = W_2' \frac{(\bar{Y}_1 - \bar{Y}_2)_a - (\bar{Y}_1 - \bar{Y}_2)_b}{\bar{Y}_a - \bar{Y}_b}. \tag{13.4.4}$$

The last expression assumes that $(W_{2a} \doteq W_{2b} \doteq W_2')$; that the proportion of nonresponses in the two classes are approximately equal. In that case, an important relative bias can occur only if the effect of nonresponse bias on one class mean is much different from that on the other. Even when separate estimates have nonresponse biases, their differences often are relatively free of bias, because of similar effects of nonresponse on the compared class means. The comparison of two periodic surveys, conducted under the "same essential conditions," may have little bias if each survey carries a similar bias.

However, we must not assume that this cancelling of biases always occurs. First, the proportion of nonresponses may be different in the two classes $(W_{2a} \neq W_{2b})$. Second, the effect of nonresponse may differ from one class to another. A large survey of the unemployed in 1937 was checked by a smaller but better sample:

> ". . . As evaluated by the check sample, approximately 67 per cent of those who should have responded actually did so, but still there were significant and important biases in the returns. The biases differed widely between men and women and between age groups and other classes. Moreover, they differed for varying characteristics within a given class. Thus, WPA workers, the fully unemployed, and the partially unemployed responded in different ways within a given age-sex group." [Hansen, Hurwitz, and Madow, 1953, 2.10.]

13.4C Some Nonresponse Results and Call-back Data

Nonresponse results depend on so many factors that a general and balanced presentation is not feasible here. I have chosen illustrations which have particular relevance for samples of persons from dwellings. I preferred unpublished and recent results to those which have already appeared in places readily available. For more data, turn to the presentations of Stephan and McCarthy [1958, Ch. 11] and Zarkovich [1963, Ch. 7].

A few points are worth noting:

1. The proportion of refusals varies greatly; it depends on the relationship of respondents to the subject matter, and it is influenced by the skill of those who write and administer the questions. That refusals can be held to reasonable limits is to the credit of patient respondents, and of researchers who have learned how far to probe and encroach on respondents' privacy and stamina without outraging their patience. Two successful examples were a 91 percent response on contraceptive practice [Freedman *et al.*, 1959], and a good return even on sexual behavior [Potter *et al.*, 1962]. The National Center of Health Statistics reports [1964] that 86.5 percent of a national sample accepted a thorough medical examination.

2. Refusals should be considered separately from not-at-homes; they are less amenable to reduction with call-backs and less tractable to generalizations. The nonrefusing portion may be regarded as the eligible population; the effect of decreasing therein the proportion of not-at-homes with increasing number of call-backs may be studied separately.

3. Characteristics of nonresponses should be considered jointly with those obtained on successive calls, since the latter throw light on the former. Otherwise, the characteristics of nonresponses are only ascertained occasionally with special effort.

4. Available data reflect the experience of the more successful organizations, I suspect. The vast bulk of low responses of the less successful may remain not only unpublished, but also untabulated.

5. To illustrate the effects of nonresponses and of responses on successive calls, we tend to select characteristics which exhibit divergences. Compared to the standard errors of ordinary samples, the great majority of items show small differences after the not-at-homes have been reduced to, say, under 10 percent with 3 to 6 calls. Nevertheless, complacency is unwarranted, since many items would show small divergences even with loose methods. Among those which do require better selection methods, a fair proportion may also require a high response rate to reduce nonresponse biases to negligible proportions.

6. Families with small children are easy to find, whereas the single and divorced are not. The old and widowed are easy to find, but often have high refusal rates; the young, especially the married young, behave in opposite ways. The employed are much harder to find than those who are not; this explains male-female differences among not-at-homes. However employed females are hard to find. (They are also missed by

quota samplers who depend on age-sex quotas.) It is harder to find the higher socio-economic classes, whether classed by income, education, occupation, or purchasing behavior; the lowest class may also be harder to find than the middle classes. Large city dwellers have higher proportions of NAH's and refusals than rural people, especially farmers. Definition of the respondent has obvious implications: randomly selected adults are twice as hard to find as "any responsible adults" and housewives, and household heads are even more elusive.

7. Our meager knowledge should not lead to inaction. The direction (sign) alone of a discrepancy is of little use; their magnitudes depend on factors we do not control. Actual differences may be much greater or smaller than expected. Furthermore, sometimes we are genuinely surprised by divergences which are contrary to our first, naive expectations; examples below are the tire dealers and the Russian peasants.

8. *Response on mail surveys* presents an elusive target for generalizations; they have been observed from less than 10 to almost 100 percent. Proper auspices employing skillful procedures have often elicited high responses from special populations, such as professional societies or business firms. (Yet results were published recently based on a 12 percent response to a single mailing to doctors.) However, the idea prevails that mail surveys cannot elicit high response rates from the general public. This belief has been successfully challenged recently by several organizations. The 1960 U.S. Census also has depended heavily on self-enumeration. Greater optimism is invited by the growing literacy of the public, by the increase of fixed and listed addresses, and by the improved skills of researchers.

The British Social Survey has obtained high responses on several mail surveys from the general public. A sample from the Electoral Register about poultry-keeping elicited a 93 percent response with three mailings [Gray, 1957]; the prevalence of poultry was 29 percent in the first mailing, but 35 percent finally; it ranged from 14 percent in the first day's answers to 43 percent at the end. Nonresponse was further decreased with interviews in half of the sample. A survey about smoking obtained a 90 percent response [Gray, 1959].

Hochstim [1962] elicited an 81 percent response with three mailings in a survey of drinking and smoking in California, and raised it to 88 percent with an interview follow-up. Sirken [1962] has had marked success with several mail surveys on health problems.

Several of these articles also contain encouraging evidence about the quality of responses in mail surveys. I have little personal experience with mail surveys, but reading the evidence leads me to the following conclusions about nonresponse to mail surveys of literate populations.

(1) High responses can be elicited with skillful, brief, simple questionnaires. (2) Three or four mailings will often raise the response over 80 and 90 percent. (3) Interview follow-ups on a subsample of nonresponses will further raise the response rate. (4) Low responses to one or even two mailings should not be accepted, because they often contain severe selection biases.

Survey Number	Total Attempts[a]	Percent Not Contacted	Total Interviews	Percent of Those Interviewed Contacted on:		
				First Call	Second Call	Third or Later Call
1	1256	7.2	1029	38.0	32.9	29.1
2	4207	13.7	3065	43.8	35.4	20.8[b]
3	736	...	610	36.9	30.8	32.3
4	25,000	6.0	...	...	...	...
5	...	...	3265	63.5	22.2	14.3
6	3006	7.0	2796[c]	72.0	19.4	8.6
7	2076	3.0	2014[c]	83.5	10.3	6.2
8	3512	4.5	...	...	...	...
9	2074	6.7	1734	77.6	16.2	6.2
10	1709	...	1530	72.2	20.8	7.0
11	1376	6.4	1183	42.3	40.8	16.8

[a] The total-attempt figures represent, insofar as it is possible to ascertain, all instances where an interview should have been obtained. They do not include attempts at vacant dwelling units, attempts where there was no person satisfying the survey definition of a respondent, and other similar types of attempts. They do include attempts classified as refusal, ill, out of town and the like.

[b] This survey, in general, required its interviewers to make only three calls at each assigned address.

[c] These values represent all households reached in the survey, and the call percentages are based on them rather than on the number of completed interviews.

TABLE 13.4.I Noncontacted Respondents and Number of Calls Required for Completed Interviews (From Stephan and McCarthy [1958, Tables 11.3 and 11.4]; look there for further references to original sources.)

1. Random adult from urban dwellings; Elmira, New York, 1948; Elmira Study.
2. Random adult from urban and rural dwellings; Cal., Ill., and N.Y., 1948; NORC.
3. Random adult from national dwellings, 1948, SRC of U. of Michigan.
4. Any responsible adult from national dwellings, 1951; CPS of Census Bureau.
5. Any responsible adult from urban national dwellings, 1943, Hilgard and Payne.
6. Any responsible adult from urban national dwellings, 1947, Market Research Company of America.
7. Any responsible adult from rural national dwellings, 1947, Market Research Company of America.

8. Head of spending unit from national dwellings, 1950; SRC of U. Michigan.
9. Housewife from urban dwellings, Milwaukee, 1946, Alfred Politz Inc.
10. Farm operators in N.Y. state, 1951; N.Y. State Farm Survey.
11. Individuals over 16 years from National Register, London, 1950; Durbin and Stuart.

Corrections by Leslie Kish: The "not contacted" of 4.5 percent for item 8 is only for not-at-homes; about 10 percent should be added for refusals and other reasons (see Table 13.4.III). I suspect this may be true for other low not-contacted rates. This demonstrates the difficulties of ascertaining comparable figures. On the other hand, the CPS nonresponse in 1954–1955 was 4.2 percent, with 1.6 percent for not-at-homes, 1.6 temporarily absent, 0.6 refusals, and 0.4 other reasons [Census Bureau, 1963]. I surmise this low nonresponse is approached by few other national samples.

Call Number	(*A*) Interviews on Three Studies			(*B*) Terminal Result				(*C*) Success Rate		
				Int.	Re-fusal	NAH	Others	Calls	Int.	Rate
1	37	37	36	40	22	14	74	1580	531	37
2	30	31	31	27	26	10	14	983	366	37
3	17	17	16	17	20	10	8	577	230	40
4	8	7	8	8	15	16	0	313	113	36
5	5	5	6	5	8	18	4	156	69	44
6+	3	3	3	3	9	32	0	129	32	25
Total Percent	100	100	100	100	100	100	100			36
Total Number	1743	1307	1360	1341	106	105	50	3738	1341	
			Percent	84	6.5	6.5	3			

TABLE 13.4.II Results on Successive Calls in Four National Samples of Persons from Dwellings (SRC of University of Michigan, 1960–1961)

In four national epsem samples of dwellings, one person interviewed: the household head or his wife was randomly designated. This population approximates closely a random sample of the adult population.

(*A*) Percentages of interviews obtained on successive calls are shown for three surveys.
(*B*) Terminal results on successive calls are shown for a fourth survey; noninterview reasons are refusals, not-at-homes, and other reasons. The latter may actually include some ineligibles, and the not-at-homes some vacant homes. Hence, the true response rate may be 1 or 2 percent higher than the 84 percent shown.
(*C*) The success rate denotes the percentage of calls that resulted in interviews, shown for the fourth study. On other studies, the success rate would be a little lower on the first call and higher on the second.

	Responses	Nonresponses				Number of Eligible Respondents
		Total	Refusal	NAH	Other reasons	
Total	86	14	7	4	3	7155
18–24 years old	94	6	2	3	1	350
25–34 years old	93	7	4	2	1	1320
With children under 18	92	8	4	2	1	3430
Negro	90	10	4	4	2	656
12 largest cities	79	21	11	4	5	1067
3+ adults in dwelling	93	7	4	1	1	954

TABLE 13.4.III Percentage of Nonresponses from Four National Dwelling Samples, of the SRC (University of Michigan, 1959–1960)

Head or wife of head randomly selected from 7155 households; this approximates the characteristics of the adult population. These subclasses had greatest divergences. Other subclasses investigated were within 3 percent of the overall 86 percent response rate. (From memorandum by I. I. Hess, January, 1962.)

	Number of Calls						
	All	1	2	3	4	5	6 or more
Number of cases	1310	427	391	232	123	77	59
Percentage of cases	100	33	30	18	9	6	4
Education of Head							
Some grade school; none	32	38	32	26	26	30	22
Some high school	19	19	19	22	26	21	7
Completed high school	23	19	26	25	21	21	30
Some college	12	10	12	12	14	14	19
Completed college	11	9	10	13	11	11	20
Family Income							
Under $3000	25	40	20	15	18	19	15
$3000 to $4999	18	16	17	23	19	19	12
$5000 to $7499	23	19	25	26	25	26	22
$7500 to $9999	14	13	14	14	15	16	17
$10,000 and over	17	11	20	19	21	17	31
Median Income	$5598	$4188	$5880	$6010	$6200	$6010	$7443
Plans to Buy a Car							
New	10	7	10	10	13	16	17
Used	8	8	8	9	8	7	12
Will not buy	81	85	80	79	77	75	69

TABLE 13.4.IV Percentages for Several Variables on Successive Calls

Each category would sum to 100 percent with the addition of "Not Ascertained," "Don't Know," and "Other" classes. From a 1963 Survey of Consumer Expectations of the Survey Research Center of the University of Michigan. Respondents were head or wife of head, randomly selected, which approximates a population of adults. Total nonresponse was 16 percent. (From a 1964 memorandum of J. B. Lansing.)

	All	Number of Calls 1	2	3	4	5	6	7+	Call not recorded
Percentage who have ever taken an air trip	29	25	27	35	36	39	35	32	24
Percentage of all adults obtained on this call	100.0	34.4	31.6	15.2	9.7	4.3	1.9	1.3	1.6

TABLE 13.4.V Percentages of Air Travelers on Successive Calls

In the 1962 Survey of Consumer Finances (SRC), the air travel experience of 5520 adults was obtained in an epsem of half as many households. The proportion of those who have ever taken an air trip increases by about half from first to later calls. (Data from a memorandum of J. B. Lansing, December, 1962.)

	Percent in the Replacement	Percent in the Cross-Section
Wife of head	37	43
Female	57	56
No children under 18	56	40
One adult in family	23	14
Income under $5000	59	51

TABLE 13.4.VI Divergences Found in Replacement Addresses

Between 1957 and 1961 the not-at-home addresses from six samples were sent out as "replacement addresses" (13.6D) in succeeding and similar surveys. Of 450 address sent out, 10 percent were vacant. Among the others, 65 percent became replacement interviews, 50 percent with the same family and 15 percent with a new family that moved into the dwelling. Of the 35 percent nonresponses, 15 were recorded as refusals, 15 as not-at-home, and 5 as others. Comparison of the 65 percent replacement interviews (about 260) was made with the cross-sectional samples. The replacement addresses may be regarded as if they constituted 6 percent not-at-homes remaining after about six calls; the replacement interviews then resemble 4 percent obtained with further calls. The table displays the salient differences we found between the cross-sectional samples and the replacements.

"In comparison with the cross-section sample, these respondents tend to be younger, have less money and fewer children under 18 years of age. Replacements are a little more likely to be the head of a one adult family. Replacement and cross-section respondents have about the same percent distributions by sex, race, geographic region, and size of place." (Memorandum, I. I. Hess, February, 1964.)

Result of Call	Percentage Distribution by Number of Calls						Mean No. of Calls
	One	Two	Three	Four	Five or more	Total sample	
Completed interview	28	36	28	31	25	87	2.6
Refusal or "too busy"	20	16	21	15	18	8	4.6
Respondent not home	20	14	15	16	13	3	6.4
No one home	30	32	34	37	43	1	7.0
Respondent ill, senile	1	1	1	1	1	1	3.2
Language problem	1	1	1	*	*	*	3.7
Total	100	100	100	100	100	100	2.9
Number of cases	2646	1888	1164	767	1278	2651	7743

* Less than 0.5 percent.

Demographic Characteristics	Number of cases	Percentage Distribution by Number of Calls					Mean No. of Calls
		One	Two	Three	Four or more	Total	
All Respondents	2313	32	30	14	24	100	2.6
Employment Status							
Employed	1357	21	31	18	30	100	3.0
Not employed	952	48	28	9	15	100	2.1
Age							
21–29 years	508	34	28	14	24	100	2.5
30–39 years	634	28	33	14	25	100	2.7
40–49 years	455	29	34	15	22	100	2.6
50–64 years	502	32	25	16	27	100	2.8
65 years or older	212	47	25	11	17	100	2.2
Marital Status							
Never married	186	28	21	14	37	100	3.3
Divorced, separated	105	26	24	16	34	100	3.2
Widowed	163	45	23	11	21	100	2.3
Married, no children present	864	31	26	16	27	100	2.8
Married, children present	990	33	36	13	18	100	2.3
Sex							
Male	1088	25	27	20	28	100	2.8
Female	1225	38	30	12	20	100	2.4
Relationship of Head of Household							
Head	1176	26	29	17	28	100	2.9
Wife	910	40	31	11	18	100	2.3
Other relative	189	33	31	14	22	100	2.6

TABLE 13.4.VII Characteristics on Several Calls of a Sample of Randomly Selected Adults from Dwellings in the Detroit Metropolitan Area

Responses were 87 percent; refusals 8, not-at-home 4, and others 1 percent. Three Detroit Area Studies of 1957, 1958, and 1959 were combined by Sharp and Feldt [1959].

	Number of Respondents	Percent with Garden	Median Size of Garden		
			All with Garden	Know Yield	Do not know Yield
First call	921	47	455	573	325
Second call	209	38	294	440	100
Third call	46	30	226	400	50
Substitutes	239	50	565	712	334

TABLE 13.4.VIII Percentage of Nonfarm Housewives with Home Gardens in a National Sample in 1944. (From a memorandum by J. A. Bayton for a sample of the Division of Program Surveys in the U.S. Department of Agriculture.)

Substitutes were made after two or three calls failed to produce a response; their nonrepresentative character was noted; they were drastically adjusted in an attempt to correct for obvious bias.

Illustration 13.4.IX

"To fill this need (administration of tire-rationing during the war), inventories of tire dealers were taken each month. . . . the '*complete count*' of September 1944 resulted in 24,015 nonresponse or 17 percent of the 140,989 questionnaires mailed out, whereas in the December 1944 and March 1945 *samples* the nonresponses were only 1.2 and 3.9 percent respectively of the 16,000 questionnaires mailed out, and it be noted that both these samples included a sample of nonresponses of September. Incidentally, the average number of tires per dealer was 50 percent higher in both December and March for this group than the average inventory of all other dealers, thus pretty well deflating the possible interpretation in which refuge is so often taken, that nonresponses can be ignored on the pretext that they average, or are composed mostly of small dealers, or represent no stock at all. These results bear out an interesting observation that had been made in tabulating the returns from the September complete count, viz., that the last 3,000 dealers that reported subsequently to the second follow-up letter sent to delinquents, actually had on hand an average of over 10 tires per dealer against an overall average of 6 tires per dealer for the 114,000 dealers that had already responded." [Deming, 1950, pp. 357–358.]

Illustration 13.4.X

The hard-to-find need not necessarily belong to the higher classes. Note this early and interesting example from the first decade of this century, presented by Jensen [1926], together with other examples of the dangers of substitutions and other selection biases:

"Many examples could be given a sample being less representative than expected, because one has rejected or passed over units which, for some reason or other, were not regarded as being suitable for inclusion in the enquiry. In an investigation into economic conditions in Russia, information was obtained from a certain number of farms in each district. If the investigator did not find the farmer at home, he substituted that farm with its nearest neighbor. It turned out that in the sample the more well-to-do farmers had for the most part been included, and thus the sample did not give a proper view of the totality. The reason was that the rejected farms were for the most part owned by peasants who, owing to poverty, had to work as day labourers on other farms, and therefore were not at home when the investigator called at their farms."

	Percentage Response	Average Number of Fruit Trees
First mailing	10	456
Second mailing	17	382
Third mailing	14	340
Nonresponse	(59)	290
Total	100	329

TABLE 13.4.XI Responses to Three Mailings to Fruit Growers. (From Finkner [1950] and Cochran [1963, 13.2].)

Mail survey of a population of 3116 fruit growers in North Carolina in 1946. The number of fruit trees was available from other sources for the population, hence for the nonresponses.

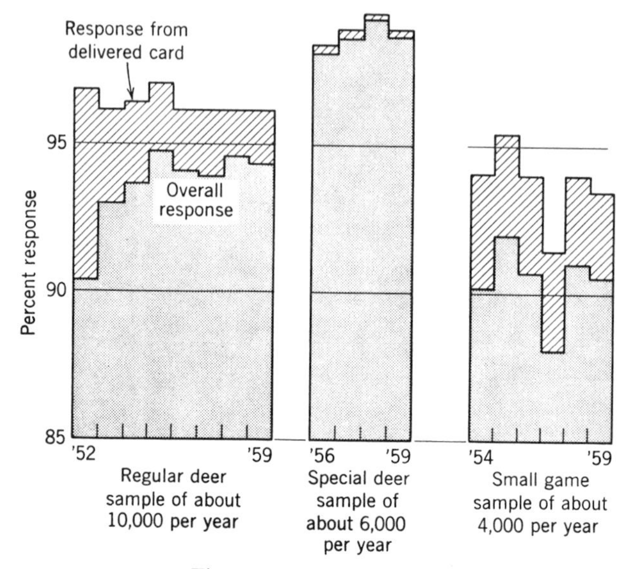

Figure 1. Survey response rates.

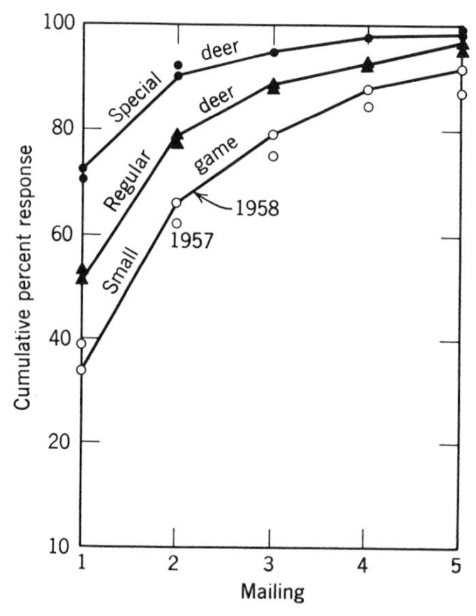

Figure 2. Cumulative response rates by mailing.

TABLE 13.4.XII Results of Periodic Mail Samples of Licensed Hunters in Michigan. (From Eberhardt and Murray [1960].)

Mailing	Regular Deer	Special Deer	Small Game
1	51%	72%	34%
2	58	66	48
3	47	52	38
4	36	46	45
5	26	28	27

Figure 3. Response rates by mailing—1958 surveys.

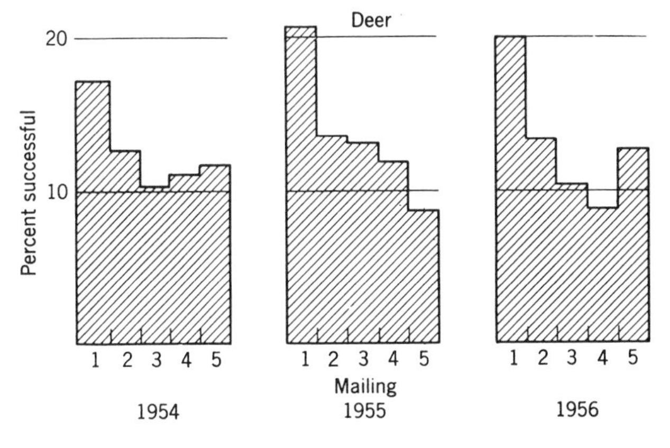

Figure 4. Reported hunting success by mailing from which response was received.

TABLE 13.4.XII *(Continued)*

The Department of Conservation had elicited, with five mailings, over 90 percent response for years. (1) The response for regular deer licensees has been about 94 percent of all mailed cards, and 97 percent of the cards actually delivered by the post office. (2) The response of special deer licensees is even higher, and of smaller game lower. The cumulative response to successive mailings are shown for 1958. (3) The success rate of mailings becomes low only on the fifth mailing. (4) The selection bias in successive mailings is strong, since successful hunters answer more readily; for other species this bias is weaker. The population consists of about 400,000 for deer licensees, and over a million for all licenses; the samples total about 25,000 each year. This remarkably successful sample replaced a previous system of a 100 percent requested returns; these averaged about 20 percent actual returns for five years; since the successful hunters reported three times as often as nonsuccessful hunters, the results were extremely biased.

13.5 CONTROL OF NONRESPONSE

13.5A Methods and Aims of Control

Few people are either always or never at home; over a survey period the probability $(1 - P_i)$ of becoming a not-at-home is neither 0 nor 1 for most individuals. For each individual assume a probability P_i of being found at home (response), defined for an interviewing period and operation. That probability lies between 0 and 1, and it varies between individuals. On first calls we obtain an overrepresentation of persons with high probabilities. The results of second, third, and later calls contain increasing proportions of persons with low probabilities. Although each call collects a mixture of persons with different probabilities of response, the mixtures vary; the earlier calls contain too small a proportion of low probabilities. More of these are harvested with later calls, and increasing the number of calls decreases the difference between the sample and population mixtures.

With only a reasonable number of calls, some discrepancy remains between sample and population mixtures. Furthermore, people with zero probabilities of being found, because of work schedule, or absence during the survey period, will remain unrepresented in the sample. This type is rare for not-at-homes, but common for refusals; the hard-core of complete refusals may consistently resist all efforts. Other refusals may resemble the not-at-homes in being partial, relative, and remediable, rather than total, absolute, and final; hence a model based on a scale of several degrees of refusals, may be more appropriate than an "all or nothing" (0 or 1) model. This model is also more rewarding; because it tends to reduce the margin of ignorance, and points to possible remedy and control. However, that model is complicated, and the assumption of $P_i = 0$ for most refusals may represent a reasonable approximation. Therefore, three remedial methods that depend on $P_i \neq 0$ are much less useful for refusals than for not-at-homes: call-backs, replacement, and the Politz plan. Four other methods, may improve both not-at-homes and refusals: improved procedures, subsampling, effect estimation, and substitution. The remarks about refusals are also pertinent to other sources of nonresponse: incapacity, not found, and destroyed schedules.

The following methods can be used in different situations to reduce either the percentage of nonresponse or its effects.

1. *Improved procedures* for collecting data is the most obvious remedy for increasing response. Improvements advocated for reducing refusals are: (*a*) guarantees of *anonymity*, perhaps with facilities for direct mailing

of responses; (*b*) *motivation* of the respondent to cooperate; (*c*) *arousing the respondents' interest* with clever opening remarks and questions; (*d*) *advance notice* to the respondent, though sometimes harmful, may increase the proportion found at home. The last may be achieved by arranging favorable hours and days, evenings and weekends; or with information, perhaps by telephone, about the respondent's time at home; or with repeated attempts while the interviewer is in the same neighborhood. It is clear that we cannot do justice here to this vast subject and that the researcher must plan for specific situations.

2. *Call-backs* are most effective for reducing not-at-homes in personal interviews, as are repeated mailings to no-returns in mail inquiries.

3. *Subsampling* the call-backs can be economical, if they are much more expensive than the first effort.

4. *Estimating the effect* of nonresponse may provide evidence of the absence of large biases due to nonresponse. Reporting the size and sources of nonresponse has become standard practice in better surveys. Sometimes formal methods can be designed to estimate the size of the bias and to reduce it. This and the following three remedies are further discussed in Section 13.6.

5. *Substitution* for the nonresponses is often suggested as a remedy. Usually this is a mistake, because the substitutes resemble the responses rather than the nonresponses.

6. Nevertheless, beneficial substitution methods can sometimes be designed, such as the *replacement procedure*.

7. The *Politz Scheme* is a procedure for obtaining differences in the probabilities of responses and weighting them accordingly.

Attempts to reduce the percentage or effects of nonresponses aim at reducing the bias caused by differences of nonrespondents from respondents. *The nonresponse bias should not be confused with the reduction of sample size due to nonresponse.* This latter effect can be easily overcome, either by anticipating the size of the nonresponse in designing the sample size, or by compensating for it with a supplement. *These adjustments only increase the size of the response* and the sampling precision, but they do not reduce the percentage nonresponse, nor the possible bias that it can cause.

If the percentages of nonresponse and noncoverage can be anticipated, use them to obtain a desired sample size n by increasing the initial sample size in the ratio: $n' = n/(\text{Response} \times \text{Coverage})$. Alternatively, we can regard the population effectively represented by the final sample as $N' = N \times \text{Response} \times \text{Coverage}$. With either plan, compute an actual working sampling fraction as

$$f = n'/N = n/(N \times \text{Response} \times \text{Coverage}) = n/N'. \quad (13.5.1)$$

Thus the sample represents that portion of the population which can be covered with the essential survey operations. Use the best available estimates or guesses for the different components of response (not-at-homes, refusals, etc.) and noncoverage. Recent experiences with similar procedures and problems provide the best sources. Seldom can we afford large enough pilot studies to provide better estimates. If the sample size remains subject to great uncertainties, it may be controlled with supplements, or with sequential release of subsamples, when circumstances permit these (8.4D).

The *coverage rate* represents the proportion of the eligible population actually included in the designated sample by the survey operations. The *response rate* represents the proportion of the designated population from which observations are obtained. The product then denotes the proportion of the eligible population represented in the sample.

If the size N of the eligible population is known, the sampling rate can be planned as $f = n'/N$, whether or not we know the numbers of sampling units involved in the selection. For example, a sample of $n' = fN$ persons aged 65 years and over can be designated if their number N is known, whether or not we know the numbers of dwellings and blocks containing them. When N is unknown, it may need to be computed from a sampling unit base K and an estimate of the *eligibility rate* (N/K). For example, a sample of dwellings containing persons aged 65 or over may be designed, if the known number of dwellings K is multiplied by the estimated eligibility rate of such dwellings. The eligibility rate may be estimated from a model or from experience with similar surveys.

13.5B Call-Backs

Call-backs denote deliberate new attempts to obtain responses from the nonresponses. This is the most common and successful means for reducing the proportion of nonresponses, particularly the not-at-homes. Because of their similarity, repeated mailings to the no-returns of mailed inquiries also fit this category. The subject has been treated in many articles, countless memoranda, appendixes to survey reports, and sampling texts. Among these, Chapter 11 of Stephan and McCarthy [1958] and Chapter 7 of Zarkovich [1963] contain much useful material. Here we shall treat only the fundamental arguments.

After obtaining a proportion of responses, the decision to make the additional effort of another call rests on two assumptions. First, the new responses must be numerous enough to justify the effort. This holds for not-at-homes, when on the preceding call the response rate was considerably different from either unity or zero. The former may indicate no need and the latter no hope for call-backs. The hard core of zero responses may be so large among refusals that call-backs would seem futile. However, the not-at-homes include many intermediate probabilities of response, because a large portion of the population is neither easy nor impossible to find. Second, decreasing the proportion of nonresponse also reduces its effect, if the relationship of the probability of response to the survey variable is regular, preferably monotonic. If there is no relationship, neither good nor harm can result. If the relationship is strong, the additional response tends to yield correspondingly large reductions of the nonresponse effects. Only in unusual situations can reduced nonresponse increase the effect of nonresponse.

The plan for call-backs should be included in computing costs and sample size, in designing the sample, and the field procedures. The schedules should carry accounts of the timing and results of each call; this information may be tabulated, and the final call number punched on the machine cards. Call-back plans may affect the sample design, especially the desirable size of the cluster: the workloads in the clusters should be figured not only for the first call, but also for the call-backs. Consider, for example, a proposed design of area segments of four dwellings, selected in the first stage. This design can serve a city, because travel for call-backs is short. It can be used to cover the U.S. only if no call-backs are needed; after obtaining the four observations designed for a day's work, the field worker drives on to the next segment; the problem of call-backs makes this design impractical for large territories.

The number of call-backs need not be the same over the entire sample, but can be varied for different parts of the sample. For example, we find that the proportion of not-at-homes is greater in the metropolitan areas, and especially in New York, than in the rural areas of the U.S.; by requiring more call-backs in the former than in the latter, we tend to equalize the not-at-home rates over the different types. This can be done with different numbers of calls fixed in advance for each type of primary area, in accord with estimates of nonresponse rates for each type. Alternatively, we can make the number of calls depend on the response rate actually found in the primary area with instructions along these lines: (*a*) make up to three calls on all not-at-homes; (*b*) compute the response rate as interviews divided by eligible respondents; (*c*) if the response rate is less than 0.85, make a fourth call on all not-at-homes; (*d*) compute the response rate again, and if it is still less than 0.85, make a fifth call on all not-at-homes; (*e*) compute the nonresponse rate again, and if it is still less than 0.85, make a sixth call on all not-at-homes.

If call-backs to remote segments are much more expensive, their number may be reduced according to optimum allocation formulas. If this introduces appreciable differences in response rates, reweighting may be needed. This topic belongs more properly to subsampling.

To avoid a common misunderstanding of the cost for call-backs, consider a simple model of 110 potential respondents in Table 13.5.I. Note that a total of 200 calls yields 98 responses from the 100 possible respondents, for an average of 2.04 calls per response. In addition, 10 other respondents gave no interviews, and 17 calls were made on these refusals and other nonproductive types, including some permanent not-at-homes (vacation, etc.). These increase the calls per response to 217/98 = 2.21. Note that 100/42 = 2.38 calls per response are needed to obtain first calls, or 110/42 = 2.62 total calls per response, including 10 refusals.

Of the 98 final responses, 91 came from the first three calls. *The first call yields the most responses, but the second and third calls have higher responses per call.* This occurs frequently because the first call yields information about the respondent that permits more efficient timing for later calls. Many refusals and other hopeless cases are also discovered on the first call. *The low response rates of the last three calls have little effect on the cumulated calls per response, because they are so few.* For example, if those three responses were 2, 2, and 1 from 10, 8, and 6 possibles (for response rates of 0.20, 0.25, and 0.17), these drastic changes would yield a final 205/95 = 2.16 calls per response instead of 2.04.

The model closely resembles actual results of samples of randomly selected adults, from the United States and from Britain (13.4*C*). More than 2.0 calls per response are needed for samples of household heads or employees. On the contrary, samples of farm operators, housewives, or "any responsible adult" need fewer calls. A model for 100 such responses yields 77 on the first call, 15 more on second calls, and 8 more from 18 more calls, for an average of 125/100 = 1.25 calls per response. Later calls may elicit somewhat lower responses than the first, but this has little effect on overall rates, since it involves a small proportion of the sample.

If any responsible adult may be interviewed, the costs will be lower than if the interviewer must search for a randomly selected adult, or for the household head. Farm operators are also easier to find than urban employees. Obviously a high rate of response per call reduces the field costs. This obvious fact leads to a common mistake about samples which need several calls to elicit a high response rate: the mistaken belief that it would be much cheaper to confine the sample to first calls.

First, the above realistic example of Table 13.5.I required *up to 6 calls*, but an *average number of* 2.04 calls per response. Second, the number of calls per response is 2.38 on first calls, higher than on six calls. If we include the calls on refusals and other futile calls, 2.62 first calls are needed per response, instead of 2.21 with 6 calls. Third, while 3 calls seem the

Wave of Call (1)	Responses (2)	Cumulated Responses (3)	Calls on Possible Respondents (4)	Cumulated Calls (5)	Response Rate (6)	Calls per Response (7)	Mean Calls of Cumulated Response (8)
i	n_i	$\sum_{r=1}^{i} n_r$	t_i	$\sum_{r=1}^{i} t_r$	n_i/t_i	t_i/n_i	$\Sigma t_r/\Sigma n_r$
First	42	42	100	100	0.42	2.38	2.38
Second	35	77	58	158	0.60	1.66	2.05
Third	14	91	23	181	0.61	1.64	1.99
Fourth	4	95	10	191	0.40	2.50	2.01
Fifth	2	97	6	197	0.33	3.33	2.03
Sixth	1	98	3	200	0.33	3.33	2.04
Refusals, etc.	0		10	17			
Total	98		110	217	0.89		

TABLE 13.5.I A Model for Responses Cumulated with 6 Calls on Not-at-Homes

Wave of Call	Responses	Cumulated Responses	Model I of Costs			Model II of Costs		
			Cost per Response	Cost for Wave of Call	Mean Cost of Cumulated Response	Cost per Response	Cost for Wave of Call	Mean Cost of Cumulated Response
(1)	(2)	(3)	(4)	(5)	(6)	(7)	(8)	(9)
i	n_i	$\sum_{r=1}^{i} n_r$	a_i'	$a_i' n_i$	$\frac{\Sigma a_r' n_r}{\Sigma n_r}$	a_i''	$a_i'' n_i$	$\frac{\Sigma a_r'' n_r}{\Sigma n_r}$
1	42	42	1.0	42.0	1.00	1.0	42.0	1.00
2	35	77	0.9	31.5	0.95	1.1	38.5	1.04
3	14	91	1.1	15.4	0.98	1.4	29.6	1.21
4	4	95	1.4	5.6	1.00	2.0	8.0	1.24
5	2	97	2.0	4.0	1.02	3.0	6.0	1.28
6	1	98	2.4	2.4	1.03	4.0	4.0	1.31
Total	98			100.9			128.1	

TABLE 13.5.II Two Models of Mean Cumulated Costs with 6 Calls on Not-at-Homes

most productive, the addition of later calls on the small remaining proportion has little effect on the overall number of calls per response.

If the mean cost per call is the same for successive calls, then the number of calls per response can measure cost directly. For example, interviewing cost within a university could depend entirely on the calls per response. But if the unit cost per call varies between the successive calls, the comparisons must include the variable unit costs.

Table 13.5.II presents cost models with cost per response (a_i) substituted for calls per response (t_i/n_i); cost of call ($a_i n_i$) for number of calls (t_i); mean cost ($\Sigma\, a_r n_r/\Sigma\, n_r$) for mean calls ($\Sigma\, t_r/\Sigma\, n_r$) of cumulated response. Model I represents reasonable unit costs for the response rates of Table 13.5.I. The results are similar to a survey by Edwards [1953], analyzed by Durbin [1954]. The unit costs per response (a_i') are higher on the fourth, fifth, and sixth calls, because both the costs per call and calls per response are higher. Nevertheless, they have only modest effects on the overall average cost, because their proportions are low. On second calls, the cost per response appears low because the call per response is generally low.

The comparison of unit costs on first call to second and third calls is important. In Model II, the unit costs go up rather sharply for later calls. This may represent a situation where travel costs rise sharply because the sample is widely spread, and the basic interview cost is relatively low. These factors (the a_i'') represent rather extreme values that should be avoided, perhaps with greater clustering of elements to reduce the unit cost of later calls.

The effect of increasing unit calls is lessened when the response rates are higher. Then the later calls have less effect on the mean cost of the cumulated response. Substituting another set of n_i(77, 15, 3, 2, 1, 2) in Table 13.5.II, we find the mean final costs of 1.03 for Model I and 1.13 for Model II.

Attention focused on travel costs can mislead us to overestimate the cost increase for later calls. If the fixed cost per interview, including coding and processing, is high, then differences in travel costs may have little effect. Moreover, *the cost of the first call should include the entire cost of selecting the sample cases, finding them, and identifying them*, plus completing and handling the "face sheet" of the schedule. It should carry a large portion of the cost for screening out vacant homes, ineligibles, refusals, and other permanent nonresponses. These may well amount to 10 percent or more of the fixed unit cost. I suspect this often raises the cost for the first call, and places it at greater disadvantage than appears in our table. If the sample also involves a larger screening operation, it can add considerably to first-call costs [Houseman, 1953].

Remark 13.5.1 An explicit model for the structure of unit costs on different calls may clarify the discussion. The cost of the ith wave of calls may be denoted with $a_i n_i = c_i t_i + K n_i$, where a_i is the unit cost, n_i the number of responses; t_i is the number of calls made, and c_i the unit cost for a call without response. The cost of a response is $c_i + K$, where K is the unit cost of interviewing and coding which does not change with the call number. The unit cost on the ith call is $a_i = c_i t_i / n_i + K$. The larger K is, the less the relative variation in unit costs. Large ratios of calls per response t_i/n_i tend to increase the unit cost of later calls, because the cost of mere calls c_i tends to increase for later calls. Furthermore, the ratio t_i/n_i tends to increase in later calls. Both c_i and t_i/n_i are often lower on first than on second calls.

13.5C Subsampling Nonresponses

Subsampling suggests itself as a ready solution for reducing the number of later calls when these are expensive. But on most surveys this solution is impractical, for two reasons. First, the unit costs of early and late calls seldom differ enough to justify introducing the complexities of subsampling with its bookkeeping and weights. The principles of optimum allocation hold here; unit costs should differ by a factor of at least 4 for small savings, and by perhaps 25 to obtain large savings. Second, the introduction of subsampling into the field procedures is unwieldy in many survey situations.

However, subsampling has been used for following up the nonresponse of mail inquiries with personal interviews [Hansen and Hurwitz, 1946]. The previous two objections are not present: the interviews cost much more, and they can be introduced readily at the end of the mailed efforts. Interviews are valuable if they can obtain responses from a considerable proportion of the nonresponses remaining after repeated mailings. Mailed questionnaires can be justified in the first phase, if they are much less expensive than interviews.

Assume that n questionnaires are mailed out for a total cost of $C_0 n$. Also that a number Rn respond and are processed for a total cost of $C_r Rn$. These can represent the results of several mailings. Together these two costs can be stated simply as $C_1 Rn$, where the unit cost for the Rn responses, $C_1 = C_0/R + C_r$, includes both the unit costs C_0 of initial mailing and the processing costs C_r. Thus we have the total cost

$$C = C_0 n + C_r Rn + C_2(1 - R)\frac{n}{k} = C_1 Rn + C_2(1 - R)\frac{n}{k}, \quad (13.5.2)$$

where C_2 is the unit cost of the interviews applied to the fraction $1/k$ of the $(1 - R)n$ nonresponses of the mailed inquiry.

The variance of the combined sample mean may be denoted as

$$R^2S_r^2/Rn + (1 - R)^2S_n^2/[(1 - R)n/k].$$

We can apply (8.5.9) to this variance and to (13.5.2), and obtain that $Rn/[(1 - R)n/k] = [RS_r/(1 - R)S_n]\sqrt{C_2/C_1}$ is optimum. Ordinarily, in our ignorance about the ratio of variances of responses S_r^2 and nonresponses S_n^2, we assume that $S_r/S_n \doteq 1$. Hence, we obtain the

$$\text{optimum } k = \left(\frac{C_2}{C_1}\right)^{1/2} = \left(\frac{C_2R}{C_0 + C_rR}\right)^{1/2}. \qquad (13.5.3)$$

Finite population corrections do not disturb this relation. For departures from simple random selections, allowance can be made with the methods of Section 8.2. Thus a factor of $C_2/C_1 = 4$ results in an optimum subsampling of $k = 2$ for the interviews; and this will produce only modest gains. If $C_2/C_1 = 25$ or 100, subsampling interviews with a value of 5 or 10 for k can become effective. However, unit costs seldom differ so much between the two parts, especially if the basic costs of interviewing, coding, and processing are the same for both.

13.6 FOUR PROPOSED REMEDIES FOR NONRESPONSE

13.6A Estimation of Effects

Estimation of nonresponse effects appears inevitably in survey results. Typically, the estimates are merely implicit in results with small nonresponses: we believe that a small nonresponse is unlikely to produce a large effect on the sample mean. Reporting the size of the nonresponse permits the reader to make a guess about the likely effects. Such reporting has become standard practice for better surveys. For estimating its effect on survey statistics, the size of the nonresponse must be linked somehow to estimates of differences between responses and nonresponses. Information about these differences can come from the sample itself, either from intensive follow-ups on a subsample, or from extrapolating the differences found on successive calls. More often we depend on vague knowledge accumulated in past surveys.

Distinct from the usual vague guesses are some formal attempts to estimate the nonresponse bias, and to reduce it by incorporating that estimate into the statistics. After plotting the accumulated averages against the percentage response of successive calls, a curve may be extrapolated to include the nonresponses. Any method must rely on some assumptions about the nonresponses, preferably an explicit mathematical model to link the means of the successive waves [Hendricks, 1949]. Little research of this kind has been published; for a good review see

Houseman [1953] and Zarkovich [1963 Ch. 7]. Good data are much rarer than good judgment, and these are amenable to a personalistic (Bayesian) approach [Schlaifer, 1959, Ch. 31].

Related statistical problems concern models for combining the bias of nonresponse with the sampling variance into the total error. Birnbaum and Sirken [1950], as well as Cochran [1963, 13,2], assume complete ignorance about the nonresponses, and investigate the effects of the extreme values of 0 or 1 for all the nonresponses of a binomial variable. This approach leads to some harsh conclusions that ordinary surveys with nonresponses of 5 to 10 percent cannot accommodate. The model proposed by Deming [1953] is more flexible: it assumes mixtures of several classes of different response probabilities within the responses of each successive call. Assuming different survey means for each response class, the biases can be computed for varying numbers of calls. For each of these, by combining bias and variance, the total errors can be found along with the optimum combinations for given cost factors.

13.6B Substitutions for Nonresponses

Although substitution is often proposed naively as a solution, it generally is of little help and may actually make matters worse. Substitution often does offer the possibility of controlling the sample size exactly, but this can be better attained approximately with supplements. Moreover, by anticipating it with increased sample size, we can often facilitate the field work considerably (8.4*D*). Entirely distinct from size control is the use of substitutes for reducing the bias of nonresponse. For this purpose substitutes are useless when they merely replace nonresponses with more elements that resemble the responses already in the sample.

Substitution for nonresponse is based on the notion of dividing the sample into subclasses which have widely divergent response rates and substantial internal homogeneity of survey characteristics. Divergence of response rates, together with differences in survey variates, create bias; internal homogeneity, if it existed, would permit its correction.

Weighting subclasses inversely to their response rates can correct for deviations between them. It avoids the problems of field substitution, and it increases the variance less than duplication. But it destroys a self-weighting sample mean, and the complications of weighting are sometimes expensive.

No method of substitution is generally free of disadvantages. These often outweigh the advantages, and cause us to avoid them. But we may choose the method with least disadvantages for a specific situation. The extent of substitution in a sample should be reported in the same manner as nonresponse, since it poses similar issues of interpretation to the reader.

The case is stronger for *imputation* or *editing* of missing characteristics ("not ascertained") of a nearly complete interview [Taeuber and Hansen, 1963]. First, the presence of several correlated variables may permit a reasonably good prediction of the missing variable; regression or "missing plot" techniques may be used. Second, the alternatives may be worse: discarding the entire interview, or using varying numbers of cases in different tables, or presenting the "not ascertained" percentages in each table. The last is often the simplest, but editing may be more useful.

13.6C The Politz Scheme

The aim of this technique is appealing: to avoid call-backs altogether by collecting only first calls, which are then corrected with information about the probability of finding the respondent. He is asked on how many of k similar periods he would have been available; if the answer is r, the interview is weighted by $(k + 1)/(r + 1)$. Typically $k = 5$ has been used. If the respondent says he has been available on $r = 1$ similar periods, he gets the weight 6/2. Thus of 600 respondents, each of whom has probability of 1/3 of being found at home, about 200 will be found; giving the answer $r = 1$, and getting the weight 6/2, they also take the places of the 400 not found at home. The procedure is described in detail by Politz and Simmons [1949].

Note two important assumptions: First, that the information r about availability is correct; or at least that errors in r are not correlated strongly with the survey variables. For example, a positive bias results if respondents with high Y_i values tend to underestimate their availability more than respondents with low values. Second, that we can enforce the random choice of the actual call period among k similar periods; or at least that deviations in the choice of call periods will not bias the results. The problems of meeting such assumptions must be justified in practical empirical results. The evidence is not favorable in two studies that compared the weighted first call results with those of several call-backs. Durbin and Stuart [1954] found that the weighted results resembled those of the first call rather than the combined call-back results. Simmons [1954] also found that the weighted first call fared badly. He also investigated a method of weighting the combined results of three calls against the results of more calls, and found it to perform well; but the unweighted results of three calls also performed well, because the nonresponse was not great.

Thus the weighted first-call procedure has serious problems of validity and practicality. Furthermore, in most situations it fails to prove more economical than call-backs. First, the unequal weights from 1 to 6 will likely increase the element variance by about 25 to 35 percent (see 11.7*C*). It seems unlikely that the cost per element is that much greater for a 6-call procedure than for a first-call procedure (13.5*B*). The increase of the variance caused by weighting is much less if the range of weights is reduced to 1 to 3 or 1 to 2, by using 2 or 3 calls. Second, the cost of complicating the analysis with weights may be substantial, particularly for an otherwise self-weighting sample. Third, the cost of obtaining, recording, and coding the necessary questions must be added. This can be considerable, particularly in relation to an otherwise inexpensive interview (8.2). Fourth, enforcing randomized calls may cost much more than ordinary procedures, which permit free planning by the interviewer, who is generally a part-time employee. I suspect that, typically, the calls are not randomized, and this can be a source of serious bias.

In most situations I doubt that the scheme will prove valid, practical, and economical. Nevertheless, if the survey situation permits only a single call, weighted results, with their bias and increased variances, may be preferable to accepting the bias of unweighted first calls. Such situations can occur when speed is paramount. Even then we may try the compromise of weighted results for two or perhaps three calls [Simmons, 1954]. If the possible bias from nonresponses remaining after several calls is great enough, weighting may reduce it. Survey evidence on this point would be welcome, especially if compared with other procedures for estimating the effects of nonresponse.

13.6D A Replacement Procedure

In this plan we include with the new survey addresses some nonresponse addresses from an earlier survey which had similar sampling procedures. Thus interviews from addresses that were nonresponses on former surveys become *replacements* for nonresponse addresses in the current survey The plan is proposed for interview surveys of dwellings, but it can be adapted to other survey procedures. It is particularly well-suited to organizations which frequently conduct surveys involving similar sampling procedures [Kish and Hess, 1959].

The procedure is suggested as an improvement for surveys now using k calls. With the new sample addresses sent out for k calls, include addresses from a similar recent survey that resulted in nonresponses after k calls. These addresses are called *replacement addresses*; and j calls made on them replace the additional j calls, that could be, but are not, made on nonresponses among the new sample addresses. The j calls made on the replacement addresses can be considered additional to the original k calls at those addresses. If, for simplicity, the replacement addresses receive the same k calls as the new survey addresses, the effect resembles a $2k$-call procedure.

For example, assume a plan to send out 1111 addresses for a survey, with the expectation, based on experience, that a 90 percent response would result in $1111 \times 0.90 = 1000$ responses after k calls. However, 103 nonresponse addresses from an earlier study are available, and a 70 percent return of $103 \times 0.70 = 72$ responses from the former nonresponse addresses is assumed. Hence, it is necessary to send out not 1111 new addresses, but only enough new addresses to yield $(1000 - 72)$ or 928 responses. The necessary number of new addresses is computed to be $928/0.90 = 1031$. With these assumed response rates, the replacement technique yields the originally planned $1031 \times 0.90 + 103 \times 0.70 = 928 + 72 = 1000$ responses. But the area of uncertainty due to nonresponse has been reduced from 10 percent to around $0.10 \times 0.30 = 3$ percent. The effect may be reduced further by weighting these 72 responses with $1/0.70$ up to the 103 nonresponses. It will not be completely eliminated since the hardest-to-get nonresponses may differ from the responses obtained.

Some nonresponse effect remains if the mean response in the replacement portion differs from the mean of nonresponses. Since a 100 percent response seems unrealistic, the replacement procedure is not aimed at the elimination of all nonresponse effects, but merely at their reduction.

In addition to two conditions necessary for successful call-backs (13.5*B*), a third condition is required for effective use of the replacement procedure. The nonresponses of the current survey must be similar to nonresponses from the earlier survey from which replacement addresses are taken. This factor depends on constancy in the behavior of the population, and similarity in survey procedures. The time between the two surveys must be brief enough, so that most people will not have moved or sharply changed their habits. Replacement addresses should be chosen from surveys using similar kinds of respondents; the type of nonresponses may depend on the kind of respondents designated for the surveys.

For not-at-homes the replacement can be chosen from surveys with different objectives, if the respondents are similar. The number (k) and timing of calls should be the same for both surveys, in order to maintain similarity of not-at-homes between the original and the replacement segment. However, the nature of refusals may also depend on survey objectives and questions; replacements should preferably be confined to surveys with similar objectives. The replacement procedure may also meet greater obstacles with refusals, because revisiting the scenes of former refusals is an unpleasant task, and because the proportion of hard-core absolute refusals may be large. On the other hand, many refusals are probably not of that kind; Stephan and McCarthy [1958, p. 238] report a response of 68 out of 92 former refusals. Attempts at the Survey Research Center to interview former refusals have been less successful. Perhaps the smaller the refusal portion, the greater the proportion of hard-core refusals among them.

For not-at-homes the procedure has proven practical, simple, and inexpensive. There may be a higher cost per interview because these addresses are more scattered and difficult to contact, but the cost of listing new addresses is saved. How is this procedure more efficient than simply making further call-backs on the nonresponses of the survey itself? First, the limits of the survey period often put an end to further recalls. Second, after k calls the remaining nonresponses are widely scattered; but the replacement addresses are surrounded by new survey addresses in the same neighborhoods, perhaps in the same blocks.

However, it would be risky to depend on a replacement scheme to avoid recalls altogether. For example, a scheme of one call on both the original addresses and the replacement addresses would be similar only to a two-call procedure, which is seldom sufficient or economical. The replacement technique should be used in addition to a recall procedure; we may try for an optimum number of calls, combining the original and replacement portions.

13.7 QUOTA SAMPLING

Questions from students and colleagues have induced me to discuss briefly these methods which are so widely practiced and publicized. My reluctance is due to the frustrations of covering so complex and ambiguous a topic in a short section. It falls outside the domain of probability sampling, because it does not attain randomization with a mechanical selection from a frame. This and the problems of judgment sampling are discussed in Sections 1.7 and 14.3. It is hard to say just what quota sampling is, what it is supposed to do, and how well it does it.

"It is not sufficient simply to state that quota sampling was used in a survey and expect any one to have more than a very general idea of how the sample was drawn." Yet most quota samples have some basic aspects in common. Sizes of subclasses in the population are estimated, based on Census or other data. *Quotas* of desired numbers of sample cases are computed proportionally to the population subclasses. The sample quotas are divided among the interviewers, who then do their best to find persons who fit the restrictions of their quota controls. These methods are better and more fully described by Stephan and McCarthy [1958, 3.6], the source of the above quotation.

Choice of the quota controls would challenge the quota sampler's ingenuity, were the controls not severely restricted by practical requirements. First, they must be available and reasonably recent. Second, they must be applicable to the interviewers' areas. For example, an ethnic or occupation group may be highly concentrated in specific areas. Third, classification by the interviewers must be reasonably easy. Fourth, to be practicable, the controls must be reduced to a small number of cells. Fifth, to serve public relations functions, the quotas should be readily and widely understood. Sixth, the quota variables should be strongly related to the survey variables. These would thereby become substantially homogeneous within the quota cells. To the degree that homogeneity can be achieved, selection bias is eliminated. This would also drastically reduce the variance.

Whereas discussions of quota sampling emphasize the last consideration, in practice the others predominate since they are necessary. Age, sex, and geographic regions are the most commonly used controls, even when their relation to survey variables is weak. For example, age and sex groups account only for 1 or 2 percent of individual variance in voting behavior. Quotas of economic levels are often represented by either income, or education, or occupation, and by color in the United States. These may often yield higher correlations than age and sex, but they are harder to attain with ordinary field procedures. Skill and care with these can, I suspect, do much to improve quota samples.

Geographic spread is assured with quota controls for sampling units, such as counties, cities and towns, neighborhoods, and even blocks. Methods of selecting these sampling units, and degrees of control over the field work, should have appreciable effects on the quality of results. Large differences exist between the practices of different organizations, and between different products of the same organization. Some may practice judgment sampling for all units. Others may select all units down to the block level with probability sampling, but permit quota sampling within blocks. They may describe their procedures as "probability-quota sampling," or some other label including the prestige word "probability." (See Problem 13.4.)

Quota sampling is not one defined scientific method. Rather, each one seems to be an artistic production, hard to define or describe. Hence, a general critique cannot be detailed, and a specific critique of one procedure may not fit another. Yet some general observations may enlighten readers who are unacquainted with quota samples.

1. The superficial resemblance of quotas to the strata of probability sampling must not obscure two basic differences. (*a*) Errors in stratum sizes and assignments can seriously bias quota samples, but not stratified probability samples. (*b*) Selection within the quotas is not randomized selection from frames within strata, but is directed by the judgment of interviewers.
2. Sample bias can be caused by inaccuracies in the computed sizes of sampling units, as well as in quota sizes. The inaccuracies may be caused by changes, growth, and mobility of the population; or because of differences between the survey population and the Census population. Probability samples are self-adjusting; changes in sampling units get reflected in changed cluster sizes.
3. Lower element costs are often attained with large clusters at the price of increased element variance. Inasmuch as interviewers receive fixed fees for each interview, as well as great freedom in obtaining them, they tend to maximize their incomes with many interviews at one location. Interviewers experiences abound with tales, such as: "At our park, I can get over a dozen fine interviews in an afternoon, from a good sample of mothers sunning their infants." Some samples do restrict the interviewers to quotas in assigned blocks. Such restrictions probably make the samples simultaneously better and more expensive.
4. Many quota samples depend essentially on first calls, since not-at-homes and reluctants are readily replaced with substitutes.
5. Typically, no attempt is made to compute the variance properly, and pq/n (or s^2/n) is boldly assumed and presented, disregarding clustering and other complexities of the sample.
6. Simple and avoidable mistakes are often added to the less avoidable problems of obtaining information cheaply. For example, equal quotas of dwellings are commonly assigned to blocks selected with equal probability; this results in unequal probabilities (b/N_α) of selection from unequal sized blocks. Or random starts for the clusters of dwellings are not attempted, even when these would cost little more.

In spite of their many faults, quota samples probably produce some good results. The bold predictions of elections have frequently come close to the election results. They often represent considerable improvements over other forms of judgment sampling they have partially replaced—such as journalists' haphazard encounters; a "typical" town, or "bell-weather" district; a poll of volunteer correspondents, or of a magazine's subscribers. Over other forms of judgment sampling, quota samples have the advantage of greater spread. This is chiefly expressed geographically, not only in many sample counties, but also with instructions and block quotas that tend to scatter interviewers into areas they would otherwise avoid. Furthermore, the quota controls also tend to spread the sample in other dimensions. I believe also that, for example, a quota sample is more likely to represent the attitude of the nation's young people, than a probability sample of a college's students.

Quota sampling is practiced mainly because its cost per element is lower than for probability sampling. But ordinary gross comparisons of element costs confound several factors, such as shorter interviews and large clusters, with the easier availability of quota respondents. The quota interview is cheapest, I suspect, when the interviewer has wide latitude. Then his clustering effect is probably large, because he selects large groups, and particularly because his individual selection bias has wide amplitude. As block and economic quotas become stricter, his choice and effect are narrowed, and the cost goes up. When the cost of quota sampling can be more properly compared to a probability sample, the latter may be made more nearly competitive in element cost. Probability sampling may be made less expensive with larger clusters, sacrificing some spread and precision.

The paramount issue still remains that nobody knows how well quota samples really perform. After all, the poorest methods can produce good results for variables which are randomized over the population (1.7). The Literary Digest poll was successful for years. In many other cases there are no outside checks. We must brush aside the naive attempts to use the quota controls, like age, sex, and region, as proofs of sample performance. Socio-economic checks on variables that are not used for quotas often reveal great discrepancies. Furthermore, election predictions often do fail, sometimes by wide margins. Many of these are quietly forgotten. A few become noisy scandals; the attempt is then made to explain these away with *ad hoc* explanations and excuses, which are ignored until they are needed.

The interested reader is advised to consult some of the more thorough investigations of quota sampling. Stephan and McCarthy [1958] have studied at length the results and methods of quota samplers. Chapter 10 describes the vexing problems of trying to assess their true variability. Chapter 13 presents comparisons of their results. The authors' critique is kinder perhaps than the methods merit.

The polls' failure in the 1948 U.S. presidential election led to an important report by a committee for the Social Science Research Council [Mosteller *et al.*, 1949] and a fine analysis by Katz [1949]. Sampling methods of the Kinsey research on sexual behavior were criticized in a remarkable report [Cochran, Mosteller, and Tukey, 1954].

Moser and Stuart's analysis [1953] of general problems is structured around the results of a field experiment. This was organized by the

Survey Techniques Unit of London University, and it involved interviewers of several British polling firms. Among interesting results was the computation of the sampling variance, based on differences between pairs of interviewers given identical instructions. It showed large increases over simple random variance. Some variances were so large they cast serious doubt on the economy of quota samples, if these were actually measured in precision per dollar, rather than merely interviews per dollar. Although this computation omits the effects of biased selection on the total accuracy, it can be used to compute the sampling variance around the biased mean. I find it disturbing that this minimum security has not been adopted by quota samplers.

*13.8 EFFECTS OF BIAS ON PROBABILITY STATEMENTS

Assume that we aim to produce sample results that permit probability statements with fixed probabilities; statements of this kind:

$$\begin{aligned} &\bar{Y} \text{ is greater than } \bar{y} - t\sigma && \text{(lower limit)},\\ &\bar{Y} \text{ is less than } \bar{y} + t\sigma && \text{(upper limit)}, \quad (13.8.1)\\ &\bar{Y} \text{ is within the limits } \bar{y} - t\sigma \text{ and } \bar{y} + t\sigma && \text{(both limits)}. \end{aligned}$$

In this section we use σ to denote briefly the standard error of the sampling distribution and its sample estimate, for which our usual symbols have been SE $(\bar{y})$ and se $(\bar{y})$, respectively. We assume (as in 1.4) that the

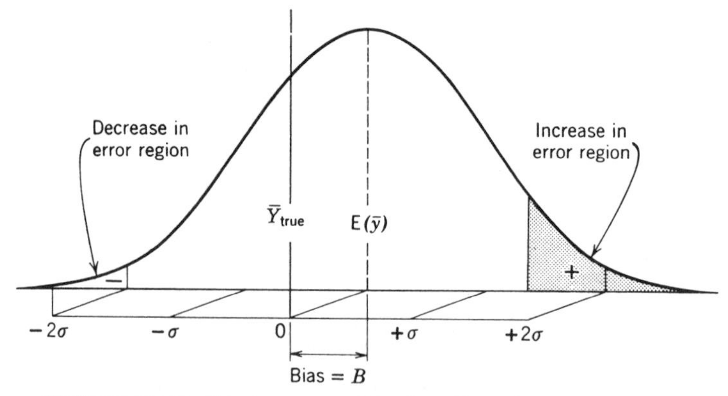

FIGURE 13.8.I Displacement of the Sampling Distribution by Bias.

We note the effects of Bias $= +B$ on the dispersion of estimates around $\bar{Y}$. The proportion of sample values lying beyond $+2\sigma$ are increased because of the positive bias, whereas those beyond -2σ are decreased. A negative bias will have the opposite effects. Although each has width B, the two areas (shown as + and −) of distortion of the error regions are not equal, and the increase side is always greater than the decrease. The amounts of these distortions vary with t and with B.

probability statements denote intervals based on the normal distributions for the sample values $\bar{y}$. What is the effect of bias in the sample design on such probability statements?

Suppose that $\bar{Y}$ is the "true value" to be estimated, and that $B = E(\bar{y}) - \bar{Y}$ is the total bias of the survey design. The entire sampling distribution is displaced from the unbiased value of $\bar{Y}$ by the amount B. If we know the value B, we can use the unbiased $(\bar{y} - B)$ to estimate $\bar{Y}$. Prior knowledge about the bias, even if subject to error, may be valuable; for example, if the bias were normally distributed with variance σ_b^2 around mean B, the combined estimate would have mean $(\bar{Y} - B)$ and variance $\sigma^2 + \sigma_b^2$. (See Example 13.1*a* and Schlaifer [1959, Chapter 31].) Suppose now that the bias is unknown and perhaps even unsuspected. Given that $\bar{y}$ is normally distributed with variance σ^2, what is the effect of the linear displacement B on the distribution's tail areas on the left of $-t\sigma$, to right of $+t\sigma$, and on the sum of both tail areas? The effects of the bias can be translated into equivalent distortions of the tail areas around unbiased estimators. Thus,

$$\begin{aligned} \bar{y} - t\sigma &= (\bar{y} - B) - \sigma(t - B/\sigma), \\ \bar{y} + t\sigma &= (\bar{y} - B) + \sigma(t + B/\sigma), \\ \bar{y} \pm t\sigma &= (\bar{y} - B) \pm \sigma(t \pm B/\sigma). \end{aligned} \tag{13.8.2}$$

A decrease of the effective t value inflates the error region, and an increase of t reduces it. A similar increase of the effective t, hence a reduction of the error region, is caused by a positive bias on the upper limit, or a negative bias on the lower limit. But a decrease in the effective t, hence an inflation of the error region, results from a negative bias on the upper limit or a positive bias on the lower limit.

The effect on two-sided limits is measured by the sum of the two error regions, outside the interval $-t\sigma(1 - B/\sigma t)$ to $+t\sigma(1 + B/\sigma t)$, instead of $-t\sigma$ to $+t\sigma$. One error region is inflated and the other reduced, but the two are not equal; the net result is an inflation of the combined error region. Roughly, the error rates double for every increase of 0.5 in the relative bias B/σ. Figure 13.8.II shows these effects in detail.

We get a different perspective from appraising the survey error not in terms of the standard error (σ) alone, but in terms of the total error $= \sqrt{\sigma^2 + B^2} = \sigma\sqrt{1 + B^2/\sigma^2}$. Assume that for a biased estimator we are able to make probability statements of this kind:

$$\begin{aligned} &\bar{Y} \text{ is greater than } \bar{y} - t\sigma\sqrt{1 + B^2/\sigma^2} && \text{(lower limit)}, \\ &\bar{Y} \text{ is less than } \bar{y} + t\sigma\sqrt{1 + B^2/\sigma^2} && \text{(upper limit)}, \\ &\bar{Y} \text{ is within the limits } \bar{y} \pm t\sigma\sqrt{1 + B^2/\sigma^2} && \text{(both limits)}. \end{aligned} \tag{13.8.3}$$

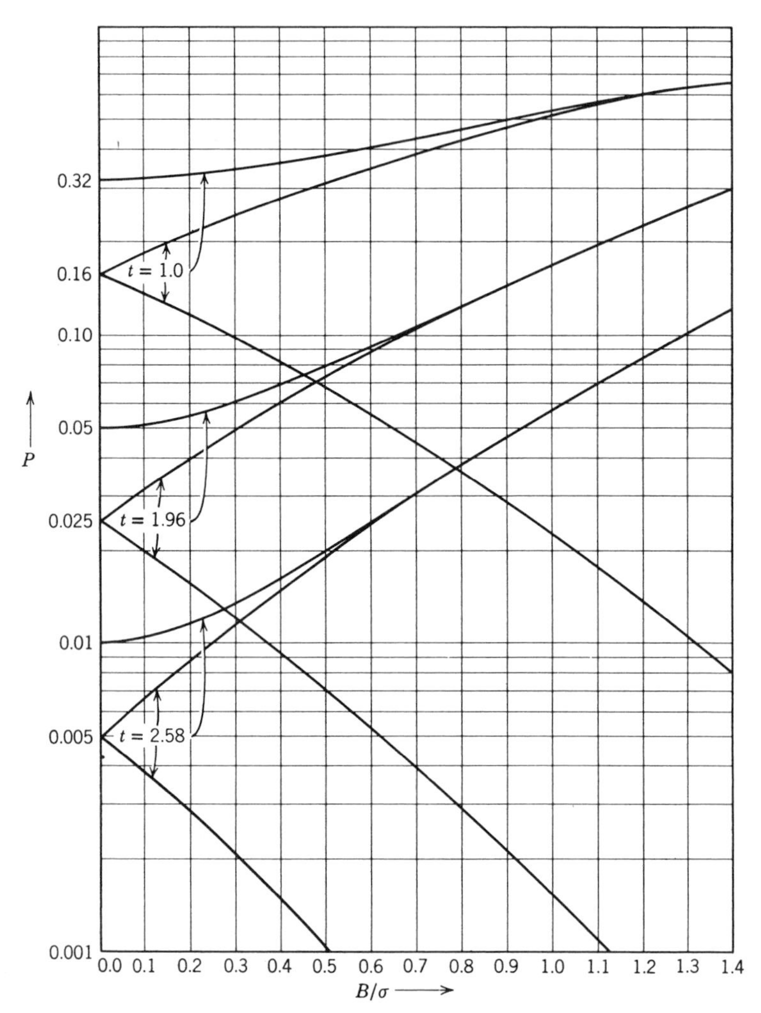

FIGURE 13.8.II Effect of Bias on Probability Statements $P\{\bar{y} \pm t\sigma\}$. Values of $P = (2\pi)^{-0.5} \int_x^\infty \exp(-y^2/2)\, dy$, for $x = t \pm B/\sigma$, shown for $t = 1.00$, 1.96, and 2.58.

The horizontal scale presents values for the ratio B/σ of the bias to the standard error. The vertical scale of P values is logarithmic, hence the slopes compare rates of relative change. Three levels of probability statements correspond to t values of 1.00, 1.96, and 2.58. At each level, three curves show the respective effects on the plus and minus error regions and on their sum.

At all three probability levels, increases in the effective t sharply reduce the error regions; an increase due to about $B/\sigma = 0.3$ will halve the error. On the other hand, decreases in the effective t sharply inflate the error rate; a decrease of about $B/\sigma = 0.3$ will double it. The two curves are almost linear on the logarithmic scale, showing about equal relative changes.

Noting that $(\bar{y} - B)$ would be unbiased, as before, the above statements can be transformed to statements about unbiased estimators by changing the nature of the limits:

$$\begin{aligned}\bar{y} - t\sigma\sqrt{1 + B^2/\sigma^2} &= (\bar{y} - B) - \sigma[t\sqrt{1 + B^2/\sigma^2} - B/\sigma],\\ \bar{y} + t\sigma\sqrt{1 + B^2/\sigma^2} &= (\bar{y} - B) + \sigma[t\sqrt{1 + B^2/\sigma^2} + B/\sigma],\\ \bar{y} \pm t\sigma\sqrt{1 + B^2/\sigma^2} &= (\bar{y} - B) \pm \sigma[t\sqrt{1 + B^2/\sigma^2} \pm B/\sigma].\end{aligned} \quad (13.8.4)$$

A decrease in the bracketed quantity, which denotes the "effective t," inflates the error rate, while an increase in the "effective t" reduces the error rate. The error rate for "both limits" is the sum of the rates for the lower and upper limits.

t	1	1.28	1.64	1.96	2.58	3.29	3.89	4.42
P_0 for $B/\sigma = 0$	0.169	0.100	0.050	0.025	5×10^{-3}	5×10^{-4}	5×10^{-5}	5×10^{-6}
P_{max}/P_0	2.96	2.11	1.92	1.84	1.76	1.72	1.70	1.69
at $B/\sigma = 1/\sqrt{t^2 - 1}$	∞	1.247	0.766	0.593	0.421	0.319	0.266	0.232

TABLE 13.8.I Maximum Increases of Error Rates Due to Intervals $(t\sqrt{1 + B^2/\sigma^2} - B/\sigma)$ for Several Fixed Values of t

The error rates P_0 correspond to fixed values of t for the normal curve $N(0, 1)$ with mean 0 and variance 1. If instead of the lower limit $-t\sigma$, the limit $x = -\sigma(t\sqrt{1 + B^2/\sigma^2} - B/\sigma)$ is used, the error rates increase with B/σ to a maximum P_{max} until x becomes minimum, and then they decrease. The minimum value is $x = \sqrt{t^2 - 1}$, and it occurs for $B/\sigma = 1/\sqrt{t^2 - 1}$. For $x > 1$, the minimum exists; for larger values of t, it occurs for $B < 0.6$, and it results in the maximum distortions P_{max}/P_0 with values about 1.7 to 2.0.

Figure 13.8.III presents the effects on error rates in detail. These curves help to answer the question: to what extent can the mean square error $\sqrt{\sigma^2 + B^2}$ represent the total survey error? For two-sided intervals the representation is good, because the curves are flat for moderate values of the relative bias B/σ. Thus we can agree with Hansen, Hurwitz, and Madow [1953, p. 57] that ". . . so long as the bias of an estimate is no greater than the standard error, and so long as one is interested only in the absolute magnitude of errors, and not in their direction, the probability that a particular estimate $\bar{x}$, will differ from the value being estimated by more than $k\sqrt{\text{MSE}(\bar{x})}$ can, in practice, be interpreted in the same way as if the estimate were unbiased and $k\sigma_{\bar{x}}$ were used to set probability limits."

But the absolute magnitudes of the inflation becomes increasingly larger than the reduction. These facts explain the curve for their sum, which denotes the size of the two-tailed error region.

Each curve for two tailed regions shows slight change to about $B/\sigma = 0.3$, hence a relative bias of this magnitude has little effect on the error rate. Doubling the error rate from $P = 0.05$ to 0.10 occurs at $B/\sigma = 0.7$, and from $P = 0.01$ to 0.02 it occurs at $B/\sigma = 0.5$. From there on, increases are logarithmically linear, doubling the error rates for increases of about 0.5 in B/σ.

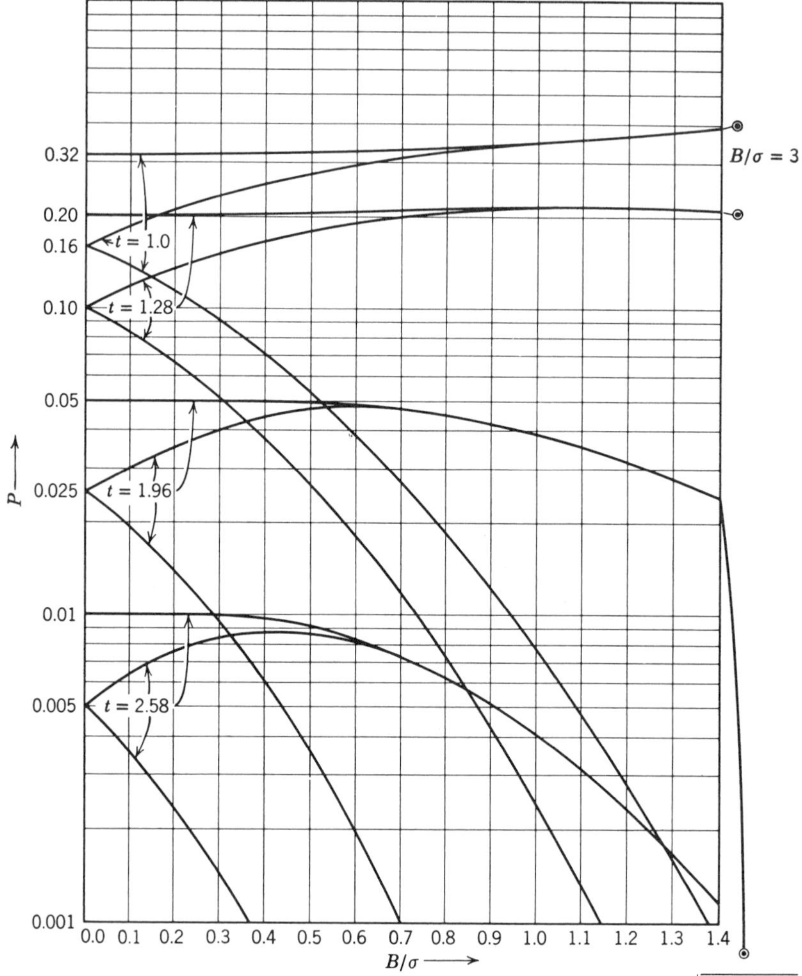

FIGURE 13.8.III Effect of Bias on Probability Statements $P\{\bar{y} \pm t\sigma\sqrt{1 + B^2/\sigma^2}\}$. Values of $P = (2\pi)^{-0.5}\int_x^\infty \exp(-y^2/2)\,dy$, for $x = [t\sqrt{1 + B^2/\sigma^2} \pm B/\sigma]$ shown for $t = 1.00$, 1.28, 1.96, and 2.58.

The horizontal scale represents the ratio B/σ of bias to standard error. The vertical scale of P values is logarithmic. At each of four levels of t values, three curves show effects on plus and minus error regions and on their sum.

At all four levels, the reduction of error rates is rapid when the second term is positive, because then B/σ increases both terms within the bracket. But when the second term is negative, the two effects of B/σ within the brackets work against each other; this results in a curvilinear relationship (for $t > 1$), with a moderate inflation of error rates followed by continuous and accelerating declines. The maximum ratios of inflation for these error rates are about 1.7 to 2.0 (See Table 13.8.I).

The sum of the two error rates yields the effect on error rates for both limits. These curves are flat for moderate values of B/σ, say, 1 for $P = 0.05$, and 0.6 for $P = 0.01$. Then they decline with accelerating speed.

Lower error rates (such as $P = 0.01$) are reduced considerably by larger values of B/σ, hence $\sqrt{B^2 + \sigma^2}$ becomes a "conservative" overestimate of the total error. For one-sided intervals, $\sqrt{\sigma^2 + B^2}$ overestimates badly on one side, and is a rough and variable estimate on the other side.

For some statistics, although the bias cannot be measured, its effect can be incorporated into a computed mean square error. According to the above results, these computations may yield either fairly good or conservative results; but seldom will they underestimate the total error.

PROBLEMS

13.1. A polling organization finds that 207 out of 576, or 36 percent, of adults interviewed in a state express preference for a certain candidate for governor in a poll conducted two months before election. A statement is made that, because $\sqrt{pq/n} = 2$ percent, the true percentage of voters favoring the candidate is between the limits of 32 percent and 40 percent. Indicate briefly several *distinct* reasons why that statement does not necessarily have a probability of 0.95 of being true. Specify what you think are the four most important reasons in practical cases of this sort.

13.2. Suppose that a sample survey needs to be repeated a year later, but with increased accuracy: the *total survey error* (the mean square error) is to be cut in half. A proposal is made to double the number of interviews (elements). Several arguments are made against this plan; decide separately for each whether the following argument is valid and true (T), or irrelevant and fallacious (F). "Doubling the number of interviews may have little effect on the total survey error, the mean square error, because . . ."

(*a*) The effect of proportionate stratification on the sample design is great.
(*b*) A return of 20,500 mailed questionnaires was obtained from a sample of 80,000 elements.
(*c*) The characteristics being measured are heterogeneous and are mostly binomial.
(*d*) The effect of clustered selection on the variance is large.
(*e*) A stratified sample was selected according to optimum allocation.
(*f*) The errors of response and nonresponse are large compared to sampling errors.

13.3. Variations between the styles of individual interviewers can result in variations in the average of responses obtained by them. This variation among their averages can be expressed usefully as the variance between interviewers; but measuring this *interviewer variance* presents special problems. Discuss briefly:

(*a*) Why we cannot usually assign the same respondents to two or more interviewers for repeated measurements; contrast this with measuring the coder variance from completed interviews in the office.

(*b*) Why simply taking the average results obtained from interviewers on an ordinary survey is not satisfactory.
(*c*) Why complete randomization of all respondents among all interviewers is often difficult.
(*d*) A design you consider feasible for measuring interviewer variance—either in general, or in a specific situation you choose and describe briefly.

13.4. For an epsem sample of dwellings in a city, 100 sample blocks are selected with equal probability in the first stage. Discuss shortcomings of each of the alternative plans proposed for subsampling the blocks:
(*a*) After a truly random start from the listed dwellings of each sample block, take the next 9 dwellings, making call-backs as needed, leading to an *average* of 8 responses from the 9 sample dwellings.
(*b*) After the random start, contact enough dwellings consecutively until 8 responses are obtained on a single visit.
(*c*) Let the interviewer pick a start and continue in strict order until 8 responses are obtained.
(*d*) Let the interviewer pick dwellings in the block until 8 responses are obtained.
(*e*) Let the interviewer find responses in a randomly chosen area of about 4 × 4 blocks.
(*f*) Let the interviewer find 8 responses in the city.

13.5. For a national interview survey you need about $n = 720$ interviews. The response rate may be as low as 80 percent, or as high as 96 percent; judging by past results the best guess seems to be about 90 percent. Several strategies (*a*) to (*g*) are suggested below for sending out an initial sample of m to obtain approximately the right final sample size n; rank these from *1* for the most desirable to *7* for the least desirable. Your choices demand subjective judgment. You may indicate ties, and write a few brief explanations. (*h*) If you prefer another strategy to any of the 7 below, describe it briefly. (*i*) Would you change the rankings if the sample were confined to a city of 7200 dwellings? (*j*) Or for a mail survey? How?
(*a*) Send out $m = 1000$ initial addresses, and accept the first 720 returned to the office.
(*b*) Send out $m = 900$, and stop when 720 are returned.
(*c*) Send out $m = 820$, and accept all you can get; then send out more only if many are needed.
(*d*) Send out 800, try hard for all you can get, and stop.
(*e*) Select a sample of 1000; from these, select and send out an initial sample of 750. Near the end send as large a sample of the 250 as are needed.
(*f*) Send out 720, and ask interviewers to substitute for nonresponses the "substitute address" sent out with each initial sample address.
(*g*) Take a pilot sample of 36 addresses, and use its response rate to decide on a strategy.

13.6. (*a*) Explain briefly why substitution of a "neighbor" element (next dwelling, or next line, or next card) does not generally correct for the bias of nonresponse.

(*b*) Under what conditions would such substitution be an effective corrective procedure for the bias of nonresponse?

(*c*) Under such conditions, what are the relative advantages and disadvantages of substitutions versus doubling the weights or machine cards of responding "neighbors"?

13.7. A survey was taken from a city's households: dwellings occupied by city residents. Compute the response rate for households from an epsem sample of 1083 sample addresses. Note that 1037 dwellings were found at 1000 dwelling addresses.

Responses	860
Responses from visitors to city (in homes, trailers, motels, hotels)	12
Nonresponses from visitors to city (in homes, trailers, motels, hotels)	4
Refusals	58
Not-at-homes	29
Other nonresponses (too ill, incapacitated)	3
Away on vacation	18
Vacant homes	53
Address not a dwelling	83

14
Some Issues of Inference from Survey Data

14.1 COMPUTATION AND PRESENTATION OF SAMPLING ERRORS

When survey statistics consist of a few means and aggregates, their standard errors can be computed and presented together with the statistics. This is also possible with standard errors of comparisons between a few pairs of statistics that measure changes and differences. However, most surveys yield hundreds or thousands of statistics. Scores of characteristics (dependent variables) can be obtained for each survey element. The characteristics are presented not only for the entire sample, but also for many subclasses, based on other variables.

Survey reports commonly contain dozens of tables, each containing perhaps $10 \times 10 = 100$ means. Moreover, in each column of such a table $9 \times 10/2 = 45$ potential comparisons lurk among pairs of means. In 10 rows and 10 columns, $20 \times 45 = 900$ such potential comparisons exist. Still another factor intrudes when analysis is three-dimensional: when two-dimensional tables are presented in sequence along a third. Computing and presenting standard errors on such a vast scale can become a formidable, even a forbidding task.

Consider first the simplest statistics: the means and aggregates, without their comparisons. We could conceivably compute standard errors for each survey mean; this is becoming more feasible with modern electronic computers. Still, standard errors are generally considerably more expensive to compute than means. Moreover, presenting all standard errors would double the number of entries, and greatly complicate the presentation of any survey report. Hence, standard errors for all survey means and aggregates are not often presented. This topic is discussed by Yates [1960, 7.26], Hansen, Hurwitz, and Madow [1953, 12.15], and Corlett [1963].

For comparisons of pairs of means the task is much more formidable. The computations are more expensive, the numbers potentially desired by readers are far greater, and problems of presentation become vastly, perhaps hopelessly, more complicated.

These difficulties force us to search for methods to reduce and pool the efforts of computing and presenting standard errors. Incidentally, such an approach may also improve the precision of variance estimates, which are subject to very large errors when they are based on few degrees of freedom. For example, a selection of 15 pairs of counties from as many strata, has only 15 degrees of freedom, or less (8.6). A sample comprised of 10 replications has only 9 degrees of freedom.

Statistical theory tells us how to compute variances for specific statistics, but little about variances for a group of statistics. The literature of "pooled variances" in the analysis of variance does not solve our problems. We cannot wait for solid theoretical foundations, because some generalized statement about standard errors in the survey report is often the only practical alternative to neglecting standard errors altogether. The following remarks, although incomplete and indefinite, may acquaint the reader with some current practices.

Solutions can be readily found if the sample results closely approximate those of simple random sampling; that is, if the *design effect* is sufficiently close to 1. Furthermore, if the survey results deal with proportions p (or aggregates Np), the reader can be encouraged to compute his own approximations $\sqrt{pq}/\sqrt{n}$. Providing a table or a graph of curves can facilitate the reader's efforts.

If the survey results are not proportions, the problem becomes less tractable, and we should look for regularities to construct some model for the variances. First, we can decide whether the standard deviations S_y or the coefficients of variation $S_y/\bar{Y}$ behave more uniformly. Second, we may find a group of variables with similar errors. Third, we may describe two or more groups of subclasses which have similar errors.

Hence furnishing relevant values of n should help the reader assess the precision of samples that can be considered approximately simple random. These may include most cases of *epsem element sampling*—for example, most of the proportionate stratified samples of elements and most of the systematic samples of elements. Stratification tends to reduce the variance, and the design effect may be less than 1; hence, the standard error will be overestimated by the simple random assumption. However, this effect will often be slight, especially for subclasses (4.5). Although clustering tends to increase the variance, having a design effect greater than 1, its effect may be slight if the clusters are small, particularly for subclasses; or if it affects only a small portion (or supplement) rather than the entire sample. It may be worth some sacrifice of efficiency to achieve a design that permits the assumption of simple random sampling.

In complex survey samples the design effect may frequently differ considerably from unity, and its effect on standard errors must be considered separately for diverse statistics. Publishing the values of n (and S or P) does not provide sufficient information for estimating the standard error, when $S/\sqrt{n}$ is not a good approximation to it. Presenting generalized sampling errors under these circumstances raises several critical questions, to which our answers currently are vague, tentative, and controversial.

What Is the Nature of Variation in Design Effect? We get the simplest results when we can assume that the design effect is a constant Deff for all survey statistics, or at least for all important survey means. Then

Reported Percentage	Sampling Error in Percent, by Size of Sample or Group					
	2000[a]	1000	700	500	300	100
50	3	4	5	6	8	14
30 or 70	3	4	5	6	7	13
20 or 80	2	4	4	5	6	11
10 or 90	2	3	3	4	5	8
5 or 95	1	2	2	3	4	

TABLE 14.1.I Example of Sampling Errors for Percentages

[a] Approximate size of entire sample in 1963 *Survey of Consumer Finances*.
Note: The chances are 95 in 100 that the value being estimated lies within a range equal to the reported percentage plus or minus the number of percentage points shown above.

The sampling errors express generalized and approximate values of 2se(p). They are based on computations of the design effect $\sqrt{\text{deff}} = \text{se}(p)/\sqrt{pq/n}$. These values were computed for many subclasses, then plotted against n and averaged. Actually the values above represent not a true average, but a "safe" or "conservative" value which exceeds most of those actually computed (see Table 14.1.IV.). (*Source:* Katona, Lininger, and Mueller [1964].

procedures suitable for simple random approximations may be extended by furnishing the reader with an estimate of $\sqrt{\text{Deff}}$ with which to multiply all estimates of $S/\sqrt{n}$. Perhaps the reader's task is made easiest if the factor $\sqrt{\text{Deff}}$ is incorporated into an "effective t_p": $t_{\text{deff}} = t_p\sqrt{\text{Deff}}$, a changed normal deviate for constructing probability intervals. Alternatively, it can be included in values of "effective n": $n_{\text{deff}} = n/\text{Deff}$ (8.2).

The last procedure can be adapted to a more complicated model, where we assume that the design effect is a linear function of n: $\text{Deff} = An + B$.

Size of Estimate	Both Sexes		Male		Female	
	Total or White	Nonwhite	Total or White	Nonwhite	Total or White	Nonwhite
10,000	6,000	6,000	8,000	6,000	6,000	6,000
50,000	14,000	13,000	18,000	13,000	12,000	13,000
100,000	19,000	18,000	25,000	18,000	18,000	18,000
250,000	30,000	28,000	39,000	28,000	28,000	28,000
500,000	42,000	38,000	55,000	38,000	39,000	38,000
1,000,000	60,000	50,000	75,000	50,000	55,000	50,000
2,500,000	90,000	65,000	110,000	65,000	90,000	65,000
5,000,000	130,000	65,000	140,000	...	120,000	...
10,000,000	170,000	...	170,000	...	170,000	...
20,000,000	230,000	...	190,000	...	220,000	...
30,000,000	260,000	...	...	...	...	...
40,000,000	280,000	...	...	...	...	...

TABLE 14.1.II Example of Generalized Sampling Errors

These values represent generalized and approximate values of se(Np). They are based on averages from many sets of computations. They serve the monthly reports of the Current Population Surveys described in Section 10.4. (*Source:* U.S. Census Bureau, Technical Report No. 7 [1963].)

This is not unreasonable since the design effect may often be approximated with $1 + \text{roh}\,(n/a - 1) = (1 - \text{roh}) + n\text{roh}/a$. This may hold well for a single characteristic, to the extent that roh remains constant for diverse subclasses. We may speculate on the extension of this to a multivariate model: $\text{Deff} = An + B_1 + B_2 + B_3 + \ldots$, where the B_j represent various factors mentioned below.

If we assume that the design effect depends only on n, we can present simple tables of approximate standard errors for proportions p, such as Table 14.1.I; or for simple aggregate counts Np, such as Table 14.1.II.

Similarly, a two-way table of standard errors for various S and n values can be constructed. This could also be used together with a set of S values for different characteristics.

What Statistics Should Be Grouped for a Single Design Effect? Even if we consider only sample means based on samples of similar sizes, this question has two dominant dimensions. First, for survey characteristics (dependent variables) that are very dissimilar, the design effects may differ widely. For example, a contagious disease, or voting behavior, or home ownership may be more homogeneous in clusters than some widespread attitudes or buying behaviors. Second, for the same characteristic, the effect may differ for diverse types of subclasses. A regional mean may involve a tenth of all primary selections ($a/10$), whereas an age-sex subclass will be fairly evenly spread over all a primaries. Some occupations (farmers, miners) and some ethnic groups (nonwhites, Mormons, Jews) may also be unevenly distributed.

How Many Separate Groups Should Be Recognized? Although caution presses us to distinguish many groups, reports are actually confined to presenting only one or two sets of standard errors. First, most readers tend to become confused by a multitude of tables. Second, often the variation of the computed variances is too great to permit good empirical separation of groups of statistics (8.6). Presenting several groups is more feasible for larger samples with many primary selections and for periodic surveys of similar data; both interest in and possibilities for precision are likely to be greater than for a small one-time survey [U.S. Census Bureau, 1963; Kish, 1963].

For Which Statistics Should Variances Be Computed? The number of statistics of interest may far exceed the number for which variance computations are feasible. Choices must be made frequently among the survey characteristics, and even more among the subclasses. We can offer little beyond some common-sense advice. Choose those categories of statistics that have most relevance in the analysis. Try to cover a good range of the likely degrees of variation. This involves anticipating a range of magnitudes of design effects, for example, if these are the functions being averaged. We may be particularly careful to include the high end of this variation. The low end may be easier to guess; a simple random model may serve this purpose.

What Function of the Variances Should Be Averaged? Instead of averaging variances or standard errors directly, we should attempt to overcome some extraneous sources of differences. Coefficients of variation may serve this purpose. I prefer averaging Deff; or better still, $\sqrt{\text{Deff}}$, which is less affected by extreme values. Unfortunately, a thorough discussion of this topic would be too obscure, and it should await a stronger theoretical foundation than is now available.

The measurement and utilization of an average standard error for a group of variables needs theoretical foundations. In addition, there are also issues of strategic presentation. I think that an average of a group of sampling errors is probably best, but other functions of the distribution are possible candidates. In past years we had published a set of "conservative" values, near the maximum of the plotted values of $\sqrt{\text{Deff}}$. This way the reader of survey reports receives "safe" values such that most standard errors would actually be lower. For any single statistic the danger of underestimating its standard error was almost eliminated. But this safety was bought at the cost of overestimating the standard error for most statistics.

Another possible strategy is to publish both "lower" and "upper" values such that most values fall between these limits. This strategy gives more information to readers than the upper values alone, but is also less easy to comprehend. The strategy received favorable comments, particularly for tables of standard errors for comparisons. We have used it at the SRC for years, but not regularly; other researchers have also borrowed this technique.

A related problem, less statistical than psychological, can be stated simply: Should we publish sampling errors of two standard errors or simply standard errors? Both procedures have been followed by some survey organizations. The standard error is the basic unit; its use should help the reader select the probability level he prefers for the specific situation. This may help to banish the practice of always using the $P = 0.95$ level (or any other rigidly fixed level), which may be either too high or too low for any actual situation.

Nevertheless some arguments favor publishing tables of sampling errors equal to two standard errors. First, since the 95 percent level is widely understood and expected, departure from it may lead to confusion. Second, the reader should be mildly skeptical about the multitude of statistics in most survey reports; the use of two standard errors covers a wider range of this "mildly but not too strongly skeptical" range of acceptance.

Probability level $1-P$	1/2	1/3	1/5	1/10	1/20	1/50	1/100	1/250	1/1000
Multiples of standard errors	2/3	1	1 1/3	1 2/3	2	2 1/3	2 2/3	3	3 1/3
Normal deviate t_p	0.68	0.97	1.28	1.64	1.96	2.33	2.58	2.88	3.29
Student t_p (20 deg. of fr.)	0.69	0.99	1.37	1.72	2.09	2.53	2.85	3.25	3.85

No. of Inter-views	No. of Interviews							
	2000	1000	700	500	400	300	200	100
				For Percentages from 35 to 65				
2000	3.2–4.0	3.9–4.9	4.4–5.5	5.0–6.2	5.5–6.9	6.2–7.8	7.4–9.2	10–12
1000		4.5–5.6	4.9–6.1	5.5–6.9	5.9–7.4	6.6–8.3	7.7–9.6	10–13
700			5.3–6.6	5.9–7.4	6.3–7.9	6.9–8.6	8.0–10	11–13
500				6.3–7.9	6.7–8.4	7.3–9.1	8.4–10	11–13
400					7.1–8.9	7.6–9.5	8.7–11	11–14
300						8.2–10	9.1–11	12–14
200							10–12	12–15
100								14–17
				For Percentages around 20 and 80				
2000	2.5–3.1	3.1–3.9	3.5–4.4	4.0–5.0	4.4–5.5	5.0–6.2	5.9–7.4	8.2–9.8
1000		3.6–4.5	3.9–4.9	4.4–5.5	4.7–5.9	5.3–6.6	6.2–7.8	8.4–10
700			4.3–5.4	4.7–5.9	5.0–6.2	5.5–6.9	6.4–8.0	8.6–10
500				5.1–6.4	5.4–6.8	5.8–7.2	6.7–8.4	8.8–11
400					5.7–7.1	6.1–7.6	6.9–8.6	9.0–11
300						6.5–8.1	7.3–9.1	9.2–11
200							8.0–10	9.8–12
100								11–14
				For Percentages around 10 and 90				
2000	1.9–2.4	2.3–2.9	2.6–3.2	3.0–3.8	3.3–4.1	3.7–4.6	4.4–5.5	
1000		2.7–3.4	3.0–3.8	3.3–4.1	3.6–4.5	4.0–5.0	4.6–5.8	
700			3.2–4.0	3.5–4.4	3.8–4.8	4.1–5.1	4.8–6.0	
500				3.8–4.8	4.0–5.0	4.4–5.5	5.0–6.2	
400					4.2–5.2	4.6–5.8	5.2–6.9	
300						4.9–6.1	5.5–6.9	
200							6.0–7.5	
				For Percentages around 5 and 95				
2000	1.4–1.8	1.7–2.1	1.9–2.4	2.2–2.8	2.4–3.0	2.7–3.4		
1000		1.9–2.4	2.1–2.6	2.4–3.0	2.6–3.2	2.9–3.6		
700			2.3–2.9	2.6–3.2	2.7–3.4	3.0–3.8		
500				2.8–3.5	2.9–3.6	3.2–4.0		
400					3.1–3.9	3.3–4.1		
300						3.6–4.5		

TABLE 14.1.III Example of Sampling Errors of Differences between Percentages

The values shown are the differences required for significance (two standard errors) in comparisons of percentages derived from *two different subgroups* of the survey. Two values—low and high—are given for each cell.

These generalized and approximate values of 2 se $(p - p')$ represent the results of many computations. The low values are merely $2[PQ(1/n + 1/n')]^{1/2}$, corresponding to two simple random samples.' The high values are about 1.25 greater. Most of the actually computed values of the standard error fell between these two boundaries. (*Source:* Freedman, Whelpton, and Campbell [1959].)

	Percent	Standard Error		se $(r-r')$	$\sqrt{\text{Deff}}$ for	
	(r)	se (r)	se $(r-r')$	se (r)	(r)	$(r-r')$
Consumer-Economic Attitudes						
1. Getting along better	34	1.52	1.84	1.21	1.20	1.02
2. Will be better off next year	34	1.54	1.83	1.19	1.20	1.01
3. Good conditions expected next year	69	1.65	2.14	1.30	1.32	1.21
4. Good conditions expected for five years	40	1.80	1.87	1.04	1.37	1.00
5. Good time now to buy durable goods	49	1.80	2.11	1.17	1.35	1.10
6. Favorable prices expected next year	30	1.28	1.78	1.39	1.03	1.01
7. Business is better now	34	1.83	2.32	1.27	1.44	1.29
8. Business will be better next year	23	1.41	1.73	1.23	1.24	1.11
9. Good time to buy car	38	1.75	2.23	1.27	1.32	1.21
Buying Intentions for Next Year						
10. Automobile	14	0.94	1.22	1.30	1.00	0.92
11. Washing machines	7	0.68	0.92	1.35	1.00	0.94
12. Refrigerators	6	0.74	0.99	1.34	1.16	1.10
13. Television set	6	0.63	0.77	1.22	0.99	0.86
14. Cooking range	5	0.58	0.75	1.29	0.99	0.90
15. Build own home	6	0.69	0.94	1.36	1.07	1.02
16. Repair own home	18	1.35	1.69	1.25	1.29	1.14
Socio-Economic Characteristics						
17. Education: grade school	36	1.73	2.19	1.27	1.33	1.19
18. Education: college degree	9	1.02	1.09	1.07	1.28	0.97
19. Occupation: professional, etc.	23	1.46	1.80	1.23	1.28	1.11
20. Occupation: laborer, craftsman	40	1.56	1.89	1.21	1.18	1.01
21. Income below $3000	29	1.80	1.82	1.01	1.47	1.05
22. Income above $10,000	8	0.89	1.13	1.27	1.24	1.11
23. Own telephone	74	1.69	1.93	1.14	1.42	1.14
24. Own home	62	1.84	2.05	1.11	1.39	1.10
25. Own automobile	76	1.46	1.74	1.19	1.26	1.05
26. Color: nonwhite	11	1.59	1.63	1.03	1.89	1.37

TABLE 14.1.IV Standard Errors from Consumer Surveys of the Survey Research Center (University of Michigan)

These computations are averages of computations on 6 similar surveys from a yearly series on Consumer Expectations and Finances [Katona and Mueller, 1956 and 1964; Kish 1962]. Each sample consists of interviews in about 1350 households from about 500 segments in 66 primary sampling areas in the U.S. Standard errors of the percentages (r) are shown, followed by standard errors of the difference $(r-r')$ between two successive surveys taken about 4 months apart. The ratios se $(r-r')$/se (r) are consistently below $\sqrt{2} = 1.41$, due to positive correlations between surveys based on the same primary areas (counties), though seldom the same segments and never the same dwellings. These ratios average 1.23, and their square 1.50 is an estimate of $(2\sigma^2 - 2R\sigma^2)/\sigma^2 = 2 - 2R$; hence, we estimate $R = 0.25$ as the average correlation due to the partial overlaps.

The correlations reduce the design effect on the variance of differences. The values of $\sqrt{\text{Deff}}$ for se $(r-r')$ are shown in the last column to be slightly greater than 1, but less than the $\sqrt{\text{Deff}}$ for se (r). This, for the first group of attitudes (1–9), averages 1.28. The buying intentions (10–16) have low percentages, and they show low values of $\sqrt{\text{Deff}}$. This occurs often with rare characteristics. Note, however, that repairs of own home (16) has higher $\sqrt{\text{Deff}}$, due probably to clustering in blocks of owned homes. Furthermore, the percentage of nonwhites (26) has much the greatest clustering effect, 1.89. The other socio-economic characteristics (17–25) average 1.32.

In samples of inventories, in business, legal or industrial uses, three standard errors are commonly employed.

"The maximum variation between the results of repeated samples all drawn from the same complete coverage, and following a prescribed sampling procedure, is usefully placed at 3 standard deviations ($3\sigma_x$) in either direction from EX. This rule is a statistical standard long used in industry.

"I believe that what the user in commerce and in law needs is limits that represent practical certainty: that is, the 3-sigma limits." [Deming, 1960, p. 55; also 1954.]

How to Present Standard Errors of Comparisons? Differences between means of similar characteristics play crucial roles in survey analysis; sometimes for comparing two separate samples, but most frequently for comparing two subclasses of the same sample. All problems of individual means are present, plus some new complications. (Variance estimates are subject to larger errors. Two subclasses may have different design effects.) If the two groups contain overlapping elements, a distinct analysis is required (12.4). If the two samples are based on entirely different sampling units, the standard error is simply the root of the sum of squares of the two: $\sigma_{12} = \sqrt{\sigma_1^2 + \sigma_2^2}$. For example, this holds for comparisons of two regions. However, two subclass means from clustered samples, although based on distinct sets of elements, tend to come from the same set of clusters. The positive correlation between cluster influences on the two means tends to reduce the variance of the difference. It seems that for most comparisons the variance is reduced below the sum of the individual variances, but probably not below the variance of two simple random samples:

$$\sigma_1^2 + \sigma_2^2 \geq \sigma_{12}^2 \geq S_1^2/n_1 + S_2^2/n_2. \tag{14.1.1}$$

Not only does this conform to our intuitive expectations, but it also tends to be verified in a large body of computations from many surveys (Table 14.1.IV) [Kish, 1962].

Variances for other analytical statistics cause severe difficulties, as discussed in Section 14.2.

14.2 ANALYTICAL STATISTICS FOR COMPLEX SAMPLES

Consistent with sampling literature, this book is organized around issues provoked by complex selection designs—in contrast with simple random sampling (srs) assumptions, common to the rest of statistics. It deals almost entirely with *descriptive statistics*, particularly estimates of aggregates (Y) and means ($\bar{Y} = Y/N$), also their ratios (X/Y) and products (XY). I have detailed special methods for subclasses and taken care to develop methods for measuring differences between pairs of means.

The difference of two means may be regarded as the simplest example of *analytical statistics.* I do not aim at a formal definition, but merely at a rough distinction from descriptive statistics. Analytical statistics measure relationships; they correspond to explanatory statistics in discussions by Wold [1956] and Dalenius [1957, Ch. 1]. Among the most valuable analytical statistics are regression and correlation, discriminant analysis, the analysis of variance and related methods of experimental design, and so on through the many topics one finds in statistics.

Standard statistical literature has been developed almost entirely on the assumption of simple random sampling of elements (mostly with replacement, though this is not germane to my discussion). The assumption is sometimes explicit, more often not. Frequently the single word "sampling" automatically implies srs, granting license for the author's resort to srs formulas, such as $\sigma/\sqrt{n}$. The ubiquity of this assumption in statistical theory can be explained by the basic nature of srs, and by its facilitation of immediate theoretical results. In applying these, the reasonable nature of the assumption can often be plausibly argued (1.7), and may be supported by "tests of randomness." Most survey sampling, especially in social research, depends on and is in fact effected by complex sample designs—almost inevitably. As sampling theory develops, probability sampling is capturing the field of respectable sampling practice with designs that are simultaneously economical and complex. Clustered and stratified samples are characteristic of large surveys, and often depart drastically from a simple random model. One result of this development is the rapidly increasing production of good sample estimates of means and totals. Another result is an increasing volume of high quality data, which researchers can subject to more complex analysis.

Theory is lacking for analytical statistics for complex samples; researchers conduct analysis without it, but with the srs formulas they find in statistical books. The situation is depicted thus by Thionet [1954] for significance tests:

> "The required sampling assumptions are, in general, extremely restrictive; the majority of tests of significance still relate only to extremely simple Bernoullian samples and the sampling procedures now in general use ('cluster' sampling, two-stage sampling, etc.) remain, in the present state of theory, outside the scope of these tests."

That this vital area of statistical theory has been largely neglected is not due entirely to the indifference of statisticians. Severe problems of distribution theory will need to be solved; this may have to be achieved separately for major sampling methods. They are bound to be considerably more formidable than the mean and its standard error. Nevertheless,

"... Statisticians have an obligation to clarify the foundation of their techniques for their clients.... Statistical methods should be tailored to the real needs of the user.... 'What should be done' is almost always more important than '*what can be done exactly*.' Hence new developments in experimental statistics are more likely to come in the form of approximate methods than in the form of exact ones." [Tukey, 1954.]

It is in this spirit that the following remarks attempt to help the reader recognize obstacles, and perhaps find a strategy for circumventing them.

Let us state two fundamental conjectures about statistics computed from large probability samples. It was shown in Section 2.8*C* that sums of functions of sample observations are unbiased estimates of the corresponding population values. For example,

$$E(\sum y_j) = f \sum Y_i, \quad E(\sum y_j^k) = f \sum Y_i^k,$$

and

$$E[\sum g(y_j)] = f \sum g(Y_i).$$

Similar average values are also unbiased if the sample size is fixed, or consistent if it is not fixed. From this it follows (although formal proof is lacking) that the distribution of elements in the sample approaches the distribution of elements in the population. This and similar consequences are derived in statistics books as asymptotic properties of simple random samples [Cramér, 1946, Ch. 28]. However, the simple random assumption is often gratuitous and unnecessary, and I believe that similar crucial properties will also be proven generally for probability samples. Hence, "primary" statistics, which measure the magnitude of relations among elements from probability samples (without having to assume independent selections) will also possess consistency and other desirable properties (asymptotically maximum likelihood also). Hence, not only means, but also medians, regression, coefficients, etc., computed from large probability samples will be proven to be good estimates. If the sample is not an epsem element sample, a large number of elements does not by itself guarantee a large probability sample. To define what is a large sample would be a complex task, but clearly the number of primary selections (a) is vital. Perhaps the best single criterion is that no single primary selection receive more than γ of the entire sample weight; γ may be 1/20 or 1/100 or some other small fraction in accord with requirements (7.1).

Variances, standard errors, and tests of significance based on the assumption of independent selections are not valid for complex samples and result in mistakes. We saw that the mean $\bar{y}$ of a probability sample is a good estimate of $\bar{Y}$, but that $s/\sqrt{n}$ can be a severe underestimate of the standard error (5.4). Similarly, a regression coefficient will be a good estimate of the corresponding population value, but the srs formula of the standard error $(1/\sqrt{n - k - 1})$ can be a poor estimate. Measures of variability of sample results depend on the sample design and are subject to "design effects." Most of our evidence on analytical statistics comes from analyses of differences between pairs of means (14.1); characteristically the design effects tend to average less than for means, but more than for simple random sampling.

One of the following methods for computing or approximating standard errors can sometimes overcome the lack of general and formal methods.

1. *Propagation of variances* refers to an asymptotic method of approximate variances for functions of random variables in large samples. Let the joint distribution of the statistics $\bar{y}_1, \bar{y}_2, \ldots, \bar{y}_k$ tend to k-variate normal form with mean values $\bar{Y}_1, \bar{Y}_2, \ldots, \bar{Y}_k$ and dispersion matrix V_{ij} (all finite). This means that the variables $(\bar{y}_1 - \bar{Y}_1), (\bar{y}_2 - \bar{Y}_2), \ldots, (\bar{y}_k - \bar{Y}_k)$ are in the limit distributed as a k-variate normal distribution with zero mean values and dispersion matrix V_{ij}. If $g(\bar{y}_1, \bar{y}_2, \ldots, \bar{y}_k)$ is a continuous function with continuous first partial derivatives (not all simultaneously zero), then the variable

$$\mu = g(\bar{y}_1, \bar{y}_2, \ldots, \bar{y}_k) - g(\bar{Y}_1, \bar{Y}_2, \ldots, \bar{Y}_k)$$

is distributed normally in the limit with zero mean and variance

$$\sum_i \sum_j V_{ij} \frac{\partial g}{\partial \bar{Y}_i} \frac{\partial g}{\partial \bar{Y}_j} . \tag{14.2.1}$$

Here $V_{ij} = \text{Cov}(\bar{y}_i, \bar{y}_j)$ and $V_{ii} = \text{Var}(\bar{y}_i)$, terms in the $k \times k$ covariance matrix of the k variates. $\partial g/\partial \bar{Y}_i$ represents a partial derivative of the function. Thus the variance of functions of variates can be expressed approximately and asymptotically in terms of the variances and covariances of the variates. Illustrations are given in Problem 14.6. Applications to complex functions that contain many covariance terms should be accompanied with investigation about the quality of the approximation. The statement above was paraphrased from a derivation by Rao [1952, 5e.1]; see also Cramér [1946, 28.4] and Deming [1960, 390–396]. They and others immediately assume simple random sampling so that the covariance matrix terms are $\sigma_{ij}/\sqrt{n}$; but the treatment of Kendall and Stuart [1958, Vol. I, 10.6] is more general. I believe it holds generally and approximately for all large probability samples.

2. *Replicated sampling* permits simple computation of the variances for practically any statistic (4.4). Thus replicated or interpenetrating samples can simply circumvent the severe theoretical obstacles which

otherwise prevent computing of sampling variability for analytical statistics. The method has practical drawbacks: few replications provide poor estimates, whereas many replications interfere with efficient design and increase the cost of variance computations.

The "*jackknife technique*" (so-called by John Tukey) has been recently advanced by several authors [Quenouille, 1956; Durbin, 1959; Mickey, 1959; Jones, 1965]. Instead of the simple replicated statistic $\bar{y}_\gamma$, the new statistic $\bar{y}_\gamma{}^*$ is formed, where $\bar{y}_\gamma{}^* = c\bar{y}_c{}' - (c - 1)\bar{y}'_{c-\gamma}$; $\bar{y}_c{}'$ is the statistic based on all c replications, and $\bar{y}'_{c-\gamma}$ omits only the γth replication. The mean of the $\bar{y}_\gamma{}^*$ statistics is $\bar{y}_c{}^* = \Sigma\, \bar{y}_\gamma{}^*/c$, and its variance is var $(\bar{y}_c{}^*) = \Sigma\,(\bar{y}_\gamma{}^* - \bar{y}_c{}^*)^2/c(c - 1)$; this is also an excellent approximation for the variance of the mean of the replications $\bar{y}_c = \Sigma\, \bar{y}_\gamma/c$. This method has two main motivations. First, when $\bar{y}_c$ may be subject to bias, $\bar{y}_c{}^*$ is much less so. Second, the statistics $\bar{y}_\gamma{}^*$ are more nearly normally distributed than the $\bar{y}_\gamma$.

3. *Random splitting* of samples can provide variance computations with increasing ease due to the development of high speed computers. Random subsamples, so-called Monte Carlo or simulation techniques, are widely employed. For complex samples the random splits must follow the selection methods. For example, if a primary clusters were selected with random choice, a random subsample of $a/2$ would provide a proper split into two halves. Stratification introduces further intricacies which must be followed in the splitting. For paired selections from $a/2$ strata, select at random one member of each pair into a random half-sample. From the two random half-samples, compute $(\bar{y}_1 - \bar{y}_2)^2$ as the variance of $\bar{y}_* = (\bar{y}_1 + \bar{y}_2)$, as shown in Section 4.3. Since

$$2\bar{y}_1 - (\bar{y}_1 + \bar{y}_2) = (\bar{y}_1 - \bar{y}_2),$$

we may compute $(2\bar{y}_1 - \bar{y}_*)^2$ for the variance; $\bar{y}_2$ need not be computed separately, because $\bar{y}_*$ is known for the entire sample. Since a single variance estimate $(\bar{y}_1 - \bar{y}_2)^2$ is subject to extremely large sampling fluctuations, we must make the split repeatedly to secure several estimates; their average can then be employed to estimate the variance of $\bar{y}_*$. The repetition can be made within the same sample, or from periodic surveys of similar data. A strategy and a large scale example are described by the Census Bureau [1963, VIII]. The method could also be combined with the "jackknife" technique (as John Tukey pointed out).

Early in 1965 we completed the first group of a series of computations: for 28 multivariate regression coefficients standard errors were computed with random halves. A balanced design of 48 replications extracted full precision from 48 paired selections. (The design follows a development by Philip J. McCarthy.) The design effect of the standard errors shows an average of $\sqrt{\text{deff}} = 1.06$. This appears consistent with an average $\sqrt{\text{deff}} = 1.07$ found for comparisons of means from similar consumer behavior data [Kish, 1962b].

4. *Design and take a simple random sample, or a plausible approximation, preferably an epsem element sample.* When listing and travel costs are not prohibitive, moderate losses of efficiency may be tolerated to permit valid error computations for analytical statistics. For example, for a mailed questionnaire, select individuals from a mailing list. A school's students can be selected individually rather than in classes. A city's dwellings may be selected from a directory rather than as a block sample (9.8).

5. *When an exact formula is not available, search for a reasonable approximation.* Choice of a good model involves statistical theory, intimacy with the actual sample design, and substantive knowledge of the likely sources of variation. From the available statistics perhaps a model can be found that fits the situation fairly well, perhaps better than an srs model. Perhaps treating the means of primary clusters as the units of statistical analysis may facilitate a solution; however, this may involve questions about weights (11.6); with careless application, severe mistakes of interpretation can easily arise [Robinson, 1950; Goodman, 1959].

6. *Translate the problem into another for which a solution is available.* Formulas for the difference of two means can help to solve several problems. For example, the difference of two separate proportions can substitute for the ordinary 2×2 chi-square test; the difference of two matched binomials can substitute for the case of a correlated 2×2 chi-square test (6.5*A*, 12.10). Standard errors for medians and other quantiles can be transformed into those of proportions (12.9).

7. *From computations on available statistics, inferences may be made by analogy to other statistics for which variances are not available.* The difference of two means may serve here also. Suppose, for example, that we want to test the null hypothesis of zero difference between $k > 2$ means. Compute variances for differences between pairs of means, and obtain the ratio (Deff) to srs variances. From these ratios we can infer an adjustment for the F ratio among the k means. I guess that similar inferences would be more valuable for multiple comparison tests [Cochran and Cox, 1957, 3.5.4a]. The technique and justification for the analogy should be carefully chosen; the discussion of 14.1 may help.

14.3 SOME REMARKS ABOUT STATISTICAL INFERENCE

It would be futile and inappropriate to aim at a comprehensive discussion here. Nevertheless, I feel obliged to add a few remarks to those in Sections 1.4 and 14.2, within the same restricted framework. I shall consider only problems of interval estimation in large samples, which are mostly based on complex designs. Small and simple samples receive the bulk of attention in statistical literature and can be safely omitted here.

Statistical intervals based on standard errors can also be translated to construct common tests of significance of null hypotheses [D. R. Cox, 1958]. However, I think that those tests are especially inappropriate when applied to the results of large sample surveys. Yates [1951], after praising Fisher's classic *Statistical Methods*, makes the following observations on the use of tests of significance:

"Second, and more important, it has caused scientific research workers to pay undue attention to the results of the tests of significance they perform on their data, particularly data derived from experiments, and too little to the estimates of the magnitude of the effects they are investigating.

"Nevertheless the occasions, even in research work, in which quantitative data are collected solely with the object of proving or disproving a given hypothesis are relatively rare. Usually quantitative estimates and fiducial limits are required. Tests of significance are preliminary or ancillary.

"The emphasis on tests of significance, and the consideration of the results of each experiment in isolation, have had the unfortunate consequence that scientific workers have often regarded the execution of a test of significance on an experiment as the ultimate objective. Results are significant or not significant and this is the end of it."

Yates' remarks about research in general apply even more forcefully to the results of large sample surveys. If the sample is large enough, the weakest and least significant relationship will appear 'statistically significant,' if tested. The results of statistical 'tests of significance' are functions of the numbers of sampling units, as well as of the magnitude of the relationships being tested. The word significance in 'tests of significance' has contributed, I think, to their common misuse for measuring meaningful relations. [Kish, 1959.]

Tests of null hypotheses of *zero* differences, of no relationships, are frequently weak, perhaps trivial, statements of the researcher's aims. In place of the test of zero difference, the nullest of null hypotheses, it is more meaningful to measure the magnitudes of the relationships, attaching proper statements of their sampling variation. The magnitudes of relationships cannot be measured by levels of significance; they can be measured by the difference of two means, or by the proportion of the total variance "explained," by coefficients of correlation and regression, by measures of association, and so on.

"Null hypotheses of no difference are usually known to be false before the data are collected; when they are, their rejection or acceptance simply reflects the size of the sample and the power of the test, and is not a contribution to science." [I. R. Savage, 1957.]

Too much of social research is planned and presented in terms of the mere existence of some relationship, such as: individuals high on variate

x are also high on variate y. Although exploratory stages of research may be well served by statements of this order, they are inconclusive, and can serve only in the primitive stages of research. Contrary to a common misconception, the more advanced stages of research should be phrased in the quantitative aspects of the relationships.

"There are normal sequences of growth in immediate ends. One natural sequence of immediate ends follows the sequence: (1) Description, (2) Significance statements, (3) Estimation, (4) Confidence statement, (5) Evaluation. . . . There are, of course, other normal sequences of immediate ends, leading mainly through various decision procedures, which are appropriate to development research and to operations research, just as the sequence we have just discussed is appropriate to basic research.

"Statistical methods should be tailored to the real needs of the user. In a number of cases, statisticians have led themselves astray by choosing a problem which they could solve exactly, but which was far from the needs of their clients. . . . The broadest class of such cases comes from the choice of significance procedures rather than confidence procedures. It is often much easier to be 'exact' about significance procedures than about confidence procedures. By considering only the most null 'null hypothesis' many inconvenient possibilities can be avoided." [Tukey, 1954.]

Probability intervals based on standard errors have been denoted by $\bar{y} \pm t_p \text{se}(\bar{y})$; we have tried to avoid a technical qualifier for the term "interval." The majority of contemporary (1965) readers will interpret these as *confidence intervals*. Frequently, however, only *approximate confidence intervals* are available. Exact confidence intervals are usually precluded by irregularities characteristic of large scale survey samples, especially by the effect on ratio means due to variations in cluster sizes.

Although the computation of these intervals is well established in many cases, their interpretation is subject to disagreement and controversy among mathematical statisticians. The Neyman-Pearson theory of confidence intervals is best established now, and is described in most current statistics books. Some of these also discuss Fisher's *fiducial intervals*. In contrast to both, but in less than complete agreement with each other, are proponents of several methods based on the primacy of the likelihood function [Jeffries, 1962; Birnbaum, 1962], and especially the growing *neo-Bayesian* or *personalistic* school of statisticians [Schlaifer, 1961; Raiffa and Schlaifer, 1961; Savage, 1954, 1962].

That the foundations of statistics are now in controversy is itself a controversial statement. However, this turbulence need not shock anybody with a broad view of science. The foundations of physics, astronomy, and of mathematics are also in turmoil, at the same time that they are advancing at an unprecedented pace. For healthy science, perhaps it is a characteristic rather than an abnormal condition when its central core, as well as its growing edges, are in dynamic stages of flux.

Space is lacking for a thorough development of this profound topic, nor do I feel fully qualified for the task. Moreover, this topic belongs in general statistical literature, rather than in this specialized volume. The only safe course would be to avoid the topic altogether. However, I feel bound to say a few words to readers who are mystified by these references to a topic they have largely overlooked. As an illustration, suppose that 10.50 percent of consumers in a survey indicate intentions to purchase new automobiles next year, and that this figure has a standard error of 0.5 percent. Thus survey results indicate, within *one* standard error, 10.0 to 11.0 percent purchasers; applied to 60 million consumer units, this means a market of 6.0 to 6.6 million new automobiles.

Second, suppose that 51.5 percent of a state sample of voters indicate preference for a Democratic candidate, and that this has a standard error of 1 percent. Thus survey results indicate a Democratic vote from 50.5 to 52.5 percent, within 1 standard error on each side of 51.5.

The two examples of standard errors above are not arbitrary choices from an academic exercise; they represent good statistics from "real-life" sample surveys. First, purchasers of new cars amount to about 0.1 of U.S. consumers; since $\sqrt{PQ/n} = \sqrt{(0.10)(0.90)/3600} = 0.30/60 = 0.0050$, an $n = 60^2 = 3600$ would be needed with simple random sampling and, say, 5000 with clustered sampling. Second, Democrats ordinarily receive *about* 0.5 of the vote; since $\sqrt{(0.50)(0.50)/2500} = 0.50/50 = 0.01$, a simple random sample of $n = 50^2 = 2500$, and a clustered sample of 4000 may be needed. Thus these are roughly the standard errors we are likely to attain with sample surveys in today's world. They are two examples of what we expect from many actual survey statistics: to provide us with valuable but not overwhelmingly convincing evidence.

In both cases, a larger sample would provide higher precision. By quadrupling all sampling units, the same span—6.0 to 6.6 million cars, or 50.5 to 52.5 percent Democrats—would represent plus and minus two standard errors instead of one. However, expanding the sample may be prohibited by lack of time or money. Or buying a larger sample may seem unwise, because of uncertainties about measurement biases. Therefore, suppose that we cannot buy higher protection with a larger sample.

We could obtain higher protection from the sample by using two or three standard errors, but that strategy might make the samples less, rather than more, useful. The range of 9.5 to 11.5 percent (or 5.7 to 6.9 millions) of auto buyers may be too wide; similarly, the range of 49.5 to 53.5 percent of the vote may also be too wide. The corresponding one-sided intervals (over 5.8 million cars, and over 49.8 percent of voters) may also be too indefinite. We suppose that those intervals are too wide and indefinite, since we can form narrower limits by *a priori* judgment, based on information available from sources outside the sample.

The *personalistic or neo-Bayesian statistician* asks You to describe Your distribution of the parameter $\bar{Y}$, based on Your own personal *a priori* opinion, without considering the sample evidence. This You is meant by personalists, to represent impersonally an arbitrary but specific decision-maker. Here we shall write about an imaginary personalistic or neo-Bayesian statistician, as if a typical specimen actually existed. Opposed to him, we must imagine a typical *objectivist* statistician. If Your opinion is unimodal, it is convenient if You state Your *a priori* distribution in a normal form around the *a priori* mean $\tilde{Y}_p$. Assume then that Your *a priori* distribution, independent of the sample evidence, has mean $\tilde{Y}_p$ and variance σ_p^2. The sample mean was $\bar{y}$, with variance $\sigma_{\bar{y}}^2$. Then Your *posterior distribution* has a mean

$$\tilde{Y}_{\text{post}} = \frac{1/\sigma_p^2}{1/\sigma_p^2 + 1/\sigma_{\bar{y}}^2}\tilde{Y}_p + \frac{1/\sigma_{\bar{y}}^2}{1/\sigma_p^2 + 1/\sigma_{\bar{y}}^2}\bar{y} \qquad (14.3.1)$$

$$= (\sigma_{\bar{y}}^2 \tilde{Y}_p + \sigma_p^2\bar{y})/(\sigma_{\bar{y}}^2 + \sigma_p^2).$$

The *posterior mean* results from combining Your *a priori* and the sample means with weights inversely proportional to their variances. The posterior distribution of $\bar{Y}$ is normally distributed with variance

$$\text{Var}(\tilde{Y}_{\text{post}}) = \frac{1}{1/\sigma_p^2 + 1/\sigma_{\bar{y}}^2}. \qquad (14.3.2)$$

We may also regard this as the variance of two normal means combined with weights inversely proportional to their variances, σ_p^2 and $\sigma_{\bar{y}}^2$.

Combining *a priori* judgment and sample information is worthwhile when their precisions are comparable. When $\sigma_{\bar{y}}^2 = \sigma_p^2$, the combination has precision twice as great, or variance half as large, as either alone. If the *a priori* variance is much greater, it may be neglected, since it does not strengthen the sample evidence. If the sample variance is much greater, the sample mean may be discarded; or preferably, this decision having been anticipated in the design, the sample would not have been taken.

Neither objectivist nor personalistic statisticians would gather sample evidence which is worthless, because Your judgment is much more reliable: when $\sigma_p^2 \ll \sigma_{\bar{y}}^2$. For this reason pilot studies, for example, are seldom taken (8.4*D*). Also, many sample plans are abortive, because it soon becomes apparent that satisfactory precision cannot be had for a feasible expense. The personalistic statistician would presumably ask You to formulate Your judgment into a distribution, to help You make a decision in a rational, systematic manner.

The action of objectivist and personalist statisticians is similar at the other extreme, when a large sample contains all useful information: when $\sigma_p^2 \gg \sigma_{\bar{y}}^2$. This situation is of prime importance, and is called *precise measurement* or *stable estimation* by some personalists [Edwards, Lindman, and Savage, 1963]. It occurs when *a priori* information is diffuse, so that it has no appreciable influence on the interpretation of sample values. Then for large samples, means and related statistics and their standard errors comprise all useful information at our disposal. This situation characterizes most results of large samples. Hence, large samples are in a fortunate situation since they often yield sample values that are normally distributed with a variance that is relatively small. These are the assumptions behind most of our presentation, as well as survey sampling literature in general.

The two examples of automobile buying and voting, where available *a priori* information is competitive with likely sample information, do not represent all situations, but neither are they unique. Another example can be found in the most famous of United States sample results: the employment and unemployment data of the United States Census Bureau's Current Population Surveys (10.4). Most economists agree, I think, that its results are valuable, but not exhaustive; they interpret, combine, and adjust them with judgment and with other data.

I would even hazard the guess that data from sample surveys are commonly "mined" until the precision of survey data becomes competitive with *a priori* knowledge. This describes methods and objectives for dividing and subdividing samples into subclasses, then analyzing and comparing their results. We should approve this practice when properly executed. To stop analysis when $\sigma_{\bar{y}}^2 \ll \sigma_p^2$ may often leave fertile areas unexplored. Unfortunately, the "mining" of data often reaches regions of unjustifiable returns: when data are grasped for support rather than illumination (in the manner of a drunk misusing a lamp-post).

Prior information and judgment were mostly neglected in this book. When mentioned, they were placed in a competitive position, as alternatives to sample information. Here we mention a few especially salient examples of the possible joint utilization of prior and sample information.

1. For skewed distributions, the sample may give poor information about the important tail end (11.4*B*). Instead of ignoring it with a "cut-off," *a priori* judgment about this portion could be combined with sample information for the rest of the distribution. Ericson [1963, 1965] has extended this idea to a formal optimum mixture of *a priori* and sample information for diverse strata.

2. Judgment estimates of measurement biases may be combined with the sampling variances to construct more realistic estimates of the total survey error (13.1).

3. The ratio mean is one example of several, where unbiased estimation leads to theoretical complications. These may be circumvented by likelihood methods, some personalists hopefully believe.

4. The design of sample size has been formulated (Chapter 8) by fixing either a required variance or cost, and solving for the other. A more realistic, though more complex, model should include curves of both cost and utility as functions of survey errors. The optimum variance would depend on the excess of utility over cost. Utility depends, generally monotonically, on the accuracy of the available *a priori* distribution.

14.4 EXPERIMENTS AND SURVEYS

"The statistician cannot evade the responsibility for understanding the processes he applies or recommends. My immediate point is that the questions involved can be disassociated from all that is strictly technical in the statistician's craft, and *when so detached*, are questions only of the right use of human reasoning powers, with which all intelligent people, who hope to be intelligible, are equally concerned, and on which the statistician, as such, speaks with no special authority. The statistician cannot excuse himself from the duty of getting his head clear on the principles of scientific inference, but equally no other thinking man can avoid a like obligation." [Fisher, 1953, p. 1.]

That correlation does not prove causation has been shown by many writers in all walks of life, including G. B. Shaw in his preface to "The Doctor's Dilemma." Searching for causal factors among data, trying to separate true explanatory variables, free them from "spurious" correlation with extraneous variables, is a vital and eternal task of all scientists. Statisticians have contributed tools to the task; and a good summary appears in Yule and Kendall [1937, Ch. 4, 15, and 16]. See also Neyman [1952], Simon [1956], Kendall [1951], Wold [1956], Wright and Tukey [in Kempthorne *et al.*, 1954], Blalock [1960, p. 337], Campbell [1957], and Kish [1959].

The researcher designates explanatory variables on the basis of substantive scientific theories. He recognizes the evidence of other *sources of variation*, and he must separate these from the explanatory variables. Sorting all sources of variation into four classes seems to me a useful simplification. Furthermore, no confusion need result from talking about sorting and treating "variables," instead of "sources of variation."

I. The *explanatory variables*, sometimes called the *experimental variables*, are the objects of research. They are the variables among which the researcher wishes to find and measure some specified relationships. They include both the *dependent* or *predictand* variables, and the *independent* or *predictor* variables [Kendall, 1951; Kendall and Buckland, 1957]. With respect to the aims of the research all other variables, of which there are three classes, are extraneous.

II. Some extraneous variables may be controlled. The control may be exercised in either the selection or the estimation procedures.

III. Other extraneous uncontrolled variables may be confounded with the Class I variables.

IV. There remain extraneous, uncontrolled variables which are treated as randomized errors. In "ideal" experiments they are actually randomized; in surveys and investigations they are only assumed to be randomized. Randomization may be regarded either as a substitute for experimental control, or as a form of control.

The aim of efficient design, both in experiments and in surveys, is to place as many extraneous variables as is feasible into the second class. The aim of randomization in experiments is to place all of the third class into the fourth class; in ideal experiments there are no variables in the third class. It is the aim of various kinds of controls in surveys to separate variables of the third class from those of the first class; these controls may involve the use of repeated cross-tabulations, regression, standardization, matching of units, and so on.

The function of statistical tests and intervals is to measure the effects found among Class I variables against the effects of variables of Class IV. In an ideal experiment this is accomplished without confusion with Class III variables. In survey comparisons there are Class III variables confounded with those of Class I; statistical tests actually contrast the effects of random variables of Class IV against the explanatory variables of Class I, which are confounded with unknown effects of Class III variables. In both ideal experiments and surveys, statistical tests serve to separate the effects of random errors of Class IV from the effects of other variables. In surveys, these are a mixture of explanatory and confounded variables; their separation poses severe problems for logic and for scientific methods; statistics is only one of the tools in the endeavor.

The scientist must make many decisions as to which variables are extraneous to his objectives, which should and can be controlled, and what methods of control he should use. He must decide where and how to introduce statistical tests into the analysis. Regarding the control of confounded variables in sample surveys, two sources of confusion should be brought into the open and flatly contradicted. First, even after controlling every possible variable, all the potentially confounding variables are not ordinarily removed. Second, neither the advance of science nor the application of statistical tests can wait for the control of all relevant, but still uncontrolled and confounded, factors.

Until now, the theory of sample surveys has been developed chiefly to provide descriptive statistics, especially estimates of means, and totals. Experimental designs have been used primarily in the analytical search for explanatory variables and for measuring relationships. However, surveys rather than experiments, must serve frequently as analytical tools in many fields, including the social sciences. Furthermore, in many research situations neither true experiments nor sample surveys are practical, and other investigations are employed.

By experiments I mean here *ideal experiments* in which the explanatory variables are deliberately introduced, and all extraneous variables are either controlled or randomized. By *sample surveys*, I mean surveys based on probability samples. By *other investigations* I mean collecting data, with considerable care and control, but without either the randomization of experiments or the probability sampling of surveys. The difference between experiments, surveys, and investigations are not consequences of statistical techniques. They result from different methods of introducing the variables and for selecting the population elements [Wold, 1956; Campbell, 1957].

The separation of sample surveys from experimental designs is an accident in the recent history of science, and it will diminish. With complete command over the research situation, we could introduce, with firm experimental controls, the desired effects and measurements into controlled and randomized portions of the entire target population. It is enlightening to think about actual and potential situations when that research ideal may be approached—also about the obstacles to that ideal. Such research situations are rare. Ordinarily only some of the factors required for inference can be brought under objective and firm control; others must be left to more or less vague and subjective—however skillful—judgment. The scientist seeks to maximize the controlled first portion, in order to minimize the tenuous second portion. After assessing the research ends, the feasible means, and the cost factors, he makes a strategic choice of methods. He is faced with the three basic problems of research: measurement, representation, and control. The problems of measurement are too vast and deep for adequate treatment here; we chiefly compare representation and control in experiments and surveys.

Experiments are strong on control through randomization; but they are weak on representation, and sometimes on the "naturalism" of measurement. Surveys are strong on representation, but they are often weak on control. Investigations are weak on control and often on representation; their use is due mainly to convenience or low cost, and to the desire for measurements in "natural settings."

Experiments have three chief advantages. (1) Through randomization of extraneous variables, the confounding variables (Class III) are eliminated. (2) Control over the introduction and variation of the predictor variables clarifies the direction of causation from predictor to predictand variables. In contrast, the direction of correlations is not clear for many survey variables; for example, between some behaviors and correlated attitudes. (3) The modern design of experiments allows for great flexibility, efficiency, and powerful statistical manipulation, whereas the analytical use of survey data presents new problems for statistical theory.

The advantages of the experimental method are so well known that we need not dwell on them here. It is the scientific method par excellence—when feasible. In many situations experiments are not feasible, and it is a common mistake to regard this handicap as separating the social from the physical and biological sciences. Such situations also occur frequently in the physical sciences, meteorology, astronomy, geology, the biological sciences, medicine, and elsewhere.

The experimental method also has some shortcomings. First, it is often difficult to choose the control variables so as to exclude all the confounding extraneous variables; that is, it may be difficult or impossible to design an ideal experiment. Thus the advantages of experiments over surveys in permitting better control are only relative, not absolute [Cornfield, 1954]. The design of proper experimental controls is not automatic; it is an art that requires scientific knowledge, foresignt in planning the experiment, and hindsight in interpreting the results. Nevertheless, the distinction in control between experiments and surveys is real and considerable; to emphasize this distinction we refer here to ideal experiments in which the control of the random variables is complete.

Second, it is difficult to design experiments so as to represent a specified important population. In fact, the questions of sampling, of making the experimental results representative of a specified population, have been largely ignored in experimental design until recently. Both in theory and in practice, experimental research has largely neglected the basic truth that causal systems, the distributions of relations—like the distributions of characteristics—exist only within specified universes. But the distributions of relationships, as of characteristics, exist only within the framework of specific populations. Probability distributions, like all mathematical models, are abstract systems; their application to the physical world must include the specification of the populations. For example, it is generally accepted that the statement of a value for mean income has meaning only with reference to a specified population; but this is not generally and clearly recognized in the case of regression of assets on income and occupation. Similarly, statistical inferences derived from the experimental testing of several treatments are restricted to the population(s) included in the experimental design [Wilk and Kempthorne, 1955 and 1956; Cornfield and Tukey, 1956; McGinnis, 1958].

Third, for many research aims, especially in the social sciences, contriving the desired "natural setting" for the measurements is not feasible in experimental design. Hence, social experiments often give clear answers to questions which have vague implications. That is, artificially contrived experimental variables may have but tenuous relationship to the variables the researcher would like to investigate.

The second and third weaknesses of experiments point to the advantages of surveys. Not only do probability samples permit clear statistical inferences to defined populations, but the measurements can often be made in the natural settings of actual populations. Thus, in practical research situations, the experimental method, like the survey method, has its distinct problems and drawbacks as well as its advantages. In practice, we generally cannot solve simultaneously all of the problems of measurement, representation, and control; rather, we must choose and compromise. In any specific situation one method may be better or more practical than the other; but neither method has all superiority in all situations. Understanding the advantages and weaknesses of both methods should lead to better choices, and to efforts for overcoming their particular weaknesses.

14.5 MULTIPLE OBJECTIVES; MULTIPURPOSE SURVEYS

For simplicity, sampling methods are developed around the framework of univariate theory. However, many—perhaps, most—samples have multiple objectives, because multipurpose surveys have overwhelming advantages along several dimensions. First, several characteristics (survey variables), often many of them, may be measured on the same set of sample elements—elicited from the same respondents on one visit. The advantages resemble those of the multifactor design of experiments, which have been so well stated by Fisher [1953, 6.39]:

"We have seen that the factorial arrangement possesses two advantages over experiments involving only single factors: (*i*) Greater efficiency, in that these factors are evaluated with the same precision by means of only a quarter of the number of observations that would otherwise be necessary; and (*ii*) Greater comprehensiveness in that, in addition to the 4 effects of single factors, their 11 possible interactions are evaluated. There is a third advantage which, while less obvious than the former two, has an important bearing upon the utility of the experimental results in their practical application. This is that any conclusion, such as that it is advantageous to increase the quantity of a given ingredient, has a wider inductive basis when inferred from an experiment in which the quantities of other ingredients have been varied, than it would have from any amount of experimentation, in which these had been kept strictly constant. The exact standardization of experimental conditions, which is often thoughtlessly advocated as a panacea, always carries with it the real disadvantage that a highly standardized experiment supplies direct information only in respect of the narrow range of conditions achieved by standardization. Standardization, therefore, weakens rather than strengthens our ground for inferring a like result, when, as is invariably the case in practice, these conditions are somewhat varied."

Second, repeated observations on the same sample elements can be profitably regarded as special cases of observing several variables. Two observations of the same variable on the jth element yields the two variables y_j and y_j', plus a third variable in the change $d_j = y_j - y_j'$. (Three observations involve three changes, and k observations $k(k - 1)/2$ changes.) Whether both observations are obtained at the same time, or on two separate occasions, depends on survey methods; timing is not relevant conceptually. Discussions about overlapping samples and panel studies (12.4 and 12.5) represent special cases of multiple objectives.

Third, the same survey yields statistics for many populations, represented by distinct subclasses. These must be determined by appropriate survey variables; the relationship of survey characteristics to subclasses is roughly that of dependent to independent, or predictand to predictor variables. For example, statistics may be given for regions, age, sex, occupation, education, and many other classes, and for combinations of them. *Domains* denote subclasses, such as regions, that are specifically controlled in the sample design—for example, either with proportionate or disproportionate sampling. Most subclasses cannot be readily allocated and are merely accepted as found; for example, sex and age classes in household samples. In some cases, a *screening* device may increase the yield of a specially desired subclass (11.4). Subclasses represent variations of the *extent* of the population. Variations in *content* and units can produce still other populations (1.2).

Fourth, the same set of variables can be combined several ways to yield a variety of statistics. We have dealt chiefly with means, aggregates, and their comparisons. There are many other valuable statistics: medians, deciles, and other quantiles; ratios and indexes; regression and correlation coefficients.

Each of the four dimensions may contribute to the multiplicity of survey objectives. The interaction of dimensions further multiplies the possible objectives; for example, the relationship of characteristics may be analyzed for diverse subclasses.

Survey economics point to two further considerable advantages for multiple objectives. First, initial contact with respondents is expensive, hence eliciting additional variables appears relatively less expensive. For this reason market surveys often incorporate several distinct surveys within one set of interviews. The cost of initial contact must cover sampling, gaining admission and confidence, and eliciting all necessary background information. Similarly, other (nondwelling, nonhuman) observational units may also require relatively large initial expense. Second, common frames and basic samples yield economies and multiply the objectives for continuing survey operations (12.6).

Of course, other factors impose limits on multiplying the objectives of a survey. Increased complexity of analysis presents obstacles. The number of survey variables must also be limited because every measurement is "destructive" in some sense; surveys should not exceed the number of valid observations that can be obtained from the subjects.

The economics of operation and the richness of multivariate explanation offer advantages, which generally overcome, even overwhelm, the disadvantages of multipurpose surveys: greater complexity and a poverty of formal theory. Actually, the answers we can offer do not measure up to the magnitude and complexity of the questions we have posed.

Remarks about multiple objectives appear in discussions of sample size (2.6, 3.5.IV, Chapter 8), formation of strata (3.6), disproportionate allocation (3.5) and (4.5C), cluster design (5.4). Good general advice is given by Cochran [1963, 4.7] and Mahalanobis [1944]. Hansen, Hurwitz, and Madow give examples of optimum allocation [1953, 5.13] and of optimum cluster sizes [1953, 6.24, 9.8], which are also treated by Sukhatme [1954, 6.a.6, 7.5]. Both problems received early attention from Jessen [1942].

If one principal statistic can be designated, that alone can determine the sample design. If a small number of principal statistics can be separated, a reasonable compromise may be found. Proportional allocation may represent such a compromise when the separate allocations are not too disparate. A common disproportionate optimum allocation may also be efficient among related characteristics; an illustrated model appears in an article [Kish, 1961] which also deals with optima for diverse subclasses. Good compromise allocations are often possible because regions of optimum allocation are broad. Compromises can be assisted by stratifying with several variables (3.6G, H, J).

Nevertheless, compromise is impossible for statistics that are too disparate. Surveys of the entire population and of a minute subclass may require vastly disparate sampling fractions. Design for a highly concentrated industry or occupation group may differ too much to be accommodated in a cross-sectional survey. Optima for medians require much less disproportionate allocation than for means of highly skewed items, such as income and wealth [Kish, 1961].

Before abandoning a joint design for conflicting purposes, we should explore the adaptability of survey resources. Sample size need not be equal for all statistics. For example, splitting the interview schedule reduces the sample size of some variables, but retains a larger size for a common core of questions. Lengthy observation may be confined to a small portion of a large sample; as an extreme, the latter may serve chiefly as a first phase for screening. A large sample for a simple survey may be drawn by attaching it to several separate surveys. Periodic surveys offer another dimension of flexibility; small subclasses can be accumulated; the sample base can be spread or contracted for diverse variables in accord with their specific needs; alternated samples may be selected for some variables and domains (11.5D).

Formal models are possible, but satisfactory ones are hard to construct. Computations can quickly become complicated; but this obstacle becomes less formidable with the growth of computer technology. However, it is hard to construct mathematical models which can adequately represent the complex requirements. These also can differ widely in diverse situations. One class of models minimizes each of a set of variances, whereas another minimizes an average function of the variances. Yates presents a method of optimum allocation to achieve required variances for several characteristics at minimum cost [1960, 10.4]. Srikantan [1963] obtains required variances for several subclasses and an entire sample with linear programming. Dalenius [1957, Chapter 9] has a thorough discussion, including the following model for minimizing an average loss function.

Let L_g represent the relative *loss of information* on variable Y_g, measured as the relative increase of variance. Then minimize a weighted average loss $\sum_g I_g L_g$, where

$$L_g = \frac{\sum_h^H W_h^2 \sigma_{gh}^2 / n_h}{\sigma_{g\,\min}^2} - 1. \tag{14.5.1}$$

$\sum_h^H W_g^2 \sigma_{gh}^2 / n_h$ is the variance of the gth variable with the compromise allocation, and $\sigma_{g\min}^2$ is its minimum variance with its own optimum allocation. The minimum $\sum_g I_g L_g$ is obtained for

$$\frac{n_h}{n} = \frac{\sqrt{\sum_g I_g W_h^2 \sigma_{gh}^2 / \sigma_{g\,\min}^2}}{\sum_h \sqrt{\sum_g I_g W_h^2 \sigma_{gh}^2 / \sigma_{g\,\min}^2}}. \tag{14.5.2}$$

PROBLEMS

14.1. Discuss briefly these two statements:

(*a*) Representing specific populations with probability sampling is important for descriptive statistics, but not for discovering fundamental relationships between variables. (Synthetic restatement of a common argument.)

(*b*) "... it is not true that one can uncover 'general' relationships by examining some arbitrarily selected population. ... There is no such thing as a completely general relationship which is independent of population, time, and space. The extent to which a relationship is constant among different populations is an empirical question which can be resolved only by examining different populations at different times in different places." [McGinnis, 1958.]

In the following four problems, sampling aspects interact with the broader problems of the research design, and uniformly best solutions are not available.

14.2. Experts in the Treasury Department are quietly discussing the possibility of a temporary tax reduction to combat a recession, by releasing such funds into the hands of consumers. The question is how much will be spent rather than saved by consumers. It is desired to obtain some interviews on buying intentions *before* any plans are released to the public, and some *after* the tax reductions are announced and publicized widely. Discuss:

(*a*) Should the *before* and *after* interviews be made on the same respondents, on different respondents, or on some of each?

(*b*) Should the samples consist of national cross sections; or be confined to a typical, homogeneous city or county; or taken in 6 or 8 counties, cities, or metropolitan areas, chosen to represent diverse regional, size, and economic classes?

14.3. A new vaccine against the common cold is available for a field trial of 5000 people over 10 years of age. Discuss these questions about the design:

(1) Should we use for control: (*a*) nothing; (*b*) before-and-after observations on the 5000 vaccines; (*c*) 5000 controls with no vaccines; (*d*) 5000 people with a placebo; (*e*) 5000 people with placebo, and before-and-after observations on all 10,000?

(2) Should the sample consist of: (*a*) 5000 individually selected persons; (*b*) 2000 families averaging 2.5 persons each?

(3) Should the sample consist of: (*a*) national probability sample; (*b*) 4 to 12 places, of contrasting characteristics on factors considered relevant by experts, such as climate, size, smoke, socio-economic status, national origin? (*c*) Should these places be large cities, city tracts, suburbs, small towns, or one Army post?

(4) How many observations should be made on each individual and over how long a period?

Cost factors are: (*a*) Cost of vaccine, cost of placebo. (*b*) Cost of reaching dwelling. (*c*) Cost of individual observation. (*d*) Cost of cooperation from county medical society.

14.4. Some of the adult population of 10 large cities, near which there are large atomic installations, were thought to be worried about radiation hazards. A sample of each of these cities is to be interviewed. Their attitudes should be contrasted with a control population, to ascertain differences in the level and prevalence of anxiety between the ten "treatment" cities and the control population. Discuss reasons for choosing one of these alternatives as controls.

(*a*) A sample of the rest of the entire country.

(*b*) One control city, paired with each of ten treatment cities, chosen by careful judgment as most resembling it.

(*c*) Three cities chosen at random from each of 10 classes of cities; each class broadly defined by size, region, etc., from the entire country and containing 1 of the 10 cities.

14.5. A large organization is divided into about 60 large units, separate plants and large divisions, with rather autonomous managements. It desires to conduct research on a moderate scale on the effects on production and satisfaction of a conversion from a hierarchical to a more democratic (decentralized) method of management. Discuss the advantages and drawbacks of three alternative research plans.

(*a*) Introduce a strong version of the new method into one entire large unit.

(*b*) Introduce necessarily weakened versions of the method into work sections, with matched controls, within several of the large units.

(*c*) In a good cross-section sample of the entire organization, conduct long and thorough interviews on the attitudes toward the possible introduction of the new methods.

14.6. Use the technique of (14.2.1) to develop the following approximate variances and relvariances for some functions of the variates y and x. Assume $E(y) = Y$ and $E(x) = X$.

(*a*) $\text{Var}(ky^2) = 4k^2Y^2\,\text{Var}(y)$, and $CV^2(ky^2) = 4CV^2(y)$.

(*b*) $\text{Var}(k\sqrt{y}) = k^2\text{Var}(y)/4Y$, and $CV^2(k\sqrt{y}) = CV^2(y)/4$.

(*c*) $\text{Var}\left(\frac{k}{y}\right) = \frac{k^2}{Y^4}\text{Var}(y)$, and $CV^2\left(\frac{k}{y}\right) = CV^2(y)$.

(*d*) $\text{Var}\left(\frac{y}{x}\right) = \frac{1}{X^2}\left[\text{Var}(y) + \frac{Y^2}{X^2}\text{Var}(x) - 2\frac{Y}{X}\text{Cov}(y, x)\right]$.

(*e*) $\text{Var}(yx) = X^2\,\text{Var}(y) + Y^2\,\text{Var}(x) + 2YX\,\text{Cov}(y, x)$.

APPENDIXES

Appendix A
Summary of Symbols and Formulas

In response to requests for a list of symbols, I submit the following descriptive remarks. They are organized to correspond to the first eight chapters, and are numbered to tally with those in appropriate sections. Since descriptions can be more readily understood, I have elected to use them rather than formal definitions, which can be found in their proper sections. These pages can also serve as a rapid review and guide for the first eight chapters; the other six did not seem readily tractable with this approach.

Chapter 2 Basic Sampling Concepts

2.1 2.2. The population mean is denoted by $\bar{Y} = Y/N$, where $Y = \sum_i^N Y_i$ is the population total (aggregate) of the Y_i variable. The sample mean is $\bar{y} = y/n$, where $y = \sum_i^N y_j$ is the sample total. Lower case letters versus upper case letters distinguish sample from population values. Elements are numbered with j in the sample and with i in the population. A bar(—) over the symbol for a total denotes a mean. These conventions are maintained throughout the book; they are common in much of sampling literature. It is impossible to be consistent with all statistical literature. Many would prefer the symbol for the sample total to be $y.$ or y_n.

The sample value for the element variance is $s_y^2 = \sum_j^n (y_j - \bar{y})^2/(n-1)$; s_y is the standard deviation. The element variance in the population has the two expressions $\sigma_y^2 = \sum_i^N (Y_i - \bar{Y})^2/N$ and $S_y^2 = N\sigma_y^2/(N-1)$, a distinction without practical importance; σ_y and S_y both denote standard deviations. Often we may omit the subscript and write simply s^2, S^2, and σ^2.

The mean of a simple random sample (srs), distinguished by the subscript 0, has variance

$$\text{Var}(\bar{y}_0) = (1-f)\frac{S^2}{n} = \frac{N-n}{N-1}\frac{\sigma^2}{n}, \quad \text{estimated by var}(\bar{y}_0) = (1-f)\frac{s^2}{n}. \tag{2.2.2}$$

The standard error is $\text{se}(\bar{y}_0) = \sqrt{1-f}\,s/\sqrt{n}$. The standard error of the estimated total is $\text{se}(N\bar{y}_0) = N\,\text{se}(\bar{y}_0) = N\sqrt{1-f}\,s/\sqrt{n}$. Here $f = n/N$, the sampling fraction, and $(1-f)$ is the *fpc*.

2.4. A *proportion p* is a mean for dichotomous variables Y_i that take values of 0 and 1 only. For simple random samples the variance is estimated by

$$\text{var}(p_0) = (1-f)\frac{p(1-p)}{n-1}. \tag{2.4.1}$$

2.5. Relative error can be measured with the coefficient of variation $C_y = S_y/\bar{Y}$, estimated by $c_y = s_y/\bar{y}$. The coefficients of variation of the mean and aggregate are equal:

$$\text{cv}(\bar{y}) = \frac{\text{se}(\bar{y})}{\bar{y}} = \frac{N\,\text{se}(\bar{y})}{N\bar{y}} = \text{cv}(N\bar{y}). \tag{2.5.2}$$

For simple random sampling

$$\text{cv}(\bar{y}_0) = \sqrt{1-f}\,c_y/\sqrt{n} = \sqrt{1-f}\,s_y/\bar{y}\sqrt{n}. \tag{2.5.4}$$

These are estimates of similar population values $\text{CV}(\bar{y})$. *Relvariances* denote relative variances, the squares of coefficients of variation; $\text{cv}^2(\bar{y}) = \text{var}(\bar{y})/\bar{y}^2 = (1-f)c_y^2/n$, the last expression being appropriate to srs only.

Chapter 3 Stratified Sampling

3.2. The population is divided into H strate with relative sizes W_h, where $\Sigma W_h = 1$. The population mean $\Sigma W_h \bar{Y}_h$ is estimated by $\bar{y}_w = \Sigma W_h \bar{y}_h$, where $\bar{y}_h$ is the sample mean in the hth stratum. The variance of the sample mean is estimated by

$$\text{var}(\bar{y}_w) = \sum W_h^2\,\text{var}(\bar{y}_h). \tag{3.2.3}$$

If proportions of elements serve as weights, $W_h = N_h/N$; then

$$\bar{y}_w = \sum N_h \bar{y}_h / N, \quad \text{and} \quad \text{var}(\bar{y}_w) = \sum N_h^2 \,\text{var}(\bar{y}_h)/N^2. \qquad (3.2.3')$$

3.3 The above are general formulas for stratified samples. When the stratum means are based on simple random samples of size n_h, with sampling fractions $f_h = n_h/N_h$, the variance becomes

$$\text{var}(\bar{y}_{w0}) = \sum (1\text{-}f_h) \frac{W_h^2 s_h^2}{n_h}, \quad \text{where } s_h^2 = \frac{1}{n_h - 1}\left(\sum_i^{n_h} y_{hi}^2 - \frac{y_h^2}{n_h}\right). \qquad (3.3.2)$$

3.4. *In proportionate element sampling*, the sampling fractions are constant, $f_h = f$, and $n_h/n = W_h$. The sample mean is self-weighting ($\bar{y} = y/n = \bar{y}_w = \Sigma W_h \bar{y}_h$), and the variance can be written as

$$\text{var}(\bar{y}_{\text{prop}}) = \frac{1\text{-}f}{n} \sum W_h s_h^2 = \frac{1\text{-}f}{n} s_w^2, \quad \text{where } s_w^2 = \sum W_h s_h^2. \qquad (3.4.2)$$

3.5. *Optimum allocation* either minimizes the variance for a fixed cost $\Sigma J_h n_h$, or minimizes this cost for fixed variance. It occurs when the sample sizes are

$$n_h = \frac{K' W_h S_h}{\sqrt{J_h}}. \qquad (3.5.3)$$

When the element costs J_h are similar in all strata, one minimizes $n = \Sigma n_h$, and the allocation becomes $n_h \doteq k' W_h S_h$. When $W_h = N_h/N$, this becomes $f_h = n_h/N_h = k S_h$. The constants of proportionality are given in 4.6.

Chapter 4 Systematic Sampling; Stratification Techniques

4.1. A *systematic* sample of n elements is selected with an interval F applied successively to a population of $N = nF$ listings. A variance formula for systematic samples utilizes the $(n - 1)$ successive differences $(y_g - y_{g+1})$, the subscript denoting order of selection:

$$\text{var}(\bar{y}_{\text{syst}}) = \frac{1\text{-}f}{2n(n-1)} \sum_g^{n-1} (y_g - y_{g+1})^2. \qquad (4.1.2)$$

Alternatively, we may use only $n/2$ comparisons from that many distinct pairs in the sample.

4.3. For *paired selections* of n elements, the population is divided into $n/2$ zones of $2F$ elements each. Two random selections from each zone yield $n/2$ pairs with a sampling fraction of $2/2F = 1/F = f = n/N$. The simple mean is also the mean of $n/2$ pairs: $\bar{y} = y/n = \sum_h (y_{ha} + y_{hb})/n$. Its variance is

$$\text{var}(\bar{y}_{\text{pair}}) = \frac{1\text{-}f}{n^2} \sum_h^{n/2} D^2 y_h, \quad \text{where } D^2 y_h = (y_{ha} - y_{hb})^2. \qquad (4.3.2)$$

The aggregate Fy has the variance $\text{var}(Fy) = F^2(1-f)\sum_h D^2 y_h$. Variable sampling fractions can be introduced with two selections from zones of size $2F_h$. The mean $\bar{y}_w = \Sigma F_h(y_{ha} + y_{hb})/(2\Sigma F_h)$ has variance

$$\text{var}(\bar{y}_w) = \frac{\sum F_h^2(1-f_h)D^2 y_h}{(2\sum F_h)^2}. \quad (4.3.5)$$

4.4. *Replicated sampling* refers to selecting c similar samples, each with the sampling fraction $1/cF$, for an overall rate of $c/cF = 1/F = f$. Within each zone of size cF, make c independent selections of similar design. Compute c estimates x_γ of the corresponding population value. Their mean, $\bar{x} = \Sigma x_\gamma/c$, has variance

$$\text{var}(\bar{x}) = \frac{1-f}{c(c-1)}\sum (x_\gamma - \bar{x})^2. \quad (4.4.2)$$

The replicates consist of n_γ elements, and $n = \Sigma n_\gamma$ is the overall size. If the n_γ differ greatly, we may use a ratio mean for c independent clusters: $\bar{y}_r = y/n = \Sigma y_\gamma/\Sigma n_\gamma$. Its variance is

$$\text{var}(\bar{y}_r) = \frac{(1-f)c}{n^2(c-1)}\sum (y_\gamma - n_\gamma \bar{y}_r)^2. \quad (4.4.5)$$

4.5. *Subclasses from stratified element samples* cause difficulties because the number m_h of subclass members found among n_h elements is a random variable. If the stratum proportions $\bar{M}_h = M_h/N_h$ of members are known in the entire population, this may be used to adjust the data. But if only $\bar{m}_h = m_h/n_h$ in the sample are known, control over the stratification is partially lost, with consequent increase in the variance. The stratum mean becomes a ratio mean $\bar{y}_h = y_h/m_h$, where $y_h = \sum_i^{m_h} y_{hi}$, the aggregate of the Y_i variable for the m_h subclass members in the hth stratum. The sample mean is

$$\bar{y}_w = \frac{\sum F_h y_h}{\sum F_h m_h} = \sum w_h \bar{y}_h, \quad (4.5.2)$$

where $w_h = F_h m_h/\Sigma F_h m_h$ and $F_h = N_h/n_h$.

The variance of this mean is

$$\text{var}(\bar{y}_w) \doteq \sum_h^H (1-f_h)\frac{w_h^2}{m_h'}[v_h^2 + (1-\bar{m}_h)(\bar{y}_h - \bar{y}_w)^2] \quad (4.5.4)$$

$$= \sum_h^H (1-f_h)\frac{w_h^2}{m_h'}[t_h^2 - \bar{m}_h(\bar{y}_h - \bar{y}_w)^2], \quad (4.5.5)$$

where $m_h' = m_h(n_h - 1)/n_h$;

$$v_h^2 = \sum_i^{m_h} (y_{hi} - \bar{y}_h)^2/m_h;$$

$$t_h^2 = \sum_i^{m_h} (y_{hi} - \bar{y}_w)^2/m_h = v_h^2 + (\bar{y}_h - \bar{y}_w)^2.$$

When $\bar{m}_h \doteq 1$, the ordinary stratified formula will do; for small subclasses, with $\bar{m}_h$ small, t_h^2 can represent the bracketed quantities. These can be denoted as b_h, and optimum allocation formulas, with f_h proportional to $b_h/\sqrt{J_h}$, can be obtained (4.5*C*).

The variance for the difference $(\bar{y}_{wa} - \bar{y}_{wb})$ of two subclasses is approximately

$$\text{var}\,(\bar{y}_{wa} - \bar{y}_{wb}) = \sum (1\text{-}f_h)\left[\left(\frac{W_h^2 t_h^2}{m_h'}\right)_a + \left(\frac{W_h^2 t_h^2}{m_h'}\right)_b\right]. \qquad (4.5.6)$$

The variance for the simple expansion estimate $\Sigma\, F_h y_h$ of the total Y_i for a subclass is

$$\text{var}\,(\sum F_h y_h) = \sum (1\text{-}f_h)\frac{N_h^2}{n_h - 1}\,\bar{m}_h[v_h^2 + (1 - m_h)\bar{y}_h^2]. \qquad (4.5.7)$$

Chapter 5 Cluster Sampling and Subsampling

5.2–5.3. *Equal clusters* refer to a population divided into A clusters of B elements each: $N = AB$. The population mean is

$$Y = \frac{Y}{N} = \frac{1}{AB}\sum_\alpha^A \sum_\beta^B Y_{\alpha\beta} = \frac{1}{A}\left(\frac{1}{B}\sum_\alpha^A Y_\alpha\right) = \frac{1}{A}\sum_\alpha^A \bar{Y}_\alpha. \qquad (5.2.1)$$

Epsem choice of *a complete clusters* yields a sample of $n = aB$ elements and a sampling fraction of $a/A = n/N = f$. *Subsampling*, using epsem selection in two stages, yields a uniform sampling fraction $(a/A)(b/B) = ab/AB = n/N = f = f_a f_b$. Complete clusters are a special case of subsampling, when $b = B$.

The sample mean is $\bar{y} = y/n = \sum_\alpha^a y_\alpha/n = \sum_\alpha^a y_\alpha/a$, where $\bar{y}_\alpha = y_\alpha/b$ represent cluster means, and $y_\alpha = \sum_\beta^b y_{\alpha\beta}$ cluster totals. When entire clusters are selected $(b = B)$, these are $\bar{y}_\alpha^* = y_\alpha/B$ and $y_\alpha^* = \sum_\beta^B y_{\alpha\beta}$; but the asterisks can usually be omitted without causing confusion.

For random selection of a clusters the variance is

$$\text{var}\,(\bar{y}) = (1\text{-}f)\frac{s_a^2}{a} = (1\text{-}f)\frac{s_y^2}{ab^2}, \qquad (5.3.3)$$

where

$s_a^2 = [1/(a-1)](\sum \bar{y}_\alpha^2 - a\bar{y}^2)$, and $s_y^2 = b^2 s_a^2 = [1/(a-1)](\sum y_\alpha^2 - y^2/a)$.

When complete clusters are selected, use $b = B$ (5.2.3).

Selection in the first stage is assumed random with replacement. Subsampling is epsem without replacement of the elements: within these restrictions diverse procedures may be freely chosen for selecting b from B elements. This *simple replicated subsampling* permits computation of the variance with only the *primary selection* values, either $\bar{y}_\alpha$ or y_α. The method leads readily to two or more stages of selection, with $f = (a/A)(b/B)(c/C)$ etc.

5.4. *Homogeneity* of elements within clusters is measured with *roh*, the coefficient of intraclass correlation. Technical definition is given in 5.6, but a simple computing formula is

$$\text{deff} = \frac{s_a^2/a}{s^2/n} = [1 + \text{roh}\,(b-1)], \tag{5.4.2}$$

where s^2 is the variance of sample elements. The *design effect*, deff, is the ratio of the actual variance of the mean to simple random variance for the same number of elements.

5.5. Stratified selection of a_h clusters has a mean

$$\bar{y}_w = \sum W_h \bar{y}_h \tag{5.5.1}$$

with variance
$$\text{var}\,(\bar{y}_w) = \sum_h^H W_h^2 \frac{(1-f_h)}{a_h} s_{ha}^2,$$

where
$$s_{ha}^2 = \frac{1}{(a_h-1)b_h^2}\left(\sum_\alpha^{a_h} y_{h\alpha}^2 - \frac{y_h^2}{a_h}\right); \tag{5.5.2}$$

a_h is the number of primary selection in the hth stratum.

Chapter 6 Unequal Clusters

6.1. The size of primary sampling units (PSU's) can be denoted either as N_α or X_α; the population size is $N = \sum_\alpha^A N_\alpha = \sum_\alpha^A X_\alpha = X$. The Y_i variable total is $Y = \sum_\alpha^A Y_\alpha$, where $Y_\alpha = \sum_\alpha^{X_\alpha} Y_{\alpha b}$ is its total for the PSU. The population mean per element is denoted $\bar{Y} = R = Y/X$.

6.2. *Simple replicated subsampling* is assumed again: selecting with $f_a = a/A$ and with replacement in the first stage; then subsampling with f_b in each of the a selected clusters. The sample mean $\bar{y} = y/n$ is a *ratio mean*:

$$r = \frac{y}{x} = \frac{\sum_\alpha^a y_\alpha}{\sum_\alpha^a x_\alpha}. \tag{6.2.1}$$

6.3. The variance of the ratio mean is

$$\text{var}(r) = \frac{1}{x^2}[\text{var}(y) + r^2 \text{var}(x) - 2r \text{cov}(y, x)]. \tag{6.3.1}$$

For random choice of a clusters, (6.3.1) becomes

$$\text{var}(r) = \frac{1-f}{x^2} a[s_y^2 + r^2 s_x^2 - 2rs_{yx}], \tag{6.3.2}$$

where
$$s_y^2 = \frac{1}{a-1}\left(\sum_\alpha^a y_\alpha^2 - \frac{y^2}{a}\right),$$

$$s_x^2 = \frac{1}{a-1}\left(\sum_\alpha^a x_\alpha^2 - \frac{x^2}{a}\right),$$

and
$$s_{yx} = \frac{1}{a-1}\left(\sum_\alpha^a y_\alpha x_\alpha - \frac{yx}{a}\right).$$

This variance may also be computed with the variable $z_\alpha = y_\alpha - rx_\alpha$:

$$\text{var}(r) = \frac{1-f}{x^2} as_z^2,$$

where
$$s_z^2 = \frac{1}{a-1}\sum_\alpha^a z_\alpha^2. \tag{6.3.5}$$

We also define $s_r^2 = s_z^2/(x/a)^2$; hence,

$$\text{var}(r) = (1-f)s_r^2/a. \tag{6.3.6}$$

6.4. *Stratified selection of a_h clusters* from the hth stratum results in the ratio mean

$$r = \frac{y}{x} = \frac{\sum_h^H y_h}{\sum_h^H x_h} = \frac{\sum_h^H \sum_\alpha^{a_h} y_{h\alpha}}{\sum_h^H \sum_\alpha^{a} x_{h\alpha}}. \tag{6.4.1}$$

For stratified samples, the variance of the ratio mean above becomes

$$\text{var}(r) = \frac{1}{x^2}\left[\sum_h^H d^2y_h + r^2 \sum_h^H d^2x_h - 2r\sum_h^H dy_h\, dx_h\right], \tag{6.4.2}$$

where $\quad \text{var}(y_h) = d^2y_h = \dfrac{1-f_h}{a_h-1}\left(a_h \sum_\alpha^{a_h} y_{h\alpha}^2 - y_h^2\right),\quad$ and

$$dy_h\, dx_h = \frac{1-f_h}{a_h-1}\left(a_h \sum_\alpha^{a_h} y_{h\alpha}x_{h\alpha} - y_h x_h\right). \tag{6.4.3}$$

Another computing formula is

$$\text{var}(r) = \frac{1}{x^2}\sum_h^H d^2z_h, \tag{6.4.4}$$

where

$$d^2z_h = \frac{1-f_h}{a_h-1}\left(a_h \sum_\alpha^{a_h} z_{h\alpha}^2 - z_h^2\right),$$

$$z_{h\alpha} = y_{h\alpha} - rx_{h\alpha}, \quad \text{and}$$

$$z_h = \sum_\alpha^{a_h} z_{h\alpha} = y_h - rx_h = \sum_\alpha^{a_h} y_{h\alpha} - r\sum_\alpha^{a_h} x_{h\alpha}.$$

6.5. The difference of two ratio means $(r - r') = (y/x - y'/x')$ has the variance

$$\begin{aligned}\text{var}(r - r') &= \text{var}(r) + \text{var}(r') - 2\,\text{cov}(r, r')\\ &= \frac{1}{x^2}\sum_h^H d^2z_h + \frac{1}{x'^2}\sum_h^H d^2z_h' - \frac{2}{xx'}\sum_h^H dz_h\, dz_h',\end{aligned} \tag{6.5.4}$$

where $dz_h\, dz_h' = \dfrac{1-f_h}{a_h-1}\left(a_h \sum_\alpha^{a_h} z_{h\alpha}z_{h\alpha}' - z_h z_h'\right).$

For paired selections, $a_h = 2$, easy computing forms are

$$\begin{aligned}Dz_h &= (z_{h1} - z_{h2}) = (y_{h1} - rx_{h1}) - (y_{h2} - rx_{h2})\\ &= (y_{h1} - y_{h2}) - r(x_{h1} - x_{h2}) = (Dy_h - r\, Dx_h)\\ &= (dy_h - r\, dx_h)/(1\text{-}f_h) = dz_h/(1\text{-}f_h).\end{aligned}$$

The dz_h forms incorporate the factors $(1\text{-}f_h)$.

Chapter 7 Selection with Probabilities Proportional to Size (PPS)

Selecting elements in two stages with a uniform overall rate f may be done in two stages with PPS:

$$\frac{\text{Mos}_\alpha}{Fb^*} \cdot \frac{b^*}{\text{Mos}_\alpha} = \frac{1}{F} = f, \tag{7.2.2}$$

where Mos_α is the measure of size of the αth primary unit, and b^* is the planned subsample size per unit.

Chapter 8 The Economic Design of Surveys

8.1. If a denotes the number of primary selections, and s_g^2 their *unit variance*, the variance of the mean $\bar{y}$ of an epsem may be written generally as

$$\text{var}(\bar{y}) = \frac{1-f}{a} s_g^2 . \tag{8.1.1}$$

8.2. The results of a past survey may be used to design the planned size $\tilde{a}$, the sampling fraction $\tilde{f}$, and the variance $\widetilde{\text{var}}(\bar{y}) = V^2$ of a future sample from a similar population:

$$\frac{\tilde{a}}{a} = \frac{\text{var}(\bar{y})}{\widetilde{\text{var}}(\bar{y})} \cdot \frac{(1-\tilde{f})}{(1-f)} . \tag{8.2.1}$$

Valuable tools for designing complex samples are the *design effect*

$$\text{Deff} = \frac{\text{Var}(\bar{y})}{(1-f)S^2/n} , \tag{8.2.3}$$

and the *element variance* $= V_v^2 = S^2 \text{ Deff} = n \text{ Var}(\bar{y})/(1-f)$. For ratio means this is $V_v^2 = S^2 \text{ Deff} = x \text{ Var}(r)/(1-\text{f})$.

8.3. A general model for cost factor is

$$T = K + C = K + K_v + nc + nc_v . \tag{8.3.1}$$

K_v depends on the nature v of the design whereas K does not; neither is affected by the size n of the sample. Both c and c_v are directly proportional to the number of elements; c_v depends on the nature of the design, c does not.

The assumption that $c_v = aC_a/n$ and $K_v = 0$ results in a more specific cost function:

$$C = nc + aC_a , \tag{8.3.5}$$

where C_a is the cost per cluster. This permits finding the

$$\text{optimum } b = \left[\frac{C_a}{c}\frac{(1-\text{roh})}{\text{roh}}\right]^{1/2} = \left(\frac{C_a}{c}\frac{S_b^2}{S_u^2}\right)^{1/2} , \tag{8.3.6'}$$

where $S_u^2 = S_a^2 - S_b^2/B$.

8.5. If cost components can be written as $C = \sum C_i m_i$ and variance components as $V^2 = \Sigma V_i^2/m_i$, then the Cauchy inequality yields a minimum V^2C at the

$$m_i = \frac{KV_i}{\sqrt{C_i}} , \tag{8.5.3}$$

where K is a constant of proportionality.

Appendix B
Unit Analysis of Statistical Formulas

I find it convenient to note the *units of statistical formulas*—a technique similar to the *dimension analysis* I learned in physics. It helps me to know formulas by understanding the relationship of components, without trying to memorize them. It also assists me to avoid and discover large computational errors. The reader may also find the technique helpful and amusing. I cannot decide whether the best name for this technique in statistics should be unit, dimension, or scale analysis; each has some advantages and disadvantages.

If $[u]$ is the unit of an element, then $\bar{y}$ and $\bar{Y}$, also s, S, and σ, are expressed in the unit $[u]$, whereas s^2, S^2, and σ^2 are all in $[u^2]$.

Var $(\bar{y})$ and var $(\bar{y})$ are in $[u^2n^{-1}]$, and SE $(\bar{y})$ and se $(\bar{y})$ in $[un^{-1/2}]$. The estimated population aggregate $N\bar{y}$ is in units of $[Nu]$, its standard error se $(N\bar{y})$ in $[Nun^{-1/2}]$, and its variance var $(N\bar{y})$ in $[N^2u^2n^{-1}]$. This is true for element samples; units for cluster samples are explained later.

The sample total y is in units $[nu]$, its variance var (y) in $[n^2u^2n^{-1}] = [nu^2]$, and its standard error se (y) in $[n^{1/2}u]$.

For example, compute the units of the variance of a simple random mean:

$$\text{var}\,(\bar{y}_0) = (1 - f)\frac{1}{n(n-1)}\left\{\sum_j^n y_j^2 - \frac{y^2}{n}\right\}$$

$$\rightarrow \left[1\,\frac{1}{n^2}\left(nu^2 - \frac{u^2n^2}{n}\right)\right] = \left[\frac{1}{n^2}\,nu^2\right] = [u^2n^{-1}].$$

The variance of the aggregate $(N\bar{y}_0)$ is simply $\{N^2\,\text{var}\,(\bar{y}_0)\}$, hence its units are $[N^2u^2n^{-1}]$; for the standard error $[\sqrt{N^2u^2n^{-1}}] = [Nun^{-1/2}]$.

Proportions are pure numbers, with the unit [1]. Hence the units of P and p are both [1], also of PQ, pq, $\sqrt{PQ}$, and $\sqrt{pq}$. The unit of var $(p_0) = (1\text{-}f)pq/(n-1)$ is $[1 \times 1 \times n^{-1}] = [n^{-1}]$. The scale of percentages $(100p)$ contains a constant 10^2, and its standard error contains the same constant. The variance of proportions, therefore, includes the constant 10^4, which may be checked against $(10^2p)(10^2q)/n \to 10^4[n^{-1}]$.

The count variate $r = np$ has the unit $[n]$. The simple random variance of p may be written as $r(n-r)/n^2(n-1) \to [n^2/n^3] = [n^{-1}]$. The variance of the sample total r should have a variance with a unit $[n^2]$ times the units $[n^{-1}]$ for the variance of p; this can be checked against var $(np) = (1\text{-}f)npq \to [n]$.

The coefficient of variation of elements, $c_y = s_y/\bar{y}$, has units of $[u/u] = [1]$; hence, it is a pure number. The coefficients of variation of a mean or an aggregate both have units of $[n^{-1/2}]$; their squares, the relvariances, have $[n^{-1}]$. For example, $\text{cv}^2(y_0) = (1\text{-}f)s_y^2/n\bar{y}^2$ has a scale $[1 \times u^2/nu^2] = [n^{-1}]$.

Note that sample values (estimates) have the same units as the corresponding population values. Numbers like $(1\text{-}f)$, $n/(n-1)$, $N/(N-1)$ have units of [1]. The terms of a sum must be homogeneous in units, all terms having the same units.

Variance of a sample mean for equal clusters has units of $[u^2a^{-1}]$, where a is the total number of primary clusters; $n = ab$ is the sample size with b elements from each primary cluster. For an unstratified sample

$$\text{var}(\bar{y}) = \frac{1-f}{a(a-1)b^2}\left[\sum_{\alpha}^{a} y_\alpha^2 - \frac{y^2}{a}\right]$$

$$\to \left[\frac{1}{a^2b^2}\left(ab^2u^2 - \frac{a^2b^2u^2}{a}\right)\right] = \left[\frac{ab^2u^2}{a^2b^2}\right] = [u^2a^{-1}].$$

Note that the cluster totals y_α had units $[bu]$, whereas cluster means $\bar{y}_\alpha = y_\alpha/b$ have the elements' unit of $[u]$; then $\Sigma\, \bar{y}_\alpha^2 \to [au^2]$. For a stratified clustered sample the variance is

$$\text{var}(\bar{y}) = \frac{1\text{-}f}{n^2}\left\{\sum_{h}^{H} \frac{a_h}{a_h - 1} \sum_{\alpha}^{a_h} y_{h\alpha}^2 - \frac{y_h^2}{a_h}\right\}$$

$$\to \left[\frac{1}{a^2b^2}\sum_{h}^{H} 1(a_h b^2u^2 - a_h b^2u^2)\right] = \left[\frac{ab^2u^2}{a^2b^2}\right] = [u^2a^{-1}].$$

Note that $s_a^2 = (\Sigma\, \bar{y}_\alpha^2 - a\bar{y}^2)/(a-1)$ has units $[u^2]$, and

$$s_y^2 = (\sum y_\alpha^2 - y^2/a)/(a-1)$$

has units $[b^2u^2]$. The variance var (y) has units $[ab^2u^2]$. The element variance V_t^2 has units $[bu^2]$.

Coefficients of variation of cluster samples se $(\bar{y})/\bar{y}$ or $\{\text{se}(N\bar{y})/N\bar{y}\}$ have units of $[ua^{-1/2}/u] = [a^{-1/2}]$. When the mean is a proportion, the elementary unit has $[u] = [1]$ and its variance is in units of $[a^{-1}]$.

For unequal clusters, consider ratio means of the general form

$$r = y/x = \sum_h^H \sum_\alpha^{a_h} y_{h\alpha} \Big/ \sum_h^H \sum_\alpha^{a_h} x_{h\alpha}.$$

This is in units of

$$\left[\sum_h^H a_h u_y \Big/ \sum_h^H a_h u_x\right] = [au_y/au_x] = [u_y/u_x],$$

where $[u_y]$ and $[u_x]$ are the units for a single primary cluster of which there are a_h in the hth stratum and a in the entire sample.

A ratio mean $r = y/x$ has units $[u_y/u_x]$. Its unit variance is

$$\text{var}(r) = \frac{1}{x^2}\left\{\text{var}(y) + r^2\,\text{var}(x) - 2r\,\text{cov}(y, x)\right\}$$

$$\rightarrow \left[\frac{1}{a^2u_x^2}\left\{au_y^2 + \frac{a^2u_y^2}{a^2u_x^2}au_x^2 - \frac{au_y}{au_x}au_yu_x\right\}\right] = \left[\frac{au_y^2}{a^2u_x^2}\right] = \left[\frac{u_y^2}{au_x^2}\right].$$

Here

$$\text{var}(y) = (1-f)\sum_h^H \frac{a_h}{a_h - 1}\left(\sum_\alpha^{a_h} y_{h\alpha}^2 - \frac{y_h^2}{a_h}\right)$$

$$\rightarrow \left[1\sum_h^H 1\left(a_hu_y^2 - \frac{a_h^2u_y^2}{a_h}\right)\right] = \left[\sum_h^H a_hu_y^2\right] = [au_y^2],$$

with similar expressions for $\text{var}(y) \rightarrow [au_x^2]$ and $\text{cov}(y, x) \rightarrow [au_yu_x]$. Note that the three terms inside the { } have the same units. Frequently the ratio mean denotes the sample mean from unequal clusters, and $x_{h\alpha}$ is the size of the αth sample cluster in the hth stratum. Adopting the notation of equal clusters, we can write $[b]$ for the units of $x_{h\alpha}$ and $[bu]$ for the units of $y_{h\alpha}$. The ratio mean r then has units $[abu/ab] = [u]$, and its variance is

$$\text{var}(r) \rightarrow \left[\frac{1}{a^2b^2}\left\{ab^2u^2 + \frac{a^2b^2u^2}{a^2b^2}ab^2 - \frac{abu}{ab}abub\right\}\right] = \left[\frac{ab^2u^2}{a^2b^2}\right] = \left[\frac{u^2}{a}\right].$$

The coefficient of variation has units $[ua^{-1/2}/u] = [a^{-1/2}]$. If the ratio is a proportion, the elementary unit has $[u] = [1]$ and the variance of the ratio mean has $[a^{-1}]$.

Appendix C
Remarks on Computations

Many statistics books contain sections on computational methods, the reader can best make his own choice among them. I found helpful remarks in Scarborough [1950, Chapter 1], Cochran and Cox [1957; 3.31–3.34] and Snedecor [1956, 5.1–5.6]. I shall merely add a few hints about sampling computations.

1. Finite summation, denoted by Σ, is a basic tool explained in most elementary statistics books; also by Hansen, Hurwitz, and Madow [1953, II, 2.2]. I have covered residual needs with detailed presentations of individual formulas, and merely add a few reminders here. See also Problem 1.14.

(*a*) $\sum_j (y_j + x_j) = \sum_j y_j + \sum_j x_j$. This illustrates the general rule that in a rectangular matrix the order of summation is reversible:

$$\sum_i^m \sum_j^n y_{ij} = \sum_j^n \sum_i^m y_{ij}.$$

(*b*) $\left(\sum_j y_j\right)^2 = \sum_i y_i^2 + \sum_{i \neq j} y_i y_j = \sum_j y_j^2 + 2 \sum_{i<j} y_i y_j$. Hence the square of a sum does not equal the sum of squares. Similarly, for products

$$\left(\sum_j y_j\right)\left(\sum_j x_j\right) = \sum_j y_j x_j + \sum_{i \neq j} y_i x_j.$$

(*c*) $\dfrac{\sum_j y_j}{\sum_j x_j} \neq \sum_j \dfrac{y_j}{x_j}$.

(*d*) $\sum_j^n \left[y_j - \frac{(\sum y_j)}{n}\right]^2 = \sum_j \left[y_j^2 - 2y_j \frac{\sum y_j}{n} + \frac{(\sum y_j)^2}{n^2}\right]$

$$= \sum_j y_j^2 - 2 \sum y_j \frac{\sum y_j}{n} + n \frac{(\sum y_j)^2}{n^2}$$

$$= \sum y_j^2 - \frac{(\sum y_j)^2}{n}.$$

This form occurs in all variance computations. It also illustrates basic summation rules.

(*e*) $\sum_j^n (ay_j + b)^2 = \sum_j (a^2 y_j^2 + 2aby_j + b^2)$

$$= a^2 \sum_j y_j^2 + 2ab \sum_j y_j + nb^2.$$

(*f*) $\sum_j (10^k x_j + y_j)^2 = 10^{2k} \sum_j x_j^2 + 2 \times 10^k \sum_j x_j y_j + \sum_j y_j^2.$

This form facilitates computing, with one operation on a desk machine, the terms needed for variances of ratio means. On the upper dial of the computer $\Sigma\,(10^k x_j + y_j) = 10^k \sum_j x_j + \sum_j y_j$ can be simultaneously cumulated.

When adding or cumulating a long list of numbers, it is wise to write down subtotals at reasonable intervals, say, every 10th or 50th number.

2. When sample variables must be weighted, the weights enter similarly into each sample moment. The weights w_j' can be standardized so that they sum to unity: $w_j = w_j'/\sum_j w_j'$; $\sum_j w_j = 1$. Thus equal weights are merely a special case of $w_j' = 1$ and $w_j = 1/n$. The mean is $\bar{y} = \sum_j w_j y_j$, the sum of squares is $\Sigma\, w_j y_j^2$, and the kth moment is $\sum_j w_j y_j^k = \sum_j w_j' y_j^k / \sum_j w_j'$. Moments of deviations may be computed also as $\sum_j w_j(y_j - \bar{y})^k$. For example, note that $\Sigma\, w_j y_j^2 - \bar{y}^2$. Similarly, $\Sigma\, w_j(y_j - a)^2 = \Sigma\, w_j y_j^2 - a^2$. When grouped observations are used, w_j' denotes the frequency and w_j the relative frequency of the jth group.

3. A few remarks on rounding may help readers unfamiliar with statistical computations.

(*a*) Rounding may be safest and most convenient on raw variables (y_j, x_j). It is seldom worth maintaining these variables with finer precision than a standard deviation (σ_y).

(*b*) Rounding a series of n numbers to the nearest k value tends to add an error of about $k^2/12n$ to the variance of the mean. This occurs because the rectangular distribution of length k has the variance $k^2/12$. Thus rounding 833 values to the nearest 0.01 increases the variance by $0.01^2/10{,}000 = (0.01/100)^2 = (0.0001)^2$.

(*c*) Keep equal *numbers of decimal places* when adding quantities.

(*d*) Keep *equal numbers of significant* places when multiplying or dividing several quantities. Employing equal numbers of decimals for products or quotients is a common mistake. For example 4.3×1114.3 or $4.3/1114.3$ are good only for two significant places.

4. Present the proper number of significant places. For example, 1.6, 1.60, and 1.600 are all distinct numbers: 1.6 means 1.6 ± 0.05; 1.60 means 1.60 ± 0.005; and 1.600 means 1.600 ± 0.0005. Thus $1.60/1.0 = 1.6$; $1.60/1.00 = 1.60$; $1.600/1.00 = 1.60$; $1.600/1.000 = 1.600$. Thus $0.0016/0.0010 = 1.6$, and writing 1.600 would be wrong. Furthermore, $1.600/100.0 = 0.01600$; writing merely 0.016 is a typical mistake, and 0.02 is even worse.

It is convenient to bring large or small numbers closer to a moderate value by using the factors 10^k. For example, $1{,}600{,}000 = 1.6 \times 10^6$ and $0.0000016 = 1.6 \times 10^{-6}$. The factors 10^{-2} for percentages and their standard errors, and 10^{-4} for their variances, are also convenient.

5. For computing quick approximations of square roots a simple useful trick depends on knowing that $\sqrt{1 \pm d} \doteq 1 \pm d/2$ when $d < 0.3$, say, because $(1 \pm d/2)^2 = 1 \pm d + d^2/4$. For example, $\sqrt{1 + 0.34} = 1.158$, whereas $1 + 0.34/2 = 1.170$; $\sqrt{1 - 0.25} = 0.866$, whereas $1 - 0.25/2 = 0.875$. Between 0.75 and 1.34 the approximation is better than 1 percent, and much better near 1, of course.

The trick can also be extended to a number far from 1, by finding a nearby square of an integer. For example, $\sqrt{0.656} = \sqrt{0.64 + 0.016} = 0.8\sqrt{1 + 16/640} = 0.8\sqrt{1 + 1/40} \doteq 0.8(1 + 1/80) = 0.810$; actually $\sqrt{0.656} = 0.80994$.

We can compute pq quickly as $pq = 0.25 - \delta^2$, where $\delta = |p - 0.5| = |q - 0.5|$. Thus $0.2 \times 0.8 = 0.16 = 0.25 - 0.3^2$. Furthermore, $\sqrt{pq} = \sqrt{0.25 - \delta^2} = 0.5\sqrt{1 - 4\delta^2} \doteq 0.5(1 - 2\delta^2) = 0.5 - \delta^2$; this approximation is within 1 percent when $\delta < 0.25$, hence, for p between 1/4 and 3/4. Thus for $\sqrt{1/4 \cdot 3/4} = 0.4330$, whereas $0.5 - 0.25^2 = 0.5 - 0.0625 = 0.4375$.

6. Helpful shortcuts and checks for standard errors can be based on the sample range: the highest minus the lowest value in a random sample

of size n. These techniques depend on the stability of the ratio R/σ of the range to standard deviation in samples from normal populations. For rectangular distributions the ratio is less, and for skewed distributions it is greater than for the normal. However, the ratio is rather stable, not greatly affected by moderate departures from normality. Hence, it affords fairly reliable grounds for approximate values. Some easy rules-of-thumb depend on the coincidence that the ratio R/σ is close to the value $\sqrt{n}$, especially for n between 3 and 13.

The estimated standard error $\hat{\sigma}/\sqrt{n}$ of the mean of n random normal variables may be approximated by range$/(R/\sigma)\sqrt{n}$. When $R/\sigma \doteq \sqrt{n}$, merely dividing the range by n can be used for the approximation. The standard error $\sqrt{n}\hat{\sigma}$ of the sum of n random variables may be approxi-

n	$\sqrt{n}$	R/σ	$R/\sigma\sqrt{n}$	n	$\sqrt{n}$	R/σ	$R/\sigma\sqrt{n}$
2	1.41	1.13	0.80	21	4.58	3.78	0.82
3	1.73	1.69	0.98	22	4.69	3.82	0.81
4	2.00	2.06	1.03	23	4.80	3.86	0.80
5	2.24	2.33	1.04	24	4.90	3.90	0.80
6	2.45	2.53	1.03	25	5.00	3.93	0.79
7	2.65	2.70	1.02	26	5.10	3.96	0.78
8	2.83	2.85	1.01	27	5.20	4.00	0.77
9	3.00	2.97	0.99	28	5.29	4.03	0.76
10	3.16	3.08	0.97	29	5.39	4.06	0.75
11	3.32	3.17	0.96	30	5.48	4.09	0.75
12	3.46	3.26	0.94	35	5.92	4.21	0.71
13	3.61	3.34	0.93	40	6.32	4.32	0.68
14	3.74	3.41	0.91	45	6.71	4.42	0.66
15	3.87	3.47	0.90	50	7.07	4.50	0.64
16	4.00	3.53	0.88	60	7.75	4.64	0.60
17	4.12	3.59	0.87	70	8.37	4.75	0.57
18	4.24	3.64	0.86	80	8.94	4.85	0.54
19	4.36	3.69	0.85	90	9.49	4.94	0.52
20	4.47	3.74	0.84	100	10.00	5.02	0.50

TABLE C.I Mean Ratios R/σ of Range to Standard Deviation in Random Samples of Size n for Normal Distributions [Tippett, 1925]

Note that R/σ is close to $\sqrt{n}$ for $n = 3$ through 13. The ratio $R/\sigma\sqrt{n}$ is well approximated by $(1 - \sqrt{n}/30)$ near $n = 20$; by $(1 - \sqrt{n}/25)$ near $n = 25$; and by $(1 - \sqrt{n}/20)$ for n from 30 to 100.

mated with range $\sqrt{n}/(R/\sigma)$; when $R/\sigma \doteq \sqrt{n}$, the sample range alone can serve for an approximation. For larger n the values of R/σ diverge from $\sqrt{n}$; the approximation can be improved by dividing the range by $R/\sigma\sqrt{n}$ for the sum, and by $\sqrt{n}R/\sigma$ for the mean.

Sample ranges are particularly helpful for quick rough checks of the sum of squares. We often need the value $\Sigma\,(y_j - \bar{y})^2$, which estimates $(n - 1)\sigma^2$. Hence $\sqrt{n - 1}$ range$/(R/\sigma)$ can be checked against $\sqrt{\Sigma\,(y_j - \bar{y})^2}$. For a rapid check against $\sqrt{\Sigma\,(y_j - \bar{y})^2}$ use the following factors for various sample sizes.

(*a*) From 3 to 15, use the range.
(*b*) Near $n = 20$, divide the range by about $(1 - \sqrt{n}/30)$.
(*c*) Near $n = 25$, divide the range by about $(1 - \sqrt{n}/25)$.
(*d*) From $n = 30$ to $n = 100$, divide the range by about $(1 - \sqrt{n}/20)$.

I think these checks are more valuable than actually using the range for estimating the standard error. The efficiency of the range is close to 100 percent only for small n. It becomes 85 percent for $n = 10$ and 70 percent for $n = 20$. Normal variables are assumed for these efficiencies. With good computers available not many researchers would care to sacrifice that much efficiency.

Appendix D
Tables of Random Numbers

Ninety-fifth Thousand

	1–4	*5–8*	*9–12*	*13–16*	*17–20*	*21–24*	*25–28*	*29–32*	*33–36*	*37–40*
1	77 66	88 40	86 61	96 70	78 75	29 77	21 94	12 37	66 11	53 42
2	74 81	53 71	16 61	59 13	33 02	25 95	92 37	03 18	46 26	37 86
3	05 88	20 12	10 45	80 22	38 70	94 11	22 02	08 37	74 87	49 04
4	05 79	76 95	69 00	48 70	60 14	53 11	06 57	06 26	60 31	06 74
5	79 98	70 98	97 94	55 99	44 04	75 89	69 50	64 03	96 68	17 89
6	55 09	79 15	11 56	65 88	08 16	96 95	33 17	60 45	81 31	50 46
7	79 19	16 49	99 08	80 01	56 35	41 42	72 58	20 39	33 53	85 26
8	28 70	12 06	71 02	34 50	30 16	83 58	39 98	84 01	27 85	17 35
9	54 44	53 59	34 44	49 93	61 75	19 87	34 93	85 16	18 79	65 94
10	93 69	31 43	93 93	77 39	72 40	66 32	90 86	65 88	41 19	36 86
11	24 94	65 41	64 64	95 13	46 97	43 12	86 02	79 50	67 90	14 19
12	04 07	67 01	59 03	27 37	83 20	17 82	11 80	46 08	32 68	60 26
13	67 24	63 38	76 53	29 14	02 47	70 31	20 88	24 31	14 65	23 35
14	69 06	90 51	48 94	89 77	41 66	54 60	66 95	46 73	76 59	20 05
15	66 56	20 91	61 48	91 73	98 80	96 94	45 09	93 21	90 40	03 01
16	36 48	02 01	88 94	20 08	07 64	08 84	26 41	25 54	43 65	82 24
17	62 93	85 57	12 06	07 88	22 37	03 84	80 69	93 29	22 34	67 88
18	94 01	05 57	71 98	47 26	58 99	72 11	69 93	22 46	72 52	75 62
19	52 94	18 97	82 49	76 84	86 83	05 27	53 27	16 40	94 34	81 86
20	27 43	78 39	71 17	16 72	43 37	60 73	83 41	31 32	61 05	37 89
21	46 00	19 71	63 06	75 27	01 57	59 61	86 70	33 35	54 77	81 38
22	29 58	01 44	39 62	83 16	97 46	31 27	27 43	67 66	35 08	86 34
23	19 31	80 79	63 47	80 56	00 71	06 17	49 70	26 75	55 43	46 84
24	02 52	31 23	74 12	16 62	21 19	76 63	33 43	17 16	96 00	42 50
25	06 00	13 63	57 37	51 83	45 58	21 01	02 89	88 07	74 32	21 87

Ninety-sixth Thousand

	1–4	*5–8*	*9–12*	*13–16*	*17–20*	*21–24*	*25–28*	*29–32*	*33–36*	*37–40*
1	71 84	75 11	67 59	58 68	58 82	31 86	05 72	67 80	07 17	27 77
2	53 03	17 77	77 20	33 26	17 76	34 97	27 38	98 29	48 87	94 10
3	46 94	37 49	80 90	79 67	68 11	05 05	46 48	80 41	97 57	61 85
4	84 19	12 26	67 68	28 64	35 48	32 54	83 89	59 06	26 64	48 31
5	71 48	58 93	09 06	11 80	17 38	48 55	84 43	19 15	72 49	29 35
6	89 50	27 14	20 08	84 94	10 97	46 38	63 23	86 62	43 32	15 52
7	79 31	14 76	36 38	41 19	19 30	55 46	46 86	50 07	10 26	66 96
8	50 50	49 02	77 68	59 39	25 70	57 03	60 62	67 20	55 65	87 94
9	24 56	90 38	34 84	87 09	25 90	40 33	84 77	06 57	78 75	06 00
10	21 16	52 91	93 82	81 36	45 27	79 55	42 23	61 78	70 26	04 20
11	01 93	80 67	91 22	77 35	12 45	28 06	03 33	82 67	15 04	42 44
12	38 38	27 05	94 29	39 24	92 73	12 94	97 10	15 80	40 41	05 20
13	90 87	61 03	96 35	90 27	11 97	36 79	91 98	40 46	18 03	71 59
14	48 49	85 86	63 34	08 92	37 83	86 68	08 96	38 08	26 83	78 69
15	45 03	39 55	51 37	89 28	46 68	47 22	07 01	50 00	05 36	78 13
16	14 71	66 70	37 56	61 38	55 05	23 47	94 51	85 65	92 49	87 31
17	02 10	51 75	02 42	44 84	51 18	18 07	19 96	95 51	62 77	18 73
18	38 93	08 89	78 98	77 29	55 49	55 55	22 51	42 53	26 64	83 23
19	17 56	97 82	02 37	27 53	67 99	92 67	34 63	88 67	84 75	22 70
20	30 95	82 49	04 20	08 91	11 46	62 60	96 57	24 75	41 58	43 25
21	96 16	76 52	88 95	49 13	21 82	85 84	19 01	03 64	74 91	50 92
22	01 22	04 38	45 59	91 92	53 20	86 75	18 12	30 15	44 28	22 73
23	44 11	38 22	82 31	01 46	05 89	36 44	14 07	25 80	80 04	06 77
24	26 87	15 33	90 55	71 13	93 31	07 30	21 59	71 41	77 03	47 04
25	49 10	33 76	70 24	35 33	19 69	41 17	60 48	78 72	21 23	44 24

(*Source:* Kendall, M. G., and Babington Smith, *Tables of Random Sampling Numbers*, Tracts for Computers No. 27, Cambridge University Press, 1954.)

Appendix E
List of a Population of 270 Blocks

This is a list of the 270 blocks in Ward 1 of Fall River, Massachusetts from its volume of Block Statistics of the 1950 U.S. Census. The blocks are arranged in 27 groups of 10 blocks each. The block numbers (i) range from 232 to 772; the 271 missing numbers among the 541 block numbers denote blocks that were either located in other wards or contained no dwellings. This illustrates irregularities that occur on many lists. One may select either directly from this list or from a compact list of integers from 1 to 270.

The X_i represent the number of dwellings, and they total $\Sigma X_i = 6786$. The Y_i denote dwellings occupied by renters, and they total $\Sigma Y_i = 4559$; the blank values mean $Y_i = 0$. When the total dwellings X_i are either 1 or 2, the Census always reports blanks for Y_i; there is a large number of these among the last 50 blocks. There should be 14 renter occupied dwellings among these, since the Census reported a total of 4573.

Further use of this population may be aided by knowing that $\Sigma X_i^2 = 338{,}694$, $\Sigma Y_i^2 = 192{,}555$, and $2 \Sigma X_i Y_i = 499{,}568$. Also $\bar{X} = 25.13$, $\bar{Y} = 16.88$, $\sigma_x^2 = 622.75$, $\sigma_x = 24.94$, $\sigma_y^2 = 428.07$, $\sigma_y = 20.69$, $\sigma_{yx} = 500.76$, and $R_{yx} = 0.971$.

The values of the cumulated number of dwellings are added in the columns marked "Cum X_i" to help students work out problems involving selections with probabilities proportional to these measures of size.

i	Y_i	X_i	Cum X_i	i	Y_i	X_i	Cum X_i	i	Y_i	X_i	Cum X_i
232	131	149	149	396	30	43	2,123	458	2	5	3,836
233	6	10	159	397	41	53	2,176	459	5	11	3,847
240	23	30	189	398	37	47	2,223	460	6	14	3,861
242	79	90	279	399	12	18	2,241	461	13	29	3,890
243	47	56	335	400	25	41	2,282	462	28	42	3,932
244	34	42	377	401	23	33	2,315	463	27	36	3,968
245	97	113	490	402	26	39	2,354	464	22	30	3,998
246	30	45	535	403	50	67	2,421	465	4	11	4,009
247	11	16	551	404	31	41	2,462	466	10	19	4,028
248	45	67	618	405	24	35	2,497	467	25	42	4,070
249	19	23	641	406	47	53	2,550	468	67	110	4,180
250	17	18	659	407	27	28	2,578	469	44	57	4,237
278	25	33	692	408	80	90	2,668	470	43	81	4,318
279	84	89	781	420	52	68	2,736	471	15	23	4,341
280	91	114	895	421	90	99	2,835	472	17	25	4,366
281	48	66	961	422	78	89	2,924	473	29	59	4,425
282	48	61	1,022	423	46	48	2,972	474	18	27	4,452
283	20	25	1,047	424	35	48	3,020	475	14	22	4,474
284	34	46	1,093	425	59	62	3,082	476	24	29	4,503
285	42	58	1,151	426	27	33	3,115	477	35	44	4,547
286	35	44	1,195	427	33	43	3,158	478	48	53	4,600
287	55	66	1,261	428	27	37	3,195	479	20	27	4,627
288	42	61	1,322	429	9	14	3,209	480	24	28	4,655
289	36	45	1,367	430	9	15	3,224	481	55	62	4,717
290	13	20	1,387	431	12	21	3,245	521	43	56	4,773
291	7	16	1,403	432	49	68	3,313	522	13	22	4,795
292	8	15	1,418	433	60	81	3,394	523	19	22	4,817
293	18	26	1,444	434	35	59	3,453	524	48	57	4,874
300	20	22	1,466	435	11	23	3,476	525	44	57	4,931
374	18	22	1,488	436	21	32	3,508	526	36	46	4,977
375		2	1,490	437	22	36	3,544	527	3	8	4,985
376	23	29	1,519	438	10	16	3,560	528	2	4	4,989
377		3	1,522	439	9	15	3,575	529	13	18	5,007
378	19	29	1,551	440	7	16	3,591	530	34	42	5,049
379	11	21	1,572	441	3	8	3,599	531	28	32	5,081
380	11	15	1,587	442	5	25	3,624	532	23	28	5,109
381	42	54	1,641	443	2	11	3,635	534	8	14	5,123
382	28	42	1,683	444	8	9	3,644	535	69	76	5,199
383	8	13	1,696	445	14	19	3,663	536		19	5,218
384		2	1,698	446	5	5	3,668	537	5	9	5,227
385	34	48	1,746	447	1	3	3,671	538	6	37	5,264
386	13	24	1,770	448	22	37	3,708	539	4	11	5,275
388	16	27	1,797	449	25	30	3,738	540	9	24	5,299
389	21	32	1,829	450	2	3	3,741	541	54	102	5,401
390	12	14	1,843	451		4	3,745	542	50	82	5,483
391	10	18	1,861	452	7	13	3,758	543	9	24	5,507
392	50	61	1,922	453	15	24	3,782	544	6	18	5,525
393	58	65	1,987	455	10	19	3,801	545	5	18	5,543
394	17	25	2,012	456	5	17	3,818	546	1	3	5,546
395	41	68	2,080	457	8	13	3,831	547		6	5,552

i	Y_i	X_i	Cum X_i	i	Y_i	X_i	Cum X_i	i	Y_i	X_i	Cum X_i
548	3	8	5,560	672	8	18	6,128	717		1	6,615
549	4	12	5,572	673		1	6,129	718	2	7	6,622
550	18	27	5,599	677	4	10	6,139	719	2	8	6,630
551	1	3	5,602	678	1	4	6,143	720	3	12	6,642
552	1	3	5,605	679	3	9	6,152	721		4	6,646
553	3	6	5,611	680		5	6,157	722	6	8	6,654
554	6	14	5,625	681	14	20	6,177	723	3	9	6,663
555	5	15	5,640	682	3	5	6,182	724	3	7	6,670
556	5	14	5,654	683	5	13	6,195	725	5	12	6,682
557	4	9	5,663	684		1	6,196	726	3	10	6,692
558		1	5,664	685	11	23	6,219	727		1	6,693
559		4	5,668	686	19	39	6,258	728		1	6,694
560	7	12	5,680	687	5	9	6,267	730		1	6,695
561	7	22	5,702	688		2	6,269	731	2	4	6,699
562	3	11	5,713	689	3	5	6,274	732		1	6,700
563	12	27	5,740	690	12	26	6,300	734		1	6,701
564	11	20	5,760	691	4	10	6,310	735		2	6,703
565	27	38	5,798	692	14	35	6,345	736		1	6,704
566	14	31	5,829	693		4	6,349	737		1	6,705
567	2	4	5,833	694	20	38	6,387	741		1	6,706
569	3	23	5,856	695	8	10	6,397	742		1	6,707
570	6	12	5,868	696	22	36	6,433	743		2	6,709
571	5	15	5,883	697	2	7	6,440	744		2	6,711
572	31	39	5,922	698	2	4	6,444	745		4	6,715
573	7	9	5,931	699		2	6,446	746		2	6,717
575	3	11	5,942	700	1	3	6,449	747		2	6,719
576	1	10	5,952	701	17	29	6,478	752		4	6,723
577	3	8	5,960	702	24	44	6,522	753		1	6,724
578	1	3	5,963	703	3	5	6,527	754		1	6,725
579	6	10	5,973	704	7	17	6,544	755	3	5	6,730
580	4	8	5,981	705		2	6,546	756	4	9	6,739
581	5	12	5,993	706	18	42	6,588	757		1	6,740
582	1	3	5,996	708	2	4	6,592	758		2	6,742
665		1	5,997	709		1	6,593	766		2	6,744
666		3	6,000	710		1	6,594	767	2	6	6,750
667	8	22	6,022	712		3	6,597	768	1	4	6,754
668	8	18	6,040	713		1	6,598	769	3	7	6,761
669	16	25	6,065	714	5	14	6,612	770	11	14	6,775
670	9	25	6,090	715		1	6,613	771	3	9	6,784
671	6	20	6,110	716		1	6,614	772		2	6,786

References

The best and most complete bibliography in sampling is now by M. N. Murthy *et al.* [1962], *Bibliography on Sampling Theory and Methods*, Calcutta: Indian Statistical Institute. The list below does not represent a comprehensive bibliography of sampling, but a list of books and articles to which I made one or more references. Many important contributions are not here because they have been thoroughly incorporated in sampling textbooks, which serve as ready references. The names of four journals are represented by their initials: *JASA* for the *Journal of the American Statistical Association; AMS* for the *Annals of Mathematical Statistics; BISI* for the *Bulletin of the International Statistical Institute;* and *JRSS* for the *Journal of the Royal Statistical Society*, with series (*A*) and (*B*) so distinguished.

Aitchison, J., and Brown, J. A. C. [1957], *The Lognormal Distribution*, Cambridge: Cambridge University Press.

Anderson, R. L., and Bancroft, T. A. [1952], *Statistical Theory in Research*, New York: McGraw-Hill Book Co.

Azorin, F. P. [1962], *Curso de Muestreo y Aplicaciones*, Madrid: Instituto Nacional de Estadistica.

Bergsten, J. W. [1958], "A nationwide sample of girls," *Journal of Experimental Education*, **50,** 197–208.

Birnbaum, A. [1962], "On the foundation of statistical inference," *JASA*, **57,** 269–306.

Birnbaum, Z. W., and Sirken, M. G. [1950], "On the total error due to random sampling," *International Journal of Opinion Attitude Research*, **4,** 179–191.

Blalock, H. M. [1960], *Social Statistics*, New York: McGraw-Hill Book Co.

Brillinger, D. R. [1964], "The asymptotic behavior of Tukey's general method of setting confidence limits (the Jackknife) when applied to maximum likelihood estimates," *Review of International Statistical Institute*, **32,** 202–206.

Bryant, E. C., Hartley, H. O., and Jessen, R. J. [1960], "Design and estimation in two-way stratification," *JASA*, **55,** 105–124.

Campbell, D. T. [1957], "Factors relevant to the validity of experiments in social settings," *Psychological Bulletin*, **54,** 297–312.

Chapman, D. G. [1954], "The estimation of biological populations," *AMS*, **25,** 1–15.

Coale, A. J. [1955], "The population of the United States in 1950 classified by age, sex, and color—a revision of census figures," *JASA*, **50,** 16–54.

Cochran, W. G. [1946], "Relative accuracy of systematic and stratified random samples for a certain class of populations," *AMS*, **17,** 164–177.

Cochran, W. G. [1950], "The comparison of percentages in matched samples," *Biometrika*, **37,** 256–266.

Cochran, W. G. [1954], "The combination of estimates from different experiments," *Biometrics*, **10,** 101–129.

Cochran W. G. [1961], "Comparison of methods for determining stratum boundaries," *Bulletin of the International Statistical Institute* (37.II), 345–58.

Cochran, W. G. [1963], *Sampling Techniques*, 2nd ed., New York: John Wiley and Sons.

Cochran, W. G., and Cox, G. M. [1957], *Experimental Designs*, 2nd ed., New York: John Wiley and Sons.

Cochran, W. G., Mosteller, F., and Tukey, J. W. [1953], "Statistical problems of the Kinsey Report," *JASA*, **48,** 673–716.

Cochran, W. G., Mosteller, F., and Tukey, J. W. [1954], "Principles of sampling," *JASA*, **49,** 1–12.

Corlett, T. [1963], "Rapid methods of estimating standard errors of stratified multi-stage samples: a preliminary investigation," *The Statistician*, **13,** 5–16.

Cornfield, J. [1954], "Statistical relationships and proof in medicine," *American Statistician*, **8,** 19–21.

Cornfield, J., and Tukey, J. W. [1956], "Average values of mean squares in factorials," *AMS*, **27,** 907–949.

Cox, D.R. [1958], "Some problems connected with statistical inference," *AMS*, **29,** 357–372.

Craig, C. C. [1953], "On the utilization of marked specimens in estimating populations of flying insects," *Biometrika*, **40,** 170–176.

Cramér, H. [1946], *Mathematical Methods of Statistics*, Princeton, N.J.: Princeton University Press.

Darroch, J. N. [1958], "The multiple recapture census I., Estimation of a closed population," *Biometrika*, **45,** 343–359; also [1959], **46,** 336–351.

Dalenius, T. [1957], *Sampling in Sweden*, Stockholm: Almquist and Wicksell.

Dalenius, T., and Gurney, M. [1951], "The problem of optimum stratification, II," *Skandinavisk Aktuarietidskrift*, **34,** 133–148.

Deming, W. E. [1943], *Statistical Adjustment of Data*, New York: John Wiley and Sons.

Deming, W. E. [1950], *Some Theory of Sampling*, New York: John Wiley and Sons.

Deming, W. E. [1953], "On a probability mechanism to attain an economic balance between the resultant error of response and the bias of nonresponse," *JASA*, **48,** 743–772.

Deming, W. E. [1954], "On the contributions of standards of sampling to legal evidence and accounting," *Current Business Studies* (New York University), **19,** 14–32.

Deming, W. E. [1960], *Sample Design in Business Research*, New York: John Wiley and Sons.

Deming, W. E. [1963], Chapter 1 in Rainer J. D. et al, *Family and Mental Health Problems in a Deaf Population*, Department of Medical Genetics, Columbia University, N.Y.

Deming, W. E., and Glasser, G. J. [1959], "On the problem of matching lists by samples," *JASA*, **54,** 403–415.

Dorfman, R. [1943], "The detection of defective numbers," *AMS*, **14,** 436–439.

Durbin, J. [1958], "Some results in sampling theory when the units are selected with unequal probabilities," *JRSS(B)*, **15,** 262–269.

Durbin, J. [1954], "Nonresponse and callbacks in surveys," *BISI*, **34/2,** 72–86.

Durbin, J. [1959], "A note on the application of Quenouille's method of bias reduction to the estimation of ratios," *Biometrika*, **46,** 477–480.

Durbin, J., and Stuart, A. [1958], "An experimental comparison between coders," *Journal of Marketing*, **19,** 54–66.

Durbin, J., and Stuart, A. [1954], "Call-backs and clustering in sample surveys," *JRSS(A)*, **117**, 387–428.

Dwyer, P. S. [1951], *Linear Computations*, New York: John Wiley and Sons.

Eberhardt, L. and Murray, R. M. [1960], "Estimating the kill of game by licensed hunters," *Proceedings of the Social Statistics Section of the ASA*, 182–188.

Ecimovic, J. P. [1956], "Three-stage sampling with varying probabilities of selection," *Indian Journal of Agricultural Statistics*, **8**, 14–44.

Eckler, A. R. [1953], "Extent and character of errors in the 1950 census," *The American Statistician*, **7**, 5, 15–21.

Eckler, A. R. [1955], "Rotation sampling," *AMS*, **26**, 664–685.

Edwards, F. [1953], "Aspects of random sampling for a commercial survey," *The Incorporated Statistician*, **4**, 9ff.

Edwards, W., Lindman, H., and Savage, L. J. [1963], "Bayesian statistical inference for psychological research, *Psychological Review*, **70**, 193–242.

Eisenhart, C. [1963], "The background and evolution of the method of least squares," *Int. Stat. Inst.*, 34th Session, Ottawa.

El-Badry, M. A., and Stephan, F. F. [1955], "On adjusting sample tabulations to census counts," *JASA*, **50**, 738–762.

Erdos, P., and Renyi, A. [1959], "On the central limit theorem for samples from a finite population," *Pub. Math. Inst.*, Hungarian Academy of Sciences, **4**, 49–57.

Ericson, W. A. [1963], "Optimum Stratified Sampling Using Prior Information." Unpublished Ph.D. Thesis, Harvard University.

Ericson, W. A. [1965], "Optimum sampling using prior information," *JASA* (in press).

Fellegi, I. P. [1963], "Sampling with varying probabilities without replacement," *JASA*, **58**, 183–201.

Fieller, E.C.[1940],"The biological standardization of insulin,"*JRSS* Supplement **7**, 1–64.

Fisher, R. A. [1953], *The Design of Experiments*, 6th ed., London: Oliver and Boyd.

Frankel, L. R., and Stock, J. S. [1942], "On the sample survey of unemployment," *JASA*, **10**, 288–293.

Freedman, R., Whelpton, P. K., and Campbell, A. A. [1959], *Family Planning, Sterility and Population Growth*, New York: McGraw-Hill Book Co.

Gales, K., and Kendall, M. G., "An inquiry concerning interviewer variability," *JRSS(A)*, **120**, 121–147.

Gibson, W. M., and Jowett, G. H. [1957], "Three-groups' regression analysis," *Applied Statistics*, **6**, 114-130 and 189–197.

Goodman, L. A. [1949], "On the estimation of the number of classes in a population," *AMS*, **20**, 572–579.

Goodman, L. A. [1952], "On the analysis of samples from *k* lists," *AMS*, **23**, 632-634.

Goodman, L. A. [1953], "Sequential sampling tagging for population size problems," *AMS*, **24**, 56–69.

Goodman, L. A. [1959], "Some alternatives to ecological correlation," *American Journal of Sociology*, **64**, 610–625.

Goodman, L. A. [1961], "Snowball sampling," *AMS*, **32**, 148–170.

Goodman, R. [1960], "Survey sampling and implementation for development programs," *Proceedings of the Social Statistics Section of the ASA*, 2–4.

Goodman, R., and Kish, L. [1950], "Controlled selection—a technique in probability sampling," *JASA*, **45**, 350–372.

Gray, P. G. [1956], "Examples of interviewer variability taken from two sample surveys," *Applied Statistics*, **5**, 73–85.

Gray, P. G. [1957], "A sample survey with both a postal and an interview stage," *Applied Statistics*, **6**, 139–153.

References

Gray, P. G., and Corlett T. [1950], "Sampling for the social survey," *JRSS(A)*, **113,** 150–206.

Gray, P. G., and Parr, E. A. [1959], "The length of cigarette stubs," *Applied Statistics*, **8,** 92–103.

Hagood, M. J., and Bernert, E. H. [1945], "Component indexes as a basis for stratification in sampling," *JASA*, **40,** 330–341.

Hájek, J. [1958], "On the theory of ratio estimates," *BISI*, **37/2,** 219–226.

Hájek, J. [1958], "Some contributions to the theory of probability sampling," *BISI*, **36/3,** 127–134.

Hansen, M. H., and Hurwitz, W. N. [1946], "The problem of nonresponse in sample surveys," *JASA*, **41,** 517–529.

Hansen, M. H., Hurwitz, W. N., and Bershad, M. [1961], "Measurement errors in censuses and surveys," *BISI*, **38/2,** 359–374.

Hansen, M. H., Hurwitz, W. N., and Madow, W. G. [1953], *Sample Survey Methods and Theory*, New York: John Wiley and Sons, Vols. I and II. When volume is not indicated, refer to Volume I.

Hansen, M. H., and Steinberg, J. [1956], "Control of error in surveys," *Biometrics*, **12,** 462–474.

Hanson, R. H., and Marks, E. S. [1958], "Influence of the interviewer on the accuracy of survey results," *JASA*, **53,** 635–655.

Hartley, H. O. [1962], "Multiple frame surveys," *Proceedings of the Social Statistics Section, American Statistical Association*, 203–206.

Hendricks, W. A. [1949], "Adjustment for bias by non-response in mailed surveys," *Agricultural Economics Research*, **1,** 52 ff.

Hendricks, W. A. [1956], *The Mathematical Theory of Sampling*, New Brunswick, N.J.: Scarecrow Press.

Hess, I., Riedel, D. C., and Fitzpatrick, T. B. [1961], *Probability Sampling of Hospitals and Patients*, Ann Arbor, Mich.: Bureau of Hospital Administration.

Hildreth, C. [1963], "Bayesian statisticians and remote clients," *Econometrica*, **31,** 422–438.

Hochstim, J. R. [1962], "Comparison of three information-gathering strategies in a population study of sociomedical variables," *Proceedings of the Social Statistics Section, American Statistical Association*, 154–159.

Horvitz, D. G., and Thompson, D. J. [1952], "A generalization of sampling without replacement from a finite universe," *JASA*, **47,** 663–685.

Houseman, E. E. [1953], "Statistical treatment of the non-response problem," *Agricultural Economics Research*, **5,** 12–18.

Hyman, H. H., Cobb, W. J., Feldman, J. J., Hart, C. W., and Stember, C. H. [1954], *Interviewing in Social Research*, Chicago: The University of Chicago Press.

Jeffries, H. R. [1962], *Theories of Probability*, 3rd ed., London: Oxford University Press.

Jensen, A. [1926], "The representative method in practice," *BISI*, **22,** 381–439.

Jessen, R. J. [1942], "Statistical Investigation of a Sample Survey for Obtaining Farm Facts," *Bull.* 304, Agricultural Experiment Station, Ames, Iowa: Iowa State College.

Jessen, R. J., Blythe, R. H., Kempthorne, O., and Deming, W. E. [1947], "On a population sample of Greece," *JASA*, **42,** 357–384.

Jessen, R. J., and Houseman, E. E. [1944], *Statistical Investigations of Farm Sample Surveys Taken in Iowa, Florida, and California*, Agricultural Experiment Station, Bulletin 329, Ames, Iowa: Iowa State College.

Jones, H. L. [1956], "Investigating the properties of a sample mean by employing random subsample means," *JASA*, **51,** 54–83.

Jones, H. L. [1965], "The jackknife method," *Proceedings of the IBM Scientific Computing Symposium on Statistics*, 1964 (in press).

Jowett, G. H. [1955], "Least squares regression analysis for time series," *JRSS(B)*, **17,** 91ff.

Kahn, R. L., and Cannell, C. F. [1957], *The Dynamics of Interviewing: Theory, Techniques, and Cases*, New York: John Wiley and Sons.

Katona, G., Kish, L., Lansing, J. B., and Dent, J. [1950], "Methods of the survey of consumer finances," *Federal Reserve Bulletin*, **36,** 795–809.

Katona, G., Lininger, C. A., and Mueller, E. [1964], 1963 *Survey of Consumer Finances*, Ann Arbor, Mich.: Institute for Social Research.

Katona, G., and Mueller, E. [1956], *Consumer Expectations* 1953–1956, Ann Arbor, Mich.: Institute for Social Research.

Katz, D. [1949], "An analysis of the 1948 polling predictions," *Journal of Applied Psychology*, **33,** 15–28.

Kempthorne, O. [1952], *The Design and Analysis of Experiments*, New York: John Wiley and Sons.

Kempthorne, O. *et al.* [1954], *Statistics and Mathematics in Biology*, Ames, Iowa: Iowa State College Press.

Kendall, M. G. [1951], "Regression, structure, and functional relationship," *Biometrika*, **38,** 12–25. Also [1952], *Biometrika*, **39,** 96–108.

Kendall, M. G., and Babington Smith [1954], *Tables of Random Sampling Numbers*, Cambridge: Cambridge University Press.

Kendall, M. G., and Buckland, W. R. [1957], *A Dictionary of Statistical Terms*, London: Oliver and Boyd.

Kendall, M. G., and Stuart, A. [1958], *The Advanced Theory of Statistics*, Vols. I and II, London: Griffin and Co.

Keyfitz, N. [1951], "Sampling with probability proportional to size; adjustment for changes in probabilities," *JASA*, **46,** 105–109.

Keyfitz, N. [1953], "A factorial arrangement of comparisons of family size," *American Journal of Sociology*, **53,** 470.

Keyfitz, N. [1957], "Estimates of sampling variance when two units are selected from each stratum," *JASA*, **52,** 503–10.

Keyfitz, N. [1960], "The design of surveys to provide experimental contrasts," *BISI*, **38,** 227–230.

King, A. J. and Jessen, R. J. [1945], "The master sample of agriculture," *JASA*, **40,** 38–56.

Kish, L. [1949], "A procedure for objective respondent selection within the household," *JASA*, **44,** 380–387.

Kish, L. [1953], "*Selection of the sample*," Chapter 5 in Festinger and Katz (eds.), *Research Methods in the Behavioral Sciences*, New York: Dryden Press, 175–240.

Kish, L. [1957], "Confidence intervals for clustered samples," *American Sociological Review*, **22,** 154–165.

Kish, L. [1959], "Some statistical problems in research design," *American Sociological Review*, **24,** 328–338.

Kish, L. [1961a], "Efficient allocation of a multi-purpose sample," *Econometrica*, **29,** 363–385.

Kish, L. [1961b], "A measurement of homogeneity in areal units," *BISI*, **33/4,** 201–209.

Kish, L. [1962a], "Studies of interviewer variance for attitudinal variables," *JASA*, **57,** 92–115.

Kish, L. [1962b], "Variances for indexes from complex samples," *Proceedings of the Social Statistics Section, American Statistical Association*, 190–199.

Kish, L. [1963], "Changing strata and selection probabilities," *Proceedings of the Social Statistics Section, American Statistical Association*, 124–131.

Kish, L. [1964], "Generalizations for complex probability sampling," *Proceedings of the Social Statistics Section, American Statistical Association.*

Kish, L., and Hess, I. [1958], "On noncoverage of sample dwellings," *JASA*, **53**, 509–524.

Kish, L., and Hess, I. [1959a], "On variances of ratios and their differences in multi-stage samples," *JASA*, **54**, 416–446.

Kish, L., and Hess, I. [1959b], "A 'replacement' procedure for reducing the bias of nonresponse," *The American Statistician*, **13**, 4, 17–19.

Kish, L., and Hess, I. [1959c], "Some sampling techniques for continuing survey operations," *Proceedings of the Social Statistics Section, American Statistical Association*, 139–143.

Kish, L., and Lansing, J. B. [1954], "Response error in estimating the value of homes", *JASA*, **49**, 520–538.

Kish, L., and Lansing, J. B. [1954], "Response error in estimating the value of homes", *JASA*, **22**, 154–165.

Kish, L., Lovejoy, W., and Rackow, P. [1961], "A multi-stage probability sample for continuous traffic surveys," *Proceedings of the Social Statistics Section, American Statistical Association*, 227–230.

Kish, L., Namboodiri, N. K., and Pillai, R. K. [1962], "The ratio bias in surveys," *JASA*, **57**, 863–876.

Klein, L. R. (Ed.) [1954], *Contributions of Survey Methods to Economics*, New York: Columbia University Press.

Lahiri, D. B. [1954], "A method of sample selection providing unbiased ratio estimates," *BISI*, **33/2**, 133–140.

Lahiri, D. B. [1958], "Observations on the use of interpenetrating samples in India," *BISI*, **36/3**, 144–152.

Lansing, J. B., and Kish, L. [1957], "Family life cycle as an independent variable," *American Sociological Review*, **22**, 512–519.

Madow, L. H. [1950], "Systematic sampling and its relation to other sampling designs," *JASA*, **45**, 30–47.

Madow, W. G. [1948], "On the limiting distributions of estimates based on samples from finite universes," *AMS*, **19**, 535–545.

Madow, W. G., and Madow, L. H. [1944], "On the theory of systematic sampling, I," *AMS*, **15**, 1-24. Also see II, **20** [1949], 333–354, and III, **24** [1953].

Mahalanobis, P. C. [1944], "On large-scale sample surveys," *Phil. Trans. Roy. Soc.*, **B231**, 329–451.

Mahalanobis, P. C. [1946], "Recent experiments in statistical sampling in the Indian Statistical Institute," *JRSS*, **109**, 325–378.

McGinnis, R. [1958], "Randomization and inference in sociological research," *American Sociological Review*, **23**, 408–414.

McNemar, Q. [1962], *Psychological Statistics*, 3rd ed., New York: John Wiley and Sons.

Mickey, M. R. [1959], "Some finite population unbiased ratio and regression estimators," *JASA*, **54**, 594–612.

Milne, A. [1959], "The centric systematic area sample treated as a random sample," *Biometrics*, **15**, 270-295.

Mood, A. M. [1950], *Introduction to the Theory of Statistics*, New York: McGraw-Hill Book Co.

Moser, C. A. [1955], "Recent developments in the sampling of human populations in Great Britain," *JASA*, **50**, 1195–1214.

Moser, C. A., and Stuart, A. [1953], "An experimental study of quota sampling," *JRSS(A)*, **116,** 349–405.
Mosteller, F. [1946], "On some useful 'inefficient' statistics," *AMS*, **17,** 377–408.
Mosteller, F. [1952], "Some statistical problems in measuring the subjective responses to drugs," *Biometrics*, **8,** 220–226.
Mosteller, F., Hyman, H., McCarthy, P. J., Marks, E. S., and Truman, D. B. [1949], *The Pre-election Polls of 1948*, Bulletin 60, New York: Social Science Research Council.
Murthy, M. N., and Sethi, V. K. [1961], "Randomized rounded-off multipliers in sampling theory," *JASA*, **56,** 328–334.
National Center for Health Statistics [1964], "Cycle I of the health examination survey: sampling and response," *Series* 11, *No.* 1, Washington: Government Printing Office.
Neyman, J. [1934], "On the two different aspects of the representative method," *JRSS*, **97,** 558–625.
Neyman, J. [1952], *Lectures and Conferences on Mathematical Statistics and Probability*, Washington: Graduate School of Department of Agriculture.
Olkin, I. [1958], "Multi-variate ratio-estimation for finite populations," *Biometrika*, **45,** 154–165.
Politz, A., and Simmons, W. R. [1949], "An attempt to get the 'Not-at-Homes' into the sample without call-backs," *JASA*, **44,** 9–31.
Patterson, H. D. [1950], "Sampling on successive occasions with partial replacement of units," *JRSS(B)*, **12,** 241–255.
Patterson, H. D. [1954], "The errors of lattice sampling," *JRSS(B)*, **16,** 140–149.
Potter, R. G., Jr., Sagi, P. P., and Westoff, C. F. [1962] "Knowledge of the ovulatory cycle and coital frequency, etc., " *Milbank Memorial Quarterly*, **40,** 46–58.
Quenouille, M. H. [1956], "Notes on bias in estimation, *Biometrika*, **43,** 353–360.
Raiffa, H., and Schlaifer, R. [1961], *Applied Statistical Decision Theory*, Boston: Grad. School of Business Adm., Harvard University.
Raj, D. [1958], "On the relative accuracy of some sampling techniques," *JASA*, **53,** 98–101.
Raj, D. [1964], "The use of systematic sampling with probability proportional to size in a large scale survey," *JASA*, **59,** 251–255.
Rao, C. R. [1952], *Advanced Statistical Methods in Biometric Research*, New York: John Wiley and Sons.
Rao, J. N. K. [1957], "Double ratio estimate in forest surveys," *Jour. Indian Society of Agricultural Stats.*, **9,** 191–204.
Rao, J. N. K. [1963], "On three procedures of unequal probability sampling without replacement," *JASA*, **58,** 202–215.
Rao, J. N. K., Hartley, H. O., and Cochran, W. G. [1962], "On a simple procedure of unequal probability sampling without replacement," *JRSS(B)*, **24,** 482–491.
Robinson, W. S. [1950], "Ecological correlations and the behavior of individuals," *American Sociological Review*, **15,** 351–357.
Roy, K. P. [1961], "A method of screening defectives by collective testing," *Calcutta Statistical Association Bulletin*, **40,** 139–146.
Savage, L. J. [1954], *The Foundation of Statistics*, New York: John Wiley and Sons.
Savage, L. J. *et al.* [1962], *The Foundations of Statistical Inference*, London: Methuen and Co.
Savage, I. R. [1957], "Nonparametric statistics," *JASA*, **52,** 331–344.
Scarborough, J. B. [1950], *Numerical Mathematical Analysis*, London: Oxford University Press.
Schlaifer, R. [1959]. *Probability and Statistics for Business Decisions*, New York: McGraw-Hill Book Company.

Schlaifer, R. [1961], *Introduction to Statistics for Business Decisions*, New York: McGraw-Hill Book Co.

Sethi, V. K. [1964], "Contributions to Stratified Sampling and Some Related Problems," Ph.D. thesis, Institute for Social Sciences, Agra, India: Agra University.

Sharp, H. [1961], "Graduate training through the Detroit area study," *American Sociological Review*, **26**, 110–114.

Sharp, H., and Feldt, A. [1959], "Some factors in a probability sample survey of a metropolitan community," *American Sociological Review*, **24**, 650–661.

Simon, H. A. [1957], *Models of Man: Social and Rational*, New York: John Wiley and Sons. Also in *JASA*, 1954.

Simmons, W. R. [1954], "A plan to account for 'not-at-homes' by combining weighting and call-backs," *Journal of Marketing*, **19**, 42–53.

Sirken, M. G., and Brown, M. L. [1962], "Quality of data elicited by successive mailings in mail surveys," *Proceedings of the Social Statistics Section, American Statistical Association.*

Sobol, M. G. [1959], "Panel mortality and panel bias," *JASA*, **54**, 52–68.

Snedecor, G. W. [1956], *Statistical Method*, Ames, Iowa: Iowa State College Press, 5th Ed.

Srikantan, K. S. [1963], "A problem in optimum allocation," *Operations Research*, **11**, 265–273.

Stephan, F. F. [1948], "History of the uses of modern sampling procedures, *JASA*, **43**, 12–39.

Stephan, F. F., and El-Badry, M. A. [1955], "On adjusting sample tabulations to census counts," *JASA*, **50**, 738–762.

Stephan, F. F., and McCarthy, P. J. [1958], *Sampling Opinions*, New York: John Wiley and Sons.

Sterrett, A. [1957], "On the detection of defective members of large populations," *AMS*, **28**, 1033–1036.

Stuart, A. [1954], "A simple presentation of optimum sampling results," *JRSS(B)*, **16**, 239–241.

Sukhatme, P. V. [1947], "The problem of plot size in large-scale surveys," *JASA*, **42**, 297–310.

Sukhatme, P. V. [1954], *Sampling Theory of Surveys with Application*, Ames, Iowa: Iowa State College Press.

Sukhatme, P. V., and Seth, G. R. [1952], "Non-sampling errors in surveys," *Jour. Indian Society of Agricultural Stats.*, **5**, 5–41.

Taeuber, C., and Hansen, M. H. [1963], "A preliminary evaluation of the 1960 censuses of population and housing," *Proceedings of the Social Statistics Section, American Statistical Association*, 56–73.

Tepping, B. J., Hurwitz, W. H., and Deming, W. E. [1943], "On the efficiency of deep stratification in block sampling," *JASA*, **38**, 93–100.

Thionet, P. [1954], "Mathematical methods in public opinion polls," *International Social Science Bulletin*, **6**, 652.

Tippett, L. H. C. [1925], "On the extreme individuals and the range of samples taken from a normal population," *Biometrika*, **17**, 364 ff.

Tippett, L. H. C. [1937], *The Methods of Statistics*, 2nd ed., London: Williams and Norgate.

Tippett, L. H. C. [1956], *Statistics*, 2nd ed., London: Oxford University Press.

Tschuprow, A. [1923], "On the mathematical expectation of the moments of frequency distributions in the case of correlated observations," *Metron*, **2**, 646–680.

Tukey, J. W. [1948], "Approximate weights," *AMS*, **19**, 91–92.

Tukey, J. W. [1954], "Unsolved problems of experimental statistics," *JASA*, **49,** 706–731.

United Nations Statistical Office [1950], *The Preparation of Sampling Survey Reports,* New York: U.N. Series C, No. 1. Also [1964], Series C, No. 1, Rev. 2.

U.S. Bureau of the Census [1963], *The Current Population Survey; A Report on Methodology*, Technical Paper No. 7, Washington: Superintendent of Documents.

U.S. Bureau of the Census [1964], "Concepts and methods used in the household survey statistics on employment and unemployment," Series P-23, Nov. 13, (Also BLS 279), Washington: Government Printing Office.

Vance, L. L., and Neter, J. [1956], *Statistical Sampling for Auditors and Accountants,* New York: John Wiley and Sons.

Walsh, J. E. [1947], "Concerning the effect of intraclass correlation on certain significance tests," *AMS*, **18,** 88–96.

Walsh, J. E. [1949], "On the 'information' lost by using a *t*-test when the population variance is known," *JASA*, **44,** 122–125.

West, Q. M. [1952], "The results of applying simple random sampling processes to farm management data," Agricultural Experiment Station, Ithaca, N.Y.: Cornell University.

Wilk, M. B., and Kempthorne, O. [1955], "Fixed, mixed, and random models," *JASA*, **50,** 1144–1167.

Wilk, M. B., and Kempthorne, O. [1956], "Some aspects of the analysis of factorial experiment in a completely randomized design," *AMS*, **27,** 950–985.

Wilkerson, M. [1960], The revised city sample for the Consumer Price Index," *Monthly Labor Review*, 1078–1083.

Wilks, S. S. [1948], "Order statistics," *Bulletin of the American Mathematical Society*, **54,** 6–50.

Wold, H. [1956], "Causal inference from observational data," *JRSS(A)*, **119,** 28–61.

Woodruff, R. S. [1952], "Confidence intervals for medians and other position measures," *JASA*, **47,** 635–646.

Woodruff, R. S. [1963], "The use of rotating samples in the Census Bureau's monthly surveys," *JASA*, **58,** 454–467.

Yates, F. [1948], "Systematic sampling," *Phil. Trans. Roy. Soc. (A)*, 241, 345–377.

Yates, F. [1951], "The influence of statistical methods for research workers on the development of the science of statistics," *JASA*, **46,** 19–34.

Yates, F. [1960], *Sampling Methods for Censuses and Surveys*, 3rd ed., London: Chas. Griffin and Company.

Yates, F., and Grundy, P. M. [1953], "Selection without replacement from within strata with probability proportional to size," *JRSS(B)*, **15,** 235–261.

You Poh Seng [1951], "Historical survey of the development of sampling theories and practice," *JRSS(A)*, **114,** 214–231.

Yule, G. U., and Kendall, M. G. [1937], *An Introduction to the Theory of Statistics*, 11th ed., London: Griffin.

Zarkovich, S. S. [1956], "Notes on the history of sampling in Russia," *JRSS(A)*, **119,** 336–338.

Zarkovich, S. S. [1961], *Sampling Methods and Censuses; Vol. I., Collecting Data and Tabulation* (Draft), Rome: FAO.

Zarkovich, S. S. [1963], *Sampling Methods and Censuses; Vol. II, Quality of Statistical Data* (Draft), Rome: FAO.

Answers to Selected Problems

Chapter 1

1.13 Yes: *a, c, d, e, f, g, j*; No: *b, h, i*.

1.14 (*a*) 33 and 3; (*b*) 131; (*c*) 1089; (*d*), (*e*), (*f*) 35.2; (*g*) 1815; (*h*) 137; (*i*), (*j*) 88; (*k*) 0.60; (*l*) indef.; (*m*) 165; (*n*) $33k$; (*o*) $1089k^2$; (*p*) 231; (*q*) -0.47.

1.15 (*a*) 55 and 5; (*b*) 385; (*c*) 3025; (*d*), (*e*), (*f*) 121.

Chapter 2

2.2 5.20.

2.3 (*a*) 0.125; (*b*) 425.

2.5 4.62 ± 0.51 and $18{,}500 \pm 2040$.

2.6 (*a*) 164; (*b*) 167; (*c*) 110.

Chapter 3

3.1 (c'') 28.64 ± 2.80; (d'') 20.34 ± 2.57; (e'') $31{,}820 \pm 3{,}110$ and 9.66×10^6.

3.2 (*a*) 3.54; (*b*) 6.51; (*c*) 3.61.

3.3 (*a*) 10.05; if $S_2/\sqrt{J_2} = S_3/\sqrt{J_3}$; (*b*) 10.05; (*c*) 37.73 ± 4.98; (*d*) 2428.

3.10 (*a*) $\sigma^2/600$; (*b*) $\sigma^2/533$; (*c*) $\sigma^2/384$; thus in ratios of 1:1.12:1.56.

3.12 588.

3.13 (*a*) $P_2/P_1 = 0.28/0.72$; (*b*) $P_2/P_1 = 0.02/0.38$; (*c*) 24%; (*d*) 30% and 47%.

3.17 (*a*) 0.022 and 0.000245; (*b*) gains of 8% and 4%; (*c*) gain of 40% and loss of 25%.

Chapter 4

4.1 0.255.

4.2 0.339.

4.6 (*a*) 1,345 and 19,247; (*b*) 0.059 and 0.048; (*c*) 19,714; (*d*) 1,363 and 19,209; (*e*) 1,640 and 19,660.

Chapter 5

5.1 (*a*) 0.385 ± 0.064 and 0.00203; (*b*) $40.7/11.90 = 3.42$ and roh $= 0.27$; (*c*) 0.00247; (*d*) $77{,}000 \pm 12{,}800$.

5.2 (*a*) $0.385 \pm .073$ and 0.0027; (*b*) $53.25/11.90 = 4.47$ and roh $= 0.39$; (*c*) 0.00303; (*d*) $77{,}000 \pm 14{,}600$.

5.5 (*a*) $s_a^2 = 0.081$, $s_b^2 = 0.177$; (*b*) 0.00104.

5.14 (*a*) roh $= 0.5$, $S_u^2 = S_b^2 = 0.5$, $S_a^2 = 0.625$; (*b*) roh $= 0.2$, $S_b^2 = 0.8$, $S_u^2 = 0.2$, $S_a^2 = 0.4$.

Chapter 6

6.3 (*a*) 0.000573; (*b*) 0.000555.

6.5 (*a*) 0.494 ± 0.060; (*b*) 2.28; (*c*) 0.00181 and 0.0030.

6.8 (*a*) 0.500; (*b*) 0.000528; (*c*) 5.28/2.50 = 2.11; (*d*) $\leq$ 0.000132; (*e*) 0.00034.

6.9 (*a*) 0.259 ± 0.029; (*b*) 8.37/9.99 = 0.838; (*c*) 0.000209; (*d*) reduce it somewhat.

6.10 (*a*) 0.427 ± 0.043; (*b*) 0.168 ± 0.053.

Chapter 7

7.1 (*a*) $\frac{\text{Mos}_\alpha}{1000} \times \frac{20}{\text{Mos}_\alpha} = \frac{1}{50}$; (*b*) $\frac{y}{n}$; (*c*) $\frac{0.98}{1740} \sum_g^{29} (\bar{y}_g - \bar{y}_{g+1})^2$ or $\frac{0.98}{900} \sum_h^{15} (\bar{y}_{h1} - \bar{y}_{h2})^2$ or $\frac{0.98}{n^2} \cdot \frac{30}{58} \sum_g^{29} (Dy_g - rDx_g)^2$ or $\frac{0.98}{n^2} \sum_h^{15} (Dy_g - rDx_g)^2$.

7.2 (*a*) 6400, since 8/6400 = 1/800; (*c*) y/n and $\frac{120}{238n^2} \sum_g^{119} (Dy_g - rDx_g)^2$ or $\frac{1}{n^2} \sum_h^{60} (Dy_h - rDx_h)^2$; (*d*) $F'y \pm F' \sqrt{\text{var}(y)}$ where $F' = 800/0.95$, or $N\bar{y} \pm N\sqrt{\text{var}(\bar{y})}$ if N is available.

Index